Ingredients Extraction by Physicochemical Methods in Food

Ingredients Extraction by Physicochemical Methods in Food

Handbook of Food Bioengineering, Volume 4

Edited by

Alexandru Mihai Grumezescu
Alina Maria Holban

ACADEMIC PRESS
An imprint of Elsevier

Academic Press is an imprint of Elsevier
125 London Wall, London EC2Y 5AS, United Kingdom
525 B Street, Suite 1800, San Diego, CA 92101-4495, United States
50 Hampshire Street, 5th Floor, Cambridge, MA 02139, United States
The Boulevard, Langford Lane, Kidlington, Oxford OX5 1GB, United Kingdom

Library of Congress Cataloging-in-Publication Data
A catalog record for this book is available from the Library of Congress

British Library Cataloguing-in-Publication Data
A catalogue record for this book is available from the British Library

ISBN: 978-0-12-811521-3

For information on all Academic Press publications visit our website at https://www.elsevier.com/books-and-journals

Publisher: Andre Gerhard Wolff
Acquisition Editor: Nina Bandeira
Editorial Project Manager: Jaclyn Truesdell
Production Project Manager: Caroline Johnson
Designer: Matthew Limbert

Typeset by Thomson Digital

Contents

List of Contributors

Cristóbal N. Aguilar Autonomous University of Coahuila, Saltillo, Coahuila, Mexico
Pedro Aguilar-Zárate Instituto Tecnológico de Ciudad Valles, Tecnológico Nacional de México, Ciudad Valles, San Luis Potosí, México
Alma Alarcon-Rojo Autonomous University of Chihuahua, Chihuahua, Mexico
Pedro Aqueveque Development of Agro industries Technology Center, University of Concepción, Chillán, Chile
Juan A. Ascacio-Valdés Autonomous University of Coahuila, Saltillo, Coahuila, Mexico
Amra Bratovcic University of Tuzla, Tuzla, Bosnia and Herzegovina
Romilson Brito MeditBio and University of Algarve, Faro, Portugal
Ramiro A. Carciochi National University of Central Buenos Aires, Olavarría, Buenos Aires, Argentina
Chung-Hung Chan Malaysian Palm Oil Board, Kajang, Selangor, Malaysia
Karina Cruz Autonomous University of Coahuila, Saltillo, Coahuila, Mexico
Rui M.S. Cruz MeditBio and University of Algarve, Faro, Portugal
Leandro G. D'Alessandro Lille University, INRA, ISA, Artois University, University of Littoral Opal Coast, Charles Viollette Institute, Lille, France
Krasimir Dimitrov Lille University, INRA, ISA, Artois University, University of Littoral Opal Coast, Charles Viollette Institute, Lille, France
Mohamed H. Abd El-Salam National Research Centre, Cairo, Egypt
Mayyada El-Sayed American University in Cairo, New Cairo; National Research Centre, Giza, Egypt
Safinaz El-Shibiny National Research Centre, Cairo, Egypt
Hanaa Essa American University in Cairo, New Cairo; Agriculture Research Centre, Giza, Egypt
Daisy Fleita American University in Cairo, New Cairo, Egypt
Nor D. Hassan University of Technology Malaysia, Johor Bahru, Johor, Malaysia
Amit K. Keshari Babasaheb Bhimrao Ambedkar University, Lucknow, Uttar Pradesh, India
Siddhartha Maity Jadavpur University, Kolkata, West Bengal, India
Siti N.H. Mamat University of Technology Malaysia, Johor Bahru, Johor, Malaysia
Diana B. Muñiz-Márquez Instituto Tecnológico de Ciudad Valles, Tecnológico Nacional de México, Ciudad Valles, San Luis Potosí, México
Ida I. Muhamad University of Technology Malaysia, Johor Bahru, Johor, Malaysia
Norazlina M. Nawi University of Technology Malaysia, Johor Bahru, Johor, Malaysia
Gek Cheng Ngoh University of Malaya, Kuala Lumpur, Malaysia

Zoe Nikolaidou Alexander Technological Educational Institute of Thessaloniki (ATEITh), Thessaloniki, Greece
Susana M. Nolasco National University of Central Buenos Aires, Olavarría, Buenos Aires, Argentina
Margarita Ocampo Development of Agro industries Technology Center, University of Concepción, Chillán, Chile
Amra Odobasic University of Tuzla, Tuzla, Bosnia and Herzegovina
Larysa Paniwnyk Coventry University, Coventry, United Kingdom
Rudi Radrigán Development of Agro industries Technology Center, University of Concepción, Chillán, Chile
Amit Rai Babasaheb Bhimrao Ambedkar University, Lucknow, Uttar Pradesh, India
Vinit Raj Babasaheb Bhimrao Ambedkar University, Lucknow, Uttar Pradesh, India
Priyanka Rao Institute of Chemical Technology, Mumbai, Maharashtra, India
Wahida A. Rashid University of Technology Malaysia, Johor Bahru, Johor, Malaysia
Virendra Rathod Institute of Chemical Technology, Mumbai, Maharashtra, India
Carlos Reyes-Luna Instituto Tecnológico de Ciudad Valles, Tecnológico Nacional de México, Ciudad Valles, San Luis Potosí, México
Dalia Rifaat American University in Cairo, New Cairo, Egypt
Raúl Rodríguez Autonomous University of Coahuila, Saltillo, Coahuila, Mexico
María M. Rodriguez National University of Central Buenos Aires, Olavarría, Buenos Aires, Argentina
José C. Rodriguez-Figueroa Autonomous University of Chihuahua, Chihuahua, Mexico
Sudipta Saha Babasaheb Bhimrao Ambedkar University, Lucknow, Uttar Pradesh, India
Indira Sestan University of Tuzla, Tuzla, Bosnia and Herzegovina
Vassilia J. Sinanoglou Technological Education Institution of Athens, Egaleo, Greece
Ashok K. Singh Babasaheb Bhimrao Ambedkar University, Lucknow, Uttar Pradesh, India
Petros Smirniotis Alexander Technological Educational Institute of Thessaloniki (ATEITh), Thessaloniki, Greece
Nuraimi A. Tan University of Technology Malaysia, Johor Bahru, Johor, Malaysia
Mihai Toma Costin D. Nenițescu–Institute of Organic Chemistry of the Romanian Academy, Bucharest, Romania
Thalia Tsiaka Institute of Biology, Medicinal Chemistry and Biotechnology, National Hellenic Research Foundation, Athens; Technological Education Institution of Athens, Egaleo; University of Athens, Athens, Greece
Siddharth Vats Shri Ram Swaroop Memorial University, Lucknow, Uttar Pradesh, India
Peggy Vauchel Lille University, INRA, ISA, Artois University, University of Littoral Opal Coast, Charles Viollette Institute, Lille, France
Margarida C. Vieira MeditBio and University of Algarve, Faro, Portugal
Elżbieta Włodarczyk Koszalin University of Technology, Koszalin, Poland
Jorge E. Wong-Paz Instituto Tecnológico de Ciudad Valles, Tecnológico Nacional de México, Ciudad Valles, San Luis Potosí, México
Rozita Yusoff University of Malaya, Kuala Lumpur, Malaysia
Paweł K. Zarzycki Koszalin University of Technology, Koszalin, Poland
Panagiotis Zoumpoulakis Institute of Biology, Medicinal Chemistry and Biotechnology, National Hellenic Research Foundation; National Hellenic Research Foundation; University of Athens, Athens, Greece

Foreword

In the last 50 years an increasing number of modified and alternative foods have been developed using various tools of science, engineering, and biotechnology. The result is that today most of the available commercial food is somehow modified and improved, and made to look better, taste different, and be commercially attractive. These food products have entered in the domestic first and then the international markets, currently representing a great industry in most countries. Sometimes these products are considered as life-supporting alternatives, neither good nor bad, and sometimes they are just seen as luxury foods. In the context of a permanently growing population, changing climate, and strong anthropological influence, food resources became limited in large parts of the Earth. Obtaining a better and more resistant crop quickly and with improved nutritional value would represent the Holy Grail for the food industry. However, such a crop could pose negative effects on the environment and consumer health, as most of the current approaches involve the use of powerful and broad-spectrum pesticides, genetic engineered plants and animals, or bioelements with unknown and difficult-to-predict effects. Numerous questions have emerged with the introduction of engineered foods, many of them pertaining to their safe use for human consumption and ecosystems, long-term expectations, benefits, challenges associated with their use, and most important, their economic impact.

The progress made in the food industry by the development of applicative engineering and biotechnologies is impressive and many of the advances are oriented to solve the world food crisis in a constantly increasing population: from genetic engineering to improved preservatives and advanced materials for innovative food quality control and packaging. In the present era, innovative technologies and state-of-the-art research progress has allowed the development of a new and rapidly changing food industry, able to bottom-up all known and accepted facts in the traditional food management. The huge amount of available information, many times is difficult to validate, and the variety of approaches, which could seem overwhelming and lead to misunderstandings, is yet a valuable resource of manipulation for the population as a whole.

The series entitled *Handbook of Food Bioengineering* brings together a comprehensive collection of volumes to reveal the most current progress and perspectives in the field of food engineering. The editors have selected the most interesting and intriguing topics, and have dissected them in 20 thematic volumes, allowing readers to find the description of basic processes and also the up-to-date innovations in the field. Although the series is mainly dedicated to the engineering, research, and biotechnological sectors, a wide audience could benefit from this impressive and updated information on the food industry. This is because of the overall style of the book, outstanding authors of the chapters, numerous illustrations, images, and well-structured chapters, which are easy to understand. Nonetheless, the most novel approaches and technologies could be of a great relevance for researchers and engineers working in the field of bioengineering.

Current approaches, regulations, safety issues, and the perspective of innovative applications are highlighted and thoroughly dissected in this series. This work comes as a useful tool to understand where we are and where we are heading to in the food industry, while being amazed by the great variety of approaches and innovations, which constantly changes the idea of the "food of the future."

Anton Ficai, PhD (Eng)
Department Science and Engineering of Oxide Materials and Nanomaterials,
Faculty of Applied Chemistry and Materials Science, Politehnica University of Bucharest,
Bucharest, Romania

Series Preface

The food sector represents one of the most important industries in terms of extent, investment, and diversity. In a permanently changing society, dietary needs and preferences are widely variable. Along with offering a great technological support for innovative and appreciated products, the current food industry should also cover the basic needs of an ever-increasing population. In this context, engineering, research, and technology have been combined to offer sustainable solutions in the food industry for a healthy and satisfied population.

Massive progress is constantly being made in this dynamic field, but most of the recent information remains poorly revealed to the large population. This series emerged out of our need, and that of many others, to bring together the most relevant and innovative available approaches in the amazing field of food bioengineering. In this work we present relevant aspects in a pertinent and easy-to-understand sequence, beginning with the basic aspects of food production and concluding with the most novel technologies and approaches for processing, preservation, and packaging. Hot topics, such as genetically modified foods, food additives, and foodborne diseases, are thoroughly dissected in dedicated volumes, which reveal the newest trends, current products, and applicable regulations.

While health and well-being are key drivers for the food industry, market forces strive for innovation throughout the complete food chain, including raw material/ingredient sourcing, food processing, quality control of finished products, and packaging. Scientists and industry stakeholders have already identified potential uses of new and highly investigated concepts, such as nanotechnology, in virtually every segment of the food industry, from agriculture (i.e., pesticide production and processing, fertilizer or vaccine delivery, animal and plant pathogen detection, and targeted genetic engineering) to food production and processing (i.e., encapsulation of flavor or odor enhancers, food textural or quality improvement, and new gelation- or viscosity-enhancing agents), food packaging (i.e., pathogen, physicochemical, and mechanical agents sensors; anticounterfeiting devices; UV protection; and the design of stronger, more impermeable polymer films), and nutrient supplements (i.e., nutraceuticals, higher stability and bioavailability of food bioactives, etc.).

The series entitled *Handbook of Food Bioengineering* comprises 20 thematic volumes; each volume presenting focused information on a particular topic discussed in 15 chapters each. The volumes and approached topics of this multivolume series are:

Volume 1: Food Biosynthesis

Volume 2: Food Bioconversion

Volume 3: Soft Chemistry and Food Fermentation

Volume 4: Ingredient Extraction by Physicochemical Methods in Food

Volume 5: Microbial Production of Food Ingredients and Additives

Volume 6: Genetically Engineered Foods

Volume 7: Natural and Artificial Flavoring Agents and Food Dyes

Volume 8: Therapeutic Foods

Volume 9: Food Packaging and Preservation

Volume 10: Microbial Contamination and Food Degradation

Volume 11: Diet, Microbiome, and Health

Volume 12: Impacts of Nanoscience on the Food Industry

Volume 13: Food Quality: Balancing Health and Disease

Volume 14: Advances in Biotechnology in the Food Industry

Volume 15: Foodborne Diseases

Volume 16: Food Control and Biosecurity

Volume 17: Alternative and Replacement Foods

Volume 18: Food Processing for Increased Quality and Consumption

Volume 19: Role of Material Science in Food Bioengineering

Volume 20: Biopolymers for Food Design

The series begins with a volume on *Food Biosynthesis*, which reveals the concept of food production through biological processes and also the main bioelements that could be involved in food processing. The second volume, *Food Bioconversion*, highlights aspects related to food modification in a biological manner. A key aspect of this volume is represented by waste bioconversion as a supportive approach in the current waste crisis and massive pollution of the planet Earth. In the third volume, *Soft Chemistry and Food Fermentation*, we aim to discuss several aspects regarding not only to the varieties and impacts of fermentative processes, but also the range of chemical processes that mimic some biological processes in the context of the current and future biofood industry. Volume 4, *Ingredient Extraction by Physicochemical Methods in Food*, brings the readers into the world of ingredients and the methods that can be applied for their extraction and purification. Both traditional and most of the modern techniques can be found in dedicated chapters of this volume. On the other hand, in volume 5, *Microbial Production of Food Ingredients and Additives*, biological methods of ingredient production, emphasizing microbial processes, are revealed and discussed. In volume 6, *Genetically Engineered Foods*, the delicate subject of genetically engineered plants and animals to develop modified foods is thoroughly dissected. Further, in volume 7, *Natural and Artificial Flavoring Agents and Food Dyes*, another hot topic in food industry—— flavoring and dyes—is scientifically commented and valuable examples of natural and artificial compounds are generously offered. Volume 8, *Therapeutic Foods*, reveals the most utilized and investigated foods with therapeutic values. Moreover, basic and future approaches for traditional and alternative medicine, utilizing medicinal foods, are presented here. In volume 9, *Food Packaging and Preservation*, the most recent, innovative, and interesting technologies and advances in food packaging, novel preservatives, and preservation methods are presented. On the other hand, important aspects in the field of *Microbial Contamination and Food Degradation* are presented in volume 10. Highly debated topics in modern society: *Diet, Microbiome, and Health* are significantly discussed in volume 11. Volume 12 highlights the *Impacts of Nanoscience on the Food Industry*, presenting the most recent advances in the field of applicative nanotechnology with great impacts on the food industry. Additionally, volume 13 entitled *Food Quality: Balancing Health and Disease* reveals the current knowledge and concerns regarding the influence of food quality on the overall health of population and potential food-related diseases. In volume 14, *Advances in Biotechnology in the Food Industry*, up-to-date information regarding the progress of biotechnology in the construction of the future food industry is revealed. Improved technologies, new concepts, and perspectives are highlighted in this work. The topic of *Foodborne Diseases* is also well documented within this series in volume 15. Moreover, *Food Control and Biosecurity* aspects, as well as current regulations and food safety concerns are discussed in the volume 16. In volume 17, *Alternative and Replacement Foods*, another broad-interest concept is reviewed. The use and research of traditional food alternatives currently gain increasing terrain and this quick emerging trend has a significant impact on the food industry. Another related hot topic, *Food Processing for Increased Quality and Consumption*, is considered in volume 18. The final two volumes rely on the massive progress made in material science and the great applicative impacts of this progress on the food industry. Volume 19, *Role of Material Science in Food Bioengineering*, offers a perspective and a scientific introduction in the science of engineered materials, with important applications in food research and technology. Finally, in the volume 20, *Biopolymers for Food Design*, we discuss the advantages and challenges related to the development of improved and smart biopolymers for the food industry.

All 20 volumes of this comprehensive collection were carefully composed not only to offer basic knowledge for facilitating understanding of nonspecialist readers, but also to offer valuable information regarding the newest trends and advances in food engineering, which is useful for researchers and specialized readers. Each volume could be treated individually as a useful source of knowledge for a particular topic in the extensive field of food engineering or as a dedicated and explicit part of the whole series.

This series is primarily dedicated to scientists, academicians, engineers, industrial representatives, innovative technology representatives, medical doctors, and also to any nonspecialist reader willing to learn about the recent innovations and future perspectives in the dynamic field of food bioengineering.

Alexandru M. Grumezescu
Politehnica University of Bucharest, Bucharest, Romania

Alina M. Holban
University of Bucharest, Bucharest, Romania

Preface for Volume 4: Ingredients Extraction by Physicochemical Methods in Food

Numerous food-related compounds have proved their additional beneficial effects, along with their nutritional properties. Some food ingredients have been utilized for centuries in traditional and preventive therapy for their health-promoting effect, while others may have an impact on various industries (i.e., food, chemical, biotechnological, and pharmaceutical) and even on our environment. A key factor in the production of such ingredients is represented by their physicochemical extraction technique. Extraction methods are variable, and great progress has been made in this field in the past decade. This book describes the most utilized methods developed for ingredients extraction and the anticipated design of future approaches. Intelligent systems have recently emerged to obtain useful and innovative ingredients from plants, exotic fruits, and spices, their impact on the quality and development of the food industry being impressive.

This book has aimed to bring together the most interesting and investigated aspects of ingredients extraction and the most important technologies, to obtain specific and valuable food-related compounds for improved food quality, health promotion, and environmental protection in the context of a sustainable food industry. Classical and newest technologies, along with their applicability spectrum and their main advantages and drawbacks, are presented within this volume.

The volume contains 15 chapters prepared by outstanding authors from France, the United Kingdom, India, Poland, Mexico, Bosnia and Herzegovina, Chile, Greece, Egypt, Portugal, and Malaysia.

The selected manuscripts are clearly illustrated and contain accessible information for a wide audience, especially food scientists, engineers, biotechnologists, biochemists, and industrial companies, but also any reader interested in learning about the most interesting and recent advances on the field of ingredients extraction and food processing.

Chapter 1, prepared by Vats, is entitled *Methods for Extractions of Value-Added Nutraceuticals From Lignocellulosic Wastes and Their Health Application*. In this work, the author introduces readers to contributions in the field of food ingredients extraction from various sources, such as medicinally valuable phytochemicals, nutraceuticals functional foods, fruit purees and powders, biochemicals, electrolytes blends, health-promoting agents, nutritive oils, antimicrobial products, bioactive compounds, commercially valuable food flavoring and additives compounds of a biochemical nature, proteins, nutritional supplements, and personal and cosmetic care, as well as drugs and pharmacophores from eukaryotic and prokaryotic cultured cells or from plants, animals, and microbes. The main extraction methods, such as standard physical extraction procedures and solvents-based approaches, are also discussed.

Saha and collaborators, in Chapter 2, *Modern Extraction Techniques for Drugs and Medicinal Agents*, reveal various physicochemical methods of extraction that comprise microwave-assisted extraction, pressurized liquid extraction, supercritical fluid extraction, liquid phase microextraction, solid phase extraction, ultrasound-assisted extraction, cloud-point extraction, enzyme-assisted extraction, membrane-based microextraction, and cooling-assisted microextraction. These are the commonly used and modern techniques in terms of isolation and separation of ingredients from both chemical and biological mixtures.

Chapter 3, entitled *Advances in Extraction, Fractionation, and Purification of Low-Molecular Mass Compounds From Food and Biological Samples*, written by Włodarczyk and Zarzycki, gives an overview concerning current extraction and quantification protocols of bioactive substances, which are recently designed for analytical and technological applications of food processing. Generally, extraction, fractionation, and purification are critical issues for both analytical applications and technological processes involving food and biological samples. The authors discuss methodological approaches depending on the expected outcomes and physicochemical properties of a given product.

In Chapter 4, *Valorization of Agrifood By-Products by Extracting Valuable Bioactive Compounds Using Green Processes*, prepared by Carciochi et al., is presented the current challenge for the food industry, related to the exploitation of various by-products as sources of new commodities using eco-friendly technologies with an optimal cost-benefit relationship. The main green technologies used to recover natural products from agrifood by-products, such as enzyme-assisted extraction, ultrasound-assisted extraction, microwave-assisted extraction, electrically assisted extraction, pressurized liquid extraction, supercritical fluid extraction, and instant controlled pressure drop, are presented here.

Wong-Paz and coworkers, in Chapter 5, *Extraction of Bioactive Phenolic Compounds by Alternative Technologies*, describe the advances in the research done on bioactive phenolic compound (BPC) extraction using alternative extraction technologies. In addition, the important parameters influencing its performance, the basic theory of reactions present, and the direct effect of alternative extraction technologies on BPCs are also included. Advantages and drawbacks of the alternative extraction technologies on BPC extraction with regard to conventional extraction technologies are summarized. Finally, a perspective and general conclusion are presented.

Chapter 6, *The Extraction of Heavy Metals From Vegetable Samples*, prepared by Odobašic´ and collaborators, describes types of extraction approaches in order to determine the origin of metals and the efficiency of their removal. It is considered that metals that are in an adsorption and exchangeable phase are more weakly bonded and more easily bioavailable and because of that have anthropogenic origins. Metals in an inert residual fraction indicate natural origin. Thus, these separation approaches bring essential information on the nature and potential impact of various heavy metals in foods.

Radrigán et al., in Chapter 7, *Extraction and Use of Functional Plant Ingredients for the Development of Functional Foods*, offer an interesting collection of technical extraction particularities of active principles of biomaterials, with emphasis on their importance in the development of functional foods, since some biomaterials are used as supplements in the diet.

In Chapter 8, prepared by Tsiaka and coworkers, *Extracting Bioactive Compounds From Natural Sources Using Green High-Energy Approaches: Trends and Opportunities in Lab- and Large-Scale Applications*, are explored the main concepts and principles of high-energy extraction techniques in selective and targeted extraction of bioactive or functional molecules with health-promoting properties. In addition, a special allusion is made to the significance of optimization strategies and experimental design models in the improvement of the extraction procedures. Furthermore, arguments presented in this review are supported by a variety of examples and peer-reviewed articles published over the past 3 years. Finally, the possibilities and economic feasibility of adopting and scaling up high-energy extraction techniques to the industrial level is investigated in order to define the framework of their implementation and the future potential.

El-Sayed et al., in Chapter 9, entitled *Assessment of the State-of-the-Art Developments in the Extraction of Antioxidants From Marine Algal Species*, critically assess the state-of-the-art methods for extracting antioxidants, with emphasis on sulfated polysaccharides (SPs), from green, red, and brown algae. The evaluation made by these authors is primarily based on the yields and antioxidant activities of the extracted SPs, in addition to other technical, economic, and environmental criteria.

Chapter 10, *The Use of Ultrasound as an Enhancement Aid to Food Extraction*, written by Paniwnyk and collaborators, covers a range of areas that have employed ultrasound to process food materials, such as extraction of plants, seeds, and fruit; the recycling of food waste; and the effect on the properties of meat and dairy products. Some discussion on scale-up processes is also included.

In Chapter 11, *Extraction of Bioactive Compounds From Olive Leaves Using Emerging Technologies*, Cruz et al. discuss extraction techniques including microwave, supercritical fluid, superheated liquid, and ultrasound that are used to extract bioactive compounds from olive leaves, as well as their antioxidant and antimicrobial properties.

El-Salam and El-Shibiny, in Chapter 12, *Separation of Bioactive Whey Proteins and Peptides*, highlight the basic principles and application of the technologies used for the isolation of casein macropeptide, lactoferrin, and lactoperoxidase from cheese whey. Also, methods for the separation of bioactive peptides from whey protein hydrolysates are presented.

Chapter 13, *Phytochemicals: An Insight to Modern Extraction Technologies and Their Applications*, prepared by Rao and Rathod, provides a holistic insight into the modern approaches for physicochemical extraction alongside conventional techniques, giving a balanced outline of the applications and latest developments of each technique.

Chapter 14, prepared by Muhamad et al., *Extraction Technologies and Solvents of Phytocompounds From Plant Materials: Physicochemical Characterization and Identification of Ingredients and Bioactive Compounds From Plant Extract Using Various Instrumentations*, aims to review several physicochemical extraction techniques, including conventional and advanced techniques, such as solvent extraction, microwave-assisted extraction, ultrasonic-assisted extraction, aqueous extraction, enzymatic extraction, and supercritical fluid extraction. These physicochemical characterizations of ingredients and bioactive compounds using various instrumentations could provide informative and scientific reference for diverse potential uses of plant extracts, especially for nutraceuticals and functional food applications.

Chan and collaborators in Chapter 15, *An Energy-Based Approach to Scale Up Microwave-Assisted Extraction of Plant Bioactives*, discuss the optimization and modeling techniques based on energy-based parameters that enable scale-up of microwave-assisted extraction (MAE) of plant-derived bioactive compounds. Energy-based parameters, namely absorbed power density (APD) and absorbed energy density (AED), are able to characterize the extraction kinetics of MAE in predicting the extraction profiles and the optimum conditions at various conditions, particularly at larger-scale operations. This chapter also discusses the applications of APD and AED in equipment design, operational flexibility, and adaptability for various types of plant extraction with the aim to commercialize MAE.

Alexandru M. Grumezescu
Politehnica University of Bucharest, Bucharest, Romania

Alina M. Holban
University of Bucharest, Bucharest, Romania

CHAPTER 1

Methods for Extractions of Value-Added Nutraceuticals From Lignocellulosic Wastes and Their Health Application

Siddharth Vats

Shri Ram Swaroop Memorial University, Lucknow, Uttar Pradesh, India

1 Introduction

Nature is the foremost and oldest inexhaustible source of chemotypes and pharmacophores (Mukherjee and Wahile, 2006). The 21st century has seen a wide and explosive surge in natural product chemistry because of new compounds with diversified structural and chemical properties. With the change in the drug discovery process, there is a need for the improvements in natural product research to maintain the cutting edge of alternative medicines (Mukherjee and Wahile, 2006). According to the International Environment Technology Centre's (United Nations Energy Program) report, the types and the volume of biomass waste generated have increased because of intensive agricultural practices to meet the needs of growing populations. In most of the agricultural practicing countries, like India and China, the major chunk of lingocellulosic waste generated remain unutilized either are burned to make way for new crops or are allowed to rot, eventually causing harm to the environment by emitting methane and generating CO_2, which causes climate change (DTIE, 2009). Obtaining value-added products from plants has gained momentum. But there is complete negligence toward forests wastes; these can also be the source of various value-added products (Vats et al., 2013). Similarly, a major share of forest waste also contributes to forest fires and other environmental changes. Forests are rich in diverse medicinal plants; the waste generated from them can be explored for extraction of medicinal valuable products, because the situations for the health care sector also have challenges from new emerging diseases. Stressful lifestyles, adulteration in the food, rising antibiotic resistance among microbes, all provide the right platform for the emergence of new global diseases. The area where we see a justifiable great scope for the natural health-care product is in metabolic diseases, immunosuppressants, and diseases caused by antibiotic resistant microbes, DNA damage, cellular injuries, and so on. Cancers and infectious diseases kill millions of people worldwide. Various reactive oxygen species (ROS), oxidize different intracellular

Ingredients Extraction by Physicochemical Methods in Food
http://dx.doi.org/10.1016/B978-0-12-811521-3.00001-6

components and lead to cancer, aging, heart failure, diabetes, neurodegenerative diseases, and so on. Alternative drugs are losing their edge against the metabolic diseases and infectious microbes with many being challenged by the resistant microbes. The rise in resistant infectious microbes and disadvantages associated with synthetic anticancerous drugs have put phytochemicals on top of the research list of microbiologists and oncologists. Providing economical anticancerous and antimicrobial drugs with negligible side effects is the only ground-level solution. The packaged food market is growing with an ever-increasing rate. The use of additives and preservatives like sodium benzoate (SB), sodium thiosulfates (ST), sodium nitrates (SN), oxalic acid (OA), sodium citrate (SC), and benzoic acids (BA), used on large scale. But it has been found that all major additives used for any purpose in food are also responsible for various health-related issues. The future for natural product-based drugs is bright as a wide range of terrestrial and marine plants and herbs found in extreme geographic and physical conditions are unexplored. Biomasses generated from trees, plants, and shrubs growing in extreme geographic conditions contain some special components with great medicinal values.

1.1 Medicines and Present Scenario

Cheap and better medicines, nutritious food, and chemical-free food preservation strategies are the most important basic necessities for mankind of the present and future centuries (Sun and Cheng, 2002). A diet rich in organic food, fruits, and vegetables may decrease the chances of deadly diseases like heart diseases, cancers (Boyer and Liu, 2004). Vegetables, fruit, and plant materials are rich sources of nonnutritive bioactive chemicals called phytochemicals, like flavonoids, alkaloids, cartenoids, phenolics, and other phytochemicals. "Eating one apple each day keeps doctor away" is also scientifically proven that there is an inhibition of cancer cell proliferation, oxidation of lipids, and lowering of blood cholesterol level by consumption of the same (Boyer and Liu, 2004). Epidemiological studies have also confirmed health benefits associated with phytochemicals. The traditional Indian medicine system *Ayurveda* utilized phytochemicals obtained from plants and herbs to impart health benefits among humans (Moon et al., 2010). Most of the economically leading nations of the world also lead the number of deaths due to cardiovascular diseases and cancers. More than 25% of total drugs available in the market are made up of plants or plant-derived substances (Ameen et al., 2011). In recent years plant-derived metabolites are analyzed and investigated on a large scale as a source of new drugs by considering the antibiotic resistance among microbes for conventional and presently available antibiotics (Ameen et al., 2011). Plants synthesize secondary metabolites as defensive molecules against predations and microbial attacks (Liu, 2004; Mallikharjuna et al., 2007). Chemoprevention and chemotherapy are two separate terms with completely different meanings. Some important issues that make them more complex are age and cancer types. Chemoprevention can be best for healthy people as it is best to prevent them for getting cancers but should be less toxic and free from side effects. Chemotherapy is a potent weapon against patients who already have tumors or cancers (Aggarwal et al., 2004).

1.2 Main Classes of Phytochemicals and Their Sources

There are various types of phytochemicals and they belong to many classes. Concentrations and quantities of different phytochemicals vary in different plants (Table 1.1).

Table 1.1: Main classes and source of phytochemicals.

S. no.	Main Group	Examples	Main Sources	References
1.	Alkaloids	Caffeine, theobromine, theophylline	Coffee, tea, onion, curly kale, green bean, broccoli, endive, celery, cranberry, orange juice, salad tomato, bell pepper, strawberry, broad bean, apple, grape, red wine, tomato juices, cabbage, carrot, mushroom, pea, spinach, peach, and white wine	Kern and Lipman (1977); Mukherjee and Menge (2000)
2.	Anthocyanin	Malvidin, cyanidin	Flowers, fruits, and vegetables	Wang et al. (1997); Bridle and Timberlake (1997); Ribereau (1974)
3.	Carotenoids	Beta-carotene, lutein lycopene	Tomatoes, watermelon, guava, and pink grapefruits	Mortensen (2006)
4.	Coumestans	Coumestrol	Split peas, pinto beans, lima beans, alfalfa and clover sprouts, soy products, cereals and breads, nuts and oilseeds, vegetables, alcoholic beverages, fruits, and nonalcoholic beverages	Thompson et al. (2006)
5.	Flavonoids	Epicatechin, hesperidin, isorhamnetin, kaempferol, myricetin, narinfin, nobiletin, oroanthocyanidins, quercetin, rutin, tangerertin	Seeds, citrus fruits, olive oil, tea, and red wine	Middleton et al. (2000)
6.	Monoterpens	Geraniol, limonene	Orange, citrus peel oils, cherry, spearmint, caraway, lemongrass	Crowell (1997)
7.	Phytosterols	Beta-sitosterol	Nuts and seeds, plants oils, fruits, and vegetables	Weihrauch and Gardner (1978)
8.	Organosulfides	Allicin, glutathione, indole-3-carbinol	Garlic	Srivastava et al. (1997)
9.	Stylbenes	Pterostilbene, resveratrol	Polygonum cuspidatum	Aggarwal et al. (2004); Kundu and Surh (2008); Ulrich et al. (2005)

(Continued)

Table 1.1: Main classes and source of phytochemicals. (*cont.*)

S. no.	Main Group	Examples	Main Sources	References
10.	Triterpenoids	Ursolic	*Rosa woodsii*, *Prosopis glandulosa*, *Phoraderndran juniperinum*, *Syzygium claviflorum*, *Hyptis capitata*, *Ternstromia gymnanthera*	Kashiwada et al. (1998)
11.	Xanthophylls	Astaxanthin, beta-crytoxanthin	Yellow corn, microbial xanthophylls, alfalafa, pimiento pepper, dehydrated lettuce meal clover meal	Swallen and Gottfried (1942); Bhosale and Bernstein (2005); Brambila et al. (1963)
12.	Isoflavones	Diadzein, genistein	Soybeans and soy foods, leguminous plants, fruits, whole grains, clover, and oilseeds	Kaufman et al. (1997); Kurzer and Xu (1997)
13.	Hydroxycinamic acids	Chicoric, coumarin, ferulic acid, scopoletin	Citrus fruits, Brassica oilseed, corn flour, raspberries, plums, and umbelliferous vegetables	Ho (1992)
14.	Lignans	Sylamarin	Nuts and oilseeds, cereals and breads, legumes, fruits, vegetables, soy products, processed foods, alcoholic and nonalcoholic beverages, and flax seed	Thompson et al. (2006)
15.	Monophenols	Hydroxytyrosol	Fruits and vegetables, seeds, cereals, berries, wine, tea, onion bulbs, aromatic plants, and olive oils	Goya et al. (2007b); Dimitrios (2006)
16.	Isothyiocynates, thiocynates		Caper, eruca sativa, wild mustard (*Brassica napus*)	Esiyok et al. (2004); Morra and Kirkegaard (2002)
17.	Polyphenols	Curcumin	Turmeric, black berry, raspberry, plum, cherry, yellow onion, apple, apricot, tomato, red wine, green tea, beans, soy flour, peach	Manach et al. (2004)
18.	Others	Capsaicin, ellagic acid, gallic acid, rosmarinic acid, tannic acid		

2 Phytochemicals and Health

Since the beginning of human civilization diseases have affected humans and their livestocks. The only medicine they had was the food they ate. Plants and animal-based food was the main source for food and medicines. Today, due to pollution, contamination, and adulteration in food, stressful hectic work routines, busy lifestyles, and heavy dependence on packaged food have led to many diseases. India is one of seven hot spots in terms of biodiversity of flora and fauna. The land has provided habitat rich in medicinal plants and 30% of the world's cattle, out of which 7500 plants species have proven medicinal values (Kirtikar and Basu 1918). Numerous studies have been published on the antimicrobial activities of plant extracts against different types of microbes (Dorman and Deans, 2000; Shan et al., 2007). Stems of *Fadogia agrestis* showed the presence of saponins, steroids, terpenoids, flavonoids, tannins, anthraquinone, glycosides, and alkaloids. Extracts demonstrated antibacterial activity against *Staphylococcus aureus*, *S.* spp., *Bacillus subtilis*, and *Escherichia coli* (Ameen et al., 2011; Yakubu et al., 2005). According to the report *US Cancer Statistics: 2007 Incidence and Mortality*, published by the Centers for Disease Control and Prevention, every year more than half a million Americans lose their lives to cancer and more than this to heart disease (Jemal et al., 2007). Cancer is the second leading cause of death in the United States of America after heart disease. Tables 1.2A and 1.2B show the major types of cancer affecting males and females in the USA.

Humans have always searched for drugs to prevent diseases. Like a secular tradition, herbal plants have always been used by every culture and country for primary health care. The ultimate source of drugs is medicinal plants and herbs, which are abundant in nature. But what matters is that part of the plant should be extracted at the right time, the right season, and the right stage of its growth (Shahid-Ud-Daula and Basher, 2009). Pathogens and diseases have affected humans and livestocks since the beginning of time. Humans have always searched and needed drugs to prevent diseases. Like a secular tradition, herbal plants have always been used by every culture and country for primary health care. Antibiotics are losing their edge in the fight against diseases and pathogens. Many antibiotic resistance microbes like quilone and ciprofloxacin resistance

Table 1.2A: Major types of cancers among men in America.

S. no.	Most Common Cancer Among Men	Description
1.	Prostate cancer	156.9; First among all races and populations + Hispanic males
2.	Lung cancer	80.5; Second among white, black, American-Indian, and Asia/Pacific island men
3.	Colorectal cancer	52.7; Second for Hispanic populations and third among white and black, American-Indian, and Asian/Pacific islander men

Table 1.2B: Major types of cancer among women in America.

S. no.	Most Common Cancers Among Women	Description
1.	Breast cancer	120.4; This is first among the women of all races + Hispanic population
2.	Lung cancer	54.5; Second for white, black, and American-Indian but third for Asian/Pacific islander + Hispanic women
3.	Colorectal cancer	39.7; Second for Asian/Pacific islander + Hispanic women, third for white, black, and American-Indian women

Pseudomonas aeruginosa (QCPRA), methicillin-resistant *Staphylococcus aureus* (MRSA), penicillin-resistant *S. aureus* (PRSA), vancomycin resistant *Enterococcus* (VRE), pose a challenge to our well being. Many food-borne pathogens, such as *E. coli*, *Salmonella*, and *Campylobacter*, are responsible for diarrhea and gastroenteritis that have resistance to antibiotics. Sexually transmitted bacteria responsible for gonorrhea, penicillin-resistant *Streptococci* causative agent for pneumonia, microbes responsible for tuberculosis, influenza, HIV, and malaria all have become antibiotic-resistant (Haydel et al., 2008).

2.1 Package Food and Health Issues

Packaged foods have become part of our daily lives and there is a huge dependence on them (Mamur et al., 2012). To add shelf life, preventing food from spoiling and achieving desired color, taste, and texture, chemical additives and preservatives are added to the packaged food (Mamur et al., 2012). Additives and preservatives like SB, ST, SN, OA, SC, and BA, are used on a very large scale in packaged food. It has been confirmed from various studies that these food preservatives have done more harm to human health than serving any good (Gamze et al., 2014). Regular consumption of food additives above the acceptable daily intake (ADI) promote cancers, aging, asthma, ulcerative colitis, kidney stones, urinary problems, hypertension, and disturb normal metabolic reactions. Many metabolic reactions generate ROS and free radicals, neutralized by the body's efficient self-defense mechanism to maintain cellular homeostasis (Halliwell and Gutteridge, 2007b). The presence of reactive species in excess causes oxidative damage to cellular biomolecules. DNA, proteins, and polyunsaturated fatty acids (PUFA) present in membranes are the most important biomolecules of any cells. Any damage to them can lead to serious problems and diseases (Rajesh et al., 2013). Cell membranes play an important role in cell adhesions, cell signaling, ion conductivity, and cell potential. Any damage to cell membranes can lead to cell death (Pagán and Mackey, 2000). DNA stores the information that governs all the functions of somatic and germ cell lines. DNA during the division of the cells undergoes replication and this information is then translated into proteins, which are part of all kinds of biochemical and physiochemical reactions occurring intracellular or extracellular. Therefore, any kind of damage to DNA molecules can alter the normal reaction and fundamental process of the both somatic and germ line cells, which are the unit of life (structural and functional). And any kind of biomolecular or organelle damage due to exposure to a number of endogenous and exogenous agents over a period of time affects the normal functioning of the cells. The aim of this study is to analyze the protective effect of phytochemicals. As the body has its own defensive mechanism to neutralize ROS, but with the consumption of packaged food containing chemical additives, oxidative stress can not be lowered by the natural defense system of body (Droge, 2002). Antioxidants must be supplied exogenously. And the phytochemicals obtained from plants can be commercially exploited because of their defensive role in maintaining cellular homeostasis. SB helps stop the fermentation or acidification of foods and can be found in sodas and many fruit juices (Saad et al., 2005), and when get mixed with vitamin C, it can create benzene, a known carcinogen (Clauson et al., 2003). Preservatives like SNs and nitrites are used in meats (ham and bacon), and gives hot dogs their red coloring. The American Cancer Society recommends avoiding consumption of processed meats containing nitrites, because it is linked to asthma, nausea, vomiting, headaches, and cancer (Kilfoy et al., 2011). Preservative benzoic acid on the other hand is associated with damage to the nervous system, asthma, and increased hyperactivity in children (Clauson et al., 2003). BA is used in processed foods like cheeses, varied sauces, margarine, fruit juices, carbonated beverages, and meats. ST and sulfites are used to prevent fungal spoilage and browning of peeled fruits and vegetables and are responsible for causing allergic reactions (Vally et al., 2009). SC should be avoided by people suffering with kidney disease, heart disease, high blood pressure, a history of heart attack, urinary problems, swelling (edema), or chronic diarrhea (such as ulcerative colitis, Crohn's disease) and should avoid the use of SC as it can increase the chances of all these mentioned diseases. OA is responsible for kidney stones in many patients. It is used in industry as a bleaching agent and for rust removal. In the body, OA can combine with calcium in the kidneys to form kidney stones in susceptible people. OA is poisonous when consumed in high quantities, so people with certain health conditions should avoid high oxalate foods.

3 Phytochemicals and Health Benefits

3.1 Alkaloids

Alkaloids are plant products that have a great impact on the social, economic, and political matters for a long time. These are major players in the field of therapy and include agents like atropine, morphine, quinine, vincristine (Fazel et al., 2008). Researchers are working on *Xylocarpus grantum* root bark containing, alkaloid *N*-methyl-flindersine and many inorganic compounds like Na^{+}, K^{+}, Ca^{++}, Cl^{-}, and Mg^{++} in leaves. Jordan et al. (1991) found anticancerous effects of *vinca* alkaloids. They took five *vinca* alkaloids and studied inhibitory activity against a proliferation of cancer. The antiproliferative activity of *vinca* alkaloids was based on the observation of the inhibition of cell growth by arresting cells at metaphase even at the lowest effective concentration with almost nil microtubule depolymerization and spindle disorganization instead by altering the dynamics of tubulin at the end of spindle microtubules. Ergot alkaloids are one type of alkaloids that find clinical use for the treatment of complicated problems like uterine atonia, postpartum bleeding, sensile cerebral insufficiency, hypertension, migraine, and so on (Kren, 1997; Mukherjee and Menge, 2000). Ergot alkaloids are produced from fungus *Claviceps purpurea.* In the field of agriculture pure alkaloid standards are used to investigate the presence of alkaloid and glycoalkaloid in agricultural products like potatoes as these also cause acute toxicity. All those new varieties of potatoes to be commercialized first and have to be ensured to be free of acute toxicity of alkaloids. Two important sources of alkaloid for standards are glycoalkaloid alpha chaconine and alkaloid solanidine (Bushway et al., 1987). Another important bioactive alkaloid is tetrahydro-beta-carbolines, which are mainly present in the mammalian tissues, fluids, and brain (Herraiz and Galisteo, 2003), but nobody is sure about their biological origin. Some fruits and juices contains 1-methyl-1,2,3,4-tetrahydro-β-carboline-3-carboxylic acid and 1,2,3,4-tetrahydro-β-carboline-3-carboxylic acid, generally in citrus fruits 1-methyl-1,2,3,4-tetrahydro-β-carboline (mainly found in tomato juice, tomatoes, and kiwi) and 6-hydroxy-1-methyl-1,2,3,4-tetrahydro-β-carboline. All the fruits that contain these alkaloids are good sources of antioxidants and show good free radical scavenger activity. Piper retrofractum fruits are sources of piperidine alkaloids, namely piperoctadecalidine and pipereicosalidine (Wong et al., 1992). Purine alkaloids are one of the main phytochemicals and caffeine, theobromine and theeophyline are the main flag bearers of this category. Caffeine is most abundant in coffee, tea, and yerba; on the other hand, cocoa seeds are the most abundant source of theobromine. Purine alkaloids caffeine and theobromine from the fruits of tea *Camellia sinensis* L. were studied by Suzuki and Waller (1985). They found that caffein amounts vary with the growth and growing season till the complete ripening of fruits. Taste, color, and flavor of coffee and tea make them good and bad. Suzuki and Waller (1985) also quantified the purine alkaloid content in the fruit. Dry fruit's pericarp contains the maximum with seed coat, fruit stalk, and the seed in decreasing level respectively.

3.2 Anthocyanins

Out of various phytochemicals being colorful anthocyanins are most attractive and widely recognized. These are most abundant in fruits and vegetables (Wang and Stoner, 2008). Anthocyanins are plants pigments that belong to the flavonoid group of phytochemicals. Main sources of anthocyanins are teas, wines, vegetables, nuts, olives, honey, cocoa, and cereals. Commercially, anthocyanins are produced from grapes, elderberry, red cabbage, roselle, and so on (Bridle et al., 1996). The red color of wines, which evolves on aging, is also because of anthocyanins present in the grapes skin (Ribereau, 1974). Anthocyanin absorbs light at 500 nm because of the presence of conjugated bonds in their structures and give purple, blue, and red colors to vegetables and fruits (Wang and Stoner, 2008). Anthocyanins are bioactive phytochemicals with strong antioxidant activity and free radical scavenging activities (Tsuda et al. 1997, 2004a,b). Anthocyanin's rich mixture of flavonoids provide protection to DNA and reduce estrogen activity, inhibition of enzymes, and improve immunological responses by enhancing the production of cytokines, reducing inflammation and lipid peroxidations, and strengthening the membrane by decreasing the permeability of capillary and their fragility (Lila, 2004). Tsuda et al. (2004b), when mice were fed with a high fat diet with anthocyanins extracted from purple corn, found effective inhibition of the increase of adipose tissue and body weight gain. Lila (2004) found that biological activity of anthocyanin pigments present in the human body are either phytochemical dependent or almost never independent. Generally, anthocyanin and other flavonoid components or nonflavonoid components provide full benefit when they work synergistically. Plants that have a rich source of anthocyanin have a complex phytochemical cocktail and these are products, which are produced by the plants as attractive agents and in their defense from pathogens and predators. Anthocyanins are present in flowers, fruits, and vegetables and show very good oxygen radical absorbing capacity. Wang et al. (1997) determined antioxidant activity of 14 anthocyanins and found kuromanin had the highest antioxidant activity among them. Today anthocyanins are a hot topic for research because of health benefits. Anthocyanins act as nutraceuticals because of their antioxidant effects and are used in therapy because of their role in treating cardiovascular diseases, cancers, and even HIV-1 (Lila, 2004; Talavéra et al., 2006; Zafra-Stone et al., 2007). In vitro experiments have proved that anthocyanins are good antioxidant with beneficial effects in humans but the real potential can be achieved when researchers understand their in vivo bioavailability with functions. If researchers look at the industrial point of view, especially the wine industries where grapes and other fruits are used, which are good sources of anthocyanins, they are influenced by the phenomena of copigmentations.

3.3 Cartenoids

Color adds value to the food and it is the color of food that gives the first impression about its quality and taste. Color makes food more tempting and helps in fulfilling expectations. There are many reasons that are associated with the addition of color to food, like to compensate color lost during processing of food, support the already existing colors, to tackle color-based quality variations, and to add color to uncolored food (Mortensen, 2006). Carotenoids are one of the most important pigments and natural colorants. Carotenoids are lipid soluble and are generally yellow, orange, and red pigments found among all higher plants and in few animals (Mortensen, 2006). Carotenoids are classified in two categories; one is made up of only carbon and hydrogen, and other has carbon, oxygen, and hydrogen and are called carotenes and xanthophyls, respectively. European legislation has set guidelines that carotenes can be obtained from plants we eat and there is specifically mention of carrots, oils obtained from plants like from oil palm fruits, alfalfa, and some grasses. The main types of carotenes are alpha carotenes and beta carotene. Lycopenes is another important carotene. Lycopene's concentration is highest in tomatoes, with 28–42 micrograms per gram, which increases up to 86 – 131 micrograms per gram of weight in juices and sauces (Rao et al., 1998). Lycopene is the precursor for the beta form of carotenes and generally found in the plants containing beta carotenes. Single lycopene molecules can neutralize oxygen molecules and one lycopene molecule can scavange more than one ROS because of the number of conjugated double bonds (Krinsky and Johnson, 2005). This makes lycopene a good candidate for chemopreventive and chemotherapeutic drugs. Lycopene has good antiprostate cancer activities based on antiproliferative and proapoptotic properties as this isomer gets accumulated within the prostate region (Krinsky and Johnson, 2005). Lutein is another common carotenoid and obtained from Aztec marigold. Annatto tree seeds are the source for the colorant annatto. Use of annatto is more restricted in the European Union than in the United States. Paprika obtained from fruits pod of capsicum annuum is a well-known spice and is used as a colorant more than as a spice. Paprika is the source of various pigments and, most importantly, one is capasnthin and is almost 50% of the total pigments present in the paprika (Minguez-Mosquera and Hornero-Mendez, 1993). Saffron is also an important source of carotenoids and whole stigma is added to the food for taste and color.

3.4 Xanthophylls

Xanthophylls are yellow pigments that are one of the important divisions of the carotenoid group. The word xanthophylls is made up of the Greek word xanthos, meaning yellow, and phyllon, meaning leaf. The major difference between xanthophylls and carotenes is that xanthophylls contain oxygen atoms in the form of a hydroxyl group or epoxides while carotenes are molecules with only hydrocarbons and no oxygen. Yellow corn contains various pigments and primarily xanthophylls and some amount of carotenes (Swallen, 1942). Xanthophylls are concentrated at leaves like all other carotenoids and modulate the light energy. Xanthophylls found in all other animals or humans or other dietary animals are only plant derived. The three main types of oxygenated carotenoids in human diets are lutein, zeaxanthin, and cryptoxanthanin and their concentration in the human blood is high. In algae and vascular plants xanthophylls pigments play various important structural, as well as functional roles (Niyogi et al., 1997). Xanthophylls are found in all photosynthetic eukaryotes in bound form generally with chlorophyll molecules and proteins present in the integral membranes.

3.5 Coumestan

Coumestan are derivatives of coumarin and forms the central core for a variety of natural compounds. Some of the phytochemicals also have oestrogenic properties and are termed as phytoestrogens and mainly belong to the flavonoids groups (Kuhnau, 1976). These are called phytoestrogen as they have same effect on the central nervous system of human as estrogen. Main sources of coumestans are split peas, pinto beans, lima beans, alfalfa, and clover sprouts. Coumstans are one of the three classes of falvonoids that are termed as phytoestrogens, namely coumestans, prenylated flavonoids, and isoflavones. Coumestans have physical and chemical properties similar to isoflavones (Humfrey, 1998). Phytoestrogens have nonsteroidal structures, which makes them resemble mammalian estrogens. They can bind to estrogen receptors (ER) in both agonists and antagonists for estrogen (Jenkins et al., 2002). According to Thompson et al. (2006), phytoestrogens like coumestan, lignans, and isoflavones may be potential candidates to treat cancers associated with hormones.

3.6 Flavonoids

Vegetables, berries, and fruits and beverages are good sources of flavonoids and are associated with reducing the risks of a number of diseases. Flavonoids have shown positive effects on the immune system both in vitro, as well as in vivo (Middleton and Kandaswami, 1992). Flavonoids are phytochemicals of low molecular weight, have three-ring structures, and are of various types based on the different substitutions (Middleton et al., 2000). Flavonoids have several important roles in plants as antimicrobials, antioxidant, attractors, light receptors, and many other biological activities (Pietta et al., 2000). The main possible mechanism is their antioxidant activity. Antioxidants evolved as an important part of natural defense mechanisms among living organisms (Jovanovic et al., 1994). These are the molecules that scavenge the free radical species and inhibit the chain reactions that can damage vital molecules of living organisms. Though they are very beneficial, one antioxidant molecule interacts with one free radical so they should be replenished to meet the constant challenge posed by various free radicals. Flavonoids have various biological activities. Today, flavonoid-based products are flushing in the market. For example, propolis is the material bees use to protect their hives. This propolis has various biological activities like antibacterial, antiviral, antiinflamatory, and also anesthetic properties. Of more than 150 components present in them, flavonoids are the major player (Chang et al., 2002). Animal and cellular studies confirm that flavonoids inhibit cancer proliferation. These studies are conducted with high concentrations of flavonoids, but will they help in the same way when tested on humans. Would humans be able to cope with the high concentration of flavonoid? These questions have to be answered. Intake of flavonols and flavones can reduce the chances of heart disease, like myocardial infarction and strokes that increase the phytochemical productivity employs the basic understanding of genes regulating and controlling the pathways responsible, and lead to the synthesis of these compounds in fruits and other vegetables. Biochemical and molecular techniques employability enhances the production of phytochemicals.

3.7 Isoflavones

Isoflavones are a subclass of flavonoinds and are scarcely distributed in nature. Soybeans are the main source of isoflavones and soy foods are consumed on a high level in Asian countries, mainly Pacific (Russo et al., 2010). Out of various isoflavones, genistein has the maximum percentage among leguminous plants and helps in fighting various kinds of cancers. Breast and prostate cancers among humans are the most common types. Genistein has a strong role as anticancerous biomolecules against breast and prostate cancers (Lampe et al., 2007). Genistein phytochemicals and their analogous molecules resemble estrogen hormones, molecular structure-wise and is the reason they are also called "phytoestrogen." ERs, like ER alpha and ER beta, receive various estrogens, like hormones, like 17-beta estradiol E2, which by the use of these receptors, act on the estrogen dependent tissues, like the uterus, ovary, and breast. ER alpha is associated with the growth effects of estrogen and found mainly in the uterus and liver while ER beta have antiproliferative properties and are found mainly in the ovary (Russo et al., 2010). Genistein inhibits growth of most types of hormone dependant and independent cancer cells (Chang et al., 2009).

3.8 Monoterpenes

Monoterpenes are phytochemicals of C_{10} representation of terpenoid family. Gershenzon et al. (1989) studied biosynthesis and catabolism of monoterpenes. Russin et al. (1989) studied the effects of monoterpenoids in inhibition of mammary carcinogenesis in rats. In this study the authors focused on limonene and oxygenated [(−)-menthol] and nonoxygenated (rf-limonene) monocyclic forms, oxygenated (1,8-cineole) and nonoxygenated [(±)-a-pinene) bicyclic forms and oxygenated [(±)-linalool and nonoxygenated 03-myrcene] acyclic forms. Dietary feed of each of the monocyclic terpenes, D-limonene or (−)-menthol resulted in a significant inhibition of mammary carcinogenesis. Out of all of them menthol was more potent even to limonene. Monoterpens are also used as fumigating agents. Lee et al. (2003) found monoterpenoids in fumes form; they show antipest effects and can be used to kill pests and insects for several stored products. Concentrations of 50 μg/mL in air caused 100% mortality in rice weevil, *Sitophilus oryzae*, the red flour beetle, *Tribolium castaneum*, the sawtoothed grain beetle, *Oryzaephilus surinamensis*, the housefly, *Musca domestica*, and the German cockroach, *Blattella germanica*, cineole, *l*-fenchone, and pulegone.

3.9 Phytosterols

Phytosterols are plant sterols. For a long time their biological role was underestimated in mammals. But in 1983 it was found that phytosterols are effective in treating patients with hypercholesterolimic (Bouic, 2001). In a clinical review written by scientist Bouic (2001), it was mentioned that phytosterols have immunological activity in animal models suffering from inflammation and colorectal and breast cancer in vivo and in vitro. Phytosterols are the main component of plant membranes and free phytosterols help in stabilizing phospholipid bilayers in plants as same as the role played by cholesterol in animal cells (Moreau et al., 2002). When consumed by humans phytosterols present in diets, they act as cholesterol-lowering agents (Moreau et al., 2002).

3.10 Organosulphides

Garlic is the most important source of organosulphides of various types (Srivastava et al., 1997). Srivastava et al. (1997) studied how organosulfides diallyl sulfide (DAS), diallyl disulfide (DADS), diallyl trisulfide (DATS), dipropyl sulfide (DPS), and dipropyl disulfide (DPDS) are effective against benzo(*a*)pyrene (BP)-induced cancer in mice by modulating enzymes involved in BP activation/inactivation pathways. Jakubikova and Sedlak (2005) also studied organosulfides and their mechanism for inducing cytotoxicity, apoptosis, arresting of cell cycle, and oxidative stress in human colon carcinoma cell lines (Caco-2 and HT-29 colon carcinoma).

3.11 Stylbenes

Stylebenes, like resveratrol, are phytochemicals responsible for antiaging properties (Baur and Sinclair, 2006). Asian medicine very commonly uses the extract of *Polygonum cuspidatum*, which is a rich source of stylbenes like resveratrol (Aggarwal et al., 2004).

3.12 Triterpenoids

Triterpenoids are secondary metabolites of plant origin and can be obtained from marine and terrestrial plants (Mahato and Sen, 1997). They can also be obtained from nonphotosynthetic bacteria and this created interest among scientists studying evolution. Androgen-associated diseases, such as prostate cancer, acne, hirsutism, benign prostatic hyperplasia (BPH), are serious problems affecting males nowadays (Culig et al. 2002; Wasser and Weis, 1999). BPH affects more than 90% of total males 80–90 years of age.

3.13 Hydroxycinnamic Acid

Other important phytochemicals are a class of polyphenols with C_6–C_3 skelton and hydroxy derivatives of cinnamic acid. Hydroxycinnamic acid compounds are widespread among plant kingdoms and are important as they are an important source for antioxidants. Chen and Ho (1997) studied antioxidant activities of caffeic acid phenethyl ester (CAPE), caffeic acid (CA), ferulic acid (FA), ferulic acid phehethyl ester (FAPE), rosmarinic acid (RA), and chlorogenic acid (CHA), compared to the antioxidant activities of alpha tocopherol and butylated hydroxyanisole (BHT). They performed the rancimat test with all these compounds and the time of lipid oxidation with these compounds was, in decreasing order, CA≈alpha tocopherol >CAPE ≈ RA >CHA > BHT > FA ≈ FAPE. But when lipid substrate changed to corn oil the order of RA > CA CAPE CHA > α-tocopherol > BHT; FA and FAPE antioxidant effect was nil in the corn oil system. Gallardo et al. (2006) studied the hydroxycinnamic acid composition and in vitro antioxidant activity of selected grain fractions. Soluble extracts from rye, buck-wheat, and wheat were used to determine the composition of hydroxycinnamic acid and their antioxidant activities. To determine soluble, insoluble, and free hydroxycinnmic acids composition HPLC-diode aray (DAD) was used. Rye bran and wheat bran fractions has the highest level of hydroxycinnamic acid whereas only traces of the same quantity were observed and noticed in the flour from buckwheat. When tested for antioxidant activities all cereal fraction water extracts have somewhat the same. Among them buckwheat- and wheatgerm-based products have the highest antioxidant acitvites while rye products were the lowest.

3.14 Lignans

Lignans are ubiquitous in angiosperms and gymnosperms. Lignans show various bioactive properties, like antitumor, antimitotic, and antimicrobial, especially against viruses, and inhibit some enzymes (MacRae et al., 1989). Olive oil is an important staple diet of Mediterranean region. It is a rich source for lignans as a major component in the phenolic fraction of olive oil (Owen et al., 2000). Phenolic antioxidants were isolated and purified and subjected to structural analysis using several spectroscopic techniques like mass spectrometery (MS) and nuclear magnetic resonance (NMR). It was found that (+)-1-acetoxypinoresinol and (+)-pinoresinol are an important lignan component. In this study they concluded that lignans may act as modulators in the cancer chemopreventive activity (Owen et al., 2000). In mammalian lignan precursors, like secoisolariciresinol diglycoside (SD), the flaxseeds are the richest source for it and have been found to help in decreasing the colon cancer (Jenab and Thompson, 1996). Similar kinds of study were conducted by Rickard (1996), where they studied the dose-dependent production of mammalian lignans in rats and in vitro, from the purified precursor secoisolariciresinol diglycoside in flaxseed. Mammals produce lignans enterodiol (ED) and enterolactaone (EL) by the action of their colonic bacteria on lignans precursors present in their diets. Good sources of lignan precursors are flaxseed and have secoiolariciresinol diglycoside (SDG) (Rickard, 1996). In his study Rickard (1996) studied the various parameters. First was focused on SDG present in flaxseed as not the only source for all the lignans produced; second, SDG produced in mammals was related to the dose intake; and the third was focused on finding any relation between in vitro production and in vivo urinary excretions. Lignan obtained from plants like lariciresinol, matairesinol, pinoresinol, syringaresinol, arctigenin, 7-hydroxymatairesinol, isolariciresinol, and secoisolariciresinol, are metabolized by human fecal microflora so their properties were studied and quantified using HPLC (Puupponen-Pimiä et al., 2005).

3.15 Monophenols

Among monophenols the most important one is hydroxytyrosol, which is also called 3,4-dihydroxyphenylethanol (DOPET). It is one of the most important phytochemicals as it is a very strong antioxidant. One of the important sources of monophenols, like hydroxytyrosol, is olive oil. In vivo studies show that hydroxytyrosol have antioxidant activities like quercetin. Human- and animal-based studies proved the bioavailability of hydroxytyrosol; it is absorbed in the intestine on ingestion and becomes metabolized (Goya et al., 2007a). Bulotta et al. (2014) found that hydroxytyrosol, which is a natural antioxidant, prevents the protein damage induced by long-wave ultraviolet radiations in melanoma cells. Long-wave ultraviolet radiations generate free radical and ROS, which in turn damages the skin. They studied the effect of hydroxytyrosol present in olive oil on the viability and redox status of

HepG2 cell lines and how hydroxytyrosol protects cell lines from oxidative stress caused by tert-butylhydroperoxide (t-BOOH); 10–40 μM HTy treatment for 2 – 20 h prevented the cell damage by tert-butylhydroperoxide (t-BOOH).

3.16 Isothiocynates

Raw food from plants like vegetables and fruits are good sources of secondary metabolites of plants origin, which supports good health by adding more nutrition than basic. Best examples are glucosinolates, from Brassicaceae family. Plants use this secondary metabolite for the synthesis of defensive molecules against herbivores by imparting bitter or sharp taste. To date, more than 120 glucosinolates have been identified among plants (Russo et al., 2010). Many studies on animals also showed anticancerous properties of isothiocynates by inhibiting cancer's initial stages (Kuroiwa et al., 2006) and have anticancerous effects (Hecht, 2000). Phytochemicals can be the potion to tackle cancers and represent very optimistic candidates for chemotherapy and chemoprevention.

3.17 Polyphenols

Curcumin is one of the most important members of the polyphenol phytochemical family. It is derived from the rhizome of the *Curcuma longa* or turmeric. This herb is cultivated mainly in Asian countries on a large scale. Each and every Indian family uses turmeric powder in their meal for flavor and color. Curcumin contains yellow-pigmented fractions of curcumin, demethoxycurcumin, bisdemethoxycurcumin, and cyclocurcumin (Yang et al., 2004).

3.18 Capsaicin

As the name suggests, capsaicin is obtained from plants belonging to the genus *capsicum* and is an active ingredient of many hot and spicy foods. It is one of the components of chili peppers and is an irritant to mammals, responsible for burning sensations to the tissues it contacts. Pain is the outcome of chemical, mechanical, or thermal stimuli, which activates peripheral subgroups of sensory neurons, which are called nociceptors (Fields et al., 1991). Nociceptors transmit information to the brain and spinal cord. These are also very sensitive to capsaicin. Generally, the body responds to pain and releases local inflammatory mediators on exposure to capsaicin to the nociceptor terminals, but prolonged exposure can lead to insensitivity of nociceptor terminals.

3.19 Health-Supporting Properties of Phytochemicals and the Mechanism Behind It

Phytochemicals show antimicrobial, antioxidant, and many health-boosting effects (Table 1.3).

Table 1.3: Different types of phytochemicals and their biochemical properties.

Molecular Formula	Phytochemical	Economical Values	HBA	HBD	Pharmacological Effect	Molecular Mechanism	References
$C_6H_{14}O_1$	3-hexenol	It occurs naturally in the flavor and aroma of plants like pineapple. Used as a food additive to add flavor.	1	1	Antimicrobial	Affects the molecular arrangements of the microbes. Maximum antimicrobial activity was observed with hydrophobic chain length from hydrophilic hydroxyl group.	Kubo and Kinst-Hori (1999)
$C_{10}H_{16}$	Alpha-thujene	Natural organic compound classified as a monoterpene.	0	0	Antimicrobial	Disturb molecular arrangement of the microbes.	Cosentino et al. (1999)
$C_{10}H_{16}$	Alpha-pinene	Monoterpenic in nature.	0	0	NK activity enhancer, antistress, anticancer	Acts on reactive oxygen species; fragrance has alleviating effects.	Kose et al. (2010); Akutsu et al. (2002)
$C_{10}H_{16}$	Camphene	Camphene is used for making fragrances and food additives and fuels. It is also an explosive agent.	0	0	Reduces plasma cholesterol and triglycerides, prevents hyperlipidemia. Important part of various essential oils, such as turpentine, cypress oil, camphor oil, citronella oil, neroli, ginger oil, and valerian.	Inhibits lipoprotein lipase, HMG-Coa reductase.	Vallianou et al. (2011)
$C_{10}H_{16}$	Myrcene	Olefinic nature organic compound part of various essential oils and intermediate in the production of various fragrances.	0	0	Antioxidative properties, intermediate of various fragrances, has analgesic effects and antiinflamatory properties, sedatives.	Increases GSH, CAT, GSH-Px, and Cuzn-SOD. Antiinflamatory actions through prostaglandin E2. Analgesic actions by blocking of nalozone and yohimbine, alpha 2-adrenoceptor stimulated release of opioids.	Ciftci et al. (2011)

$C_{10}H_{16}$	Delta 3 carene	A sweet and pungent smell, water-insoluble, bicyclic monoterpene. CNS depressant.	0	0	Dry excess fluids, tears, running noses, excess menstrual flow and perspiration,	CNS depressant.	—
					Antibacterial.	Affects molecular arrangement of the microbes.	Massumi et al. (2007)
					Larvicidal, mosquito repellents, insecticides, repellents, and antifeedants.	Multiple and novel target sites, affects the growth rate, reproduction, and behavior of insects.	Giatropoulos et al. (2013)
$C_{10}H_{16}$	Limonene	Colorless cyclic terpene with smell of oranges. It is used in the preparation of renewable cleansers. Biologically D-limonene is more available and commercially obtained by centrifugal separation and steam distillation. It is common in parts of various cosmetic products and medicines and natural falvoring agent. Biofuels.	0	0	Antifungal and mosquito deterrent.	Multiple and novel target sites, affects the growth rate,	Bhat et al. (2011)
						reproduction, and behavior of insects.	Giatropoulos et al. (2013)
						Inhibits gamma-interferon and IL-4.	Aggarwal et al. (2009)
					Remedies for cerebral disease.	Inhibition of acetylcholinesterase	Bhat et al. (2011)
					Remedy for colds, coughs, pleurisy, wind, colic, rheumatism, and diseases of the urinary organs, skin rashes, wounds, rheumatism, and toothaches.	Diffused through the entire pulmonary region.	Bhat et al. (2011)
					Cosmetic products, used as flavoring agent. Natural reliever to gastroesophageal reflux disease and heart-burn.	Masks the bitter taste of flavonoids.	Sun (2007)
					Hepatoprotective.	Inhibits the malondialdehyde formation.	Bhat et al. (2011)
					Carminative, diaphoretic, diuretic, antiseptic, and antidepressant.		Bhat et al. (2011)

(*Continued*)

Table 1.3: Different types of phytochemicals and their biochemical properties. (*cont.*)

Molecular Formula	Phytochemical	Economical Values	HBA	HBD	Pharmacological Effect	Molecular Mechanism	References
$C_{10}H_{16}$	Gama-terpinene	Commonly used in perfumes and cosmetics, and also as flavoring agent in food.	0	0	Antifungal, antioxidative, Cosmetics, flavoring agent.		Singh et al. (2004) Wang et al. (2009)
$C_{10}H_{14}O$	Verbenone	It is used for insect control, especially pine beetle, which is a major threat to pine trees. It does have pleasant smell and used in perfumery, aromatherapy, and herbal remedies.	1	0	Significant enhancement of performance and overall quality of memory.	Antioxidant and antiacetylcholinesterase activities.	Santoyo et al. (2005); Mata et al. (2007)
$C_{10}H_{18}O_1$	Linalool	Natural terpene alcohol found in spices and herbs. It is used in toiletries, cleaning agents, perfumes, soaps, shampoos. Linalool is also used for insecticides and in the synthesis of vit-E.	1	1	Linalool is anxiolytic. Antimicrobial and good smell.	Acts upon GABA/ benzodiazepine receptor. Soothing smell and effects.	Lopez et al. (2006) Maia and Moore (2011)
$C_{10}H_{18}O_1$	Sabinene hydrate	Sabinene is natural bicyclic monoterpene. Sabinene contributes to the spiciness of black pepper.	1	2	Remedies for cerebral disease, antifungal, and mosquito deterrent. Antioxidant.	Inhibition of acetylcholinesterase. Multiple and novel target sites; affect the growth rate, reproduction, and behavior of insects. DPPH free-radical scavenging activity	Bhat et al. (2011) Bhat et al. (2011); Giatropoulos et al. (2013) Bhat et al. (2011)
$C_{10}H_{18}O_1$	Endo-fenchol	Its terpenic in nature and isomer of borneol. Has great use in perfumery.	1	1	Broad-spectrum antimicrobial activity.	Affects proteins and lipids of microbial cells.	Kotan et al. (2007)

$C_{10}H_{18}O_1$	Terpinen-4-ol	Common ingredient in perfumes, cosmetics, and flavors.	1	1	Antifungal, antioxidative,		Singh et al. (2004)
					pleasant smelling agent, common ingredient in perfumes, cosmetics, and flavors.		Yao et al. (2005)
$C_{15}H_{28}$	Germacrene	Volatile organic sesquiterpenes	0	0	Carminative, diaphoretic, diuretic, antiseptic, and antidepressant.	Affect the growth rate, reproduction, and behavior of insects.	Giatropoulos et al. (2013)
					Produced by large numbers of plants for antimicrobial and insecticides actions.	Act as pheromones.	Adio (2009)
$C_{15}H_{24}O_1$	8-Hydroxy-alphahumulene	Alpha humulene or alpha caryophyllene is monocyclic sequisterpene and found in the essential oil of humulus lupulus. It gives beer a hoppy aroma and has medicinal properties.	0	0	Humulene has antiinflamotory properties, and can decrease edema. Anticancerous.	It inhibits the cancer by inhibiting the inflammation caused by TNF alpha and IL 1beta.	Fernandes et al. (2007)
$C_{13}H_{15}NO_3$	Ethyl-12-dimethyl-5-5-Hydroxylinndole-3-coarboxylate	Used in medicinal purposes.	2	1	Medicinal uses against viral diseases like dengue and West Nile viral diseases.	Inhibits virus protease actions.	Nitsche et al. (2011)
$C_{14}H_{12}N_2O_2$	Beta-carboline-1-propionic acid		4	2	It is used in preparation of psychedelic drug brews obtained by infusion of various plants.	It prevents the breakdown of dimethyltryptamine in the mammalian gut by blocking the action of monoamine oxidase.	Venault and Chapouthier (2007)
					It increases the memory, acts as anxiogenic, and has convulsive affects.	By acting inverse agonist for benzodiazepine receptor.	Venault and Chapouthier (2007)

(*Continued*)

Table 1.3: Different types of phytochemicals and their biochemical properties. (*cont.*)

Molecular Formula	Phytochemical	Economical Values	HBA	HBD	Pharmacological Effect	Molecular Mechanism	References
$C_{11}H_{12}O_6$	Eucomic acid	It is found in louts and *Vanda teres*, *Eucomis puncata*, and is a secondary metabolite involved in leaves rolling and opening in plants.	6	4	It is used in traditional medicines to treat sore throat and bronchitis and also as paste applied externally on scorpion stings.	Antimicrobial properties and soothing effect.	Sahakitpichan et al. (2012)
					Eucomic acid and eucomate derivatives, vandaterosides I (2), II (3), and III (4), cellular antiaging properties were evaluated on human immortalized keratinocyte cell line (HaCaT).	It shows their effect on cytochrome c oxidase and increased enzymatic activity or expression without increasing the cellular mitochondrial content.	Simmler et al. (2011)
$C_{10}H_{30}O_1$	Farnesyl acetone	A terpene ketone in which an (*E*,*E*)-farnesyl group is bonded to one of the α-methyls of acetone.	1	0	Antifungal properties.	Nonselective activity against cyclooxygenase-1 (COX-1) and cyclooxygenase-2 (COX-2) whereas 3 (IC(50) 1.88 microg/mL) preferentially inhibited the enzyme COX-2.	Duru et al. (2003)
					Antidiabetic.	Antihyperglycemic and insulin secretory activity.	Jose and Reddy (2010)
					Antibacterial against *Propionibacterium acnes*.	Hydrophobic alkyl groups of this long chain alcohol have antimicrobial activity.	Kubo et al. (1994)
$C_{15}H_{10}O_5$	Apigenin	Coloring agent for wool has chemopreventive role against leukemia, important in metabolism of various drugs as prevent inhibition of various CYP2C9 enzymes.	5	3	Strong antioxidant, antiinflammatory, and anticarcinogenic properties.	Able with differential effects in normal versus cancer cells, inhibits IL-1beta and TNF.	Patel et al. (2007)
					Antihepatitis activity.	Inhibits hepatitis virus.	Cushnie and Lamb (2005)
					Anti Lassa viral activity, antiviral activity.	Inhibits binding and entry of Lassa virus into the cells.	Liu et al. (2008)
					Antimicrobial activity against bacteria.	Affects cellular molecular arrangements.	Konstanti-nopoulou et al. (2003)
					Metabolisms of various drugs.	Prevents inhibition of various YP2C9 enzymes.	Si Dayong et al. (2009)

$C_{16}H_{10}O_5$	Damnacanthal	It is an anthraquinone and has anticancerous properties.	5	1	Acts as anticancerous substance.	p56lck tyrosine kinase inhibitor.	Faltynek et al. (1995)
$C_{17}H_{14}O_4$	Dihydronortanshi-none	Broad spectrum antibacterial.	3	0	Antibacterial activity.	DNA, RNA, and protein syntheses in microbes were nonselectively inhibited by these compounds.	Lee et al. (1999)
					Used in neural system.	Natural product inhibitors of AChE.	Beri et al. (2013)
$C_{15}H_{10}O_6$	Kaempferol	Natural flavonoid found in many plants and is a good source for anticancerous drugs and cardiovascular diseases.	6	4	Antimicrobial activity against bacteria.	Affects cellular molecular arrangements.	Konstanti-nopoulou et al. (2003)
					Anticancerous and reduces the chances for cardiovascular diseases, neuroprotective, antiestrogenic, anxiolytic, analgesic, antiallergic, antidiabetic, antiinflamatory.	Glycosides of kaempferol like kaempferitrin and astragalin.	
$C_{15}H_{10}O_6$	Luteolin	Luteolin is a flavonoid, with antiskin cancerous and antiallergic activity.	6	4	Antiallergic and anticancerous.	Inhibition of degranulation and release of TNF-alpha and IL-4 in RBL-2H3 cells.	Sethi et al. (2009)
					Antiinflamatory activity.		
$C_{17}H_{23}NO_4$	Anisodamine	It is also a naturally occurring tropane alkaloid found in some plants of the Solanaceae family.	5	2	For the treatment of acute circulatory shock.	Anticholinergic and α_1-adrenergic receptor antagonist.	
$C_{18}H_{23}NO_4$	Capsaicin	Capsaicinoid and dihydrocapsaicin potent taste and nerve modulators.	4	2	Anticancerous and antiinflamatory.	Acts on STAT-3 pathway, down regulate cyclin-D1, VEGF.	Aggarwal et al. (2009)
					Antifungal, topical analgesic, ointments used in making dermal patches, antipsoriasis, antidiabetic, anticancerous, prostate cancer and lung cancer.	Binds to a protein known as TRPV1.	Glinski et al. (1991); Lejeune et al. (2003)

(*Continued*)

Table 1.3: Different types of phytochemicals and their biochemical properties. (*cont.*)

Molecular Formula	Phytochemical	Economical Values	HBA	HBD	Pharmacological Effect	Molecular Mechanism	References
$C_{15}H_{10}O_7$	Quercetin	Quercetin is a flavonoid, used in beer and beverages industries. It is used in making medicines that are antiviral, antiasthamtic, anticancer, antiexzemic, antiinflamatory.	7	5	Antiinflammatory, antioxidant, antiviral (HIV), antitoxic, free radical scavenging, cardioprotectant, hepatoprotectant, antitussive, antihemorrhagic.	Strong inhibitor of HBsAg and HBeAg secretion. Inhibit HIV-1 reverse transcriptase.	Yu et al. (2007); Wu et al. (2007)
					Antiviral, antiasthamtic, anticancer, antiexzemic, antiinflamatory.	Inhibit reverse transcriptase.	Jung et al. (2010)
					Anticancerous and antiinflamatory.	PMACI-induced activation of NF-katta B and p38 MAPK in human mast cell line HMC-1.	Aggarwal et al. (2009)
					Antimicrobial activity against bacteria.	Affects cellular molecular arrangements.	Konstanti-nopoulou et al. (2003)
$C_{18}H_{23}NO_4$	Lycoremine	Lycorine is a toxic alkaloid.	5	1	Antiherpes activity.	Kill the herpes-infected cells. Inhibitory effect on the cytopathogenic effect of viral DNA and RNA.	
					Antimicrobial activity	Anti bacterial activity by affecting ascorbic acid synthesis.	Evidente et al. (1985)
$C_{17}H_{19}NO_5$	Piplartine	Natural phytochemicals obtained from pepper obtained in south India and southeast Asia.	6	0	Anticancerous and antitumoric.	Selectively kill the cancerous cells, inhibits tumor growth with no toxicity toward healthy cells.	
					Very selectively kills the cancer cells.	Blocks tumor growth and metastasis.	Raj et al. (2011)

$C_{20}H_{32}O_3$	Ginkgoneolic acid	It is obtained from Chinese herb *Ginkgo biloba* and commercially has a strong presence in the herbal market in USA.	3	2	Anticancerous	PI-PLCgamma1 inhibitory activity. These compounds inhibit the multiplication and growth among various cancer cell lines of human origin, but do not affect normal colon cells.	Lee et al. (2004b)
					Used in the treatment of the dementia, anxiety, schizophrenia, and insufficient flow to the brain conditions.		Singh et al. (2010); Ahlemeyer et al. (2001)
					Antimicrobial.		He et al. (2013)
$C_{15}H_{12}O_8$	Dihydromyricetin	It is a flavanonol used in making hepatoprotective drugs and drugs to counter the affect of alcohol on brain.	6	8	Hepatoprotective effects. Has ability to counter effects of alcohol on brain and fight hangover.		Hase et al. (1997); Hänsel and Klaffenbach (1961)
$C_{18}H_{13}NO_5$	Decumbenine B	It's an alkaloid, used in medicines, alkaloid antitumor agent.	1	6	Antitumor	Exhibits topoisomerase II poison activity, as well as catalytic inhibition activity.	Xu (2000)
$C_{16}H_{12}O_7$	Rahmnetin	It is O-methylated flavonol and is commercially obtained from cloves and used as flavoring agent and in tooth pain relievers.	4	7	Broad spectrum. Antimicrobial.	Targets the beta hydroxyacyl-acyl carrier proteins.	Zhang et al. (2008); Konstantinopoulou et al. (2003)
					Antimutagenic.	Affects indirect mutagens like I&, B(alP, AFB1, Trp-P-1, and Glu-P-1).	
$C_{19}H_{14}NO_4$	Coptisine	An alkaloid and produces bitter taste used in making medicines for digestive disorders. Antidepressant.	0	5	Antidepressants.	Inhibition of monoamine oxidase A.	Ro et al. (2001)
					Used to treat digestive disorders.	Inhibitory effects in the proximal and the excitatory effects in the distal stomach.	Schemann et al. (2006)
$C_{21}H_{22}O_4$	Bavachinin	Its flavonoid with medicinal valuable activity against cancers.	1	4	Anticancerous.	By blocking tumor angiogenesis, as bavachinin, has potent antiangiogenic activity in vitro and in vivo	Nepal et al. (2012)

(*Continued*)

Table 1.3: Different types of phytochemicals and their biochemical properties. (*cont.*)

Molecular Formula	Phytochemical	Economical Values	HBA	HBD	Pharmacological Effect	Molecular Mechanism	References
$C_{15}H_{11}ClO_7$	Delphinidin cloride	It's an anthochanidin primary pigment from plants with strong antioxidant activities.	—	—	Antioxidant.	Protects human HaCaT Keratinocytes against UVB-mediated oxidative stress and apoptosis.	Faq et al. (2007)
$C_{20}H_{22}N_2O_3$	Perivine	It's alkaloidic in nature. It's used in treatment of infectious diseases and cancers.	2	5	Antimicrobial.	Microbial molecular arrangement disturbances.	Vijayalakshmidevi et al. (2011)
					Anticancerous activity.	Inhibit proliferation and growth of human breast cancer	Spelman et al. (2006)
$C_{20}H_{18}O_5$	Methyltanshionate	Phenanthrene quinone derivatives.	0	3	Antimicrobial and bacteriostatic only towards gram positive and	Effects on microbial molecular distrurbances.	Baricevic and Bartol (2000)
$C_{20}H_{22}N_2O_3$	Picrinine	Alkaloidic in nature, commercially used in African nations.	1	5	Spermicidal activity.	Inhibits sperm motility.	Chattopadhyay et al. (2005)
					Antiinflammatory and analgesic effect.	Inhibited proinflammatory enzymes COX-1, COX-2, or 5-LOX in vitro.	Meena et al. (2001)
$C_{22}H_{33}NO_2$	Denudatine	It's alkaloidic in nature obtained from plant *Delphinium denudatum* grown in Himalyan region. Has commercial value for treatment of piles, brain diseases, and fungal infections.	2	3	Used in treatment of aconite poisoning, piles, toothaches, fungal infections, brain diseases.	Acts on dopaminergic-D2 receptor; brain monoamines.	Rahman et al. (2002); Ahmad et al. (2006)
					Vaso relaxing effects.	Is an antiplatelet agent.	
$C_{20}H_{25}NO_4$	Wilsonine	Alkaloidic phytochemicals used in preparing antitumor medicines.	0	5	Anti-Hepa-3B hepatoma cells and also acts against oral epidermoid carcinoma.	Important agent in fight against against (KB) oral epidermoid carcinoma, as well as (Hepa-3B) hepatoma cells.	Kuo et al. (2002)

$C_{20}H_{16}N_2O_4$	Camptothecin	Quinolinic alkaloidic compounds that inhibit the DNA topoisomerase and used for treatement of cancers.	2	6	Anticancer activity.	Inhibits the DNA topoisomerase activity.	Takimoto and Calvo (2008)
$C_{21}H_{18}NO_4$	Chelerythrine	A benzophenanthridine alkaloid used as antimicrobial agent.	0	5	Antimicrobial activity.	Potent, selective, and cell permeable protein kinase C inhibitor.	Gibbons et al. (2003)
$C_{22}H_{26}N_2O_2$	Vinpocetine	Semisynthetic derivative of alkaloid vincamine. It has found use as a neuroprotective and cerebrovascular disorders treating agent.	0	4	Neuroprotective and to treat age-related memory impairment. Antiinflammatory.	Enhances the cerebral blood flow, which helps memory impairment. Inhibits the upregulation of NF-κB by tumor necrosis factor alpha.	McDaniel et al. (2003)
					Helps in treating Parkinson's disease and Alzheimer's disease.	Antiinflammatory effect, vasodilation and nootropic.	Jeon et al. (2010)
$C_{20}H_{19}NO_5$	Chelidonine	Alkaloidic in nature used in treatment of nephrotoxicity.	1	6	Antinephrotoxicity; antispasmodic and relaxant activity of chelidonine, acetylcholinesteras, and butyrylcholinesterase inhibitory compounds.	Inhibits activity against HuAChE and HuBuChE.	Koriem et al. (2013)
$C_{20}H_{19}NO_5$	Protopine	Alkaloid obtained from opium poppy and commercially used for medical analgesic.	0	6	Analgesic.	Inhibit histamine H1 receptors and platelet aggregation.	Saeed et al. (1997)
$C_{20}H_{18}O_6$	Asarinin	It is furofuran lignans and has been important part of food in the Western countries. Used in graft immune modulation and anticancerous properties.	0	6	Graft coating can enhance graft survival.	Increases the immune tolerance induced by donor splenocytes or bone marrow cells combined with asarinin on graft.	Guifang et al. (1999)
					Beneficial effects of these lignans in breast, colon, and prostate cancer.	Hormonal metabolism or availability, angiogenesis, antioxidation, and gene suppression.	Landete (2012)

(*Continued*)

Table 1.3: Different types of phytochemicals and their biochemical properties. (*cont.*)

Molecular Formula	Phytochemical	Economical Values	HBA	HBD	Pharmacological Effect	Molecular Mechanism	References
$C_{16}H_{18}O_9$	Caffeoylquinic acid	Natural polyphenolic compound with chemopreventive activities.	6	9	Chemopreventive effect and laxative effects.	Prevents and treats estrogen independent breast cancer cells.	Noratto et al. (2009); Stacewicz-Sapuntzakis et al. (2001)
					Antiviral activity, antiviral activity against respiratory syncytial virus.	Could inhibit RSV directly, extracellularly, inhibition of virus–cell fusion in the early stage, and the inhibition of cell–cell fusion at the end of the RSV replication cycle.	Li et al. (2005)
$C_{16}H_{18}O_9$	1-Chlorogenic acid	Chlorogenic acid (CGA) is a natural chemical compound, which is the ester of caffeic acid and (−)-quinic acid.	6	9	Antidiabetic and antioxidant.	Slows the release of glucose into the bloodstream after a meal.	Johnston et al. (2003)
$C_{16}H_{18}O_9$	Cryptochlorogenic acid	Its a natural polyphenolc compound chemopreventive	6	9	Food, flavouring agents and in pharmaceuticals and antimicrobial.	Taste modulators and affects microbial molecular arrangements.	Omar (1992)
$C_{16}H_{18}O_9$	Neo chlorogenic acid	It's a natural polyphenolc compound chemopreventive.	6	9	Chemopreventive and laxative effects.	Kills cancer cells.	Stacewicz-Sapuntzakis et al. (2001)
$C_{16}H_{18}O_9$	Scopolin	Scopolin is a glucoside of scopoletin by the action of enzyme scopoletin glucosyltransferase. Used in treating infection among plants.	4	9	Antimicrobial, Plant Antiviral kilss TMV	Antibiotic potentiators or virulence attenuators, reinforcement of the cell wall, biosynthesis of the lytic enzymes, and also production of secondary metabolites like scopolin.	Kuć and Currier (1975); González-Lamothe et al. (2009)
$C_{21}H_{26}N_2O_3$	Vinacamine	Vincamine is a monoterpenoid indole alkaloid used as peripheral vasodilator, which increases the blood flow to the brain.	1	5	Used to combat aging and stress.	Vasodilator, which increases the blood flow to the brain. Used as nootropic agent.	Cook and James (1981)

$C_{21}H_{26}N_2O_3$	Yohimbine		2	5	Stimulant has aphrodisiac effects, antitype two diabetics.	Acts polymorphic to as antagonist to α_{2A}-adrenergic receptor gene for the treatment of the diabetes.	Rosengren et al. (2009); Verwaerde et al. (1997)
					Antidepressants.	Blocks the function of monoamine oxidase enzymes.	Millan et al. (2000)
					Sexual dysfunctionality treatment.	Yohimbine blocks the pre- and postsynaptic α_2 receptors.	Adeniyi et al. (2007)
$C_{18}H_{16}O_8$	Centaureidine	An ortho methylated flavonol obtained from many plants like *Brickellia veronicaefolia, Polymnia fruticosa.* Commercially used in cosmetics all over the world in skin creams, skin whitening.	3	8	Protects from large number of skin diseases, including melasma, postinflammatory hyperpigmentation, and lentigo.	Inhibited the dendrite outgrowth in melanocytes and reduced epidermal pigmentation also inhibited melanogenesis and reduced the total amount of tyrosinase.	Solano et al. (2006); Park et al. (2010); Saeki et al. (2003)
$C_{21}H_{28}O_5$	Prednisolone	Derivative of cortisol and synthetically produced as glucocorticoid. Commcercially used for treatment of various autoimmune disorders and inflammatory conditions. Patients with hepatic failure also treated with this.	3	5	Antiinflamatory and antiautoimmune diseases like uveitis, pyoderma gangrenosum, rheumatoid arthritis, ulcerative colitis, pericarditis, temporal arteritis, and Crohn's disease, Bell's palsy, multiple sclerosis, cluster headaches, vasculitis, acute lymphoblastic leukemia, and autoimmune hepatitis.	Prednisolone is a corticosteroid drug with predominant glucocorticoid and low mineralocorticoid activity, and antiinflammatory and antiautoimmune conditions.	Fiel and Vincken (2006); Thrower (2009); Lambrou et al. (2009); Czock et al. (2005)
					For the treatment of the patients of the hepatic failure.	Metabolize prednisone to prednisolone as it is active metabolite of the commercially available drug prednisone.	

(*Continued*)

Table 1.3: Different types of phytochemicals and their biochemical properties. (*cont.*)

Molecular Formula	Phytochemical	Economical Values	HBA	HBD	Pharmacological Effect	Molecular Mechanism	References
$C_{18}H_{16}O_8$	Rosmarinic acid	Caffeic acid ester with strong antioxidant and medicinal values. Potential anxiolytic and antiviral agent.	5	8	Antioxidant, antiviral, antibacterial, anticanerous properties.	Inhibits the formation and decomposition of hydroperoxides, superoxide molecules, inhibit COX-2, reduces the levels of proinflammatory cytokines and a chemokine, downstream inhibitor of IK kinase-β activity, increased levels of pNF-κB and Cox-2.	Frankel et al. (1996); Awad et al. (2009); Pedro et al. (2011)
					Anxiolytic.	Acts as GABA transaminase inhibitors.	Awad et al. (2009)
					Neural acetylcholinesterase inhibitor.	Unconjugated rosmarinic acid and its metabolites remain in the bloodstream, bound to human serum albumin and lysozyme.	Pedro et al. (2011)
$C_{24}H_{21}O_5.HCL.H_2O$	Leonurine Hcl	Commercially obtained from South African plant *Leonotis leonurus* and is mildly psychoactive alkaloid.	5	8	Antimicrobial against respiratory tract infections, bone infections, skin infections.	Acts against *S. pneumonia* and haemolytic streptococci, staphylococci, *P. mirabilis*.	Pingale et al. (2013); Im et al. (2012)
$C_{20}H_{24}O_7$	Diosbulbin A	It is norclerodane diterpenoid.	1	7	Antimicrobial.	Affects molecular arrangement of microbes.	Prakash and Hosetti (2012)
					Anticancerous.	Inhibits proliferation and induce apoptosis in human colon cancer.	Liu et al. (2011)
$C_{22}H_{20}N_2O_4$	Ethofenprox	A pyrethroid used in making commercial insecticides.	0	3	Insecticides.	Neurophysiological effect, affects the sodium channels.	Becker et al. (2010); Nishimura et al. (1986)

$C_{25}H_{28}O_3$	Epitriptolide	An antiinflammatory isolate of *Tripterygium wilfordii*.	2	7	Antiinflamatory and anticancerous.	Epitriptolide derivatives as potential anticancer agents were synthesized and tested for their cytotoxicity against SKOV-3 and PC-3 tumor cell lines.	Ma et al. (2007); Xu et al. (2014)
$C_{20}H_{24}O_7$	Ailanthone	Allelopathic chemical used as weedicides, inhibits the growth of other plants.	3	7	Growth inhibitors of other plants, weedicides, Herbicidal effects.	Inhibits caretonoids synthesis by inhibiting HPPD enzymes (hydroxyphenyl-puruvatedioxygenase) used in carotenoid synthesis and which protects the chlorophyll from damage from sunlight.	Heisey and Heisey (2003)
$C_{27}H_{41}NO_2$	Cyclopamine	Steroidal jerveratrum alkaloid used for the treatment of cancers.	2	3	Anticancerous activity, treatment agent for multiple myeloma, basal cell carcinoma, sarcoma, and so on.	Lowers the Hh activity, which caused tumor.	Beachy et al. (2000)
$C_{29}H_{48}O$	Fucosterol	Sterol with antidiabetic and antioxidant properties.	1	1	Antidiabetic	Inhibition of blood glucose level and degradation of glycogen; fucosterol-inhibited aldose reductase (AR).	Lee et al. (2004a); Jung et al. (2013)
					Antioxidant properties and hepatoprotective.	It increases the antioxidant enzymes hepatic cytosolic superoxide dismutase, catalase, and glutathione peroxidise.	Lee et al. (2003)
$C_{27}H_{45}NO_2$	Hupehenine	Bioactive alkaloid with lowering blood pressure and antiviral properties.	2	3	Anti HIV-1.	HIV integrase strand transfer inhibition	Zhang et al. (2010)
					Lowers blood pressure.	Inhibition of angiotensin convering enzyme and angitensin II receptor blocking.	Patten et al. (2013)
$C_{27}H_{41}NO_3$	Peimisine	Bioactive alkaloid.	2	4	Antibacterial.	Molecular arrangements disturbances	Yang et al. (1996)

(*Continued*)

Table 1.3: Different types of phytochemicals and their biochemical properties. (*cont.*)

Molecular Formula	Phytochemical	Economical Values	HBA	HBD	Pharmacological Effect	Molecular Mechanism	References
$C_{27}H_{43}NO_3$	Peiminine	Alkaloidic phytochemicals.	2	4	Treatment of cough.	TRPV1 (transient receptor potential vanilloid 1) and transient receptor potential ankyrin 1 (TRPA1).	Zhang et al. (2011)
$C_{24}H_{31}NO_6$	Guan-Fu Base B	Bioactive alkaloids	2	7	Antiarrhythmic effects, anticardiac arrhythmias, and myocardial contractility.	Anticholinesterase activity.	Dong and Chen (1994)
$C_{21}H_{20}O_{10}$	Apigenin 8-C glucoside (Vitexin)	Vitexin is an apigenin flavone glucoside, a chemical coupound found in the passion flower.	7	10	Antioxidant and antiradical activities.	Acts against oxidative reactive species.	Burda and Oleszek (2001)
$C_{21}H_{20}O_{10}$	Apigenin 6-C-glucoside (Isovitexin)	Isovitexin flavonoids is used commercially for the production of antioxidant and antiradicals medicines.	7	10	Carbonyl products, including glyoxal, are reportedly hazardous to human health because of its genotoxicity.	Inhibitory effect of 2 ″-O-glycosyl isovitexin on genotoxic glyoxal formation in a lipid peroxidation system, inhibited xanthine oxidase.	Nishiyama et al. (1994); Ramarathnam et al. (1989)
$C_{21}H_{20}O_1$	Homoorientin	Homoorientin is a flavonoid glycosides.	8	11	Antimycotic properties.	Inhibit their growth.	Turchetti et al. (2005)
					Antiinflamatory and anticancerous activity.	Reduces proinflamatory responses and kills oxidant species and shows antiinfllmatory activity.	Schauss (2012)
$C_{21}H_{20}O_{11}$	Cynaroside	Flavones 7- -O-glucoside of luteolin and can be found, and have phytopharmaceutical applications.	7	11	Flavonoids increase NOS expression in endothelial cells, inhibit platelet aggregation.	Phytopharmaceutical applications.	Negro et al. (2012)
					Hypocholesterolemic potentials	Decreases plasma cholesterol levels.	Mukherjee (2003)

$C_{21}H_{20}O_{11}$	Cyanidine 3-O-glucoside	Cyanidin 3-O-glucoside, is an anticancer agent.	8	11	Anticancer, antiproliferative, antioxidant.	Inhibits ROS production, morphological changes, and alterations in oligonucleosomal DNA fragments.	Moongkarndi et al. (2004)
					Antiacetylcholinesterase and antioxidant.	Inhibitory effects on mutagenesis and carcinogenesis in human.	
$C_{24}H_{25}NO_7$	8-O-acetyl-excelsine	Alkaloid used in medicines, vitamins, proteins.	—	—	Have anaphylaxis effect, highly nutritious part of Brazil nuts.	Therapeutic effects.	Nagashiro et al. (2001)
$C_{24}H_{39}NO_7$	Fuziline	It is a diterpenoid alkaloid.	4	8	Antidepressant.	Enhanced the ratio of phospho-CREB/CREB (cAMP response element-binding) and BDNF (brain-derived neurotrophic factor) protein level in the frontal cortex and hippocampus.	Liu et al. (2012)
$C_{23}H_{23}N_3O_5$	Topotecan hydrchloride	It is a water-soluble derivative of camptothecin, also called hycamtin. It is used in treatment of cancers.	2	8	Ovarian, lung, and cervical anticancerous agent.	Inhibit topoisomerase I, Topotecan intercalates DNA bases and disrupts the DNA duplication machinery; this disruption prevents DNA replication, and ultimately leads to cell death.	Staker et al. (2002)
$C_{20}H_{27}NO_{11}$	Amygdalin (D/R)	Anticancerous glycoside with other name as vitamin B-17.	7	12	Anticancer.	Induces apoptosis in the prostate cancer cell lines by regulating Bax and Bcl-2 expression.	Moertel et al. (1982); Newmark et al. (1981)
$C_{22}H_{26}O_{11}$	Agnuside	Terpenic phytochemicals with anticancer and antiallergic properties.	6	11	Anticancer and antiallergic.	COX-1; antiinflammatory.	Bellik et al. (2012)

(*Continued*)

Table 1.3: Different types of phytochemicals and their biochemical properties. (*cont.*)

Molecular Formula	Phytochemical	Economical Values	HBA	HBD	Pharmacological Effect	Molecular Mechanism	References
$C_{22}H_{26}O_{11}$	Curculigoside	Curculigosides are natural phenols that could be useful against β-amyloid aggregation in Alzheimer's disease.	5	11	Reduces cerebral ischemia injury.	Curculigoside A reduced the oxygen–glucose deprivation-induced cytotoxicity and apoptosis, blocked TNF-α-induced NF-κB and IκB-α phosphorylation, and decreased HMGB1 expression.	Jiang et al. (2011)
					Prevents damages in osteoblasts.	H_2O_2-induced reduction of differentiation markers, such as alkaline phosphatase, calcium deposition, and Runx2 level was significantly recovered due to CUR. CUR prevents from H_2O_2-induced stimulation of extracellular signal-regulated kinase 1/2, and nuclear factor-κB signaling and the inhibition of p38 mitogen-activated protein kinase activation.	Wang et al. (2012)
					Circuligoside attenuates human umbilical veins endothelial cell injury.	Curculigoside can inhibit H_2O_2-induced injury in human umbilical vein endothelial cells, pretreatment with curculigoside decreased the activity of caspase-3 and p53 mRNA expression, which was known to play a key role in H_2O_2-induced cell apoptosis.	
$C_{30}H_{54}O_4$	25-Hydroxyprot-opanaxatriol.		4	4			

$C_{22}H_{22}O_{12}$	Isorhamnetin-3-O-galactoside	Natural flavonoid used to tackle inflammation, ROS, phagocytosis.	7	12	Inhibitory effect on ROS in PMNs, prevents oxidative burst of PMNs and also protect membrane damage by lipid peroxidations.	Used to tackle dangerous agents released during the inflammations, ROS. Tackeling phagocytosis caused by PMNs, poly morphonuclear neutrophils, archidonic acid metabolites.	Zielinska et al. (2001)
$C_{22}H_{22}O_{12}$	Isorhamnetin-3-O-glucoside	Flavonoidic in nature used in various medicines.	7	12	Treatment of injured or inflamed arteries.	Quercetin acts on the inflamed and injured arteries by activated macrophages. Macrophages are the targets in human atherosclerotic arteries.	Kawai et al. (2008)
$C_{22}H_{22}O_{12}$	Neptrin	Important components of various Chinese medicines.	—	—	Used as anticold and anticough.	It resolves the phlegm and prevents coughing and wheezing.	Xia (1998)
$C_{28}H_{46}O_6$	Brassinolide	A steroidal lactone used as growth promoters.	4	6	Responses similar to IAA were elicited by Brassinolide in plants.	Has role in hypocotyl hook opening, elongation of maize mesocotyl, fresh weight increase in Jerusalem artichoke (2,4-D used pea epicotyls, azuki bean epicotyl sections) and pea epicotyl sections.	Mandava et al. (1981)
$C_{35}H_{52}O_4$	Hyperforin	Hyperforin is a prenylated phloroglucinol derivative, produced by some of the members of the plant genus *Hypericum*. It accumulates in oil glands, pistils, and fruits, probably as a plant defense against herbivory.	1	4	Antidepressant, antixiolytic.	Reuptake inhibitor of monoamines like norepinephrine, serotonin,, dopamine. Reuptake inhibitors of GABA and glutamate. Modulates release of acetylcholine release in himppocampal region. Also help in release of acetylcholine release from striatum.	Buchholzer et al. (2002); Kiewert et al. (2004)

(*Continued*)

Table 1.3: Different types of phytochemicals and their biochemical properties. (*cont.*)

Molecular Formula	Phytochemical	Economical Values	HBA	HBD	Pharmacological Effect	Molecular Mechanism	References
					Topical antibiotic.	Affect molecular arrangements of microbes like MRSA.	Reichling et al. (2001)
					Anti-Alzheimer's disease.	Procognotive, find use in the treatment of Alzheimer's disease.	
					Sexual problems.	Effective against the premature ejaculation.	Kim and Chancellor (2008)
					Anticancerous effects.	Acts as antianjgiogenic and proapoptopic, also anticlastogenic.	Quiney et al. (2006); Martínez-Poveda et al. (2010); Sun et al. (2011)
					Antidepressant.	Acts upon sigma and D2 receptor.	
					Anti-inflamatory effects.	Inhibits COX1 and 5-LO.	
$C_{40}H_{56}$	Lycopene	Carotenoid found in tomatoes and many fruits. Carotene with no Vitamin-A.	0	0	Anticancerous properties and prevent skin damage.	PSA level.	Ilic et al. (2011); Rizwan et al. (2011)
$C_{29}H_{44}O_9$	Rhodexin A	Carotenoid in nature used in anticancerous medicines.	5	9	Anticancerous.	Cytotoxic activity against leukemia cell K562.	Umebayashi et al. (2003)
$C_{40}H_{56}$	Beta-carotene	Natural lipid antioxidant.	0	0	Lipid antioxidant, anticancerous (in on smokers).	Good radical-trapping antioxidant behavior.	Mayne (1996); Burton and Ingold (1984); Peto et al. (1981)

$C_{30}H_{18}O_{10}$	Amentoflavon	Biflavonoid, found in various plants used in various antimalarial and metabolism of various drugs.	6	10	Antimalarial and leishmanicidal activity.	Targets the M1 alanyl and M17 leucyl aminopeptidase.	Thivierge et al. (2012); Kunert et al. (2008)
					Helps in drug metabolisms.	Inhibitior of CYP3A4 and CYP2C9 enzymes involved in drug metabolisms.	
					Anticancer activity.	Inhibits the fatty acid synthatase enzymes.	Lee et al. (2013); Wilsky et al. (2012)
$C_{31}H_{24}O_{10}$	Sikokianin A	Biflavonoid used in various medicinal uses.	5	10	Antimicrobial.	Antimitotic and antifungal.	Yang et al. (2005)
					Antimalarial.	Inhibitory affect against malaria.	
					Antiviral activity.	Shows anti-HBV effect HBsAg secretion.	Yang and Chen (2008)
$C_{32}H_{44}O_{8}$	Cucurbitacin E	It is an oxidated steroid consisting of tetracyclic triterpenes.	3	8	Antiinflamtory activity.	Inhibition of cyclo-oxygenase, ROS and RNS macrophages, produces cytokines, reactive nitrogen species, growth factors by the activation by chemical mediators.	Abdelwahab et al. (2011)
					Inhibition of the tumor.	Inhibits tumor angiogenesis.	Chen et al. (2012)
					Antioxidant properties.	Scavanges the free radicals.	Tannin-Spitz et al. (2007)
					Cytostatic activity.	Inhibits the cell mitosis during the S and M phase. Also disrupts the cytoskeleton actin.	
					Hepatoprotective.	Reduces the GPT, GOT, ALP, TP, and TBIL.	Chen et al. (2012)
					Antimetastatis.	Inhibits adhesions of cancer cell acts upon collagen type 1.	Chen et al. (2012)

(*Continued*)

Table 1.3: Different types of phytochemicals and their biochemical properties. (*cont.*)

Molecular Formula	Phytochemical	Economical Values	HBA	HBD	Pharmacological Effect	Molecular Mechanism	References
$C_{32}H_{22}O_{10}$	Ginkgetin	It is biflavone. Used in making medicines.	4	10	Potent antiarthritic activity, analgesic activity and inhibit arthiritis.	Inhibitor of group II phospholipase A 2.	Kwak et al. (2002)
					Antiherpes.	Inhibition of virus replication.	Hayashi et al. (1992)
$C_{32}H_{22}O_{10}$	Isoginkgetin	A natural biflavonoid used in various medicinal formulations for antidibeties, antioxidants, and hair growth properties.	4	10	Fight baldness.	Stimulates hair growth.	Gallwitz et al. (2010)
					Antidiabetic.	Enhances adiponectin secretion from differentiated adiposarcoma cells via a novel pathway involving AMP-activated protein kinase.	Liu et al. (2007)
					Antioxidant.	Scavenges the free radicals.	Ellnain-Wojtaszek et al. (2003)
					Antitumor.	Splicing inhibition is the mechanistic basis of the antitumor activity of isoginkgetin.	O'Brien et al. (2008)
$C_{15}H_{12}O_5$ $C_3H_{23}N_3O_5$	Naringin	Citrus bioflavonoid, flavanone glycoside.	8	14	Affects the drug uptake.	Naringin inhibits some drug-metabolizing cytochrome P450 enzymes, including CYP3A4 and CYP1A2.	Bailey et al. (1998)
					Antineuropathic.	Reduced diabetes-induced neuropathy.	Kandhare et al. (2012)
					Protective effect.	Protective effects on cognition and oxidative.	Kumar et al. (2010)
	Rhamnosylvitexin	Flavonoid used in medicines.	9	14	Anticancer.	Antiproliferative and apoptotic effects in human breast cancer.	Way et al. (2009)
$C_{27}H_{30}O_{14}$	Rhoifolin	Flavone glycoside, or flavones, a type of flavonoid found in canton lemon, grapefruit.	8	14	Anticancerous and antiproliferative.	Anticancerous effect on hepatocellular (Hep G2), colon (HCT-116), and fetal human lung fibroblast (MRC-5).	Eldahshan (2013)

$C_{27}H_{30}O_{10}$	Vitexin-4	Apigenein flavones glucoside found in pearl millet, passion flower and so on.	9	14	Responsible for causing goiter.	It inhibits enzyme thyroid peroxidise and contributes goiter.	Gaitan (1990)
$C_{27}H_{30}O_{15}$	Vitexin-2″-O-rhamnoside	These are O-methylated flavonoid	9	14	Anticancer	Strongly inhibited DNA synthesis in MCF-7 cells.	Ninfali et al. (2007)
$C_{33}H_{52}O_{8}$	Diosgenin glucoside	Plant phytochemicals used in various medicines.	4	8	Antioxidant.	Neutralizes ROS.	Araghiniknam et al. (1996)
					Antitumor.	Antiproliferative activities.	Carvalho et al. (2010)
					Antihypercholesterolemia and atherosclerosis.	Inhibits intestinal absorptions of cholesterol.	Malinow (1986)
$C_{33}H_{52}O_{8}$	B3-procyanidin	Used in medicines for hair growth.	10	12	Antibald activity.	Hair growth stimulant.	Benavides et al. (2006)
$C_{32}H_{45}NO_{10}$	O-Benzoylmesaco-nine	Aconitine alkaloids.	4	11	Antiburn infections caused by HSV type 1 and candida albicans.	Inhibits the cytokine production by various type-2 T cells.	Kobayashi et al. (1998)
					Analgesic/antiinflammatory/antipyretic activity.	Prevent growth of cells.	Murayama (1998)
$C_{27}H_{30}O_{15}$	Kaempferol 7-neohesperidoside	Phenolic compounds used in various medicines.	9	15	Antiinflammatory	TNF, IL-1beta, caspases, and specifically targets JAK-3 of hemopoietic cells.	Aggarwal et al. (2009)
$C_{30}H_{26}O_{13}$	Procyanidin	Member of proanthocyanidin class of flavonoids.	10	13	It has antioxidant capacity, that is, oxygen radical absorbance capacity (ORAC). Have antioxidant properties higher than vitamins.	Neutralizes free radicals.	
					Help in smoothing blood vessels.	Suppress production of protein endothelin-1	Corder et al. (2006)
					Antidepressants.	MAO inhibitory.	Xu et al. (2010)
$C_{37}H_{42}N_{2}O_{6}$	Dauricinoline	Alkaloids used in various psychotropic medicines. One of important crude drugs exported by China.	2	8	Find use for treatment of allergy, arrhythmia and inflammation.	Ability to modulate ca^{2+} uptake and several k+ channels. It also affects human-ether-a-go-go-related gene (HERG) channels.	Zhao et al. (2012)

(Continued)

Table 1.3: Different types of phytochemicals and their biochemical properties. (*cont.*)

Molecular Formula	Phytochemical	Economical Values	HBA	HBD	Pharmacological Effect	Molecular Mechanism	References
$C_{37}H_{42}N_2O_6$	Daurinoline	Alkaloids used in various psychotropic medicines. One of important crude drugs exported by China.	2	8	Find use for treatment of allergy, arrhythmia, and inflammation.	Ability to modulate ca^{2+} uptake and several k+ channels. It also affects human-ether-a-go-go-related gene (HERG) channels.	Zhao et al. (2012)
$C_{37}H_{42}N_2O_6$	Daurisoline	Alkaloids used in various psychotropic medicines. One of important crude drugs exported by China.	2	8	Used in the treatment of inflammation, allergy, and arrhythmia.	Ability to modulate ca^{2+} uptake and several k+ channels. It also affects human-ether-a-go-go-related gene (HERG) channels.	Zhao et al. (2012)
$C_{37}H_{42}N_2O_6$	N-desmethyldau-ricine	A new phenolic dauricine alkaloids.	2	8	Antitumor.	Immunoregulatory activity of PE2.	Shan et al. (2006)
$C_{23}H_{23}N_3O_5$	Hesperdine	Flavanones used in various human disorders treatment.	8	8	Anti-inflammatory.	Inhibits both acute and chronic inflammations.	Guardia et al. (2001)
					Enhances the bioavailability of flavonoids	Interact with gut microflora and undergo better absorptions.	Selma et al. (2009); Yao et al. (2004)
					Anticancerous.	Apoptotic effects on the human colon cancer cells by acting upon caspase3.	Yao et al. (2004)
$C_{37}H_{42}N_2O_6$	Liensinine	Alkaloids with antioxidant effects and phytochemicals effects.	2	8	Antioxidant.	Antioxidant effects because of vasodilating effects.	Lee et al. (2005)
$C_{28}H_{34}O_{15}$	Neohesperidin	Flavones with high health benefits.	2	8	Increases antioxidant capacity, acts as antioxidant,	Prevents phospholipid hydroper-oxidations.	Yao et al. (2004)
					lipolytic affects inhuman body fat adipocites.	Inhibition of cAMP-phosphodiesterase (PDE).	Dallas et al. (2008)
$C_{27}H_{30}O_{16}$	Rhodiosin	Flavonol glycoside.	2	8	Antioxidant and helps in treatment of diabetes mellitus, cancer, Alzheimer's disease.	Neutralizing the ROS.	Kwon et al. (2009)
$C_{23}H_{23}N_3O_5$	Quercetin 3-rutinoside	Flavonoids used in various human medicines.	2	8	Cancer inhibitory.	Growth inhibitory effects of Q was due to a blocking action in the G_0/G_1 phase	Chen et al. (2006); Ranelletti et al. (1992)
$C_{23}H_{23}N_3O_5$	Cycleanine	Alkaloids with anticancer activity.	0	8	Anticancer activity against various human cancer cell lines.	Inhibits cell growth.	De Wet et al. (2009); Ono et al. (1994)

$C_{36}H_{62}O_8$	Ginsenoside Ck or Rh2	Steroid glycosides, and triterpene saponins.	6	8	Anticancerous and immunomodulating.	Inhibition of tumor angiogenesis and metastatis.	Sato et al. (1994)
$C_{29}H_{34}O_{15}$	Pectolinarin	Flavonoid glycosides.	7	15	Anticancerous and prevent hepatic injury.	It has analgesic, antiinflammatory activity and prevents spreading of cancer.	Martinez-Vazquez et al. (1998); Lim et al. (2008)
						Prevents injury to hepatic cells by being antioxidant.	Yoo et al. (2008)
$C_{38}H_{42}N_2O_6$	Tetrandrine	A *bis*-benzylisoquinoline alkaloid.	0	8	It does have antiinflamatory properties.	Reduces the inflammation induced by Interlukin-1, TNF and PAF	Wong et al. (1992)
					Anticancerous.	Induces Cell growth arrest and apoptosis.	Lee et al. (2002)
$C_{31}H_{40}O_{15}$	Martynoside	Phenylpropanoid glycosides.	7	15	Helps in retardation of skeletal muscle fatigue.	Effects muscle contractility.	Liao et al. (1999)
					Anticancerous, cytotoxic, and antimetastatic.	Acts on cancer cells.	Liao et al. (1999)
					Estrogenic and antietrogenic properties.	Activate estrogen receptor isoforms ERα and ERβ	Papoutsi et al. (2006)
$C_{32}H_{46}O_{16}$	Secoisolariciresinol diglucoside	Plant lignin.	10	16	Reduces serum cholesterol and hypercholesterolemic atherosclerosis. Antioxidant activities.	Lignans are platelet activating factor receptor antagonists and lowers the production of oxygen free radicals. Antioxidant property is also responsible for reduction of hypercholesterolemic atherosclerosis. There was reduction in serum cholesterol, LDL-C, and lipid peroxidation product while an increase in HDL-C and antioxidant reserve.	Prasad (1999);
$C_{42}H_{72}O_{14}$	Ginsenoside Rf	Found in traditional oriental medicinal plants. It is a type of saponins.	2	2	Antinociception activity.	Ginsenosides Rf inhibits Ca^{2+} channels in mammalian sensory neurons.	Mogil et al. (1998)

(Continued)

Table 1.3: Different types of phytochemicals and their biochemical properties. (*cont.*)

Molecular Formula	Phytochemical	Economical Values	HBA	HBD	Pharmacological Effect	Molecular Mechanism	References
					Expectorant, antiinflammatory, vasoprotective, hypocholesterolemic, immunomodulatory, hypoglycaemic, molluscicidal, antifungal, antiparasitic and many others properties.	—	Podolak et al. (2010)
$C_{45}H_{72}O_{17}$	Gracillin	It is a type of saponin and is used in medicinal products.	9	17	Methyl protogracillin is good in the treatment of cancer of urinary bladder, cervical carcinoma, carcinoma of and renal tumor. Pancreatic lipase inhibitor, suppresses blood triacylglycerol level, inhibits adipogenesis.	Induces ERα expression, inhibits p38 MAPK, decreases expression of C/EBPα, LPL, PPARγ and leptin.	Hu and Yao (2001); Vermaak et al. (2011)
$C_{45}H_{73}NO_{16}$	Solasonine	It is a type of glycoalkaloid.	10	17	Antiproliferative activities against human colon (HT29) and liver (HepG2) cancer cells.	Inhibition of growth of cancer cells.	Lee et al. (2004a,b)
$C_{46}H_{56}N_4O_{10.}H_2SO_4$	Vincristine sulfate	Antileukaemic alkaloids and has huge market for leukemic patients. Generation of gian spermatogonial cells, which can be controlled by hormone.	5	18	Antileukaemic alkaloids.	By probably affecting spermato-genic mitosis.	Stanley and Akbarsha (1992); Steinberger (1971)
$C_{48}H_{78}O_{17}$	Saikosaponin C	It's a triterpenoid glycoside.	11	18	Saponins are antiinflamatory, immunomodulatory, hepatoprotective, antitumor and antiviral activities, antimicrobial, antihepatitis B virus.	Induction of apoptosis through the activation of caspases 3 and 7, which subsequently resulted in poly-ADP-ribose-polymerase (PARP) cleavage.	WHO (1998); Shao et al. (1999); Uckan et al. (2005); Chiang et al. (2003)
$C_{46}H_{72}O_{19}$	Saundersioside A	Steroidal glycosides or cholestane glycosides hold potential as phyto-chemicals based drugs.	3	3	Antileukemic.	Potent cytostatic activity, induction of apoptosis by cell morphology and DNA fragmentation.	Kuroda et al. (1997)

HBA, Hydrogen bond acceptor; HBD, hydrogen bond donor

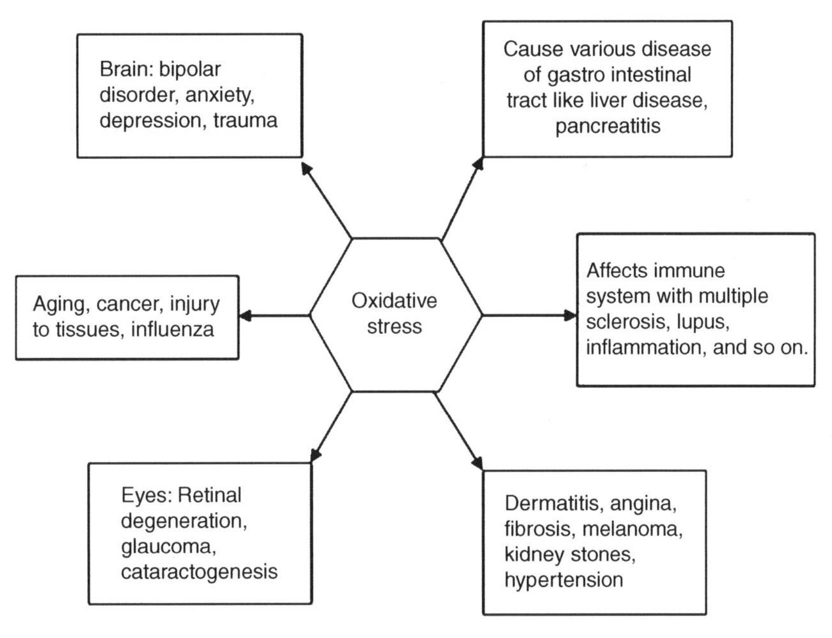

Figure 1.1: Oxidative Stress Circle (Ilie and Margină, 2012).

Oxidative stress (Fig. 1.1) is the outcome of imbalance between the amount of prooxidant and antioxidant; the amount of prooxidant is high, which leads to the generation of ROS (Sies, 1991). There are many free radical productions during normal biochemical reaction that a cell undergoes. Most of them are centered on oxygen and halogens and are sulfur-based (Halliwell and Gutteridge, 2007a). Most of the endogeneous free radical species are nitric monoxide (NO), singlet oxygen ($^1\Sigma g^+O_2$), hydroxyl (OH), and superoxide (O_2^-). Others are peroxynitrite ($ONOO^-$) and hydrogen peroxide (H_2O_2), which are responsible for generating free radicals in the living cells by getting involved in various chemical reactions (Halliwell, 2006).

Various studies have indicated that the formation of direct DNA oxidants that are the derivative of hydrogen peroxide is a cell metabolism-dependent process (Imlay et al., 1988). This DNA oxidant is generated in the presence of reducing equivalents and any source of iron ions ($FeCl_3$, $FeSO_4$), and initiates a Fenton reaction, where hydrogen peroxide is reduced by ferrous iron to a reactive radical and this then affects various biomolecules of the cells (Fig. 1.2).

4 Disturbing the Lingocellulosic Arrangement

Special arrangements of lingocellulosic components make plant biomass resistant for degradation to biological and chemical agents. Lingocellulosic component pretreatment strategies and conditions also vary with the type and part of plants used. The pretreated bark

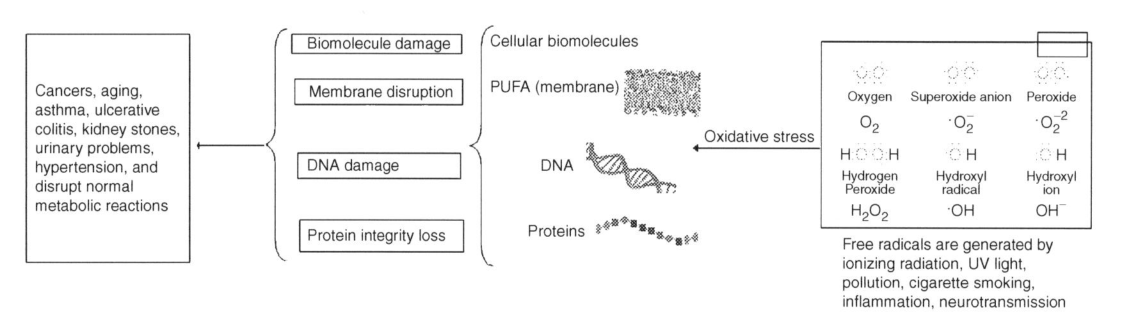

Figure 1.2: Damage to Cells and Biomolecules by Free Radicals.

of poplar trees and corn leaves with dilute acid but with the same method was not suitable for corn stalks and sweet-gum bark; instead, the enzymatic pretreatment strategy was more suitable (Donghai et al., 2006). For biological degradation of lingocellulosic materials, crystallinity, hemicelluloses-lignin percentage, free accessible area, degree of acetylation and polymerization of hemicelluloses, and celluloses are major factors (Wyman, 1996). There are various kinds of interactions that take place between the components of lignocelluloses. The main types of bonds that exist are hydrogen bonds, ester bonds, carbon-to-carbon interaction, and ether bonds, details of which are provided in Table 1.4.

4.1 Pretreatment of Lingocellulosic Biomass

Pretreatment of plant biomass can be done by physical, chemical, and biological methods. Pretreatment of biomass like *Pinus robxburghii* forest wastes helps in the bioavailability to enzymes and biodigestibility in bioethanol production.

4.1.1 Physical pretreatment

Physical pretreatment is generally done to reduce the size of lingocellulosic waste material or biomass by involving grinders, millers, or chippers either singularly or in combinations. Reduction of size also reduces the crystallinity of the biomass. Millers are used when the size

Table 1.4: Different types of bonding among lingocellulosic components (Faulon and Hatcher, 1994).

S. no.	Types of Bonds	Components	Linkages Type
1.	Hydrogen bond	Cellulose	Intrapolymeric
2.	Carbon to carbon bond	Lignin	Intrapolymeric
3.	Ether bond	Lignin and hemicelluloses	Interapolymeric
4.	Ester bond	Hemicelluloses	Intrapolymeirc
5.	Ether bond	Cellulose-lignin, hemicelluloses-lignin,	Interpolymeric
6.	Ester bond	Hemicellulose-lignin	Interplymeric
7.	Hydrogen bond	Cellulose-hemicellulose, hemicelluloses-lignin	Interpolymeric

of biomass is to be reduced up to 0.2–2.0 mm, similar to the effect of grinders. Chippers reduce the size up to 10–30 mm (Sun and Cheng, 2002). But it is important to note that the power consumptions are related to the size reduction of the biomass (Cadoche and López, 1989).

4.1.1.1 Pyrolysis

In this process biomasses are thermochemically decomposed at very high temperatures in the absence of oxygen. It is the process that is irreversible and involves both chemical and physical changes. Pyrolysis of biomass rich in organic components generates gas and liquid byproducts while a solid is left richer in organic carbon, called char, or sygnas and biochar. Pyrolysis of wood or plant biomaterial is done at temperatures between 200 and 300°C, because at low temperatures the products are less volatile (Kumar et al., 2009). Fan et al. found that mild acid hydrolysis of the product obtained from pyrolysis resulted in more than 80% conversion of cellulose to reducing sugar while 50% glucose generation, finally (Jia et al., 1996). Though the process of pyrolysis was carried out in the absence of oxygen, the same was enhanced in the presence of oxygen in the case of biomass.

4.1.1.2 Steam explosion

Steam explosion of biomass is the most common and preferred method for pretreatment of biomasses. Steam explosion involves first the increase of steam pressure up to or more than 15 psi at 120–260°C for a few seconds, and then pressure is reduced at once, which leads to the explosion of the biomass material. Steam explosion improves hydrolysis of hemicelluloses. Steam explosion causes conversion of lignin and degradation of hemicellulose, benefiting the overall conversion. It is believed that hemicellulose undergoes hydrolysis by the action of acetic acid and other acids produced due to high temperature and pressure. Compared to steam exploded and untreated poplar chips, enzymatic hydrolysis was better up to 90% in the case of pretreated chips while it was only 15% for untreated chips (Sun and Cheng, 2002). It is believed that lignin is not fully removed from the fiber; instead it gets redistributed on the biomass or fiber surface because of depolymerization after being melted at high temperatures, but sudden lowering of temperature due to the release of steam (Li et al., 2009). This leads to an increase in the surface area of action for the enzymes (Duff and Murray, 1996). It's also well known that water at high temperature can act as acid (Wyman et al., 2005). In the addition of acid, like H_2SO_4, or CO_2, or SO_2 up to 3% in the steam explosion, the time period and temperature can be reduced and less inhibitor formations and more hemicelluloses removal (Ballesteros et al., 2006). The factors that affect the efficacy of steam explosion pretreatment are the size of biomass, residence, or holding time, moisture content and temperature (Duff and Murray, 1996). Steam explosion is a cost-effective technology as tested on pilot scale for hard wood and agricultural waste but not for soft wood (Sun and Cheng, 2002). While a limitation of this technology is that xylan gets destroyed, incomplete destruction of matrix-holding lignin and carbohydrates, and, most importantly, generation of inhibitors, inhibit growth of microbes that are used in further downstream processing (Mackie et al., 1985).

4.1.1.3 Ammonia fiber explosion

In this strategy biomass is physicochemically treated, where biomass or lignocellulosic waste are exposed to liquid ammonia at temperatures above 90°C. In this process biomass is treated with ammonia solutions up to 1–2 kg/kg weight of dry biomass at temperatures of 90°C or more for 30 min holding time (Kumar et al., 2009).

Ammonia recycle percolation is another technique in which ammonia solution 10%–15% in water is allowed to pass from biomass at temperatures of 140–180°C, with fluid moving with a velocity of approximately 1–2 cm/min and has a holding time of 14 min. The ammonia is obtained and separated, which is then further recycled. Liquid ammonia solution reacts with lignin and is responsible for breaking the polymer linkages of lignin and carbohydrates. This process holds the advantage of not producing inhibitors, which is a water-saving method (Galbe and Zacchi, 2002; Kumar et al. 2009).

4.1.1.4 Carbon dioxide explosion

Carbon dioxide explosion is cheaper than ammonia explosion and requires less temperature for treatment compared to steam explosion. Supercritical fluids are gases compressed to have liquid like density at temperature more than their critical point (Kumar et al., 2009). The advantages of supercritical fluids are many, like CO_2 releases carbonic acid, which increases the rate of hydrolysis of substrate. The molecules of carbon dioxide are the same size as that of ammonia and water so it can easily penetrate the same way. Low temperature lowers the decomposition of monomeric sugars by the acids. Supercritical carbon dioxide explosion also increases the surface and are accessible to enzymes of microbes.

4.2 Chemical Treatment

4.2.1 Ozonolysis

Lignin content can be reduced by the use of ozone and leads to an increase of digestibility of the biomass. Ozone does not produce inhibitors so it's better than other chemical methods. Ozone has been used to degrade various kinds of biomass, like wheat straw, bagasse, peanut, pines (Ben-Ghedalia and Miron, 1981). When used to pretreat poplar sawdust it was found that enzymatic hydrolysis increased up to 50% and lignin content decreased to 8% from 29% (Vidal and Molinier, 1988). Less inhibitor generation at room temperature and atmospheric pressure all make this process advantageous over other processes. To prevent harm to the environment used ozone is then treated with a catalytic bed or simply by increasing the temperature it can decompose the ozone (Quesada et al., 1999). Now, most of the ozone-based pretreatment of biomass is done with hydrated fixed bed because it leads to more oxidation of the same better than H_2O solution suspension or 45% acetic acid solutions (Vidal and Molinier, 1988).

4.2.2 Acid hydrolysis

Acids like H_2SO_4 and HCl are used in treating lignocellulosic biomass for enhancing enzymatic hydrolysis and the release of more sugar. Acids have toxic, corrosive, hazardous natures so special types of reactors are required, which are resistant to corrosion. So, to make the process feasible at a commercial level, recovery of the acid used is required (Sun and Cheng, 2002). For acid pretreatment of biomass either dilute or strong acids are used. Dilute acid hydrolysis is done with H_2SO_4 (4%), as it is economical and effective. Furfural is the product treatment of cellulose with dilute acid (Zeitsch, 2000). Dilute acid acts on the biomass to convert hemicelluloses to xylose and other sugars and then to furfural (Weil et al., 1997). H_2SO_4 recovers almost 100% of hemicelluloses and high temperature favors the digestibility of cellulose in the residual solids (Mosier et al., 2005). There are two types of dilute acid pretreatment methods involving high temperature and others at low temperature. The high temperature process works at ($T > 160°C$), with continuous flow for biomass in a 5%–10% of the substrate to weight of overall mixture ratio. While temperatures below 160°C and substrate loading to overall reaction mixture in a ratio of 10%–40% in a batch process makes the process of low temperature dilute acid pretreatment (Brennan et al., 1986). Acid pretreatment can be used on a wide range of substrates like hard woods to agricultural residues and grasses (Kumar et al., 2009). But there are some drawbacks associated with acid pretreatment involving high-energy consumption. Equipment configurations, corrosion, negative impact on enzymes on later-on stages, and a large amount of water consumption in upstream and downstream processing makes the process costly even more than steam explosion and AFEX (Kumar et al., 2009).

4.2.3 Alkaline hydrolysis

Depending on the percentage of lignin in the lingocellulosic biomass some masses can also be used (McMillan, 1994). Alkali pretreatment is carried out at low temperatures and pressure compared to other technologies but the order of pretreatment is of hours or maybe days. Alkali pretreatment produces fewer sugars compared to acid pretreatment. Basic reagents like hydroxides of calcium, potassium, sodium, and ammonium are used for alkali pretreatment. Out of these four, though sodium hydroxide is studied more, but on the basis of cost, hydroxides of calcium are more effective and the least expensive of all. Recovery of calcium from calcium carbonate from the used solution is also very easy by using carbon dioxide and regenerates it to calcium hydroxide by using lime-kiln technology. Lime pretreatment removes amorphous portions like lignin and hemicelluloses and increases the crystallinity index. Enzymatic pretreatments are affected by the structural changes occur due to the pretreatment (Kim and Holtzapple, 2006).

4.2.4 Oxidative delignification

Oxidative delignification can be achieved by use of peroxidase enzyme in the presence of hydrogen peroxide (Azzam, 1989). Azzam (1989) found 50% lignin removal and almost all hemicelluloses solubilization at 2% hydrogen peroxide. Bjerre and coworkers used wet oxidative alkaline hydrolysis-based delignification at 170°C for 5–10 min and noted 85% of total cellulose being converted to glucose (Ahring et al., 1996). Polysaccharide can be made more susceptible to enzymatic hydrolysis after performing wet oxidative pretreatment with the addition of base and making maximum hydrolysis of polysaccharides (Kumar et al., 2009).

4.2.5 Organosolv process

Organosolvation process utilizes either an organic solvent or an aqueous organic solvent mixed with inorganic acid catalysts like HCl or H_2SO_4. This pretreatment strategy breaks down the bonds between internal lignin and the hemicelluloses. Solvents preferred in this process are methanol, ethanol, acetone, triethylene glycol, acetone, and tetrahydrofurfuryl alcohol (Chum et al., 1988).

4.3 Biological Pretreatment

Each process has its own requirement of energy and cost. Processes like physical, mechanical treatment along with thermochemical pretreatment require abundant energy for processing the biomass. Pretreatment processes involving the use of whole microbes or enzymes are called biological pretreatment processes. Bacteria or fungi are involved, which are less energy consuming and are safe to the environment but efficient in lignin removal from the plant biomass (Okano et al., 2005). Biological degradations and hydrolysis of lingocellulosic biomass for pretreatment and posttreatment can be done by the use of cocktails of enzymes, like keratinase, cellulase, amylase, protease, and lipase. Pretreatment of food wastes for sugar production and then for fermentation of that sugar for ethanol production was carried out. Ethanol production from food waste was produced by using a lab scale fermenter after the pretreatment by using a mixture of enzymes like carbohydrase, glucoamylase, cellulase, and protease. Fermentation was carried by using enzyme carbohydrase and microbes *Saccharomyces cerevisiae*.

4.3.1 Enzyme-based biological hydrolysis

The use of alternate substrates as the raw material for the production of enzymes and then produced enzymes are employed for the production of alcohol is a good idea. Kitchen waste was used for the production of ethanol and for rearing of enzymes used in the biofuel production by Wang et al. (2010). To obtain enzyme cocktails different substrates were used and one of them was forest wastes and kitchen wastes. But the problem of low porosity of kitchen waste was overcome by the addition of paddy husk and corn stover. Corn stover was found to be more effective than paddy waste and was used with kitchen waste with a ratio of 1:3.75 (Wang et al., 2010). The production of enzyme cocktails can be done by growing all enzyme producers together. There have been a number of reports supporting cocultivation of microbes for the release lignocellulosic waste degrading enzymes (Vats et al., 2013). Basidomycetes consortium capable of producing cellulase, xylanase, and peroxidases has been employed for lignocellulosic waste degradation (Baldrian et al., 2005, 2011). Consortia prepared by Haruta et al. (2002) also degraded many cellulosic substances, like filter paper, printing paper, cotton, and rice straws, which were 60% degraded by their consortia within 4 days. Microbial consortia prepared by Sarkar et al. (2011), capable of producing enzyme with concomitant activity, were used to degrade kitchen wastes. In their study they noted a 55%–65% decrease in the volume of the kitchen wastes, with less time of degradation and reduced foul smell.

5 Mathematical Modeling

To disturb the lingocellulosic structural framework, pretreatment of lignocellulosic biomass is a must to ease the extraction of value-added products. Biological pretreatment methods for lignocellulosic biomass wastes generated from industries are better than chemical and physical treatment methods, being ecofriendly, specific, and producing less side products (Table 1.5). Physicobiological pretreatment methods like steam explosion-based enzymatic pretreatment process is significantly influenced by factors like steam pressure, ratios at which each enzyme used and incubation time with solvent system, incubation time during steam explosion, steam explosion pressure, and solvent system volume. Practically, optimizing "one variable at a time" approach disregards the complex interactions among parameters. Out of various new statistical optimization methods, response surface methodology (RSM) combines mathematical and statistical techniques for analyzing problems with several independent variables having control on a dependent variable. One parameter at one time, based optimization cannot study the combined effect of all variables.

6 Conclusions

History has witnessed high numbers of loss to human life either by war or to diseases. Havoc created by infectious diseases like plague, tuberculosis, cholera, and malaria have pivotally shaken the course of history. Challenges posed by new emerging diseases is no stranger to humans living in any part of world. The present world is a global village and millions of people travel across one country to another, where large populations provide all possible opportunities for contagious diseases to spread as epidemics and pandemics. Stressful lifestyles, adulteration in the food, rising antibiotic resistance among microbes,

Table 1.5: Comparative analysis of various extraction methods employed[a].

S. No.	Method	Solvents Used	Temperature (°C)	Pressure (ATM)	Time Required	Volume of Solvent Required (mL)
1.	Soxhlet extraction	Methanol, ethanol, mixture of water or alcohol	Based on solvent employed	NA	Depends on cycle, 3–20 h	100–200
2.	Sonication	Methanol, ethanol, mixture of water or alcohol	Only heating	NA	30 min–1 h	Depends on the matter, 50–100
3.	Maceration	Methanol, ethanol, mixture of water or alcohol	RT	NA	3–4 days with respect to sample size	Depends on the size of the matter
4.	Super critical fluid extraction	CO_2 mixture of CO_2 and methanol	35–105	245–450	Half an hour to 2 h	NA
5.	Microwave-assisted extraction	Methanol; ethanol; mixture of water and methanol; water, methanol, and alcohol	75–155	Depends on the type of system used closed or open	5–40 min	15–100
6.	Pressure liquid extraction	Methanol	80–200	10–20 bar	15 min to 1 h	20–30
7.	Microwave-assisted, alkali extraction, and enzymatic hydrolysis	1%–2% NaOH	RT for pretreatment and 50 for enzymatic saccharification	1 bar	1 min for microwave	2,4,6, and 8 mL

[a]Zygmunt and Namieśnik (2003); Huie (2002); Cunha et al. (2011); Arora et al. (2010)

all provide the right platform for the emergence of new global diseases. The World Health Organization has asked world nations to be ready for a postantibiotic era. Providing medicines and healthy food is a global challenge. Some medicines and nutritious food are of very high cost, like cancer drugs and nutraceuticals, and are considered as a costly commodity. In the case of food preservation, additives and preservatives are used on a very large scale in packaged food. These food additives above the ADI enhance the generation of ROS and free radical species, which when in excess cause oxidative damage to the cellular biomolecules, like DNA, proteins, and PUFA of membranes, and result in cancers, aging, asthma, ulcerative colitis, kidney stones, urinary problems, and hypertension by disturbing the metabolism of the body and disturbing the metabolic reactions. With the production of value-added products from the biomass wastes, it would be a giant step toward solving the problems related to the health-care system.

References

Abdelwahab, S.I., Hassan, L.E.A., Sirat, H.M., Yagi, S.M.A., Koko, W.S., Mohan, S., Taha, M.M.E., Ahmad, S., Chuen, C.S., Narrima, P., Rais, M.M., Hadi, A.H.A., 2011. Anti-inflammatory activities of cucurbitacin E isolated from *Citrullus lanatus* var. Citroides: role of reactive nitrogen species and cyclooxygenase enzyme inhibition. Fitoterapia 82 (8), 1190–1197.

Adeniyi, A.A., Brindley, G.S., Pryor, J.P., Ralph, D.J., 2007. Yohimbine in the treatment of orgasmic dysfunction. Asian J. Androl. 9 (3), 403–407.

Adio, A.M., 2009. Germacrenes A–E and related compounds: thermal, photochemical and acid induced transannular cyclizations. Tetrahedron 65 (8), 1533–1552.

Aggarwal, B.B., Bhardwaj, A., Aggarwal, R.S., Seeram, N.P., Shishodia, S., Takada, Y., 2004. Role of resveratrol in prevention and therapy of cancer. Anticancer Res. 24 (5A), 2783–2840.

Aggarwal, B.B., Van Kuiken, M.E., Iyer, L.H., Harikumar, K.B., Sung, B., 2009. Molecular targets of nutraceuticals derived from dietary spices: potential role in suppression of inflammation and tumorigenesis. Exp. Biol. Med. 234 (8), 825–849.

Ahlemeyer, B., Selke, D., Schaper, C., Klumpp, S., Krieglstein, J., 2001. Ginkgolic acids induce neuronal death and activate protein phosphatase type-2C. Eur. J. Pharmacol. 430 (1), 1–7.

Ahmad, M., Yousuf, S., Khan, M.B., Ahmad, A.S., Saleem, S., Hoda, M.N., Islam, F., 2006. Protective effects of ethanolic extract of *Delphinium denudatum* in a rat model of Parkinson's disease. Hum. Exp. Toxicol. 25 (7), 361–368.

Ahring, B.K., Jensen, K., Nielsen, P., Bjerre, A.B., Schmidt, A.S., 1996. Pretreatment of wheat straw and conversion of xylose and xylan to ethanol by thermophilic anaerobic bacteria. Bioresour. Biotechnol. 58 (2), 107–113.

Akutsu, H., Kikusui, T., Takeuchi, Y., Sano, K., Hatanaka, A., Mori, Y., 2002. Alleviating effects of plant-derived fragrances on stress-induced hyperthermia in rats. Physiol. Behav. 75 (3), 355–360.

Ameen, O.M., Olatunji, G.A., Atata, R.F., Usman, L.A., 2011. Antimicrobial activity, cytotoxic test and phytochemicals screening of the extracts of the stem of *Fadogia agrestis*. Nigerian Soc. Exp. Biol. J. 11 (1), 79–84.

Araghiniknam, M., Chung, S., Nelson-White, T., Eskelson, C., Watson, R., 1996. Antioxidant activity of dioscorea and dehydroepiandrosterone (DHEA) in older humans. Life Sci. 59 (11), PL147–PL157.

Arora, R., Manisseri, C., Li, C., Ong, M.D., Scheller, H.V., Vogel, K., Singh, S., 2010. Monitoring and analyzing process streams towards understanding ionic liquid pretreatment of switchgrass (*Panicum virgatum* L.). BioEnergy Res. 3 (2), 134–145.

Awad, R., Muhammad, A., Durst, T., Trudeau, V.L., Arnason, J.T., 2009. Bioassay-guided fractionation of lemon balm (*Melissa officinalis* L.) using an in vitro measure of GABA transaminase activity. Phytother. Res. 23 (8), 1075–1081.

Azzam, A.M., 1989. Pretreatment of cane bagasse with alkaline hydrogen peroxide for enzymatic hydrolysis of cellulose and ethanol fermentation. J. Environ. Sci. Health B 24 (4), 421–433.

Bailey, D.G., Malcolm, J., Arnold, O., David Spence, J., 1998. Grapefruit juice–drug interactions. Br. J. Clin. Pharmacol. 46 (2), 101–110.

Baldrian, P., Valášková, V., Merhautová, V., Gabriel, J., 2005. Degradation of lignocellulose by *Pleurotus ostreatus* in the presence of copper, manganese, lead and zinc. Res. Microbiol. 156 (5), 670–676.

Baldrian, P., Voříšková, J., Dobiášová, P., Merhautová, V., Lisá, L., Valášková, V., 2011. Production of extracellular enzymes and degradation of biopolymers by saprotrophic microfungi from the upper layers of forest soil. Plant Soil 338 (1–2), 111–125.

Ballesteros, I., Negro, M.J., Oliva, J.M., Cabañas, A., Manzanares, P., Ballesteros, M., 2006. Ethanol production from steam-explosion pretreated wheat straw. Twenty-Seventh Symposium on Biotechnology for Fuels and Chemicals. Humana Press, pp. 496–508.

Baricevic, D., Bartol, T., 2000. The biological/pharmacological activity of the salvia genus. In: Kintzios, S.E. (Ed.), The Genus Salvia. CRC Press, pp. 143–184.

Baur, J.A., Sinclair, D.A., 2006. Therapeutic potential of resveratrol: the in vivo evidence. Nat. Rev. Drug Discov. 5 (6), 493–506.

Beachy, P.A., Taipale, J., Chen, J.K., Cooper, M.K., Wang, B., Mann, R.K., Milenkovic, L., Scott, M.P., 2000. Effects of oncogenic mutations in smoothened and patched can be reversed by cyclopamine. Nature 406 (6799), 1005–1009.

Becker, N., Petric, D., Zgomba, M., Boase, C., 2010. Mosquitoes and Their Control. Springer, Berlin, Heidelberg, p. 463.

Bellik, Y., M Hammoudi, S., Abdellah, F., Iguer-Ouada, M., Boukraa, L., 2012. Phyto-chemicals to prevent inflammation and allergy. Rec. Pat. Inflam. Allerg. Drug Disc. 6 (2), 147–158.

Benavides, A., Montoro, P., Bassarello, C., Piacente, S., Pizza, C., 2006. Catechin derivatives in Jatropha macrantha stems: characterisation and LC/ESI/MS/MS quali–quantitative analysis. J. Pharm. Biomed. Anal. 40 (3), 639–647.

Ben-Ghedalia, D., Miron, J., 1981. The effect of combined chemical and enzyme treatments on the saccharification and in vitro digestion rate of wheat straw. Biotechnol. Bioeng. 23 (4), 823–831.

Beri, V., Wildman, S.A., Shiomi, K., Al-Rashid, Z.F., Cheung, J., Rosenberry, T.L., 2013. The natural product dihydrotanshinone I provides a prototype for uncharged inhibitors that bind specifically to the acetylcholinesterase peripheral site with nanomolar affinity. Biochemistry 52 (42), 7486–7499.

Bhat, Z.A., Kumar, D., Shah, M.Y., 2011. Angelica archangelica Linn. is an angel on earth for the treatment of diseases. Int. J. Nutr. Pharmacol. Neurol. Dis. 1 (1), 36.

Bhosale, P., Bernstein, P.S., 2005. Microbial xanthophylls. Appl. Microbiol. Biotechnol. 68 (4), 445–455.

Bouic, P.J., 2001. The role of phytosterols and phytosterolins in immune modulation: a review of the past 10 years. Curr. Opin. Clin. Nutr. Metab. Care 4 (6), 471–475.

Boyer, J., Liu, R.H., 2004. Apple phyto-chemicals and their health benefits. Nutr. J. 12 (3), 5.

Brambila, S., Pino, J.A., Mendoza, C., 1963. Studies with a natural source of xanthophylls for the pigmentation of egg yolks and skin of poultry. Poult. Sci. 42 (2), 294–300.

Brennan, A.H., Hoagland, W., Schell, D.J., Scott, C.D., 1986, January. High temperature acid hydrolysis of biomass using an engineering-scale plug flow reactor. Results of low testing solids. In: Biotechnol. Bioeng. Symp., (United States) (Vol. 17, No. CONF-860508). Solar Energy Research Institute, Golden, CO 80401, USA.

Bridle, P., Bakker, J., Gracia-Viguera, C., Picinelli, A., 1996. Polymerisation reactions in red wines and the effect of sulphur dioxide. In: Fenwick, G.R., Hedley, C., Richards, R.L., Khokhar, S. (Eds.), Agri-Food Quality: An Interdisciplinary Approach. Royal Society of Chemistry, Cambridge, pp. 258–260.

Bridle, P., Timberlake, C.F., 1997. Anthocyanins as natural food colours—selected aspects. Food Chem. 58 (1), 103–109.

Buchholzer, M.L., Dvorak, C., Chatterjee, S.S., Klein, J., 2002. Dual modulation of striatal acetylcholine release by hyperforin, a constituent of St. John's wort. J. Pharmacol. Exp. Ther. 301 (2), 714–719.

Bulotta, S., Celano, M., Lepore, S.M., Montalcini, T., Pujia, A., Russo, D., 2014. Beneficial effects of the olive oil phenolic components oleuropein and hydroxytyrosol: focus on protection against cardiovascular and metabolic diseases. J. Transl. Med. 12 (1), 219.

Burda, S., Oleszek, W., 2001. Antioxidant and antiradical activities of flavonoids. J. Agr. Food Chem. 49 (6), 2774–2779.

Burton, G.W., Ingold, K.U., 1984. Beta-carotene: an unusual type of lipid antioxidant. Science 224 (4649), 569–573.

Bushway, R.J., Savage, S.A., Ferguson, B.S., 1987. Inhibition of acetyl cholinesterase by solanaceous glycoalkaloids and alkaloids. Am. Potato J. 64 (8), 409–413.

Cadoche, L., López, G.D., 1989. Assessment of size reduction as a preliminary step in the production of ethanol from lignocellulosic wastes. Biol. Wastes 30 (2), 153–157.

Carvalho, M., Silva, B.M., Silva, R., Valentao, P., Andrade, P.B., Bastos, M.L., 2010. First report on *Cydonia oblonga* Miller anticancer potential: differential antiproliferative effect against human kidney and colon cancer cells. J. Agr. Food Chem. 58 (6), 3366–3370.

Chang, K.L., Cheng, H.L., Huang, L.W., Hsieh, B.S., Hu, Y.C., Chih, T.T., Su, S.J., 2009. Combined effects of terazosin and genistein on a metastatic, hormone-independent human prostate cancer cell line. Cancer letters 276 (1), 14–20.

Chang, C.C., Yang, M.H., Wen, H.M., Chern, J.C., 2002. Estimation of total flavonoid content in propolis by two complementary colorimetric methods. J. Food Drug Anal. 10 (3).

Chattopadhyay, D., Dungdung, S.R., Mandal, A.B., Majumder, G.C., 2005. A potent sperm motility-inhibiting activity of bioflavonoids from an ethnomedicine of Onge, Alstonia macrophylla Wall ex A. DC, leaf extract. Contraception 71 (5), 372–378.

Chen, J.H., Ho, C.T., 1997. Antioxidant activities of caffeic acid and its related hydroxycinnamic acid compounds. J. Agr. Food Chem. 45 (7), 2374–2378.

Chen, P.N., Chu, S.C., Chiou, H.L., Kuo, W.H., Chiang, C.L., Hsieh, Y.S., 2006. Mulberry anthocyanins, cyanidin-3-rutinoside and cyanidin-3-glucoside, exhibited an inhibitory effect on the migration and invasion of a human lung cancer cell line. Cancer Lett. 235, 248–259.

Chen, X., Bao, J., Guo, J., Ding, Q., Lu, J., Huang, M., Wang, Y., 2012. Biological activities and potential molecular targets of cucurbitacins. Anticancer Drugs 23 (8), 777–787.

Chiang, L.C., Ng, L.T., Liu, L.T., Shieh, D.E., Lin, C.C., 2003. Cytotoxicity and anti-hepatitis B virus activities of saikosaponins from Bupleurum species. Planta Med. 69 (08), 705–709.

Chum, H.L., Johnson, D.K., Black, S., Baker, J., Grohmann, K., Sarkanen, K.V., Schroeder, H.A., 1988. Organosolv pretreatment for enzymatic hydrolysis of poplars: I. enzyme hydrolysis of cellulosic residues. Biotechnol. Bioeng. 31 (7), 643–649.

Ciftci, O., Ozdemir, I., Tanyildizi, S., Yildiz, S., Oguzturk, H., 2011. Antioxidative effects of curcumin, β-myrcene and 1,8-cineole against 2,3,7,8-tetrachlorodibenzo-p-dioxin-induced oxidative stress in rats liver. Toxicol. Ind. Health 27 (5), 447–453.

Clauson, K.A., Shields, K.M., McQueen, C.E., 2003. Safety issues associated with commercially available energy drinks College of Pharmacy-West Palm Beach, Nova Southeastern University, Palm Beach Gardens, Florida. J. Am. Pharmacol. Assoc. 48 (3), 55–63.

Cook, P., James, I., 1981. Cerebral vasodilators. N. Engl. J. Med. 305 (26), 1560–1564.

Corder, R., Mullen, W., Khan, N.Q., Marks, S.C., Wood, E.G., Carrier, M.J., Crozier, A., 2006. Oenology: red wine procyanidins and vascular health. Nature 444 (7119), 566.

Cosentino, S., Tuberoso, C.I.G., Pisano, B., Satta, M., Mascia, V., Arzedi, E., Palmas, F., 1999. In-vitro antimicrobial activity and chemical composition of *Sardinian thymus* essential oils. Lett. Appl. Microbiol. 29 (2), 130–135.

Crowell, P.L., 1997. Monoterpenes in breast cancer chemoprevention. Breast Cancer Res. Treat. 46 (2-3), 191–197.

Culig, Z., Klocker, H., Bartsch, G., Hobisch, A., 2002. Androgen receptors in prostate cancer. Endocr. Rel. Cancer 9 (3), 155–170.

Cunha, J.A., Pereira, M.M., Valente, L.M., De la Piscina, P.R., Homs, N., Santos, M.R.L., 2011. Waste biomass to liquids: low temperature conversion of sugarcane bagasse to bio-oil: the effect of combined hydrolysis treatments. Biomass Bioenergy 35 (5), 2106–2116.

Cushnie, T.P., Lamb, A.J., 2005. Antimicrobial activity of flavonoids. Int. J. Antimicrob. Agents 26 (5), 343–356.

Czock, D., Keller, F., Rasche, F.M., Häussler, U., 2005. Pharmacokinetics and pharmacodynamics of systemically administered glucocorticoids. Clin. Pharmacokinet. 44 (1), 61–98.

Dallas, C., Gerbi, A., Tenca, G., Juchaux, F., Bernard, F.X., 2008. Lipolytic effect of a polyphenolic citrus dry extract of red orange, grapefruit, orange (SINETROL) in human body fat adipocytes. Mechanism of action by inhibition of cAMP-phosphodiesterase (PDE). Phytomedicine 15 (10), 783–792.

De Wet, H., Fouche, G., Van Heerden, F.R., 2009. In vitro cytotoxicity of crude alkaloidal extracts of South African menispermaceae against three cancer cell linese. Afr. J. Biotechnol. 8 (14), 3332–3335.

Dimitrios, B., 2006. Sources of natural phenolics antioxidants. Trends Food Sci. Technol. 17, 505–512.

Dong, Y.L., Chen, W.Z., 1994. Effects of guan-fu base a on experimental cardiac arrhythmias and myocardial contractility. Acta Pharm. Sin. 30 (8), 577–582.

Donghai, S.U., Junshe, S., Ping, L.I.U., Yanping, L.Ü., 2006. Effects of different pretreatment modes on the enzymatic digestibility of corn leaf and corn stalk. Chin. J. Chem. Eng. 14 (6), 796–801.

Dorman, H.J.D., Deans, S.G., 2000. Antimicrobial agents from plants: antibacterial activity of plant volatile oils. J. Appl. Microbiol. 88 (2), 308–316.

Droge, W., 2002. Free radicals in the physiological control of cell function. Physiol. Rev. 282, 47–95.

DTIE, U., 2009. Converting waste agricultural biomass into a resource. Compendium of Technologies. Osaka, United Nations Environment Programme.

Duff, S.J., Murray, W.D., 1996. Bioconversion of forest products industry waste cellulosics to fuel ethanol: a review. Biores. Technol. 55 (1), 1–33.

Duru, M.E., Cakir, A., Kordali, S., Zengin, H., Harmandar, M., Izumi, S., Hirata, T., 2003. Chemical composition and antifungal properties of essential oils of three Pistacia species. Fitoterapia 74 (1), 170–176.

Eldahshan, O.A., 2013. Rhoifolin: a potent antiproliferative effect on cancer cell lines. Br. J. Pharm. Res. 3 (1), 46–53.

Ellnain-Wojtaszek, M., Kruczyński, Z., Kasprzak, J., 2003. Investigation of the free radical scavenging activity of *Ginkgo biloba* L. leaves. Fitoterapia 74 (1), 1–6.

Esiyok, D., Otles, S., Akcicek, E., 2004. Herbs as a food source in Turkey. Asian Pac. J. Cancer Prev. 5 (3), 334–339.

Evidente, A., Randazzo, G., Surico, G., Lavermicocca, P., Arrigoni, O., 1985. Degradation of lycorine by *Pseudomonas* species strain ITM 311. J. Nat. Prod. 48 (4), 564–570.

Faltynek, C.R., Schroeder, J., Mauvais, P., Miller, D., Wang, S., Murphy, D., Maycock, A., 1995. Damnacanthal is a highly potent, selective inhibitor of p56lck tyrosine kinase activity. Biochemistry 34 (38), 12404–12410.

Faq, F., Syed, D.N., Malik, A., Hadi, N., Sarfaraz, S., Kweon, M.-H., Khan, N., Zaid, M.A., Mukhtar, H., 2007. Delphinidin, an anthocyanidin in pigmented fruits and vegetables, protects human HaCaT keratinocytes and mouse skin against UVB-mediated oxidative stress and apoptosis. J. Invest. Dermatol. 127 (1), 222–232.

Faulon, J.L., Hatcher, P.G., 1994. Is there any order in the structure of lignin? Energy Fuels 8 (2), 402–407.

Fazel, S., Hamidreza, M., Rouhollah, G., Mohammadreza, V.R., 2008. Spectrophotometric determination of total alkaloids in some Iranian medicinal plants. Thai J. Pharm. Sci. 32, 17–20.

Fernandes, E.S., Passos, G.F., Medeiros, R., da Cunha, F.M., Ferreira, J., Campos, M.M., Pianowski, L.F., Calixto, J.B., 2007. Anti-inflammatory effects of compounds alpha-humulene and (-)-trans-caryophyllene isolated from the essential oil of *Cordia verbenacea*. Eur. J. Pharmacol. 569 (3), 228–236.

Fiel, S.B., Vincken, W., 2006. Systemic corticosteroid therapy for acute asthma exacerbations. J. Asthma 43 (5), 321–331.

Frankel, E.N., Huang, S.W., Aeschbach, R., Prior, E., 1996. Antioxidant activity of a rosemary extract and its constituents, carnosic acid, carnosol, and rosmarinic acid, in bulk oil and oil-in-water emulsion. J. Agric. Food Chem. 44 (1), 131–135.

Gaitan, E., 1990. Goitrogens in food and water. Ann. Rev. Nutr. 10, 21–39.

Galbe, M., Zacchi, G., 2002. A review of the production of ethanol from softwood. Appl. Microbiol. Biotechnol. 59 (6), 618–628.

Gallardo, C., Jimenez, L., Garcia-Conesa, M.T., 2006. Hydroxycinnamic acid composition and in vitro antioxidant activity of selected grain fractions. Food Chem. 99 (3), 455–463.

Gallwitz, W.E., Garrett, I.R., Gutierrez, G., 2010. US Patent No. 7,678,395. Washington, DC: US Patent and Trademark Office.

Gamze, Y., Pandir, D., Bas, H., 2014. Protective role of catechin and quercetin in sodium benzoate-induced lipid peroxidation and the antioxidant system in human erythrocytes in vitro. The Scientific World Journal, 2014.

Giatropoulos, A., Pitarokili, D., Papaioannou, F., Papachristos, D.P., Koliopoulos, G., Emmanouel, N., Michaelakis, A., 2013. Essential oil composition, adult repellency and larvicidal activity of eight Cupressaceae species from Greece against Aedes albopictus (Diptera: Culicidae). Parasitol. Res., 1–11.

Gibbons, S., Leimkugel, J., Oluwatuyi, M., Heinrich, M., 2003. Activity of Zanthoxylum clava-herculis extracts against multi-drug resistant methicillin-resistant *Staphylococcus aureus* (mdr-MRSA). Phytother. Res. 17 (3), 274–275.

González-Lamothe, R., Mitchell, G., Gattuso, M., Diarra, M.S., Malouin, F., Bouarab, K., 2009. Plant antimicrobial agents and their effects on plant and human pathogens. Int. J. Mol. Sci. 10 (8), 3400–3419.

Goya, L., Mateos, R., Bravo, L., 2007a. Effect of the olive oil phenol hydroxytyrosol on human hepatoma HepG2 cells: protection against oxidative stress induced by tert-butylhydroperoxide. Eur. J. Nutr. 46 (2), 70–78.

Goya, L., Delgado-Andrade, C., Rufián-Henares, J.A., Bravo, L., Morales, F.J., 2007b. Effect of coffee Melanoidin on human hepatoma HepG2 cells: protection against oxidative stress induced by tert-butylhydroperoxide. Mol. Nutr. Food Res. 51 (5), 536–545.

Guardia, T., Rotelli, A.E., Juarez, A.O., Pelzer, L.E., 2001. Anti-inflammatory properties of plant flavonoids: effects of rutin, quercetin and hesperidin on adjuvant arthritis in rat. Il Farmaco 56 (9), 683–687.

Guifang, W., Shouyao, Z., Suqin, Z., 1999. Determination of L-sesamin and L-asarinin in asarumheterotropoides Fr. var. mandshuricum (Maxim) Kitag. by HPLC. Chin. J. Pharma. Anal. 4, 251–253.

Halliwell, B., 2006. Reactive species and antioxidants: redox biology is a fundamental theme of aerobic life. Plant Physiol. 141 (2), 312–322.

Halliwell, B., Gutteridge, J.M.C., 2007a. Cellular responses to oxidative stress: adaptation, damage, repair, senescence and death. Free Rad. Biol. Med. 4.

Halliwell, B., Gutteridge, J.M.C., 2007b. Free Radicals in Biology and Medicine, fourth ed. Oxford University Press, New York.

Haruta, S., Cui, Z., Huang, Z., Li, M., Ishii, M., Igarashi, Y., 2002. Construction of a stable microbial community with high cellulose-degradation ability. Appl. Microbiol. Biotechnol. 59 (4–5), 529–534.

Hase, K., Ohsugi, M., Xiong, Q., Basnet, P., Kadota, S., Namba, T., 1997. Hepatoprotective effect of *Hovenia dulcis* THUNB: on experimental liver injuries induced by carbon tetrachloride or D-galactosamine/lipopolysaccharide. Biol. Pharm. Bull. 20 (4), 381–385.

Hayashi, K.Y.O.K.O., Hayashi, T., Morita, N., 1992. Mechanism of action of the antiherpesvirus biflavone ginkgetin. Antimicrob. Agents Chemother. 36 (9), 1890–1893.

Haydel, S.E., Remenih, C.M., Williams, L.B., 2008. Broad-spectrum in vitro antibacterial activities of clay minerals against antibiotic-susceptible and antibiotic-resistant bacterial pathogens. J. Antimicrob. Chemother. 61 (2), 353–361.

He, J., Wang, S., Wu, T., Cao, Y., Xu, X., Zhou, X., 2013. Effects of ginkgoneolic acid on the growth, acidogenicity, adherence, and biofilm of *Streptococcus* mutans in vitro. Folia Microbiol. 58 (2), 147–153.

Hecht, S.S., 2000. Inhibition of carcinogenesis by isothiocyanates. Drug Metab. Rev. 32 (3–4), 395–411.

Heisey, R.M., Heisey, T.K., 2003. Herbicidal effects under field conditions of *Ailanthus altissima* bark extract, which contains ailanthone. Plant Soil 256 (1), 85–99.

Herraiz, T., Galisteo, J., 2003. Tetrahydro-β-carboline alkaloids occur in fruits and fruit juices. Activity as antioxidants and radical scavengers. J. Agric. Food Chem. 51 (24), 7156–7161.

Ho, C.T., 1992. Phenolic compounds in food. In: Phenolic Compounds in Food and Their Effects on Health I (vol. 506). Chapter 1 ACS Symposium Series, American Chemical Society, Washington, DC, pp. 2–7.

Hu, K., Yao, X., 2001. Methyl protogracillin (NSC-698792): the spectrum of cytotoxicity against 60 human cancer cell lines in the National Cancer Institute's anticancer drug screen panel. Anticancer Drugs 12 (6), 541–547.

Huie, C.W., 2002. A review of modern sample-preparation techniques for the extraction and analysis of medicinal plants. Anal. Bioanal. Chem. 373 (1–2), 23–30.

Humfrey, C.D., 1998. Phytoestrogens and human health effects: weighing up the current evidence. Nat. Toxins 6 (2), 51–59.

Ilic, D., Forbes, K.M., Hassed, C., 2011. Lycopene for the prevention of prostate cancer. Cochrane Database Syst. Rev. (11).

Ilie, M., Margină, D., 2012. Trends in the evaluation of lipid peroxidation processes. In: Catala, A. (Ed.), Lipid peroxidation. InTech, Rijeka, Croatia, (Chapter 5).

Im, A., Han, L., Kim, E., Kim, J., Kim, Y.S., Park, Y., 2012. Enhanced antibacterial activities of Leonuri herba extracts containing silver nanoparticles. Phytother. Res. 26 (8), 1249–1255.

Imlay, J.A., Chin, S.M., Linn, S., 1988. Toxic DNA damage by hydrogen peroxide through the Fenton reaction in vivo and in vitro. Science 240 (4852), 640–642.

Jakubikova, J., Sedlak, J., 2005. Garlic-derived organosulfides induce cytotoxicity, apoptosis, cell cycle arrest and oxidative stress in human colon carcinoma cell lines. Neoplasma 53 (3), 191–199.

Jemal, A., Siegel, R., Ward, E., Murray, T., Xu, J., Thun, M.J., 2007. Cancer statistics. CA Cancer J. Clin. 57 (1), 43–66.

Jenab, M., Thompson, L.U., 1996. The influence of flaxseed and lignans on colon carcinogenesis and β-glucuronidase activity. Carcinogenesis 17 (6), 1343–1348.

Jenkins, D.J., Kendall, C.W., Connelly, P.W., Jackson, C.J.C., Parker, T., Faulkner, D., Vidgen, E., 2002. Effects of high-and low-isoflavone (phytoestrogen) soy foods on inflammatory biomarkers and proinflammatory cytokines in middle-aged men and women. Metabolism 51 (7), 919–924.

Jeon, K.-I., Xu, X., Aizawa, T., Lim, J.H., Jono, H., Kwon, D.-S., Abe, J.-I., Berk, B.C., Li, J.-D., Yan, C., 2010. Vinpocetine inhibits NF-B-dependent inflammation via an IKK-dependent but PDE-independent mechanism. Proc. Natl. Acad. Sci. 107 (21), 9795–9800.

Jia, X.S., Furumai, H., Fang, H.H., 1996. Yields of biomass and extracellular polymers in four anaerobic sludges. Environ. Technol. 17 (3), 283–291.

Jiang, W., Fu, F., Tian, J., Zhu, H., Hou, J., 2011. Curculigoside A attenuates experimental cerebral ischemia injury in vitro and vivo. Neuroscience 192, 572–579.

Johnston, K.L., Clifford, M.N., Morgan, L.M., 2003. Coffee acutely modifies gastrointestinal hormone secretion and glucose tolerance in humans: glycemic effects of chlorogenic acid and caffeine. Am. J. Clin. Nutr. 78 (4), 728–733.

Jordan, M.A., Thrower, D., Wilson, L., 1991. Mechanism of inhibition of cell proliferation by *Vinca* alkaloids. Cancer Res. 51 (8), 2212–2222.

Jose, B., Reddy, L.J., 2010. Analysis of the essential oils of the stems, leaves and rhizomes of the medicinal plant *Costus pictus* from southern India. Int. J. Pharm. Pharm. Sci. 2, 163–173.

Jovanovic, S.V., Steenken, S., Tosic, M., Marjanovic, B., Simic, M.G., 1994. Flavonoids as antioxidants. J. Am. Chem. Soc. 116 (11), 4846–4851.

Jung, M.K., Hur, D.Y., Song, S.B., Park, Y., Kim, T.S., Bang, S.I., Kim, S., Song, H.K., Park, H., Cho, D.H., 2010. Tannic acid and quercetin display a therapeutic effect in atopic dermatitis via suppression of angiogenesis and TARC expression in Nc/Nga Mice. J. Invest. Dermatol. 130 (5), 1459–1463.

Jung, H.A., Islam, M.N., Lee, C.M., Oh, S.H., Lee, S., Jung, J.H., Choi, J.S., 2013. Kinetics and molecular docking studies of an anti-diabetic complication inhibitor fucosterol from edible brown algae *Eisenia bicyclis* and *Ecklonia stolonifera*. Chem. Biol. Interact. 206 (1), 55–62.

Kandhare, A.D., Raygude, K.S., Ghosh, P., Ghule, A.E., Bodhankar, S.L., 2012. Neuroprotective effect of naringin by modulation of endogenous biomarkers in streptozotocin induced painful diabetic neuropathy. Fitoterapia 83 (4), 650–659.

Kashiwada, Y., Wang, H.K., Nagao, T., Kitanaka, S., Yasuda, I., Fujioka, T., Yamagishi, T., Cosentino, L.M., Kozuka, M., Okabe, H., Ikeshiro, Y., Hu, C.Q., Yeh, E., Lee, K.H., 1998. Anti-AIDS agents. 30. Anti-HIV activity of oleanolic acid, pomolic acid, and structurally related *triterpenoids*. J. Nat. Prod. 61 (9), 1090–1095.

Kaufman, P.B., Duke, J.A., Brielmann, H., Boik, J., Hoyt, J.E., 1997. A comparative survey of leguminous plants as sources of the isoflavones, genistein and daidzein: implications for human nutrition and health. J. Alter. Complement. Med. 3 (1), 7–12.

Kawai, Y., Nishikawa, T., Shiba, Y., Saito, S., Murota, K., Shibata, N., Terao, J., 2008. Macrophage as a target of quercetin glucuronides in human atherosclerotic arteries implication in the anti-atherosclerotic mechanism of dietary flavonoids. J. Biol. Chem. 283 (14), 9424–9434.

Kern, J.W., Lipman, A.G., 1977. Rational theophylline therapy: a review of the literature with a guide to pharmacokinetics and dosage calculation. drug Intell. Clin. Pharm. 11 (3), 144–153.

Kiewert, C., Buchholzer, M.L., Hartmann, J., Chatterjee, S.S., Klein, J., 2004. Stimulation of hippocampal acetylcholine release by hyperforin, a constituent of St. John's Wort. Neurosci. Lett. 364 (3), 195–198.

Kilfoy, B.A., Zhang, Y., Park, Y., 2011. Dietary nitrate and nitrite and the risk of thyroid cancer in the NIH-AARP diet and health study. Int. J. Cancer 129 (1), 160–172.

Kim, D.K., Chancellor, M.B., 2008. Men reporting lasting longer with hyperforin. Int. Braz. J. Urol. 34 (3), 370–371.

Kim, S., Holtzapple, M.T., 2006. Effect of structural features on enzyme digestibility of corn stover. Bioresour. Technol. 97 (4), 583–591.

Kirtikar, K.R., Basu, B.D., 1918. Indian medicinal plants. Indian Med. Plants.

Kobayashi, M., Kobayashi, H., Mori, K., Pollard, R.B., Suzuki, F., 1998. The regulation of burn-associated infections with herpes simplex virus type 1 or *Candida albicans* by a non-toxic aconitine-hydrolysate, benzoylmesaconine part 2: mechanism of the antiviral action. Immunol. Cell Biol. 76 (3), 209–216.

Konstantinopoulou, M., Karioti, A., Skaltsas, S., Skaltsa, H., 2003. Sesquiterpene lactones from anthemis a ltissima and their anti-helicobacter pylori activity. J. Nat. Prod. 66 (5), 699–702.

Koriem, K.M., Arbid, M.S., Asaad, G.F., 2013. *Chelidonium majus* leaves methanol extract and its chelidonine alkaloid ingredient reduce cadmium-induced nephrotoxicity in rats. J. Nat. Med. 67 (1), 159–167.

Kose, E.O., Deniz, I.G., Sarikurkcu, C., Aktas, O., Yavuz, M., 2010. Chemical composition, antimicrobial and antioxidant activities of the essential oils of *Sideritis erythrantha* Boiss. and Heldr. (var. erythrantha and var. cedretorum P.H. Davis) endemic in Turkey. Food Chem. Toxicol. 48, 2960–2965.

Kotan, R., Kordali, S., Cakir, A., 2007. Screening of antibacterial activities of twenty-one oxygenated monoterpenes. Zeitschrift Nat. C 62 (7/8), 507.

Kren, V.C., 1997. New method of preparation of ß-D-O-galactosides of ergot alkaloids. Top Curr. Chem. 186, 45.

Krinsky, N.I., Johnson, E.J., 2005. Carotenoid actions and their relation to health and disease. Mol. Asp. Med. 26 (6), 459–516.

Kubo, I., Kinst-Hori, I., 1999. Flavonols from saffron flower: tyrosinase inhibitory activity and inhibition mechanism. J. Agric. Food Chem. 47 (10), 4121–4125.

Kubo, I., Muroi, H., Kubo, A., 1994. Naturally occurring antiacne agents. J. Nat. Prod. 57 (1), 9–17.

Kuć, J., Currier, W., 1975. Phytoalexins, plants and human health. Adv. Chem.

Kuhnau, J., 1976. Flavonoids: a class of semi-essential food components: their role in human nutrition. World Rev. Nutr. Diet..

Kumar, Y.B., Babu, G.S., Bhaskar, P.U., Raja, V.S., 2009. Effect of starting-solution pH on the growth of Cu2ZnSnS4 thin films deposited by spray pyrolysis. Phys. Status Solidi A 206 (7), 1525–1530.

Kumar, A., Dogra, S., Prakash, A., 2010. Protective effect of naringin, a citrus flavonoid, against colchicine-induced cognitive dysfunction and oxidative damage in rats. J. Med. Food 13 (4), 976–984.

Kundu, J.K., Surh, Y.J., 2008. Cancer chemopreventive and therapeutic potential of resveratrol mechanistic perspectives. Cancer Lett. 269 (2), 243–261.

Kunert, O., Swamy, R.C., Kaiser, M., Presser, A., Buzzi, S., Appa Rao, A.V.N., Schühly, W., 2008. Antiplasmodial and leishmanicidal activity of biflavonoids from Indian *Selaginella bryopteris*. Phytochem. Lett. 1 (4), 171–174.

Kuo, Y.H., Hwang, S.Y., Yang Kuo, L.M., Lee, Y.L., Li, S.Y., Shen, Y.C., 2002. A novel cytotoxic C-methylated biflavone, taiwanhomoflavone-B from the twigs of *Cephalotaxus wilsoniana*. Chem. Pharm. Bull. 50 (12), 1607–1608.

Kuroda, M., Mimaki, Y., Sashida, Y., Hirano, T., Oka, K., Dobashi, A., Harada, N., 1997. Novel cholestane glycosides from the bulbs of *Ornithogalum saundersiae* and their cytostatic activity on leukemia HL-60 and MOLT-4 cells. Tetrahedron 53 (34), 11549–11562.

Kuroiwa, Y., Nishikawa, A., Kitamura, Y., Kanki, K., Ishii, Y., Umemura, T., Hirose, M., 2006. Protective effects of benzyl isothiocyanate and sulforaphane but not resveratrol against initiation of pancreatic carcinogenesis in hamsters. Cancer Lett. 241 (2), 275–280.

Kurzer, M.S., Xu, X., 1997. Dietary phytoestrogens. Ann. Rev. Nutr. 17, 353–381.

Kwak, W.J., Han, C.K., Son, K.H., Chang, H.W., Kang, S.S., Park, B.K., Kim, H.P., 2002. Effects of Ginkgetin from *Ginkgo biloba* leaves on cyclooxygenases and in vivo skin inflammation. Planta Med. 68 (04), 316–321.

Kwon, H.J., Ryu, Y.B., Jeong, H.J., Kim, J.H., Park, S.J., Chang, J.S., Lee, W.S., 2009. Rhodiosin, an antioxidant flavonol glycoside from *Rhodiola rosea*. J. Korean Soc. Appl. Biol. Chem. 52 (5), 486–492.

Lambrou, G.I., Vlahopoulos, S., Papathanasiou, C., Papanikolaou, M., Karpusas, M., Zoumakis, E., Tzortzatou-Stathopoulou, F., 2009. Prednisolone exerts late mitogenic and biphasic effects on resistant acute lymphoblastic leukemia cells: Relation to early gene expression. Leuk. Res. 33 (12), 1684–1695.

Lampe, C.A., Ellison, N., Steinfield, C., 2007. A familiar face (book): profile elements as signals in an online social network. In: Proceedings of the SIGCHI conference on Human factors in computing systems. ACM, pp. 435–444.

Landete, J.M., 2012. Plant and mammalian lignans: a review of source, intake, metabolism, intestinal bacteria and health. Food Res. Int. 46 (1), 410–424.

Lee, D.S., Lee, S.H., Noh, J.G., Hong, S.D., 1999. Antibacterial activities of cryptotanshinone and dihydrotanshinone I from a medicinal herb, *Salvia miltiorrhiza* Bunge. Bio. Biotechnol. Biochem. 63 (12), 2236–2239.

Lee, J.H., Kang, G.H., Kim, K.C., Kim, K.M., Park, D.I., Choi, B.T., Choi, Y.H., 2002. Tetrandrine-induced cell cycle arrest and apoptosis in A549 human lung carcinoma cells. Int. J. Oncol. 21 (6), 1239–1244.

Lee, S., Peterson, C.J., Coats, J.R., 2003. Fumigation toxicity of monoterpenoids to several stored product insects. J. Stored Prod. Res. 39 (1), 77–85.

Lee, J.S., Kim, J., Yu, Y.U., Kim, Y.C., 2004a. Inhibition of phospholipase Cγ1 and cancer cell proliferation by lignans and flavans from *Machilus thunbergii*. Arch. Pharm. Res. 27 (10), 1043–1047.

Lee, K.R., Kozukue, N., Han, J.S., Park, J.H., Chang, E.Y., Baek, E.J., Friedman, M., 2004b. Glycoalkaloids and metabolites inhibit the growth of human colon (HT29) and liver (HepG2) cancer cells. J. Agr. Food Chem. 52 (10), 2832–2839.

Lee, H.K., Choi, Y.M., Noh, D.O., Suh, H.J., 2005. Antioxidant effect of Korean traditional lotus liquor (Yunyupju). Int. J. Food Sci. Technol. 40 (7), 709–715.

Lee, J.S., Sul, J.Y., Park, J.B., Lee, M.S., Cha, E.Y., Song, I.S., Kim, J.R., Chang, E.S., 2013. Fatty acid synthase inhibition by amentoflavone suppresses HER2/neu(erbB2) oncogene in SKBR3 human breast cancer cells. Phytother. Res. 27 (5), 713–720.

Lejeune, M.P., Kovacs, E.M., Westerterp-Plantenga, M.S., 2003. Effect of capsaicin on substrate oxidation and weight maintenance after modest body-weight loss in human subjects. Br. J. Nutr. 90 (3), 651–659.

Li, Y., But, P.P., Ooi, V.E., 2005. Antiviral activity and mode of action of caffeoylquinic acids from *Schefflera heptaphylla* (L.) Frodin. Antiviral Res. 68 (1), 1–9.

Liao, F., Zheng, R.L., Gao, J.J., Jia, Z.J., 1999. Retardation of skeletal muscle fatigue by the two phenylpropanoid glycosides: verbascoside and martynoside from *Pedicularis plicata* Maxim. Phytother. Res. 13 (7), 621–623.

Lila, M.A., 2004. Anthocyanins and human health: an in vitro investigative approach. BioMed Res. Int. 2004 (5), 306–313.

Lim, H., Son, K.H., Chang, H.W., Bae, K., Kang, S.S., Kim, H.P., 2008. Anti-inflammatory activity of pectolinarigenin and pectolinarin isolated from *Cirsium chanroenicum*. Biol. Pharm. Bull. 31 (11), 2063–2067.

Liu, R.H., 2004. Potential synergy of phyto-chemicals in cancer prevention: mechanism of action. J. Nutr. 134 (12), 3479S–3485S.

Liu, G., Grifman, M., Macdonald, J., Moller, P., Wong-Staal, F., Li, Q.X., 2007. Isoginkgetin enhances adiponectin secretion from differentiated adiposarcoma cells via a novel pathway involving AMP-activated protein kinase. J. Endocrinol. 194 (3), 569–578.

Liu, A.L., Liu, B., Qin, H.L., Lee, S.M.Y., Wang, Y.T., Du, G.H., 2008. Anti-influenza virus activities of flavonoids from the medicinal plant *Elsholtzia rugulosa*. Planta Med. 74 (08), 847–851.

Liu, S.L., Chen, G., Liu, P., Tian, D.Z., Wang, P., 2011. Analysis on the properties of 146 Chinese traditional medicines for lung channel tropism. Lishizhen Med. Mater. Res. 10, 095.

Liu, L., Li, B., Zhou, Y., Wang, L., Tang, F., Shao, D., Li, Y., 2012. Antidepressant-like effect of fuzi total alkaloid on ovariectomized mice. J. Pharm. Sci. 120 (4), 280–287.

Lopez, R., Pina, M.B., Estrada, R.R., Heinze, G., Martinez, V.M., 2006. Anxiolytic effect of hexane extract of the leaves of *Annona cherimolia* in two anxiety paradigms: possible involvement of the GABA/benzodiazepine receptor complex. Life Sci. 78, 730–737.

Ma, J., Dey, M., Yang, H., Poulev, A., Pouleva, R., Dorn, R., Lipsky, P.E., Kennelly, E.J., et al., 2007. Anti-inflammatory and immunosuppressive compounds from *Tripterygium wilfordii*. Phytochemistry 68 (8), 1172–1178.

Mackie, K.L., Brownell, H.H., West, K.L., Saddler, J.N., 1985. Effect of sulphur dioxide and sulphuric acid on steam explosion of aspenwood. J. Wood Chem. Technol. 5 (3), 405–425.

MacRae, W.D., Hudson, J.B., Towers, G.H., 1989. The antiviral action of lignans. Planta Med. 55 (6), 531–535.

Mahato, S.B., Sen, S., 1997. Advances in triterpenoid research, 1990–1994. Phytochemistry 44 (7), 1185–1236.

Maia, M.F., Moore, S.J., 2011. Plant-based insect repellents: a review of their efficacy, development and testing. Malar J. 10 (Suppl. 1), S11.

Malinow, M.R., 1986. US Patent No. 4,602,005. Washington, DC: US Patent and Trademark Office.

Mallikharjuna, P.B., Rajanna, L.N., Seetharam, Y.N., Sharanabasappa, G.K., 2007. Phyto-chemicals studies of *Strychnos potatorum* Lf-A medicinal plant. J. Chem. 4 (4), 510–518.

Mamur, S., Yüzbaşıoğlu, D., Ünal, F., Aksoy, H., 2012. Genotoxicity of food preservative sodium sorbate in human lymphocytes in vitro. Cytotechnology 64 (5), 553–562.
Manach, C., Scalbert, A., Morand, C., Rémésy, C., Jiménez, L., 2004. Polyphenols: food sources and bioavailability. Am. J. Clin. Nutr. 79 (5), 727–747.
Mandava, N.B., Sasse, J.M., Yopp, J.H., 1981. Brassinolide, a growth-promoting steroidal lactone. Physiol. Plant. 53 (4), 453–461.
Martínez-Poveda, B., Verotta, L., Bombardelli, E., Quesada, A.R., Medina, M.A., 2010. Tetrahydrohyperforin and octahydrohyperforin are two new potent inhibitors of angiogenesis. PLoS One 5 (3), e9558.
Martinez-Vazquez, M., Ramirez Apan, T.O., Lastra, A.L., Bye, R., 1998. A comparative study of the analgesic and anti-inflammatory activities of pectolinarin isolated from *Cirsium subcoriaceum* and linarin isolated from *Buddleia cordata*. Planta Med. 64 (2), 134–137.
Massumi, M.A., Fazeli, M.R., Alavi, S.H.R., Ajani, Y., 2007. Chemical constituents and antibacterial activity of essential oil of *Prangos ferulacea* (L.) Lindl. fruits. Iranian J. Pharm. Sci. 3 (3), 171–176.
Mata, A.T., Proença, C., Ferreira, A.R., Serralheiro, M.L.M., Nogueira, J.M.F., Araújo, M.E.M., 2007. Antioxidant and antiacetylcholinesterase activities of five plants used as Portuguese food spices. Food Chem. 103 (3), 778–786.
Mayne, S.T., 1996. Beta-carotene, carotenoids, and disease prevention in humans. FASEB J. 10 (7), 690–701.
McDaniel, M.A., Maier, S.F., Einstein, G.O., 2003. "Brain-specific" nutrients: a memory cure? Nutrition 19 (11–12), 957–975.
McMillan, J.D., 1994. Pretreatment of lignocellulosic biomass (vol. 566). In: ACS Symposium Series, National Renewable Energy Laboratory, CO, USA, pp. 292–324 (Chapter 15).
Meena, A.K., Nitika, G., Jaspreet, N., Meena, R.P., Rao, M.M., 2001. Review on ethnobotany, phyto-chemicals and pharmacological profile of alstonia scholaris. Int. Res. J. Pharm. 2 (1), 49–54.
Middleton, E.J., Kandaswami, C., 1992. Effects of flavonoids on immune and inflammatory cell functions. Biochem. Pharmacol. 43 (6), 1167–1179.
Middleton, Jr., E., Kandaswami, C., Theoharides, T.C., 2000. The effects of plant flavonoids on mammalian cells: implications for inflammation, heart disease, and cancer. Pharmacol. Rev. 52 (4), 673–751.
Millan, M.J., Newman-Tancredi, A., Audinot, V., et al., 2000. Agonist and antagonist actions of yohimbine as compared to fluparoxan at alpha(2)-adrenergic receptors (AR)s, serotonin (5-HT)(1A), 5-HT(1B), 5-HT(1D) and dopamine D(2) and D(3) receptors. Significance for the modulation of frontocortical monoaminergic transmission and depressive states. Synapse 35 (2), 79–95.
Minguez-Mosquera, M.I., Hornero-Mendez, D., 1993. Separation and quantification of the carotenoid pigments in red peppers (*Capsicum annuum* L.), paprika, and oleoresin by reversed-phase HPLC. J. Agr. Food Chem. 41 (10), 1616–1620.
Moertel, C.G., Fleming, T.R., Rubin, J., 1982. A clinical trial of amygdalin (Laetrile) in the treatment of human cancer. N. Engl. J. Med. 306 (4), 201–206.
Mogil, J.S., Shin, Y.H., McCleskey, E.W., Kim, S.C., Nah, S.Y., 1998. Ginsenoside Rf, a trace component of ginseng root, produces antinociception in mice. Brain Res. 792 (2), 218–228.
Moon, K., Khadabadi, S.S., Deokate, U.A., Deore, S.L., 2010. *Caesalpinia bonducella* F: an overview, report and opinion. Sci. Pub. J. 2 (3), 83–90.
Moongkarndi, P., Kosem, N., Kaslungka, S., Luanratana, O., Pongpan, N., Neungton, N., 2004. Antiproliferation, antioxidation and induction of apoptosis by *Garcinia mangostana* (mangosteen) on SKBR3 human breast cancer cell line. J. Ethnoparmacol. 90 (1), 161–166.
Moreau, R.A., Whitaker, B.D., Hicks, K.B., 2002. Phytosterols, phytostanols, and their conjugates in foods: structural diversity, quantitative analysis, and health-promoting uses. Prog. Lipid Res. 41 (6), 457–500.
Morra, M.J., Kirkegaard, J.A., 2002. Isothiocyanate release from soil-incorporated Brassica tissues. Soil Biol. Biochem. 34, 1683–1690.
Mortensen, A., 2006. Carotenoids and other pigments as natural colorants. Pure Appl. Chem. 78 (8), 1477–1491.
Mosier, N., Wyman, C., Dale, B., Elander, R., Lee, Y.Y., Holtzapple, M., Ladisch, M., 2005. Features of promising technologies for pretreatment of lignocellulosic biomass. Bioresour. Technol. 96 (6), 673–686.
Mukherjee, P.K., 2003. Plant products with hypocholesterolemic potentials. Adv. Food Nutr. Res. 47, 277–338.

Mukherjee, P.K., Wahile, A., 2006. Integrated approaches towards drug development from ayurveda and other Indian system of medicines. J. Ethnopharmacol. 103 (1), 25–35.
Murayama, M., 1998. US Patent No. 5,770,604. Washington, DC: US Patent and Trademark Office.
Nagashiro, C.W., Saucedo, A., Alderson, E., Wood, C.D., Nagler, M.J., 2001. Chemical composition, digestibility and aflatoxin content of Brazil nut (*Bertholletia excelsa*) cake produced in north-eastern Bolivia. Livestock Res. Rural Dev. 13, 2.
Negro, D., Montesano, V., Grieco, S., Crupi, P., Sarli, G., De Lisi, A., Sonnante, G., 2012. Polyphenol compounds in artichoke plant tissues and varieties. J. Food Sci. 77 (2), C244–C252.
Nepal, M., Jung Choi, H., Choi, B.Y., Lim Kim, S., Ryu, J.H., Hee Kim, H., Soh, Y., 2012. Anti-angiogenic and anti-tumor activity of Bavachinin by targeting hypoxia-inducible factor-1α. Eur. J. Pharmacol..
Newmark, J., Brady, R.O., Grimley, P.M., Gal, A.E., Waller, S.G., Thistlethwaite, J.R., 1981. Amygdalin (Laetrile) and prunasin beta-glucosidases: distribution in germ-free rat and in human tumor tissue. Proc. Natl. Acad. Sci. 78 (10), 6513–6516.
Ninfali, P., Bacchiocca, M., Antonelli, A., Biagiotti, E., Di Gioacchino, A.M., Piccoli, G., Brandi, G., 2007. Characterization and biological activity of the main flavonoids from Swiss Chard (*Beta vulgaris* subspecies cycla). Phytomedicine 14 (2), 216–221.
Nishimura, K., Kobayashi, T., Fujita, T., 1986. Symptomatic and neurophysiological activities of new synthetic non-ester pyrethroids, ethofenprox, MTI-800, and related compounds. Pest. Biochem. Physiol. 25 (3), 387–395.
Nishiyama, T., Hagiwara, Y., Hagiwara, H., Shibamoto, T., 1994. Inhibitory effect of 2″-O-glycosyl isovitexin and α-tocopherol on genotoxic glyoxal formation in a lipid peroxidation system. Food Chem. Toxicol. 32 (11), 1047–1051.
Nitsche, C., Steuer, C., Klein, C.D., 2011. Arylcyanoacrylamides as inhibitors of the dengue and West Nile virus proteases. Bioorg. Med. Chem. 19 (24), 7318–7337.
Niyogi, K.K., Bjorkman, O., Grossman, A.R., 1997. Chlamydomonas xanthophyll cycle mutants identified by video imaging of chlorophyll fluorescence quenching. Plant Cell 9 (8), 1369–1380.
Noratto, G., Porter, W., Byrne, D., Cisneros-Zevallos, L., 2009. Identifying peach and plum polyphenols with chemopreventive potential against estrogen-independent breast cancer cells. J. Agric. Food Chem. 57 (12), 5219–5226.
O'Brien, K., Matlin, A.J., Lowell, A.M., Moore, M.J., 2008. The biflavonoid isoginkgetin is a general inhibitor of Pre-mRNA splicing. J. Biol. Chem. 283 (48), 33147–33154.
Okano, K., Kitagawa, M., Sasaki, Y., Watanabe, T., 2005. Conversion of Japanese red cedar (*Cryptomeria japonica*) into a feed for ruminants by white-rot basidiomycetes. Anim. Feed Sci. Technol. 120 (3), 235–243.
Omar, M.M., 1992. Phenolic compounds in botanical extracts used in foods, flavors, cosmetics, and pharmaceuticals. In: ACS Symposium Series (vol. 506). American Chemical Society, NJ, United States, pp. 154 (Chapter 12).
Ono, M., Tanaka, N., Orita, K., 1994. Positive interactions between human interferon and cepharanthin against human cancer cells in vitro and in vivo. Cancer Chemother. Pharmacol. 35 (1), 10–16.
Owen, R.W., Giacosa, A., Hull, W.E., Haubner, R., Würtele, G., Spiegelhalder, B., Bartsch, H., 2000. Olive-oil consumption and health: the possible role of antioxidants. Lancet Oncol. 1 (2), 107–112.
Pagán, R., Mackey, B., 2000. Relationship between membrane damage and cell death in pressure-treated *Escherichia coli* cells: differences between exponential-and stationary-phase cells and variation among strains. Appl. Environ. Microbiol. 66 (7), 2829–2834.
Papoutsi, Z., Kassi, E., Mitakou, S., Aligiannis, N., Tsiapara, A., Chrousos, G.P., Moutsatsou, P., 2006. Acteoside and martynoside exhibit estrogenic/antiestrogenic properties. J. Steroid Biochem. Mol. Biol. 98 (1), 63–71.
Park, K.C., Huh, S.Y., Choi, H.R., Kim, D.S., 2010. Biology of melanogenesis and the search for hypopigmenting agents. Dermatol. Sin. 28 (2), 53–58.
Patel, D., Shukla, S., Gupta, S., 2007. Apigenin and cancer chemoprevention: progress, potential and promise (review). Int. J. Oncol. 30 (1), 233.
Patten, G.S., Abeywardena, M.Y., Bennett, L.E., 2013. Inhibition of angiotensin converting enzyme, angiotensin II receptor blocking and blood pressure lowering bioactivity across plant families. Crit. Rev. Food Sci. Nutr. 56 (2), 181–214.

Pedro, L.V., Falé, P.J., Madeira, M.A., Florêncio, H., Ascensão, L., Serralheiro, M.L.M., 2011. Function of *Plectranthus barbatus* herbal tea as neuronal acetylcholinesterase inhibitor. Food Funct. 2, 130–136.

Peto, R., Doll, R., Buckley, J.D., Sporn, M.B., 1981. Can dietary beta-carotene materially reduce human cancer rates? Nature.

Pietta, P., Simonetti, P., Gardana, C., Mauri, P., 2000. Trolox equivalent antioxidant capacity (TEAC) of *Ginkgo biloba* flavonol and *Camellia sinensis* catechin metabolites. J. Pharm. Biomed. Anal. 23 (1), 223–226.

Pingale, R., Pokharkar, D., Phadtare, S., 2013. A review on ethnopharmacolgy, phytochemistry and bioactivity of *Leonitis nepatofolia*. Int. J. Pharm. Tech. Res. 5 (3), 1161–1164.

Podolak, I., Galanty, A., Sobolewska, D., 2010. Saponins as cytotoxic agents: a review. Phytochem. Rev. 9 (3), 425–474.

Prakash, G., Hosetti, B.B., 2012. Bio-efficacy of *Dioscorea pentaphylla* from mid-Western Ghats, India. Toxicol. Int. 19 (2), 100.

Prasad, K., 1999. Reduction of serum cholesterol and hypercholesterolemic atherosclerosis in rabbits by secoisolariciresinol diglucoside isolated from flaxseed. Circulation 99 (10), 1355–1362.

Puupponen-Pimiä, R., Nohynek, L., Hartmann-Schmidlin, S., Kähkönen, M., Heinonen, M., Määttä-Riihinen, K., Oksman-Caldentey, K.M., 2005. Berry phenolics selectively inhibit the growth of intestinal pathogens. J. applied microbiology 98 (4), 991–1000.

Quesada, J., Rubio, M., Gómez, D., 1999. Ozonation of lignin rich solid fractions from corn stalks. J. Wood Chem. Technol. 19 (1–2), 115–137.

Quiney, C., Billard, C., Salanoubat, C., Fourneron, J.D., Kolb, J.P., 2006. Hyperforin, a new lead compound against the progression of cancer and leukemia? Leukemia 20 (9), 1519–1525.

Rahman, S., Khan, R.A., Kumar, A., 2002. Experimental study of the morphine de-addiction properties of *Delphinium denudatum* Wall. BMC Compl. Alter. Med. 2 (1), 6.

Raj, L., Ide, T., Gurkar, A.U., Foley, M., Schenone, M., Li, X., Tolliday, N.J., Golub, T.R., Carr, S.A., Shamji, A.F., Stern, A.M., Mandinova, A., Schreiber, S.L., Lee, S.W., 2011. Selective killing of cancer cells by a small molecule targeting the stress response to ROS. Nature 475 (7355), 231.

Rajesh, K.P., Manjunatha, H., Krishna, V., Kumara Swamy, B.E., 2013. Potential in vitro antioxidant and protective effects of *Mesua ferrea* Linn. bark extracts on induced oxidative damage. Ind. Crops Prod. 47, 186–198.

Ramarathnam, N., Osawa, T., Namiki, M., Kawakishi, S., 1989. Chemical studies on novel rice hull antioxidants. 2. Identification of isovitexin, a C-glycosyl flavonoid. J. Agric. Food Chem. 37 (2), 316–319.

Ranelletti, F.O., Ricci, R., Larocca, L.M., Maggiano, N., Capelli, A., Scambia, G., Piantelli, M., 1992. Growth-inhibitory effect of quercetin and presence of type-II estrogen-binding sites in human colon-cancer cell lines and primary colorectal tumors. Int. J. Cancer 50 (3), 486–492.

Rao, A.V., Waseem, Z., Agarwal, S., 1998. Lycopene content of tomatoes and tomato products and their contribution to dietary lycopene. Food Res. Int. 31 (10), 737–741.

Reichling, J., Weseler, A., Saller, R., 2001. A current review of the antimicrobial activity of *Hypericum perforatum* L. Pharmacopsychiatry 34 (Suppl. 1), S116–S118.

Ribereau, G.P., 1974. The chemistry of red wine color. In: Webb, D.A. (Ed.), Chemistry of Winemaking. American Chemical Society, Washington, pp. 50–87.

Rickard, S.E., 1996. Dose-dependent production of mammalian lignans in rats and in vitro from the purified precursor secoisolariciresinol diglycoside in flaxseed. J. Nutr. 126 (8), 2012.

Rizwan, M., Rodriguez-Blanco, I., Harbottle, A., Birch-Machin, M.A., Watson, R.E.B., Rhodes, L.E., 2011. Tomato paste rich in lycopene protects against cutaneous photodamage in humans in vivo: a randomized controlled trial. Br. J. Dermatol. 164 (1), 154–162.

Ro, J.S., Lee, S.S., Lee, K.S., Lee, M.K., 2001. Inhibition of type a monoamine oxidase by coptisine in mouse brain. Life Sci. 70 (6), 639–645.

Rosengren, A.H., Jokubka, R., Tojjar, D., Granhall, C., Hansson, O., Li, D.-Q., Nagaraj, V., Reinbothe, T.M., et al., 2009. Overexpression of alpha2A-adrenergic receptors contributes to type 2 diabetes. Science 327 (5962), 217–220.

Russin, W.A., Hoesly, J.D., Elson, C.E., Tanner, M.A., Gould, M.N., 1989. Inhibition of rat mammary carcinogenesis by monoterpenoids. Carcinogenesis 10 (11), 2161–2164.

Russo, M., Spagnuolo, C., Tedesco, I., Russo, G.L., 2010. Phyto-chemicals in cancer prevention and therapy: truth or dare? Toxins 2 (4), 517–551.

Saad, B., Bari, M.F., Saleh, M.I., Ahmad, K., Talib, M.K.M., 2005. Simultaneous determination of preservatives (benzoic acid, sorbic acid, methylparaben and propylparaben) in foodstuffs using high-performance liquid chromatography. J. Chromatogr. A 1073 (1), 393–397.

Saeed, S.A., Gilani, A.H., Majoo, R.U., Shah, B.H., 1997. Anti-thrombotic and anti-inflammatory activities of protopine. Pharmacol. Res. 36 (1), 1–7.

Saeki, Y., Kanamaru, A., Matsumoto, K., Tada, A., 2003. PP-01 The inhibitory effects of centaureidin on the outgrowth of dendrites, melanosome transfer and melanogenesis in normal human melanocyte. Pigment Cell Res. 16 (5), 593.

Sahakitpichan, P., Mahidol, C., Disadee, W., Chimnoi, N., Ruchirawat, S., Kanchanapoom, T., 2012. Glucopyranosyloxybenzyl derivatives of (*R*)-2-benzylmalic acid and (*R*)-eucomic acid, and an aromatic glucoside from the pseudobulbs of *Grammatophyllum speciosum*. Tetrahedron.

Santoyo, S., Cavero, S., Jaime, L., Ibañez, E., Señoráns, F.J., Reglero, G., 2005. Chemical composition and antimicrobial activity of *Rosmarinus officinalis* L.: essential oil obtained via supercritical fluid extraction. J. Food Prot. 68 (4), 790–795.

Sarkar, P., Meghvanshi, M., Singh, R., 2011. Microbial consortium: a new approach in effective degradation of organic kitchen wastes. Int. J. Environ. Sci. Dev. 2 (3), 171–174.

Sato, K., Mochizuki, M., Saiki, I., Yoo, Y.C., Samukawa, K., Azuma, I., 1994. Inhibition of tumor angiogenesis and metastasis by a saponin of *Panax ginseng*, ginsenoside-Rb2. Biol. Pharm. Bull. 17 (5), 635–639.

Schauss, A.G., 2012. Polyphenols and Inflammation. In: Watson, R.R. (Ed.), Bioactive Food as Dietary Interventions for Arthritis and Related Inflammatory Diseases: Bioactive Food in Chronic Disease States. Academic Press, Cambridge, MA, p. 379.

Schemann, M., Michel, K., Zeller, F., Hohenester, B., Rühl, A., 2006. Region-specific effects of STW 5 (Iberogast) and its components in gastric fundus, corpus and antrum. Phytomedicine 13, 90–99.

Selma, M.V., Espin, J.C., Tomas-Barberan, F.A., 2009. Interaction between phenolics and gut microbiota: role in human health. J. Agr. Food Chem. 57 (15), 6485–6501.

Sethi, G., Sung, B., Kunnumakkara, A.B., Aggarwal, B.B., 2009. Targeting TNF for treatment of cancer and autoimmunity. Therapeutic Targets of the TNF Superfamily. Springer, New York, pp. 37–51.

Shahid-Ud-Daula, A.F.M., Basher, M.A., 2009. Phyto-chemicals screening, plant growth inhibition, and antimicrobial activity studies of *Xylocarpus granatum*. Malays. J. Pharm. Sci. 7 (1), 9–21.

Shan, B.E., Liu, D.Q., Liang, W.J., Zhang, J., Li, Q.X., 2006. The anti-tumor effect and immunoregulatory activity of PE2 from *Rhizoma menipermi* extracts in vivo. Carc. Terat. Mutag. 5, 007.

Shan, B., Cai, Y.Z., Brooks, J.D., Corke, H., 2007. The in vitro antibacterial activity of dietary spice and medicinal herb extracts. Int. J. Food Microbiol. 117 (1), 112–119.

Shao, F., Hu, Z., Xiong, Y.M., Huang, Q.Z., Wang, C.G., Zhu, R.H., Wang, D.C., 1999. A new antifungal peptide from the seeds of *Phytolacca americana*: characterization, amino acid sequence and cDNA cloning. Biochem. Biophys. Acta 1430, 262–268.

Si Dayong, Wang, Y., Zhou, Y.-H., Guo, Y., Wang, J., Zhou, H., Li, Z.-S., Fawcett, J.P., 2009. Mechanism of CYP2C9 inhibition by flavones and flavonols. Drug Metabol. Dispos. 37 (3), 629–634.

Sies, H., 1991. Oxidative stress: from basic research to clinical application. Am. J. Med. 91 (3), S31–S38.

Simmler, C., Antheaume, C., André, P., Bonté, F., Lobstein, A., 2011. Glucosyloxybenzyl eucomate derivatives from *Vanda teres* stimulate HaCaT cytochrome C oxidase. J. Nat. Prod. 74 (5), 949–955.

Singh, G., Maurya, S., Catalan, C., De Lampasona, M.P., 2004. Chemical constituents, antifungal and antioxidative effects of ajwain essential oil and its acetone extract. J. Agric. Food Chem. 52 (11), 3292–3296.

Singh, V., Singh, S.P., Chan, K., 2010. Review and meta-analysis of usage of ginkgo as an adjunct therapy in chronic schizophrenia. Int. J. Neuropsychopharmacol. 13 (02), 257–271.

Solano, F., Briganti, S., Picardo, M., Ghanem, G., 2006. Hypopigmenting agents: an updated review on biological, chemical and clinical aspects. Pigment Cell Res. 19 (6), 550–571.

Spelman, K., Duke, J.A., Bogenschultz-Godwin, M.J., 2006. The Synergy Principle Atwork With Plants, Pathogens, Insects, Herbivores, and Humans. In: Cseke, L.J. et al., (Ed.), Natural Products from Plants. 2nd ed. CRC Taylor & Francis, Boca Raton, pp. 475–501.

Srivastava, S.K., Hu, X., Xia, H., Zaren, H.A., Chatterjee, M.L., Agarwal, R., Singh, S.V., 1997. Mechanism of differential efficacy of garlic organosulfides in preventing benzo (a) pyrene-induced cancer in mice. Cancer Lett. 118 (1), 61–67.
Stacewicz-Sapuntzakis, M., Bowen, P.E., Hussain, E.A., Damayanti-Wood, B.I., Farnsworth, N.R., 2001. Chemical composition and potential health effects of prunes: a functional food? Crit. Rev. Food Sci. Nutr. 41 (4), 251–286.
Staker, B.L., et al., 2002. The mechanism of topoisomerase I poisoning by a camptothecin analog. PNAS 99 (24), 15387–15392.
Stanley, A., Akbarsha, M.A., 1992. Giant spermatogonial cells generated by vincristine and their uses. Curr. Sci. 63 (3), 144–147.
Steinberger, E., 1971. Hormonal control of mammalian spermatogonial cells generated by vincristine and their uses. Curr. Sci. 63, 144.
Sun, J., 2007. D-limonene: safety and clinical applications. Altern. Med. Rev. 12 (3), 259–264.
Sun, Y., Cheng, J., 2002. Hydrolysis of lignocellulosic materials for ethanol production: a review. Bioresour. Technol. 83 (1), 1–11.
Sun, F., Liu, J.Y., He, F., Liu, Z., Wang, R., Wang, D.M., Wang, Y.F., Yang, D.P., 2011. In-vitro antitumor activity evaluation of hyperforin derivatives. J. Asian Nat. Prod. Res. 13 (8), 688–699.
Suzuki, T., Waller, G.R., 1985. Purine alkaloids of the fruits of *Camellia sinensis* L. and *Coffea arabica* L. during fruit development. Ann. Botany 56 (4), 537–542.
Swallen, L.C., 1942. US Patent No. 2,287,649. Washington, DC: US Patent and Trademark Office.
Swallen, L.C., Gottfried, J.B., 1942. Purification and uses of as xanthophyl-containing oil derived from corn. J. Am. Oil Chem. Soc. 19 (3), 58–59.
Takimoto, C.H., Calvo, E., 2008. Principles of oncologic pharmacotherapy. In: Pazdur, R., Wagman, L.D., Camphausen, K.A., Hoskins, W.J. (Eds.), Cancer Management: A Multidisciplinary Approach. eleventh ed. CMP United Business Media, Manhasset, NY
Tannin-Spitz, T., Bergman, M., Grossman, S., 2007. Cucurbitacin glucosides: antioxidant and free-radical scavenging activities. Biochem. Biophys. Res. Commun. 364 (1), 181–186.
Thivierge, K., Mathew, R.T., Nsangou, D.M., Da Silva, F., Cotton, S., Skinner-Adams, T.S., Dalton, J.P., 2012. Anti-malaria drug development targeting the M1 alanyl and M17 leucyl aminopeptidases. ARKIVOC 4, 330–346.
Thompson, L.U., Boucher, B.A., Liu, Z., Cotterchio, M., Kreiger, N., 2006. Phytoestrogen content of foods consumed in Canada, including isoflavones, lignans, and coumestan. Nutr. Cancer 54 (2), 184–201.
Thrower, B.W., 2009. Relapse management in multiple sclerosis. Neurologist 15 (1), 1–5.
Tsuda, T., Kojima, M., Harada, H., Nakajima, A., Aoki, S., 1997. Acute toxicity, accumulation and excretion of organophosphorous insecticides and their oxidation products in killifish. Chemosphere 35 (5), 939–949.
Tsuda, H., Ohshima, Y., Nomoto, H., Fujita, K.I., Matsuda, E., Iigo, M., Moore, M.A., 2004a. Cancer prevention by natural compounds. Drug Metab. Pharm. 19 (4), 245–263.
Tsuda, T., Ueno, Y., Aoki, H., Koda, T., Horio, F., Takahashi, N., Osawa, T., 2004b. Anthocyanin enhances adipocytokine secretion and adipocyte-specific gene expression in isolated rat adipocytes. Biochem. Biophys. Res. Commun. 316 (1), 149–157.
Turchetti, B., Pinelli, P., Buzzini, P., Romani, A., Heimler, D., Franconi, F., Martini, A., 2005. In vitro antimycotic activity of some plant extracts towards yeast and yeast-like strains. Phyto. Res. 19 (1), 44–49.
Uckan, F.M., Rustamova, L., Vassilev, A.O., Tibbles, H.E., Petkevich, A.S., 2005. CNS activity of pokeweed anti-viral protein (PAP) in mice infected with lymphocytic choriomeningitis virus (LCMV). BMC Inf. Dis. 5, 9.
Umebayashi, C., Yamamoto, N., Nakao, H., Toi, Y., Chikahisa-Muramatsu, L., Kanemaru, K., Oyama, Y., 2003. Flow cytometric estimation of cytotoxic activity of Rhodexin A isolated from *Rhodea japonica* in human leukemia K562 cells. Biol. Pharm. Bull. 26 (5), 627–630.
Vallianou, I., Peroulis, N., Pantazis, P., Hadzopoulou-Cladaras, M., 2011. Camphene, a plant-derived monoterpene, reduces plasma cholesterol and triglycerides in hyperlipidemic rats independently of HMG-CoA reductase activity. PloS ONE 6 (11), e20516.
Vally, H., Misso, N.L., Madan, V., 2009. Clinical effects of sulfite additives National Centre for Epidemiology and Population Health, ANU College of Medicine and Health Services, the Australian National University, Canberra, Australia. Clin. Exp. Allergy 139 (11), 1643–1651.

Vats, S., Maurya, D.P., Shaimoon, M., Agarwal, A., Negi, S., 2013. Development of a microbial consortium for the production of blend enzymes for the hydrolysis of agricultural waste into sugars. J. Sci. Ind. Res. 72, 585–590.

Venault, P., Chapouthier, G., 2007. From the behavioral pharmacology of beta-carbolines to seizures, anxiety, and memory. Sc. World J. 7, 204–223.

Vermaak, I., Viljoen, A.M., Hamman, J.H., 2011. Natural products in anti-obesity therapy. Nat. Prod. Rep. 28 (9), 1493–1533.

Verwaerde, P., Tran, M.A., Montastruc, J.L., Senard, J.M., Portolan, G., 1997. Effects of yohimbine, an α_2 receptors antagonist, on experimental neurogenic orthostatic hypotension. Fund. Clin. Pharmacol. 11 (6), 567–575.

Vidal, P.F., Molinier, J., 1988. Ozonolysis of lignin: improvement of in vitro digestibility of poplar sawdust. Biomass 16 (1), 1–17.

Vijayalakshmidevi, S., Anju, S., Melchias, G., 2011. In vitro screening of *C. Roseus* alkaloids for antibacterial activity. In Vitro 6 (2), 245–248.

Wang, L.S., Stoner, G.D., 2008. Anthocyanins and their role in cancer prevention. Cancer Lett. 269 (2), 281–290.

Wang, H., Cao, G., Prior, R.L., 1997. Oxygen radical absorbing capacity of anthocyanins. J. Agric. Food Biochem..

Wang, Lu, Wang, Z., Zhang, H., Li, X., Zhang, H., 2009. Ultrasonic nebulization extraction coupled with headspace single drop microextraction and gas chromatography–mass spectrometry for analysis of the essential oil in *Cuminum cyminum* L. Anal. Chim. Acta 647 (1), 72–77.

Wang, C.Z., Du, G.J., Zhang, Z., Wen, X.D., Calway, T., Zhen, Z., Yuan, C.S., 2012. Ginsenoside compound K, not Rb1, possesses potential chemopreventive activities in human colorectal cancer. Int. J. Oncol. 40 (6), 1970.

Way, T.D., Lin, H.Y., Hua, K.T., Lee, J.C., Li, W.H., Lee, M.R., Lin, J.K., 2009. Beneficial effects of different tea flowers against human breast cancer MCF-7 cells. Food Chem. 114 (4), 1231–1236.

Weihrauch, J.L., Gardner, J.M., 1978. Sterol content of foods of plant origin. J. Am. Diet. Assoc. 73 (1), 39–47.

Weil, J., Sarikaya, A., Rau, S.L., Goetz, J., Ladisch, C.M., Brewer, M., Ladisch, M.R., 1997. Pretreatment of yellow poplar sawdust by pressure cooking in water. Appl. Biochem. Biotechnol. 68 (1–2), 21–40.

WHO, 1998. WHO Regional publication Western Pacific series No.21. Medicinal Plants in the Republic of Korea. WHO Regional Office, Manila, Phillipines, p. 316.

Wilsky, S., Sobotta, K., Wiesener, N., Pilas, J., Althof, N., Munder, T., Wutzler, P., Henke, A., 2012. Inhibition of fatty acid synthase by amentoflavone reduces coxsackievirus B3 replication. Arch. Virol. 157 (2), 259–269.

Wong, C.W., Seow, W.K., Ocallaghan, J.W., Thong, Y.H., 1992. Comparative effects of tetrandrine and berbamine on subcutaneous air pouch inflammation induced by interleukin 1, tumour necrosis factor and platelet-activating factor. Agents Act. 36, 112–118.

Wu, L.L., Yang, X.B., Huang, Z.M., Liu, H.Z., Wu, G.X., 2007. In vivo and in vitro antiviral activity of hyperoside extracted from *Abelmoschus manihot* (L) medik. Acta Pharmacol. Sin. 28 (3), 404–409.

Wyman, C., 1996. Handbook on Bioethanol: Production and Utilization. CRC Press.

Wyman, C.E., Dale, B.E., Elander, R.T., Holtzapple, M., Ladisch, M.R., Lee, Y.Y., 2005. Coordinated development of leading biomass pretreatment technologies. Bioresour. Technol. 96 (18), 1959–1966.

Xia, B., 1998. 17 herbs that resolve phlegm and stop coughing and wheezing. Chin. Mater. Med., 477.

Xu, R.S., 2000. Some bioactive natural products from Chinese medicinal plants. Studies Nat. Prod. Chem. 21, 729–772.

Xu, Y., Li, S., Chen, R., et al., 2010. Antidepressant-like effect of low molecular proanthocyanidin in mice: involvement of monoaminergic system. Pharmacol. Biochem. Behav. 94 (3), 447–453.

Xu, H., Tang, H., Feng, H., Li, Y., 2014. Design, synthesis and anticancer activity evaluation of novel C14 heterocycle substituted epi-triptolide. Eur. J. Med. Chem. 73, 46–55.

Yakubu, M.T., Akanji, M.A., Oladiji, A.T., 2005. Aphrodisiac potentials of the aqueous extract of *Fadogia agrestis* (Schweinf. Ex Hiern) stem in male albino rats. Asian J. Androl. 7 (4), 399–404.

Yang, G., Chen, D., 2008. Biflavanones, flavonoids, and coumarins from the roots of *Stellera chamaejasme* and their antiviral effect on hepatitis B virus. Chem. Biodiv. 5 (7), 1419–1424.

Yang, L., Xueru, L., Ya, Z., Ningping, W., 1996. Study on antibacterial activity of total alkaloia from *F. taitaiensis* L. and *Shedan chuanbeiye* in vitro. Ning. Med. J. 3, 147–148.

Yang, H., Li, J., Tang, C., Gong, L., Wang, H., 2004. Role of active components of curcumin in anti-angiogenesis in vitro. Acta Acad. Med. Militar. Tertiae 27 (11), 1068–1070.

Yang, G., Liao, Z., Xu, Z., Zhang, H., Chen, D., 2005. Antimitotic and antifungal C-3/C-3″-biflavanones from *Stellera chamaejasme*. Chem. Pharm. Bull. 53 (7), 776–779.

Yao, L.H., Jiang, Y.M., Shi, J., Tomas-Barberan, F.A., Datta, N., Singanusong, R., Chen, S.S., 2004. Flavonoids in food and their health benefits. Plant Foods Hum. Nutr. 59 (3), 113–122.

Yao, S.S., Guo, W.F., Lu, Y., Jiang, X.Y., 2005. Flavor characteristics of lapsang souchong and smoked lapsang souchong, a special chinese black tea with pine smoking process. J. Agri. Food Chem. 53 (22), 8688–8693.

Yoo, Y.M., Nam, J.H., Kim, M.Y., Choi, J., Park, H.J., 2008. Pectolinarin and pectolinarigenin of *Cirsium setidens* prevent the hepatic injury in rats caused by D-galactosamine via an antioxidant mechanism. Biol. Pharm. Bull. 31 (4), 760–764.

Yu, Y.B., Miyashiro, H., Nakamura, N., Hattori, M., Park, J.C., 2007. Effects of triterpenoids and flavonoids isolated from *Alnus firma* on HIV-1 viral enzymes. Arch. Pharm. Res. 30 (7), 820–826.

Zafra-Stone, S., Yasmin, T., Bagchi, M., Chatterjee, A., Vinson, J.A., Bagchi, D., 2007. Research Article Berry anthocyanins as novel antioxidants in human health and disease prevention. Mol. Nutr. Food Res. 51, 675–683.

Zeitsch, K.J., 2000. The Chemistry and Technology of Furfural and Its Many by-ProductsElsevier.

Zhang, L., Kong, Y., Wu, D., Zhang, H., Wu, J., Chen, J., Shen, X., 2008. Three flavonoids targeting the β-hydroxyacyl-acyl carrier protein dehydratase from *Helicobacter pylori*: crystal structure characterization with enzymatic inhibition assay. Prot. Sci. 17 (11), 1971–1978.

Zhang, X.J., Yang, G.Y., Wang, R.R., Pu, J.X., Sun, H.D., Xiao, W.L., Zheng, Y.T., 2010. 7, 8-Secolignans from Schisandra wilsoniana and their Anti-HIV-1 Activities. Chem. biodiver. 7 (11), 2692–2701.

Zhang, Y., Sreekrishna, K., Lin, Y., Huang, L., Eickhoff, D., Degenhardt, C., Xu, T., 2011. Modulation of transient receptor potential (TRP) channels by Chinese herbal extracts. Phyto. Res. 25 (11), 1666–1670.

Zhao, J., Lian, Y., Lu, C., Jing, L., Yuan, H., Peng, S., 2012. Inhibitory effects of a bisbenzylisoquinline alkaloid dauricine on HERG potassium channels. J. Ethnopharmcol. 141 (2), 685–691.

Zielinska, M., Kostrzewa, A., Ignatowicz, E., Budzianowski, J., 2001. The flavonoids, quercetin and isorhamnetin 3-O-acyl-glucosides diminish neutrophil oxidative metabolism and lipid peroxidation. Acta Biochim. Polon. 48 (1), 183–190.

Zygmunt, B., Namieśnik, J., 2003. Preparation of samples of plant material for chromatographic analysis. J. Chroma. Sci. 41 (3), 109–116.

Further Reading

Arma, D.R., Yue, T.L., 1986. Adrenoceptor blocking properties of atropine-like agents anisodamine and anisodine on brain and cardiovascular tissues of rats. Br. J. Pharmacol. 87 (3), 587–594.

Bharat, G., Parrabia, M.H., 2010. Pharmacognostic: evaluation of bark and seeds of *Mimusops elengi* l. Int. J. Pharm. Pharm. Sci. 2 (4), 110–113.

Cheng, K.F., Yip, C.S., Yeung, H.W., Kong, Y.C., 1979. Leonurine, an improved synthesis. Experientia 35 (5), 571–572.

Davis, K.D., Meyer, R.A., Campbell, J.N., 1993. Chemosensitivity and sensitization of nociceptive afferents that innervate the hairy skin of monkey. J. Neurophysiol. 69 (4), 1071–1081.

Dézsi, L., Kis-Varga, I., Nagy, J., Komlódi, Z., Kárpáti, E., 2002. Neuroprotective effects of vinpocetine in vivo and in vitro: apovincaminic acid derivatives as potential therapeutic tools in ischemic stroke. Acta Pharm. Hungarica 72 (2), 84–91.

Fields, H.L., Heinricher, M.M., Mason, P., 1991. Neurotransmitters in nociceptive modulatory circuits. Ann. Rev. Neurosci. 14 (1), 219–245.

Gershenzon, J., Maffei, M., Croteau, R., 1989. Biochemical and histochemical localization of monoterpene biosynthesis in the glandular trichomes of spearmint (*Mentha spicata*). Plant Physiol. 89 (4), 1351–1357.

Glinski, W., Glinska-Ferenz, M., Pierozynska-Dubowska, M., 1991. Neurogenic inflammation induced by capsaicin in patients with psoriasis. Acta Derm. Venereol. 71 (1), 51–54.

Gupta, M., Shaw, B.P., Mukherjee, A., 2010. A new glycosidic flavonoid from *Jwarhar mahakashay* (antipyretic) ayurvedic preparation. Int. J. Ayurv. Res. 1, 106–111.

Hänsel, R., Klaffenbach, J., 1961. Optisch aktives dihydromyricetin aus *Erythrophleum africanum*. Archiv. Pharm. 294 (3), 158–172.

Iacobellis, N.S., Lo, C.P., Capasso, F., Senatore, F., 2005. Antibacterial activity of *Cuminum cyminum* L. and *Carum carvi* L. essential oils. J. Agr. Food Chem. 53 (1), 57–61.

Ieven, M., Vlietinick, A.J., Berghe, D.V., Totte, J., Dommisse, R., Esmans, E., Alderweireldt, F., 1982. Plant antiviral agents. III. Isolation of alkaloids from *Clivia miniata* Regel (Amaryl-lidaceae). J. Nat. Prod. 45 (5), 564–573.

Infante, R., Contador, L., Rubio, P., Aros, D., Peña-Neira, Á., 2011. Postharvest sensory and phenolic characterization of "Elegant Lady" and "Carson" peaches. Chilean J. Agr. Res. 71 (3), 445–451.

Jong-Woong, A., Mi-Ja, A., Ok-Pyo, Z., Eun-Joo, K., Sueg-Geun, L., Hyung, J.K., Kubo, I., 1992. Piperidine alkaloids from Piper retrofractum fruits. Phytochemistry 31 (10), 3609–3612.

Khory, R.N., Katrak, N.N., 1985. Materia Medica of India and Their Therapeutics. Neeraj Publishing House, Delhi, pp. 10-11.

Lee, J.S., Lee, M.S., Oh, W.K., Sul, J.Y., 2009. Fatty acid synthase inhibition by amentoflavone induces apoptosis and antiproliferation in human breast cancer cells. Biol. Pharm. Bull. 32 (8), 1427–1432.

Li, Q., He, Y.C., Xian, M., Jun, G., Xu, X., Yang, J.M., Li, L.Z., 2009. Improving enzymatic hydrolysis of wheat straw using ionic liquid 1-ethyl-3-methyl imidazolium diethyl phosphate pretreatment. Biores. Technol. 100 (14), 3570–3575.

Liu, R.H., 2003. Health benefits of fruit and vegetables are from additive and synergistic combinations of phyto-chemicals. Am. J. Clin. Nutr. 78 (3), 517s–520s.

Mukherjee, J., Menge, M., 2000. Progress and prospects of ergot alkaloid research. Adv. Biochem. Eng. Biotechnol. 68, 1–20.

Novel compound selectively kills cancer cells by blocking their response to oxidative stress. Science Daily, July 2011.Park, H.J., Kim, M.J., Ha, E., Chung, J.H., 2008. Apoptotic effect of hesperidin through caspase3 activation in human colon cancer cells, SNU-C4. Phytomedicine 15 (1), 147–151.

Policegoudra, R.S., Aradhya, S.M., Singh, L., 2011. Mango ginger (*Curcuma amada* Roxb.): a promising spice for phyto-chemicals and biological activities. J. Biosci. 36 (4), 739–748.

Pretorius, J.C., Magama, S., Zietsman, P.C., 2003. Growth inhibition of plant pathogenic bacteria and fungi by extracts from selected South African plant species. S. Afr. J. Bot. 20, 188–192.

Riju, A., Sithara, K., SUJA, S.N., Shamina, A., EAPEN, S.J., 2009. In silico screening major spice phyto-chemicals for their novel biological activity and pharmacological fitness. J. Bioequival. Bioavail. 1 (2), 63–73.

Rossi, F., Jullian, V., Pawlowiez, R., Kumar-Roiné, S., Haddad, M., Darius, H.T., Gaertner-Mazouni, N., Chinain, M., Laurent, D., 2012. Protective effect of *Heliotropium foertherianum* (Boraginaceae) folk remedy and its active compound, rosmarinic acid, against a Pacific ciguatoxin. J. Ethnopharmacol. 143 (1), 33–40.

Shamsa, F., Monsef, H., Ghamooshi, R., Verdian-rizi, M., 2008. Spectrophotometric determination of total alkaloids in some Iranian medicinal plants. Thai J. Pharm. Sci. 32, 17–20.

Talavéra, S., Felgines, C., Texier, O., Besson, C., Mazur, A., Lamaison, J.L., Rémésy, C., 2006. Bioavailability of a bilberry anthocyanin extract and its impact on plasma antioxidant capacity in rats. J. Sci. Food Agr. 86 (1), 90–97.

Ulrich, S., Wolter, F., Stein, J.M., 2005. Molecular mechanisms of the chemopreventive effects of resveratrol and its analogs in carcinogenesis. Molecular nutrition & food research 49 (5), 452–461.

Vila, F.C., Colombo, R., de Lira, T.O., Yariwake, J.H., 2008. HPLC microfractionation of flavones and antioxidant (radical scavenging) activity of *Saccharum officinarum* L. J. Braz. Chem. Soc. 19 (5), 903–908.

Wasser, S.P., Weis, A.L., 1999. Therapeutic effects of substances occurring in higher Basidiomycetes mushrooms: a modern perspective. Crit. Rev. Immun. 19 (1).

Wolter, F., Ulrich, S., Stein, J., 2005. Molecular mechanisms of the chemopreventive effects of resveratrol and its analogs in colorectal cancer: key role of polyamines? J. Nutr. 134 (12), 3219–3222.

CHAPTER 2

Modern Extraction Techniques for Drugs and Medicinal Agents

Sudipta Saha*, Ashok K. Singh*, Amit K. Keshari*, Vinit Raj*, Amit Rai*, Siddhartha Maity**

**Babasaheb Bhimrao Ambedkar University, Lucknow, Uttar Pradesh, India;*
***Jadavpur University, Kolkata, West Bengal, India*

1 Introduction

The physicochemical properties of drugs and medicinal agents derived from natural origins mostly rely on the selection of appropriate extraction techniques (Sasidharan et al., 2011). Extraction is the pioneer step of the study of any medicinal plant, playing a crucial role in the final outcomes. Extraction techniques are, therefore, occasionally referred as "sample preparation techniques" (Hennion et al., 1998).

The traditional extraction techniques are time-consuming, lack efficiency in extracting the target molecule, and need large volumes of nonenvironmentally benign organic solvents, such as dichloromethane and methanol, sorbents, and sample. Traditional solid–liquid extraction (SLE) techniques consist of maceration, soxhlet extraction, percolation, soaking, turbo-extraction (high-speed mixing), and sonication. Apart from this, the current extraction techniques provide fast processing of samples, easy automation, high reproducibility, and use of low volumes of organic toxic solvents in compliance with green analytical chemistry. "Green extraction" depends on the improvement of extraction techniques that minimize energy consumption, allocate the utilization of environmentally benign solvents and inexhaustible natural products, and ensure a safe and high-quality extract (Vazquez-Roig and Picó, 2015). Furthermore, when considering the extraction of drugs and medicinal agents that are sensitive, thermolabile, and found in low concentrations, traditional extraction techniques might not be the best option, probably due to very low product yields. To acquire those drugs and medicinal agents in an environmental friendly way with adequate yields, a green extraction approach is required (Mustafa and Turner, 2011). These newer and resourceful extraction techniques, robust for the need of green analytical chemistry, are pressurized liquid extraction (PLE), microwave-assisted extraction (MAE), different types of liquid-liquid microextraction (LLME), and supercritical fluid extraction (SFE) (Vazquez-Roig and Picó, 2015).

Ingredients Extraction by Physicochemical Methods in Food
http://dx.doi.org/10.1016/B978-0-12-811521-3.00002-8

However, between both the traditional and newer methods established to date, no single method is considered a standard for the extraction of biochemical ingredients. The assortment of extraction methods frequently rely on the critical input parameters, an understanding of the nature of the matrix embedded with the extracting compounds, the chemistry of biochemical ingredients to be extracted, and the scientific skill (Azmir et al., 2013).

Until now, numerous extraction techniques and their utilization in diverse analytical fields have been increasing day by day, usually to progress compatibility with modern analytical instruments or to consent to utilize an environmental friendly way of green analytical chemistry. In this chapter, the latest developments toward modern extraction techniques will be briefly explained, and the variations in previously existing techniques for the extraction of different bioactive compounds in recent years will be described. The main aim of this chapter is to provide for the reader a brief layout to the essentials and utilization of various recently developed extraction method for the analysis of bioactive ingredient.

2 Various Extraction Procedures

2.1 Pressurized Liquid Extraction

PLE combines elevated pressure and temperature with liquid solvents to attain rapid and proficient extraction of the analytes from the solid matrix. It applies maximum temperatures with a decrease in solvent viscosity, which increases the solvent's capability to wet the matrix and solubilize the target component. Temperature also functions in the distribution of analyte–matrix bonds and inspires analyte diffusion to the matrix surface. However, PLE has significant importance over challenging techniques with regards to solvent use, time saving, automation, and efficiency. For example, PLE shows an advantage over MAE in that no extra filtration step is essential since the matrix ingredients that are not solubilize in the extraction solvent may be received within the sample extraction cell. This is very expedient for the determination of automation and on-line coupling of the separation and extraction techniques (Carabias-Martínez et al., 2005).

The technique is used for extraction with solvents at a temperature and high pressure without their critical point being reached; it has received different names, such as pressurized fluid extraction (PFE), accelerated solvent extraction (ASE), pressurized-hot-water extraction (PHWE), high-pressure high-temperature solvent extraction (HPHTSE), and subcritical solvent extraction (SSE). If water is used as a solvent for the extraction, then the technique is called PHWE, subcritical water extraction, or superheated water extraction (Mustafa and Turner, 2011). More specifically, it depends on which water solvent at atmospheric pressure below the critical point of water (374°C/647 K, 22.1 MPa) and above the boiling point of water (100°C/273 K, 0.1 MPa) is used (Plaza and Turner, 2015).

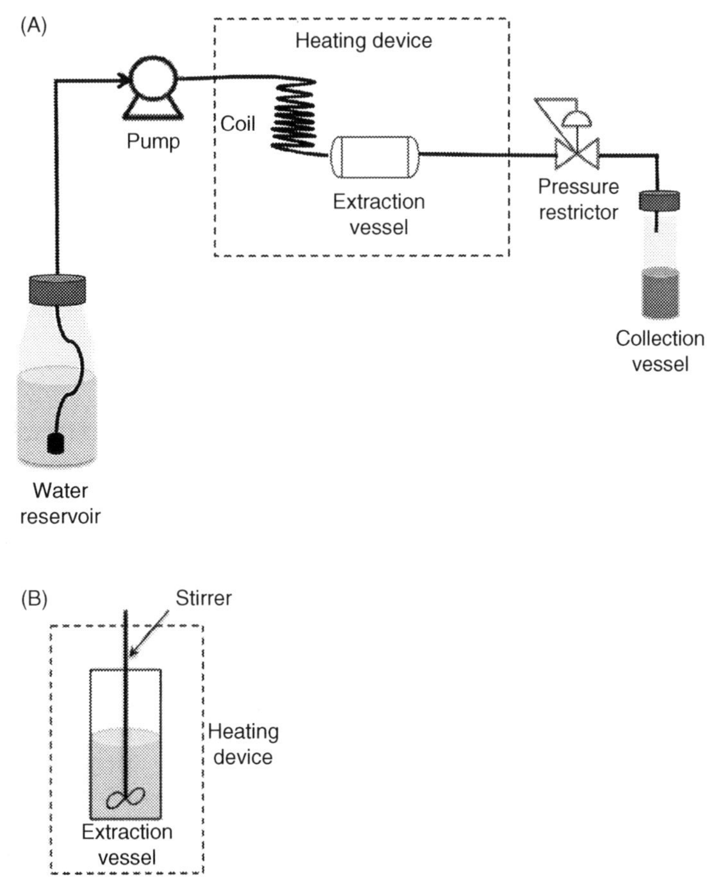

Figure 2.1: The main parts of a dynamic PLE system (A) and a static PLE system (B) (Plaza and Turner, 2015).

2.1.1 Method of pressurized liquid extraction technique

Here we address how to perform extraction practically using water as extractant. First, either tap or deionized water is used as the extractant. The water might be oxygen-free for avoiding oxidation of the analytes. The methods most commonly used for obtaining oxygen-free water are a helium purge or ultrasound (sonication), of which the former is more economic if an ultrasound bath is available. There are two types of equipment: dynamic (continuous-flow) systems and static (batch) systems, and combinations of the two, (Fig. 2.1) (Plaza and Turner, 2015).

Dynamic PLE basically needs a pump, an extraction vessel, a heating device, a pressure restrictor, and a collection vial. The pump delivers water to the extraction vessel, and via the pressure restrictor to the collection vial. The pump should be able to achieve the pressure necessary to keep the water in a liquid state during the extraction process (normally 3.5–20 MPa).

Heating of the water is by use of an oven, heating tape, or a heating jacket. The extraction vessel is usually made of stainless steel, and should have frits at both ends in order to avoid sample losses. For a typical extraction of bioactive compounds at a moderate temperature and pressure, 100–200°C and 5–10 MPa, an empty high-performance liquid chromatography (HPLC) column may be used as an extraction vessel. However, in case the temperatures are near the critical point, above 250°C, the extraction vessel might be of a corrosion-resistant metal alloy (e.g., Hastelloy). Pressure restriction is used to control the pressure within the extraction vessel and to prevent boiling-off effects of water at the exit of the extraction vessel. In static PLE, a pump is unnecessary and convection is accomplished using a stirrer to speed up the mass transfer. The extraction vessel in static PLE is usually of wider diameter than in dynamic PLE, for fit in the stirrer. For heating, an oven, a heating jacket, or heating tape is appropriate. A pressure restrictor is not needed, unless the speed of removing the extract from the vessel is to be controlled (Plaza and Turner, 2015).

There are advantages and disadvantages for using both types of system. Static PLE is simpler and easier to use than a pump and does not need pressure restrictor. The residence time of the analytes is greater than in dynamic PLE, which may cause thermally labile analytes to degrade (Liu et al., 2013a). Furthermore, in static PLE, equilibrium for distribution of analytes from the sample matrix to the extractant will stabilize after some time, since the volume of the extractant is constant. However, in dynamic PLE, the residence time of analytes in the high-temperature water is shorter, since fresh extractant is continuously being pumped into the extraction vessel and out to the collection vial. In this case, the flow rate of the extraction will control the residence time, so extraction and degradation kinetics in PLE are easier to control using a dynamic extraction setup. The disadvantage of dynamic PLE is that it is more costly, and there is always a risk of clogging inside the tubing during the extraction. Downstream of the extraction vessel, water with the extracted analytes is cooled to temperatures at which some of the extracted analytes are no longer soluble, so they precipitate and block the tubing. There are two ways to avoid this problem. One is to use heating tape around the tubing from the exit of the oven to the collection vial. Another is to use an additional pump to wash the lines after the extraction vessel and before the pressure restrictor (Monrad et al., 2012).

2.1.2 Method optimization

Extraction of analytes from solid and semisolid samples can be described by the following five steps:

1. wetting the sample matrix with solvent;
2. to initial desorption from the sample matrix;
3. to diffuse inside the pores of the sample matrix;
4. to partition between the extractant and the sample matrix;
5. to diffuse through the sluggish extractant layer until the zone of convection is reached (Pawliszyn, 2003).

All these steps happen more or less in parallel. In PHWE, the temperature is the key parameter to optimize, since it affects the efficiency of all these five steps, as described earlier. In addition, extraction time and/or flow rate are important variables to optimize.

2.1.3 Influential parameters

The *higher temperature* of the water leads to improved wetting of the sample matrix [see preceding step (1)]. Further, increasing the temperature also favors mass-transfer kinetics by disrupting analyte-matrix interactions, especially hydrogen bonding and other dipole-dipole forces, by way of promoting initial desorption of the analytes from the sample matrix [step (2)]. A higher temperature also results in faster diffusivity [steps (3) and (5)] as altered (usually higher) solubility, the latter leading to a shift in Z of the analytes between the extractant and the sample matrix [step (4)]. In summary, an elevated temperature in PHWE brings several advantages in terms of improved extraction kinetics. There are three main drawbacks in using elevated temperatures in PHWE: decreasing selectivity of the extraction, pertinent degradation of the analytes, and other chemical reactions in the sample matrix (Plaza and Turner, 2015).

A *sustain flow system* with a high enough flow rate of the extractant minimizes chemical reactions during PLE. A higher flow rate will not only decrease the residence time for the analytes in the elevated temperature water but also enhance the extraction rate of the analytes (Liu et al., 2013a).

Pressure has very little influence on the properties of water, as long as the water remains in the liquid state. Hence, a pressure of 5–10 MPa is usually employed unless the saturation pressure of water is used (Plaza and Turner, 2015).

The *poring of some inorganic and organic modifiers, surfactants, and additives* may increase the solubility of the analytes in the extractant, and affect the physical properties of the desorption of analytes and the sample matrix from the sample. For example, 5% ethanol and 1% formic acid in water favored the extraction of anthocyanins from red cabbage (Arapitsas and Turner, 2008).

The *particle size* of the sample influences the extraction kinetics since a smaller particle size leads to increasing contact surface between the sample and the extractant. The *solvent-to-sample ratio* is an important parameter in PHWE. An increase in the ratio of extractant to sample results in a larger fraction of the analytes being extracted without replacing the extractant with fresh solvent. The maximum of solvent to sample ratio need more water to be heated (Rezaei et al., 2013).

The *moisture* content of the sample is another parameter that may influence the extraction yield. Some studies show that crude samples with high moisture content give better extraction yields of polyphenols than dried samples (Monrad et al., 2014).

2.1.4 Applications

During recent years, different bioactive compounds have been extracted through either static or dynamic PHWE techniques, which include polysaccharides (from *Lycium barbarum, Chlorella vulgaris, Himanthalia elongate, Haematococcus pluvialis, and Dunaliella salina*), phenolic compounds (from lemon balm, grape pomace, and potato peel), antioxidants (cow cockle seed, oregano, and olive leaves), flavonols (apple by-products, grape pomace, and *Moringa oleifera* leaf), and anthocyanins (red cabbage, red onion, and grape pomace) (Plaza and Turner, 2015).

2.2 Microwave-Assisted Extraction

MAE has drawn significant research attention in various fields for medicinal plant research, moderate capital cost, special heating mechanism, and its good efficacy under atmospheric conditions. Microwaves consist of an electric field and a magnetic field oscillating at frequency ranged from 0.3 to 300 GHz. Microwaves may enter into calm materials and interact with the polar components to generate heat. The principle of MAE depends on ionic conduction and dipole rotation via a direct effect of microwaves on molecules of the extracted system and also works because only targeted and selective materials may be heated on the basis of their dielectric constant (Sparr Eskilsson and Björklund, 2000). The heating process and efficiency of the microwave are based on dissipation factor of the material tan δ. However, the measurement of the sample to absorb microwave energy and disperse heat to the neighboring molecules is given by Eq. (2.1).

$$tan\ \delta = \varepsilon'' / \varepsilon' \tag{2.1}$$

where ε'' is the dielectric loss that expresses the efficiency of changing microwave energy into heat energy, whereas ε' is the dielectric constant, denoting the measurement of the ability of the material to absorb microwave energy. The rate of conversion of electrical energy into thermal energy in the material is described by Eq. (2.2).

$$P = K \cdot f\ \varepsilon' E^2\ tan\ \delta \tag{2.2}$$

where P indicates the microwave power dissipation per unit volume, f is the applied frequency, K is a constant, E is the electric field strength, ε' is the material's absolute dielectric constant, and tan δ is the dielectric loss tangent (Chan et al., 2011).

MAE may happen in any one or several of the following three heating mechanisms:

1. Extracting analyte into a single solvent or mixture of solvents that absorb microwave energy strongly.
2. Extracting analyte into a combined solvent containing both high and low dielectric losses mixed in various proportions.
3. Extracting analyte with a microwave transparent solvent from a sample of high dielectric loss. MAE can be carried out in a closed or open microwave transparent vessels; thereby sample and solvent are placed and then exposed to microwave energy (Madej, 2009).

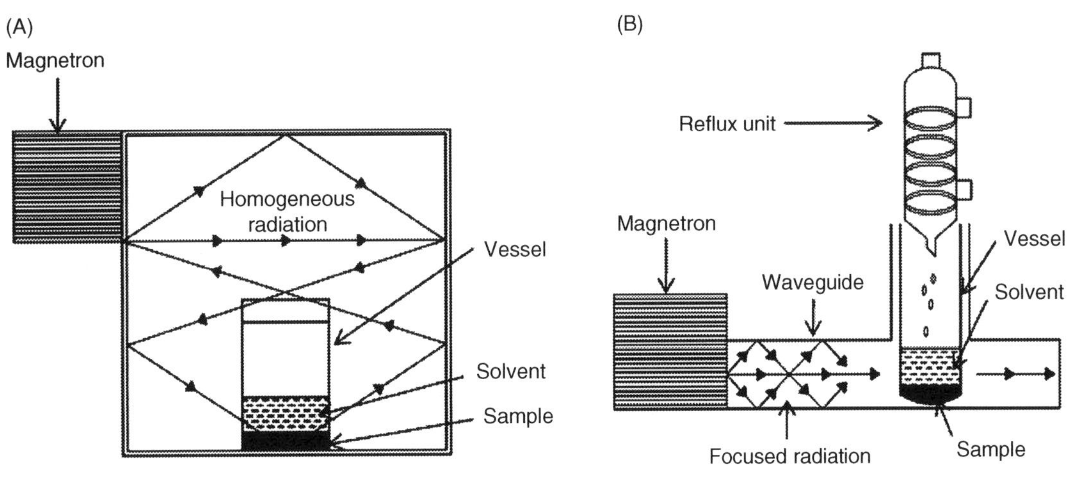

Figure 2.2: (A) Closed type microwave system and (B) open type microwave system (Chan et al., 2011).

2.2.1 Method of microwave-assisted extraction technique

MAE may be classified into "closed system" and "open system" on the basis of operating above and under atmospheric pressure, respectively. Further, consider the closed system and open system, for which schematic diagrams are shown in Fig. 2.2 (Chan et al., 2011).

In a closed MAE system, the different mode of microwave radiation is used for extraction in a sealed vessel with sustained microwave heating. High working temperature and pressure of the system permit fast and competent extraction. The pressure is essentially controlled inside the extraction vessel in such a way that it cannot go beyond the working temperature and pressure of the vessel and can be regulated above the normal boiling point of the extraction solvent. Recent upgrading of the closed system has led to the improvement of high-pressure microwave-assisted extraction (HPMAE). The increase in pressure and temperature expedite microwave-assisted extraction due to the capability of extraction solvent to absorb microwave energy. The closed system makes fast and proficient extraction with less solvent spending, easily preventing the loss of volatile compounds with limited sample throughput (Wang et al., 2008).

The thermolabile compounds are safely extracted out using an open system to counter the shortcomings of a closed system. The productivity of extraction has greater volumes of solvent that might be used any time during the process. Basically, the open system operates at more mild conditions. The open MAE system is frequently used in the extraction of active compounds and also used in analytical chemistry. This system transports at atmospheric conditions and only part of the vessel is directly uncovered to the promulgation of microwave radiation (monomode). The condensation of any vaporized solvent that happens via the upper part of the vessel is connected to a reflux. Besides that, multimode radiation may also be engaged in an open MAE system with the reflux unit (Luque-García and Luque de Castro, 2003).

Poor extraction yield due to oxidation and thermal degradation of some active compounds has led to the enlargement of more proficient MAE that requires additional instruments on top of the commercial system. These are vacuum microwave-assisted extraction (VMAE), nitrogen-protected microwave-assisted extraction (NPMAE), dynamic microwave-assisted extraction (DMAE), ultrasonic microwave-assisted extraction (UMAE), and solvent-free microwave-assisted extraction (SFME).

In the cases of NPMAE and VMAE, additional vacuum pump and nitrogen sources are supplemented. The vacuum pump is used to grant vacuum pressure for VMAE and it is also used to remove oxygen before nitrogen is pressurized into the vessel for NPMAE. Further, a reflux system is installed to avoid any additional pressure built up during the extraction process. In some NPMAE, the inert gas is pressurized straight inside the extraction vessel containing the sample–solvent mixture—and put into the closed-type microwave cavity. For UMAE, the ultrasonic sound transducer might be installed so that the wave can proliferate directly into the extraction vessel of the alert-type microwave system. Further, in the case of DMAE, most of the instrument setup is custom made. The system made up MAE extraction and HPLC analysis in a single step. The extraction step begins by introduction of the sample vessel into the resonance cavity and the solvent is circulated through the extraction loop. The heating of the microwave is activated once the solvent flow rate reaches steady state. The regulation screws in the microwave resonance are attuned to decrease the reflected power. When the extraction is complete, the extract is determined by the sample loop. The solvent is then mixed with the mobile phase and proceeds to the analytical step (Chan et al., 2011).

Apart from these, the SFME method involves insertion of vegetable material in a microwave reactor without the addition of any solvent or water. The inside heating of the in situ water within the plant material distends the plant cells and leads to breakdown of the cells. Thus, this process releases essential oils containing bioactive compounds, which are evaporated by the in situ water of the plant material. The continuous condensation of the distillate outside the microwave oven allows the cooling of the system, which comprises essential oils and water. The extreme water is refluxed to the reactor to maintain the proper humidity rate of the plant materials (Li et al., 2013).

2.2.2 Influential parameters

The main parameters influencing MAE performance, include solvent nature, the solvent-to-feed ratio, microwave power, extraction time, and temperature.

In the extraction of most bioactive compounds, organic solvents are used. When selecting the solvent, consideration should mainly be focused on the microwave-absorbing properties of the solvent, selectivity toward the analyte, and interaction of the solvent with the matrix. Generally, in conservative extractions, a maximum volume of solvent has augmented the revival of the analyte, but, in MAE, the same move toward may give lower recoveries, which may be due to the inadequate stirring of the solvent by the microwaves. The solvent volume depends on the type and the size of the sample, but, on average, the amounts of solvent may be about 10 times less than those used in classical extractions. The microwave power and the corresponding time depend on the type of sample and solvent used. In theory, one should use high-power microwaves to reduce exposure time as much as possible. However, in some cases, a very high-power microwave decreases the extraction efficiency (EE) by degrading the sample or rapidly boiling the solvent in open-vessel systems, which hinders contact with the sample. Generally, extraction times in MAE are much shorter than those of classical extraction techniques. Usually, elevating extraction times above the optimal range does not improve EE, and, in some cases, may even decrease analyte recoveries (e.g., thermolabile compounds). In most cases of the MAE, high temperatures result in enhanced EE, but particular deliberation should be specified to applications dealing with thermolabile substances, which may be decomposed at high temperatures. Further, by introducing inspiring in MAE, the negative effect of low solvent-to-feed ratio of extraction yield may be reduced (Madej, 2009).

2.2.3 Applications

Standard MAE is commonly employed either in open or closed systems to extract thermostable compounds. For extraction of degradable active compounds, there are various modified MAE techniques that are suitable for the application. DMAE is suitable to extract degradable compounds that require multiple extraction cycles as the technique performs under mild conditions and in a continuous manner. This technique promotes efficient and fast analytical step, as it may be coupled on-line with HPLC analysis system. Moreover, for highly brittle compounds, which forecast high risks of oxidation and thermal degradation, VMAE is appropriate because the extraction is done at low temperature and in a vacuum condition. Alternatively, extraction of thermally degradable compounds can also be achieved by NPMAE. It gives faster extraction than VMAE but requires additional extraction steps. Besides, SFME is preferable to for use in essential-oil extraction and it is more efficient than the traditional HD method. In some circumstances in which the associated active compounds have low diffusion and are difficult to extract, UMAE can be employed because it improves the mass-transfer mechanism and reduces the extraction time. This technique can provide high activation energy or the impact energy required for the extraction to proceed (Chan et al., 2011).

2.3 Supercritical Fluid Extraction

SFE is the process of extraction using supercritical fluids as the extracting solvent. Extraction is usually from a solid matrix, but can also be from liquids. SFE provides several operational advantages since it uses supercritical solvents, with different physicochemical properties, such as density, diffusivity, viscosity, and dielectric constant. The extraction speed is mainly dependent on the viscosity and diffusivity of the mobile phase. With a low viscosity and high diffusivity, the component that is to be extracted can pass through the mobile phase easily. The lower viscosity and higher diffusivity of supercritical fluids, as compared to regular extraction liquids, help the components to be extracted faster than through other techniques. Thus, an extraction method may take just 10–60 min with SFE, whereas it would take hours or even days with classical methods. Altering its pressure and/or temperature can modify the density of the supercritical fluid. Since density is related to solubility, by varying the extraction pressure, the solvent strength of the fluid can be adapted. Other advantages, compared to other extraction techniques, are the use of solvents generally known as safe (GRAS) and the higher efficiency of the extraction process in terms of lower extraction times and increasing yields (da Silva et al., 2016).

2.3.1 Fluid materials used in supercritical fluid extraction

There are several material compounds that may be used as supercritical fluids; the one most frequently used is carbon dioxide as a solvent. For practice, more than 90% of all analytical SFE is carried out with carbon dioxide (CO_2) for numerous practical reasons. Besides having a relatively low temperature (32°C) and critical pressure (74 bar), CO_2 is comparative, nonflammable, nontoxic, existing in high purity at rather a low cost, and is easily removed from the extract. In the supercritical state, the CO_2 shows similar polarity with liquid pentane and is, therefore, superlatively suitable for lipophilic compounds. The major disadvantage of CO_2 is its lack of polarity for the extraction of polar analytes. In the 1990s, a number of reports were published about the alternative of N_2O as an extraction fluid for analytical SFE. This fluid was measured to be more suitable for polar compounds because of its enduring dipole moment. One of the applications in which N_2O exhibited major progression when compared to CO_2 is, for example, the extraction of polychlorinated dibenzodioxins from fly ash. Unfortunately, this fluid has been shown to cause violent explosions when used for samples having high organic content and should, therefore, be used only when absolutely necessary. Other more exotic supercritical fluids that have been used for environmental SFE are freons and SF_6. The SF_6 is a nonpolar molecule (although easily polarizable) and as a supercritical fluid, it has been revealed to selectively extract aliphatic hydrocarbons up to around C-24 from a mixture containing both aliphatic and aromatic hydrocarbons. Freons, particularly $CHClF_2$ (Freon-22), have on a number of occasions been shown to augment the EE compared to conducting extractions with CO_2 (Pourmortazavi and Hajimirsadeghi, 2007).

Although supercritical H_2O has often been used for the obliteration of hazardous organics, the high pressure and temperature needed ($P > 221$ bars and $T > 374$°C) in concert with the corrosive nature of H_2O in these circumstances, has limited the promising practical applications in plant-oil analysis. H_2O at subcritical conditions is used as an effective fluid for the extraction of several classes of essential oil. Propane, ethane, dimethyl ether, ethylene, and so forth have also been suggested as solvents under sub- and supercritical circumstances for extraction (Illés et al., 1997). A substitute to CO_2 in supercritical extractions is the exploit of propane. Although propane does not recommend many of the characters that are generally associated with CO_2, this reasonably inexpensive solvent can be a better choice for the extraction of oils and natural products. Propane does not leave a toxic filtrate just as CO_2 but the required extraction pressures are inferior to those applied with CO_2 (Sparks et al., 2006).

2.3.2 Method of supercritical fluid extraction techniques

Only dry samples might be preferred for the extraction through SFE. When a fresh plant material is extracted, its high moisture content may cause mechanical difficulties, such as restrictor clogging due to ice formation. One uncomplicated yet efficient way to avoid such problems is to mix the sample with anhydrous Na_2SO_4. Anhydrous Na_2SO_4 can improve SFE results because: (1) it can make available improved contact between SFs and samples, (2) it can decrease the dead volume effects, and (3) it can effectively retain the moisture. Further researchers, however, assumed that silica gel was a better choice in retaining the moisture for SFE of fresh samples. Enhanced SFE results were also experiential as fresh ginger samples were blended with coarse granulated celite (30–60 mesh) before loading the samples into the SFE cell for extraction.

Sample particle size is a critical factor for a suitable SFE process. Large particles may result in lengthened extraction because the process can become diffusion controlled. Therefore, pulverizing a sample into fine powder can speed up the extraction and progress the efficiency, but it may also introduce complexity in maintaining an appropriate flow rate. One effective way to prevail over the flow rate difficulty is to pack the sample with glass beads or other rigid inert materials, such as sea sand (Lang and Wai, 2001).

The required apparatus for a basic SFE setup is simple. Fig. 2.3 depicts the basic elements of an SFE instrument, a pressure tuning injection unit, composed of a reservoir of supercritical fluid, two pumps (to take the components in the mobile phase in and to throw them out of

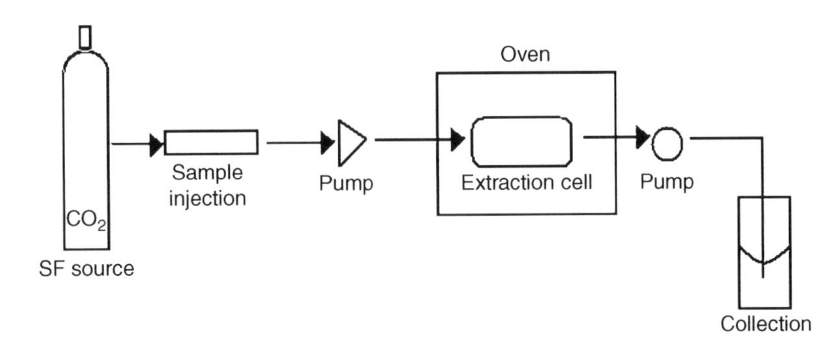

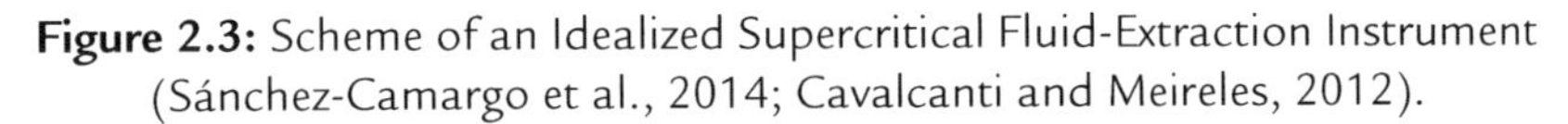

Figure 2.3: Scheme of an Idealized Supercritical Fluid-Extraction Instrument (Sánchez-Camargo et al., 2014; Cavalcanti and Meireles, 2012).

the extraction cell), and a collection chamber. There are two principle modes to run the instrument: static extraction and dynamic extraction.

In the dynamic extraction, the second pump transfers the materials out to the collection chamber and is always open during the extraction process. However, the mobile phase reaches the extraction cell and extracts components to take them out consistently.

In the static extraction experiment, there are two different steps in the process:

1. The mobile phase fills the extraction cell and interacts with the sample.
2. The second pump is opened and the extracted substances are taken out at once.

In order to choose the mobile phase for SFE, parameters taken into consideration, include the polarity and solubility of the samples in the mobile phase. Carbon dioxide is the most common mobile phase for SFE. It has the capability to dissolve nonpolar materials, such as alkanes. For semipolar compounds (such as polycyclic aromatic hydrocarbons, aldehydes, esters, alcohols, etc.) carbon dioxide can be used as a single-component mobile phase. However, for compounds that have polar characteristic, supercritical carbon dioxide must be modified by the addition of polar solvents, such as methanol. These extra solvents can be introduced into the system through a separate injection pump. There are two modes in terms of collecting and detecting the components: off-line and on-line extraction. Off-line extraction occurs by captivating the mobile phase out with the extracted components and directing them into the collection chamber. At this point, the supercritical fluid phase is evaporated and free to the atmosphere and the components are captured in a convenient adsorption surface or a solution. Then the extracted fragments are processed and prepared for a separation method. This extra manipulation step between extractor and chromatography instrument can cause errors. In the on-line method, all extracted materials are directly transferred to a separation unit due to a more sensitive, usually a chromatography instrument, without taking them out of the mobile phase. In this extraction/detection type, there is no extra sample preparation after extraction for the separation process. This minimizes the errors coming from manipulation steps. Additionally, sample loss does not occur and sensitivity increases (Cavalcanti and Meireles, 2012; Sánchez-Camargo et al., 2014).

2.3.3 Influential parameters

For successful SFE, a number of factors must be in use on reflection prior to the experiments. These factors, include the type of sample, type of fluid; method of sample preparation; choice of modifiers; method of fluid feeding; and extraction conditions including temperature, pressure, flow rate, and extraction time. To optimize the SFE conditions, a statistical experimental design based on a "second order central composite design" was carried and reported by Adas¸og˘lu et al. (1994). The diffusion rate of a compound from the sample matrix may be exaggerated by the following three factors: (1) occupation of the matrix sites by the SF molecules, which could decrease the affinity of the matrix for the solutes; (2) dissolution of the solutes in the SF, which is directly related to the fluid density; and (3) temperature effects, which can influence the volatility of the solutes, particularly for those with high boiling points. Apart from this, the solubility of a target compound in an SF is a major factor determining its EE. The solubility is controlled by the sum of two factors: the volatility of the substance, which is a function of temperature; and the salvation effect of the SFs, which is a function of fluid density [8]. To achieve a good selectivity for an SFE process, careful control of the fluid density is essential. By controlling the fluid density, fractionation of the extracts could be achieved (Lang and Wai, 2001).

2.3.4 Applications

SFE can be applied to a broad range of materials, such as carbohydrates, polymers, oils, lipids, pesticides, organic pollutants, volatile toxins, polyaromatic hydrocarbons, biomolecules, foods, flavors, explosives, pharmaceutical metabolites, and organometallics, and so forth. Common industrial applications consist of the biochemical industry and pharmaceutical; industrial synthesis in the polymer industry; and extraction, natural product chemistry, and the food industry.

Examples of materials analyzed in environmental applications: oils and fats, pesticides, alkanes, organic pollutants, volatile toxins, herbicides, nicotine, phenanthrene, aromatic surfactants, fatty acids, in samples from clay to petroleum waste, from soil to river sediments. In food analyses: caffeine, peroxides, oils, acids, cholesterol, and so forth are extracted from samples, such as coffee, olive oil, lemon, cereals, wheat, potatoes, and dog feed. Through industrial applications, the extracted materials vary from additives to different oligomers, and from petroleum fractions to stabilizers. Samples analyzed are plastics, PVC, paper, wood, and so forth. Drug metabolites, enzymes, and steroids are extracted from plasma, urine, serum, or animal tissues in biochemical applications (Herrero et al., 2010).

2.4 Liquid Phase Microextraction

Liquid-liquid extraction (LLE) is a process to transfer the analyte from an aqueous sample to a large amount of water-immiscible solvent based on their relative solubilities. LLE is almost certainly the oldest separation technique in analytical chemistry and residue between the best-known and commonly used techniques. Although LLE offers many advantages, it has been overshadowed by the extensive use of solid-phase extraction (SPE). The most significant merits of both SPE, which have made them so popular among analytical chemists, mainly include: (1) the efficient use of time and labor; (2) the significant reduction in a number of organic solvents used, thus leading to the lower cost for analysis; (3) the reduction in the amount of waste generated, which contributes to their environmentally safe character; and (4) the possibility of their being automated (Andruch et al., 2012). However, to minimize the known disadvantages of LLE while also preserving the advantages it offers, a great variety of miniaturized pretreatment techniques based on LLE have been developed, termed as liquid phase microextraction (LPME)/solvent microextraction.

LPME emerged from liquid-liquid extraction; it is a widely used sample extraction and separation procedure in spite of its clear disadvantages, such as high consumption of time and strong toxicity of solvent, as well as its tedious application (Asensio-Ramos et al., 2011). LPME, in general, takes place between a number of microliters of water-immiscible solvent extraction phase or acceptor phase (AP) and an aqueous sample phase or donor phase (DP), which contains the target analytes of interest. LPME can be classified into three main categories: single-drop liquid-phase microextraction (SD–LPME), dispersive liquid-liquid microextraction (DLLME), and hollow fiber liquid-phase microextraction (HF–LPME) (Yan et al., 2014).

2.4.1 Single-drop liquid-phase microextraction

Single-drop microextraction (SDME), as first advanced by Frederick F. Cantwell in the late 1990s, was in the beginning shared with gas chromatography. Presently, seven dissimilar modes of solvent microextraction fall under the category of SDME. However, the headspace (HS) and direct immersion (DI) modes are the ones that are used repeatedly. A microliter drop of water-immiscible organic solvent is either in the HS above the sample or immersed directly into the sample by an ordinary chromatography syringe. The aqueous sample is stirred and after the extraction, the drop is retracted back into the syringe and transferred into the chromatography system for detection.

Despite the information that the favored technique for the analysis of extracts after SDME is gas chromatography, to date SDME has also been collective with high-performance liquid chromatography, graphite furnace atomic absorption spectrometry, inductively coupled plasma mass spectrometry, capillary electrophoresis, and mass spectrometry (Andruch et al., 2012).

To decrease evaporation risk during the process to obtain the desired results and extraction period, a number of significant factors of extraction solvent, such as relatively low vapor pressure or high boiling point, density, suitable chromatographic behavior, high viscosity, and high EE or EF for the target analytes must be considered (Bai et al., 2008). On the basis of these facts, frequent extraction solvents used are toluene, hexane, octane, dodecane, and xylene. Except that, certain ionic liquids (ILs), such as 1-butyl-3-methylimidazollium hexafluorophosphate ([BMIM][PF6]), 1-hexyl-3-methylimidazolium hexafluorophosphate ([HMIM][PF6]), and 1-octyl-3-methylimidazolium hexafluorophosphate ([OMIM][PF6]) can also provide satisfactory results with better reproducibility. This production is due to the ILs, with surface tension and high viscosity, which helps to form a stable drop of a very much larger volume to prolong the extraction time. In addition, β-cyclodextrine, surfactants, and supramolecular solvents as extractants were also planned in the field of SD-LPME (Yan et al., 2014).

In short, SD-LPME has to be converted into one of the most popular sample preparation techniques because of its: (1) simplicity and speed, (2) low cost and low environmental pollution, (3) extensive range of solvent types being selected and used, (4) applicability to different complex matrices, (5) extraction and separation of both inorganic and organic compounds, and (6) compatibility with chromatographic or electrophoretic injection systems. However, HS-SDME is more often applied to analytes with relatively high vapor pressure, such as the volatile components of essential oil (Deng et al., 2005).

2.4.2 Dispersive liquid-phase microextraction

Dispersive liquid-liquid microextraction is a miniaturized version of conventional LLE and requires only microliter volumes of solvents. As the name suggests, DLLME is equivalent to a miniaturized type of LLE that is generally established on a ternary component solvent system, in which an appropriate disperser solvent is introduced to help the dispersion of an organic extraction solvent into an aqueous sample and further achieve a highly efficient extraction. Briefly, a DLLME procedure may be outlined as follows: (1) the injection of extraction and disperser solvents into sample solution, (2) the formation of so-called cloudy state due to the solvency of the disperser solvent with two other solvents, (3) the achievement of extraction equilibrium in a short time based on the extensive surface contact between the droplets of the extraction solvent and the sample, and (4) centrifugation to completely separate the two phases and force the organic phase with the extracted analyte to the bottom (Yan and Wang, 2013).

The mixture of dispersive solvent and extraction solvent is rapidly injected into a sample solution with a glass syringe. At the moment of injection, fine droplets of the extraction solvent are dispersed in the aqueous phase, forming a so-called cloudy state, which should remain stable for a while. In this regard, the very small size of the extraction droplets, the big surface area thus produced enables a very rapid mass transfer of the analyte from one phase into another. The cloudy solution is then uncovered to centrifugation to completely separate the two phases and force the organic phase with the extracted analyte to the bottom of the test tube (Andruch et al., 2012).

Obviously, the advantages of DLLME are mainly the following: (1) the use of only microliter volumes of extraction solvent, which makes the procedure environmentally friendly, (2) the short extraction time as a result of the rapid achieving of the equilibrium state, and (3) the high enrichment factor as a result of the high phase ratio. Accordingly, the DLLME technique is simple, quick, efficient, and simultaneously meets the development requirement of green chemistry.

Nevertheless, this technique has certain limitations, which primarily result from requirements related to the extraction and disperser solvents—namely: (1) the extraction solvent has to be immiscible with water, have a high extraction potential for the target analyte, and has to have a density higher than that of water due to simple phases separation by centrifugation; (2) the disperser solvent, on the other hand, has to be miscible with both the extraction solvent and water (the sample solution) to enable the dispersion of fine particles of the extraction solvent into the aqueous phase containing the analyte. For this explanation, chloroform, 1,2-dichlorobenzene, carbon tetrachloride, and dichloromethane are most frequently apply as an extraction solvent, and ethanol, methanol, acetonitrile, and acetone are usually used as disperser solvent (Yan and Wang, 2013).

The DLLME technique uses somewhat bigger volumes of organic solvents than are general in SDME, but this still only involves microliters. Therefore, DLLME, much like SDME, is also most often combined with GC, HPLC, GF-AAS, and FAAS detection. However, articles in which DLLME is collective with UV–vis spectrophotometry—are being published less often (Andruch et al., 2012).

Furthermore, ILs, known as "green solvents," are a group of nonmolecular solvents that can be defined as organic salts that remain in a liquid state at room temperature (RTILs). These solvents have numerous exclusive physicochemical properties, such as variable viscosity, negligible vapor pressure, and high thermal stability. On the basis of the diverse combinations of organic cations and various anions, ILs can be structurally tailored to be hydrophilic or hydrophobic, such as miscible or immiscible with the disperser solvent in DLLME. In addition, the high density and the low volatility of ILs are also significant features; the former facilitates phase separation, and the latter offers stable droplets. ILs are consequently regarded as environmentally kind replacements for traditional toxic organic solvents and can be potentially used in DLLME (termed as IL-DLLME) (Table 2.1). In the years that followed, many applications based on IL-DLLME technique were carried, focusing on the determination of pesticides, metal ions, pharmaceuticals and other organic pollutants in water, food, urine, or even cosmetics (Yan and Wang, 2013).

2.4.3 Hollow-fiber liquid-phase microextraction

To augment single-drop stability and decrease AP pollution from the substrate in DP or impurities, mentioned earlier, SD-LPME, hollow-fiber liquid-phase microextraction (HF-LPME) was introduced. HF-LPME is a technique that allows extraction and preconcentration of analytes from complex samples in both a simple and inexpensive way (Pedersen-Bjergaard and Rasmussen, 1999). In the two-phase LPME sampling mode (HF-LPME), the analyte is extracted from an aqueous sample to a water-immiscible extractant immobilized in the pores of a hollow fiber, typically made of polypropylene and supported by a microsyringe. In this sampling mode, the AP is organic—that is, attuned with atomic detectors for total resolve, such as HPLC and GC for the coupling of chromatographic separation techniques to atomic detectors (Psillakis and Kalogerakis, 2003).

Table 2.1: Applications of SD-LPME and ionic liquid D-LPME for the extraction of drugs.

Analyte	Matrix	LPME Type	Detection Techniques	References
Salinomycin, gramicidin D	Water/human urine and plasma	SD-LPME	MALDI-MS	Wu et al. (2011)
Berberine, palmatine, tetrahydropalmatine	Human urine	SD-LPME	MECK	Gao et al. (2011a)
Gatifloxacin, lomefloxacin, enoxacin, ciprofloxacin, ofloxacin, pefloxacin	Human urine	SD-LPME	CE	Gao et al. (2011b)
Mizolastine, chlorpheniramine, pheniramine	Human urine	SD-LPME	MECK	Gao et al. (2012)
Illicit drugs	Horse urine	SD-LPME	GC-ECD	Stege et al. (2011)
Statins	Serum sample/ water	SD-LPME	CEC	Jahan et al. (2015)
Growth hormones	Bovine urine	SD-LPME	GC-MS	George et al. (2015)
Sufentanil alfentanil	Human urine, wastewater	SD-LPME	GC-FID	Fakhari et al. (2011)
Amitriptyline, nortriptyline	Human urine and plasma	SD-LPME	GC-FID	Yazdi et al. (2008)
Nortriptyline, imipramine, trimeprazine, promethazine, imino dibenzyl	Human urine	SD-LPME	CE	Wu et al. (2014)
Clozapine, desmethyl clozapine, or clozapine	Human urine and serum	IL-D-LPME, [EMIM] [NtfO2]	CE	Breadmore (2011)
Ephedrine, ketamine	Human urine	IL-D-LPME, [BMIM] [PF6]	CE	Liu et al. (2013b)
Rifaximin	Rat serum	IL-D-LPME, [BMIM] [PF6]	HPLC-UV	Nageswara Rao et al. (2012)
Eprosartan, valsartan, irbesartan, losartan, telmisartan	Rat serum	IL-D-LPME, [BMIM] [PF6]	HPLC-UV	He et al. (2009)
Balofloxacin	Rat serum	IL-D-LPME, [BMIM] [PF6]	HPLC-UV	Rao et al. (2014)
Emodin, metabolites	Rat urine	IL-D-LPME, [C6MIM][PF6]	HPLC-UV	Tian et al. (2012)

In HF-LPME, limited to analytes with ionizable functionalities, the analyte is extracted from an aqueous sample through the water-immiscible extractant immobilized in the pores of the hollow fiber and ultimately into an aqueous phase inside the lumen of the hollow fiber. Because the AP is aqueous in this microextraction mode, the technique might be attuned not only with atomic detectors for determination of totals but also with hyphenated techniques involving HPLC separations for speciation (Pena-Pereira et al., 2009).

Table 2.2: Comparison of advantages and drawbacks of SDME, HFME, and DLLME microextraction techniques.

SDME	HFME	DLLME
Inexpensive, simple, easy to operate, nearly solvent free, more suitable for volatile and semivolatile analytes, environmental friendly, various extraction modes, ease of automation, high extraction efficiency	Inexpensive, simple, environmental friendly, high versatility and selectivity, headspace and immersion modes	Simple, high enrichment factors, rapid, inexpensive, environmental friendly, enormous contact area between acceptor phase and sample, fast reaction kinetics, instantaneous extraction, complete analyte recovery, DLLME can be coupled with SPE, SFE, SBSE, nano techniques
Problem of drop dislodgment, time-consuming, incomplete equilibrium	Poor reproducibility, time-consuming, formation of air bubble	Minor restrictions in solvent selection and automation

DLLME, Dispersive liquid-liquid microextraction; SDME, single-drop microextraction; SFE, supercritical fluid extraction; SBSE, stir-bar sorptive extraction; SPE, solid-phase extraction.

Solvent impregnation of the fiber is necessary since extraction occurs on the surface of the immobilized solvent. The pores of a porous hydrophobic polymer membrane are packed with an organic liquid, which is apprehended by capillary forces. The extractant should have a polarity matching that of the hollow fiber to be easily immobilized within its pores. Usually, the extraction effectiveness gained with HF-LPME is greater than with direct-SDME, since hydrophobic hollow fibers permit the use of vigorous stirring rates to increase speed of the extraction kinetics. Moreover, the contact area between the aqueous sample and the extractant phase is higher than in the case of SDME, favoring the mass transfer rate. The use of the hollow fiber provides protection of the extractant phase and hence, the analysis of dirty samples is feasible. Moreover, the small pore size allows microfiltration of the sample, thus yielding very clean extracts (Table 2.2) (Rasmussen and Pedersen-Bjergaard, 2004).

2.5 Solid Phase Extraction

SPE is a sample extraction technique normally used in laboratories for the extraction of a complex matrix, such as urine, blood, food samples, water, etc. Traditionally, liquid-liquid extraction (LLE) was developed and employed to screen for general unknowns. However, SPE is becoming more popular than LLE for analyte preconcentration and matrix removal, due to its simplicity and economy in terms of time and solvent (Picó et al., 2007). SPE has gained wide acceptance because of the inherent disadvantages of LLE, whose drawbacks include (Płotka-Wasylka et al., 2015):

1. incapability to extract polar compounds,
2. being laborious and time-consuming,
3. expense,
4. tendency to form emulsions,

5. need for evaporation of huge volumes of solvents,
6. discarding of toxic or flammable chemicals.

By contrast, SPE is a more efficient separation process than LLE and is becoming one of the most routinely used procedures for the separation and preconcentration of a variety of compounds and elements from complex samples due to their well-known advantages, which include the high enrichment factor, easier to obtain a higher recovery of the analyte by using a reduced volume of solvents and the possibility of automation (off- or on-line) of the whole process. Moreover, modern regulations pertaining to the use of organic solvents have made LLE techniques undesirable. LLE procedures that need a number of consecutive extractions to make progress greater than 99% of the analyte can often be replaced by SPE methods. Furthermore, SPE does not need the phase separation required for LLE, and that eliminates errors associated with a variable or inaccurately measured extract volumes. In recent decades, the use of SPE has increased due to the progress of a variety of new materials that may be engaged as solid sorbents (Płotka-Wasylka et al., 2015).

2.5.1 Method of solid-phase extraction technique

SPE is widely accepted for analyte extraction and preconcentration as a substitute to time consuming and laborious liquid–liquid extraction (LLE) procedures. SPE uses the distinction of attraction between an analyte and interferents present in a liquid matrix, for a solid phase (sorbent). This affinity allows the partition of the target analyte from the interferents. There are four steps involved in solid phase extraction (Fig. 2.4):

1. First, the cartridge is conditioned or equilibrated with a solvent to wet the sorbent.
2. The loading solution containing the analyte is percolated through the solid phase. Ideally, the analyte and some impurities are retained on the sorbent.

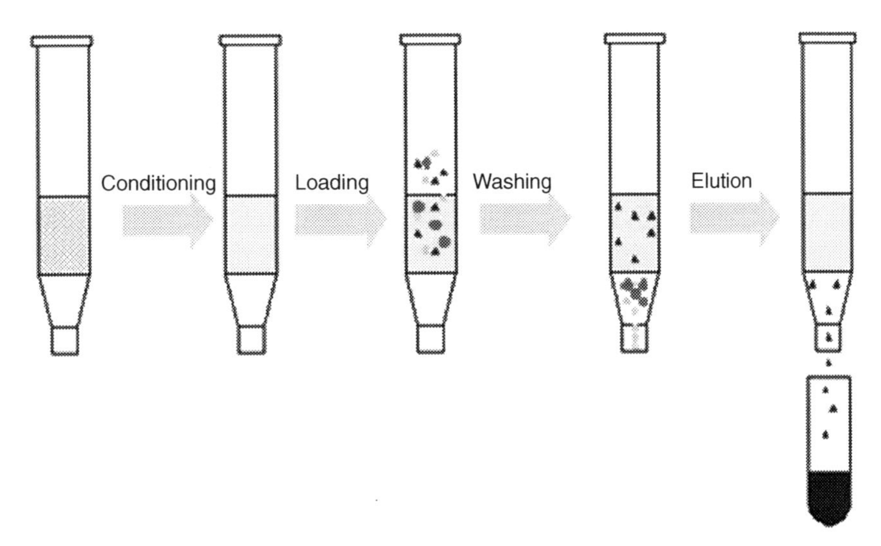

Figure 2.4: The Fundamental Method of SPE Technique (Herrero-Latorre et al., 2015).

3. The sorbent is then washed to eradicate impurities.
4. The analyte is collected during this elution step.

However, this technique suffers from some drawbacks, depending on the type of SPE applied, the sorbent used, and the characteristics of the sample. The most frequently used techniques make use of column immobilized SPE and these may result in long treatment times, high back-pressure in the packing method and low extraction efficiencies in convinced cases when compared to other SPE methodologies. Therefore, in the past few years, other SPE approaches, such as solid-phase microextraction (SPME), dispersive solid-phase extraction (DSPE), magnetic solid-phase extraction (MSPE), molecularly imprinted solid-phase extraction (MISPE), and matrix solid-phase dispersion extraction (MSPDE), have been applied in an effort to overcome these problems (Herrero-Latorre et al., 2015).

2.6 Carbon Nanotubes Solid Phase Extraction

Nanotechnology is currently one of the largest significant trends in science, which includes the production of novel and revolutionary materials of the size of 100 nm or even smaller. Carbon nanotubes (CNTs) are part of these novel materials. Their discovery was a direct consequence of the synthesis of fullerenes, especially the buckminsterfullerene, C60. CNTs are molecular-scale tubes of graphitic carbon that can be consequently considered as a graphene sheet in the shape of a cylinder (Gouda, 2014). CNTs can be visualized as sheets of graphite rolled into a tube; they are known to have strong interactions with molecules or ions. Their great surface areas show them to have potential to be used for solid-sorbent for preconcentration procedures. CNTs can be further divided into multiwalled carbon nanotubes (MWCNTs) and single-walled carbon nanotubes (SWCNTs) according to the carbon atom layers in the wall of the nanotubes. Multiwalled CNTs (MWCNTs) better reflect their structure. The key predicament when applying SPE always remains the method development and the choice of the most appropriate sorbent, which depends on the physicochemical properties of the analytes. In this sense, CNT's high surface area; ability to establish π–π interactions; excellent chemical, mechanical and thermal stability; and so forth make them very attractive as SPE materials for either nonpolar (in the case of nonfunctionalized CNTs) and polar compounds for which functionalization of the tubes plays a key role in selectivity. Because of its elevated extraction effectiveness, the ease of method development, the lower amount of organic solvents used, and the possibilities of automation, SPE has been increasingly used regarding the extraction of various organic analytes (pesticides, pharmaceuticals, phthalate esters, and phenolic compounds), as well as inorganic analytes (Table 2.3) (Ravelo-Pérez et al., 2010).

2.7 Magnetic Solid-Phase Extraction

Magnetic separation and preconcentration using magnetic carbon nanotubes (M-CNTs) give optimum and discriminating sample pretreatment measures that minimized the mandatory

Table 2.3: Applications of CNTs as SPE sorbents for the extraction of drugs.

Analyte	Matrix	Detection Techniques	CNTs Characteristics	Remarks	References
Atrazine simazine	Water	GC/MS	MWCNTs	–	Katsumata et al. (2010)
Benzodiazepine residues: diazepam, estazolam, alprazolam, and triazolam	Pork	GC-MS	MWCNTs	Analytes were extracted by ultrasonic-assisted extraction	Wang et al. (2006)
Sulfonamides: sulfadiazine, sulfamerazine, sulfadimidine, sulfathiazole, sulfamoxole, sulfamethizole, sulfamethoxypyridazine, sulfachlorpyridazine, sulfadoxine, and sulfisoxazole	Eggs and pork	HPLC-UV	MWCNTs	Ultrasonic-assisted extraction and online SPE	Fang et al. (2006)
NSAIDs: tolmetin, ketoprofen, and indomethacin	Urine	CE-MS	Carboxylated SWCNTs immobilized on an inert porous glass	–	Suárez et al. (2007b)
Barbiturates: barbital, amobarbital, and phenobarbital	Pork	GC-MS/MS	MWCNTs	Barbiturates were extracted by ultrasonic-assisted extraction and derivatized after SPE procedure	Zhao et al. (2007)
Tetracyclines: oxytetracycline, tetracycline, and doxycycline	Surface water	CE-MS	SWCNTs	Only MWCNTs provided adequate results	Suárez et al. (2007a)
Cephalon, cephalexin, cephradine, benzoic acid, cefaclor, sulfathiazole, sulfadiazine, sulfamethazine, phenol, hydroxyquinone, guaiacol, 1,3,5-trihydroxybenzene, and 3,5-dihydroxybenzoic acid	Tap and well water	HPLC-UV	MWCNTs	Retention abilities of C18 and graphitized carbon black were also investigated	Niu et al. (2007)
Antidepressants: imipramine, nortriptyline, desipramine, amitryptiline, clomipramine, trimipramine, trazodone, fluoxetine, and mianserin	Urine	HPLC-UV	MWCNTs	Use of ionic liquids to improve the HPLC chromatographic the behavior of the analytes	Cruz-Vera et al. (2008)

(*Continued*)

Table 2.3: Applications of CNTs as SPE sorbents for the extraction of drugs. (*cont.*)

Analyte	Matrix	Detection Techniques	CNTs Characteristics	Remarks	References
Thiamine	Serum and urine; pharmaceutical and foodstuffs	Spectrofluorimetry	MWCNTs	–	Daneshvar Tarigh and Shemirani (2014)
Diethylstilbestrol estrone estriol	Tap, river, and mineral water; honey	MEKC	MWCNTs	–	Guan et al. (2010)
Estradiol ethinyloestradiol hexestrol	Milk	HPLC	MWCNTs	–	Ding et al. (2011)
Bovine serum albumin	Bovine calf serum	HPLC	MWCNTs	–	Zhang et al. (2011)
Quercetin luteolin kaempferol	Human urine	HPLC	MWCNTs	–	Xiao et al. (2014)
Gatifloxacin	Serum	HPLC	MWCNTs	–	Xiao et al. (2013)
Naproxen	Human urine	Spectrofluorimetry	MWCNTs	–	Madrakian et al. (2013)
Doxorubicin	Rat tissues	HPLC	PEGylatede MWCNTs	–	Shen et al. (2011)
Methylprednisolone	Rat plasma	HPLC	MWCNTs	–	Yu et al. (2014a)
Puerarin	Rat plasma	HPLC	PEGylatede MWCNTs	–	Yu et al. (2014b)

sample preparation time, reduced the number of matrix manipulations and the sample-preparation steps, and avoided the introduction of sources of uncertainty in the analytical process.

M-SPE involves the accumulation of magnetic sorbent particles to the sample solution. The external magnetic field is applied for the separation of sample when the target compound is adsorbed onto the magnetic material and the magnetic particle (containing the analyte). Finally, the analyte is recovered from the adsorbent by elution with the suitable solvent and it is subsequently analyzed. A general scheme for the M-SPE procedure is shown in Fig. 2.5. This access has a number of advantages over long-established SPE (Herrero-Latorre et al., 2015):

1. It averts time-consuming and tedious on-column SPE procedures.
2. It provides a rapid and simple analyte separation that avoids the need for centrifugation or filtration steps.
3. The magnetic sorbents have high selectivity, even when complex matrices from environmental or biological fields were analyzed.

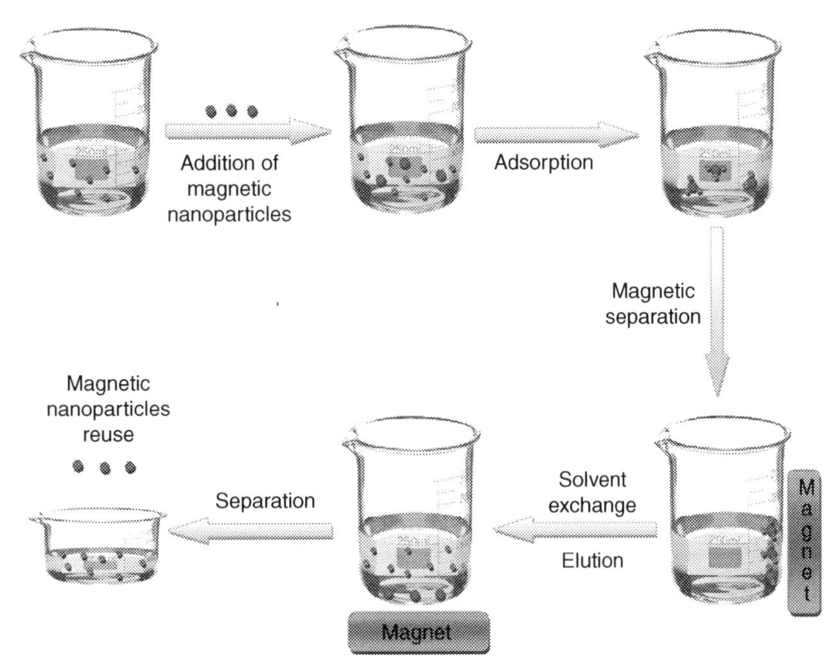

Figure 2.5: The Procedure Used for Magnetic Solid-Phase Extraction (M-SPE) (Płotka-Wasylka et al., 2015).

4. Sample impurities are diamagnetic; they do not get in the way with magnetic particles during the magnetic separation step.
5. Automation of the whole process is possible with flow injection analysis and other related techniques, which leads to rapid, selective, sensitive, and repeatable methods for routine determinations.

MSPE has been extensively used in many fields, including (Chen et al., 2012):

1. biomedicine, to separate cells and to isolate enzymes, proteins;
2. environmental science, for the isolation of pesticides, metal ions, dyes, PAHs, drugs, antibiotics, and carcinogenic, surfactants, and mutagenic compounds in water and sewage samples;
3. food analysis, to extract pesticides, antibiotics, metals, and drugs from different kinds of food samples.

2.8 Molecularly Imprinted Solid Phase Extraction (MISPE)

Molecular imprinting is a method used for producing synthetic polymers with predetermined molecular recognition properties. Molecularly imprinting technology depends on the arrangement of a complex between an analyte (template) and a functional monomer. A huge surplus of a cross-linking agent is required to form a three-dimensional polymer network.

After the polymerization process, the template is removed from the polymer leaving specific recognition sites complementary in shape, size, and chemical functionality to the template molecule. Mainly between the template molecules, intermolecular interaction, such as dipole–dipole, hydrogen bonds, and ionic interactions occurred. These interactions act as functional groups present in the polymer matrix and drive the molecular appreciation phenomena. Thus, the resultant polymer perceives and binds selectively only to the template molecules (Vasapollo et al., 2011).

The principle of MISPE is based on the same main steps as conventional SPE: habituation of the sorbent, loading the sample, washing away interferences, and elution of the target analytes.

In the second step, the sample is percolated through the molecularly imprinted polymer. It is significant to provide similar solvent polarity to that used in the polymerization process since it enhances the number of relations between the analyte and specific binding sites in the MIP sorbents. Acrylate-based MIPs became a popular choice as selective/specific sorbents for SPE and nowadays, applications of these conventional MIPs are focused on the large and complex molecules (Płotka-Wasylka et al., 2016).

Recently, MIPs attracted much attention due to their outstanding advantages (He et al., 2007):

1. predetermined recognition ability;
2. mechanical and chemical stability;
3. relative ease and simplicity of preparation;
4. low cost of preparation;
5. potential application to a wide range of target molecules.

Recently, MIPs have been attracting widespread interest in many fields of science, especially to mimic natural recognition entities, such as antibodies and biological receptors; they are also useful to separate and analyze complex materials, such as biological fluids, food, drug, and environmental samples (Płotka-Wasylka et al., 2016).

2.9 Solid-Phase Microextraction

SPME is a class of SPE, in which SPME disposes of two of the largest substantial faults of SPE (i.e., the length of time for extraction and, more importantly, the need to use organic solvents). Sample preparation using SPME gained appreciation among a large group of analytical scientists because of the following benefits (Duan et al., 2011; Souza Silva et al., 2013; Spietelun et al., 2013).

1. The possibility of simultaneous download, concentration, and analyte determination, which significantly shortened the time to make an analysis.
2. High sensitivity (possibly to determine the substance at the ppt level).
3. Small sample size.

4. Simplicity and speed of analysis, where use of complicated equipment, tools, and devices or precise operations was not required.
5. Cost minimization by eliminating organic solvents and expensive toxic and multiple uses of SPME fibers.
6. Small fibers, which consent to the device to download samples in in situ conditions.
7. The prospect of automation.
8. The possibility of joining with other instrumental techniques—most often with liquid chromatography (LC), gas chromatography (GC), high-performance liquid chromatography (HPLC), and capillary electrophoresis (CE) in the off-line or on-line modes.

The numerous advantages of SPME mean that it is almost universal, because it allows analysis of many kinds of the sample in different physical states—liquid, gas, and solid—often with very complex matrixes, and it contributes determination of analytes at trace and ultratrace levels. In SPME, thin fibers containing melted silica, coated with an appropriate sorption material, are engaged in catching the target analytes. Compounds that exist in the sample are divided into the matrix and the coated fiber. The quantity of absorbed analyte depends on the partition coefficients between the sorbent layer coating the fiber and the matrix of the samples and on the analyte affinity, the time of contact, and other variables (Lord and Pawliszyn, 2000).

2.10 Stir-Bar Sorptive Extraction

Stir-bar sorptive extraction (SBSE) emerged as a novel application involving the use of polydimethylsiloxane (PDMS) polymer as a sorbent for SPE. SBSE is depended on the similar principles as SPME, but, as a substitute of a polymer-coated fiber, stir bars are coated with PDMS, and a polar polymeric phase is used for hydrophobic relations with target molecules (commercially available as Twister, Gerstel GmbH), whereas, the retention process in the PDMS phase is based on van der Waals forces and the hydrogen bonds that may be produced with oxygen atoms of PDMS, depending on the molecular structure of the target analytes. Due to its elimination of solvents and reduction of the labor-intensive and time-consuming sample preparation step, SBSE also fulfills the requirements of green analytical chemistry (Nogueira, 2012).

Sampling is done by introducing the SBSE device into the aqueous sample. While it is stirred, the bar absorbs analytes to be extracted. The bar is removed from the sample, rinsed with deionized water, and dried. Afterward, the analytes are desorbed from the enriched sorbent phase by thermal desorption (TD) but, in the case of analytes that decompose at low temperatures, the analytes are desorbed by liquid desorption (LD).

The greatest applicable improvement to expand the significant SBSE would be new coatings, which would allow the analysis of polar compounds. Unfortunately, these compounds generally cannot be analyzed using gas chromatography (GC), and TD cannot be used. These new coating materials include (Płotka-Wasylka et al., 2015):

1. Polyurethane foams
2. Silicone materials
3. Poly (ethylene glycol)-modified silicone (EG Silicone Twister)
4. Poly(dimethylsiloxane)/polypyrrole
5. Polyvinyl alcohol
6. Poly(phthalazine ether sulfone ketone)
7. Polyacrylate (acrylate twister)
8. Carbon nanotube-polydimethylsiloxane) (CNT-PDMS)
9. Alkyl-diol-silica (ADS) constrained right of entry materials
10. Cyclodextrin
11. Monolithic materials
12. Sorbents obtained with sol–gel techniques

The SBSE technique offers several advantages; however, since a single polar polymer covers the stir bar, it may only be applied to semivolatile, thermo-stable compounds. Moreover, SBSE has fewer disadvantages, such as (1) limited spectrum of analyte polarities for the offered stationary phases, (2) the existence of strong matrix effects, and (3) the need for high control of extraction conditions. SBSE is effectively applied to many analytical fields (e.g., clinical, environmental, and food analysis) and to different kinds of the matrix, including environmental water, wastewater, soils, biological fluids, and gaseous samples. SBSE-based methods have developed in the field of clinical analysis and pharmaceuticals (Table 2.4). However, the number of applications within this field is lower than in the field of environment and food analysis (Popp et al., 2001).

2.11 Electrical-Field-Induced Solid-Phase Extraction

The electrical-field-induced solid-phase extraction (EF-SPE) technique is a topic that is less interrogate. Electrically or electrochemically induced solid-phase extraction techniques have originated significant applications in SPE and SPME techniques with the synthesis of sorbents and augmentation of extraction. Generally, an electrical field has two effects in the solid-based extraction techniques; it directly affects the surface of solid sorbents and provides manipulation possibility of extraction, or it has an indirect effect that only provides an electrokinetic migration of charged analytes toward solid sorbent or facilitates elution of the analytes (Yamini et al. 2014). The substantial escalating movement in application of the electrical field in solid depends on extraction techniques, which may be recognized for several reasons:

1. Properties of the conducting polymer can be modified by varying the conditions during the electropolymerization step to enable extraction of analytes with different sizes and charges.

Table 2.4: Applications of SBSE method for the extraction of drugs.

Analyte	Matrix	Detection Techniques	SBSE-Coating Materials	References
Benzothiazole	Wastewater	GC–MS	PA	Camino-Sánchez et al. (2014)
Ractopamine, isoxsuprine, clenbuterol, and fenoterol	Pork, liver, and feed	HPLC-UV HPLC-FID	MIP with ractopamine	Xu et al. (2010a)
Ractopamine	Pork meat	ECL	MIP	Wang et al. (2012)
Ketamine	Urine	HPLC-UV	Titania-OH-TSO	Xu et al. (2010b)
Paracetamol, caffeine, antipyrine, propranolol, carbamazepine, ibuprofen, diclofenac, methylparaben, ethylparaben, and propylparaben	River water, effluent and influent wastewater	LC–MS/MS	Hydrophilic polymer based on poly(*N*-vinylpyrrolidone-*co*-divinyl benzene)	Bratkowska et al. (2011)
Paracetamol, caffeine, antipyrine, propranolol, carbamazepine, naproxen, and diclofenac	Environmental water	LC–MS/MS	Poly(MAA-*co*-DVB)	Bratkowska et al. (2012)

2. Compared to conventional solid-based extraction techniques, in which a material with a fixed number of exchange sites is employed, the electrically assisted technique offers higher flexibility. This is because the properties of the material and number of exchange sites may be on the outside prohibited by electrochemically controlling the charge of the material.
3. The use of polymer on the fiber films in solid-based extraction techniques may be comprehensive to the analysis of neutral, electro-inactive analytes by engaging the advantage of electrochemically controlled hydrophilic or hydrophobic "switching".
4. Electrically assisted solid-based extraction techniques may be used for extractions and analytes that usually need to be derivatized prior to traditional solid-based extractions.
5. In electrically assisted solid based extraction techniques, the extraction and desorption steps are performed merely by changing the potential of the conducting polymer-coated electrode. In this way, there is no need for changes of the solvent to for facilitate desorption of the compounds.
6. By altering the electrochemical potential of the polymer, desorption of electrostatically held analytes may also be faster compared to the desorption techniques normally used in traditional solid-based extractions. This makes the technique mainly attractive for utilizing in conjunction with miniaturized analytical systems.

2.12 Ultrasound-Assisted Extraction

Ultrasound-assisted extraction (UAE) is regularly fetching a matter of routine practice in analytical chemistry, which uses this energy for a multiplicity of purposes in relation to sample preparation, mostly sample extraction. Solvent extraction of organic compounds enclosed within the seeds and plants are extensively enhanced by using the power of ultrasound. The mechanical property of ultrasound affords a greater solvent penetration into cellular materials and progress mass transfer due to the property of microstreaming. This is combined with an additional benefit of using ultrasound in extractive processes—namely, disruption of biological cell walls to release the cell contents. Overall, UAE is renowned as a competent extraction technique that dramatically reduces working times, increasing yields, and repeating the quality of the extract (Awad et al., 2012).

Ultrasound consists of mechanical waves that require an elastic medium to spread. The difference between ultrasound and sound is the frequency of the wave; sound waves are at human-hearing frequencies (16 Hz to 16–20 kHz), whereas ultrasound has frequencies above human hearing but below microwave frequencies (from 20 kHz to 10 MHz). For the classification of ultrasound applications, the amount of energy generated, characterized by sound power (W), sound intensity (W/m^2), or sound energy density (W/m^3) is the key criterion. The uses of ultrasound is broadly distinguished into two groups: high intensity and low intensity (Chemat et al., 2011).

Low-intensity ultrasound—high frequency (100 kHz–1 MHz), low power (typically $<1\ W/cm^2$) —is involved in nondestructive analysis, particularly for quality assessment. On the other hand, high-intensity ultrasound—low frequency (16–100 kHz) high power (typically 10–1000 W/cm^2)—is used, among other applications, to speed up and to progress the effectiveness of sample preparation (Awad et al., 2012).

During the sonication process, when a sonic wave meets longitudinal waves, a liquid medium is produced, as a result of generating regions of exchanging compression and rarefaction waves induced on the molecules of the medium (Soria and Villamiel, 2010). In these regions of altering pressure, cavitation bubbles are generated close to the plant material surface; then, during a compression cycle, this bubble collapses and a microjet directed toward the plant matrix is created. The high temperature and pressure involved in this process demolish the cell walls of the plant matrix, and its content can be released into the medium. This is an extremely attractive tool for extraction of ingredients from natural products. This technique requires a liquid medium, an energy generator, and a transducer, which transforms the electric, magnetic, or kinetic energy into acoustic energy (Chemat et al., 2011).

Ultrasound-assisted solvent extraction is also considered a good option for organic-compound extraction from different matrices, because it provides more efficient contact between solid and solvent due to an increase of pressure (which favors penetration and transport) and temperature (which improves solubility and diffusivity). Several extractions can be performed simultaneously, and, as no specialized laboratory equipment is required, the technique is relatively inexpensive compared to most modern extraction methods. Several classes of food components (e.g., aroma, pigments, antioxidants, and other organic and mineral compounds), additives, and environmental contaminants have been extracted and analyzed efficiently from a variety of matrices (mainly animal tissues, food, plant materials, water, soil, and sediment). Due to its characteristics, UAE can be used as a pretreatment prior to more sophisticated extraction, such as solvent-phase extraction/solid-phase extraction/SFE/or MSE (Gómez-González et al., 2010).

2.13 Cloud-Point Extraction

Cloud-point extraction (CPE), an easy, environmental friendly, rapid, safe, and inexpensive methodology for preconcentration and disassembly of trace metals from aqueous solutions has recently become an attractive area of research and an alternative to liquid–liquid extraction.

Aiming at greener analytical chemistry, several strategies have been adopted, involving replacement of toxic reagents, minimized reagent consumption, recycling, and waste treatment. One of them is the use of surfactants. Micellar systems are also extensively utilized in analytical chemistry, particularly in separation/preconcentration procedures based on spectral methods. Also, the clouding behavior of micellar solutions is extensively oppressed as cloud point extraction for the extraction and preconcentration of various organic, metal ions, and inorganic industrial pollutants, pharmaceuticals, pesticides, and proteins.

Aqueous solutions of many zwitterionic surfactant micellar and nonionic systems become turbid over a slight temperature range when the experimental conditions have been changed. This temperature is named "lower consolute temperature" (LCT) or "cloud-point temperature" (CPT). The value of CPT depends on the structure and concentration of the surfactant, and the presence of additives, such as salts, alcohols, other surfactants, polymers, and some inorganic or organic compounds, which can cause a decrease or increase of CPT. Therefore, in micellar systems, two types of phase separation can be experimental: lower consolute type (occurring below CPT) and upper consolute type (occurring above CPT). The lower consolute type of phase separation is rarely obtained in micellar systems. On the other hand, in some situations, the cloud point was attained under CPT and even at room temperature, by altering pressure or via various additives, together with electrolytes (Nascentes and Arruda, 2003). In micellar media, the upper consolute type of phase separation dominates. In this case, clouding phenomenon is observed especially with polyoxyethylene surfactants and can be ascribed to the efficient dehydration of hydrophilic portion of micelles at higher temperature conditions and due to the interaction of nonionic surfactant micelles via an attractive potential, whose well depth increases with temperature. These micelles attract each other and form clusters with the approach of the cloud point. However, the mechanism that stops after the lower consolute behavior of nonionic surfactant systems remains obscure (Mukherjee et al., 2011).

More simply, CPE may be defined as a progression of transferring a nonionic surfactant from one liquid phase to a further one by heating. As the temperature of the solution rises, the

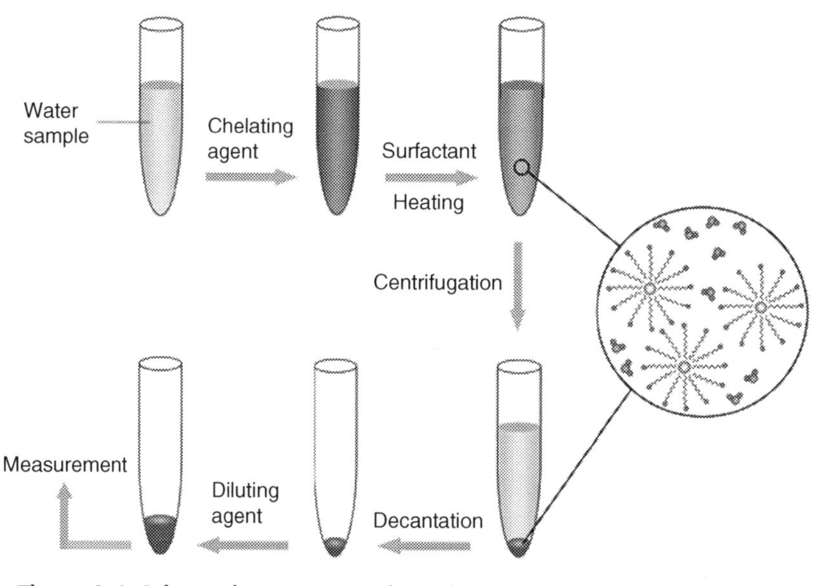

Figure 2.6: Schematic Representation of a Basic Cloud-Point Extraction (Pytlakowska et al., 2013).

surfactant molecules form micelles, and if the temperature increases above the cloud point (CPT), the micelles become dehydrated and aggregate. This leads to macroscopic phase separation of the solution into a solvent phase and surfactant-rich phase. The typical cloud point methodology is used for extraction of metal ions is given in Fig. 2.6 (Pytlakowska et al., 2013).

CPE is an efficient separation/preconcentration technique that has been established to have the separate merits of simplicity, low cost, speed, lower toxicity to the environment, and a high ability to concentrate an extensive variety of analytes with high recoveries and high concentration factors. In addition, the simple combination with chromatographic, atomic absorption, spectral, and electrochemical analyses allows cloud-point extraction for elaborating highly sensitive and convenient analytical methods.

In the CPE technique, the surfactant is one of the key reagents. The surfactants that are used in cloud point extraction are frequently of nonionic types, such as polyoxyethylene-9.5-octylphenoxy ether (Triton X-100), polyoxyethylene-7.5-octyl-phenoxy ether (Triton X-114), or PONPE (poly nonyl phenyl ether). The technique is extensively used for the extraction and preconcentration of various organic and inorganic industrial pollutants, metal ions, pharmaceuticals, pesticides, and proteins (Madej, 2009).

2.14 Enzyme-Assisted Extraction

Enzyme-based extraction of bioactive compounds from plants is a prospective substitute to conventional solvent-dependent extraction methods. Enzymes are ideal catalysts to assist in the modification, extraction, or synthesis of complex bioactive compounds of natural origin. Enzyme-based extraction depends on the inherent capability of enzymes to catalyze reactions with regioselectivity, exquisite specificity, and a capacity to function under mild processing conditions in aqueous solutions (Gardossi et al., 2010). This method also offers the opportunity of greener chemistry as pressure mounts on pharmaceutical companies and the food industry to identify cleaner routes for the extraction of new compounds. Enzymes have the capability to degrade or disrupt membranes and cell walls, thus enabling enhanced release and extra proficient extraction of bioactive (Pinelo et al., 2006).

Enzyme-assisted extraction (EAE) methods are a fast consideration because of the need for ecofriendly extraction technologies. A particularly useful application of enzymes increases the result of solvent pretreatment and either reduces the amount of solvent needed for extraction or increases the yield of extractable compounds. Enzymes, such as cellulases, pectinases, and hemicellulases are extensively used in juice extraction and beer clarification to degrade cell walls and increase juice extractability. The interruption of the cell wall matrix also releases components, such as phenolic compounds into the juice, thus enhancing product quality (Puri et al., 2012).

EAE methods have been exposed to accomplish high extraction yields for compounds including natural pigments, polysaccharides, flavors, oils, and medicinal compounds (Sowbhagya and Chitra, 2010). Recent studies on enzyme-assisted extraction have exposed more rapid extraction, reduced solvent usage, higher recovery, and lower energy consumption when compared to nonenzymatic methods.

2.14.1 Method of enzyme-assisted extraction technique

Enzymes have been used particularly for the treatment of plant material prior to conventional methods for extraction. A prior acquaintance of the cell wall composition of the raw materials helps in the collection of an enzyme or enzymes useful for pretreatment. Various enzymes, such as cellulases, pectinases, and hemicellulose are frequently mandatory to interrupt the structural integrity of the plant cell wall, thus enhancing the extraction of bioactive materials from plants. These enzymes hydrolyze cell wall components thereby increasing cell wall permeability, which results in higher extraction yields of bioactive materials. To utilize enzymes efficiently for extraction applications, it is important to understand their catalytic properties and modes of action, optimal operational conditions, and which enzyme or enzyme combination is suitable for the plant material chosen. Enzymes most frequently increase the yield of extraction of carotenoids, lycopene, phenolic compounds, polyphenols, and anthocyanins (Puri et al., 2012).

2.14.2 Benefits and limitations of enzyme-assisted extraction technique

Enzyme pretreatment of raw material normally results in a reduction in extraction time, minimizes usage of solvents, and provides increased yield and quality of the product. The enzyme-assisted extraction of natural compounds can save processing time and energy, and potentially provide a more reproducible extraction process at the commercial scale (Sowbhagya and Chitra, 2010).

Enzyme-assisted extraction of bioactive compounds from plants has potential commercial and technical limitations: (1) the cost of enzymes is relatively expensive for dispensation of bulky volumes of raw material; (2) currently, available enzyme preparations cannot completely hydrolyze plant cell walls, limiting the extraction yields of compounds, including the extraction of stevioside; and (3) enzyme-assisted extraction can be difficult to scale up to industrial scale because enzymes behave in a different way from environmental conditions, such as the percentage of dissolved oxygen, temperature, and nutrient availability vary. However, if the aforementioned limitations can be overcome, then enzyme-based extraction could provide an opportunity not only to increase extraction yields but also to enhance product quality by enabling the use of milder processing conditions, such as lower extraction temperatures (Puri et al., 2012).

2.15 Membrane-Based Microextraction

A membrane is a selective barrier, through which different vapors, gasses, and liquids move at varying rates. The membrane facilitates the two phases approaching contact with each other without direct mixing. Molecules shift through membranes via the progression of diffusion and are driven by a concentration, pressure, or electrical potential gradient. The diffusion-based transport can be expressed by Flick's first law of diffusion, according to Eq. (2.3).

$$J = -D(\mathrm{dc}/\mathrm{dx}) \tag{2.3}$$

where J is the rate of transfer (or flux) (g/cm^2/s), D is the diffusion coefficient (cm^2/s), and dc/dx is the concentration gradient.

The integration of the above equation gives Eq. (2.4).

$$J = D(C_{is} - C_{il})/L \tag{2.4}$$

where C_{is} is the concentration of i at the outer membrane surface, C_{il} is the concentration of i in the lumen, and L is the membrane thickness. The flux of analytes across the membrane is also affected by temperature since this determines the diffusion coefficient. In the case of liquids, this can best be described by the Stokes–Einstein equation, shown in Eq. (2.5).

$$D = kT/6\pi\alpha\eta \tag{2.5}$$

where k is the Boltzmann constant, T corresponds to absolute temperature, π is the pi number, α is the radius of the solute, and η is the solution viscosity. The size, quantity, and distribution of pores throughout the membrane structure represent important characteristics of the membrane morphology. Membranes that have no pores in their structure are recognized as nonporous, whereas those that have pores are referred to as porous. The shape, size, and distribution of pores in a membrane are mainly reliant on the processes through which they are completed and play a significant role in their mode of separation. Selected molecules pass through the openings in porous membranes, and movement through these membranes involves size exclusion. Therefore, these membranes are useful in applications, such as nanofiltration and dialysis. Nonporous membranes are solid (pore-free) structures and the molecules move through them via diffusion, and, therefore, their compatibility with the analyte is critical (Hylton and Mitra, 2007).

Relating to the chemical nature and the variety of analyte, membranes can be classified by the source of their hydrophilic or hydrophobic nature. Regenerated cellulose, cellulose acetate, polysulfone, polyamide, polyethersulfone, polycarbonate, and ion exchange membranes are well suited for the isolation of ionic or polar species from aqueous/organic solutions. Additionally, hydrophobic membranes (e.g., polytetrafluoroethylene, polypro- pylene, silicone rubber, latex, polyvinylchloride, or polyvinylidene difluoride) are commonly exploited for the selective on-line separation of gasses and volatile organic or inorganic compounds (Miró and Frenzel, 2004). The membrane mainly used in microextraction procedures is Acurel 3/2 Q, consisting of a hollow microporous polypropylene tube (Membrana GmbH, Wuppertal, Germany) with an inner diameter of 600 mm, a wall thickness of 200 mm, a pore size of 0.2 mm, and a wall porosity (by volume) of around 70%. Other types of capillary membranes, such as those composed of polysulfone and cellulose, have also been used for sample preparation purposes. For the thin-film microextraction approach, membranes composed of polydimethylsiloxane, carboxen/poly-dimethylsiloxane, and polydimethylsiloxane/divinylbenzene have been produced and applied; in this case, the thin polymeric membranes sheets are used not only to sample cleanup but also as extraction phases (preconcentration) (Vora-adisak and Varanusupakul, 2006; Zhang et al., 1996).

The SPME procedure represents a great advance in the miniaturization of sample preparation techniques. In SPME, a little amount of the extracting phase connected with a solid support is located to make contact with the sample matrix or in the HS above the matrix for a prearranged time. If this time is long enough, concentration equilibrium is recognized between the sample matrix and the extraction phase. When equilibrium circumstances are reached, exposure of the fiber for a longer amount of time does not lead to a greater gathering of analytes. The constant for the distribution of the amount of analyte contained in the matrix [A(matrix)] and the amount extracted by the SPME fiber [A(fiber)] can be represented as shown in Eq. (2.6).

$$\mathrm{A(matrix)} \rightleftarrows \mathrm{A(fiber)} K_{\mathrm{mf}} = C_f / C_m \tag{2.6}$$

where K_{mf} represents the matrix–fiber distribution constant; C_{f} represents the analyte concentration in the SPME fiber, and C_{m} represents the analyte concentration in the matrix.

The total analyte concentration can be described as the sum of the amount of analyte present in the matrix, in the HS, and in the SPME fiber, as described in Eq. (2.7).

$$n_0 = n_f + n_{hs} + n_m \tag{2.7}$$

where n_0 represents the total amount of analyte, n_f represents the amount of analyte in the SPME fiber, n_{hs} represents the amount of analyte in the HS, and n_m represents the amount of analyte present in the matrix. Using Eq. (2.7) with some mathematical adjustments, the total amount of analyte mass extracted by the SPME fiber can be described by Eq. (2.8).

$$n^f = C_0 \times V_m \times V_f \times K_{mf} / V_m + V_f \times K_{mf} \tag{2.8}$$

where n_f represents the total mass of analyte extracted by the SPME fiber, V_f represents the fiber coating volume, K_{mf} represents the fiber coating–sample matrix distribution constant, C_0 represents initial concentration, and V_m represents the matrix volume (Pawliszyn, 2000).

In the membrane-protected SPME mode, a membrane is placed around the SPME fiber to protect it. This is a useful precaution when matrices containing solids in suspension or particulate material, such as soil, milk, urine, and blood, are analyzed [16]. The same SPME extraction mode was useful successfully [36] for the determination of the free concentration of the medicament paclitaxel in liposome formulations. The end of the membrane was sealed and great care was taken to avoid air entering the space between the fiber and the membrane. The membrane blocked the admittance of large particles, such as liposomes to the coating surface, while target analytes with low molecular weight diffused through the membrane and reached the extraction phase. A dialysis membrane was used, which inhibited the direct interaction between the SPME extraction phase and the liposome content of the liposome formulations. The setup for the membrane-protected SPME used in this lesson is shown in Fig. 2.7 (Carasek and Merib, 2015).

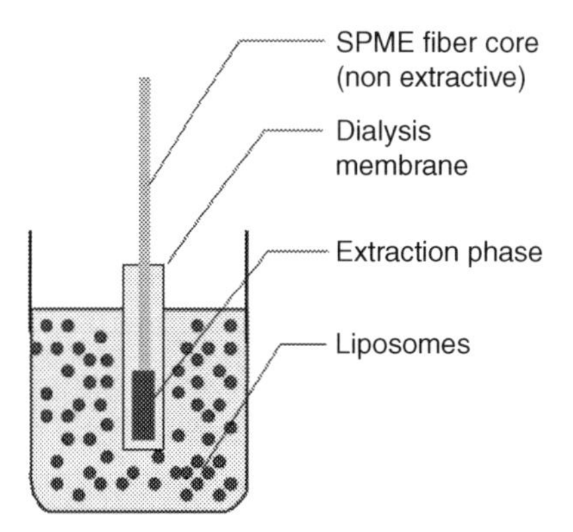

Figure 2.7: Scheme of the Membrane-Protected Solid-Phase Microextraction (SPME) (Carasek and Merib, 2015).

This microextraction form was applied successfully to the resolve of the free paclitaxel in liposome formulations, with good linearity over the range of concentrations of interest, this being a very useful alternative to the application of SPME for complex samples.

The most important advantage of this approach is the possibility of using SPME to analyze complex matrices or matrices containing particulate or suspended material. The polymeric membrane acts as an obstruction to compounds of high molecular weight allowing the dispersal of compounds with low molecular weight through it, facilitating the determination of several analytes in complex matrices. Apart from membrane-protected solid-phase microextraction, some other membrane-based microextraction (MME) techniques, such as thin film microextraction, hollow-fiber liquid-phase microextraction, electro membrane extraction have been developed for its more specificity for different types of analysis (Carasek and Merib, 2015).

2.16 Cooling-Assisted Microextraction

Recently, a special consideration was specified by the scientists to the cooling-assisted approach in many analytical methods, including liquid-phase microextraction (LPME) and solid-phase microextraction (SPME). One of the few succeeded endeavors made to raise the efficiency of SPME was cold-fiber solid-phase microextraction (CF-SPME). This system can simultaneously provide heating sample matrix and cool fiber coating. The first encouragement in SPME effectiveness was prepared by introducing internally cooled solid-phase microextraction (IC-SPME) device. The IC-SPME device was successfully evaluated for the quantitative extraction of benzene, ethylbenzene, toluene, and xylene (BTEX) in clay soil sample. In general, using the IC-SPME device was tedious. Nevertheless, it was the initial point for improving the microextraction methods by cooling process (Zhang and Pawliszyn, 1995).

The project remained inactive until 2006, when an adapted version of the previous design named cold-fiber HS solid-phase microextraction (CF-SPME) device was introduced. The CF-SPME design used a 33-gauge stainless steel tubing to deliver liquid carbon dioxide for cooling the fiber. It has a number of separate parts, such as solenoid valve, CO_2 tank, thermocouple wires, temperature controller, and stainless steel tubing, which make it complex and not feasible to apply in field studies (Ghiasvand et al., 2006).

A thermoelectric cooler (TEC), which has been used in miniaturized analytical instruments, is a proper alternative-cooling tool for CF-SPME. For instance, TEC has low cost, small size, low weight, no moving parts, and can precisely control the temperature. However, the major factor necessary to attain higher efficiencies is the capability to directly cool the extraction phase (Jiemin et al., 2003).

In addition to liquid CO_2 and thermoelectric cooler, various circulating cooled fluids, such as ice, alcohol, and cold water have also been used to cool the extraction phase in SPME and

LPME methods. In one of these studies, methyl *tert*-butyl ether (MTBE) was extracted at sub-ppb levels in surface water by a cooled-fiber SPME, hyphenated to GC-MS. A cooling cylinder packed with ice was located around a commercial fiber holder to cool a profitable SPME fiber to 5°C and the sample solution over the range of 5–30°C during the extraction.

The highest extraction competence was achieved when the fiber was cooled to 5°C and the sample solution was kept at 18–19°C. In this way, advantages of cold fiber SPME (CF-SPME) and thin film microextraction (TFME) were merged in a new setup—namely, a cooled membrane device (CMD). The cooling-assisted solid-phase microextraction (CA-SPME) techniques are generally carried out to extract organic compounds in various matrices, such as soil, sand, water, air particulate, and medicinal herbs (Ghiasvand et al., 2016).

3 Conclusions

The ever-growing demand to extract drugs and medicinal agents encourages continuous search to achieve convenient, rapid, and environmental-friendly extraction methods. The increasing economic significance of bioactive compounds may lead to finding out more sophisticated extraction methods in the future. As an unconventional to traditional procedures as liquid-liquid or solid-phase extraction, microextraction methods have become a well-established field of analytical studies, with a growing number of applications. The initially proposed procedures have had a continuous evolution; new methods and procedures have been proposed every year to develop the effectiveness of these techniques or to overcome some of their traditional disadvantages. Nevertheless, the most recent developments in this field have aimed to miniaturize and automate the microextraction procedures to reduce the required sample volume and manual operations and the development of more efficient and obtainable materials. Such miniaturization and automation of microextraction techniques remain an attractive and demanding task for analytical chemists. The understanding of every aspect of the nonconventional extraction process is essential as most of these methods are based on different mechanism and extraction enhancement results from the different process.

References

Adas¸og˘lu, N., Dinçer, S., Bolat, E., 1994. Supercritical-fluid extraction of essential oil from Turkish lavender flowers. J. Supercritic. Fluids 7, 93–99.

Andruch, V., Kocúrová, L., Balogh, I.S., Škrlíková, J., 2012. Recent advances in coupling single-drop and dispersive liquid-liquid microextraction with UV–vis spectrophotometry and related detection techniques. Microchem. J. 102, 1–10.

Arapitsas, P., Turner, C., 2008. Pressurized solvent extraction and monolithic column-HPLC/DAD analysis of anthocyanins in red cabbage. Talanta 74, 1218–1223.

Asensio-Ramos, M., Ravelo-Pérez, L.M., González-Curbelo, M.Á., Hernández-Borges, J., 2011. Liquid phase microextraction applications in food analysis. J. Chromatogr. A 1218, 7415–7437.

Awad, T.S., Moharram, H.A., Shaltout, O.E., Asker, D., Youssef, M.M., 2012. Applications of ultrasound in analysis, processing and quality control of food: a review. Food Res. Int. 48, 410–427.

Azmir, J., Zaidul, I.S.M., Rahman, M.M., Sharif, K.M., Mohamed, A., Sahena, F., Jahurul, M.H.A., Ghafoor, K., Norulaini, N.A.N., Omar, A.K.M., 2013. Techniques for extraction of bioactive compounds from plant materials: a review. J. Food Eng. 117, 426–436.

Bai, X.-H., Li, J.-W., Chen, X., 2008. Determination of clomifene citrate in biological matrices by liquid-phase microextraction combined with photochemical fluorescence high performance liquid chromatography. Chinese J. Pharm. Anal. 28, 49–53.

Bratkowska, D., Marcé, R.M., Cormack, P.A.G., Borrull, F., Fontanals, N., 2011. Development and application of a polar coating for stir bar sorptive extraction of emerging pollutants from environmental water samples. Anal. Chim. Acta 706, 135–142.

Bratkowska, D., Fontanals, N., Cormack, P.A.G., Borrull, F., Marcé, R.M., 2012. Preparation of a polar monolithic stir bar based on methacrylic acid and divinylbenzene for the sorptive extraction of polar pharmaceuticals from complex water samples. J. Chromatogr. A 1225, 1–7.

Breadmore, M.C., 2011. Ionic liquid-based liquid phase microextraction with direct injection for capillary electrophoresis. J. Chromatogr. A 1218, 1347–1352.

Camino-Sánchez, F.J., Rodríguez-Gómez, R., Zafra-Gómez, A., Santos-Fandila, A., Vílchez, J.L., 2014. Stir bar sorptive extraction: recent applications, limitations and future trends. Talanta 130, 388–399.

Carabias-Martínez, R., Rodríguez-Gonzalo, E., Revilla-Ruiz, P., Hernández-Méndez, J., 2005. Pressurized liquid extraction in the analysis of food and biological samples. J. Chromatogr. A 1089, 1–17.

Carasek, E., Merib, J., 2015. Membrane-based microextraction techniques in analytical chemistry: a review. Anal. Chim. Acta 880, 8–25.

Cavalcanti, R.N., Meireles, M.A.A., 2012. Fundamentals of supercritical fluid extraction A2. In: Pawliszyn, J. (Ed.), Comprehensive Sampling and Sample Preparation. Academic Press, Oxford.

Chan, C.-H., Yusoff, R., Ngoh, G.-C., Kung, F.W.-L., 2011. Microwave-assisted extractions of active ingredients from plants. J. Chromatogr. A 1218, 6213–6225.

Chemat, F., Zill, E.H., Khan, M.K., 2011. Applications of ultrasound in food technology: processing, preservation and extraction. Ultrason. Sonochem. 18, 813–835.

Chen, X.-W., Mao, Q.-X., Liu, J.-W., Wang, J.-H., 2012. Isolation/separation of plasmid DNA using hemoglobin modified magnetic nanocomposites as solid-phase adsorbent. Talanta 100, 107–112.

Cruz-Vera, M., Lucena, R., Cárdenas, S., Valcárcel, M., 2008. Combined use of carbon nanotubes and ionic liquid to improve the determination of antidepressants in urine samples by liquid chromatography. Anal. Bioanal. Chem. 391, 1139–1145.

Daneshvar Tarigh, G., Shemirani, F., 2014. Simultaneous in situ derivatization and ultrasound-assisted dispersive magnetic solid phase extraction for thiamine determination by spectrofluorimetry. Talanta 123, 71–77.

da Silva, R.P.F.F., Rocha-Santos, T.A.P., DuarteF A.C., 2016. Supercritical fluid extraction of bioactive compounds. Trends Anal. Chem. 76, 40–51.

Deng, C., Yao, N., Wang, A., Zhang, X., 2005. Determination of essential oil in a traditional Chinese medicine, *Fructus amomi* by pressurized hot water extraction followed by liquid-phase microextraction and gas chromatography–mass spectrometry. Anal. Chim. Acta 536, 237–244.

Ding, J., Gao, Q., Li, X.-S., Huang, W., Shi, Z.-G., Feng, Y.-Q., 2011. Magnetic solid-phase extraction based on magnetic carbon nanotube for the determination of estrogens in milk. J. Sep. Sci. 34, 2498–2504.

Duan, C., Shen, Z., Wu, D., Guan, Y., 2011. Recent developments in solid-phase microextraction for on-site sampling and sample preparation. Trends Anal. Chem. 30, 1568–1574.

Fakhari, A.R., Tabani, H., Nojavan, S., 2011. Immersed single-drop microextraction combined with gas chromatography for the determination of sufentanil and alfentanil in urine and wastewater samples. Anal. Methods 3, 951–956.

Fang, G.-Z., He, J.-X., Wang, S., 2006. Multiwalled carbon nanotubes as sorbent for on-line coupling of solid-phase extraction to high-performance liquid chromatography for simultaneous determination of 10 sulfonamides in eggs and pork. J.Chromatogr. A 1127, 12–17.

Gao, W., Chen, G., Chen, Y., Li, N., Chen, T., Hu, Z., 2011a. Selective extraction of alkaloids in human urine by on-line single drop microextraction coupled with sweeping micellar electrokinetic chromatography. J. Chromatogr. A 1218, 5712–5717.

Gao, W., Chen, G., Chen, Y., Zhang, X., Yin, Y., Hu, Z., 2011b. Application of single drop liquid-liquid-liquid microextraction for the determination of fluoroquinolones in human urine by capillary electrophoresis. J. Chromatogr. B 879, 291–295.

Gao, W., Chen, Y., Chen, G., Xi, J., Chen, Y., Yang, J., Xu, N., 2012. Trace analysis of three antihistamines in human urine by on-line single drop liquid-liquid-liquid microextraction coupled to sweeping micellar electrokinetic chromatography and its application to pharmacokinetic study. J. Chromatogr. B 904, 121–127.

Gardossi, L., Poulsen, P.B., Ballesteros, A., Hult, K., Švedas, V.K., Vasic´-Racˇki, Đ., Carrea, G., Magnusson, A., Schmid, A., Wohlgemuth, R., Halling, P.J., 2010. Guidelines for reporting of biocatalytic reactions. Trends Biotechnol. 28, 171–180.

George, M.J., Marjanovic, L., Williams, D.B.G., 2015. Picogram-level quantification of some growth hormones in bovine urine using mixed-solvent bubble-in-drop single drop micro-extraction. Talanta 144, 445–450.

Ghiasvand, A.R., Hosseinzadeh, S., Pawliszyn, J., 2006. New cold-fiber headspace solid-phase microextraction device for quantitative extraction of polycyclic aromatic hydrocarbons in sediment. J. Chromatogr. A 1124, 35–42.

Ghiasvand, A.R., Hajipour, S., Heidari, N., 2016. Cooling-assisted microextraction: comparison of techniques and applications. Trends Anal. Chem. 77, 54–65.

Gómez-González, S., Ruiz-Jiménez, J., Priego-Capote, F., Luque, D.E., Castro, M.A.D., 2010. Qualitative and quantitative sugar profiling in olive fruits, leaves, and stems by gas chromatography−tandem mass spectrometry (GC-MS/MS) after ultrasound-assisted leaching. J. Agric. Food Chem. 58, 12292–12299.

Gouda, A.A., 2014. Solid-phase extraction using multiwalled carbon nanotubes and quinalizarin for preconcentration and determination of trace amounts of some heavy metals in food, water and environmental samples. Int. J.Environ. Anal. Chem. 94, 1210–1222.

Guan, Y., Jiang, C., Hu, C., Jia, L., 2010. Preparation of multi-walled carbon nanotubes functionalized magnetic particles by sol–gel technology and its application in extraction of estrogens. Talanta 83, 337–343.

He, C., Long, Y., Pan, J., Li, K., Liu, F., 2007. Application of molecularly imprinted polymers to solid-phase extraction of analytes from real samples. J. Biochem. Biophys. Methods 70, 133–150.

He, L., Luo, X., Xie, H., Wang, C., Jiang, X., Lu, K., 2009. Ionic liquid-based dispersive liquid-liquid microextraction followed high-performance liquid chromatography for the determination of organophosphorus pesticides in water sample. Anal. Chim. Acta 655, 52–59.

Hennion, M.C., Cau-Dit-Coumes, C., Pichon, V., 1998. Trace analysis of polar organic pollutants in aqueous samples: tools for the rapid prediction and optimisation of the solid-phase extraction parameters. J. Chromatogr. A 823, 147–161.

Herrero, M., Mendiola, J.A., Cifuentes, A., Ibáñez, E., 2010. Supercritical fluid extraction: recent advances and applications. J. Chromatogr. A 1217, 2495–2511.

Herrero-Latorre, C., Barciela-García, J., García-Martín, S., Peña-Crecente, R.M., Otárola-Jiménez, J., 2015. Magnetic solid-phase extraction using carbon nanotubes as sorbents: a review. Anal. Chim. Acta 892, 10–26.

Hylton, K., Mitra, S., 2007. Automated, on-line membrane extraction. J. Chromatogr. A 1152, 199–214.

Illés, V., Szalai, O., Then, M., Daood, H., Perneczki, S., 1997. Extraction of hiprose fruit by supercritical CO_2 and propane. J. Supercrit. Fluids 10, 209–218.

Jahan, S., Xie, H., Zhong, R., Yan, J., Xiao, H., Fan, L., Cao, C., 2015. A highly efficient three-phase single drop microextraction technique for sample preconcentration. Analyst 140, 3193–3200.

Jiemin, L., Guibin, J., Jingfu, L., Qunfang, Z., Ziwei, Y., 2003. Development of cryogenic chromatography using thermoelectric modules for the separation of methyltin compounds. J. Sep. Sci. 26, 629–634.

Katsumata, H., Kojima, H., Kaneco, S., Suzuki, T., Ohta, K., 2010. Preconcentration of atrazine and simazine with multiwalled carbon nanotubes as solid-phase extraction disk. Microchem. J. 96, 348–351.

Li, Y., Fabiano-Tixier, A.S., Vian M.A., Chemat, F., 2013. Solvent-free microwave extraction of bioactive compounds provides a tool for green analytical chemistry. Trends Anal. Chem. 47, 1–11.

Lang, Q., Wai, C.M., 2001. Supercritical fluid extraction in herbal and natural product studies—a practical review. Talanta 53, 771–782.

Liu, J., Sandahl, M., Sjöberg, P.J.R., Turner, C., 2013a. Pressurised hot water extraction in continuous flow mode for thermolabile compounds: extraction of polyphenols in red onions. Anal. Bioanal. Chem. 406, 441–445.

Liu, X., Fu, R., Li, M., Guo, L.-P., Yang, L., 2013b. Ionic liquid-based dispersive liquid-liquid microextraction coupled with capillary electrophoresis to determine drugs of abuse in urine. Chinese J. Anal. Chem. 41, 1919–1922.
Lord, H., Pawliszyn, J., 2000. Evolution of solid-phase microextraction technology. J. Chromatogr. A 885, 153–193.
Luque-García, J.L., Luque de Castro, M.D., 2003. Where is microwave-based analytical equipment for solid sample pre-treatment going? Trends Anal. Chem. 22, 90–98.
Madej, K., 2009. Microwave-assisted and cloud-point extraction in determination of drugs and other bioactive compounds. Trends Anal. Chem. 28, 436–446.
Madrakian, T., Ahmadi, M., Afkhami, A., Soleimani, M., 2013. Selective solid-phase extraction of naproxen drug from human urine samples using molecularly imprinted polymer-coated magnetic multi-walled carbon nanotubes prior to its spectrofluorometric determination. Analyst 138, 4542–4549.
Miró, M., Frenzel, W., 2004. Flow-through sorptive preconcentration with direct optosensing at solid surfaces for trace-ion analysis. Trends Anal. Chem. 23, 11–20.
Monrad, J.K., Srinivas, K., Howard, L.R., King, J.W., 2012. Design and optimization of a semicontinuous hot–cold Extraction of polyphenols from grape pomace. J. Agric. Food Chem. 60, 5571–5582.
Monrad, J.K., Suárez, M., Motilva, M.J., King, J.W., Srinivas, K., Howard, L.R., 2014. Extraction of anthocyanins and flavan-3-ols from red grape pomace continuously by coupling hot water extraction with a modified expeller. Food Res. Int. 65 (Part A), 77–87.
Mukherjee, P., Padhan, S.K., Dash, S., Patel, S., Mishra, B.K., 2011. Clouding behaviour in surfactant systems. Adv. Colloid Interf. Sci. 162, 59–79.
Mustafa, A., Turner, C., 2011. Pressurized liquid extraction as a green approach in food and herbal plants extraction: a review. Anal. Chim. Acta 703, 8–18.
Nageswara Rao, R., Mastan Vali, R., Vara Prasada Rao, A., 2012. Determination of rifaximin in rat serum by ionic liquid based dispersive liquid-liquid microextraction combined with RP-HPLC. J. Sep. Sci. 35, 1945–1952.
Nascentes, C.C., Arruda, M.A.Z., 2003. Cloud point formation based on mixed micelles in the presence of electrolytes for cobalt extraction and preconcentration. Talanta 61, 759–768.
Niu, H., Cai, Y., Shi, Y., Wei, F., Liu, J., Mou, S., Jiang, G., 2007. Evaluation of carbon nanotubes as a solid-phase extraction adsorbent for the extraction of cephalosporins antibiotics, sulfonamides and phenolic compounds from aqueous solution. Anal. Chim. Acta 594, 81–92.
Nogueira, J.M.F., 2012. Novel sorption-based methodologies for static microextraction analysis: a review on SBSE and related techniques. Anal. Chim. Acta 757, 1–10.
Pawliszyn, J., 2000. Theory of solid-phase microextraction. J. Chromatogr. Sci. 38, 270–278.
Pawliszyn, J., 2003. Sample preparation: quo vadis? Anal. Chem. 75, 2543–2558.
Pedersen-Bjergaard, S., Rasmussen, K.E., 1999. Liquid-liquid-liquid microextraction for sample preparation of biological fluids prior to capillary electrophoresis. Anal. Chem. 71, 2650–2656.
Pena-Pereira, F., Lavilla, I., Bendicho, C., 2009. Miniaturized preconcentration methods based on liquid-liquid extraction and their application in inorganic ultratrace analysis and speciation: a review. Spectrochim. Acta 64, 1–15.
Picó, Y., Fernández, M., Ruiz, M.J., Font, G., 2007. Current trends in solid-phase-based extraction techniques for the determination of pesticides in food and environment. J. Biochem. Biophys. Methods 70, 117–131.
Pinelo, M., Arnous, A., Meyer, A.S., 2006. Upgrading of grape skins: significance of plant cell-wall structural components and extraction techniques for phenol release. Trends Food Sci. Technol. 17, 579–590.
Plaza, M., Turner, C., 2015. Pressurized hot water extraction of bioactives. Trends Anal. Chem. 71, 39–54.
Płotka-Wasylka, J., Szczepańska, N., de La Guardia, M., Namieśnik, J., 2015. Miniaturized solid-phase extraction techniques. Trends Anal. Chem. 73, 19–38.
Płotka-Wasylka, J., Szczepańska, N., de La Guardia, M, Namieśnik, J., 2016. Modern trends in solid phase extraction: new sorbent media. Trends Anal. Chem. 77, 23–43.
Popp, P., Bauer, C., Wennrich, L., 2001. Application of stir bar sorptive extraction in combination with column liquid chromatography for the determination of polycyclic aromatic hydrocarbons in water samples. Anal. Chim. Acta 436, 1–9.

Pourmortazavi, S.M., Hajimirsadeghi, S.S., 2007. Supercritical fluid extraction in plant essential and volatile oil analysis. J. Chromatogr. A 1163, 2–24.

Psillakis, E., Kalogerakis, N., 2003. Developments in liquid-phase microextraction. Trends Anal. Chem. 22, 565–574.

Puri, M., Sharma, D., Barrow, C.J., 2012. Enzyme-assisted extraction of bioactives from plants. Trends Biotechnol. 30, 37–44.

Pytlakowska, K., Kozik, V., Dabioch, M., 2013. Complex-forming organic ligands in cloud-point extraction of metal ions: a review. Talanta 110, 202–228.

Rao, R.N., Naidu, C.G., Suresh, C.V., Srinath, N., Padiya, R., 2014. Ionic liquid based dispersive liquid-liquid microextraction followed by RP-HPLC determination of balofloxacin in rat serum. Anal. Methods 6, 1674–1683.

Rasmussen, K.E., Pedersen-Bjergaard, S., 2004. Developments in hollow fibre-based, liquid-phase microextraction. Trends Anal. Chem. 23, 1–10.

Ravelo-Pérez, L.M., Herrera-Herrera, A.V., Hernández-Borges, J., Rodríguez-Delgado, M.Á., 2010. Carbon nanotubes: solid-phase extraction. J.Chromatogr. A 1217, 2618–2641.

Rezaei, S., Rezaei, K., Haghighi, M., Labbafi, M., 2013. Solvent and solvent to sample ratio as main parameters in the microwave-assisted extraction of polyphenolic compounds from apple pomace. Food Sci. Biotechnol. 22, 1–6.

Sánchez-Camargo, A.P., Mendiola, J.A., Ibáñez, E., Herrero, M., 2014. Supercritical Fluid Extraction. Elsevier, The Netherlands.

Sasidharan, S., Chen, Y., Saravanan, D., Sundram, K.M., Yoga Latha, L., 2011. Extraction, isolation and characterization of bioactive compounds from plants' extracts. Afr. J. Tradit., Complement., Altern. Med. 8, 1–10.

Shen, S., Ren, J., Chen, J., Lu, X., Deng, C., Jiang, X., 2011. Development of magnetic multiwalled carbon nanotubes combined with near-infrared radiation-assisted desorption for the determination of tissue distribution of doxorubicin liposome injects in rats. J. Chromatogr. A 1218, 4619–4626.

Soria, A.C., Villamiel, M., 2010. Effect of ultrasound on the technological properties and bioactivity of food: a review. Trends Food Sci. Technol. 21, 323–331.

Souza Silva, E.A., Risticevic, S., Pawliszyn, J., 2013. Recent trends in SPME concerning sorbent materials, configurations and in vivo applications. Trends Anal. Chem. 43, 24–36.

Sowbhagya, H.B., Chitra, V.N., 2010. Enzyme-assisted extraction of flavorings and colorants from plant materials. Crit. Rev. Food Sci. Nutr. 50, 146–161.

Sparks, D., Hernandez, R., Zappi, M., Blackwell, D., Fleming, T., 2006. Extraction of rice brain oil using supercritical carbon dioxide and propane. J. Am. Oil Chem. Soc. 83, 885–891.

Sparr Eskilsson, C., Björklund, E., 2000. Analytical-scale microwave-assisted extraction. J. Chromatogr. A 902, 227–250.

Spietelun, A., Marcinkowski, Ł., de La Guardia, M., Namieśnik, J., 2013. Recent developments and future trends in solid phase microextraction techniques towards green analytical chemistry. J. Chromatogr. A 1321, 1–13.

Stege, P.W., Lapierre, A.V., Martinez, L.D., Messina, G.A., Sombra, L.L., 2011. A combination of single-drop microextraction and open tubular capillary electrochromatography with carbon nanotubes as stationary phase for the determination of low concentration of illicit drugs in horse urine. Talanta 86, 278–283.

Suárez, B., Santos, B., Simonet, B.M., Cárdenas, S., Valcárcel, M., 2007a. Solid-phase extraction-capillary electrophoresis-mass spectrometry for the determination of tetracyclines residues in surface water by using carbon nanotubes as sorbent material. J. Chromatogr. A 1175, 127–132.

Suárez, B., Simonet, B.M., Cárdenas, S., Valcárcel, M., 2007b. Determination of non-steroidal anti-inflammatory drugs in urine by combining an immobilized carboxylated carbon nanotubes minicolumn for solid-phase extraction with capillary electrophoresis-mass spectrometry. J. Chromatogr. A 1159, 203–207.

Tian, J., Chen, X., Bai, X., 2012. Comparison of dispersive liquid-liquid microextraction based on organic solvent and ionic liquid combined with high-performance liquid chromatography for the analysis of emodin and its metabolites in urine samples. J. Sep. Sci. 35, 145–152.

Vasapollo, G., Sole, R.D., Mergola, L., Lazzoi, M.R., Scardino, A., Scorrano, S., Mele, G., 2011. Molecularly imprinted polymers: present and future prospective. Int. J. Mol. Sci. 12, 5908–5945.

Vazquez-Roig, P., Picó, Y., 2015. Pressurized liquid extraction of organic contaminants in environmental and food samples. Trends Anal. Chem. 71, 55–64.

Vora-adisak, N., Varanusupakul, P., 2006. A simple supported liquid hollow fiber membrane microextraction for sample preparation of trihalomethanes in water samples. J. Chromatogr. A 1121, 236–241.

Wang, L., Zhao, H., Qiu, Y., Zhou, Z., 2006. Determination of four benzodiazepine residues in pork using multiwalled carbon nanotube solid-phase extraction and gas chromatography–mass spectrometry. J. Chromatogr. A 1136, 99–105.

Wang, Y., you, J., Yu, Y., Qu, C., Zhang, H., Ding, L., Zhang, H., Li, X., 2008. Analysis of ginsenosides in *Panax ginseng* in high pressure microwave-assisted extraction. Food Chem. 110, 161–167.

Wang, S., Wei, J., Hao, T., Guo, Z., 2012. Determination of ractopamine in pork by using electrochemiluminescence inhibition method combined with molecularly imprinted stir bar sorptive extraction. J. Electroanal. Chem. 664, 146–151.

Wu, H.-F., Kailasa, S.K., Lin, C.-H., 2011. Single drop microextraction coupled with matrix-assisted laser desorption/ionization mass spectrometry for rapid and direct analysis of hydrophobic peptides from biological samples in high salt solution. Rapid Comm. Mass Spectrom. 25, 307–315.

Wu, H.-F., Kailasa, S.K., Yan, J.-Y., Chin, C.-C., Ku, H.-Y., 2014. Comparison of single-drop microextraction with microvolume pipette extraction directly coupled with capillary electrophoresis for extraction and separation of tricyclic antidepressant drugs. J.Indust. Eng. Chem. 20, 2071–2076.

Xiao, D., Dramou, P., Xiong, N., He, H., Li, H., Yuan, D., Dai, H., 2013. Development of novel molecularly imprinted magnetic solid-phase extraction materials based on magnetic carbon nanotubes and their application for the determination of gatifloxacin in serum samples coupled with high performance liquid chromatography. J. Chromatogr. A 1274, 44–53.

Xiao, D., Yuan, D., He, H., Pham-Huy, C., Dai, H., Wang, C., Zhang, C., 2014. Mixed hemimicelle solid-phase extraction based on magnetic carbon nanotubes and ionic liquids for the determination of flavonoids. Carbon 72, 274–286.

Xu, Z., Hu, Y., Hu, Y., Li, G., 2010a. Investigation of ractopamine molecularly imprinted stir bar sorptive extraction and its application for trace analysis of beta2-agonists in complex samples. J. Chromatogr. A 1217, 3612–3618.

Xu, Z., Hu, Y., Hu, Y., Li, G., 2010b. Investigation of ractopamine molecularly imprinted stir bar sorptive extraction and its application for trace analysis of β2-agonists in complex samples. J. Chromatogr. A 1217, 3612–3618.

Yamini, Y., Seidi, S., Rezazadeh, M., 2014. Electrical field-induced extraction and separation techniques: promising trends in analytical chemistry - a review. Anal. Chim. Acta. 814C, 1–22.

Yan, H., Wang, H., 2013. Recent development and applications of dispersive liquid-liquid microextraction. J. Chromatogr. A 1295, 1–15.

Yan, Y., Chen, X., Hu, S., Bai, X., 2014. Applications of liquid-phase microextraction techniques in natural product analysis: a review. J. Chromatogr. A 1368, 1–17.

Yazdi, A.S., Razavi, N., Yazdinejad, S.R., 2008. Separation and determination of amitriptyline and nortriptyline by dispersive liquid-liquid microextraction combined with gas chromatography flame ionization detection. Talanta 75, 1293–1299.

Yu, P., Ma, H., Shang, Y., Wu, J., Shen, S., 2014a. Polyethylene glycol modified magnetic carbon nanotubes as nanosorbents for the determination of methylprednisolone in rat plasma by high performance liquid chromatography. J. Chromatogr. A 1348, 27–33.

Yu, P., Wang, Q., Ma, H., Wu, J., Shen, S., 2014b. Determination of puerarin in rat plasma using PEGylated magnetic carbon nanotubes by high performance liquid chromatography. J. Chromatogr. B 959, 55–61.

Zhang, Z., Pawliszyn, J., 1995. Quantitative extraction using an internally cooled solid phase microextraction device. Anal. Chem. 67, 34–43.

Zhang, Z., Poerschmann, J., Pawliszyn, J., 1996. Direct solid phase microextraction of complex aqueous samples with hollow fibre membrane protection. Anal. Comm. 33, 219–221.

Zhang, Z., Yang, X., Chen, X., Zhang, M., Luo, L., Peng, M., Yao, S., 2011. Novel magnetic bovine serum albumin imprinted polymers with a matrix of carbon nanotubes, and their application to protein separation. Anal. Bioanal. Chem. 401, 2855–2863.

Zhao, H., Wang, L., Qiu, Y., Zhou, Z., Zhong, W., Li, X., 2007. Multiwalled carbon nanotubes as a solid-phase extraction adsorbent for the determination of three barbiturates in pork by ion trap gas chromatography–tandem mass spectrometry (GC/MS/MS) following microwave assisted derivatization. Anal. Chim. Acta 586, 399–406.

CHAPTER 3

Advances in Extraction, Fractionation, and Purification of Low-Molecular Mass Compounds From Food and Biological Samples

Elżbieta Włodarczyk, Paweł K. Zarzycki
Koszalin University of Technology, Koszalin, Poland

1 Problem Overview

Extraction, fractionation, and/or purification procedures are the key and critical steps for both quantitative analysis and technological processes concerning food and biologically complex materials. The methodological approach concerning isolation of components of interest depends on the expected outcomes and physicochemical properties of given samples or product. Effective extraction of target low-molecular-mass compounds from multicomponent matrices generate a number of problems; therefore, several different approaches allowing maximization of the extraction efficiency and purification level were developed and invented (Self, 2005). This chapter summarizes the literature search concerning recent advances in extraction protocols of bioactive substances, which are designed for analytical and technological applications of food processing as well as biological materials and environmental samples. Particularly, detailed experimental conditions and the sequence of the key analytical steps required for efficient isolation of target bioanalytes, biomarkers, and micropollutants from complex materials form the main focus of this review. As it is visible from the graphs presented within Fig. 3.1, during recent years the number of research papers dealing with extraction of for example, food products, is rapidly increasing. Most recently (2015 and 2016), there are increasing numbers of publications concerning isolation of the specific low-molecular-mass compounds using various extraction methodological approaches, at analytical and preparative scales, related to the following:

- Natural phytochemicals and probiotics from functional food using advanced encapsulation techniques, as well as classical Soxhlet, ultrasound-assisted, supercritical fluid, accelerated solvent, and shake type extractions (Silva et al., 2016),

Ingredients Extraction by Physicochemical Methods in Food
http://dx.doi.org/10.1016/B978-0-12-811521-3.00003-X

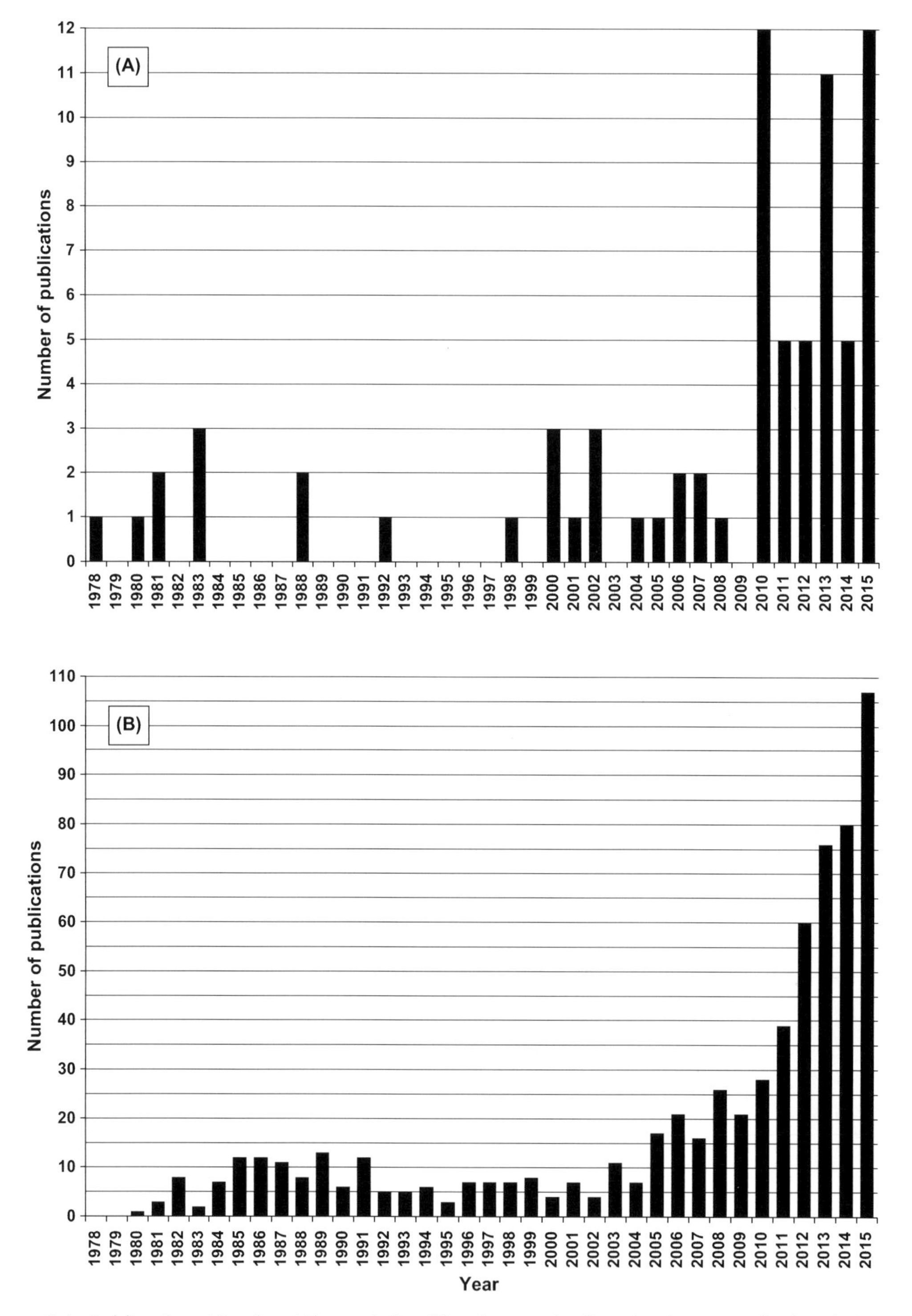

Figure 3.1: Publications Number (A) consisting "food extraction" topic phrase, and related citations (B), which were published up to 2015, according to Web of Science database.

- Organic micropollutants from food and environmental samples *via* microwave-assisted extraction (Wang et al., 2016),
- Polyphenols enriched extracts from pomegranate using classical Soxhlet extraction with different organic and organic-water based solvents (Masci et al., 2016),
- Anthocyanins from purple sweet potatoes involving conventional, ultrasound-assisted, and accelerated-solvent extractions (Cai et al., 2016),
- Assessment of nutritional properties of legumes fatty acids using Folch method and liquid solid extraction involving various extraction liquids (Caprioli et al., 2016),
- Quantification of organic contaminants from complex environmental/food samples using automated pressurized liquid extraction including pressurized hot-water extraction (Vazquez-Roig and Pico, 2015),
- Isolation of bioactive chemicals from food-processing residues involving high-pressure extraction techniques (Duba and Fiori, 2015).

It should be noted that even relatively simple materials—for example, dry cyanobacteria cells (spirulina), which are considered as "superfood" and are commonly distributed on the market in forms of various drug formulations and food supplements—may generate a number of problems concerning effective extraction of given bioactive substances (Zarzycki and Zarzycka, 2008). Therefore, several different approaches allowing to maximize the extraction efficiency and purity level of given compounds were also developed, for example, using low-parachor organic liquids (Zarzycki et al., 2011).

2 Recent Advances in Development of Extraction Protocols From Complex Matrices

In general, there are two major modes of analytical protocols designed for extraction/isolation/purification of target components from complex matrices. In the first approach, the whole sample is passing through or contacting with the entire mass of the adsorbing material (liquid or solid) allowing isolation of all the target molecules that are present in the sample. For example, in solid-phase extraction (SPE), the liquid sample is pumped through porous solid material that is located inside a tube or cartridge catching, more or less selectively, all target substances. This methodology allows quantitative extraction, especially at preparative scale and, therefore is commonly applied for extraction/fractionation of biocomponents from the food or pharmaceutical products. The second approach is based on an idea that the extraction process concerns only a small fraction of the molecules available within the sample—for example, in solid-phase microextraction (SPME) involving small bars, wires, or fibers coated with the adsorbing phase and placed within large volume of gas or liquid sample, the mass of the analyte extracted is proportional to its concentration in the sample, considering that the appropriate equilibrium was established. Such an approach is very useful for quantitative analytical applications, because it eliminates or at least significantly reduces the use of additional solvents that must be applied for components of interest elution, which are usually necessary to perform the aforementioned SPE methodology. However, the second approach is strongly limited as the extraction tool in preparative scale.

This chapter concerns selected experimental works that were published within the past few years and illustrates the main targets (bioactive substances), as well as directions in development of the extraction equipment and methodologies. Based on literature search, predominantly performed *via* Web of Science database, the number of classical and newly invented extraction protocols, which are currently applied for food and environmental samples/materials, were identified. The outline of this search, describing all steps and analytical details of the extraction procedures for given components of interest, are presented in table form, including the following:

- Molecularly imprinted solid-phase extraction (MISPE)—Table 3.1.
- Solid-phase extraction (SPE)—Table 3.2.
- Solid-phase microextraction (SPME)—Table 3.3.
- Headspace sorptive extraction (HSSE)—Table 3.4.
- Petrol ether extraction (PEE)—Table 3.5.
- Stir-bar sorptive extraction (SBSE)—Table 3.6.
- Simultaneous distillation-extraction (SDE)—Table 3.7.
- Membrane-assisted solvent extraction (MASE)—Table 3.8.
- Supercritical fluid extraction (SFE)—Table 3.9.
- Pressurized liquid extraction (PLE)—Table 3.10.
- Ultrasound-assisted extraction (UAE)—Table 3.11.
- Microwave-assisted extraction (MAE)—Table 3.12.
- Solid-liquid extraction (SLE)—Table 3.13.
- Dispersive liquid–liquid microextraction (DLLME)—Table 3.14.
- Maceration (MAC)—Table 3.15.
- Soxhlet extraction (SOX)—Table 3.16.
- Accelerated solvent extraction (ASE)—Table 3.17.
- Pressurized hot water extraction (PHWE)—Table 3.18.
- High hydrostatic pressure extraction (HHP)—Table 3.19.
- Conventional solvent extraction (CSE)—Table 3.20.
- Matrix solid-phase dispersion (MSPD)—Table 3.21.
- Quick, Easy, Cheap, Effective, Rugged, and Safe protocols (QuEChERS)—Table 3.22.

The ability to discover the new materials for extraction seems to be critical for development of high throughput protocols, especially for quantitative analysis of complex

Table 3.1: Molecularly imprinted solid-phase extraction (MISPE).

Analytes/ Compounds	Sample/ Matrix	Sample Volume (g)	Synthesis of MIP	Polymer	Extraction From Food	Elution	Target Components Analysis	References
Sudan I	Red chili peppers	2	Template: sudan I; functional monomer: methacrylic acid (MAA), 4-vinylpyridine; crosslinker: ethylene glycol dimethacrylate (EGDMA); initiator: azobisisobutyronitrile (AIBN); solvent: chloroform thermo-polymerization; extraction of MIP: 200 mL of acetic acid:methanol (1:1) for 48 h, 200 mL methanol for 48 h in a Soxhlet apparatus, dried in an oven at 60°C overnight	MIP, NIP	Sample extracted with 30 mL chloroform, filtered, 12 mL solution loaded the MISPE column, column washed with 0.3 mL dichlorometane	4 mL methanol	HPLC	Puoci et al. (2005)
α-Tocopherol	Bay leaves	2	Template: α-tocopherol; functional monomer: methacrylic acid (MAA); crosslinker: ethylene glycol dimethacrylate (EGDMA); initiator: azobisisobutyronitrile (AIBN); solvent: chloroform thermopolymerization, photopolymerization; extraction of MIP: 200 mL methanol:acetic acid (8:2, v/v) for 48 h, 200 mL methanol for 48 h in a soxhlet apparatus, dried in an oven at 60°C overnight	MIP, NIP	Sample extracted with 200 mL of ethanol, water added to 60 mL of mix to raise the loading solution (ethanol/water, 6/4, v/v; 100 mL), 5mL solution used to load the MISPE coloumn	Step 1: 3 mL ethanol/water (7/3, v/v), 5 mL ethanol step 2: 4 mL ethanol with 5% acetic acid	HPLC	Puoci et al. (2007)

(Continued)

Table 3.1: Molecularly imprinted solid-phase extraction (MISPE). (*cont.*)

Analytes/ Compounds	Sample/ Matrix	Sample Volume (g)	Synthesis of MIP	Polymer	Extraction From Food	Elution	Target Components Analysis	References
Cholesterol	Cheese	25	Template: cholesterol; functional monomer: methacrylic acid (MAA); cross-linker: ethyleneglycol dimethacrylate (EGDMA); initiator: azobisisobutyronitrile (AIBN); solvent: chloroform photopolymerization; extraction of MIP: 200 mL acetic acid:tetrahydrofuran (1:1) for 48 h, 200 mL tetrahydrofuran for 48 h in a Soxhlet apparatus, dried in an oven at 60°C overnight	MIP, NIP	Sample extracted with 150 mL acetonitrile, filtered, water added to raise the loading solution concentration (acetonitrile/water 7/3 v/v), 8 mL solution used to load the MISPE column	Step 1: 7 mL acetonitrile/ water (7:3 v/v), 5 mL acetonitrile/ water (9:1 v/v); step 2: 4 mL hot acetonitrile (50°C)	HPLC	Puoci et al. (2008)
Fenitrothion (insecticide)	Tomatoes	5	Template: fenitrothion; functional monomer: methacrylic acid (MAA); cross-linker: ethylene glycol dimethacrylate (EGDMA); initiator: azobisisobutyronitrile (AIBN); solvent: dichloromethane thermo-polymerization; extraction of MIP: washed several times with ACN/water/methanol (7:2:1, v/v/v)	MIP, NIP	Added to sample 5 mL of Britton–Robinson buffer, pH 7.75:ACN 60:40 (v/v), mixture was triturated, spiked with 500 μL of FNT, sonicated, centrifuged (980 × g) for 5 min, supernatant applied to the previously conditioned MISPE cartridge	5 mL ACN:acetic acid 90:10 (v/v)	HPLC	Pereira and Rath (2009)

17β-Estradiol	Milk powder	25	Template: 17β-estradiol; functional monomer: 2-(trifluoromethyl) acrylic acid (TFMAA); cross-linker: ethylene glycol dimethacrylate (EGDMA); initiator: azobisisobutyronitrile (AIBN); solvent: anhydrous CAN thermopolymerization, photopolymerization; extraction of MIP: 12 h on-line flushing using MeOH containing 10% HAc (v/v) at flow 0.1 mL/min, flushing 100% ACN for 2 h	MIP, NIP	Sample mixed with 75 mL MeOH, stirred, left to stand for 12 h, centrifuged at 3000 rpm for 20 min, supernatant collected into a 100 bottle, 3×1 mL loaded to the MISPE column	3 mL ACN/H_2O (3/7, v/v), n × 0.5 mL MeOH/ACN (1/1, v/v)	HPLC, GC-MS	Zhu et al. (2009)
17β-Estradiol	Dairy and meat (milk, yogurt, beef, pork, chicken)	500	Template: 17β-estradiol; functional monomer: 4-vinylpyridine (4-VP); cross-linker: ethylene; glycol dimethacrylate (EGDMA); initiator: azobisisobutyronitrile (AIBN); solvent: mineral oil–toluene 2:3, v/v photopolymerization extraction of MIP: methanol–acetic acid 9:1 (v/v) for 24 h in a soxhlet apparatus, washed by ethanol for 5 times, dried in vacuo overnight at 25°C	MIP, NIP	Sample homogenized with 150 mL methanol-acetone 1:1 (v/v), vortex-mixed for 2 min, ultrasonic for 5 min, centrifuged at 5000 g for 5 min, supernatant collected, procedure repeated thrice; combined supernatant volatilised to dry at 35°C, residue redissolved into 3 mL methanol, diluted with water to 15 mL	After MISPE, the elution concentrated to 100 μL	HPLC	Shi et al. (2011)

(Continued)

Table 3.1: Molecularly imprinted solid-phase extraction (MISPE). (*cont.*)

Analytes/ Compounds	Sample/ Matrix	Sample Volume (g)	Synthesis of MIP	Polymer	Extraction From Food	Elution	Target Components Analysis	References
Florfenicol (FF) (antibiotic)	Fish, chicken meat, honey	5, 10	Template: FF; functional monomer: 4-vinyl pyridine (4-VP); cross-linker: ethylene glycol dimethacrylate (EGDMA); initiator: azobisisobutyronitrile (AIBN); solvent: tetrahydrofuran photo-polymerization; extraction of MIP: methanol for 48 h in a Soxhlet apparatus	MIP, NIP	Fish, chicken: 3 × 10 g of sample transferred to 100 mL polyethylene containers, spiked with a standard FF solution, diluted to 50 mL with methanol, shaken, sonicated for 30 min, centrifuged at 5000 rpm for 5 min, evaporated, residues dissolved in 5 mL of methanol, 1 mL of sample diluted to 10 mL by distilled water at pH 8.0 and passed through MISPE/NISPE cartridge; honey: 5 g of sample spiked with FF aqueous solution, mixed, added 5 mL distilled water, shaken for 5 min, added 5 mL of ethylacetate, shaken for 5 min, centrifuged at 5000 rpm for 5 min, organic layer transferred to vessel, repeated two times, combined two organic layers, organic phase evaporated to dryness at 40°C, reconstituted, pH adjusted to 8.0	MeCN–water solution (30%, v/v)	HPLC	Sadeghi and Jahani (2013)

Histamine	Canned fish	1	Template: histamine; functional monomer: methacrylic acid (MAA); crosslinker: ethylene glycol dimethacrylate (EGDMA); initiator: azobisisobutyronitrile (AIBN); solvent: chloroform thermopolymerization; extraction of MIP: methanol–acetic acid (80:20, v/v), centrifuged at 3,000 rpm for 10 min	MIP, NIP	2 mL trichloroacetic acid–water (3:97, w/v) added to sample, hemogenized, centrifuged 13,000 rpm for 10 min (three times), to 1 mL supernatant added 275 μL NaOH (1 M), shaken, 0.5 mL of mixture loaded onto the column	2 mL ACN-acetic acid (95:5 v/v), 5 mL methanol-acetic acid (50:50, v/v)	HPLC	Sahebnasagh et al. (2014)
Sulfonylurea herbicides	Rice grain	5	Template: pyrazosulfuron ethyl (PS); functional monomer: methacrylic acid (MAA); crosslinker: trimethylolpropane trimethacrylate (TRIM); initiator: azobisisobutyronitrile (AIBN); solvent: dichloromethane thermopolymerization; extraction of MIP: methanol/acetonitrile/acetic acid (2:1:1, v/v/v) in a water bath at 97°C for 12 h, wash solvent replaced once every 4 h	MIP, NIP	20 mL acetonitrile added to sample, ultrasonic-assisted extraction performed for 30 min, centrifuged at 3550g for 10 min, supernatant evaporated, cleaned up with the MIP and NIP cartridges after dissolving in 1 mL of acetonitrile containing 10 μg/mL mixed sulfonylureas.		HPLC	Tang et al. (2014)
Fenarimol (fungicide)	Apple, banana, tomato	10	Template: fenarimol; functional monomer: methacrylic acid (MAA); cross-linker: ethylene glycol dimethacrylate (EGDMA); initiator: azobisisobutyronitrile (AIBN); solvent: acetonitrile thermo-olymerization; extraction of MIP: polymers separated, centrifuged, washed by ethyl acetate and acetic acid (90:10, v/v)	MIP, NIP	Macerated samples homogenized with 10 mL acetonitrile followed by salting out step with anhydrous $MgSO_4$ and NaCl, centrifuged		UPLC	Khan et al. (2016)

Table 3.2: Solid-phase extraction (SPE).

Analytes/ Compounds	Sample/ Matrix	Sorbents	Conditioning	Flow Rate (mL/min)	Washing	Elution	Target Components Analysis	References
Phenolic compounds	Virgin olive oil	Octadecyl C_{18}, octadecyl C_{18EC}	2 × 10 mL methanol, 2 × 10 mL hexane			4 × 10 mL hexane	GC-MS	Liberatore et al. (2001)
Pesticides	Tea	Envi—Carb SPE	6.0 mL acetone–ethyl acetate–*N*-hexane (1:2:1 v/v/v)			6.0 mL acetone–ethyl acetate (1:2, v/v)	GC–MS	Huang et al. (2007)
Patulin	Apple juice concentrate	C_{18}-SPE	10 mL methanol, 3 mL 10% methanol, 10 mL water	2–3	5 mL hexane, dried with a strong stream of air for 15 min	3 × 5 mL hexane:ethyl acetate:acetone (1:5:4, 1:4:5, 1:3:6)	HPLC	Li et al. (2007)
Synthetic colors	Drinks, candies	Polyamide	20 mL 1% acetic acid		20 mL 1% acetic acid, 20 mL water	15 mL 1% ammonia solution/ ethanol (1:1, v/v)	HPLC	Yoshioka and Ichihashi (2008)
Polycyclic aromatic hydrocarbons (PAHs)	Coffee brew	C_{18}-silica	5 mL MeOH, 5 mL water			2 mL eluting solvent	GC–MS	Orecchio et al. (2009)
Melamine	Liquid milk	PCX	3 mL methanol, 3 mL water		5 mL water, 5 mL methanol	6 mL 25% ammonia solution–methanol (1:20, v/v)	HPLC	Sun et al. (2010)
Insecticides	Honey	Oasis HLB cartridge [poly(divinylbenzene-*co*-*N*-pyrrolidone)]	5 mL methanol, 5 mL Milli-Q water	10	5 mL Milli-Q water	10 mL methanol–dichloromethane (30:70)	LC–IT-MS	Blasco et al. (2011)
α- and β-thujone	Alcoholic beverages	SepraC_{18}-E	5 mL *N*-hexane/ ethyl acetate (9:1)		1 mL 20% ethanol/water	5 mL *N*-hexane/ ethyl acetate	GC	Dawidowicz and Dybowski (2012)

Anthocyanins	Tea	Reverse phase (RP) C_{18}	10% MeOH/HCl		0.01% HCl in distilled water	Acidified methanol (10% formic acid v/v)	HPLC-UV-Visible	Kerio et al. (2012)
Cephalosporins (cephalosporin antibiotics)	Milk	Molecularly imprinted polymer (MIP) [Tributylammonium cefadroxil (TBA-CFD), TFMAA as monomer]; nonimprinted polymer (NIP)	7.5 mL H_3PO_4 (1.2 mol/L)		1 mL ACN	4 mL MeOH/AcOH 90/10 (v/v)	HPLC	Quesada-Molina et al. (2012)
Mycotoxins	Barley	Oasis HLB	Column prewashed with 10 mL ACN, conditioned 10 mL 5% ACN in deionized water		10 mL 5% ACN in water	5 mL ACN	UHPLC-Orbitrap^R MS	Rubert et al. (2012b)
Aflatoxin M1	Milk, milk powder	OASIS HLB	5 mL ACN, 5 mL water		10 mL 20% ACN aqueous solution	5 mL ACN	HPLC-FLD	Wang et al. (2012)
Malachite green, leucomalachite green	Fish	Graphene-based SPE	3 mL MeOH, 3 mL ultrapure	1	2 mL ultrapure water	4 mL ammoniac methanol	UPLC-MS/MS	Chen et al. (2013)
Mycotoxins	Beers	Oasis HLB	5 mL ACN/MeOH(50:50, v/v), 5 mL water		5 mL water	4 mL ACN/MeOH (50:50, v/v)	HPLC-QqQ-MS/MS	Rubert et al. (2013a)
Furan derivatives	Apple cider, wine	LiChrolut EN cartridges	4 mL methanol, 4 mL water	2	2.5 mL 40% methanol solution (apple cider) 4 mL water (wine)	4 mL dichloromethane	HPLC/DAD	Hu et al. (2013)

(Continued)

Table 3.2: Solid-phase extraction (SPE). (*cont.*)

Analytes/ Compounds	Sample/ Matrix	Sorbents	Conditioning	Flow Rate (mL/min)	Washing	Elution	Target Components Analysis	References
Synthetic food colorings	Soft drinks	Sep-Pack C_{18}	2 mL isopropyl alcohol		5 mL acetic acid	10.0 mL of 18% (v/v) isopropyl alcohol	Ion-pair-HPLC, TLC	de Andrade et al. (2014)
Plasticizer	Tea	Oasis HLB glass (5 cc; 0.2 g) cartridges	6 mL MeOH, 6 mL water	1	2 × 20 mL water	6 mL MeOH	GC-MS	Lo Turco et al. (2015)
Isoflavones, Resveratrols	Soy bean and peanut oils	Oasis[R] WCX, Hicapt SAX	3 mL MeOH, 3 mL *N*-hexane	1	3 mL 60% isopropyl alcohol/*N*-hexane	2 × 1.5 mL MeOH	LC–MS/MS	Zhao et al. (2015)
Pesticides	Red wine	Oasis HLB	3 mL ACN, 3 mL water	5	3 mL MeOH/ water (50%/50% v/v)	2 mL *N*-hexane, 4 × 2 mL ACN	GC-MS	Pelajić et al. (2016)
Organic acids	Peach fruit	Reversed phase-weak anion exchange sorbent (Strata X-AW)	3 mL MeOH, 3 mL ultrapure water		3 mL water, 3 mL MeOH	Acetonitrile, methanol, water (1% HCl), water (5% NH_4OH)	LC–ESI-MS	Sandín-España et al. (2016)

Table 3.3: Solid-phase microextraction (SPME).

Analytes/ Compounds	Sample/ Matrix	Sample Weight	Fiber Conditioning	Fiber Coatings	Temperature (°C)	Extraction Time (min)	Salt	Agitation (rpm)	Desorption	Target Components Analysis	References
Pesticides	Wine, fruit juices	5 mL		85 μm PA	Room temperature	30			250°C 5 min	GC-MS	Zambonin et al. (2004)
Pesticides (HS-SPME)	Radish		270°C 1 h	C[4]/OH-TSO	40–90 *70*	10–60 *30*		600	270°C 2 min	GC–ECD	Dong et al. (2005)
Volatile aroma compounds	Chili (*Capsicum annuum* var *annuum* cv. Kulai)	5 g		50/30 μm DVB/CAR/PDMS	25, 30, 40, 50, 60, 70 *60*	5, 10, 20, 30, 60 *30*			250°C 2 min	GC-MS	Mazida et al. (2005)
Sorbic and benzoic acids (HS-SPME)	Beverages		Manufacturer's recommended procedures	65 μm PDMS–DVB	40–90 *50*	10–60 *40*	1.0, 1.5, 2.0, *2.5*, 3.0, 4.0 g anhydrous Na_2SO_4	600	260°C 5 min	GC-FID	Dong and Wang (2006)
Aroma compound (HS-SPME)	French beans	10 g	270°C 1 h	50/30 μm DVB/CAR/PDMS	40	60			250°C 5 min	GC, GC/MS	Barra et al. (2007)
Formaldehyde	Fish		300°C under helium flow for 1.5 h	75 μm CAR-PDMS	80	30			310°C 3 min	GC–MS	Bianchi et al. (2007)
Volatiles	Vinegar	25 mL		75 μm CAR-PDMS	25	120	6.14 g NaCl	1250	40–300°C 10 min	GC	Guerrero et al. (2007b)
Vitamin K	Green tea	25 mL		7 μm *PDMS*, 85 μm PA, PDMS/DVB, PDMS/DVB/CAR	45	10, 20, 30, *40*, 60	0, 8, *10%* NaCl	1000, *1300*	5 min 250°C	GC-FID	Reto et al. (2007)
Pesticide residues	Fruit juices	1 mL	MeOH 30 min	50 μm *CW/TPR*, 60 μm *PDMS/DVB*, 85 μm PA	*20*, 30, 40, 50, 60	30, 60, *90*, 120	30% NaCl	1000	MeOH/water (70:30 v/v) 15 min	LC-MS, LC/MS/MS	Sagratini et al. (2007)

(Continued)

Table 3.3: Solid-phase microextraction (SPME). (*cont.*)

Analytes/ Compounds	Sample/ Matrix	Sample Weight	Fiber Conditioning	Fiber Coatings	Temperature (°C)	Extraction Time (min)	Salt	Agitation (rpm)	Desorption	Target Components Analysis	References
N-nitrosodimethylamine	Beer	5 mL	Manufacturer's recommended procedures	765 μm PDMS/DVB	42	40	20% NaCl		225°C 3 min	GC-MS	Pérez et al. (2008)
Volatile components	Roasted Mini-pig pork	15 g	280°C 40 min	75 μm CAR/PDMS	80	60	0,22 g salt		280°C 4 min	GC-MS, GC-O, GC-FID	Xie et al. (2008)
Pesticide residues (HS-SPME)	Fruits, vegetables			100 μm PDMS	60	30	10% NaCl	800	240°C 10 min	GC-ECD	Chai and Tan (2009)
Tetramethylene disulfotetramine (HS-SPME)	Food (water, juice, applesauce, yogurt, peanut butter, peas, tuna, potato chips)	3–5 g		70 μm CW/DVB	100	90			260°C 3 min	GC–MS	De Jager et al. (2009a)
Tetramine (HS-SPME)	Food (water, juice, applesauce, yogurt, peanut butter, peas, tuna, potato chips)	3–5 g		70 μm CW/DVB	100	90			260°C 3 min	GC-MS	De Jager et al. (2009b)
Furan	Baby food	2.5 g		75 μm CAR/PDMS	45	20	20% NaCl			GC–MS	Jestoi et al. (2009)
Flavor compounds	Extruded potato snacks	10 g	Manufacturer's recommended procedures	PDMS, CW/DVB, DVB/CAR/PDMS, PDMS/DVB CAR/PDMS	20, 50	5, 15, 30, 60, 90			5 min	GCO/AEDA, GC/MS	Majcher and Jeleń (2009)

Strobilurin fungicides	Baby foods	2 g	Manufacturer's recommended procedures	100 µm PDMS, 65 µm PDMS-DVB, 85 µm PA, 75 µm CAR-PDMS, 70 µm CW-DVB, 50/30 µm DVB-CAR-PDMS, 60 µm PEG	30–90 *60*	10–60 *40*		0–2000 *1700* rpm	240°C 4 min	GC–MS	Viñas et al. (2009)
Volatile constituents (HS-SPME)	Ginger (*Zingiber officinale*)	0.1 g	Manufacturer's recommended procedures	7 µm PDMS, *100 µm PDMS*, 85 µm PA, 75 µm CAR/PDMS	40, 50, *60*, 80	10, 20, *30*, 40			250°C 3 min	GC-MS	Yang et al. (2009)
Pesticides	Mangoes	3 g		PA, PDMS, PDMS/DVB; PDMS/CAR, DVB/CAR/PDMS	30, 40, *50*, 60	10, 20, *30*, 40	5% NaCl pH 3	250	280°C 5 min	GC–MS	Filho et al. (2010)
Esters (HS-SPME)	Beer	10 mL	Preconditioned 250°C 0.5 h	65 µm DVB/CAR/PDMS	25, *40*, 60	5, 15, 30, *60*, 90	5 g Na Cl	800	250°C 5 min	GC	Horák et al. (2010)
Pesticides (HS-SPME)	Cucumber, strawberry	30 g		100 µm PDMS	60	30	10% NaCl	800	240°C 10 min	GC–ECD	Kin and Huat (2010)
Insecticides	Honey	5 g	MeOH 30 min, cleaned MeOH 15 min	50 µm CW/TPR		120			Static MeOH/water (70:30, v/v) 15 min	LC–IT-MS	Blasco et al. (2011)

(*Continued*)

Table 3.3: Solid-phase microextraction (SPME). (*cont.*)

Analytes/ Compounds	Sample/ Matrix	Sample Weight	Fiber Conditioning	Fiber Coatings	Temperature (°C)	Extraction Time (min)	Salt	Agitation (rpm)	Desorption	Target Components Analysis	References
Volatiles	Chili pepper	1 g	Manufacturer's recommended procedures	50/30 μm DVB/CAR/PDMS, 65 μm DVB/PDMS, 70 μm CW/DVB, 75 μm CAR/PDMS, 100 μm PDMS	40	30			250°C 1 min	GC–MS	Bogusz Junior et al. (2011)
Trihalomethanes (HS-SPME)	Soft drink	20 mL	300°C 1 h	85 μm PA, 100 μm PDMS, 65 μm CW/DVB, 65 μm PDMS/DVB, 50/30 μm DVB/CAR/PDMS, 75 μm *CAR/PDMS*	10–80 *20*	10	20% NaCl	500	280°C 3 min	GC	Dos Santos et al. (2011)
Volatile components	Fermented camel milk	5 mL	Preconditioned in the injection port of GC	75 μm CAR/PDMS, 65 μm PDMS/DVB, 50/30 μm DVB/CAR/PDMS	40	20	1 g NaCl		3 min	GC/MS, GC–O	Ning et al. (2011)
Aroma compounds	Tartary buckwheat tea	1 g		65 μm PDMS/DVB	60	60			10 min	GC-MS	Qin et al. (2011)

Volatile composition (HS-SPME)	White salsify (*Tragopogon porrifolius*)	5, 6, *7* g	Manufacturer's recommended procedures	50/30 μm DVB/CAR/PDMS	50, 60, 70, *80*	45, 60, *75*		700		GC–MS	Riu-Aumatell et al. (2011)
Organic volatile flavor compounds	Fermented stinky tofu	20 g	Manufacturer's recommended procedures	*75 μm CAR/PDMS*, 65 μm PDMS/DVB, 50/30 μm DVB/CAR/PDMS, 100 μm PDMS	50	15, 30, 45, 60, 75				GC-MS	Liu et al. (2012c)
Fungicides	Wine	19 mL	Manufacturer's recommended procedures	100 μm PDMS	35	60		250		GC-MS/MS	Martins et al. (2012)
Pesticides	Lettuce	2 mL aliquot lettuce extract,	MeOH 30 min	100 μm PDMS, 60 μm PDMS/DVB, 50 μm *CW/TPR*, 85 μm PA,	22	30	18% NaCl	1000 rpm	MeOH/water (9:1) 10 min	HPLC-DAD	Melo et al. (2012)
Volatile compounds (HS-SPME)	Oolong tea (*Camellia sinensis*)	10 g	220°C 5 min	65 μm PDMS/DVB	50	40				GC-MS	Lin et al. (2013)
Volatile compounds (HS-SPME)	Banana fruit	3 g	Manufacturer's recommended procedures	100 μm PDMS, 65 μm PDMS/DVB, 50/30 μm DVB/CAR/PDMS, 85 μm CAR/PDMS	40	30	1 g NaCl		250°C 2 min	GC–O	Pino and Febles (2013)

(Continued)

Table 3.3: Solid-phase microextraction (SPME). (*cont.*)

Analytes/ Compounds	Sample/ Matrix	Sample Weight	Fiber Conditioning	Fiber Coatings	Temperature (°C)	Extraction Time (min)	Salt	Agitation (rpm)	Desorption	Target Components Analysis	References
Volatile compounds	Pu-erh tea	4 g	15 min	100 μm PDMS, 75 μm *CAR/PDMS*, 65 μm PDMS/DVB, 50/30 μm DVB/CAR/PDMS	50, 60, *70*, 80	40, 50, *60*, 70	4.8 g NaCl		250°C 5 min	GC–MS	Du et al. (2014)
Aroma	Fermented red pepper paste (gochujang)	4 g		100 μm PDMS, 65 μm PDMS/DVB, 85 μm PA, 75 μm CAR/PDMS, 65 μm CW/DVB, 50/30 μm DVB/CAR/PDMS	60	30			250°C 1 min	GC–O	Kang and Baek (2014)
Volatile compounds (HS-SPME)	Papaya cv. Red Maradol	3 g		100 μm PDMS, 65 μm PDMS/DVB, 50/30 μm *DVB/CAR/PDMS*, 85 μm CAR/PDMS	40	30	1 g NaCl			GC-O	Pino (2014)
Medium and long-chain free fatty acids volatiles	Cooked ham exudates	2 mL		50/30 μm CAR/PDMS/DVB	40	30			250°C 10 min	GC–MS	Benet et al. (2015)

Volatile compounds	Beer	2 mL		65 µm PDMS/DVB	40	30	0.6 g/mL NaCl	250	250°C 15 min	GC/FID	Castro and Ross (2015)
Methylpyrazines	Cocoa liquors	0.5 g	250°C 10 min	DVB/CAR/PDMS	60	65			250°C 5 min	GC–MS	Di Carro et al. (2015)
Volatile compounds (HS-SPME)	Wheat bread	0.25	Manufacturer's recommended procedures	50/30 µm DVB/CAR/PDMS 85 µm CAR/PDMS	50	60	20% NaCl	700		GC-MS	Raffo et al. (2015)
Salicylic acids, benzoic acid	Fruits, legumes, beverages			60 µm PDMS/DVB	20	30	0.3 g/mL NaCl		Static desorption 5 min	LC-UV/DAD	Aresta and Zambonin (2016)
Volatile flavor compound	Codfish	6 g	Thermally conditioned 250°C for 30 min, 300°C for 1.5 h, 300°C for 2 h	60 µm PDMS/DVB, 75 µm CAR/PDMS, 85 µm PA	60	40	8 mL saturated NaCl	150	10 min	GC-MS	Chang et al. (2016)
Furan, 2-methylfuran, 2-pentylfuran (HS-SPME)	Juice	5 mL diluted fruit juice (1:4) sample	Manufacturer's recommended procedures	85 µm CAR/PDMS	23, *32*, 40	5, 10, 15, *20*, 30	0%, 5%, 10%, *15%*, 20% NaCl	200, 400, *600*, 800	250°C 5 min	GC–FID	Hu et al. (2016)
Pesticides (HS-SPME)	Fruits (apple, pineapple)	2 g of		100 µm PDMS, 85 µm PA, 65 µm PDMS/DVB, 50/30 µm DVB/CAR/PDMS, MIP-coated	70	30	1.6 g NaCl	600	250°C 8 min	GC, LLE	Li et al. (2016a)

(*Continued*)

Table 3.3: Solid-phase microextraction (SPME). (*cont.*)

Analytes/ Compounds	Sample/ Matrix	Sample Weight	Fiber Conditioning	Fiber Coatings	Temperature (°C)	Extraction Time (min)	Salt	Agitation (rpm)	Desorption	Target Components Analysis	References
Acrylamide (HS-SPME)	Fried foods	1 g	Manufacturer's recommended procedures	100 μm PDMS, 85 μm PA, 65 μm CW/DVB, 50/30 μm *CAR/DVB/PDMS*	30–65 *60*	30			230°C 2 min	GC-FID	Ghiasvand and Hajipour (2016)
Aroma-active compounds	Pu-erh tea	6 g	Manufacturer's recommended procedures	65 μm PDMS/DVB, 85 μm PA, 75 μm *CAR/PDMS*, 50/30 μm DVB/CAR/PDMS, 100 μm PDMS	40, 50, 60, *70*, 80	30, 40, *50*, 60, 70	4 g KCl		At GC injector port for 3.5 min in splitless mode	GC-O, GC-MS	Xu et al. (2016)

Table 3.4: Headspace sorptive extraction (HSSE).

Analytes/ Compounds	Sample/Matrix	Sample	Extraction	Desorption	Target Components Analysis	Reference
Volatile fraction	Pasta sauce	4 g	10 mL headspace vial, PDMS stir bar, incubated 25°C 2 h	Stir bar thermodesorption, extracted compounds desorbed from 20 to 40°C at 30°C/min, held for 20 min, helium flow, volatiles trapped and focused on cooled surface of the injector liner, cryo-focalized at −120°C	GC-MS	Salvadeo et al. (2007)
Flavor compounds	Mountain cheese	2 g	20 mL headspace vial, PDMS-coated stir bar, heated 50°C in water bath 60 min	Stir bar thermodesorption	GC-MS	Panseri et al. (2008)
Volatile compounds	Red wine	5 mL	20 mL headspace vial, PDMS-coated stir bar, heated at 62°C in thermostatic bath 60 min	Compounds desorbed from 35°C to 250°C at 60°C/min, held for 5 min; injector placed in liquid nitrogen at −35°C, raised at 10°C/s to 290°C, held for 4 min	GC-MS	Callejón et al. (2009)
Volatile fraction	Pasta sauce	4 g	10 mL headspace vial, PDMS stir bar, incubated at 25°C 2 h	Stir bar thermodesorption	GC-MS	Zunin et al. (2009)
Volatile compounds	Buckwheat	2 g	20 mL headspace vial, PDMS-coated magnetic stir bar, heated at 50°C in water bath 30 min	Stir bar thermodesorption in thermal desorption unit with analyte cryotrapping in PTV injector of GC	GC-MS	Prosen et al. (2010)

Table 3.5: Petrol ether extraction (PEE).

Analytes/ Compounds	Sample/ Matrix	Sample Weight	Extraction Medium	Extraction Time	Extraction Method	Collected Over	Target Components Analysis	References
Volatile constituents	Ginger (*Z. officinale*)	100 g	Petrol ether	72 h	Soxhlet	Anhydrous sodium sulfate	GC-MS	Yang et al. (2009)
Hopanoids	Sediments, rocks, coals, crude oils, plant fossils, stromatolites, crude oils, microbial biomass	From mg to kg	Petrol ether, *N*-alkanes	From minutes to days	Soxhlet; pressurized extractions	Various protocols	GC-MS/HPLC-MS	Zarzycki and Portka (2015)

Table 3.6: Stir-bar sorptive extraction (SBSE).

Analytes/ Compounds	Sample/ Matrix	Sample Volume	Stir Bar Material	Temperature (°C)	Extraction Time (min)	Stirring Speed (rpm) (italicised numbers indicate optimal conditions)	Bar Cleaning Protocol	Desorption	Target Components Analysis	Reference
Preservatives	Beverages, vinegar, aqueous sauces, quasidrug drinks	10 mL	PDMS	*25*, 45, 70	5–360, *120*	1000	Dried with tissue, placed in a glass tube	Thermally desorbed by programming the TDS 2 from 20°C (held for 1 min) to 250°C (held for 5 min) at 60°C/min	GC–MS	Ochiai et al. (2002)
Pesticides	Pear pulp	1 g	PDMS	Room temperature	Homogenized 10 min, equilibrated 60 min		Dried with filter paper, inserted into glass tube	From 0°C to 280°C (8 min) at 60°C/min; carrier gas He, constant flow 1.0 mL/min	GC/ECD	Bicchi et al. (2003)
Organo-phosphorus pesticides (OPPs)	Cucumbers, potatoes	10 g	PDMS prepared by sol–gel coating	30	10–40, *30* *(repeated twice)*	600	Water and methylene chloride, 1 mol/L NaOH and 0.1 mol/L HCl, dried at 100°C for 2 h in GC oven	260°C 5 min	GC–TSD	Liu et al. (2005)
Pesticides	Vinegar	40 mL	PDMS	25	20–150, *150*	500–1500, *1000*	300°C 15 min	30–300°C under helium flow (75 mL/min), analytes cryofocused with liquid nitrogen at −150°C	GC-MS	Guerrero et al. (2007a)

(*Continued*)

Table 3.6: Stir-bar sorptive extraction (SBSE). (*cont.*)

Analytes/ Compounds	Sample/ Matrix	Sample Volume	Stir Bar Material	Temperature (°C)	Extraction Time (min)	Stirring Speed (rpm) (italicised numbers indicate optimal conditions)	Bar Cleaning Protocol	Desorption	Target Components Analysis	Reference
Volatiles	Vinegar	25 mL	PDMS	25	120	1250	Rinsed with distilled water, dried with lint-free tissue	40–300°C under helium flow (75 mL/min), analytes cryofocused with liquid nitrogen at −140°C	GC	Guerrero et al. (2007b)
Tetramethylene disulfotetramine	Food (water, juice, applesauce, yogurt, peanut butter, peas, tuna, potato chips)	3–5 g	PDMS		15, 30, *60*, 90	1000, *1200*, 1500	Rinsed with water, dried with lint-free tissue	Helium flow rate 50 mL/min (solvent vent mode), initial temperature 40°C with a 0.4-min delay followed by a temperature ramp at 200°C/min to 300°C and 4 min hold	GC–MS	De Jager et al. (2009a)
Tetramine	Food (water, juice, applesauce, yogurt, peanut butter, peas, tuna, potato chips)	3–5 g			60	1200	Rinsed with water, blotted with lint-free tissue, placed in desorption tube		GC-MS	De Jager et al. (2009b)

Pesticides	Vegetables (lettuce, spinach, green bean, green pepper, tomato, broccoli, potato, carrot, onion)	30 mL	PDMS	20	180	1000	Dried with lint-free tissue, placed into 2 mL glass vial filled with 1.5 mL ACN	Thermal desorption	GC–MS(SIM)	Barriada-Pereira et al. (2010)
Fungicides	Fruits (apple, pear, grape, orange, lemon, peach, plum)	5 g	PDMS	45	20	2000	Rinsed with distilled water, residual water removed with tissue paper	Extracted with 100 μL of 50/50 (v/v) ACN/water for 15 min	LC–DAD	Campillo et al. (2010)
Volatile compounds	Sherry brandy	35 mL	PDMS	25	30–120, *100*	500–1500, *1100*	Rinsed with distilled water, dried with lint-free tissue	Thermally	GC-MS	Delgado et al. (2010)
Volatile compounds	Blackberry	10 mL	PDMS		120	1000	Rinsed with distilled water, placed into sample holder for GC–MS analysis	Thermal desorption	GC–MS	Du et al. (2010)
Esters	Beer	10 mL	PDMS	Ambient temperature	60	1200	Rinsed with distilled water, dried with lint-free tissue		GC	Horák et al. (2010)
Triazine herbicides	Rice, apple, lettuce	5 g	MIP-coated		60	500	Dried with lint-free tissue	5 min at ultrasonic treatment	HPLC-UV	Hu et al. (2010)
Furan	Water, coffee, jarred baby food	10 mL	PDMS	Tempera-ture (ap-proximately 20°C)	60	500	Rinsed with small amount of ultrapure water, patted dry on clean tissue	300°C 90 min	GC–MS	Ridgway et al. (2010)

(*Continued*)

Table 3.6: Stir-bar sorptive extraction (SBSE). (*cont.*)

Analytes/ Compounds	Sample/ Matrix	Sample Volume	Stir Bar Material	Temperature (°C)	Extraction Time (min)	Stirring Speed (rpm) (italicised numbers indicate optimal conditions)	Bar Cleaning Protocol	Desorption	Target Components Analysis	Reference
Volatiles	Chinese soy sauce type liquors		PDMS	Room temperature	15, 30, 45, 60, *90*	5.55 × g	Rinsed with distilled water, dried with odorless tissue	270°C 5 min	GC-MS	Fan et al. (2011)
Volatiles	Cantaloupe, honeydew melon	1 mL	Polydimeth-ylsiloxane	37.5	60	850	Rinsed with deionized water, blotted dry on lint-free tissue, placed in glass thermodesorption tubes	After a 0.4 min delay, the temperature increased from 35 to 280°C at 200°C/min, held 5 min	GC-MS	Amaro et al. (2012)
Pesticides	Tea	2 g	PDMS		0–*60*	1000	Rinsed with small amount of distilled water	300°C 30 min, 0°C inlet temperature of cold trap	GC	Li et al. (2012)
Resveratrol, piceatannol, oxyresvera-trol isomers	Wine	10 mL	PDMS		0.5–6 h, *3 h*	600	Rinsed with water, dried with lint-free tissue, placed in glass desorption tube	50–260°C at 210°C/min, held 9.6 min	GC-MS	Cacho et al. (2013)
Chlorophe-nols, chloro-anisoles	Wine	10 mL	PA, *EG–Silicone*		0.5–6 h, *2 h*	600	Rinsed with Milli-Q water, dried with lint-free tissue, plased in glass desorption tube	Thermal desorption	GC-MS	Cacho et al. (2014)

Farnesol	Makgeolli (rice wine)	20 mL	PDMS	Room temperature	1 h	450		Thermal desorption	GC–MS	Ha et al. (2014)
Medium and long-chain free fatty acids	Cooked ham	10 mL	PDMS		15 h	300	Rinsed with distilled water, dried with clean tissues, placed in desorption tubes	250°C 3 min	GC–MS	Benet et al. (2015)
Volatile compounds	Beer	10 mL	PDMS	Room temperature	60	1200	Rinsed with distilled water, dried with tissues, placed in desorption tube	From 25°C to 240°C	GC/FID	Castro and Ross (2015)

Table 3.7: Simultaneous distillation-extraction (SDE).

Analytes/ Compounds	Sample/Matrix	Sample Weight	Addition	Solvent/ Extraction Solvent	Time (h)	Temperature (°C)	Target Components Analysis	References
Aroma compound	French beans	300 g	Water	Dichloromethane	2		GC, GC/MS	Barra et al. (2007)
Volatile compounds	Roots of licorice (*Glycyrrhiza uralensis*)	50 g	Distilled water, pH at 6.5 with 1% NaOH, IS (*N*-butylbenzene)	Redistilled *N*-pentane: diethylether (1:1, v/v)	2		GC-MS	Gyawali et al. (2008)
Volatile components	Roasted mini-pig pork	120 g	Water	Dichloromethane	1.5		GC-MS, GC-O, GC-FID	Xie et al. (2008)
Volatile compounds	Fermented soybean pastes	500 g	1 L distilled water, 5 mL IS [2- methyl-1-pentanol (10 lg/mL in methanol)]	Redistilled dichloromethane	2		GC-MS	Lee and Ahn (2009)
Flavor compounds	Extruded potato snacks	300 g	Distilled water, IS (d_4-pyrazine)	Ethyl eter/ pentane (1:1 v:v)	2	100°C	GCO/AEDA, GC/MS	Majcher and Jeleń (2009)
Volatile compounds	Almond	30 g	IS (1-penten-3-one)	Dichloromethane	2	Oil bath 160°C, water bath 55°C	GC–MS	Vázquez-Araújo et al. (2009)
Volatile constituents	Ginger (*Z. officinale*)	100 g	Deionized water	*N*-hexane	4		GC-MS	Yang et al. (2009)
Volatiles	Cooked Korean nonaromatic rice	100 g	Antifoam solution, deodorized distilled water	Dichloromethane	2		GC-MS	Park et al. (2010)
Volatile compounds	Chinese fish sauce (Yu Lu)	600 mL		Diethyl ether	1	65–70°C	GC-MS	Jiang et al. (2011)
Volatile components	Fermented camel milk	200 g	Sodium chloride, with 100 μl IS (1,2-dichloro-benzene)	Diethyl ether, *dichloromethane*	3	Water bath 60°C, oil bath (saxol) 120°C	GC/MS, GC–O	Ning et al. (2011)
Flavor volatiles	Tea	100 g	Distilled water	Dichloromethane	5	Water bath 50°C, paraffin oil bath 200°C	GC–MS	Pripdeevech and Machan (2011)
Volatile composition	White salsify (*T. porrifolius*)	16.32 g	Distilled water	Pentane-dichloromethane (3:1)	4		GC–MS	Riu-Aumatell et al. (2011)
Aroma compounds	Tartary buckwheat tea	75 g	Distilled water	Petroleum ether	8		GC-MS	Qin et al. (2011)

Volatile compounds	Foxtail millet (*Setaria italica*): brown millet, milled millet, millet bran	200 g	Distilled water	Diethyl ether	2	Water bath 40°C	GC-MS	Liu et al. (2012b)
Aroma and aroma-active compounds	Olive oil	40 mL	Pure water, 30% NaCl, IS (4-nonanol)	Dichloromethane	3		GC–MS–O	Kesen et al. (2013)
Volatile compounds	Yerba mate	50 g	Distilled water, 0.1 mL of internal standard (1-heptanol at 823 ppm in ethanol)	Dichloromethane	1		GC–MS, GC-O	Márquez et al. (2013)
Volatile compounds	Banana fruit	200 g	Distilled water, IS (methyl nonanoate)	Diethyl ether	1		GC–FID, GC–MS	Pino and Febles (2013)
Volatile compounds	Pu-erh tea	20 g	Distilled water, *N*-pentane	*N*-pentane	4		GC–MS	Du et al. (2014)
Aroma	Fermented red pepper paste (gochujang)	600 g	Deodorized distilled water	Dichloromethane	1.5		GC–O	Kang and Baek (2014)
Volatile compounds	Papaya cv. Red Maradol	200 g	Distilled water, IS (methyl nonanoate)	Diethyl ether	1		GC-O	Pino (2014)
Volatile compounds	Fermented flour paste	200 g	Distilled water	Ethyl ether	2	Water bath at 20-25°C	GC-MS, GC-O	Zhang et al. (2014)
Phthalate esters (PAEs)	Teas and tea infusions	10 g	Ultrapure water	*N*-hexane	3		GC–MS	Du et al. (2016)
volatile flavor compound	Codfish	50 g	Deionized water	Dichloromethane	2 h	65°C	GC–MS	Chang et al. (2016)
Aroma-active compounds	Pu-erh tea	30 g	Potassium chloride, distilled water	Dichloromethane	3		GC-O, GC-MS	

Table 3.8: Membrane-assisted solvent extraction (MASE).

Analytes/ Compounds	Sample/ Matrix	Membrane Conditioning	Device	Filling of Membrane Bag/ Acceptor Phase	Temperature (°C)	Extraction Time (min)	Agitation (rpm)		Target Components Analysis	References
Pesticides	Sugarcane juice	10 mL cyclohexane room temperature, 1 h, 60 rpm	20 mL vial, 15 mL NaCl saturated aqueous sample	800 μL cyclohexane	45	30	750	Solvent extract withdrawn with syringe, poured into 2 mL autosampler vial	LVI–GC–MS(SIM)	Zuin et al. (2006)
Polycyclic aromatic hydrocarbons (PAHs)	Water, beverages	50 mL cyclohexane, room temperature, 1 h, 70 rpm; solvent exchanged, extraction repeated overnight	20 mL vial, spiked aqueous samples at concentration level of 5 μg/L of each compound	400 μL ethyl acetate	50	60	750	Withdrawn with syringe from membrane bag, transferred to a 2 mL autosampler vial	LVI–GC–MS	Rodil et al. (2007)
Fungicide	Wines, juices		15 mL vials, 5 mL solution containing standards, 5 mL 0.2 M acetate/ acetic acid buffer (pH 5), 2 g sodium chloride	800 μL hexane	30	30	1700	Extract removed using microliter syringe, transferred into 2 mL sampler vial, vaporated using stream of argon, sample reconstituted with 0.5 mL water	UPLC	Viñas et al. (2008)
Tetramine	Food (water, juice, applesauce, yogurt, peanut butter, peas, tuna, potato chips)		20 mL vials, tetramine, mixed 1 min, placed in rotating end-over-end mixer 10 min, sand 2 h, 10 g water, mixed 1 min, stand 1 h	900 μL benzene	Room temperature	120	1000	Organic phase removed from using pasture pipette, transferred into 2 mL autosampler vial	GC-MS	De Jager et al. (2009b)

Table 3.9: Supercritical fluid extraction (SFE).

Analytes/ Compounds	Sample/ Matrix	(g)	SFE Extraction Fluid/Extraction Medium/Solvent/ Cosolvent	Pressure (MPa)	Temperature (°C)	Extraction Time (min)	Flow Rate ($kgCO_2/h$)	Target Components Analysis	References
PCBs, DDTs	Fish	1	SC-CO_2	35.5	150	30	3 mL/min	GC	Hale et al. (1996)
Oleoresin	Onion	750	SC-CO_2	10–30	45–65			GC	Simándi et al. (2000)
Vitamins A and E	Milk powder	0.5	5% MeOH in CO_2	37	80	15 min static, 15 min dynamic	1 mL/min	HPLC	Turner and Mathiasson (2000)
Pesticides	Baby foods	2	100% CO_2, 15% ACN in CO_2	17.24	70	15 min static, 40 min dynamic		GC/MS	Chuang et al. (2001)
Fat-soluble vitamin, fatty acids	Rice by-products	20	SC-CO_2	34.47; 51.71; 68.95	40, 60, 80	1, 2, 4 h	1.082 g/min	HPLC	Perretti et al. (2003)
Fatty acids	Pumpkin seed oil	20	SC-CO_2	18, *19*, 20	*35*; 40; 45	120		GC	Bernardo-Gil et al. (2004)
Fat-soluble vitamins	Cheese, salami	2	5% MeOH in SC-CO_2	36.60	80	15 min static; 60 min dynamic	1.082 g/min	HPLC	Perretti et al. (2004)
Lycopene	Tomato	100	10% hazel-nut oil in SC-CO_2	33.5–*45*	45–70 *66*	2–5 h	8–*20* kg/h	HPLC	Vasapollo et al. (2004)
Carotenoids	Pulp of pitanga fruits	4	SC-CO_2	10, 15, 20, *25*, 30, 35, 40	40; *60*	120 min	6.8×10^{-5} kg/s	HPLC–PDA	Filho et al. (2008)
Lycopene	Tomato pomace	2	Ethanol-modified supercritical CO_2 (16%)	40	*57*	1,8 h		HPLC	Huang et al. (2008)
Pigments (chlorophylls carotenoids), volatile constituents, fatty acids, tocopherols	Cardamom oil	45	SF-CO_2 subcritical CO_2, SC-propane, 25% EtOH in CO_2	10–30 8–10 2–5	35–55 *25*	3–4 h	1.0–1.5 mL/min	HPLC, GC–MS	Hamdan et al. (2008)

(*Continued*)

Table 3.9: Supercritical fluid extraction (SFE). (*cont.*)

Analytes/ Compounds	Sample/ Matrix	(g)	SFE Extraction Fluid/Extraction Medium/Solvent/ Cosolvent	Pressure (MPa)	Temperature (°C)	Extraction Time (min)	Flow Rate (kgCO_2/h)	Target Components Analysis	References
Lycopene	Watermelon	10	10–*15%* EtOH in CO_2	20.7; *27.6*; 34.5; 41.4	*70*, 80, 90	5 min static, 30 min dynamic	1.5 mL/min	HPLC	Katherine et al. (2008)
Total phenol content (TPC)	Guava seeds (*Psidium guajava*)	30	CO_2, EtAc in CO_2, *10% EtOH in* CO_2	*10*, 20, 30	40, 50, *60*	2 h		Follin–Ciocalteu method	Castro-Vargas et al. (2010)
Bioactive compounds (total phenolic compounds, antioxidants, total anthocyanins)	Grape peel	3	5, 6, *7*, 8% EtOH in CO_2	15.69–16.18	37, 40, 43, 46	30 min	2 mL/min		Ghafoor et al. (2010)
Flavonoids	Spearmint (*Mentha spicata*) leaves		absolute EtOH in CO_2	10, *20*, 30	40, 50, *60*	30, *60*, 90 min	15 g/min	HPLC	Bimakr et al. (2011)
Fatty acids, tocopherols	Pomegranate seeds	250	SC-CO_2	15, 30, 45	35, 50, 65	2 h		GC–FID, HPLC	Liu et al. (2012a)
Lycopene	Tomato by-product	4	SC-CO_2, cosolvents: EtOH, water, oil (5-15%)	20–*40*	70–*90*	180 min	2–4 mL/min	GC–MS, GC–FID	Machmudah et al. (2012)
Vitamins (vitamin E, provitamin A)	Red pepper (*Ca. annum*) by-products	30	CO_2, EtOH in CO_2	20, *24*, 30	45, *60*	90, 120 min	300 mL/h	HPLC	Romo-Hualde et al. (2012)
Capsaicinoids	Peppers (*Ca. frutescens*)	4	SC-CO_2	*15*, 25, 35	*40*, 50, 60	320 min	1.98 10^{-4} kg/s	U-HPLC-DAD-MS/MS	de Aguiar et al. (2013)
Total phenolic content (TPC)	Date seeds	10	EtOH in SC-CO_2	25, 30, *35*, 40, 45	40, *50*, 60	1.0, 1.5, *2.0*, 2.5, 3.0 h	4 mL/min	HPLC	Liu et al. (2013)
Fatty acids	Pomegranate seed oil	250	CO_2	38	47	120 min	21 L/h	GC–MC	Tian et al. (2013)

Total phenolic content (TPC), total flavonoids (TF), total carotenoids	Peach palm pulp (Bactris gasipaes)	10	SC-CO_2	10, 20, *30*	*40*, 50, 60	91	3 L/min	Spectrophotometric method	Espinosa-Pardoa et al. (2014)
capsaicinoids, capsinoids	Biquinho pepper	2.5	CO_2	*15*, 20, 25	*40, 50*, 60	90	2.15×10^{-4} kg/s	UPLC-DAD	de Aguiar et al. (2014)
Polyunsaturated fatty acids	Carp (*Cyprinus carpi*) viscera	20	CO_2	20, 30, 35, *40*	40, 50, *60*	30, 60, 120, *180*	0.194; 0.277; *0.354 kg/h*	GC-FID	Lisichkov et al. (2014)
Phenolic compounds, antioxidants, anthocyanins	Blueberry residues		CO_2, water and EtOH, in CO_2	15, *20*, 25	40		1.05×10^{-4}; *1.4×10^{-4} kg/s*	UPLC-QTOF-MS, UPLC-UV-vis	Paes et al. (2014)
Anthocyanins	Blackberry bagasse	5	5%, 10% EtOH, 5, 10% water; in CO_2	*15*, 20, 25	40, 50, *60*	54–57	2.77×10^{-4} kg/s	UPLC	Reátegui et al. (2014)
Glucosinolates, phenols, lipids	Rocket salad (*Eruca sativa*)		SC-CO_2 MeOH, EtOH, water in CO_2	15–30	45–75	60	0.3-0.7 kg/h	HPLC-MS	Solana et al. (2014)
Carotenoids and phenolic compounds	Spinach	500	Pure supercritical CO_2	25, 35	40, 70	6 h	60 g/min	HPLC, colorimetric method	Jaime et al. (2015)
Triglycerides and polyphenols	Hazelnut, coffee, grape wastes	16; 20	pure CO_2, 5%–25% w/w of ethanol *10%*	35–*50*	40–60	0–180	1 g/min	GC	Manna et al. (2015)
Capsaicinoids	Malagueta pepper (*Ca. frutescens*)		CO_2 99% purity SFE SFE + US (200, 280, *360* W)	15	40	*60*, 150, 240	1.673×10^{-4} kg/s	HPLC-PDA	Santos et al. (2015)

Table 3.10: Pressurized liquid extraction (PLE).

Analytes/ Compounds	Sample/ Matrix	Sample Weight (g)	Solid Sorbent/ Dispersant/ Dispersion Matrices	Solvent	Temperature (°C)	Pressure (MPa)	Cycles	Extraction Time (min)	Target Components Analysis	References
Tocopherols	Seeds, nuts	1	Hydromatrix celite	Acetonitrile	50	11	2	5 min (static time)	LC-ED	Delgado-Zamarreño et al. (2004)
Sulfonamides	Pork meat	10	Diatomaceous earth	Hot water	160	10.34	1	5 min (static time)	CE-MS/MS	Font et al. (2007)
Macrolide antibiotics	Meat, fish	5	Aluminum oxide	100% MeOH	80	10.34	2	15 min	LC-(ESI)MS	Berrada et al. (2008)
Fumonisins B1, B2, B3	Baby food	3		Ethyl acetate, toluene, ACN, *MeOH,* solvent mixtures	30, *40*, 80, 120	*3.45*; 6.89; 10.34; 13.78	*1*; 2; 3	5 min (static time)	LC-MS/MS	D'Arco et al. (2008)
Oleoresin	Turmeric leaves	2	Diatomaceous earth	*N*-hexane	60, 90, 130, 170, 200, *147*	6895, 8274, 10343, 12411, 13790, *7150 kPa*	1	17 min (static time)	GC-MS	Zaibunnisa et al. (2009)
Pesticides	Fruits	1	Sea sand, diathomeus earth	Water	60	10.34	1	12	CE-MS	Juan-García et al. (2010)
Ochratoxin A	Breakfast and infants cereals	10		MeOH, ACN, toluene, ethyl acetate, isopropanol, MeOH/water (80:20 v/v), MeOH/ACN (50:50 v/v), MeOH/ACN (80:20 v/v), *ACN/water (80:20 v/v)*, ACN/water (75:25 v/v), ACN/water (90:10 v/v) and MeOH/ACN/water (40/40/20)	40	3.45	1	5 min (static time)	LC	Zinedine et al. (2010)

Insecticides	Honey	1.5	Silica	100% ethyl acetate	75	10.34	2	7 min (static time)	LC-IT-MS	Blasco et al. (2011)
Sulfonamides	Foods of animal origin (muscle, liver, kidney of swine, bovine and muscle, liver of chicken)	5		Acetonitrile	50–90, *70*	8.27–11, *9.65*	2	2 min (static time)	HPLC LC-MS/MS	Yu et al. (2011b)
Volatile	Coffee bean	10	Diatomaceous earth	Methanol, hexane, dichloromethane	50–100	6.89–13.79	1	5-15 min (static time)	GC-MS/FID	Cheong et al. (2013)
Organophosphorus flame retardants	Fish	1	Acid-washed silica gel	Acetonitrile, methanol, acetone; *water/acetonitrile (90:10, v/v)*	105–200, *150*	10.34	*1–5*	1-10 *5 min*	GC-FPD	Gao et al. (2014)
Phenolic compounds, antioxidants, anthocyanins	Blueberry residues	20 g fresh, 5 g freeze-dried		100% ethanol, 50% ethanol and 50% water, 100% acidified water (pH 2.0), 50% acidified water (pH 2.0), 50% ethanol, 100% acetone	25–180, *40*	0.5–40, *20*		15 min, in duplicates	UPLC-QTOF-MS, UPLC-UV-vis	Paes et al. (2014)
Caffeine catechins	Green coffee beans, green tea leaves	1	Sea sand	Ethyl lactate, water, ethyl lactate/water (75:25, 50:50, 25:75, 0:100)	50; 100; 125; 150; 200	Atmospheric pressure	2	20	HPLC-DAD triple quadrupole mass spectrometer with ESI	Bermejo et al. (2015)
Aflatoxins ochratoxin A	Dried fruit	1	Diatomaceous earth	30% MeOH	110	10.34	3	5 min (static time)	UHPLC-MS/MS	Campone et al. (2015)
Phenolic compounds, anthocyanins	Blackberry residues	5	Glass wool	Pure water, pure ethanol, *water/ethanol (50% v/v)*, acidified water (pH = 2.5, adjusted by the direct addition of citric acid)	60, 80, *100*	7.5		30	UHPLC-QToF-MS, UHPLC-UV-Vis	Machado et al. (2015)

(Continued)

Table 3.10: Pressurized liquid extraction (PLE). (*cont.*)

Analytes/ Compounds	Sample/ Matrix	Sample Weight (g)	Solid Sorbent/ Dispersant/ Dispersion Matrices	Solvent	Temperature (°C)	Pressure (MPa)	Cycles	Extraction Time (min)	Target Components Analysis	References
Antioxidant compounds	Mango leaves	15		ethanol (0, 50, 100%)	40, 60, 80, 100	4, 12, 20		3 h	HPLC-ESI-MS	Fernández-Ponce et al. (2015)
Carotenoids, phenolic compounds	Spinach	5		Hexane, ethanol, water, *50:50 ethanol/water*	60, *80*, *100*	7.5		30	HPLC, colorimetric method	Jaime et al. (2015)
Phenolics, anthocyanins	Blackberry residues	5		Pure water, pure ethanol, water/ethanol (50% v/v), acidified water (pH = 2.5, by addition of citric acid)	60, 80, 100	7.5		30	UHPLC	Machado et al. (2015)
Polycyclic aromatic hydrocarbons (PAHs)	Roasted coffee	10	Activated silica gel	Hexane/dichloromethane (85:15, v/v)	100	10.34	2	5	GC-MS	Pissinatti et al. (2015)
19 congeners of PCBs	Chicken meat, clam meat, pork	1	Copper(II) isonicotinate	Hexane	120	10.34	1	15	GC-MS	Jiao et al. (2016)
carbohydrates (inositols, inulin)	Artichoke	0.3	Sand	Milli-Q water	40, *80*, 120	10	*1*; 2; 3	3, *17*, 30	GC-FID	Ruiz-Aceituno et al. (2016)
Phenolic compounds	Rice (*Oryza sativa*) grains	2.5	Sea sand	*60*, 70, 80, 90% EtOAc/ MeOH	100–200, *190*	10.31–20.27	1, 2, *3*	Static extraction cycle (5–*10* min)	HPLC	Setyaningsih et al. (2016b)

Table 3.11: Ultrasound-assisted extraction (UAE).

Analytes/ Compounds	Sample/ Matrix	Sample Weight (g)	Solvents	Ultrasonic Power	Extraction Time (min)	Temperature (°C)	Frequency	Target Components Analysis	References
Polyphenols	Orange (*Citrus sinensis* L.) peel		80% (v/v) ethanol:water	150 W	30	40	25 kHz	HPLC UV–Visible	Khan et al. (2010)
Melanin	*Auricularia auricula* fruit bodies	2 g		100–*250* W	20, 30, 40, *36*	45, 60, 75, *63*	40 kHz		Zou et al. (2010)
Antioxidants	Pomegranate (*Punica granatum*) peel	0.2 g	70% ethanol	140 W	30	60	35 kHz	Folin–Ciocalteu (FC) reagent assay	Tabaraki et al. (2012)
β-glucans	Barley	20	Water	400 W	3–10	55	24 kHz	HPLC-SEC	Benito-Román et al. (2013)
Capsaicinoids, capsinoids	Biquinho pepper	5	Hexane, ethanol, acetone, methanol	360 W	10		20 kHz	UPLC-DAD	de Aguiar et al. (2014)
Vanillin	Vanilla beans	4	70% (v/v) ethanol/ water		15–90	25	40 kHz	HPLC	Dong et al. (2014)
Mangiferin and lupeol	Mango peels (*Mangifera indica* L.)	10 g	Ethanol/water (8:2 v/v)–mangiferin, hexane–lupeol		30	25	42 kHz	HPLC	Ruiz-Montañez et al. (2014)
Capsaicinoids	Chili pepper	1	Methanol/acetone (0–100%), *100% methanol*		0–20, *10*	Room temperature (24 ± 1°C)	40 kHz	UHPLC-DAD-MS/MS	Sganzerla et al. (2014)
Flavonoids, terpene lactones, ginkgolic acids, phenylpropanols	Ginkgo seeds	0.5	Aq. *methanol* (50%, 60%, *70%*, 80%, 90% and 100%; methanol, v/v) and aq. ethanol (50%, 60%, 70%, 80%, 90% and 100% ethanol, v/v)	80, 100, 120, 140, 160, 180, 200 W	20, 30, 45, 60, 75, 90		25 kHz	UAE–UHPLC–TQ/MS^2	Zhou et al. (2014)

(*Continued*)

Table 3.11: Ultrasound-assisted extraction (UAE). (*cont.*)

Analytes/ Compounds	Sample/ Matrix	Sample Weight (g)	Solvents	Ultrasonic Power	Extraction Time (min)	Temperature (°C)	Frequency	Target Components Analysis	References
Phenolic compounds (phenols, antioxidant capacity, chlorogenic acid, caffeic acid, catechin and epicatechin)	Carrot pomace	1 g	Ethanol concentration (13–97%)	750 W	3–37	10–60	20 kHz	HPLC	Jabbar et al. (2015)
Lycopene	Papaya (*Carica papaya*) waste	2	Ethanol/ethyl acetate (20, 30, 40, 50, 60, 70%, v/v)	Electric power of 600 W, heating power of 800 W	15, 20, 25, 30, 35, 40	20, 30, 40, 50, 60, 70	40 kHz	HPLC	Li et al. (2015)
Anthocyanins	*Purple Majesty* potato	5 g	Ethanol/water (50:50 and *70:30* v/v)	23, 28, 35 W	*5*, 30, 60, 120	33	20 kHz in the continuous mode at 30%, 50% and 70% amplitude		Mane et al. (2015)
Polyphenols	*C. sinensis* peels	1	75.79% acetone		8.33	27	20 kHz	HPLC-DAD	Nayak et al. (2015)
Phenolic compounds (phenolic acids, flavonoids)	Mandarin and lime peels	1 g	Methanol, ethanol, *acetone* 20, 50, *80%* v/v	50.93 W	30	40	38.5 kHz	HPLC	Singanusong et al. (2015)
flavonoids	Red grape skin	250 mg	Acetonitrile/water/ formic acid (26:73:1, v/v/v)	320W	3-90 *15*	50	35 kHz	HPLC	Tomaz et al. (2015)
Taurine	Red algae *Porphyra yezoensis*	10	water	100, 200, *300* W	15, 30, 45, *38.3 min*	20, 40, 60, *40.5*		HPLC	Wang et al. (2015)

Anthocyanins, phenolics, flavonoids	Purple sweet potatoes	10g 0.1% (v/v) HCl	80%, *90%*, 100% (v/v) ethanol	*200*, 240, 280 W	*45*, 60, 75	40, *50*, 60		LC-MS	Cai et al. (2016)
Total monomeric anthocyanin (TMA) and total phenolic content (TPC)	Eggplant peel	3	Methanol and 2-propanol: 50% (v/v), 70% (v/v), 90% (v/v) using water		15, 30, 45	50, 60, 70	12.5; 25; 37.5 kHz	Folin–Ciocalteu (FC) method, RSM	Dranca and Oroian (2016)
Resveratrol	Cookies, jams	1	10%–70% and 30%–90% methanol in water		5–30			UPLC-FD	Guamán-Balcázar et al. (2016)
Pectin	Tomato waste		Ammonium oxalate/ oxalic acid		*15*, 30, 45, 60, 90	60, 80	37 kHz	NMR and FTIR spectroscopy	Grassino et al. (2016)
Total anthocyanins (TA) and phenolics (TP)	Blueberry wine pomace		Ethanol (70%, v/v), hydrochloric acid (0.01%, v/v)	400 W	15–35, *30*	50–70, *61.03*		UPLC-DAD-MS/MS	He et al. (2016)
Antioxidant compounds	Iranian basil (*Ocimum basilicum*)	0.2 g	0%–100% methanol/ water *65.2% (v/v)*		20–70, *20*	20–60, *59*	37 kHz		Izadiyan and Hem-mateenejad (2016)
Natural pigment (betacyanin, betaxanthin)	Waste red beet stalks	10 g		60–120, *89 W*	35	40–60, *53*			Maran and Priya (2016)
Melatonin	Red rice (*O. sativa*) grains	2	0–*50%* methanol in water, pH 3–7 *(3.5)*	200 W	5–30, *10*	10–70, *18.5*	24 kHz	UPLC-FD	Setyaningsih et al. (2016a)
Polysaccharide	Hazelnut skin	5	Ethanol	400 W	15, 30, 45, 60, 90, 120	15	24 kHz	FTIR	Yılmaz and Tavman (2016)

Table 3.12: Microwave-assisted extraction (MAE).

Analytes/ Compounds	Sample/ Matrix	Sample Weight (g)	Temperature (°C)	Solvent	Time (min)	Microwave Power	Microwave Frequency	Target Components Analysis	References
Total phenolic content	Longan (*Dimocarpus Longan*) peel	5	80	Ethanol (95%)	30	500 W	2450 MHz	Folin–Ciocalteu reagent	Pan et al. (2008)
Phenolics	Bean		25, 50, 100, 150	Water, 50% ethanol/water, 100% ethanol	15		Stirring rate of 320 rpm	Colorimetric method	Sutivisedsak et al. (2010)
Polychlorinated biphenyls, polybrominated diphenyl ethers	Fish	2	115	*N*-hexane/ acetone (1:1, v/v)	10 min, held for 15 min, cooled down in 20 min	1200 W		HRGC/HRMS	Wang et al. (2010)
total phenolics, flavonoids	Broccoli	300 mg	50–90, P: 71.51, F: 68.35	Methanol/water, 50%–90%, v/v; P: 72.06% methanol; F: 80% methanol	1–27, P: 16.94, F: 15.93	100–200 W, P:159.33, F: 175		P: Folin–Cicalteu method F: Spectrophotometric method	Jokić et al. (2012)
Melatonin	Rice grains	2.5	125–175, *195*	10%–90% EtOAc/MeOH, *100% MeOH*	5–15, *20*	500–*1000* W		HPLC-FD	Setyaningsih et al. (2012)
β-carotene carotenoids	Carrots	2	58	50% (v/v) hexane, 25% (v/v) acetone, 25% (v/v) ethanol	1, *3*, 5	100, 180, 300	2450 MHz	HPLC	Hiranvarachat et al. (2013)
Polyphenols	Ginger (*Z. officinale*)	10	50–70, 60	Ethanol, methanol, aqueous ethanol *50%*, 80%	0–5, 3	*100*, 200, 500 W	50 Hz	HPLC	Kubra et al. (2013)
Anthocyanin	Black currant marc	28	80, 69.7	Aqueous hydrochloric solution	*10*, 20, 30	140, *700* W	2450 MHz	HPLC	Pap et al. (2013)
Sulforaphane	Cabbages	5		Dichloromethane, water	1, 2, 3, 4, 5	130, 260, 390 W		HPLC	Tanongkankit et al. (2013)
Polyphenols	Mulberry fruits (*Morus alba*)	2		20%–60%, *40%* ethanol	1–9, *8*	120–240 W, *210* W	2450 MHz	HPLC	Teng and Lee (2013)

Vanillin	Vanilla beans	4	36.7–55	40, 70, 100% ethanol/water	3, 6, 9, 12, 15, 18	50, 150, 250 W		HPLC	Dong et al. (2014)
Mangiferin, lupeol	Mango peels (*M. indica*)	10	25	Ethanol/water (8:2 v/v)–mangiferin, hexane–lupeol	1 min in 30 s irradiation cycles and 10 min of cooling	600 W		HPLC	Ruiz-Montañez et al. (2014)
Polycyclic aromatic hydrocarbons (PAHs)	Grilled meat	1	300	50:50 ratio of KOH (2 mol/L), ethanol	1.5	1800 W	500 MHz	GC–MS	Kamankesh et al. (2015)
Phenolic compounds	Licorice (*G. glabra*) root			*Ethanol 80%*, methanol 80%, water	2–6, *5–6*			Folin-Ciocalteu assay	Karami et al. (2015)
Lipids	Meat (chicken leg, chicken thigh, fresh ham, pork loin, beef hump)	230–670 mg, *300*	30–60, *54*	Ethyl acetate/methanol (2:1, v/v)	2–8, *16*	400 W		GC-FID	Medina et al. (2015)
Polyphenols	*C. sinensis* peels	1	80	Acetone/water (20%–80%, v/v)	90–240 s, *122*	300–600 W, *500*	2450 MHz	HPLC-DAD	Nayak et al. (2015)
Phenolic compounds	Rice grains	2.5	125–175, *185*	10%–90% EtOAc/MeOH, *100% MeOH*	5–15 *20*	500–*1000*		HPLC	Setyaningsih et al. (2015)
Polysaccharides	Mulberry leaves	6-24 *20*		Distilled water	5–15 *10*	50–250 W *170*		FT-IR	Thirugnanasambandham et al. (2015)
Heterocyclic aromatic amines	Hamburger patties	1		Sodium hydroxide (1 mol/L), ethanol, acetone 70:10:20	1.5		500 MHz	HPLC	Aeenehvand et al. (2016)
Pectic polysaccharide	Tangerine peel		52.2	Distilled water	41.8	704 W		HPLC, HP-GPC, FTIR	Chen et al. (2016)

(*Continued*)

Table 3.12: Microwave-assisted extraction (MAE). (*cont.*)

Analytes/ Compounds	Sample/ Matrix	Sample Weight (g)	Temperature (°C)	Solvent	Time (min)	Microwave Power	Microwave Frequency	Target Components Analysis	References
Hydrophilic (H) and lipophilic (L) antioxidants	Surplus tomato crop		60–180, *149.2 (H), 60 (L)*	0%–100%, *99.1%* ethanol/water (H); *33.0%* ethanol/water (L)	0–20; *2.25 (H); 15.4 (L)*	200 W	600 rpm		Pinela et al. (2016a)
Phenolic acids, flavonoids, antioxidant ingredients	Tomato		180	Ethanol 0%	20	200 W	600 rpm	HPLC	Pinela et al. (2016b)
Carbohydrates (inositols, inulin)	Artichoke	0.1–0.5	50-120	Ultrapure water	3-30	900 W		GC-FID	Ruiz-Aceituno et al. (2016)

Table 3.13: Solid–liquid extraction (SLE).

Analytes/ Compounds	Sample/ Matrix	Sample Weight (g)	Solvents	Mode of Extraction	Temperature (°C)	Extraction Time (min)	Target Components Analysis	References
Chemo-therapeutant residues	Salmon	2	Acetonitrile (0.1% acetic acid)	0.2 g NaCl, centrifugated 5 min at 3700 rpm, 0.2 g Bondesil-H2, vortex 2 min (800 rpm), centrifugated 3 min at 3700 rpm		3	LC-TOF-MS	Hernando et al. (2006)
Caffeine	Tea waste	50	Water, chloroform		20, 97	1, 3–6 h		Senol and Aydin (2006)
Avermectin residues	Salmon muscle, pepper	3	Acetonitrile (0.1% acetic acid)	0.2 g NaCl, vortex 2 min, ultrasonication 5 min, centrifugation 5 min at 3700 rpm, for 5 min			LC–MS/MS	Hernando et al. (2007)
Pesticides	Tomato	4	8.0 mL ACN, 0.5 mL water, 1.5 mL ethyl acetate	Agitated, chilled in a freezer −20°C 6 h, filtered through layer containing 1.5 g of anhydrous sodium sulfate			GC–MS	de Pinho et al. (2010)
Antioxidant	Rosemary, oregano, marjoram	0.5	80% MeOH	Shaken at 1500 rpm in room temperature, centrifuged 15 min at 2000g	Room temperature (~23°C)		HPLC	Hossain et al. (2011)
Flavonoids	Onion by-products	5	50 mL 80% MeOH	Homogenized in ultrahomogenizer at 8000 rpm for 5 min			HPLC	Zill-e-Huma et al. (2011)
Mycotoxins	Barley	2	Acetonitrile/water/acetic acid (79:20:1, v/v/v)	Shaken 90 min, centrifuged 5 min, 11,000 rpm, 20°C		90 min	UHPLC-Orbitrap^{R} MS	Rubert et al. (2012b)
Fumonisins	Cereal-based products	25	Methanol/water (80:20, v/v)	Centrifuged 15 min at 2500g, extracted twice with 30 mL methanol/water (80:20, v/v)			HPLC-MS/MS	Rubert et al. (2013b)
Pesticides	Pineapple		Acetonitrile/water/ethyl acetate; (7.0 mL/1.5 mL/1.5 mL); (5.5 mL/3.0 mL/1.5 mL); (7.0 mL/3.0 mL/0.0 mL)	Stirred 10 min at 25°C, 200 oscillations/min, cooled at −20°C for 12 h				da Costa Morais et al. (2014)
Pesticides	Lettuce	4	ACN, *ACN/ethyl acetate (6.5:1.5, v/v)*	Stirred 10 min on a shaker table at 25°C, 200 rpm, centrifuged 1, 3, 5 min at 1200 g, frozen 3, 6, 12, 24 h, −20°C			GC/ECD	Costa et al. (2015)

(Continued)

Table 3.13: Solid–liquid extraction (SLE). (*cont.*)

Analytes/ Compounds	Sample/ Matrix	Sample Weight (g)	Solvents	Mode of Extraction	Temperature (°C)	Extraction Time (min)	Target Components Analysis	References
Steroidal alkaloids	Potato peels	5	Methanol	Shaken 60 min at room temperature (~23°C) at 1700 rpm, centrifuged 10 min at 3000 g		60	UPLC–MS/MS	Hossain et al. (2015)
Carotenoids, phenolic compounds	Spinach	1	Hexane, ethanol, water	After extraction supernatant filtered	50	24 h	HPLC, colorimetric method	Jaime et al. (2015)
Pesticides	Carrot	4	Acetonitrile	Homogenized in vortex 60 s at 2500 rpm, agitated at 200 rpm, centrifugated 10 min at 3000 rpm, stored in the freezer at −20°C			GC-MS	Araújo et al. (2016)

Table 3.14: Dispersive liquid–liquid microextraction (DLLME).

Analytes/Compounds	Sample/Matrix	Sample Weight	Type and Volume of Extractant	Type and Volume of Disperser	Salinity	pH	Time (min)	Target Components Analysis	References
Cholesterol	Food samples (milk, yolk, olive oil)	4 mL	35 μL carbon tetrachloride	0.8 mL ethanol		8.5	Shaken 1 min, centrifuged 5 min 5000 rpm	HPLC–UV	Daneshfar et al. (2009)
Cadmium, copper	Food samples (rice, defatted milk powder, tea)		100 μL CCl_4	500 μL methanol for Cd, ethanol for Cu		3.0 Cd, 9.2 Cu	Shaken manually, centrifuged 3 min 3000 rpm	UV–vis spectrophotometry	Wen et al. (2011)
Biogenic amines	Beer	5 mL	325 μL toluene	1 mL acetonitrile			Shaken by hand 5 min, centrifuged 2 min 5000 rpm	GC-MS	Almeida et al. (2012)
Ochratoxin A	Wine	5 mL	660 μL chloroform	940 μL acetonitrile	0.25 g NaCl (5%; w/v)		Shaken, centrifuged 1 min 5000 rpm	Capillary-HPLC-LIF	Arroyo-Manzanares et al. (2012)
Pesticides	Honey	0.5 g	100 μL chloroform	450 μL acetone			Shaken 5 s, centrifuged 5 min 4000 rpm	GC–MS	Kujawski et al. (2012)
Pesticides	Honey	5 mL	50 μL chloroform	750 μL acetonitrile		4.4 no pH	Shaken by hand 1 min, centrifuged 3 min 2500 rpm	GC-MS	Zacharis et al. (2012)
Insecticides	Cucumber	5.0 mL aliquot	100 μL chloroform	0.8 mL acetonitrile			Vortexing 1 min, centrifuged 5 min 3500 rpm	MEKC	Zhang et al. (2012)
Nonsteroidal antiinflammatory drugs	Milk, dairy products		150 μL chloroform	2.0 mL acetonitrile			Vortex-mixed 1 min, centrifuged 3 min 5000 rpm	FASS-CE	Alshana et al. (2013)
Deoxynivalenol (DON)	Wheat flour		0.25 mL chloroform	0.8 mL acetonitrile		11	Vortex-mixed a few seconds, centrifuged 3 min 3500 rpm		Karami-Osboo et al. (2013)
Copper	Cereals, vegetable food (maize, millet, rice, wheat, gram, lentils, kidney beans, green beans)	1 mL	0.2 mL chloroform containing PBITU	0.5 mL absolute ethanol	1.0% NaCl	6.0	Shaken a few minutes, centrifuged 2 min 754.6 g	FAAS	Shrivas and Jaiswal (2013)

(*Continued*)

Table 3.14: Dispersive liquid–liquid microextraction (DLLME). (*cont.*)

Analytes/ Compounds	Sample/Matrix	Sample Weight	Type and Volume of Extractant	Type and Volume of Disperser	Salinity	pH	Time (min)	Target Components Analysis	References
Patulin	Apple juices	5 mL	1000 μL chloroform	1000 μL 2-propanol			Shaken by hand, centrifuged 5 min 5000 rpm	MEKC	Víctor-Ortega et al. (2013)
Polycyclic aromatic hydrocarbons (PAHs)	Grilled meat	10 mL	Chloroform, ethylene tetrachloride, carbon tetrachloride; *80 μL ethylene tetrachloride*	300 μL acetone	15% NaCl		Shaken for 1 min, centrifuged 5 min 2683.2g	GC–MS	Kamankesh et al. (2015)
Cobalt, cadmium	Food (lettuce, chickpea, apple, rice, lentil fish)	0.3–1.0 g	0.5 mL chloroform	0.5 mL ethanol		5.4	Centrifuged 5 min 3800 rpm	FAAS	Bosch Ojeda and Sánchez Rojas (2014)
Phenolic compounds (phenolic acids, flavonoids)	Honey	10 g (10 mL diluted honey sample)	450 μL chloroform	750 μL Me_2CO		2	Shaken by hand 10 s, centrifuged 5 min 6000 rpm	HPLC–UV HPLC–HRMS	Campone et al. (2014)
Ochratoxin A	Rice wines	5 mL	100 μL [HMIM] [PF6]	0.1 mL ethanol		3.0	Vortex-mixed 2 min 2800 rpm, centrifuged 5 min 3000 rpm	HPLC-FLD	Lai et al. (2014)
Caffeine	Teas, coffees, beverages	1 mL	20 μL chloroform	200 μL ethanol	10% NaCl (w/v)		Centrifuged 2 min 4000 rpm	GC–NPD	Sereshti and Samadi (2014)
Ochratoxin A	Raisin		*$CHCl_3$*, CCl_4, CS_2 0.2 mL chloroform	0.8 mL 80% methanol			Centrifuged 5 min 1,132 g	HPLC-FLD	Karami-Osboo et al. (2015)
Iron	Foods (bottled mineral water, milk, banana, potato, carrot, soybean)	5 mL	100 μL carbon tetrachloride	200 μL ethanol		5.5	Shaken 3 min, centrifuged 3 min 3500 rpm	Microvolume UV–vis spectrophotometry	Peng et al. (2015)
Heterocyclic aromatic amines	Hamburger patties	10 mL	100 μL 1-octanol	600 μL methanol	1 g NaCl	11	Centrifuged 5 min 4000 rpm	HPLC	Aeenehvand et al. (2016)
Triazole fungicides	Fruit juices (peach, apple, orange)	5 mL	70 μL [C6MIM] [PF6]	0.25 mL acetonitrile	4% NaCl (w/v)		Vortexed 2 min, centrifuged 5 min 4000 rpm	HPLC	Zhang et al. (2016)

Table 3.15: Maceration (MAC).

Analytes/ Compounds	Sample/Matrix	Sample Weight (g)	Solvents	Extraction Time	Temperature (°C)	Target Components Analysis	References
Natural antioxidants	Hazelnut skin, shell by-products		80% (v/v) aqueous methanol, ethanol, acetone; *80% aqueous acetone*	20 h	Room temperature (20°–22°C)	Folin–Ciocalteau method, UV spectra	Contini et al. (2008)
Vanillin	Vanilla beans	4	70% (v/v) ethanol/ water	120–720 min	25	HPLC	Dong et al. (2014)
Phenolics	Cumin (*Cuminum cyminum*) seeds	2.5	80% aqueous acetone (v/v)	30 min	Room temperature	RP-HPLC	Rebey et al. (2014)
Mangiferin, lupeol	Mango peels (*M. indica*)	10	Ethanol/water (8:2 v/v)–mangiferin, hexane–lupeol	24 h	25	HPLC	Ruiz-Montañez et al. (2014)
Phenolic compounds, anthocyanins	Blackberry residues	5	Acidified methanol (0.01% v/v HCl)	24 h	Room temperature	UHPLC-QToF-MS UHPLC-UV-Vis	Machado et al. (2015)
Polyphenols	Dried chokeberry		25 mL distilled water, *50%*, 70%, 96% ethanol/water	15, 30, *60*, and *90 min*, 2, 2.5, 5, 10, and 18 h	Ambient temperature	HPLC	Ćujić et al. (2016)

Table 3.16: Soxhlet extraction (SOX).

Analytes/Compounds	Sample/Matrix	Sample Weight (g)	Solvents	Extraction Time (h)	Temperature (°C)	Target Components Analysis	References
Isoflavones	Soy beans	3	Dimethyl sulfoxide:acetonitrile:water (5:58:37, v/v/v)	3	Hot boiling	HPLC	Luthria et al. (2007)
Lycopene	Tomato pomace	2	Chloroform	7		HPLC	Huang et al. (2008)
Total phenolic content	Longan (*D. Longan*) peel	5	Ethanol (95%)	2		Folin–Ciocalteu reagent	Pan et al. (2008)
Oleoresin	Turmeric leaves	2	*N*-hexane	16		GC/MS	Zaibunnisa et al. (2009)
Total phenol content (TPC)	Guava seeds (*P. guajava*)	30	Acetate (EtAc) and ethanol (EtOH)	10		Follin-Ciocalteu method	Castro-Vargas et al. (2010)
Polychlorinated biphenyls, polybrominated diphenyl ethers	Fish	2	*N*-hexane/acetone (1:1, v/v)	24		HRGC/HRMS	Wang et al. (2010)
Favonoids	Spearmint (*M. spicata*) leaves	3	Methanol, ethanol (99.5%), ethanol:water (70:30), petroleum ether	6		HPLC	Bimakr et al. (2011)
Capsaicinoids	*Ca. frutescens* peppers	5	Hexane	6		U-HPLC- DAD-MS/MS	de Aguiar et al. (2013)
β-carotene carotenoids	Carrots	2	50% (v/v) hexane, 25% (v/v) acetone, 25% (v/v) ethanol	6	58	HPLC	Hiranvarachat et al. (2013)
Fatty acids	Pomegranate seed oil	5	*N*-hexane	3	60	GC–MC	Tian et al. (2013)
Phenolic compounds, antioxidants, anthocyanins	Blueberry residues	5	Methanol, ethanol, acetone	6		UPLC-QTOF-MS, UPLC–UV–vis	Paes et al. (2014)
Capsaicinoids, capsinoids	Biquinho pepper	5	Hexane, ethanol, acetone, methanol	6		UPLC-DAD	de Aguiar et al. (2014)
Phenolics	Cumin (*C. cyminum*) seeds	20	80% aqueous acetone (v/v)	6	85	RP-HPLC	Rebey et al. (2014)
Mangiferin, lupeol	Mango peels (*M. indica*)	10	Ethanol–water (8:2 v/v)–mangiferin; hexane–lupeol	8		HPLC	Ruiz-Montañez et al. (2014)
Glucosinolates, phenols, lipids	Rocket salad (*E. sativa*)		Methanol:water 7:3	18	100	HPLC-MS	Solana et al. (2014)
Capsaicinoids	Chili pepper	0.1	Methanol	1-60		HPLC-APCI-MS	Bajer et al. (2015)

Phenolic compounds, anthocyanins	Blackberry residues	5	Ethanol, methanol	5	80	UHPLC-QToF-MS e UHPLC-UV-Vis	Machado et al. (2015)
Phenolic compounds	Mango leaves	40	Water, ethanol, mixture ethanol/water (50:50)	12		HPLC-ESI-MS	Fernández-Ponce et al. (2015)
Phenolic compounds	Licorice (*G. glabra*) root	25	*Ethanol 80%, m*ethanol 80% or water	2, 4, *6*		Folin-Ciocalteu assay	Karami et al. (2015)
Lycopene	Papaya (*C. papaya*) waste	6	40% ethanol in ethyl acetate	4	95	HPLC	Li et al. (2015)
Phenolics, anthocyanins	Blackberry residues	5	Ethanol, methanol	5	80	UHPLC	Machado et al. (2015)
Triglycerides, polyphenols	Hazelnut, coffee, grape wastes	20, 16, 12	160 gN *n*-hexane or 170 g ethanol	18		GC	Manna et al. (2015)
Isoflavones	Soybeans	20	ethanol 80% in water (v/v)	9	70–80	HPLC	Nemitz et al. (2015)
Capsaicinoids	Malagueta pepper (*Ca. frutescens*)	5	hexane, dichloromethane, ethyl ether, ethyl acetate	6		HPLC-PDA	Santos et al. (2015)
Carotenoids	Kaki, peach, apricot	2	MeOH:THF, 1:1, v/v	6	66	LC–MS	Zaghdoudi et al. (2015)
Flavonoids	Orange fruit peel	1	80 mL petroleum ether	2–3		UPLC	Cao et al. (2016)
Fatty acid	Lentils	5	Acetone/hexane (1:4), CH_2Cl_2/hexane (1:4)	5		GC	Caprioli et al. (2016)
19 congeners of PCBs	Chicken meat, clam meat, pork	5	Hexane	8 h with a reflux cycle time of approximately 10 min		GC-MS	Jiao et al. (2016)
Antioxidants, fatty acids, sterols	Fig achenes' oil	4	*N*-hexane and methanol/water (60:40, v/v)	5		GC-FID, colorimetric assay	Soltana et al. (2016)

Table 3.17: Accelerated solvent extraction (ASE).

Analytes/Compounds	Sample/Matrix	Sample Weight	Solvent	Temperature [°C]	Pressure (MPa)	Time (min)	Cycles	Target Components Analysis	References
Genistin, daidzin	Soybean food	0.2 g (mixed with flavone as IS)	90% (v/v) aqueous methanol	145	15 kPa	5 min	2	HPLC/UV/MS	Klejdus et al. (2004)
Lipids, extractable organochlorine	Fish	5–15 g	Hexane/acetone (3:1)	50, *55*, 75, *100*	10.3	50 min	2		Zhuang et al. (2004)
Polychlorinated biphenyls, polybrominated diphenyl ethers	Fish	2 g (mixed with anhydrous sodium sulfate powder)	*N*-hexane/acetone (1:1, v/v)	150	10.3	7 min heating, 8 min static state	3	HRGC/HRMS	Wang et al. (2010)
Amitraz and its metabolite	Liver, kidney of swine, bovine, sheep	5 g	*N*-hexane/methanol 1:9 (v/v)	60	12	2 min	3	GC-ECD	Yu et al. (2010)
Antioxidant	Rosemary, oregano, marjoram	0.5 g (mixed with diatomaceous earth)	32%–88% methanol, *56% and 57%*	66–*129*	10.3	5 min		HPLC	Hossain et al. (2011)
Malachite green, Gentian violet, Leucomalachite green, Leucogentian violet	Shrimp, salmon	2 g (6.0 g basic alumina)	McIlvaine buffer (pH 3)/acetonitrile/hexane (2/10/2, v/v)	60	10.3	5 min	1	LC-ESI-MS/MS	Tao et al. (2011)
Pesticides	Foods of animal origin (pork, beef, chicken, fish)	10 g (mixed with 5 g Extrelut 20)	*Acetonitrile*, hexane–acetone (2:1, v/v), cyclohexane/ethyl acetate (1:1, v/v)	80	10.3	5 min heating, 5 min static time	2	GC–MS	Wu et al. (2011)
Tetracyclines	Muscle and liver of porcine, chicken, bovine	2 g	Trichloracetic acid/acetonitrile (1:2) adjusted to pH 4.0	40–80, *60*	4.5–8.5, *6.5*	5 min static time	2	HPLC–UV	Yu et al. (2011a)
Antibiotics, avermectins	Meat (muscle, kidney, liver of swine, bovine)	2 g (with macrolide antibiotics and avermectins)	Acetonitrile/methanol (1/1, v/v)	60	10.3	10 min	2	LC–MS/MS	Tao et al. (2012)
Fluoroquinolones	Food of animal origin (muscle, liver, kidney of swine, bovine, chicken, fish)	5 g (mixed with diatomaceous earth)	Acetonitrile	65	7	2 min static time	2	HPLC-UV, LC–MS/MS	Yu et al. (2012)

Caffeine	Green coffee beans, green tea leaves	1 g (tea), 3 g (coffee)	*Ethyl lactate*, ethanol, ethyl acetate	100, 150, *200*	10	10 min and 20 min static time		HPLC	Bermejo et al. (2013)
Lipids	*Amaranthus* spp. seeds	10 g (mixed with diatomaceous earth)	Hexane	40–110, *90*	10	5–25 min *17*	3	HPLC, GC	Kraujalis et al. (2013)
Sulfonamides	Aquatic products	10 g (mixed with diatomite)	Acetonitrile	70	10.3	5 min	2	CE	Sun et al. (2013)
Polychlorinated biphenyls (PCBs)	Fish (sea bass, black sea bass, sea trout, sole)	1 g (mixed with diatomaceous earth)	Acetone/*N*-hexane (1:1, v/v)	100	10	5 min heating, 5 min static time	1	GC–MS	Ottonello et al. (2014)
Pentachlorophenol	Meat, fish	5 g	Methanol/2% trichloroacetic acid (3/1, v/v)	110	10.3	10 min	1	GC–ECD, GC–MS	Zhao (2014)
Pharmaceuticals	Vegetables (celery, lettuce)	0.5 g	Acetonitrile/ methanol/water 72:8:20 (v/v/v)	80	10.3	5 min static time	2	LC-MS/MS	Chuang et al. (2015)
Antibiotics	Food samples (pork, chicken meat, clam meat)	1 g (mixed with diatomaceous earth)	Methanol	70	10.3	15 min	1	HPLC	Jiao et al. (2015)
Polyphenols	*C. sinensis* peels	1 g (mixed with diatomaceous earth)	50% acetone	120	10.3	6 min heating, 5 min	3 static cycles	HPLC-DAD	Nayak et al. (2015)
Carotenoids	Orange carrot	25 g	Acetone/Hexane (3:5 v/v); ethanol/ hexane (4:3 v/v), *ethanol:hexane/acetone (2:3:1 v/v/v)*	40, 50, *60*	10.3	5, 10, *15 min* static time	3	LC–ESI MS	Saha et al. (2015)
Carotenoids	Kaki, peach, apricot	2 g	20:80 (v:v) methanol/ tetrahydrofuran	40	10.3	5 min	5 static cycles	LC–MS	Zaghdoudi et al. (2015)
Anthocyanins, phenolics Flavonoids	Purple sweet potatoes	10g	80% (v/v) aqueous ethanol containing 0.1% (v/v) HCl	90		15 min	2 static cycles	LC-MS	Cai et al. (2016)
Lutein	Paprika leaves	1	60%, 70%, 80%, 90%, 100% ethanol/water; *79.63% ethanol*	60, 90, 120, 150, 180 *93.26*	10	1, 2, 3, 4, *5 min* static time		UPLC	Kang et al. (2016)

Table 3.18: Pressurized hot water extraction (PHWE).

Analytes/ Compounds	Sample/ Matrix	Sample	Extraction Cell	Extraction Solvents	Pressure [MPa]	Temperature (°C)	Mode	Extraction Time (min)	Flow Rate	Target Components Analysis	References
Carbamate insecticides	Bovine milk	3 mL	8.1 cm × 8.3 mm i.d. stainless steel column used as extraction cell	Water		90	Dynamic	5	1 mL/min	LC-MS	Bogialli et al. (2004)
Anthocyanins	Red cabbage	1–3 g *2.5 g*	11 mL extraction cells, cellulose paper filter at bottom	Water/ ethanol/ formic acid (94/5/1, v/v/v)	5	80–120, *99*	Static	6–11, *7*		HPLC/DAD	Arapitsas and Turner (2008)
Stevioside, rebaudioside A	*Stevia rebaudiana*	0.2 g	Stainless steel extraction cells (10 mm id 22 × 250 mm)	Water	1.1–1.3	60, 80, 100, 120	Dynamic	50	1.5 mL/min	HPLC	Teo et al. (2009)
Antioxidants, polyphenols	Grape skin	0.5, 1.0, 1.5 g	11 mL extraction cell containing inert material	Purified water	15	40–120, *80*	Static	3 × 5 min		HPLC	St’avíková et al. (2011)
Antioxidants	Onion	2 g	10 mL extraction cells	Water	8	120	Static	Total extraction time 60 min	0.3 mL/min	HPLC-UV	Andersson et al. (2012)
Isoxanthohumol	Hop (*Humulus lupulus*)	1 g	11 mL extraction cells, 2 g of sea sand	Water	10.68	50, 100, *150*, 200	Static	30 min (6 cycles of 5 min each)		HPLC-DAD	Gil-Ramírez et al. (2012)
Polyphenols	Red onions	2 g	Stainless steel extraction cell	80 mL solution of water, ethanol, formic acid (94:5:1, v/v/v)	1.5	110	Static	120 min	2, 3, *4* mL/ min	HPLC-DAD-MS	Liu et al. (2014)
Capsaicinoids	Chili pepper	0.1 g	11 mL extraction cartridge, inert glass sand	Deionized water	20	120–240, *200*	Static	5–60 min *10–20 min*		HPLC-APCI-MS	Bajer et al. (2015)
Flavonols	*Moringa oleifera* leaf	0.5 g	Stainless steel extraction cell, 0.5 g diatomaceous earth	Deionized water		100	Dynamic	20 min	1.0 mL/min	HPLC-UV	Matshediso et al. (2015)
Flavonoids	*Momordica foetida*	1; 4 g	20 mL extraction cell (70 × 30 mm)	Ultrapure water	6.89	100, 150, 200, *250*, 300			5 mL/s	UHPLC-qTOF-MS	Khoza et al. (2016)

Table 3.19: High hydrostatic pressure extraction (HHP).

Analytes/ Compounds	Sample/ Matrix	Sample Weight (g)	Solvent	Temperature (°C)	Compressor Air Speed	Pressure Fluid	Pressure (MPa)	Time (min)	Target Components Analysis	References
Anthocyanins	Grape skins		20%–100% ethanol	20–70		Water–glycol (20:80, v/v)	200, 400, 600	30, 60, 90	HPLC-DAD/ ESI-MS	Corrales et al. (2009)
Mangiferin and lupeol	Mango peels (*M. indica* L.)	10	ethanol–water (8:2 v/v)–mangiferin, hexane–lupeol	25	125 PSI	Water:anticorrosive lubricant (5:1)	150	20	HPLC	Ruiz-Montañez et al. (2014)

Table 3.20: Conventional solvent extraction (CSE).

Analytes/Compounds	Sample/Matrix	Sample Weight (g)	Solvents	Temperature (°C)	Time (min)	Target Components Analysis	References
Dyes, low-molecular mass bioanalites	*Spirulina maxima*	0,2	Methanol, ethanol, 1-propanol, 2-propanol, acetonitrile, acetone, tetrahydrofuran, dichloromethane, toluene, *N*-hexane, water, 10 mM cyclodextrin solutions in water	Room temperature	120	Reversed-phase micro TLC	Zarzycki and Zarzycka (2008); Zarzycki et al. (2010, 2011)
Quercetin	Onion skin	1	59.3% ethanol	59.2	16.5	HPLC	Jin et al. (2011)
Chlorogenic acids	Green coffee beans		Water	50	5	UV spectrophotometric method	Upadhyay et al. (2012)
Polyphenols	Ginger (*Z. officinale*)	10	Ethanol, methanol and aqueous ethanol (50%, 80%)	60	10	HPLC	Kubra et al. (2013)
Anthocyanin	Black currant marc	100	Aqueous HCl at pH2, and citric acid solutions at pH2, and a solution containing 50 ppm SO 2 and 1% citric acid	80	300	HPLC	Pap et al. (2013)
Sulforaphane	Cabbages	5	Dichloromethane with sodium sulfate anhydrous	30	30	HPLC	Tanongkankit et al. (2013)
Polyphenols	*C. sinensis* peels	1	50 mL of 50% aqueous acetone (v/v)	60	2	HPLC-DAD	Nayak et al. (2015)
Anthocyanins, phenolics, flavonoids	Purple sweet potatoes	10	70, *80*, 90% (v/v) ethanol	*60*, 70, 80	*90*, 120, 150	LC-MS	Cai et al. (2016)
Total anthocyanins (TA) and phenolics (TP)	Blueberry wine pomace		Ethanol (70%, v/v) and hydrochloric acid (0.01%, v/v)	61	35	UPLC–DAD–MS/MS	He et al. (2016)
Pectin	Tomato waste		Ammonium oxalate/oxalic acid	60, 80	24 h in the first step and 12 h in the second step	NMR and FTIR spectroscopy	Grassino et al. (2016)

Table 3.21: Matrix solid-phase dispersion (MSPD).

Analytes/ Compounds	Sample/ Matrix	Sample Weight	Sorbent Material	Column	Samples Preparation	Target Components Analysis	References
Pesticides	Bovine samples	0.5 g	2 g C_{18}	6 mL polypropylene filtration tube with polyethylene frit in the bottom	Elution: 20 mL deionized water, 3 mL acetonitrile/water (25:75 v/v), 1 mL acetonitrile/water (60:40 v/v), 5 mL acetonitrile at flow rate of 1 mL/min; cleanup of acetonitrile eluate (fraction d) was performed using 0.5 g silica gel column (previously washed with 15 mL ACN); evaporated in rotary evaporator (40°C, 200 mbar); residue redissolved in 250 μL acetonitrile	HPLC-DAD GC-MS	García de Llasera and Reyes-Reyes (2009)
Patulin	Apple products	0.5 g	2 g C_{18}	10 mL empty cartridge constructed from syringe barrel, containing 0.4 g sodium sulfate anhydrous	Elution: 3×3 mL dichloromethane, column packing dried with a strong stream of air, flow of each portion stopped for 1 min, combined solutions added to one drop of acetic acid glacial; extracts evaporated to dryness in water bath at 40°C under a gentle stream of nitrogen; residue dissolved in 0.5 mL acetic acid buffer solution	HPLC	Wu et al. (2009)
Pesticides	Olives	1 g	2 g aminopropyl sorbent (Bondesil-NH_2)	12 mL SPE cartridge containing 2 g florisil	Elution: 2×5 mL acetonitrile, first aliquot used to backwash both mortar and pestle; extracts evaporated until near dryness; residue dissolved in 1:1 acetonitrile/water, reaching final volume of 1 mL	LC–MS/MS	Gilbert-López et al. (2010)
Fungicides	Inseng extract	5 mL (1 g ginseng extract dissolved in 10 mL water)	10 g Florisil, 200 mg anhydrous Na_2SO_4	Small chromatographic column which Whatman No. 1 filters	Extracted twice with 10 mL ethyl acetate/hexane (70:30, v/v) 15 min in ultrasonic bath at room temperature; columns placed in vacuum manifold, eluting solution collected in 25 mL graduated glass tubes; concentrated to 1 mL with gentle stream of nitrogen	GC-ECD	Qi (2010)
Sudan dyes	Sauces, condiments	0.5 g	0.50 g sodium sulfate, 0.50 g Florisil, 0.50 g washed sea sand	6 mL solid phase extraction glass tube containing polyethylene frit and a filter paper disk at the bottom	Elution: 3 mL acetonitrile, elute dropwise by applying slight vacuum; extracts evaporated to dryness over 5 min under stream of argon (outlet pressure, 1–2 bar); residue reconstituted with 0.5 mL methanol/acetonitrile/water (65/20/15, v/v/v)	HPLC–DAD	Enríquez-Gabeiras et al. (2012)

(Continued)

Table 3.21: Matrix solid-phase dispersion (MSPD). (*cont.*)

Analytes/ Compounds	Sample/ Matrix	Sample Weight	Sorbent Material	Column	Samples Preparation	Target Components Analysis	References
Antibiotics	Sheep's milk	0.5 mL	2 g washed sea sand, 2 g Na_2SO_4	Solid-phase extraction column with plug of silanized glass wool at the bottom	Eluted with 5 mL 50:50 (v:v) methanol/ethyl acetate, dropwise by gravity, in five 1 min static extraction steps, flow 1 mL/min; extracts evaporated to dryness under a stream of argon at room temperature; residue reconstituted in 0.5 mL of dilution mixture	LC–UV–DAD	García-Mayor et al. (2012)
Pesticides	Seaweeds	1 g	4 g anhydrous sodium sulfate	Polypropylene syringe containing polyethylene frit at the bottom, filled with 0.4 g GCB and 3.6 g of Florisil (as cleanup adsorbents)	Gravity flow with hexane/ethyl acetate, containing 40% of ethyl acetate; extract evaporated using a gentle stream of nitrogen; adjusted to 1 mL with ethyl acetate, filtered by syringe filter	GC-MS	García-Rodríguez et al. (2012)
Mycotoxins	Baby food	1 g	1 g C_{18}	Glass column (100 mm × 9 mm i.d.)	Mixture eluted with 15 mL acetonitrile/methanol (50:50) (v/v), 1 mM ammonium formate; extract evaporated under gentle nitrogen stream at 35°C; residue reconstituted to 1 mL with methanol/water (50:50) (v/v), filtered using nylon filter	LC–MS/MS	Rubert et al. (2012a)
Mycotoxins	Barley	1 g	1 g C_{18}	Glass column (100 mm × 9 mm i.d.)	Mixture eluted with 1 mM ammonium formate in 10 mL acetonitrile/methanol (50/50, v/v); filtered through a 22 mm nylon filter	UHPLC-Orbitrap[R] MS	Rubert et al. (2012b)
Pesticides	Fruits	1 g	1 g diatomaceous earth	6-mL SPE tube	Extraction: 10 mL dichloromethane at 0.5 mL/min; extract evaporated to dryness under gentle nitrogen stream; residue reconstituted with 0.4 mL methanol	LC–MS^2	Radišić et al. (2013)
Mycotoxins	Wheat grain	1 g	1 g C_{18}	Glass column (100 mm–9 mm i.d.)	Mixture eluted with15 mL acetonitrile/methanol (50:50) (v/v), 1 mM ammonium formate; extract evaporated to dryness under gentle nitrogen stream at 45°C; residue reconstituted to 1 mL with methanol:water (80:20) (v/v)	LC-ESI-MS/MS	Blesa et al. (2014)

Dechlorane compounds	Fish	1 g	2 g silica gel	Polypropylene SPE cartridge containing 1.0 g Florisil as cleanup cosorbent	20 mL *N*-hexane by gravity flow, slight vacuum; extracts evaporated under gentle stream of nitrogen to dryness; residue redissolved in 20 μL acetone containing 50 pg/μL of IS	GC–ECNICI-MS	Chen et al. (2014a)
Herbicides	Food crops (wheat, rice, soybea)	0.02 g	0.08 g CN-silica, 0.02 g anhydrous sodium sulfate	1-mL syringe with 0.05 g anhydrous sodium sulfate	3 mL *N*-hexane, 2.5 mL dichloromethane (containing 5 μL of acetic acid); extracts evaporated to dryness under gentle flow of nitrogen; residue reconstituted with 0.2 mL methanol	HPLC	Liang et al. (2014)
Fenicols	Shrimp, fish	1 g	2 g C_{18}	Glass column (300 × 15 mm i.d.) with cotton at the bottom	Extraction solvents mixture: ethyl acetate/ACN, (20/80, v/v); ethyl acetate/ACN (10/90, v/v); ethyl acetate/ACN/25% ammonium hydroxide (10/88/2, v/v/v); ethyl acetate/ACN/25% ammonium hydroxide (10/85/5, v/v/v); eluate dried on the nitrogen dryer at 50°C under a mild stream of nitrogen; residue reconstituted into 1.0 mL 5% MeOH in 0.1% formic acid-5 mmol/L ammonium acetate solution	LC–MS/MS	Tao et al. (2014)
Pesticides	Tea	0.5 g	100 μL IS, 0.75 g C_{18}, 0.75 g FLS	Column with glass filter paper at the bottom	20 mL acetonitrile (8 mL the first time, 10 mL for the second time, 2.0 mL additional eluent was adopted to wash the mortar and pestle, and then transferred into the column); extracts evaporated under gentle stream of nitrogen at 35°C; residue dissolved in 1.0 mL initial mobile phase, filtered through 0.22 μm PTFE filter	LC–MS/MS	Cao et al. (2015)
Chloramphenicol, thiamphenicol, florfenicol	Fish	2 g	3 g C_{18}	Glass column (300 × 15 mm i.d.) with two filter paper disks	10 mL acetonitrile/water (50:50, v/v), 6 mL ethyl acetate, stirred 1 min, centrifuged at 3000 × g for for 10 min, dried under stream of nitrogen at 50°C; residue reconstituted into 1 mL methanol/water (10:90, v/v)	UPLC-MS/MS	Pan et al. (2015)
Triazines	Mussel	0.5 g	2 g Envi-C_{18}	20-mL SPE cartridge containing a triple sorbent layer of 1.5 g Supelclean Envi-Carb-II/SAX/PSA (500/500/500 mg)	20 mL ethyl acetate, 5 mL ACN; extracts evaporated in rotary-evaporator, dried by gentle nitrogen stream; residue reconstituted in 1 mL methanol, filtered through syringe filter of PTFE	HPLC-DAD	Rodríguez-González et al. (2015)

(*Continued*)

Table 3.21: Matrix solid-phase dispersion (MSPD). (*cont.*)

Analytes/ Compounds	Sample/ Matrix	Sample Weight	Sorbent Material	Column	Samples Preparation	Target Components Analysis	References
Tocotrienols, tocopherols	Barley	100 mg	0.5 g activated neutral alumina, 0.1 g sea sand	Glass syringe barrel column with Whatman GF/A glass fiber filter fitted at the bottom	Elute dropwise; evaporated under gentle stream of nitrogen (45°C); redissolved in ~100 μL methanol, vortexed 30 s	RP-HPLC-FLD	Tsochatzis and Tzimou-Tsitouridou (2015)
Flavonoids	Orange fruit peel	25 mg	25 mg SBA-15	SPE cartridge containing sieve plates	500 μL organic solvent, centrifuged at 13,000 rpm for 5 min	UPLC	Cao et al. (2016)
β-lactam Antibiotics	Pork	2 g	3 g Oasis HLB	Glass column (150 × 15 mm i.d.) with two frit discs	8 mL acetonitrile/water (50:50, v/v) containing 0.1% formic acid into a tube, 6 mL ethyl acetate, stirred 1 min, centrifuged at 3000 × g for 10 min; combined supernatant dried on the nitrogen dryer at 50°C under mild stream of nitrogen; residue reconstituted into 2 mL acetonitrile/water (10:90, v/v)	UPLC-MS/MS	Huang et al. (2016)

Table 3.22: Quick, easy, cheap, effective, rugged, and safe protocols (QuEChERS).

Analytes/ Compounds	Sample/Matrix	Sample Weight	Extraction	Cleanup	Target Components Analysis	References
Pesticides	Olives	10 g	10 mL ACN, 4 g anhydrous $MgSO_4$, 1 g NaCl, shaken 1 min, centrifuged [3700 rpm, 1377 g (rcf)] for 1 min	250 mg PSA, 250 mg C_{18} sorbent, 250 mg GCB, 750 mg $MgSO_4$, manually shaken for 30 s, centrifuged [3700 rpm, 1377 g (rcf)] for 1 min, 1 mL of extract evaporated to near dryness, taken up with 500 μL with MeOH and 500 μL mQ H_2O	LC–MS/MS	Gilbert-López et al. (2010)
Pesticides	Sugarcane juice	10 mL	10 mL ACN, mixed 1 min, 4 g anhydrous $MgSO_4$, 1 g NaCl, mixed 1 min, centrifuged for 10 min at 1200 rpm	4 mL upper layer, 200 mg PSA, 600 mg anhydrous $MgSO_4$, mixed 30 s, centrifuged for 3 min at 3500 rpm, 2 mL upper layer transferred, evaporated in a water bath at 40°C under nitrogen flow until total dryness, extract diluted in 500 μL toluene	GC–ECD	Furlani et al. (2011)
Mycotoxins	Barley	2 g	10 mL of 0.1% CH_2O_2 in deionized water, mixed 3 min, waited 10 min, 10 mL ACN, shaken 3 min	4 g $MgSO_4$, 1 g of NaCl, shaken 3 min, centrifuged (5 min, 11.000 rpm, 20°C)	UHPLC-Orbitrap[R] MS	Rubert et al. (2012b)
Isoflavone	Legumes (chickpeas, lentils, white beans)	5–7 g	10 mL ACN:H_2O (70:30, v/v), shaken 5 min, 5 mL ACN, shaken 5 min	4 g $MgSO_4$, 1 g NaCl, shaken 1 min, centrifuged at 3000 rpm for 5 min	LC–MS/MS	Delgado-Zamarreño et al. (2012)
Ochratoxin A	Bread	5 g	15 mL ACN, added to QuEChERS (6 g $MgSO_4$, 1.5 g NaCl, 1.5 g $C_6H_5Na_3O_7 \times 2H_2O$, 0.750 g sodium citrate sesquihydrate), shaken 3 min, centrifugation 10 min at 2575g		LC–FLD	Paíga et al. (2012)
Bisphenol A, Bisphenol B	Canned vegetables and fruits	10 g (IS $BPAd_{16}$)	5 mL deionized water, vortex 5 min, centrifuge at 3500 g for 2 min	10 mL MeCN, 4 g anhydrous $MgSO_4$, 1 g NaCl, shaken 15 min, centrifuged at 3500 g for 2 min	GC-MS	Cunha and Fernandes (2013)

(Continued)

Table 3.22: Quick, easy, cheap, effective, rugged, and safe protocols (QuEChERS). (*cont.*)

Analytes/ Compounds	Sample/Matrix	Sample Weight	Extraction	Cleanup	Target Components Analysis	References
Ochratoxin A	Wine	4 mL	4 mL of ACN: CH_3COOH (99:1 v/v), 2.6 g QuEChERS mixture (anhydrous $MgSO_4$, NaCl, sodium citrate tribasic dihydrate, sodium citrate dibasic sesquihydrate, 4:1:1:0.5), shaken 10 s, centrifuged at 1489 g, 5 min	1 mL acetonitrile phase transferred into PTFE dSPE cleanup tubes containing 150 mg $MgSO_4$, 50 mg different sorbents: primary and secondary amine (PSA), C_{18}, florisil, alumina, silica, cleanup by dSPE	LC–ESI-MS/MS	Fernandes et al. (2013)
Pesticides	Rice	5 g	10 mL water, 10 mL MeCN kept in refrigerator for 30 min, vortexed 1 min, QuEChERS extraction bag (4 g anhydrous MgSO4, 1 g NaCl), vortexed 1 min, centrifuged 20 min at 4000 rpm −10°C	5 mL MeCN layer transferred into centrifuge tube, cleaned up using d-SPE with 375 mg PSA, 750 mg anhydrous $MgSO_4$ vortexed 1 min, centrifuged 20 min at 4000 rpm at −10°C	GC–MS/MS	Hou et al. (2013)
Pesticide	Tea	4 g	20 mL ACN, 12,000 rpm for 2 min, 5 g NaCl, 4 g $MgSO_4$, hand-shaken for 1 min, centrifuged at 1000 × g for 10 min, 10 mL upper layer evaporated at 40°C, 3 mL ACN	Solution transferred into tube containing 200 mg PSA, 200 mg C_{18}, 100 mg GCB, 100 mg MWCNTs, 200 mg anhydrous $MgSO_4$, shaken 1 min, centrifuged at 300 × g for 10 min	GC-MS/MS	Chen et al. (2014b)
PAHs	Mussels	10 g	10 mL ACN, mixed 1 min 2,850 rpm, 4 g anhydrous $MgSO_4$, 1 g NaCl, mixed, centrifuged 5 min at 5000 rfc at 15°C	Acetonitrile layer transferred into tube containing 900 mg $MgSO_4$, 150 mg PSA, mixed 30 s, centrifugated 1 min, 5,000 rfc		Madureira et al. (2014)
Pesticides	Fruit samples (tomato, papaya, watermelon)	10 g	2 mL ACN, shaken, 4 g $MgSO_4$, 1 g NaCl, shaken 1 min, centrifuged at 3000 rpm for 10 min	Acetonitrile extract transferred to d-SPE tube containing 75 mg PSA, 75 mg C_{18}, 450 mg $MgSO_4$, d-SPE tubes sealed, shaken for 30 s, centrifuged at 3000 rpm for 3 min	HPLC	Bedassa et al. (2015)
Pharmaceuticals	Vegetables (celery, lettuce)	500 mg	2 mL of 150 mg/L of Na_2EDTA solution, 5 mL ACN/MeOH (ACN/MeOH/Na_2EDTA 46.4/25.0/28.6), shaked 1 min, 2 g anhydrous Na_2SO_4, 0.5 g NaCl, vortexed 1.5 min, centrifuged at 2990 × g for 10 min	Supernatants transferred to d-SPE tube containing 12.5 mg C_{18}, 12.5 mg PSA, 225 mg Na_2SO_4, mixed for 1 min, centrifuged at 9240 × g for 10 min	LC-MS/MS	Chuang et al. (2015)

Pesticides	Orange	15 g	15 mL of ACN/CH_3COOH (99:1, v/v), 6 g $MgSO_4$, 1.5 g $C_2H_3NaO_2$, vortexed, centrifuged 1 min at 5000 rpm	Upper layer transferred, 0.6 g $MgSO_4$, 0.2 g PSA, shaken 1 min, centrifuged at 5000 rpm for 1 min	LC–MS/MS	Golge and Kabak (2015)
Pesticides	Food of animal origin	5 g	20 mL ACN/water (1:1v/v), 25 μL triphenyl phosphate solution for procedural control, shaked 30 min, 5 mL *N*-hexane, extraction 10 min, QuEChERS salt (61.5 wt.% anhydrous magnesium sulfate, 15.4 wt.% sodium chloride, 15.4 wt.% trisodium citrate dihydrate, 7.7 wt.% disodium citrate sesquihydrate), shaken 1 min, centrifuged 5 min	Acetonitrile phase transferred to a SPE C_{18} cartridge (conditioned with 5 mL acetonitrile), eluted 6 mL ACN	GC-MS/MS, LC-MS/MS	Lichtmannegger et al. (2015)
Quinolones	Fish (bass, trout, panga)	2 g	10 mL 5% CH_2O_2 in ACN, shaken 30 s, QuEChERS kit (4 g $MgSO_4$, 1 g NaCl, 1 g sodium citrate, 0.5 g sodium citrate dibasic sesquihydrate), shaken 2 min, centrifuged at 9000 rpm for 5 min	Acetonitrile layer transferred to dSPE tube comprising C_{18}, $MgSO_4$, shaken 2 min, centrifuged at 5000 rpm for 5 min	UHPLC-FL	Lombardo-Agüí et al. (2015)
Pesticides	Peppermint	5 g	10 mL ACN, 10 mL distilled water, shaken 1 min, mixture of 1 g NaCl, 0.5 g disodium hydrogen citrate sesquihydrate, 1 g sodium citrate dehydrate, 4 g $MgSO_4$, shaken 1 min, centrifuged at 3000 rpm for 5 min	Upper layer transferred into centrifuge tube with 150 mg PSA, 45 mg GCB, 900 mg anhydrous $MgSO_4$, shaken 2 min, centrifuged at 3000 rpm for 5 min	GC/ECD/NPD	Słowik-Borowiec (2015)
Nicotine	Tea	2 g	10 mL water, vortexed 3 s, hydration 30 min, 100 μL of IS, 400 μL 5% ammonia solution, 10 mL ACN, shaken 15 min, placed in ultrasonic bath for 15 min, 4 g anhydrous $MgSO_4$, 1 g NaCl, vortexed, centrifuged for 5 min (3500 rpm)	Supernatant transferred into dSPE tube containing mg PSA, 150 mg anhydrous $MgSO_4$, vortexed 30 s, centrifuged for 5 min (3500 rpm), supernatant acidified with 50 μL 1% CH_2O_2	LC–ESI–MS/MS	Thräne et al. (2015)
PCB congeners	Catfish	3 g	5 mL water, 30 mL acetonitrile, shaken at 1000 stroke/min for 30 min, 6 g $MgSO_4$, 1.5 g NaCl, centrifuged at 5000 rpm for 10 min	Acetonitrile extract transferred into tube containing 150 mg anhydrous $MgSO_4$, 50 mg of PSA, 50 mg C_{18}, capped, vortexed for 1 min, centrifuged at 2000 rpm for 10 min	GC–MS/MS	Chamkasem et al. (2016)

(*Continued*)

Table 3.22: Quick, easy, cheap, effective, rugged, and safe protocols (QuEChERS). (*cont.*)

Analytes/ Compounds	Sample/Matrix	Sample Weight	Extraction	Cleanup	Target Components Analysis	References
Tocopherols, sitosterols	Seeds and nuts (almonds, hazelnuts, peanuts, sunflower seeds,pistachios)	1–3 g	10 mL MeOH, shaken 5 min at room temperature, 4 g $MgSO_4$, 1 g NaCl, shaken 1 min, centrifuged at 3000 rpm for 5 min	Methanol fraction transferred into SPE containing 450 mg $MgSO_4$, 75 mg PSA, shaken, centrifuged for 5 min at 3000 rpm	HPLC–DAD	Delgado-Zamarreño et al. (2016)
Chlorothalonil	Cabbage	5 g	10 mL 5% AcOH/toluene (1:1, v/v), shaken 30 min in an air bath at 22°C, 2 g anhydrous $MgSO_4$, 0.5 g anhydrous sodium acetate, vortexed 1 min, centrifuged at 2500 rcf for 5 min	Supernatant transferred centrifuge tube containing 30 mg PSA, 150 mg $MgSO_4$ for d-SPE procedure, vortexed 1 min, centrifuged at 8800 rcf for 5 min	GC-MS	Hou et al. (2016)
Thiacloprid, spirotetramat, spirotetramat's four metabolites	Pepper	10 g	10 mL ACN, 0.68 mL formic acid, shaken 3 min, 2.5 g NaCl, shaken 1 min, centrifuged 2811 × g for 5 min	Aliquot transferred into centrifuge tube containing 30 mg PSA, 100 mg C_{18}, 60 mg GCB, 150 mg anhydrous $MgSO_4$, shaken 1 min, centrifuged at 2400 × g for 5 min	UHPLC–MS/MS	Li et al. (2016b)
Amphenicol antibiotics (chlorampheni-col, thiampheni-col)	Milk, honey	2 g	IS, 15 mL 1% acetic acid/ACN (v/v), homogenized for 1 min, extraction kit (4 g anhydrous Na_2SO_4, 1 g NaCl), mixed 1 min, shaken at 1500 rpm for 10 min, centrifuged for 5 min at 4000 rpm	Pack of cleanup kit (0.9 g of anhydrous Na_2SO_4, 0.15 g C_{18} powder), 0.35 g C_{18} endcapped sorbents, 0.5 g QuE Z-Sepþ powders, mixed 1 min, shaken at 1500 rpm for 10 min, centrifuged for 30 min at 4000 rpm (1600 g, at 4°C)	LC-MS/MS	Liu et al. (2016)
Pesticides	Milk	10 mL	4 g $MgSO_4$, 1 g NaCl, 10 mL ACN, centrifuged (6000 rpm) 2 min	upper layer transferred to QuEChERS tubes, vortexed at 8000 rpm for 2 min	GC-µECD	Jawaid et al. (2016)

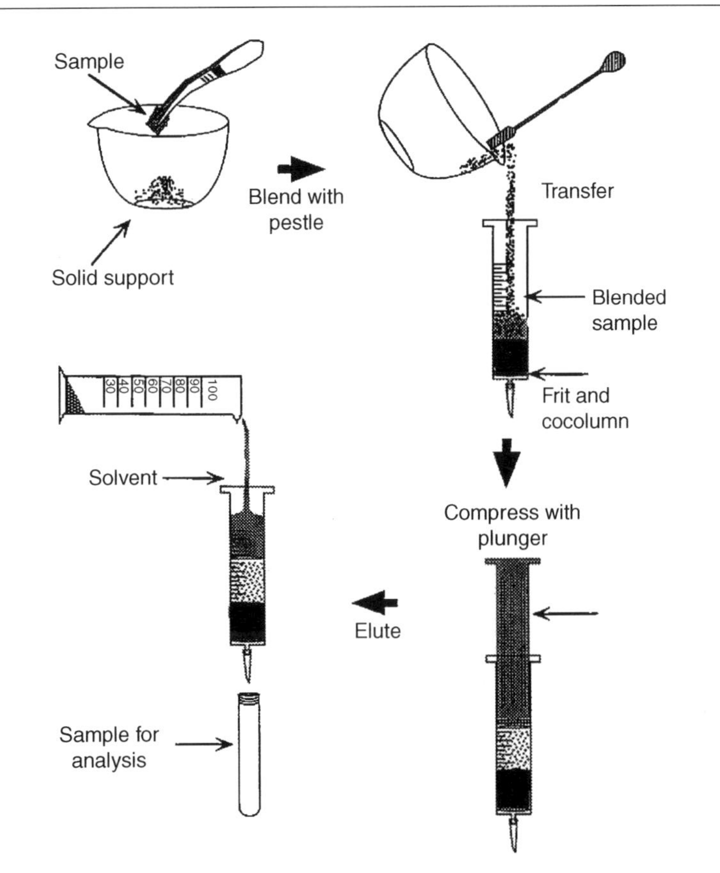

Figure 3.2: Main Analytical Steps Concerning Matrix Solid Phase Dispersion Protocol (MSPD).
Reproduced from the reference Barker, S.A., 2007. Matrix solid phase dispersion (MSPD). J. Biochem. Biophys. Method 70, 151–162, with permission, according to Fig. 1, in original publication.

samples requiring fast and simple analytical procedures (Fig. 3.2). For example, there is an increasing number of papers dealing with matrix solid-phase dispersion (MSPD) involving new extraction materials and enabling fast and simple preparation, isolation, and fractionation of solid, semisolid, and/or highly viscous biological samples. This approach significantly reduce the solvent usage (around 95% less in comparison to classical protocols) and also quantification can be performed much faster (in 90% less time when compared to alternative classical methods). In the case of solid or semisolid biological materials, this methodology can be highly competitive with different classical protocols (Barker, 2007).

There is an increasing trend in combining of classical methods with microwave heating and involving advanced multivariate optimization of the extraction process (Fig. 3.3). This approach allows decreased extraction time and increased extraction efficiency, which was

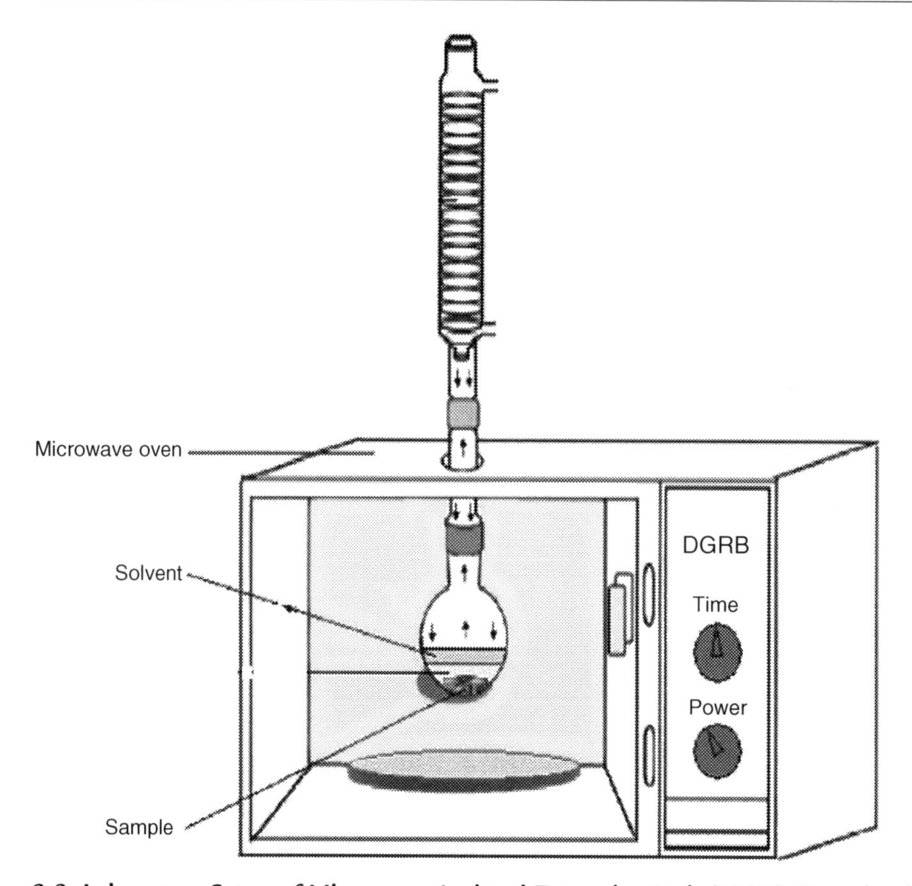

Figure 3.3: Laboratory Setup of Microwave-Assisted Extraction Unit (MAE). *Reproduced from the reference Thirugnanasambandham, K., Sivakumar, V., Maran, J.P., 2015. Microwave-assisted extraction of polysaccharides from mulberry leaves. Int. J. Biol. Macromol. 72, 1–5, with permission, according to Fig. 1, in original publication.*

well documented by isolation of thermally stable polysaccharides from mulberry leaves (Thirugnanasambandham et al., 2015). Extraction of thermally unstable and oxygen sensitive substances require different methodology involving, for example, supercritical carbon dioxide and/or pressurized liquids (Fig. 3.4). Supercritical carbon dioxide and pressurized liquids including water, ethanol, and acetone can be considered as interesting alternatives for toxic organic solvents as well as extraction methods performed at high temperature regions. This methodology was successfully applied for processing of multicomponent samples consisting of phenolic compounds, antioxidants, and anthocyanins from residues of blueberry (*Vaccinium myrtillus L.*) (Paes et al., 2014). Described technologies enable recovery of high-value components that are present in various wastes and also open a new possibility for the production of food ingredients for which cost is significantly reduced. It is noteworthy to say that pressurized hot-water extraction (Fig. 3.5) can be applied for different extraction

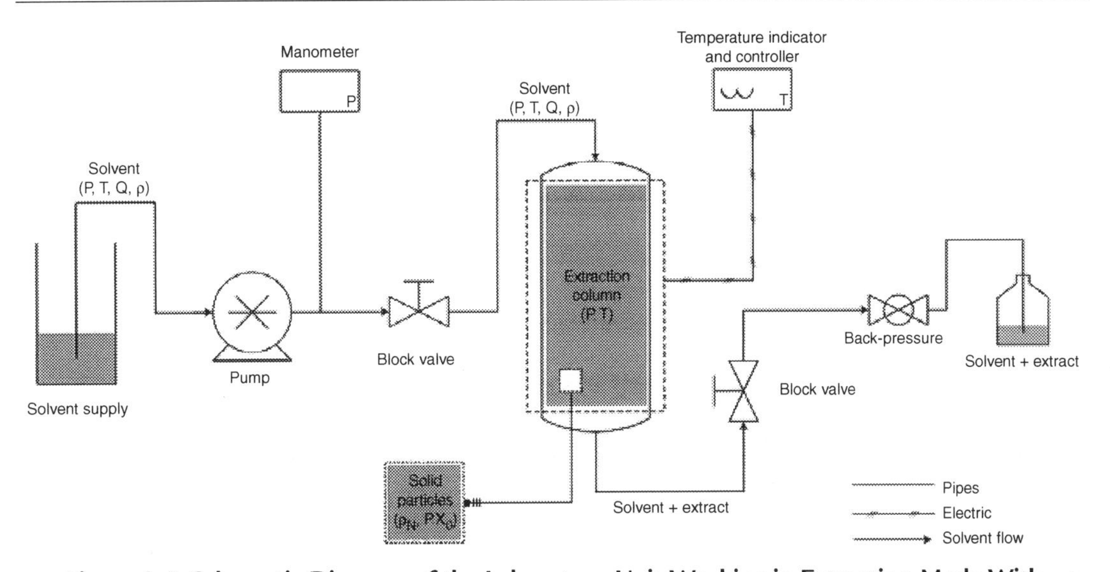

Figure 3.4: Schematic Diagram of the Laboratory Unit Working in Extraction Mode With Pressurized Liquids (PLE). *Reproduced from the reference Paes, J., Dottaa, R., Barbero, G.F., Martínez, J., 2014. Extraction of phenolic compounds and anthocyanins from blueberry (*Vaccinium myrtillus *L.) residues using supercritical CO_2 and pressurized liquids. J. Supercri. Fluid 95, 8–16, with permission, according to Fig. 2, in original publication.*

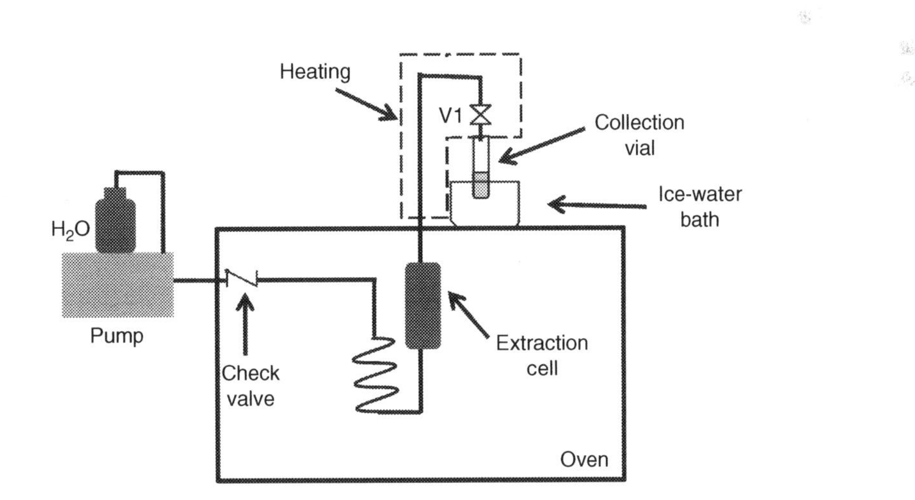

Figure 3.5: General Scheme of Pressurized Hot Water Extraction (PHWE). *Reproduced from the reference Andersson, J.M., Lindahl, S., Turner, C., Rodriguez-Meizoso, I., 2012. Pressurised hot water extraction with on-line particle formation by supercritical fluid technology. Food Chem. 134, 1724–1731, with permission, according to Fig. 1, in original publication.*

materials like nondried extracts and vegetable tissues. This on-line and one-step extraction/drying process allows efficient extraction of flavonoids and various antioxidant-acting low-molecular-mass compounds (Andersson et al., 2012).

3 Conclusions

Bioactive chemicals described in this chapter form a nonhomogenic group of the key substances that may be considered as micropollutants but also as components of interest for pharmaceutical and food industries, particularly lipids, polychlorinated biphenyls, phenolic compounds, phthalate esters, polycyclic aromatic hydrocarbons, steroids, terpenes, and may others. These chemicals may act as, for example, antibiotics, antioxidants, aroma/flavor compounds, food colorings, fungicides, insecticides, pesticides, plasticizers, preservatives, or vitamins.

Generally, isolation protocols may involve single or mixed extraction mechanisms based on partition, solvation, exclusion, distillation, adsorption, or diffusion phenomena. Supramolecular interactions (enabling, e.g., creation of highly selective host–guest complexes) allows efficient enantioselective extraction of bioactive components. Such methodology requires specific host molecules—or example, macrocyclic compounds that are covalently immobilized within solid support. Obviously, any extraction protocol must be optimized and this step is usually time consuming.

Recently, the main issue has been to eliminate the toxic organic liquids, particularly, from large-scale preparative extraction protocols, due to environment protection and green chemistry idea. Therefore, water-based extraction liquids are strongly preferred and appropriate procedures extensively developed.

References

Aeenehvand, S., Toudehrousta, Z., Kamankesh, M., Mashayekh, M., Tavakoli, H.R., Mohammadi, A., 2016. Evaluation and application of microwave-assisted extraction and dispersive liquid–liquid microextraction followed by high-performance liquid chromatography for the determination of polar heterocyclic aromatic amines in hamburger patties. Food Chem. 190, 429–435.

Almeida, C., Fernandes, J.O., Cunha, S.C., 2012. A novel dispersive liquideliquid microextraction (DLLME) gas chromatography-mass spectrometry (GC-MS) method for the determination of eighteen biogenic amines in beer. Food Control. 25, 380–388.

Alshana, U., Göğer, N.G., Ertaş, N., 2013. Dispersive liquid–liquid microextraction combined with field-amplified sample stacking in capillary electrophoresis for the determination of non-steroidal anti-inflammatory drugs in milk and dairy products. Food Chem. 138, 890–897.

Amaro, A.L., Beaulieu, J.C., Grimm, C.C., Stein, R.E., Almeida, D.P.F., 2012. Effect of oxygen on aroma volatiles and quality of fresh-cut cantaloupe and honeydew melons. Food Chem. 130, 49–57.

Andersson, J.M., Lindahl, S., Turner, C., Rodriguez-Meizoso, I., 2012. Pressurised hot water extraction with on-line particle formation by supercritical fluid technology. Food Chem. 134, 1724–1731.

Arapitsas, P., Turner, C., 2008. Pressurized solvent extraction and monolithic column-HPLC/DAD analysis of anthocyanins in red cabbage. Talanta 74, 1218–1223.

Araújo, E.A., Lara, M.C.R., dos Reis, M.R., Viriato, R.L.S., Rocha, R.A.R., Gonçalves, R.G.L., Heleno, F.F., de Queiroz, M.E.L.R., Tronto, J., Pinto, F.G., 2016. Determination of haloxyfop-methyl, linuron, and procymidone pesticides in carrot using SLE-LTP extraction and GC-MS. Food Anal. Methods 9, 1344–1352.

Aresta, A., Zambonin, C., 2016. Simultaneous determination of salicylic, 3-methyl salicylic, 4-methyl salicylic, acetylsalicylic and benzoic acids in fruit, vegetables and derived beverages by SPME–LC–UV/DAD. J. Pharmaceut. Biomed. 121, 63–68.

Arroyo-Manzanares, N., Gámiz-Gracia, L., García-Campaña, A.M., 2012. Determination of ochratoxin A in wines by capillary liquid chromatography with laser induced fluorescence detection using dispersive liquid–liquid microextraction. Food Chem. 135, 368–372.

Bajer, T., Bajerová, P., Kremr, D., Eisner, A., Ventura, K., 2015. Central composite design of pressurised hot water extraction process for extracting capsaicinoids from chili peppers. J. Food Compos. Anal. 40, 32–38.

Barker, S.A., 2007. Matrix solid phase dispersion (MSPD). J. Biochem. Biophys. Method 70, 151–162.

Barra, A., Baldovini, N., Loiseau, A.-M., Albino, L., Lesecq, C., Cuvelier, L.L., 2007. Chemical analysis of French beans (*Phaseolus vulgaris* L.) by headspace solid phase microextraction (HS-SPME) and simultaneous distillation/extraction (SDE). Food Chem. 101, 1279–1284.

Barriada-Pereira, M., Serôdio, P., González-Castro, M.J., Nogueira, J.M.F., 2010. Determination of organochlorine pesticides in vegetable matrices by stir bar sorptive extraction with liquid desorption and large volume injection-gas chromatography–mass spectrometry towards compliance with European Union directives. J. Chromatogr. A. 1217, 119–126.

Bedassa, T., Gure, A., Megersa, N., 2015. Modified QuEChERS method for the determination of multiclass pesticide residues in fruit samples utilizing high-performance liquid chromatography. Food Anal. Methods 8, 2020–2027.

Benet, I., Ibañez, C., Guàrdia, M.D., Solà, J., Arnau, J., Roura, E., 2015. Optimisation of stir-bar sorptive extraction (SBSE), targeting medium and long-chain free fatty acids in cooked ham exudates. Food Chem. 185, 75–83.

Benito-Román, Ó., Alonso, E., Cocero, M.J., 2013. Ultrasound-assisted extraction of β-glucans from barley. LWT Food Sci. Technol. 50, 57–63.

Bermejo, D.V., Luna, P., Manic, M.S., Najdanovic-Visak, V., Reglero, G., Fornari, T., 2013. Extraction of caffeine from natural matter using a bio-renewable agrochemical solvent. Food Bioprod. Process 91, 303–309.

Bermejo, D.V., Mendiola, J.A., Ibáñez, E., Reglero, G., Fornari, T., 2015. Pressurized liquid extraction of caffeine and catechins from green tea leaves using ethyl lactate, water and ethyl lactate + water mixtures. Food Bioprod. Process 96, 106–112.

Bernardo-Gil, M.G., Cardoso, Lopes, L.M., 2004. Supercritical fluid extraction of *Cucurbita ficifolia* seed oil. Eur. Food Res. Technol. 219, 593–597.

Berrada, H., Borrull, F., Font, G., Marcé, R.M., 2008. Determination of macrolide antibiotics in meat and fish using pressurized liquid extraction and liquid chromatography–mass spectrometry. J. Chromatogr. A 1208, 83–89.

Bianchi, F., Careri, M., Musci, M., Mangia, A., 2007. Fish and food safety: determination of formaldehyde in 12 fish species by SPME extraction and GC–MS analysis. Food Chem. 100, 1049–1053.

Bicchi, C., Cordero, C., Rubiolo, P., Sandra, P., 2003. Stir bar sorptive extraction (SBSE) in sample preparation from heterogeneous matrices: determination of pesticide residues in pear pulp at ppb (ng/g) level. Eur. Food Res. Technol. 216, 449–456.

Bimakr, M., Rahman, R.A., Taip, F.S., Ganjloo, A., Salleh, L.M., Selamat, J., Hamid, A., Zaidul, I.S.M., 2011. Comparison of different extraction methods for the extraction of major bioactive flavonoid compounds from spearmint (*Mentha spicata* L.) leaves. Food Bioprod. Process 89, 67–72.

Blasco, C., Vazquez-Roig, P., Onghena, M., Masia, A., Picó, Y., 2011. Analysis of insecticides in honey by liquid chromatography–ion trap-mass spectrometry: comparison of different extraction procedures. J. Chromatogr. A 1218, 4892–4901.

Blesa, J., Moltó, J.C., El Akhdari, S., Mañes, J., Zinedine, A., 2014. Simultaneous determination of *Fusarium* mycotoxins in wheat grain from Morocco by liquid chromatography coupled to triple quadrupole mass spectrometry. Food Control. 46, 1–5.

Bogialli, S., Curini, R., Di Corcia, A., Laganà, A., Nazzari, M., Tonci, M., 2004. Simple and rapid assay for analyzing residues of carbamate insecticides in bovine milk: hot water extraction followed by liquid chromatography–mass spectrometry. J. Chromatogr. A 1054, 351–357.

Bogusz Junior, S., De Melo, A.M.T., Zini, C.A., Godoy, H.T., 2011. Optimization of the extraction conditions of the volatile compounds from chili peppers by headspace solid phase micro-extraction. J. Chromatogr. A 1218, 3345–3350.

Bosch Ojeda, C., Sánchez Rojas, F., 2014. Evaluation of dispersive liquid-liquid microextraction for the determination of cobalt and cadmium by flame atomic absorption spectrometry: application in water and food samples. Sample Perp. 2, 13–20.

Cacho, J.I., Campillo, N., Viñas, P., Hernández-Córdoba, M., 2013. Stir bar sorptive extraction with gas chromatography–mass spectrometry for the determination of resveratrol, piceatannol and oxyresveratrol isomers in wines. J. Chromatogr. A 1315, 21–27.

Cacho, J.I., Campillo, N., Vias, P., Hernández-Córdoba, M., 2014. Stir bar sorptive extraction polar coatings for the determination of chlorophenols and chloroanisoles in wines using gas chromatography and mass spectrometry. Talanta 118, 30–36.

Cai, Z., Qu, Z., Lan, Y., Zhao, S., Ma, X., Wan, Q., Jing, P., Li, P., 2016. Conventional, ultrasound-assisted, and accelerated-solvent extractions of anthocyanins from purple sweet potatoes. Food Chem. 197, 266–272.

Callejón, R.M., Tesfaye, W., Torija, M.J., Mas, A., Troncoso, A.M., Morales, M.L., 2009. Volatile compounds in red wine vinegars obtained by submerged and surface acetification in different woods. Food Chem. 113, 1252–1259.

Campillo, N., Viñas, P., Aguinaga, N., Férez, G., Hernández-Córdoba, M., 2010. Stir bar sorptive extraction coupled to liquid chromatography for the analysis of strobilurin fungicides in fruit samples. J. Chromatogr. A 1217, 4529–4534.

Campone, L., Piccinelli, A.L., Pagano, I., Carabetta, S., Di Sanzo, R., Russo, M., Rastrelli, L., 2014. Determination of phenolic compounds in honey using dispersive liquid–liquid microextraction. J. Chromatogr. A 1334, 9–15.

Campone, L., Piccinelli, A.L., Celano, R., Russo, M., Valdés, A., Ibáñez, C., Rastrelli, L., 2015. A fully automated method for simultaneous determination of aflatoxins and ochratoxin A in dried fruits by pressurized liquid extraction and online solid-phase extraction cleanup coupled to ultra-high-pressure liquid chromatography–tandem mass spectrometry. Anal. Bioanal. Chem. 407, 2899–2911.

Cao, Y., Tang, H., Chen, D., Li, L., 2015. A novel method based on MSPD for simultaneous determination of 16 pesticide residues in tea by LC–MS/MS. J. Chromatogr. B 998-999, 72–79.

Cao, W., Hu, S.-S., Ye, L.-H., Cao, J., Pang, X.-Q., Xu, J.-J., 2016. Trace matrix solid phase dispersion using a molecular sieve as the sorbent for the determination of flavonoids in fruit peels by ultra-performance liquid chromatography. Food Chem. 190, 474–480.

Caprioli, G., Giusti, F., Ballini, R., Sagratini, G., Vila-Donat, P., Vittori, S., Fiorini, D., 2016. Lipid nutritional value of legumes: evaluation of different extraction methods and determination of fatty acid composition. Food Chem. 192, 965–971.

Castro, L.F., Ross, C.F., 2015. Determination of flavour compounds in beer using stir-bar sorptive extraction and solid-phase microextraction. J. Inst. Brew. 121, 197–203.

Castro-Vargas, H.I., Rodríguez-Varela, L.I., Ferreira, S.R.S., Parada-Alfonso, F., 2010. Extraction of phenolic fraction from guava seeds (*Psidium guajava* L.) using supercritical carbon dioxide and co-solvents. J. Supercrit. Fluid. 51, 319–324.

Chai, M.K., Tan, G.H., 2009. Validation of a headspace solid-phase microextraction procedure with gas chromatography-electron capture detection of pesticide residues in fruits and vegetables. Food Chem. 117, 561–567.

Chamkasem, N., Lee, S., Harmon, T., 2016. Analysis of 19 PCB congeners in catfish tissue using a modified QuEChERS method with GC–MS/MS. Food Chem. 192, 900–906.

Chang, Y., Hou, H., Li, B., 2016. Identification of volatile compounds in codfish (*Gadus*) by a combination of two extraction methods coupled with GC-MS analysis. J. Ocean Univ. China 15 (3), 509–514.

Chen, L., Lu, Y., Li, S., Lin, X., Xu, Z., Dai, Z., 2013. Application of graphene-based solid-phase extraction for ultra-fast determination of malachite green and its metabolite in fish tissues. Food Chem. 141, 1383–1389.

Chen, C.-L., Tsai, D.-Y., Ding, W.-H., 2014a. Optimisation of matrix solid-phase dispersion for the determination of Dechlorane compounds in marketed fish. Food Chem. 164, 286–292.

Chen, H., Yin, P., Wang, Q., Jiang, Y., Liu, X., 2014b. A modified QuEChERS sample preparation method for the analysis of 70 pesticide residues in tea using gas chromatography-tandem mass spectrometry. Food Anal. Methods 7, 1577–1587.

Chen, R., Jin, C., Tong, Z., Lu, J., Tan, L., Tian, L., Chang, Q., 2016. Optimization extraction, characterization and antioxidant activities of pectic polysaccharide from tangerine peels. Carbohyd. Polym. 136, 187–197.

Cheong, M.-W., Tan, A.A.-A., Liu, S.-Q., Curran, P., Yu, B., 2013. Pressurised liquid extraction of volatile compounds in coffee bean. Talanta 115, 300–307.

Chuang, J.C., Hart, K., Chang, J.S., Boman, L.E., Van Emon, J.M., Reed, A.W., 2001. Evaluation of analytical methods for determining pesticides in baby foods and adult duplicate-diet samples. Anal. Chim. Acta. 444, 87–95.

Chuang, Y.-H., Zhang, Y., Zhang, W., Boyd, S.A., Li, H., 2015. Comparison of accelerated solvent extraction and quick, easy, cheap, effective, rugged and safe method for extraction and determination of pharmaceuticals in vegetables. J. Chromatogr. A 1404, 1–9.

Contini, M., Baccelloni, S., Massantini, R., Anelli, G., 2008. Extraction of natural antioxidants from hazelnut (*Corylus avellana* L.) shell and skin wastes by long maceration at room temperature. Food Chem. 110, 659–669.

Corrales, M., García, A.F., Butz, P., Tauscher, B., 2009. Extraction of anthocyanins from grape skins assisted by high hydrostatic pressure. J. Food Eng. 90, 415–421.

Costa, A.I.G., Queiroz, M.E.L.R., Neves, A.A., de Sousa, F.A., Zambolim, L., 2015. Determination of pesticides in lettuce using solid–liquid extraction with low temperature partitioning. Food Chem. 181, 64–71.

Ćujić, N., Šavikin, K., Janković, T., Pljevljakušić, D., Zdunić, G., Ibrić, S., 2016. Optimization of polyphenols extraction from dried chokeberry using maceration as traditional technique. Food Chem. 194, 135–142.

Cunha, S.C., Fernandes, J.O., 2013. Assessment of bisphenol A and bisphenol B in canned vegetables and fruits by gas chromatographyemass spectrometry after QuEChERS and dispersive liquideliquid microextraction. Food Control. 33, 549–555.

D'Arco, G., Fernández-Franzóna, M., Font, G., Damiani, P., Manes, J., 2008. Analysis of fumonisins B1, B2 and B3 in corn-based baby food by pressurized liquid extraction and liquid chromatography/tandem mass spectrometry. J. Chromatogr. A 1209, 188–194.

da Costa Morais, E.H., Rodrigues, A.A.Z., de Queiroz, M.E.L.R., Neves, A.A., Morais, P.H.D., 2014. Determination of thiamethoxam, triadimenol and deltamethrin in pineapple using SLE-LTP extraction and gas chromatography. Food Control. 42, 9–17.

Daneshfar, A., Khezeli, T., Lotfi, H.J., 2009. Determination of cholesterol in food samples using dispersive liquid–liquid microextraction followed by HPLC–UV. J. Chromatogr. B. 877, 456–460.

Dawidowicz, A.L., Dybowski, M.P., 2012. Fast determination of α- and β-thujone in alcoholic beverages using solid-phase extraction and gas chromatography. Food Control. 25, 197–201.

de Aguiar, A.C., Sales, L.P., Coutinho, J.P., Barbero, G.F., Godoy, H.T., Martínez, J., 2013. Supercritical carbon dioxide extraction of *Capsicum* peppers: global yield and capsaicinoid content. J. Supercrit. Fluid 81, 210–216.

de Aguiar, A.C., dos Santos, P., Coutinho, J.P., Barbero, G.F., Godoy, H.T., Martínez, J., 2014. Supercritical fluid extraction and low pressure extraction of Biquinho pepper (*Capsicum chinense*). LWT Food Sci. Technol. 59, 1239–1246.

de Andrade, F.I., Guedes, M.I.F., Vieira, Í.G.P., Mendes, F.N.P., Rodrigues, P.A.S., Maia, C.S.C., Ávila, M.M.M., de Matos Ribeiro, L., 2014. Determination of synthetic food dyes in commercial soft drinks by TLC and ion-pair HPLC. Food Chem. 157, 193–198.

De Jager, L.S., Perfetti, G.A., Diachenko, G.W., 2009a. Stir bar sorptive extraction–gas chromatography–mass spectrometry analysis of tetramethylene disulfotetramine in food: method development and comparison to solid-phase microextraction. Anal. Chim. Acta. 635, 162–166.

De Jager, L.S., Perfetti, G.A., Diachenko, G.W., 2009b. Comparison of membrane assisted solvent extraction, stir bar sorptive extraction, and solid phase microextraction in analysis of tetramine in food. J. Sep. Sci. 32, 1081–1086.

de Pinho, G.P., Neves, A.A., de Queiroz, M.E.L.R., Silvério, F.O., 2010. Pesticide determination in tomatoes by solid–liquid extraction with purification at low temperature and gas chromatography. Food Chem. 121, 251–256.

Delgado, R., Durán, E., Castro, R., Natera, R., Barroso, C.G., 2010. Development of a stir bar sorptive extraction method coupled to gas chromatography-mass spectrometry for the analysis of volatile compounds in Sherry brandy. Anal. Chim. Acta. 672, 130–136.

Delgado-Zamarreño, M.M., Bustamante-Rangel, M., Sánchez-Pérez, A., Carabias-Martínez, R., 2004. Pressurized liquid extraction prior to liquid chromatography with electrochemical detection for the analysis of vitamin E isomers in seeds and nuts. J. Chromatogr. A 1056, 249–252.

Delgado-Zamarreño, M.M., Pérez-Martín, L., Bustamante-Rangel, M., Carabias-Martínez, R., 2012. A modified QuEChERS method as sample treatment before the determination of isoflavones in foods by ultra-performance liquid chromatography–triple quadrupole mass spectrometry. Talanta 100, 320–328.

Delgado-Zamarreño, M.M., Fernández-Prieto, C., Bustamante-Rangel, M., Pérez-Martín, L., 2016. Determination of tocopherols and sitosterols in seeds and nuts by QuEChERS-liquid chromatography. Food Chem. 192, 825–830.

Di Carro, M., Ardini, F., Magi, E., 2015. Multivariate optimization of headspace solid-phase microextraction followed by gas chromatography–mass spectrometry for the determination of methylpyrazines in cocoa liquors. Microchem. J. 121, 172–177.

Dong, C., Wang, W., 2006. Headspace solid-phase microextraction applied to the simultaneous determination of sorbic and benzoic acids in beverages. Anal. Chim. Acta. 562, 23–29.

Dong, C., Zeng, Z., Li, X., 2005. Determination of organochlorine pesticides and their metabolites in radish after headspace solid-phase microextraction using calix[4]arene fiber. Talanta 66, 721–727.

Dong, Z., Gu, F., Xu, F., Wang, Q., 2014. Comparison of four kinds of extraction techniques and kinetics of microwave-assisted extraction of vanillin from *Vanilla planifolia* Andrews. Food Chem. 149, 54–61.

Dos Santos, M.S., Martendal, E., Carasek, E., 2011. Determination of THMs in soft drink by solid-phase microextraction and gas chromatography. Food Chem. 127, 290–295.

Dranca, F., Oroian, M., 2016. Optimization of ultrasound-assisted extraction of total monomeric anthocyanin (TMA) and total phenolic content (TPC) from eggplant (*Solanum melongena* L.) peel. Ultrason. Sonochem. 31, 637–646.

Du, X., Finn, C.E., Qian, M.C., 2010. Volatile composition and odour-activity value of thornless "Black Diamond" and "Marion" blackberries. Food Chem. 119, 1127–1134.

Du, L., Li, J., Li, W., Li, Y., Li, T., Xiao, D., 2014. Characterization of volatile compounds of pu-erh tea using solid-phase microextraction and simultaneous distillation–extraction coupled with gas chromatography–mass spectrometry. Food Res. Int. 57, 61–70.

Du, L., Ma, L., Qiao, Y., Lu, Y., Xiao, D., 2016. Determination of phthalate esters in teas and tea infusions by gas chromatography–mass spectrometry. Food Chem. 197, 1200–1206.

Duba, K.S., Fiori, L., 2015. Extraction of bioactives from food processing residues using techniques performed at high pressures. Curr. Opin. Food Sci. 5, 14–22.

Enríquez-Gabeiras, L., Gallego, A., Garcinuño, R.M., Fernández-Hernando, P., Durand, J.S., 2012. Interference-free determination of illegal dyes in sauces and condiments by matrix solid phase dispersion (MSPD) and liquid chromatography (HPLC–DAD). Food Chem. 135, 193–198.

Espinosa-Pardoa, F.A., Martinez, J., Martinez-Correa, H.A., 2014. Extraction of bioactive compounds from peach palm pulp (*Bactris gasipaes*) using supercritical CO_2. J. Supercrit. Fluid. 93, 2–6.

Fan, W., Shen, H., Xu, Y., 2011. Quantification of volatile compounds in Chinese soy sauce aroma type liquor by stir bar sorptive extraction and gas chromatography–mass spectrometry. J. Sci. Food Agric. 91, 1187–1198.

Fernandes, P.J., Barros, N., Câmara, J.S., 2013. A survey of the occurrence of ochratoxin A in Madeira wines based on a modified QuEChERS extraction procedure combined with liquid chromatography–triple quadrupole tandem mass spectrometry. Food Res. Int. 54, 293–301.

Fernández-Ponce, M.T., Casas, L., Mantell, C., de la Ossa, E.M., 2015. Use of high pressure techniques to produce *Mangifera indica* L. leaf extracts enriched in potent antioxidant phenolic compounds. Innov. Food Sci. Emerg. 29, 94–106.

Filho, G.L., De Rosso, V.V., Meireles, M.A.A., Rosa, P.T.V., Oliveira, A.L., Mercadante, A.Z., Cabral, F.A., 2008. Supercritical CO_2 extraction of carotenoids from pitanga fruits (*Eugenia uniflora* L.). J. Supercrit. Fluid. 46, 33–39.

Filho, A.M., dos Santos, F.N., de Paula Pereira, P.A., 2010. Development, validation and application of a methodology based on solid-phase micro extraction followed by gas chromatography coupled to mass spectrometry (SPME/GC–MS) for the determination of pesticide residues in mangoes. Talanta 81, 346–354.

Font, G., Juan-García, A., Picó, Y., 2007. Pressurized liquid extraction combined with capillary electrophoresis–mass spectrometry as an improved methodology for the determination of sulfonamide residues in meat. J. Chromatogr. A 1159, 233–241.

Furlani, R.P.Z., Marcilio, K.M., Leme, F.M., Tfouni, S.A.V., 2011. Analysis of pesticide residues in sugarcane juice using QuEChERS sample preparation and gas chromatography with electron capture detection. Food Chem. 126, 1283–1287.

Gao, Z., Deng, Y., Yuan, W., He, H., Yang, S., Sun, C., 2014. Determination of organophosphorus flame retardants in fish by pressurized liquid extraction using aqueous solutions and solid-phase microextraction coupled with gas chromatography-flame photometric detector. J. Chromatogr. A 1366, 31–37.

García de Llasera, M.P., Reyes-Reyes, M.L., 2009. A validated matrix solid-phase dispersion method for the extraction of organophosphorus pesticides from bovine samples. Food Chem. 114, 1510–1516.

García-Mayor, M.A., Gallego-Picó, A., Garcinuño, R.M., Fernández-Hernando, P., Durand-Alegría, J.S., 2012. Matrix solid-phase dispersion method for the determination of macrolide antibiotics in sheep's milk. Food Chem. 134, 553–558.

García-Rodríguez, D., Cela-Torrijos, R., Lorenzo-Ferreira, R.A., Carro-Díaz, A.M., 2012. Analysis of pesticide residues in seaweeds using matrix solid-phase dispersion and gas chromatography–mass spectrometry detection. Food Chem. 135, 259–267.

Ghafoor, K., Park, J., Choi, Y.-H., 2010. Optimization of supercritical fluid extraction of bioactive compounds from grape (*Vitis labrusca* B.) peel by using response surface methodology. Innov. Food Sci. Emerg. 11, 485–490.

Ghiasvand, A.R., Hajipour, S., 2016. Direct determination of acrylamide in potato chips by using headspace solid-phase microextraction coupled with gas chromatography-flame ionization detection. Talanta 146, 417–422.

Gilbert-López, B., García-Reyes, J.F., Lozano, A., Fernández-Alba, A.R., Molina-Díaz, A., 2010. Large-scale pesticide testing in olives by liquid chromatography–electrospray tandem mass spectrometry using two sample preparation methods based on matrix solid-phase dispersion and QuEChERS. J. Chromatogr. A 1217, 6022–6035.

Gil-Ramírez, A., Mendiola, J.A., Arranz, E., Ruíz-Rodríguez, A., Reglero, G., Ibáñez, E., Marín, F.R., 2012. Highly isoxanthohumol enriched hop extract obtained by pressurized hot water extraction (PHWE). Chemical and functional characterization. Innov. Food Sci. Emerg. 16, 54–60.

Golge, O., Kabak, B., 2015. Determination of 115 pesticide residues in oranges by high-performance liquid chromatography–triple-quadrupole mass spectrometry in combination with QuEChERS method. J. Food Compos. Anal. 41, 86–97.

Grassino, A.N., Brnčić, M., Vikić-Topić, D., Roca, S., Dent, M., Brnčić, S.R., 2016. Ultrasound assisted extraction and characterization of pectin from tomato waste. Food Chem. 198, 93–100.

Guamán-Balcázar, M.C., Setyaningsih, W., Palma, M., Barroso, C.G., 2016. Ultrasound-assisted extraction of resveratrol from functional foods: cookies and jams. Appl. Acoust. 103, 207–213.

Guerrero, E.D., Mejías, R.C., Marín, R.N., Barroso, C.G., 2007a. Optimization of stir bar sorptive extraction applied to the determination of pesticides in vinegars. J. Chromatogr. A 1165, 144–150.

Guerrero, E.D., Marín, R.N., Mejías, R.C., Barroso, C.G., 2007b. Stir bar sorptive extraction of volatile compounds in vinegar: validation study and comparison with solid phase microextraction. J. Chromatogr. A 1167, 18–26.

Gyawali, R., Seo, H.-Y., Shim, S.-L., Ryu, K.-Y., Kim, W., You, S.G., Kim, K.-S., 2008. Effect of γ-irradiaton on the volatile compounds of licorice (*Glycyrrhiza uralensis* Fischer). Eur. Food Res. Technol. 226, 577–582.

Ha, J., Wang, Y., Jang, H., Seog, H., Chen, X., 2014. Determination of E,E-farnesol in *Makgeolli* (rice wine) using dynamic headspace sampling and stir bar sorptive extraction coupled with gas chromatography–mass spectrometry. Food Chem. 142, 79–86.

Hale, R.C., Gaylor, M.O., Thames, J.F., Smith, C.L., Mothershead, I.I., R.F., 1996. Robustness of supercritical fluid extraction (SFE) in environmental studies: analysis of chlorinated pollutants in tissues from the osprey (*pandion haliaetus*) and several fish species. Int. J. Environ. Anal. Chem. 64 (1), 11–19.

Hamdan, S., Daood, H.G., Toth-Markus, M., Illes, V., 2008. Extraction of cardamom oil by supercritical carbon dioxide and sub-critical propane. J. Supercrit. Fluid. 44, 25–30.

He, B., Zhang, L.-L., Yue, X.-Y., Liang, J., Jiang, J., Gao, X.-L., Yue, P.-X., 2016. Optimization of ultrasound-assisted extraction of phenolic compounds and anthocyanins from blueberry (*Vaccinium ashei*) wine pomace. Food Chem. 204, 70–76.

Hernando, M.D., Mezcua, M., Suárez-Barcena, J.M., Fernández-Alba, A.R., 2006. Liquid chromatography with time-of-flight mass spectrometry for simultaneous determination of chemotherapeutant residues in salmon. Anal. Chim. Acta. 562, 176–184.

Hernando, M.D., Suárez-Barcena, J.M., Bueno, M.J.M., Garcia-Reyes, J.F., Fernández-Alba, A.R., 2007. Fast separation liquid chromatography–tandem mass spectrometry for the confirmation and quantitative analysis of avermectin residues in food. J. Chromatogr. A 1155, 62–73.

Hiranvarachat, B., Devahastin, S., Chiewchan, N., Raghavan, G.S.V., 2013. Structural modification by different pretreatment methods to enhance microwave-assisted extraction of β-carotene from carrots. J. Food Eng. 115, 190–197.

Horák, T., Čulík, J., Kellner, V., Jurková, M., Čejka, P., Hašková, D., Dvořák, J., 2010. Analysis of selected esters in beer: comparison of solid-phase microextraction and stir bar sorptive extraction. J. Inst. Brewing. 116, 81–85.

Hossain, M.B., Barry-Ryan, C., Martin-Diana, A.B., Brunton, N.P., 2011. Optimisation of accelerated solvent extraction of antioxidant compounds from rosemary (*Rosmarinus officinalis* L.), marjoram (*Origanum majorana* L.) and oregano (*Origanum vulgare* L.) using response surface methodology. Food Chem. 126, 339–346.

Hossain, M.B., Aguiló-Aguayo, I., Lyng, J.G., Brunton, N.P., Rai, D.K., 2015. Effect of pulsed electric field and pulsed light pre-treatment on the extraction of steroidal alkaloids from potato peels. Innov. Food Sci. Emerg. 29, 9–14.

Hou, X., Han, M., Dai, X.H., Yang, X.F., Yi, S., 2013. A multi-residue method for the determination of 124 pesticides in rice by modified QuEChERS extraction and gas chromatography–tandem mass spectrometry. Food Chem. 138, 1198–1205.

Hou, F., Zhao, L., Liu, F., 2016. Determination of chlorothalonil residue in cabbage by a modified QuEChERS-based extraction and gas chromatography – Mass spectrometry. Food Anal. Method. 9, 656–663.

Hu, Y., Li, J., Hu, Y., Li, G., 2010. Development of selective and chemically stable coating for stir bar sorptive extraction by molecularly imprinted technique. Talanta 82, 464–470.

Hu, G., Hernandez, M., Zhua, H., Shaoa, S., 2013. An efficient method for the determination of furan derivatives in apple cider and wine by solid phase extraction and high performance liquid chromatography—diode array detector. J. Chromatogr. A 1284, 100–106.

Hu, G., Zhu, Y., Hernandez, M., Koutchma, T., Shao, S., 2016. An efficient method for the simultaneous determination of furan, 2-methylfuran and 2-pentylfuran in fruit juices by headspace solid phase microextraction and gas chromatography–flame ionisation detector. Food Chem. 192, 9–14.

Huang, Z., Li, Y., Chen, B., Yao, S., 2007. Simultaneous determination of 102 pesticide residues in Chinese teas by gas chromatography–mass spectrometry. J. Chromatogr. B 853, 154–162.

Huang, W., Li, Z., Niu, H., Li, D., Zhang, J., 2008. Optimization of operating parameters for supercritical carbon dioxide extraction of lycopene by response surface methodology. J. Food Eng. 89, 298–302.

Huang, Z., Pan, X.-D., Huang, B., Xu, J.-J., Wang, M.-L., Ren, Y.-P., 2016. Determination of 15 β-lactam antibiotics in pork muscle by matrix solid-phase dispersion extraction (MSPD) and ultra-high pressure liquid chromatography tandem mass spectrometry. Food Control. 66, 145–150.

Izadiyan, P., Hemmateenejad, B., 2016. Multi-response optimization of factors affecting ultrasonic assisted extraction from Iranian basil using central composite design. Food Chem. 190, 864–870.

Jabbar, S., Abid, M., Wu, T., Hashim, M.M., Saeeduddin, M., Hu, B., Lei, S., Zeng, X., 2015. Ultrasound-assisted extraction of bioactive compounds and antioxidants from carrot pomace: a response surface approach. J. Food Process. Preserv. 39, 1878–1888.

Jaime, L., Vázquez, E., Fornari, T., López-Hazas, M.C., García-Risco, M.R., Santoyoa, S., Reglero, G., 2015. Extraction of functional ingredients from spinach (*Spinacia oleracea* L.) using liquid solvent and supercritical CO_2 extraction. J. Sci. Food Agric. 95, 722–729.

Jawaid, S., Talpur, F.N., Nizamani, S.M., Khaskheli, A.A., Afridi, H.I., 2016. Multipesticide residue levels in UHT and raw milk samples by GC- μ ECD after QuEChER extraction method. Environ. Monit. Assess. 188, 230.

Jestoi, M., Järvinen, T., Järvenpää, E., Tapanainen, H., Virtanen, S., Peltonen, K., 2009. Furan in the baby-food samples purchased from the Finnish markets—determination with SPME–GC–MS. Food Chem. 117, 522–528.

Jiang, J.-J., Zeng, Q.-X., Zhu, Z.-W., 2011. Analysis of volatile compounds in traditional Chinese fish sauce (Yu Lu). Food Bioprocess Technol. 4, 266–271.

Jiao, Z., Zhu, D., Yao, W., 2015. Combination of accelerated solvent extraction and micro-solid-phase extraction for determination of trace antibiotics in food samples. Food Anal. Methods 8, 2163–2168.

Jiao, Z., Jiang, Z., Zhang, N., 2016. Determination of polychlorinated biphenyls in food samples by selective pressurized liquid extraction using copper(II) isonicotinate as online cleanup adsorbent. Food Anal. Meth. 9, 88–94.

Jin, E.Y., Lim, S., Kim, S.O., Park, Y.-S., Jang, J.K., Chung, M.-S., Park, H., Shim, K.-S., Choi, Y.J., 2011. Optimization of various extraction methods for quercetin from onion skin using response surface methodology. Food Sci. Biotechnol. 20 (6), 1727–1733.

Jokić, S., Cvjetko, M., Božić, ð., Fabek, S., Toth, N., Vorkapić-Furač, J., Redovniković, I.R., 2012. Optimisation of microwave-assisted extraction of phenolic compounds from broccoli and its antioxidant activity. Int. J. Food Sci. Tech. 47, 2613–2619.

Juan-García, A., Font, G., Juan, C., Picó, Y., 2010. Pressurised liquid extraction and capillary electrophoresis–mass spectrometry for the analysis of pesticide residues in fruits from Valencian markets. Spain. Food Chem. 120, 1242–1249.

Kamankesh, M., Mohammadi, A., Hosseini, H., Tehrani, Z.M., 2015. Rapid determination of polycyclic aromatic hydrocarbons in grilled meat using microwave-assisted extraction and dispersive liquid–liquid microextraction coupled to gas chromatography–mass spectrometry. Meat Sci. 103, 61–67.

Kang, K.-M., Baek, H.-H., 2014. Aroma quality assessment of Korean fermented red pepper paste (*gochujang*) by aroma extract dilution analysis and headspace solid-phase microextraction–gas chromatography–olfactometry. Food Chem. 145, 488–495.

Kang, J.-H., Kim, S., Moon, B.K., 2016. Optimization by response surface methodology of lutein recovery from paprika leaves using accelerated solvent extraction. Food Chem. 205, 140–145.

Karami, Z., Emam-Djomeh, Z., Mirzaee, H.A., Khomeiri, M., Mahoonak, A.S., Aydani, E., 2015. Optimization of microwave assisted extraction (MAE) and soxhlet extraction of phenolic compound from licorice root. J. Food Sci. Technol. 52 (6), 3242–3253.

Karami-Osboo, R., Maham, M., Miri, R., AliAbadi, M.H.S., Mirabolfathy, M., Javidnia, K., 2013. Evaluation of dispersive liquid–liquid microextraction–HPLC–UV for determination of deoxynivalenol (DON) in wheat flour. Food Anal. Method. 6, 176–180.

Karami-Osboo, R., Miri, R., Javidnia, K., Kobarfard, F., AliAbadi, M.H.S., Maham, M., 2015. A validated dispersive liquid–liquid microextraction method for extraction of ochratoxin A from raisin samples. J. Food Sci. Technol. 52 (4), 2440–2445.

Katherine, L.S.V., Edgar, C.C., Jerry, W.K., Luke, R.H., Julie, K.D., 2008. Extraction conditions affecting supercritical fluid extraction (SFE) of lycopene from watermelon. Biores. Technol. 99, 7835–7841.

Kerio, L.C., Wachira, F.N., Wanyoko, J.K., Rotich, M.K., 2012. Characterization of anthocyanins in Kenyan teas: extraction and identification. Food Chem. 131, 31–38.

Kesen, S., Kelebek, H., Sen, K., Ulas, M., Selli, S., 2013. GC–MS–olfactometric characterization of the key aroma compounds in Turkish olive oils by application of the aroma extract dilution analysis. Food Res. Int. 54, 1987–1994.

Khan, M.K., Abert-Vian, M., Fabiano-Tixier, A.S., Dangles, O., Chemat, F., 2010. Ultrasound-assisted extraction of polyphenols (flavanone glycosides) from orange (*Citrus sinensis* L.) peel. Food Chem. 119, 851–858.

Khan, S., Bhatia, T., Trivedi, P., Satyanarayana, G.N.V., Mandrah, K., Saxena, P.N., Mudiam, M.K.R., Roy, S.K., 2016. Selective solid-phase extraction using molecularly imprinted polymer as a sorbent for the analysis of fenarimol in food samples. Food Chem. 199, 870–875.

Khoza, B.S., Dubery, I.A., Byth-Illing, H.-A., Steenkamp, P.A., Chimuka, L., Madala, N.E., 2016. Optimization of pressurized hot water extraction of flavonoids from *Momordica foetida* using UHPLC-qTOF-MS and multivariate chemometric approaches. Food Anal. Methods 9, 1480–1489.

Kin, C.M., Huat, T.G., 2010. Headspace solid-phase microextraction for the evaluation of pesticide residue contents in cucumber and strawberry after washing treatment. Food Chem. 123, 760–764.

Klejdus, B., Mikelová, R., Adam, V., Zehnálek, J., Vacek, J., Kizek, R., Kubáň, V., 2004. Liquid chromatographic–mass spectrometric determination of genistin and daidzin in soybean food samples after accelerated solvent extraction with modified content of extraction cell. Anal. Chim. Acta. 517, 1–11.

Kraujalis, P., Venskutonis, P.R., Pukalskas, A., Kazernavičiūtė, R., 2013. Accelerated solvent extraction of lipids from *Amaranthus* spp. seeds and characterization of their composition. LWT Food Sci. Technol. 54, 528–534.

Kubra, I.R., Kumar, D., Rao, L.J.M., 2013. Effect of microwave-assisted extraction on the release of polyphenols from ginger (*Zingiber officinale*). Int. J. Food Sci. Technol. 48, 1828–1833.

Kujawski, M.W., Pinteaux, E., Namieśnik, J., 2012. Application of dispersive liquid–liquid microextraction for the determination of selected organochlorine pesticides in honey by gas chromatography–mass spectrometry. Eur. Food Res. Technol. 234, 223–230.

Lai, X., Ruan, C., Liu, R., Liu, C., 2014. Application of ionic liquid-based dispersive liquid–liquid microextraction for the analysis of ochratoxin A in rice wines. Food Chem. 161, 317–322.

Lee, S.J., Ahn, B., 2009. Comparison of volatile components in fermented soybean pastes using simultaneous distillation and extraction (SDE) with sensory characterisation. Food Chem. 114, 600–609.

Li, J., Wu, R., Hu, Q., Wang, J., 2007. Solid-phase extraction and HPLC determination of patulin in apple juice concentrate. Food Control. 18, 530–534.

Li, B., Zeng, F., Dong, Q., Cao, Y., Fan, H., Deng, C., 2012. Rapid determination method for 12 pyrethroid pesticide residues in tea by stir bar sorptive extraction–thermal desorption-gas chromatography. Phys. Procedia 25, 1776–1780.

Li, A.-N., Li, S., Xu, D.-P., Xu, X.-R., Chen, Y.-M., Ling, W.-H., Chen, F., Li, H.-B., 2015. Optimization of ultrasound-assisted extraction of lycopene from papaya processing waste by response surface methodology. Food Anal. Method 8, 1207–1214.

Li, J.-W., Wang, Y.-L., Yan, S., Li, X.-J., Pan, S.-Y., 2016a. Molecularly imprinted calixarene fiber for solid-phase microextraction of four organophosphorous pesticides in fruits. Food Chem. 192, 260–267.

Li, S., Liu, X., Dong, F., Xu, J., Xu, H., Hu, M., Zheng, Y., 2016b. Chemometric-assisted QuEChERS extraction method for the residual analysis of thiacloprid, spirotetramat and spirotetramat's four metabolites in pepper: application of their dissipation patterns. Food Chem. 192, 893–899.

Liang, P., Wang, J., Liu, G., Guan, J., 2014. Determination of sulfonylurea herbicides in food crops by matrix solid-phase dispersion extraction coupled with high-performance liquid chromatography. Food Anal. Method 7, 1530–1535.

Liberatore, L., Procida, G., d'Alessandro, N., Cichelli, A., 2001. Solid-phase extraction and gas chromatographic analysis of phenolic compounds in virgin olive oil. Food Chem. 73, 119–124.

Lichtmannegger, K., Fischer, R., Steemann, F.X., Unterluggauer, H., Masselter, S., 2015. Alternative QuEChERS-based modular approach for pesticide residue analysis in food of animal origin. Anal. Bioanal. Chem. 407, 3727–3742.

Lin, J., Zhang, P., Pan, Z., Xu, H., Luo, Y., Wang, X., 2013. Discrimination of oolong tea (*Camellia sinensis*) varieties based on feature extraction and selection from aromatic profiles analysed by HS-SPME/GC–MS. Food Chem. 141, 259–265.

Lisichkov, K., Kuvendziev, S., Zekovic, Z., Marinkovski, M., 2014. Influence of operating parameters on the supercritical carbon dioxide extraction of bioactive components from common carp (*Cyprinus carpio* L.) viscera. Sep. Purif. Technol. 138, 191–197.

Liu, W., Hu, Y., Zhao, J., Xu, Y., Guan, Y., 2005. Determination of organophosphorus pesticides in cucumber and potato by stir bar sorptive extraction. J. Chromatogr. A 1095, 1–7.

Liu, G., Xu, X., Gong, Y., He, L., Gao, Y., 2012a. Effects of supercritical CO_2 extraction parameters on chemical composition and free radical-scavenging activity of pomegranate (*Punica granatum* L.) seed oil. Food Bioprod. Proc. 90, 573–578.

Liu, J., Tang, X., Zhang, Y., Zhao, W., 2012b. Determination of the volatile composition in brown millet, milled millet and millet bran by gas chromatography/mass spectrometry. Molecules 17, 2271–2282.

Liu, Y., Miao, Z., Guan, W., Sun, B., 2012c. Analysis of organic volatile flavor compounds in fermented stinky tofu using SPME with different fiber coatings. Molecules 17, 3708–3722.

Liu, H., Jiao, Z., Liu, J., Zhang, C., Zheng, X., Lai, S., Chen, F., Yang, H., 2013. Optimization of supercritical fluid extraction of phenolics from date seeds and characterization of its antioxidant activity. Food Anal. Method 6, 781–788.

Liu, J., Sandahl, M., Sjöberg, P.J.R., Turner, C., 2014. Pressurised hot water extraction in continuous flow mode for thermolabile compounds: extraction of polyphenols in red onions. Anal. Bioanal. Chem. 406, 441–445.

Liu, H.-Y., Lin, S.-L, Fuh, M.-R., 2016. Determination of chloramphenicol, thiamphenicol and florfenicol in milk and honey using modified QuEChERS extraction coupled with polymeric monolith-based capillary liquid chromatography tandem mass spectrometry. Talanta 150, 233–239.

Lo Turco, V., Di Bella, G., Potortì, A.G., Fede, M.R., Dugo, G., 2015. Determination of plasticizer residues in tea by solid phase extraction–gas chromatography–mass spectrometry. Eur. Food Res. Technol. 240, 451–458.

Lombardo-Agüí, M., García-Campaña, A.M., Cruces-Blanco, C., Gámiz-Gracia, L., 2015. Determination of quinolones in fish by ultra-high performance liquid chromatography with fluorescence detection using QuEChERS as sample treatment. Food Control 50, 864–868.

Luthria, D.L., Biswas, R., Natarajan, S., 2007. Comparison of extraction solvents and techniques used for the assay of isoflavones from soybean. Food Chem. 105, 325–333.

Machado, A.P.D.F., Pasquel-Reátegui, J.L., Barbero, G.F., Martínez, J., 2015. Pressurized liquid extraction of bioactive compounds from blackberry (*Rubus fruticosus* L.) residues: a comparison with conventional methods. Food Res. Int. 77, 675–683.

Machmudah, S., Zakaria, Winardi, S., Sasaki, M., Goto, M., Kusumoto, N., Hayakawa, K., 2012. Lycopene extraction from tomato peel by-product containing tomato seed using supercritical carbon dioxide. J. Food Eng. 108, 290–296.

Madureira, T.V., Velhote, S., Santos, C., Cruzeiro, C., Rocha, E., Rocha, M.J., 2014. A step forward using QuEChERS (quick, easy, cheap, effective, rugged, and safe) based extraction and gas chromatography-tandem mass spectrometry—levels of priority polycyclic aromatic hydrocarbons in wild and commercial mussels. Environ. Sci. Pollut. Res. 21, 6089–6098.

Majcher, M., Jeleń, H.H., 2009. Comparison of suitability of SPME, SAFE and SDE methods for isolation of flavor compounds from extruded potato snacks. J. Food Compos. Anal. 22, 606–612.

Mane, S., Bremner, D.H., Tziboula-Clarke, A., Lemos, M.A., 2015. Effect of ultrasound on the extraction of total anthocyanins from *Purple Majesty* potato. Ultrason. Sonochem. 27, 509–514.

Manna, L., Bugnone, C.A., Banchero, M., 2015. Valorization of hazelnut, coffee and grape wastes through supercritical fluid extraction of triglycerides and polyphenols. J. Supercrit. Fluids. 104, 204–211.

Maran, J.P., Priya, B., 2016. Multivariate statistical analysis and optimization of ultrasound-assisted extraction of natural pigments from waste red beet stalks. J. Food Sci. Technol. 53 (1), 792–799.

Márquez, V., Martínez, N., Guerra, M., Fariña, L., Boido, E., Dellacassa, E., 2013. Characterization of aroma-impact compounds in yerba mate (*Ilex paraguariensis*) using GC–olfactometry and GC–MS. Food Res. Int. 53, 808–815.

Martins, J., Esteves, C., Limpo-Faria, A., Barros, P., Ribeiro, N., Simões, T., Correia, M., Delerue-Matos, C., 2012. Analysis of six fungicides and one acaricide in still and fortified wines using solid-phase microextraction-gas chromatography/tandem mass spectrometry. Food Chem. 132, 630–636.

Masci, A., Coccia, A., Lendaro, E., Mosca, L., Paolicelli, P., Cesa, S., 2016. Evaluation of different extraction methods from pomegranate whole fruit or peels and the antioxidant and antiproliferative activity of the polyphenolic fraction. Food Chem. 202, 59–69.

Matshediso, P.G., Cukrowska, E., Chimuka, L., 2015. Development of pressurised hot water extraction (PHWE) for essential compounds from *Moringa oleifera* leaf extracts. Food Chem. 172, 423–427.

Mazida, M.M., Salleh, M.M., Osman, H., 2005. Analysis of volatile aroma compounds of fresh chilli (*Capsicum annuum*) during stages of maturity using solid phase microextraction (SPME). J. Food Compos. Anal. 18, 427–437.

Medina, A.L., da Silva, M.A.O., de Sousa Barbosa, H., Arruda, M.A.Z., Marsaioli, Jr., A., Bragagnolo, N., 2015. Rapid microwave assisted extraction of meat lipids. Food Res. Int. 78, 124–130.

Melo, A., Aguiar, A., Mansilha, C., Pinho, O., Ferreira, I.M.P.L.V.O., 2012. Optimisation of a solid-phase microextraction/HPLC/diode array method for multiple pesticide screening in lettuce. Food Chem. 130, 1090–1097.

Nayak, B., Dahmoune, F., Moussi, K., Remini, H., Dairi, S., Aoun, O., Khodir, M., 2015. Comparison of microwave, ultrasound and accelerated-assisted solvent extraction for recovery of polyphenols from *Citrus sinensis* peels. Food Chem. 187, 507–516.

Nemitz, M.C., Teixeira, H.F., von Poser, G.L., 2015. A new approach for the purification of soybean acid extract: simultaneous production of an isoflavone aglycone-rich fraction and a furfural derivative-rich by-product. Ind. Crop. Prod. 67, 414–421.

Ning, L., Fu-ping, Z., Hai-tao, C., Si-yuan, L., Chen, G., Zhen-yang, S., Bao-guo, S., 2011. Identification of volatile components in Chinese Sinkiang fermented camel milk using SAFE, SDE, and HS-SPME-GC/MS. Food Chem. 129, 1242–1252.

Ochiai, N., Sasamoto, K., Takino, M., Yamashita, S., Daishima, S., Heiden, A.C., Hoffmann, A., 2002. Simultaneous determination of preservatives in beverages, vinegar, aqueous sauces, and quasi-drug drinks by stir-bar sorptive extraction (SBSE) and thermal desorption GC–MS. Anal. Bioanal. Chem. 373, 56–63.

Orecchio, S., Ciotti, V.P., Culotta, L., 2009. Polycyclic aromatic hydrocarbons (PAHs) in coffee brew samples: analytical method by GC–MS, profile, levels and sources. Food Chem. Toxicol. 47, 819–826.

Ottonello, G., Ferrari, A., Magi, E., 2014. Determination of polychlorinated biphenyls in fish: optimisation and validation of a method based on accelerated solvent extraction and gas chromatography–mass spectrometry. Food Chem. 142, 327–333.

Paes, J., Dottaa, R., Barbero, G.F., Martínez, J., 2014. Extraction of phenolic compounds and anthocyanins from blueberry (*Vaccinium myrtillus* L.) residues using supercritical CO_2 and pressurized liquids. J. Supercri. Fluid. 95, 8–16.

Paíga, P., Morais, S., Oliva-Teles, T., Correia, M., Delerue-Matos, C., Duarte, S.C., Pena, A., Lino, C.M., 2012. Extraction of ochratoxin A in bread samples by the QuEChERS methodology. Food Chem. 135, 2522–2528.

Pan, Y., Wang, K., Huang, S., Wang, H., Mu, X., He, C., Ji, X., Zhang, J., Huang, F., 2008. Antioxidant activity of microwave-assisted extract of longan (*Dimocarpus Longan* Lour.) peel. Food Chem. 106, 1264–1270.

Pan, X.-D., Wu, P.-G., Jiang, W., Ma, B., 2015. Determination of chloramphenicol, thiamphenicol, and florfenicol in fish muscle by matrix solid-phase dispersion extraction (MSPD) and ultra-high pressure liquid chromatography tandem mass spectrometry. Food Control. 52, 34–38.

Panseri, S., Giani, I., Mentasti, T., Bellagamba, F., Caprino, F., Moretti, V.M., 2008. Determination of flavour compounds in a mountain cheese by headspace sorptive extraction-thermal desorption-capillary gas chromatography-mass spectrometry. LWT Food Sci. Technol. 41, 185–192.

Pap, N., Beszédes, S., Pongrácz, E., Myllykoski, L., Gábor, M., Gyimes, E., Hodúr, C., Keiski, R.L., 2013. Microwave-assisted extraction of anthocyanins from black currant Marc. Food Bioprocess Technol. 6, 2666–2674.

Park, J.S., Kim, K.-Y., Baek, H.H., 2010. Potent aroma-active compounds of cooked Korean non-aromatic rice. Food Sci. Biotechnol. 19 (5), 1403–1407.

Pelajić, M., Peček, G., Pavlovič, D.M., Čepo, D.V., 2016. Novel multiresidue method for determination of pesticides in red wine using gas chromatography–mass spectrometry and solid phase extraction. Food Chem. 200, 98–106.

Peng, B., Shen, Y., Gao, Z., Zhou, M., Ma, Y., Zhao, S., 2015. Determination of total iron in water and foods by dispersive liquid–liquid microextraction coupled with microvolume UV–vis spectrophotometry. Food Chem. 176, 288–293.

Pereira, L.A., Rath, S., 2009. Molecularly imprinted solid-phase extraction for the determination of fenitrothion in tomatoes. Anal. Bioanal. Chem. 393, 1063–1072.

Pérez, D.M., Alatorre, G.G., Álvarez, E.B., Silva, E.E., Alvarado, J.F.J., 2008. Solid-phase microextraction of *N*-nitrosodimethylamine in beer. Food Chem. 107, 1348–1352.

Perretti, G., Miniati, E., Montanari, L., Fantozzi, P., 2003. Improving the value of rice by-products by SFE. J. Supercrit. Fluid. 26, 63–71.

Perretti, G., Marconi, O., Montanari, L., Fantozzi, P., 2004. Rapid determination of total fats and fat-soluble vitamins in Parmigiano cheese and salami by SFE. LWT Food Sci. Technol. 37, 87–92.

Pinela, J., Prieto, M.A., Barreiro, M.F., Carvalho, A.M., Oliveira, M.B.P.P., Vázquez, J.A., Ferreira, I.C.F.R., 2016a. Optimization of microwave-assisted extraction of hydrophilic and lipophilic antioxidants from a surplus tomato crop by response surface methodology. Food Bioprod. Process 98, 283–298.

Pinela, J., Prieto, M.A., Carvalho, A.M., Barreiro, M.F., Oliveira, M.B.P.P., Barros, L., Ferreira, I.C.F.R., 2016b. Microwave-assisted extraction of phenolic acids and flavonoids and production of antioxidant ingredients from tomato: a nutraceutical-oriented optimization study. Sep. Purif. Technol. 164, 114–124.

Pino, J.A., 2014. Odour-active compounds in papaya fruit cv. red maradol. Food Chem. 146, 120–126.

Pino, J.A., Febles, Y., 2013. Odour-active compounds in banana fruit cv. Giant Cavendish. Food Chem. 141, 795–801.

Pissinatti, R., Nunes, C.M., de Souza, A.G., Junqueira, R.G., de Souza, S.V.C., 2015. Simultaneous analysis of 10 polycyclic aromatic hydrocarbons in roasted coffee by isotope dilution gas chromatography-mass spectrometry: optimization, in-house method validation and application to an exploratory study. Food Control 51, 140–148.

Pripdeevech, P., Machan, T., 2011. Fingerprint of volatile flavour constituents and antioxidant activities of teas from Thailand. Food Chem. 125, 797–802.

Prosen, H., Kokalj, M., Janeš, D., Kreft, S., 2010. Comparison of isolation methods for the determination of buckwheat volatile compounds. Food Chem. 121, 298–306.

Puoci, F., Garreffa, C., Iemma, F., Muzzalupo, R., Spizzirri, U.G., Picci, N., 2005. Molecularly imprinted solid phase extraction for detection of Sudan I in food matrices. Food Chem. 93, 349–353.

Puoci, F., Cirillo, G., Curcio, M., Iemma, F., Spizzirri, U.G., Picci, N., 2007. Molecularly imprinted solid phase extraction for the selective HPLC determination of α-tocopherol in bay leaves. Anal. Chim. Acta. 593, 164–170.

Puoci, F., Curcio, M., Cirillo, G., Iemma, F., Spizzirri, U.G., Picci, N., 2008. Molecularly imprinted solid-phase extraction for cholesterol determination in cheese products. Food Chem. 106, 836–842.

Qi, X., 2010. Development of a matrix solid-phase dispersion-sonication extraction method for the determination of fungicides residues in ginseng extract. Food Chem. 121, 758–762.

Qin, P., Ma, T., Wu, L., Shan, F., Ren, G., 2011. Identification of tartary buckwheat tea aroma compounds with gas chromatography-mass spectrometry. J. Food Sci. 76 (6), 401–407.

Quesada-Molina, C., Claude, B., García-Campaña, A.M., del Olmo-Iruela, M., Morin, P., 2012. Convenient solid phase extraction of cephalosporins in milk using a molecularly imprinted polymer. Food Chem. 135, 775–779.

Radišić, M.M., Vasiljević, T.M., Dujaković, N.N., Laušević, M.D., 2013. Application of matrix solid-phase dispersion and liquid chromatography–ion trap mass spectrometry for the analysis of pesticide residues in fruits. Food Anal. Method 6, 648–657.

Raffo, A., Carcea, M., Castagna, C., Magrì, A., 2015. Improvement of a headspace solid phase microextraction-gas chromatography/mass spectrometry method for the analysis of wheat bread volatile compounds. J. Chromatogr. A 1406, 266–278.

Reátegui, J.L.P., Machado, A.P.F., Barbero, G.F., Rezende, C.A., Martínez, J., 2014. Extraction of antioxidant compounds from blackberry (*Rubus* sp.) bagasse using supercritical CO_2 assisted by ultrasound. J. Supercrit. Fluid 94, 223–233.

Rebey, I.B., Kefi, S., Bourgou, S., Ouerghemmi, I., Ksouri, R., Tounsi, M.S., Marzouk, B., 2014. Ripening stage and extraction method effects on physical properties, polyphenol composition and antioxidant activities of cumin (*Cuminum cyminum* L.) seeds. Plant Foods Hum. Nutr. 69, 358–364.

Reto, M., Figueira, M.E., Helder, M., Filipe, H.M., Almeida, C.M.M., 2007. Analysis of vitamin K in green tea leafs and infusions by SPME–GC-FID. Food Chem. 100, 405–411.

Ridgway, K., Lalljie, S.P.D., Smith, R.M., 2010. The use of stir bar sorptive extraction—a potential alternative method for the determination of furan, evaluated using two example food matrices. Anal. Chim. Acta 657, 169–174.

Riu-Aumatell, M., Vargas, L., Vichi, S., Guadayol, J.M., López-Tamames, E., Buxaderas, S., 2011. Characterisation of volatile composition of white salsify (*Tragopogon porrifolius* L.) by headspace solid-phase microextraction (HS-SPME) and simultaneous distillation–extraction (SDE) coupled to GC–MS. Food Chem. 129, 557–564.

Rodil, R., Schellin, M., Popp, P., 2007. Analysis of polycyclic aromatic hydrocarbons in water and beverages using membrane-assisted solvent extraction in combination with large volume injection–gas chromatography–mass spectrometric detection. J. Chromatogr. A 1163, 288–297.

Rodríguez-González, N., González-Castro, M.J., Beceiro-González, E., Muniategui-Lorenzo, S., 2015. Development of a matrix solid phase dispersion methodology for the determination of triazine herbicides in mussels. Food Chem. 173, 391–396.

Romo-Hualde, A., Yetano-Cunchillos, A.I., González-Ferrero, C., Sáiz-Abajo, M.J., González-Navarro, C.J., 2012. Supercritical fluid extraction and microencapsulation of bioactive compounds from red pepper (*Capsicum annum* L.) by-products. Food Chem. 133, 1045–1049.

Rubert, J., Soler, C., Mañes, J., 2012a. Application of an HPLC–MS/MS method for mycotoxin analysis in commercial baby foods. Food Chem. 133, 176–183.

Rubert, J., Dzuman, Z., Vaclavikova, M., Zachariasova, M., Soler, C., Hajslova, J., 2012b. Analysis of mycotoxins in barley using ultra high liquid chromatography high resolution mass spectrometry: comparison of efficiency and efficacy of different extraction procedures. Talanta 99, 712–719.

Rubert, J., Soler, C., Marín, R., James, K.J., Mañes, J., 2013a. Mass spectrometry strategies for mycotoxins analysis in European beers. Food Control 30, 122–128.

Rubert, J., Soriano, J.M., Mañes, J., Soler, C., 2013b. Occurrence of fumonisins in organic and conventional cereal-based products commercialized in France, Germany and Spain. Food Chem. Toxicol. 56, 387–391.

Ruiz-Aceituno, L., Jesús García-Sarrió, M., Alonso-Rodriguez, B., Ramos, L., Sanz, M.L., 2016. Extraction of bioactive carbohydrates from artichoke (*Cynara scolymus* L.) external bracts using microwave assisted extraction and pressurized liquid extraction. Food Chem. 196, 1156–1162.

Ruiz-Montañez, G., Ragazzo-Sánchez, J.A., Calderón-Santoyo, M., Velázquez-de la Cruz, G., Ramírez de León, J.A., Navarro-Ocaña, A., 2014. Evaluation of extraction methods for preparative scale obtention of mangiferin and lupeol from mango peels (*Mangifera indica* L.). Food Chem. 159, 267–272.

Sadeghi, S., Jahani, M., 2013. Selective solid-phase extraction using molecular imprinted polymer sorbent for the analysis of Florfenicol in food samples. Food Chem. 141, 1242–1251.

Sagratini, G., Mañes, J., Giardiná, D., Damiani, P., Picó, Y., 2007. Analysis of carbamate and phenylurea pesticide residues in fruit juices by solid-phase microextraction and liquid chromatography–mass spectrometry. J. Chromatogr. A 1147, 135–143.

Saha, S., Walia, S., Kundu, A., Sharma, L., Paul, R.K., 2015. Optimal extraction and fingerprinting of carotenoids by accelerated solvent extraction and liquid chromatography with tandem mass spectrometry. Food Chem. 177, 369–375.

Sahebnasagh, A., Karimi, G., Mohajeri, S.A., 2014. Preparation and evaluation of histamine imprinted polymer as a selective sorbent in molecularly imprinted solid-phase extraction coupled with high performance liquid chromatography analysis in canned fish. Food Anal. Method. 7, 1–8.

Salvadeo, P., Boggia, R., Evangelisti, F., Zunin, P., 2007. Analysis of the volatile fraction of "pesto genovese" by headspace sorptive extraction (HSSE). LWT Food Sci. Technol. 41, 185–192.

Sandín-España, P., Mateo-Miranda, M., López-Goti, C., De Cal, A., Alonso-Prados, J.L., 2016. Development of a rapid and direct method for the determination of organic acids in peach fruit using LC–ESI-MS. Food Chem. 192, 268–273.

Santos, P., Aguiar, A.C., Barbero, G.F., Rezende, C.A., Martínez, J., 2015. Supercritical carbon dioxide extraction of capsaicinoids from malagueta pepper (*Capsicum frutescens* L.) assisted by ultrasound. Ultrason. Sonochem. 22, 78–88.

Self, R., 2005. Extraction of Organic Analytes from Foods. The Royal Society of Chemistry, Cambridge, UK.

Senol, A., Aydin, A., 2006. Solid–liquid extraction of caffeine from tea waste using battery type extractor: process optimization. J. Food Eng. 75, 565–573.

Sereshti, H., Samadi, S., 2014. A rapid and simple determination of caffeine in teas, coffees and eight beverages. Food Chem. 158, 8–13.

Setyaningsih, W., Palma, M., Barroso, C.G., 2012. A new microwave-assisted extraction method for melatonin determination in rice grains. J. Cereal Sci. 56, 340–346.

Setyaningsih, W., Saputro, I.E., Palma, M., Barroso, C.G., 2015. Optimisation and validation of the microwave-assisted extraction of phenolic compounds from rice grains. Food Chem. 169, 141–149.

Setyaningsih, W., Duros, E., Palma, M., Barroso, C.G., 2016a. Optimization of the ultrasound-assisted extraction of melatonin from red rice (*Oryza sativa*) grains through a response surface methodology. Appl. Acoust. 103, 129–135.

Setyaningsih, W., Saputro, I.E., Palma, M., Barroso, C.G., 2016b. Pressurized liquid extraction of phenolic compounds from rice (*Oryza sativa*) grains. Food Chem. 192, 452–459.

Sganzerla, M., Coutinho, J.P., de Melo, A.M.T., Godoy, H.T., 2014. Fast method for capsaicinoids analysis from *Capsicum chinense* fruits. Food Res. Int. 64, 718–725.

Shi, Y., Peng, D.-D., Shi, C.-H., Zhang, X., Xie, Y.-T., Lu, B., 2011. Selective determination of trace 17b-estradiol in dairy and meat samples by molecularly imprinted solid-phase extraction and HPLC. Food Chem. 126, 1916–1925.

Shrivas, K., Jaiswal, N.K., 2013. Dispersive liquid–liquid microextraction for the determination of copper in cereals and vegetable food samples using flame atomic absorption spectrometry. Food Chem. 141, 2263–2268.

Silva, B.V., Barreira, J.C.M., Oliveira, M.B.P.P., 2016. Natural phytochemicals and probiotics as bioactive ingredients for functional foods: extraction, biochemistry and protected-delivery technologies. Trends Food Sci. Tech. 50, 144–158.

Simándi, B., Sass-Kiss, A., Czukor, B., Deák, A., Prechl, A., Csordás, A., Sawinsky, J., 2000. Pilot-scale extraction and fractional separation of onion oleoresin using supercritical carbon dioxide. J. Food Eng. 46, 183–188.

Singanusong, R., Nipornram, S., Tochampa, W., Rattanatraiwong, P., 2015. Low power ultrasound-assisted extraction of phenolic compounds from mandarin (*Citrus reticulata* Blanco cv. Sainampueng) and lime (*Citrus aurantifolia*) peels and the antioxidant. Food Anal. Meth. 8, 1112–1123.

Słowik-Borowiec, M., 2015. Validation of a QuEChERS-based gas chromatographic method for multiresidue pesticide analysis in fresh peppermint including studies of matrix effects. Food Anal. Method 8, 1413–1424.

Solana, M., Boschiero, I., Dall'Acqua, S., Bertucco, A., 2014. Extraction of bioactive enriched fractions from *Eruca sativa* leaves by supercritical CO_2 technology using different co-solvents. J. Supercrit. Fluid. 94, 245–251.

Soltana, H., Tekaya, M., Amri, Z., El-Gharbi, S., Nakbi, A., Harzallah, A., Mechri, B., Hammami, M., 2016. Characterization of fig achenes' oil of *Ficus carica* grown in Tunisia. Food Chem. 196, 1125–1130.

St'avíková, L., Polovka, M., Hohnová, B., Karáseka, P., Roth, M., 2011. Antioxidant activity of grape skin aqueous extracts from pressurized hot water extraction combined with electron paramagnetic resonance spectroscopy. Talanta 85, 2233–2240.

Sun, H., Wang, L., Ai, L., Liang, S., Wu, H., 2010. A sensitive and validated method for determination of melamine residue in liquid milk by reversed phase high-performance liquid chromatography with solid-phase extraction. Food Control 21, 686–691.

Sun, H., Qi, H., Li, H., 2013. Development of capillary electrophoretic method combined with accelerated solvent extraction for simultaneous determination of residual sulfonamides and their acetylated metabolites in aquatic products. Food Anal. Method 6, 1049–1055.

Sutivisedsak, N., Cheng, H.N., Willett, J.L., Lesch, W.C., Tangsrud, R.R., Biswas, A., 2010. Microwave-assisted extraction of phenolics from bean (*Phaseolus vulgaris* L.). Food Res. Int. 43, 516–519.

Tabaraki, R., Heidarizadi, E., Benvidi, A., 2012. Optimization of ultrasonic-assisted extraction of pomegranate (*Punica granatum* L.) peel antioxidants by response surface methodology. Sep. Puri. Technol. 98, 16–23.

Tang, K., Gu, X., Luo, Q., Chen, S., Wu, L., Xiong, J., 2014. Preparation of molecularly imprinted polymer for use as SPE adsorbent for the simultaneous determination of five sulphonylurea herbicides by HPLC. Food Chem. 150, 106–112.

Tanongkankit, Y., Sablani, S.S., Chiewchan, N., Devahastin, S., 2013. Microwave-assisted extraction of sulforaphane from white cabbages: effects of extraction condition, solvent and sample pretreatment. J. Food Eng. 117, 151–157.

Tao, Y., Chen, D., Chao, X., Yu, H., Yuanhu, P., Liu, Z., Huang, L., Wang, Y., Yuan, Z., 2011. Simultaneous determination of malachite green, gentian violet and their leuco-metabolites in shrimp and salmon by liquid chromatographyetandem mass spectrometry with accelerated solvent extraction and auto solid-phase clean-up. Food Control 22, 1246–1252.

Tao, Y., Yu, G., Chen, D., Pan, Y., Liu, Z., Wei, H., Peng, D., Huang, L., Wang, Y., Yuan, Z., 2012. Determination of 17 macrolide antibiotics and avermectins residues in meat with accelerated solvent extraction by liquid chromatography–tandem mass spectrometry. J. Chromatogr. B 897, 64–71.

Tao, Y., Zhu, F., Chen, D., Wei, H., Pan, Y., Wang, X., Liu, Z., Huang, L., Wang, Y., Yuan, Z., 2014. Evaluation of matrix solid-phase dispersion (MSPD) extraction for multi-fenicols determination in shrimp and fish by liquid chromatography–electrospray ionisation tandem mass spectrometry. Food Chem. 150, 500–506.

Teng, H., Lee, W.Y., 2013. Optimization of microwave-assisted extraction of polyphenols from mulberry fruits (*Morus alba* L.) using response surface methodology. J. Korean Soc. Appl. Biol. Chem. 56, 317–324.

Teo, C.C., Tan, S.N., Yong, J.W.H., Hew, C.S., Ong, E.S., 2009. Validation of green-solvent extraction combined with chromatographic chemical fingerprint to evaluate quality of *Stevia rebaudiana* Bertoni. J. Sep. Sci. 32, 613–622.

Thirugnanasambandham, K., Sivakumar, V., Maran, J.P., 2015. Microwave-assisted extraction of polysaccharides from mulberry leaves. Int. J. Biol. Macromol. 72, 1–5.

Thräne, C., Isemer, C., Engelhardt, U.H., 2015. Determination of nicotine in tea (*Camellia sinensis*) by LC–ESI–MS/MS using a modified QuEChERS method. Eur. Food Res. Technol. 241, 227–232.

Tian, Y., Xu, Z., Zheng, B., Lo, Y.M., 2013. Optimization of ultrasonic-assisted extraction of pomegranate (*Punica granatum* L.) seed oil. Ultrason. Sonochem. 20, 202–208.

Tomaz, I., Maslov, L., Stupi, D., Preiner, D., Ašperger, D., Kontić, J.K., 2015. Multi-response optimisation of ultrasound-assisted extraction for recovery of flavonoids from red grape skins using response surface methodology. Phytochem. Anal 27, 13–22.

Tsochatzis, E.D., Tzimou-Tsitouridou, R., 2015. Validated RP-HPLC method for simultaneous determination of tocopherols and tocotrienols in whole grain barley using matrix solid-phase dispersion. Food Anal Methods 8, 392–400.

Turner, C., Mathiasson, L., 2000. Determination of vitamins A and E in milk powder using supercritical fluid extraction for sample clean-up. J. Chromatography A. 874, 275–283.

Upadhyay, R., Ramalakshmi, K., Jagan Mohan Rao, L., 2012. Microwave-assisted extraction of chlorogenic acids from green coffee beans. Food Chem. 130, 184–188.

Vasapollo, G., Longo, L., Rescio, L., Ciurlia, L., 2004. Innovative supercritical CO_2 extraction of lycopene from tomato in the presence of vegetable oil as co-solvent. J. Supercr. Fluid 29, 87–96.

Vázquez-Araújo, L., Verdú, A., Navarro, P., Martínez-Sánchez, F., Carbonell-Barrachina, A.A., 2009. Changes in volatile compounds and sensory quality during toasting of Spanish almonds. Int. J. Food Sci. Tech. 44, 2225–2233.

Vazquez-Roig, P., Pico, Y., 2015. Pressurized liquid extraction of organic contaminants in environmental and food samples. Trends Analyt. Chem. 71, 55–64.

Víctor-Ortega, M.D., Lara, F.J., García-Campaña, A.M., del Olmo-Iruela, M., 2013. Evaluation of dispersive liquideliquid microextraction for the determination of patulin in apple juices using micellar electrokinetic capillary chromatography. Food Control 31, 353–358.

Viñas, P., Aguinaga, N., Campillo, N., Hernández-Córdoba, M., 2008. Comparison of stir bar sorptive extraction and membrane-assisted solvent extraction for the ultra-performance liquid chromatographic determination of oxazole fungicide residues in wines and juices. J. Chromatogr. A 1194, 178–183.

Viñas, P., Campillo, N., Martínez-Castillo, N., Hernández-Córdoba, M., 2009. Method development and validation for strobilurin fungicides in baby foods by solid-phase microextraction gas chromatography–mass spectrometry. J. Chromatogr. A 1216, 140–146.

Wang, H., Ding, J., Ren, N., 2016. Recent advances in microwave-assisted extraction of trace organic pollutantsfrom food and environmental samples. Trends Anal. Chem. 75, 197–208.

Wang, F., Guo, X.-Y., Zhang, D.-N., Wu, Y., Wu, T., Chen, Z.-G., 2015. Ultrasound-assisted extraction andpurification of taurine from the red algae Porphyra yezoensis. Ultrason. Sonochem. 24, 36–42.

Wang, P., Zhang, Q., Wang, Y., Wang, T., Li, X., Ding, L., Jiang, G., 2010. Evaluation of Soxhlet extraction, accelerated solvent extraction and microwave-assisted extraction for the determination of polychlorinated biphenyls and polybrominated diphenyl ethers in soil and fish samples. Anal Chim. Acta 663, 43–48.

Wang, Y., Liu, X., Xiao, C., Wang, Z., Wang, J., Xiao, H., Cui, L., Xiang, Q., Yue, T., 2012. HPLC determination of aflatoxin M1 in liquid milk and milk powder using solid phase extraction on OASIS HLB. Food Control 28, 131–134.

Wen, X., Yang, Q., Yan, Z., Deng, Q., 2011. Determination of cadmium and copper in water and food samples by dispersive liquid–liquid microextraction combined with UV–vis spectrophotometry. Microchem. J. 97, 249–254.

Wu, R.-N., Han, F.-L., Shang, J., Hu, H., Han, L., 2009. Analysis of patulin in apple products by liquid–liquid extraction, solid phase extraction and matrix solid-phase dispersion methods: a comparative study. Eur. Food Res. Technol. 228, 1009–1014.

Wu, G., Bao, X., Zhao, S., Wu, J., Han, A., Ye, Q., 2011. Analysis of multi-pesticide residues in the foods of animal origin by GC–MS coupled with accelerated solvent extraction and gel permeation chromatography cleanup. Food Chem. 126, 646–654.

Xie, J., Sun, B., Zheng, F., Wang, S., 2008. Volatile flavor constituents in roasted pork of mini-pig. Food Chem. 109, 506–514.

Xu, Y.-Q., Wang, C., Li, C.-W., Liu, S.-H., Zhang, C.-X., Li, L.-W., Jiang, D.-H., 2016. Characterization of aroma-active compounds of pu-erh tea by headspace solid-phase microextraction (HS-SPME) and simultaneous distillation-extraction (SDE) coupled with GC-olfactometry and GC-MS. Food Anal. Method 9, 1188–1198.

Yang, Z., Yang, W., Peng, Q., He, Q., Feng, Y., Luo, S., Yu, Z., 2009. Volatile phytochemical composition of rhizome of ginger after extraction by headspace solid-phase microextraction, petroleum ether extraction and steam distillation extraction. Bangladesh J. Pharmacol. 4, 136–143.

Yılmaz, T., Tavman, S., 2016. Ultrasound assisted extraction of polysaccharides from hazelnut skin. Food Sci. Technol. Int. 22 (2), 112–121.

Yoshioka, N., Ichihashi, K., 2008. Determination of 40 synthetic food colors in drinks and candies by high-performance liquid chromatography using a short column with photodiode array detection. Talanta 74, 1408–1413.

Yu, H., Tao, Y., Le, T., Chen, D., Ishsan, A., Liu, Y., Wang, Y., Yuan, Z., 2010. Simultaneous determination of amitraz and its metabolite residue in food animal tissues by gas chromatography-electron capture detector and gas chromatography–mass spectrometry with accelerated solvent extraction. J. Chromatogr. B. 878, 1746–1752.

Yu, H., Tao, Y., Chen, D., Wang, Y., Huang, L., Peng, D., Dai, M., Liu, Z., Wang, X., Yuan, Z., 2011a. Development of a high performance liquid chromatography method and a liquid chromatography–tandem mass spectrometry method with the pressurized liquid extraction for the quantification and confirmation of sulfonamides in the foods of animal origin. J. Chromatogr. B 879, 2653–2662.

Yu, H., Tao, Y., Chen, D., Wang, Y., Yuan, Z., 2011b. Development of an HPLC–UV method for the simultaneous determination of tetracyclines in muscle and liver of porcine, chicken and bovine with accelerated solvent extraction. Food Chem. 124, 1131–1138.

Yu, H., Tao, Y., Chen, D., Pan, Y., Liu, Z., Wang, Y., Huang, L., Dai, M., Peng, D., Wang, X., Yuan, Z., 2012. Simultaneous determination of fluoroquinolones in foods of animal origin by a high performance liquid chromatography and a liquid chromatography tandem mass spectrometry with accelerated solvent extraction. J. Chromatogr. B 885–886, 150–159.

Zacharis, C.K., Rotsias, I., Zachariadis, P.G., Zotos, A., 2012. Dispersive liquid–liquid microextraction for the determination of organochlorine pesticides residues in honey by gas chromatography-electron capture and ion trap mass spectrometric detection. Food Chem. 134, 1665–1672.

Zaghdoudi, K., Pontvianne, S., Framboisier, X., Achard, M., Kudaibergenova, R., Ayadi-Trabelsi, M., Kalthoum-cherif, J., Vanderesse, R., Frochot, C., Guiavarc'h, Y., 2015. Accelerated solvent extraction of carotenoids from: Tunisian Kaki (*Diospyros kaki* L.), peach (*Prunus persica* L.) and apricot (*Prunus armeniaca* L.). Food Chem. 184, 131–139.

Zaibunnisa, A.H., Norashikin, S., Mamot, S., Osman, H., 2009. An experimental design approach for the extraction of volatile compounds from turmeric leaves (*Curcuma domestica*) using pressurised liquid extraction (PLE). LWT Food Sci. Technol. 42, 233–238.

Zambonin, C.G., Quinto, M., De Vietro, N., Palmisano, F., 2004. Solid-phase microextraction—gas chromatography mass spectrometry: a fast and simple screening method for the assessment of organophosphorus pesticides residues in wine and fruit juices. Food Chem. 86, 269–274.

Zarzycki, P.K., Portka, J.K., 2015. Recent advances in hopanoids analysis: quantification protocols overview, main research targets and selected problems of complex data exploration. J. Steroid Biochem. Mol. Biol. 153, 3–26.

Zarzycki, P.K., Zarzycka, M.B., 2008. Evaluation of the water and organic liquids extraction efficiency of *Spirulina maxima* dyes using thermostated micro thin-layer chromatography. J. AOAC Int. 91, 1196–1202.

Zarzycki, P.K., Zarzycka, M.B., Ślączka, M.M., Clifton, V.L., 2010. Acetonitrile, the polarity chameleon. Anal. Bioanal. Chem. 397, 905–908.

Zarzycki, P.K., Zarzycka, M.B., Clifton, V.L., Adamski, J., Głód, B.K., 2011. Low-parachor solvents extraction and thermostated micro-thin-layer chromatography separation for fast screening and classification of spirulina from pharmaceutical formulations and food samples. J. Chromatogr. A 1218, 5694–5704.

Zhang, S., Yang, X., Yin, X., Wang, C., Wang, Z., 2012. Dispersive liquid–liquid microextraction combined with sweeping micellar electrokinetic chromatography for the determination of some neonicotinoid insecticides in cucumber samples. Food Chem. 133, 544–550.

Zhang, Y., Huang, M., Tian, H., Sun, B., Wang, J., Li, Q., 2014. Preparation and aroma analysis of Chinese traditional fermented flour paste. Food Sci. Biotechnol. 23 (1), 49–58.

Zhang, Y., Zhang, Y., Zhao, Q., Chen, W., Jiao, B., 2016. Vortex-assisted ionic liquid dispersive liquid-liquid microextraction coupled with high-performance liquid chromatography for the determination of triazole fungicides in fruit juices. Food Anal Method 9, 596–604.

Zhao, D., 2014. Determination of pentachlorophenol residue in meat and fish by gas chromatography–electron capture detection and gas chromatography–mass spectrometry with accelerated solvent extraction. J. Chromatogr. Sci. 52, 429–435.

Zhao, X., Ma, F., Li, P., Li, G., Zhang, L., Zhang, Q., Zhang, W., Wang, X., 2015. Simultaneous determination of isoflavones and resveratrols for adulteration detection of soybean and peanut oils by mixed-mode SPE LC–MS/MS. Food Chem. 176, 465–471.

Zhou, G., Yao, X., Tang, Y., Qian, D., Su, S., Zhang, L., Jin, C., Qin, Y., Duan, J., 2014. An optimized ultrasound-assisted extraction and simultaneous quantification of 26 characteristic components with four structure types in functional foods from ginkgo seeds. Food Chem. 158, 177–185.

Zhu, Q., Wang, L., Wu, S., Joseph, W., Gu, X., Tang, J., 2009. Selectivity of molecularly imprinted solid phase extraction for sterol compounds. Food Chem. 113, 608–615.

Zhuang, W., McKague, B., Reeve, D., Carey, J., 2004. A comparative evaluation of accelerated solvent extraction and Polytron extraction for quantification of lipids and extractable organochlorine in fish. Chemosphere 54, 467–480.

Zill-e-Huma, Abert-Vian, M., Elmaataoui, M., Chemat, F., 2011. A novel idea in food extraction field: study of vacuum microwave hydrodiffusion technique for by-products extraction. J. Food Eng. 105, 351–360.

Zinedine, A., Blesa, J., Mahnine, N., El Abidi, A., Montesano, D., Mañes, J., 2010. Pressurized liquid extraction coupled to liquid chromatography for the analysis of ochratoxin A in breakfast and infants cereals from Morocco. Food Control. 21, 132–135.

Zou, Y., Xie, C., Fan, G., Gu, Z., Han, Y., 2010. Optimization of ultrasound-assisted extraction of melanin from *Auricularia auricula* fruit bodies. Innov. Food Sci. Emerg. 11, 611–615.

Zuin, V.G., Schellin, M., Montero, L., Yariwake, J.H., Augusto, F., Popp, P., 2006. Comparison of stir bar sorptive extraction and membrane-assisted solvent extraction as enrichment techniques for the determination of pesticide and benzo[a]pyrene residues in Brazilian sugarcane juice. J. Chromatogr. A 1114, 180–187.

Zunin, P., Salvadeo, P., Boggia, R., Lanteri, S., 2009. Study of different kinds of "esto genovese" by the analysis of their volatile fraction and chemometric methods. Food Chem. 114, 306–309.

CHAPTER 4

Valorization of Agrifood By-Products by Extracting Valuable Bioactive Compounds Using Green Processes

Ramiro A. Carciochi*, Leandro G. D'Alessandro, Peggy Vauchel**, María M. Rodriguez*, Susana M. Nolasco*, Krasimir Dimitrov****

**National University of Central Buenos Aires, Olavarría, Buenos Aires, Argentina; **Lille University, INRA, ISA, Artois University, University of Littoral Opal Coast, Charles Viollette Institute, Lille, France*

1 Introduction

The Food and Agricultural Organization (FAO) estimates that each year, approximately one-third of all food produced for human consumption in the world is lost or wasted (FAO, 2013). The global volume of food wastage (i.e., both food loss and food waste) is estimated to be 1.6 gigatons of "primary product equivalents," whereas the total waste for the edible part of food is about 1.3 gigatons. This amount can be weighed against total agricultural production (for food and nonfood uses), which is about 6 gigatons per year (FAO, 2013). Thus, very significant amounts of edible food mass is lost, discarded, or degraded in different stages of the food supply chain: production, postharvest handling, processing, distribution, and consumption. Disposal of these wastes represents both the cost to the food processor and potential negative impact on the environment (Wijngaard et al., 2012b). At the same time, the volume of crop residues (nonedible plant parts that are left in the field or orchard after the main crop has been harvested) is also very important. Recently, much attention has been directed to the recovery of bioactive compounds from different residues: leaves, peels, barriers, seeds, wood, culls, rinds, pits, pulp, press cakes, marc, malts, hops, hulls, husks, spent grain, and so forth (Angiolillo et al., 2015). Many of these agrifood wastes could be valorized as animal feed, compost, or transformed into biomass-based energy fuel and a wide variability of industrial products, such as wood-based panels, bio-fertilizers, biofibers among others (Santana-Meridas et al., 2012). Research on these agrifood matrices has also revealed a wide range of natural bioactive compounds including a broad diversity of structures and functionalities providing an excellent pool of molecules for the production of nutraceuticals, functional foods, and food additives. The maximal valorization of bioresources by extracting first the valuable biomolecules (vitamins, minerals,

Ingredients Extraction by Physicochemical Methods in Food
http://dx.doi.org/10.1016/B978-0-12-811521-3.00004-1

proteins, alkaloids, antioxidant polyphenols, lipids, sugars, pectin, dietary fibers, aroma, and volatile compounds) from agrifood wastes before proceeding to another valorization of the vegetal matrix is closely related to the biorefinery concept requiring maximum utilization of the natural resources during their processing and minimum wastes. The biorefinery is an industrial facility, or network of facilities, that covers an extensive range of combined technologies aimed at full sustainable transformation of biomass into building blocks with the concomitant production of biofuels, energy, commodity, or specialty chemicals and materials, preferably of added value (Morais and Bogel-Lukasik, 2013). This concept should be applied also to the extraction of natural products used as ingredients (food, personal and home care, pharmaceutics), food supplements (nutrition), or active compounds (pharmaceutics) (Rombaut et al., 2014). It means that from a given agro-resource and even from a given agrifood waste, more than one product could be obtained by applying several processes adapted for recovery of different types of valuable biomolecules. For example, Kammerer et al. (2014) have described a technology allowing attainment of one natural extract rich in proteins and another extract rich in antioxidant polyphenols from sunflower press cake, and Boukroufa et al. (2015) have proposed a schema for the valorization of orange peels by recovering essential oil and two extracts: one rich in polyphenols and other in pectin. It should be mentioned that the term "food by-products" is increasingly used to signify that "food wastes" are ultimate substrates for the recapture of functional compounds and the development of new products with a market value (Galanakis, 2012).

The potentially marketable components present in foods wastes and by-products need to be separated from the vegetal matrix through combined approaches (biochemical, chemical, and physical) for selective extraction and modification of the targeted components and changed into higher-value food products or additives (Baiano, 2014). These operations must be performed avoiding microbiological hazards and ensuring that the final products comply with the existing food regulations and meet consumer liking. The processes used to recover the valuable biomolecules from agrifood by-products are the same as those used in the extraction from raw vegetal sources. In most cases, an appropriate solvent extraction is used to recover target substances and separate them from the vegetal matrix. Vegetable oils could also be extracted by mechanical pressing, with volatile compounds by steam or hydrodistillation. To enhance extraction efficiency, the extraction can be preceded by various pretreatment processes, such as milling, mixing, homogenization, extrusion, dehydration, drying, and so on. Enzymatic hydrolysis of the solid matrix could facilitate the access of solvent into the vegetal matrix and help the diffusion of the extractible molecules into the solvent. In some cases thermal pretreatment could also affect strongly extraction efficiency, since depending on the temperature, key enzymes could be activated or inactivated (Galanakis, 2012). Usually, solvent extraction is not very selective and many other molecules are coextracted with the target molecules. If the content of valuable species in the extract is not satisfactory, other separation processes can be applied to enrich the extracts in these molecules, such as membrane processes, adsorption, liquid–liquid extraction, and so forth. Finally, to obtain the final product, the liquid extract is usually dried (freeze drying or spray draying). Since it is very important to preserve the functional properties of the extracted biomolecules, in some cases it is suitable to emulsify the aqueous extracts in organic solvents (vegetable oils, for example) or encapsulate the dried extract to protect the extracted target molecules from the environment (Pittia and Gharsallaoui, 2015).

2 Green Extraction of Natural Products From Agrifood By-products

The solvent extraction of valuable compounds from agrifood by-products can be carried out by using conventional extraction processes, such as maceration, infusion, decoction, or Soxhlet extraction. To overcome the problems encountered when using conventional methods and, namely, to reduce the energy consumption and negative impact on the environment and human beings, several alternative extraction processes [ultrasound-assisted extraction (UAE), microwave-assisted extraction, pressurized-liquid extraction, supercritical-fluid extraction (SFE), etc.] have been developed. These emerging extraction processes are considered "green extraction processes" and attract more and more the interest of researchers and industrials. In 2012, Chemat et al. defined the following "6 principles of Green Extraction of Natural Products:"

- Principle 1: Innovation by selection of varieties and use of renewable plant resources;
- Principle 2: Use of alternative solvents and principally water or agrosolvents;
- Principle 3: Reduce energy consumption by energy recovery and using innovative technologies;
- Principle 4: Production of by-products instead of waste to include the bio- and agrorefining industry;
- Principle 5: Reduce unit operations and favor safe, robust and controlled processes;
- Principle 6: Aim for a nondenatured and biodegradable extract without contaminants.

The listing of these six principles should be viewed by industry and scientists as a direction to establish an innovative and green label, charter, and standard, and as a reflection to innovate not only in process but in all aspects of extraction. The principles have been identified and described not as rules but more as innovative examples to follow, discovered by scientists and successfully applied by industry (Chemat et al., 2012). The fourth principle is directly related to the valorization of agrifood by products by extracting valuable molecules.

Taking into account the large diversity of agrifood by-products, the factors that must be considered include not only the types and the amounts of valuable molecules in these wastes but also the types of the vegetal matrixes, different extraction processes, solvents, and operating conditions that could be best adapted to extract a target substance from a given by-product efficiently. Actually, many researchers work on the establishment of the optimal processes and conditions for the recovery of different biomolecules from agrifood by-products. Some examples of recent research works considering conventional and some green extraction processes are given in Table 4.1. Moure et al. (2001) and Schieber et al. (2001) have reviewed the existing studies on

Table 4.1: Some examples of extracted biomolecules from agrifood by-products.

Sources	By-Products	Target Compounds	Extraction Methods	References
Apple	Pomace	Flavan-3-ols, procyanidins, flavanols, phenolic acids	UAE	Pingret et al. (2012); Virot et al. (2010)
Apple	Pomace	Flavonoids, phenolic acids	CE, MAE, UAE, PLE	Grigoras et al. (2013)
Apple	Pressing residue	Phenolic compounds	CE	Peschel et al. (2006)
Aronia	Pressing residue	Anthocyanins	UAE	Galván D'Alessandro et al. (2014)
Black currant	Pressing residue	Anthocyanins	CE	Lapornik et al. (2005)
Black currant	Pressing residue	Anthocyanins	UAE	Holtung et al. (2011)
Broccoli	Stems, stalks	Phenolic compounds	CE	Peschel et al. (2006)
Cherry	Kernel	β-kerotene	CE, SFE	Yilmaz and Gokmen (2013)
Chesnut tree	Wood waste	Phenolic compounds	CE	Gironi and Piemonte (2011)
Chicory	Grounds	Phenolic compounds	UAE	Pradal et al. (2016)
Chicory	Stems, stalks	Phenolic compounds	CE	Peschel et al. (2006)
Citrus (*Citrus inshiu*)	Peels	Phenolic acids	CE, UAE	Ma et al. (2009)
Citrus (*C. limon*)	Peels	Phenolic compounds	CE, MAE, UAE	Dahmoune et al. (2013)
Coffee	Spent coffee grounds	Phenolic compounds, caffeine	CE, SFE, UAE	Andrade et al. (2012)
Cucumber	Stems, stalks	Phenolic compounds	CE	Peschel et al. (2006)
Endive	Stems, stalks	Phenolic compounds	CE	Peschel et al. (2006)
Grape	Mark	Anthocyanins, flavan-3-ols, flavonols, tannins	CE	Makris et al. (2008); Sant'Anna et al. (2012)
Grape	Mark	Flavonoids, catechins, phenolic acids	UAE, PLE	Paini et al. (2016)
Grape	Pomace	Anthocyanins, catechins, phenolic acids, phenolic alcohols, stilbenes, flavonol glycosides	CE	Spigno and de Faveri (2007)
Grape	Seeds	Phenolic compounds	CE	Bucic-Kojic et al. (2007)
Grape	Seeds and skins	Flavonoids, phenolilc acids, resveratrol, hydroxymethylfurfural	CE, MAE, UAE, PLE	Casazza et al. (2010)
Grape	Skins	Resveratrol	CE	Liu et al. (2013)
Grapefruit	Peels	Pectin	CE, UAE, MAE	Bagherian et al. (2011)
Mandarine	Peels	Phenolic compounds	CE	Karcheva et al. (2013)
Melon	Peels	Pectin	MAE	Prakash Maran et al. (2014)
Navel orange	Peels	Pectin	CE, PLE, MAE	Guo et al. (2012)
Olive	Leaves	Oleurepein	PLE	Xynos et al. (2014)
Olive	Leaves	Phenolic compounds	UAE	Sahin and Samli (2013)

Table 4.1: Some examples of extracted biomolecules from agrifood by-products. (*cont.*)

Sources	By-Products	Target Compounds	Extraction Methods	References
Olive	Pomace	Flavonoids, oleuropein, phenolic acids	UAE, PLE	Paini et al. (2016)
Orange	Peels	Pectin	MAE	Prakash Maran et al. (2013)
Peanut	Defatted flour	Proteins	PLE	Dong et al. (2011)
Pear	Pressing residue	Phenolic compounds	CE	Peschel et al. (2006)
Plum	Seeds	Peptides	UAE	González-García et al. (2014)
Pomegranate	Peel	Phenolic compounds	UAE	Kaderides et al. (2015); Tabaraki et al. (2012)
Pomegranate	Seeds	Vegetable oil	UAE	Kalamara et al. (2015)
Potato	Peels	Alkaloids	UAE	Hossain et al. (2014)
Potato	Peels	Glycoalkaloids, phenolic acids	CE	Maldonado et al. (2014)
Potato	Peels	Phenolic acids	PLE	Singh and Saldaña (2011)
Red beet	Pressing residue	Phenolic compounds	CE	Peschel et al. (2006)
Red currant	Pressing residue	Anthocyanins	CE	Lapornik et al. (2005)
Rosemary	Residue after SFE	Rosmarinic acid	CE, UAE	Zibetti et al. (2013)
Strawberry	Pressing residue	Phenolic compounds	CE	Peschel et al. (2006)
Tomato	Peels	Phenolic compounds	CE	Peschel et al. (2006)
Tomato	Seeds	Vegetable oil	SFE, PLE	Eller et al. (2010)

CE, Conventional extraction; MAE, microwave-assisted extraction; PLE, pressurized liquid extraction; UAE, ultrasound-assisted extraction.

extraction of natural products from food by-products at the beginning of 21st century, and very recently Galanakis (2015) has edited a special book on food-waste recovery considering different aspects of food waste management and valorization; classification of the target compounds; universal recovery strategy; conventional and emerging technologies for pretreatment, extraction, purification, and product formation; and some commercialization aspects and applications (Galanakis, 2015). Recently, some reviews have been published to present the state of the art of conventional and green extraction processes applied on some food industry by-products, such as winery by-products (grape mark) (Kammerer et al., 2014; Teixeira et al., 2014), tomato juice (Strati and Oreopoulou, 2014) and fruit-juice-industry pressing residues (Kammerer et al., 2014), sunflower oil (Kammerer et al., 2014), and olive oil production by-products (cakes, wastewaters) (Araujo et al., 2015; Roselló-Soto et al., 2015a).

In this chapter the main green extraction processes applied for recovery of bioactive compounds from agrifood by-products, such as enzyme-assisted extraction (EAE), UAE, microwave-assisted extraction (MAE), pressurized liquid extraction (PLE), and SFE will be presented. Some processes, such as enzymatic hydrolysis, ultrasound assistance, pulsed electric fields (PEF), and high-voltage electrical discharges (HVED) could be applied both in pretreatment stages and during the extraction process. Another pretreatment process will also be described, which is the instant controlled pressure drop (DIC process), which allows the considerable enhancement of the extraction process. For each process, the principle will be described, then examples of its application for recovery of valuable biomolecules from agrifood by-products will be presented. The green impact of the process will be related to the six principles of the green extraction. Finally, some examples of simultaneous application of at least two processes (hybrid extraction processes) and process integration allow the enhancement of the extraction yields and attainment of enriched natural extracts at reduced time and energy (integrated processes) will be presented.

3 Enzyme-Assisted Extraction

Enzyme-assisted extraction (EAE) is a nontraditional and environmentally friendly technology, that is, becoming popular due to its ability to improve the yield of target compounds while it reduces the use of solvents in the process. The enzymatic treatment of the plant material is usually used prior to the conventional solvent extraction process. This pretreatment of the raw material generally results in a reduction in the extraction time and offers the possibility of processing with low temperature, thereby decreasing the energy consumption (Puri et al., 2012). This is feasible because enzymes can catalyze reactions with specificity and regioselectivity operating in aqueous solutions under mild conditions (Meyer, 2010), thereby reducing the use of organic solvents and energy consumption in accordance with the second and third principles of Green Extraction of Natural Products. The use of enzymes in the extraction process is based on their ability to degrade the cell walls and membranes, thereby increasing cell wall permeability, and enabling targeted bioactive compounds release into the medium.

The composition of cell walls depends on the source, but cell walls mainly consist of interconnecting polysaccharides, such as cellulose, hemicellulose, lignin, and pectin. Therefore, to improve the extraction of bioactive compounds, the use of various common food-grade enzymes, such as cellulases, pectinases, and hemicellulases often is required to disrupt the structural integrity of the plant cell wall. Enzymes can be derived from bacteria, fungi, animal organs, or vegetable/fruit extracts, microbial enzymes being the most frequently used. Selection of the appropriate hydrolytic enzyme or enzyme combination is essential. For this purpose, it is important to know the cell wall composition of the raw material, as well as to understand the catalytic property, mode of action, and optimum operational conditions of enzymes (Puri et al., 2012). After selection of the suitable enzymes, various process conditions can be modified, to find the optimal conditions in which the recovery of desirable compounds is maximized. There are several factors directly conditioning the effect of enzymes on the release of valuable bioactive compounds. These parameters include enzyme concentration, time, and temperature of incubation, pH, enzyme-to-substrate ratio, type of extraction solvent, and solvent-to-substrate ratio, which should be optimized for each specific process. In addition, agitation, as well as substrate particle size, are also other important factors to be considered. In fact, it has been mentioned that a particle size reduction of the substrate prior to enzymatic treatment provides better accessibility of the enzyme into the plant cell, increasing significantly the extraction yields (Pinelo et al., 2006). An advantage of the EAE process is that although enzymes normally function at an optimal temperature, they can still be used over a range of temperatures, providing flexibility for both cost and product quality (Puri et al., 2012).

EAE can easily be tested and optimized on the laboratory scale, and the use of common food-grade enzymes makes it a relatively low-cost technique for extraction purposes. However, some technical limitations on large-scale application include: (1) higher relative cost for processing large volumes of raw material (associated with the price of enzymes); (2) the current availability of enzyme preparations that cannot completely hydrolyze plant cell walls; and (3) the difficulty to optimize the process at industrial scale because enzymes behave differently as the environmental conditions change (percentage of dissolved oxygen, temperature, and nutrient availability) (Baiano, 2014; Puri et al., 2012).

Table 4.2 summarizes some studies applying enzymatic treatment for the extraction of valuable compounds from agrifood by-products. Popular applications of EAE in the food industry include phenolic compound extraction from apple and grape by-products. These phytochemicals are often embedded in the plant matrix, bound to polysaccharide fraction through ether linkages and cellulosic microfibrils via hydrophobic interactions and hydrogen bonding (Pinelo et al., 2006). Moreover lignin and pectin materials further shield bound phenolics. Thus, conventional solvent extraction only extracts the most accessible and weakly bonded phenols from these sources. Hence, the application of exogenous enzymes can enhance the extraction of phenolic compounds by (1) hydrolytic degradation of the cell wall polysaccharides that retain the phenolics in the polysaccharide-lignin/pectin network by hydrogen or hydrophobic bonding or (2) direct hydrolysis of the ether and/or ester cleavage between phenolic compounds and plant cell wall polymers (Pinelo et al., 2008). However, a special remark should be taken into account for phenolic compounds extraction. Phenolic compounds are a heterogeneous group, whose differences in the chemical structure confer them differences in relation to their physicochemical properties, such as solubility, thermal and chemical stability, and ability to link to other compounds. Thus, the various polyphenol groups can react differently to the enzymatic treatment. In fact, Arnous and Meyer (2010) showed the difference in the behavior of various polyphenols to enzymatic treatment, and hence the importance of monitoring the various groups (or a particular target group) of polyphenols. The study proved that anthocyanins from grape skins were extracted in the early stage of the enzymatic (pectinolytic and cellulolytic activities) hydrolysis and were degraded after further enzymatic treatment, whereas the release of phenolic acids was

Table 4.2: Examples of EAE of valuable compounds from agrifood by-products.

Target Compounds	Agrifood By-Product Sources	Recovery Yields	Treatment Conditions	References
Pectin	Chicory root	34,600 mg/100 g	*Em* = Commercial cellulases (Celluclast 1.5 L, Cellulyve TR400, Maxazyme CL2000), pectinase (Pectinex AR), and protease (Neutrase) mixture; *E/F* = 5% (w/w); *T* = 40°C; *pH* = 5.5	Panouillé et al. (2006)
Lycopene	Tomato peel	42.9 mg/100 g for cellulase; 110.4 mg/100 g for pectinase	*Em* = Cellulase (Celluclast-1.5 L), 3% (w/w); *T* = 55°C; *t* = 15 min; *pH* = 4.5; *Em* = Pectinase (Pectinex Ultra SP-L); *E/F* = 2% (w/w); *T* = 60°C; *t* = 20 min; pH 5.0 (optimized conditions)	Choudhari and Ananthanarayan (2007)
Carotene	Carrot pomace	6.4 mg/100 g	*Em* = pectinase and cellulase; *E/F* = 0.15% (w/w); *T* = 50°C; *t* = 1 h; pH 4 (optimized conditions on pilot scale)	Stoll et al. (2003)
Phenolic compounds	Citrus peel	90–162 mg GAE/100 g	*Em* = Cellullase MX ; *E/F* = 1.5% (w/w); *T* = 50°C; *t* = 3 h	Li et al. (2006)
Phenolic compounds	Apple skin	104.94 mg GAE/L	*Em* = Pectinex Smash, Celluclast 1.5 L and Sumizyme AP; *E/F* = 0.1 % (v/w); *T* = 50°C; *t* = 10 min, *F/S* = 1 g/40 mL; *D* = 1500 μm	Pinelo et al. (2008)
Phenolic compounds	Grape pomace	6055 mg GAE/L	*Em* = Pectinase (Grindamyl); *E/F* = 10% (w/w); *T* = 40°C; *t* = 8 h; pH 5.02; *F/S* = 1 g/20 mL; *D* = 125–250 μm	Meyer et al. (1998)
Phenolic compounds	Grape pomace	~78,000 mg phenolic acids/100 g DW, ~2,000 mg anthocyanins/100 g DW	*Em* = Cellulases (Cellubrix L and Celluclast 1.5 L) and several pectinases; *E/F* = 0.5 (w/w) and 0.25% (w/w), respectively; *T* = 50°C; *t* = 2 h (pilot scale)	Kammerer et al. (2005)
Flavonoids (anthocyanins, flavan-3-ols, and flavonol glycosides)	Grape skin	4190 mg anthocyanins/100 g, 373 mg flavan-3-ols/100 g, 34 mg flavonol glycosides/100 g	*Em* = Pectinase with cellulase and hemicellulase side activities (Lallzyme EX-V); *E/F* = 0.01 % (w/w); *T* = 45°C; *t* = 3 h; *pH* = 2.0	Tomaz et al. (2016)
Phenolic compounds	Apple pomace	908 mg GAE/100 g	*Em* = Commercial pectinase; *E/F* = 12% (w/w); *T* = 37°C; t = 11 h; pH 3.6 (optimal conditions)	Zheng et al. (2008)

D, Particle size; EAE, enzyme-assisted extraction; E/F, enzyme/feed ratio; Em, enzyme mixture; F/S, feed/solvent ratio; t, time.

directly correlated with cell wall degradation, and flavonols were mainly hydrolyzed to their respective aglycones (Arnous and Meyer, 2010).

In summary, EAE is an environment-friendly technology that offers a valuable alternative for the recovery of target compounds reducing the use of solvents and saving processing time and energy. A prior knowledge of raw material composition and how target compounds are inserted in the matrix helps in the selection of the appropriate hydrolytic enzyme or enzyme combination. Nevertheless, EAE presents some technical limitations to apply optimal conditions on an industrial scale. However, if the scaling up limitations can be overcome, then EAE could provide an opportunity to increase the extraction yields using mild processing conditions.

4 Ultrasound-Assisted Extraction

Ultrasound-assisted extraction (UAE) has attracted much interest in recent years due to their many advantages in the recovery of valuable compounds from different matrices compared to conventional extraction methods. The main benefits are a shorter and a more effective extraction, thus reducing energy consumption (thus in compliance with the third principle of Green Extraction of Natural Products), and also using moderate temperatures, which is beneficial for heat-sensitive valuable compounds.

Ultrasounds (US) are mechanical waves with frequencies above human earring (from 20 kHz to 10 MHz). Particularly in a solid-liquid extraction assisted by US, the sample is immersed in the solvent and submitted to ultrasound using a US probe or US bath. When the waves are transmitted through the liquid medium, they induce a longitudinal displacement of particles, whereas the source of the sound wave acts as a piston, resulting in a succession of compression and rarefaction phases on the medium (Chemat et al., 2011). If the rarefaction cycle is strong enough, the distance between contiguous molecules of the liquid can reach or even exceed the critical molecular distance. The voids created into the medium are cavitation bubbles. These incipient bubbles are able to grow during rarefaction phases and decrease in size during compression cycles (Fig. 4.1). When the size of the bubbles reaches a critical point (at the end of a rarefaction cycle), they collapse during the subsequent compression cycle, resulting in a violent implosion that release large amounts of energy (Esclapez et al., 2011). The temperature and the pressure at the moment of collapse have been estimated to be up to 5000 K and up to

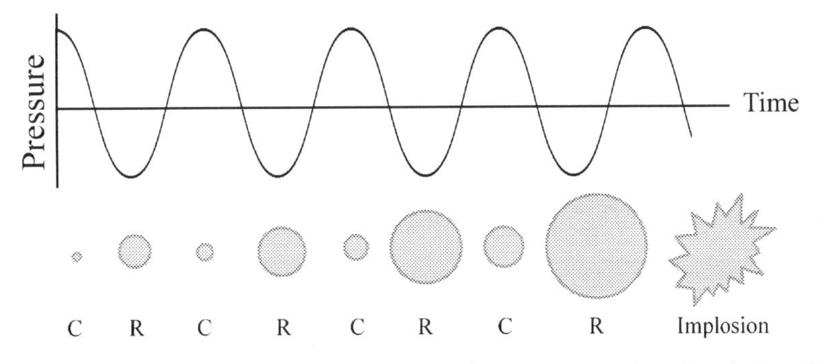

Figure 4.1: *Acoustic cavitation phenomenon, showing formation, growth and implosion of a bubble:* *C*, Compression; *R*, rarefaction.

50 MPa, respectively (Barba et al., 2016; Chemat et al., 2011). The mechanical effects induced during the collapse of the cavitation bubbles in the solid–liquid interface of a heterogeneous medium include shockwave-induced damages and microjet impacts (Esclapez et al., 2011). In this latter case, an asymmetrical collapse takes place, in which the potential energy of the expanded bubble is converted into kinetic energy of a liquid microjet that extends inside the bubble and penetrates the opposite bubble wall, reaching high velocities. In UAE from vegetal sources, when the bubbles collapse onto the surface of the solid material, the microjets directed toward the solid surface cause the breakdown of the cell walls, improving the solvent penetration into the plant matrix and releasing its content into the medium (Toma et al., 2001). Thus, the use of US can enhance the extraction process by increasing the mass transfer between the plant material and the solvent. The aforementioned statements could suggest dramatic changes in the parameters as temperature or pressure of the surrounding medium, but this is not the case because the timescale for these microreactions is really small to affect the macroscopic system. So, UAE is a potential alternative to preserve heat-sensitive compounds (Roselló-Soto et al., 2015b).

UAE technology is considered a green process, since the extraction of natural bioactive compounds from vegetal sources is carried out using biosolvents (generally water and/or water–ethanol mixtures) in accordance with the second principle of Green Extraction of Natural Products. Following this principle, the extraction of antioxidants, such as anthocyanins, flavonols, or phenolic acids has been specially addressed in the last several years (Galván D'Alessandro et al., 2014; Pingret et al., 2012; Virot et al., 2010). The properties exhibited by polyphenols are of interest in the production of functional foods increasing their antioxidant activity and enhancing their biological benefits. Table 4.3 summarizes some studies applying US to extract valuable bioactive compounds from agrifood by-products. In most cases, ultrasonic assistance allowed an improvement of extraction yields in short extraction times, as well as simplifications of handling and work-up conditions. In this way, several works have studied the suitability of UAE on apple pomace, the solid waste resulting from industrial processing of apple juice or cider production. For example, Virot et al. (2010) showed that under optimized conditions (40.1°C, 45 min, and ultrasonic power of 0.142 W/g) using 50% aqueous ethanol as solvent, the yields of extracted phenolic compounds increased 25% (964 and 769 mg of catechin equivalent per 100 g DW, respectively) in comparison to untreated samples. In a similar study on the same matrix, Pingret et al. (2012) optimized the same operational conditions but using water as solvent. The authors demonstrated that under optimal conditions (40°C, 40 min and ultrasonic power of 0.764 W/cm^2) the recovery of polyphenols was 32% higher than that those obtained by conventional solvent extraction (555 and 420 mg of catechin equivalent per 100 g DW, respectively). In addition, UAE has been mentioned as a versatile and low-cost technology, since it can be easily integrated with already existing devices as a part of the technological plant, or when upscaling is necessary (Barba et al., 2016).

Table 4.3: Examples of UAE of valuable compounds from agrifood by-products.

Target Compounds	Agrifood By-Product Sources	Recovery Yield	Treatment Conditions	References
Pectin	Tomato waste	18,500 mg/100 g DW	*S* = ammonium oxalate/oxalic acid aqueous solution; *M* =UAE bath; *T* = 80°C; *t* = 30 min; *F* = 37 kHz	Grassino et al. (2016)
Pectin	Grapefruit peel	27,340 mg/100 g	*S* = HCl aqueous solution (pH 1.5); *M* = UAE probe; *T* = 66.71°C; *t* = 27.95 min; *F* = 20 kHz; *Pw* = 12.56 W/cm; *F/S* = 21 g/50 mL (optimum conditions)	Wang et al. (2015)
Fructans (inulin + oligofructose)	Serish (*Eremurus spectabilis*) root	61,830 mg/100 g	*S* = water; *M* = UAE probe; *T* = 60°C; *t* = 29.27 min; *F* = 24 kHz; *Pw* = 200 W; *A* = 80%	Pourfarzad et al. (2014)
Phenolic compounds	Coconut shell	2244 mg GAE/100 g	*S* = 50% ethanol-water; *M* = UAE bath; *T* = 30°C; *t* = 15 min; *F* = 25 kHz; *Pw* = 150 W; *F/S* = 1.5 g/75 mL (optimum conditions)	Rodrigues et al. (2008)
Phenolic compounds	Wheat bran	312 mg GAE/100 g	*S* = 64% ethanol-water; *M* = UAE bath; *T* = 60°C; *t* = 25 min; *F* = 40 kHz; *Pw* = 250 W; *F/S* = 5 g/100 mL	Wang et al. (2008)
Phenolic compounds	Apple pomace	555 mg catechin eq./100 g	*S* = water; *M* = UAE bath; *T* = 40°C; *t* = 40 min; *F* = 25 kHz; *Pw* = 150 W; *F/S* = 15 g/100 mL	Pingret et al. (2012)
Phenolic compounds	Apple pomace	964 mg catechin eq./ 100 g	*S* = 50% ethanol-water; *M* = UAE bath; *T* = 40°C; *t* = 40 min; *F* = 25 kHz; *Pw* = 150 W; *F/S* = 15 g/100 mL (optimum conditions)	Virot et al. (2010)
Phenolic compounds, α-Tocopherol	Olive leaves	41.43 mg oleuropein eq./100 g oil, 1200 mg/100 g oil of enriched α-tocopherol	*S* = olive oil; *M* = UAE bath; *T* = 16°C; *t* = 45 min; *F* = 25 kHz; *Pw* = 60 W; *F/S* = 15 g/100 mL oil (optimum conditions)	Achat et al. (2012)
Phenolic compounds	Cauliflower waste	730 mg GAE/100 g DW	*S* = 2 M NaOH aqueous solution; *M* = UAE bath; *T* = 60°C; *t* = 30 min ; *F* = 37 kHz; *Pw* = 180 W (optimum conditions)	Gonzales et al. (2014)
Phenolic compounds	Chicory by-products	723 mg GAE/100 g DW	*S* = 37.5% ethanol-water; *M* = UAE probe; *T* = 60°C; *t* = 9.2 min; *F* = 30.8 kHz; *Pw* = 100 W [energy consumption at time 1.17 (kWh)]	Pradal et al. (2016)
Phenolic compounds, anthocyanins	Black chokeberry (*A. melanocarpa*) wastes	~7000 mg GAE/100 g DW, ~1300 mg cyanidin-3-glucoside eq./100 g DW	*S* = 50% ethanol-water; *M* = UAE probe; *T* = 40°C; t = 10 min ; *F* = 30.8 kHz; *Pw* = 100 W; *F/S* = 1 g/20 mL; *D* = <500 μm	Galván D'Alessandro et al. (2014)

A, Amplitude; F, frequency; M, method; Pw, power; S, solvent.

However, for a successful application of the UAE, it is necessary to optimize the extraction process considering the influence of several process variables, such as applied ultrasonic power and frequency, extraction temperature, solvent used, particle size, and sample-solvent ratio (Carciochi et al., 2015). It is especially important to study the kinetics of the extraction to optimize extraction process, to decrease processing residence time, and to reduce energy consumption. Recently, a kinetic mathematical model considering time, temperature of extraction, solvent composition, and ultrasound power, has been proposed for the optimization of UAE of antioxidant polyphenols from black chokeberry wastes (Galván D'Alessandro et al., 2014). This model was then upgraded for UAE of polyphenols from chicory by-product, integrating also the energy consumption, thus providing an original multicriteria optimization tool, considering both extraction yield and energy consumption (Pradal et al., 2016).

In summary, UAE is considered a clean technology that diminishes the extraction times with high reproducibility, reducing the consumption of solvent, simplifying manipulation and work-up, giving higher purity of the final product, allowing the use of green solvents, and consuming only a fraction of the fossil energy normally needed for a conventional extraction methods (Chemat et al., 2011).

5 Microwave-Assisted Extraction

Microwave-assisted extraction (MAE) is another alternative process that has recently received much attention in the food industry due to its many advantages compared with conventional extraction techniques, such as higher extraction rates with lower costs being a nondestructive method and saving time, complying with environmental and economic requirements to ensure safe and high quality extract/products (Baiano et al., 2014; Dahmoune et al., 2015; Song et al., 2011). The fact that MAE involves volumetric heating and is a rather controllable process also makes it an attractive alternative for extraction, especially when dealing with the extraction of heat-sensitive bioactive compounds from plant materials where rapid heating and hence shorter extraction time is desired (Baiano et al., 2014; Chumnanpaisont et al., 2014). Consequently, solvent consumption and extraction time using microwave can be significantly reduced and energy saved as compared to conventional heating methods (Ballard et al., 2010; Hayat et al., 2009; Wu et al., 2012).

The extraction of biomolecules using microwave assistance was first applied in the 1980s of 20th century. Since then, this technique has attracted growing interest, and it has been extensively used in analytical chemistry (Wang et al., 2016). Microwave energy is a nonionizing radiation (frequency 300 MHz–300 GHz) used to heat solvents in contact with solid or liquid samples and to extract the bound compounds inside the walls of a food material (Ahmad and Langrish, 2012). The principle of MAE is based on the direct effect of microwaves on molecules by ionic conduction and dipole rotation. Generally, polar solvents having a permanent dipole moment (e.g., water and methanol) and ionic solutions (usually acids) strongly absorb microwave energy, whereas nonpolar solvents (e.g., hexane) do not heat when exposed to microwave radiation (Madej, 2009; Wu et al., 2012). The enhancement of extract recovery by MAE is generally attributed to its quickly heating effect on solvent, which increases the solubility for the compound of interest (Ahmad and Langrish, 2012). MAE can also be used without adding water or organic solvents in a process called solvent-free microwave extraction, mainly used to recover essential oils (Chemat et al., 2012). This method is based on the combination of microwave heating and distillation performed at atmospheric pressure. When microwaves are used to heat water in the fresh matrix, glands containing essential oil are broken, thus releasing volatile oils, which are then condensed in a cooling system placed outside the microwave oven. Thus, the obtained extracts have no traces of toxic solvents and the products could be considered as nondenatured and without contaminants (the sixth principle of Green Extraction of Natural Products). In addition, as it has been mentioned earlier, MAE can be performed using either polar biobased solvents or no solvent, therefore it can be considered as a green extraction process (second principle).

The main parameters influencing MAE performance include: solvent type, solvent volume, composition of solvent, solid–to–solvent ratio, microwave power, exposure time, and temperature (Madej, 2009; Wu et al., 2012). When selecting the solvent, consideration should mainly be given to the microwave-absorbing properties of the solvent, to the selectivity toward the target molecule, and to the interaction of the solvent with the matrix. However, most MAE applications have involved mixtures of nonpolar solvent and water, including the humidity of biological matrices themselves. The solvent volume depends on the type and the size of sample, but, on average, the amounts of solvent may be about 10 times less than those used in classical extractions (Madej, 2009). The microwave power and the corresponding time depend on the type of sample and solvent used. In theory, one should use high-power microwaves to reduce exposure time as much as possible. However, in some cases, a very high-power microwave decreases the extraction efficiency through degrading the sample or rapidly boiling the solvent in open-vessel systems, which hinders contact with the sample. Generally, extraction times in MAE are much shorter than those of classical extraction techniques. Usually, elevating extraction times above the optimal range does not improve extraction efficiency, and in some cases, may even decrease target molecules recoveries (e.g., thermolabile compounds). In most cases of the MAE, elevated temperatures result in improved extraction efficiency, but particular consideration should be given to applications dealing with thermolabile substances, which may be decomposed at high temperatures (Madej, 2009).

MAE has been accepted as a potential and powerful alternative for the recovery of valuable species from food industrial wastes (Angiolillo et al., 2015). Table 4.4 show some examples of MAE applications in the extraction of bioactive molecules from agrifood by-products. In recent years, MAE has been successfully used for the recovery of phenolic compounds from citrus peels. For instance, Ahmad and Langrish (2012) have optimized the extraction process

Table 4.4: Examples of microwave-assisted extraction (MAE) of valuable compounds from agrifood by-products.

Target Compound	Agrifood By-Product Source	Recovery Yield	Treatment Conditions[a]	References
Phenolic compounds	Potato downstream wastes	1,100 mg GAE/100 g DW	*S* = 60% ethanol; *T* = 80°C; *t* =2 min; *F/S* = 1 g/40 mL	Wu et al. (2012)
Lignin	Triticale straw	91,000 mg/100 g	*S* = 92% ethanol, H_2SO_4 0.64 N; *T* = 148°C	Monteil-Rivera et al. (2012)
Phenolic compounds	*Myrtus communis* L. leaves	16,249 mg GAE/100 g DW 502 mg quercetin Eq./100 g	*S* = 42% ethanol; *t* = 62 s; *Pw* = 500 W; *F/S* =1 g/32 mL	Dahmoune et al. (2015)
Phenolic compounds	*Citrus sinensis* peels	1,220 mg GAE/100 g DW	*S* = 51% acetone; *t* = 122 s; *Pw* = 500 W; *F/S* = 1 g/25 mL	Nayak et al. (2015)
Pectin	*Carica papaya* L. peel	25,410 mg/100 g	*t* = 140 s; *Pw* = 512 W; *F/S* = 1 g/15 mL; pH = 1.8	Prakash Maran et al. (2014)
Phenolic compounds	Mandarin peels	2,320 mg GAE/100 g	*S* = water; *T* = 135°C; *t* = 3 min; *Pw* = 400 W; *F/S* = 1 g/2 mL; pH = 1.5–2.5	Ahmad and Langrish (2012)
Phenolic compounds	Peanut skins	14,360 mg GAE/g	*S* = ethanol 30%; *t* = 30 s; *Pw* = 90% of nominal	Ballard et al. (2010)
Lycopene	Tomato peels	13.592 mg/100 g	*S* = hexane:ethyl acetate; *Pw* = 400 W	Ho et al. (2015)
Phenolic acids	Mandarin peels	116.28 mg/100 g DW (phenolic acids)	*S* = 66% methanol; *t* = 49 s; *Pw* = 152 W; *F/S* = 1 g/16 mL	Hayat et al. (2009)
Phenolic compounds	*Ipomoeabatatas* leaves	6126 mg GAE/100 g DW	*S* = 53% ethanol; *t* = 123 s; *Pw* = 302 W	Song et al. (2011)
Carbohydrate	Wheat straw	Xylose: 100 g/100 g Glucose: 65 g/100 g	*S* = water; *T* = 195°C; *t* = 34 min; *F/S* = 1 g/90 mL; pH = 1.77	Yemis¸ and Mazza (2012)
Phenolic compounds	Vegetable solid wastes	Cauliflower: 45, Celery: 15.5, Chicory: 39.2 Asparagus: 24.5 mg GAE/100 g	*S* = water; *t* = 4 min; *Pw* = 750 W; *F/S* = 1 g/2 mL	Baiano et al. (2014)

[a]*F/S*, Feed/solvent ratio; *Pw*, power; *S*, solvent; *T*, temperature; *t*, time.

for mandarin peels using acidified water as solvent. Total phenolic content recovered from this source was 23.2 mg GAE/g FW at 135°C for 3 min and a microwave power of 400 W. In another study, the extraction conditions to maximize the recoveries of phenolic compound from orange peels have been optimized (Nayak et al., 2015). Under optimal MAE conditions (51% aqueous acetone, 500 W microwave power, 80°C, 2 min) the maximum level of polyphenols obtained was 12.09 mg GAE/g DW, representing a rise of 18% more than conventional solvent extraction.

Briefly, MAE has a great potential as an alternative to conventional extraction processes for the recovery of valuable substances from agrifood by-products, considering the possibility of obtaining high extraction yields with quite low costs within reduced extraction times, which enables the limitations of degradation due to temperature increase while extracting thermolabile substances. MAE also offers the possibility to work with polar biobased solvents or even without solvent.

6 Extractions Assisted by Pulsed Electric Technologies

Two promising methods to extract valuable compounds from vegetal and food by-products using electrical assistance are PEF and HVED. Typical contactors to perform extractions using these techniques are shown in Fig. 4.2.

PEF-assisted extraction basically consists in the application of direct electric pulses on the vegetal source, which is placed between two plate electrodes in a batch or in a continuous treatment chamber. The source is exposed to a pulsed voltage (usually up to 10 kV/cm, even though typically 0.1–5 kV/cm with pulses of 50–1000 μs are used), which provokes the formation of pores in the cell structures based on an increase of the transmembrane potential (Koubaa et al., 2015; Vorobiev and Lebovka, 2010). When the intensity of an applied electric field increases, the potential difference across a cell membrane also increases. If this transmembrane potential exceeds a stated threshold value (typically 0.2–1 V), a temporary loss of membrane semipermeability occurs. This phenomenon of cell damage is called electroporation (or electropermeabilization) and depending on the intensity of the electric field, it leads to the formation of temporary (reversible) or permanent (irreversible) pores. Thus, the process assists the electrophoretic movement of the intracellular charged compounds between cellular compartments and release of them without any significant

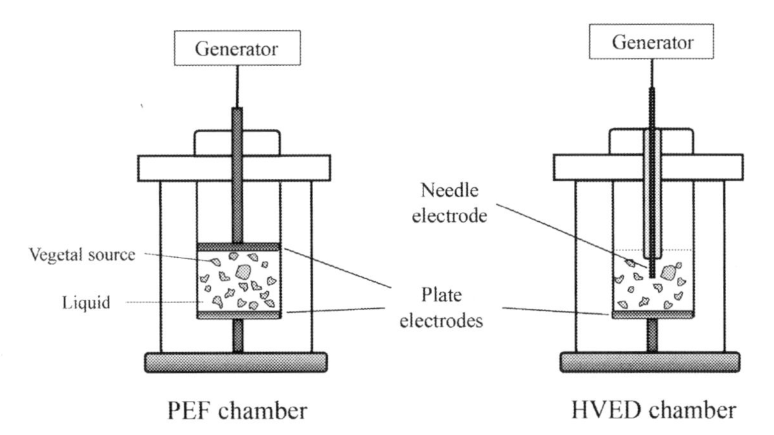

Figure 4.2: Typical reactors for PEF and HVED-assisted extraction.

increase of temperature. The fact that PEF is able to extract selectively the intracellular molecules without fragmenting the treated tissue makes it an interesting green extraction process, because it reduces the subsequent purification steps and also agrosolvents are generally used in the following diffusion step (second and fifth principles of Green Extraction of Natural Products).

The main parameters that have to be controlled in PEF-assisted extraction experiments are the intensity of the electric fields applied to the material being processed, which is related to the gap fixed between the electrodes, the delivered voltage, the electrode geometry, and their disposition in the reactor. Besides electric field strength, pulse width, and repetition rate, other process parameters should be controlled, such as the number of pulses, treatment time, and total specific energy (kJ/kg), which is generally below 20 kJ/kg (Puértolas et al., 2012). In addition, efficiency of PEF-assisted extraction also depends on (1) extraction parameters (temperature, pH, solvent type, and concentration); (2) physicochemical properties of the treated matrix (size, shape, electric conductivity, cell structure, and membrane characteristics); and (3) nature and cell location (cytoplasm or vacuoles) of the targeted molecules being extracted (Puértolas et al., 2012; Vorobiev and Lebovka, 2010).

The effect of PEF on the recovery of valuable compounds from several by-products has been recently evaluated in different works as shown in Table 4.5. For instance, the application of 1–7 kV/cm (15–150 μs) and subsequent pressing step (30 min; 5 bar) on orange peel by-product yielded increases up to 159% when the source had been PEF treated at the highest electric field strength assayed, in comparison to only pressed (non-PEF treated) samples (Luengo et al., 2013). More recently, Bobinaitė et al. (2016) have reported an increase in anthocyanins yield (95%) when blueberry pressed cake were subjected to a PEF treatment (1–5 kV/cm, 10 kJ/kg) and a pressing step (8 min, 1.32 bar), compared with the untreated sample. Both studies mentioned that recovery of phenolic compounds was improved significantly by increasing the electric field strength.

HVED technology has been used to recover bioactive compounds from several plant food materials, based on electrical breakdown phenomenon produced in cell tissues. When HVED is performed directly in extraction solvent, it injects energy directly into an aqueous solution through a plasma channel formed by a high-current/high-voltage electrical discharge (usually 40 kV and 10 kA) between two submersed electrodes (Boussetta and Vorobiev, 2014). The discharge of electrons produced from the high-voltage needle electrode to the grounded one leads to the generation of hot, localized plasma that emits high-intensity UV light, producing shock waves, cavitation bubbles, creation of liquid turbulence and formation of hydroxyl radicals during water photo dissociation in the liquid medium (Barba et al., 2016). All these phenomena lead to particle fragmentation and cell structure damage, consequently facilitating the release of intracellular compounds.

One of the most important factors in HVED-assisted extraction is the treatment energy input, which depends on the conductivity of the medium. However, the aforementioned parameters

Table 4.5: Examples of pulsed electric field (PEF) and high voltage electrical discharges (HVED) assisted extraction of valuable compounds from agrifood by-products.

Target Compound	Agrifood By-Product Source	Recovery Yield	Treatment Conditions[a]	References
Pulsed electric fields (PEF)				
Anthocyanins	Blueberry press cake	245.8 mg/L	*S* = 50% ethanol-water (v/v); *Ef* = 3 kV/cm; Sie = 10 kJ/kg; *F* = 10 Hz; *Pl* = 20 μs	Bobinait et al. (2016)
Phenolic compounds	Borage leaves	115.8 mg/100 g	*S* = water; *Ef* = 5 kV/cm; *Sie* = 0.04–61.1 kJ/kg; *F* = 1 Hz; *Pl* = 60 μs	Segovia et al. (2015)
Phenolic compounds	Orange peels	34.8 mg GAE/100 g	*S* = water; *Ef* = 20 kV/cm; *Sie* = 24 kJ/kg; *F* = 0.5 Hz; *Pl* = 60 μs	Luengo et al. (2013)
Inulin	Chicory roots	121.8 g/L	*S* = water; *Ef* = 0.6 kV/cm; *Pl* = 100 μs	Zhu et al. (2012)
Colorants	Red prickly pear peels	81.3 mg/100 g	*S* = water; *Ef* = 20 kV/cm; *Sie* = 24 kJ/kg; *F* = 0.5 Hz; *Pl* = 10 μs	Koubaa et al. (2016)
Glycoalkaloids	Potato peels	135.3 mg solanidine /100 g DW	*S* = methanol; *Ef* = 0.75 kV/cm; *Sie* = 0.05215 kJ/kg; *Pl* = 1500 μs	Hossain et al. (2015)
High voltage electrical discharges (HVED)				
Phenolic compounds Flavan-3-ols Flavonols, Stilbenes	Grape stems	7600 mg GAE/100 g 888.6 mg catechin Eq./100 g, 55.5 mg quercetin Eq./100 g 19.29 mg resveratrol Eq./100 g	*S* = water at different pH values; *Ef* = 40 kV/cm; *Sie* = 188 kJ/kg; *F* = 0.5 Hz; *Pl* = 4 ms	Brianceau et al. (2016)
Phenolic compounds	Grape pomace	2800 mg GAE/100 g DW	*S* = water; *Ef* = 40 kV/cm; *Sie* = 80 kJ/kg; *F* = 0.5 Hz; *Pl* = 10 μs	Boussetta et al. (2011)
Phenolic compounds	Olive kernel	626.6 mg GAE/L	*S* = water (pH: 2.5); *Ef* = 40 kV/cm; *Sie* = 66 kJ/kg; *F* = 0.5 Hz; *Pl* = 10 μs	Roselló-Soto et al. (2015a)
Secoisolariciresinol	Flaxseed cake	300 mg/100g DW	*S* = 25% ethanol in water; *Ef* = 40 kV/cm; Sie = 181 kJ/kg; *F* = 0.5 Hz; *Pl* = 10 μs	Boussetta et al. (2013)
Phenolic compounds	Grape seeds	8600 mg GAE/100 g DW	*S* = 30% ethanol in water; *Ef* = 40 kV/cm; *Sie* = 69 kJ/kg; *F* = 0.33 Hz; *Pl* = 10 μs	Boussetta et al. (2012a)

[a]*Ef*, Electrical field; *F*, frequency; *Pl*, pulse; *S*, Solvent; *Sie*, specific input energy.

influencing the extraction process for PEF-assisted extraction also should be taken into account here.

Several examples of HVED-assisted extraction of nutritionally valuable compounds from vegetal and food by-products are listed in Table 4.5. For instance, Boussetta et al. (2011) have mentioned an increase of tenfold in the amount of extracted phenolic compounds when grape pomace was treated by HVED in water (80 kJ/kg, 150 pulses) compared to untreated samples. In another study, HVED (40 kV, 188 kJ/kg) applied on grape stems, following a subsequent diffusion step (120 min in 50% ethanol solvent) improved significantly the extraction of flavan-3-ols and flavonols in 21 and 12%, respectively, compared to conventional hydroalcoholic extraction. Thus, HVED can be considered as a green extraction process, since water is the solvent principally used in the extractions (according to second principle of Green Extraction of Natural Products).

In comparison, although HVED is more efficient for cell disintegration than PEF, the application of the former can be less selective, since small particles produced by HVED technology lead to a more difficult solid-to-liquid separation. Concerning scaling up, both techniques are relevant for an industrial use; however, the down side effect of using these technologies for industrial purposes are mainly the high investment cost. Although different industrial prototypes have been developed for PEF over the past few years, which have reached continuous product flow rates up to 10,000 kg/h (Puértolas et al., 2012), the application of HVED at industrial levels was not currently evaluated, since it was only tested at a pilot scale (Boussetta et al., 2012b).

In summary, both pulsed electric technologies are able to enhance cell disruption combined with mild temperatures, representing an environmental green alternative to recover valuable compounds from plant food by-products. The principal benefits against conventional methods of extraction are: better extraction yields with minimal thermal degradation, while extraction time and temperature, as well as the amount of solvent, are decreased, thereby reducing the energy cost and the environmental impact (Corrales et al., 2008).

7 Pressurized Liquid Extraction

Pressurized liquid extraction (PLE) is a technique that involves extraction using liquid solvents at elevated temperature and pressure (but below their critical point values), thus enhancing solubility and mass transfer properties in comparison with techniques performed at or near room temperature and atmospheric pressure. This technique is also known as pressurized solvent extraction, accelerated solvent extraction, and enhanced solvent extraction. When 100% water is used as solvent, PLE is generally called superheated water extraction, subcritical water extraction, pressurized low polarity water extraction, or pressurized hot water extraction (Pronyk and Mazza, 2009).

PLE can be considered as a green extraction process, especially when a nontoxic solvent (water and/or water-ethanol mixtures) is used (second principle of Green Extraction of Natural Products).

Briefly, to perform a PLE, sample matrix is put into the extraction cell. From here, there are two main set-ups for PLE, static and dynamic modes. In static mode PLE, the solvent is pressurized in the extractor, keeping the outlet valve closed. After the extraction, this valve is opened to release the obtained extract. This batch process could be carried out using several cycles with addition of fresh solvent at each cycle. In dynamic mode, the solvent is continuously pumped through the extractor containing vegetal source, keeping both inlet and outlet valves open during the extraction (Carabias-Martinez et al., 2005). Static extraction is considered as more efficient because of the greater penetration of the solvent into the pores of the vegetal source (Nieto et al., 2010).

Concerning solvents, a wide range of organic nonpolar to polar solvents and their mixtures have been used in PLE for the extraction of bioactive compounds from foods and herbs. However, PLE is a technology that modifies solvent properties by increasing the temperature (e.g., up to 374°C for water), while the pressure is kept high enough to maintain the solvent in the liquid state. In this sense, it has been showed that an increase in the temperature of water during the extraction resulted in a substantial decrease of the dielectric constant (polarity), as well as in its viscosity and its surface tension (Hawthorne and Miller, 1994). This fact makes water a suitable solvent to extract selectively (at different temperatures) polar to nonpolar organic compounds from different biomass matrices (Alvarez et al., 2014). In addition, the high temperature used during the PLE increases the capacity of the solvent to solubilize the solutes, increasing the diffusion, disrupting of the solute-matrix bonds, and decreasing the viscosity and surface tension of the solvent (Ramos et al., 2002; Richter et al., 1996). The pressure is usually maintained in the range of 4 to 20 MPa, which ensures the solvent is maintained in a liquid state (Ramos et al., 2002). Thus, the solvent can drive into the pores of the matrix, enhancing the solubility of compounds (Mustafa and Turner, 2011; Ramos et al., 2002). Additionally, the equipment set-up for the PLE also provides the protection of oxygen and light sensitive compounds, such as phenolics and essential oils (Mustafa and Turner, 2011).

In Table 4.6, some examples of application of PLE for the extraction of valuable species from agrifood by-products are presented. For example, Wijngaard et al. (2012a) have reported on the optimization of the subcritical water extraction of phenolic compounds and glycoalkaloids from industrial potato peels, and Ho et al. (2007) have used pressurized low-polarity water extraction to recover proteins from defatted flaxseed meal.

Due to interest in the extraction of natural compounds from vegetal sources using green approaches, PLE extraction is gaining a lot of attention. The advantages of PLE in comparison with conventional solvent extraction are the short extraction time, the reduced usage of solvents, and the higher yields obtained. However, this method is not suitable for thermolabile compounds, because high temperatures can affect their structure and functional activity (Piasek et al., 2011).

Table 4.6: Examples of PLE of valuable compounds from agrifood by-products.

Target Compounds	Agri-sood By-Product Sources	Recovery Yield	Treatment Conditions	References
Phenolic compounds	Winery by-products	582 mg GAE/100 g	*S* = water; *T* = 150°C; *P* = 10.3 MPa	Marino et al. (2006)
Phenolic compounds	Winery waste	3169 mg GAE /100 g	*S* = water; *T* = 140°C; *P* = 11.6 MPa; *t* = 30 min	Aliakbarian et al. (2012)
Phenolic compounds	Agave Americana waste	2380 mg GAE /100 g	*S* = water; *T* = 150°C; *P* = 1 MPa; *t* = 240 min	Ben Hamissa et al. (2012)
Flavonones	Citrus unshiu peel	8370 mg GAE /100 g	*S* = water; *T* = 160°C; *P* = 5.3 MPa; *t* = 10 min	Cheigh et al. (2012)
Phenolic compounds	Mango leaves	1774 mg GAE /100 g	*S* = water; *T* = 100°C; *P* = 4 MPa; *t* = 180 min	Fernández-Ponce et al. (2012)
Phenolic compounds	Potato peel	82 mg GAE /100 g	*S* = water; *T* = 180°C; *P* = 6 MPa; *t* = 60 min	Singh and Saldaña (2011)
Phenolic compounds	Pomegranate peels	26,430 mg GAE/100 g	*S* = water; *T* = 40°C; *P* = 10.3 MPa; *t* = 10 min	Çam and His¸il (2010)
Carbohydrates	Potato peels	61,000 mg/100 g	*S* = water; *T* = 190°C; *P* = 4 MPa; *t* = 30 min	Alvarez et al. (2014)
Proteins	Defatted flax seed meal	22,500 mg/100 g	*S* = water; *T* = 130–190°C; *P* = 5.2 MPa; *t* = 400 min	Ho et al. (2007)

8 Supercritical Fluid Extraction

Supercritical fluid extraction (SFE) could be an environmentally beneficial alternative to the conventional organic solvent extraction of biological compounds. SFE is a technology that uses the ability of certain substances to become excellent solvents for certain solutes under a specific combination of pressure and temperature. A substance at a pressure higher than its critical pressure and at a temperature higher than its critical temperature is known

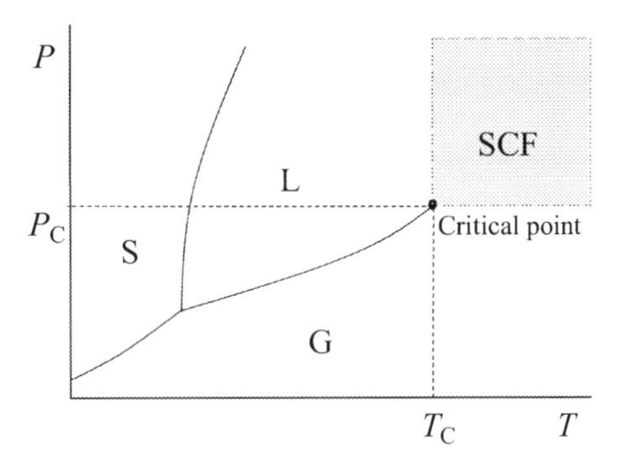

Figure 4.3: Phase Diagram for a Pure Compound in a Close System, as a Function of Temperature (T) and Pressure (P); T_C, Critical Temperature; P_C, Critical Pressure.

as a supercritical fluid (Fig. 4.3). The critical point is the pressure and temperature at which a liquid and its gas are indistinguishable. Above the critical point, the substance is a single-phase fluid, noncondensing and showing some typical physicochemical properties of gases and liquids, where the density approaches the levels of a liquid while its viscosity and diffusivity are similar to those of gases.

The SFE process consists mainly of four stages:

1. Pressurization step: a compressor or a pump is used to reach the required pressure of the solvent, above its critical pressure.
2. Temperature adjustment stage: in this stage the compressed fluid reaches the required temperature (above its critical temperature). Thermal energy is added or removed from the fluid using a heat exchanger, thermal bath or electrical resistances.
3. Extraction stage: the process is carried out in solid matrixes by means of the continuous contact between the solvent and the solid phase. In most cases, the solid is placed in a fixed bed (extractor) and the solvent flows through it.
4. Separation step: the precipitation of the solute is made by simple reduction of the pressure below the critical point, in the separator.

The low viscosity and high diffusivity of a supercritical fluid enhance its penetrating power based on the high-mass-transfer rate of the solutes into the fluid, enabling an efficient extraction of the compounds from the raw material. Likewise, a low viscosity contributes to lower fluid transportation costs. Moreover, supercritical fluids are very sensible because their properties undergo major changes with small variations. The solvation capacity of a fluid in its supercritical state depends on its density, which can be modified continuously and moderately by varying the pressure and/or temperature because in the process no phase changes occur. The highest density values are attained by combining low temperatures (no lower than the critical temperature) and high pressures. Another advantage of using supercritical fluids in extraction is that the separation of solute and solvent is achieved efficiently by a simple expansion. This avoids having to increase the temperature to remove the solvent by evaporation, as with traditional solvents. In addition, SFE processes are selective and the products are free of residual solvents.

From an industrial point of view, it is of interest that fluids do not require a very high pressure or temperature, and those levels can be achieved at relatively low cost. However, its practical application is limited to processes that are not affected by the relatively high temperatures that must be used due to high T_c. Carbon dioxide (CO_2) is the solvent most commonly used to extract bioactive compounds from natural sources using SFE since it is nontoxic, nonflammable and has high solubilization power, which provides high mass-transfer rates, and it is an environmentally friendly solvent, that is, generally recognized as safe (GRAS) for use in the food industry. It is also widely available at high purity and is inexpensive. Carbon dioxide can be easily separated after the extraction, leaving virtually no trace in the processed matrix. In addition, given its low critical temperature (31.05°C) and mild critical pressure (7.38 MPa), it is particularly recommended for the extraction of thermolabile compounds (for example, natural essences).

So, SFE can be considered as a green extraction process, since a nontoxic alternative solvent (such as supercritical CO_2) is usually used (second principle of Green Extraction of Natural Products) and the possibility to separate the extracted molecules from the solvent by simple pressure release detent allows reduction of the number of unit operations (fifth principle).

The application of supercritical CO_2 shows good performance in the extraction of nonpolar substances, such as oil, fatty acids, carotenoids, and tocopherols (Ekinc and Gürü, 2014; Espinosa-Pardo et al., 2014; Liu et al., 2009; Prado et al., 2012; Viganó and Martinez, 2015), thus it does not present good performance in the recovery of compounds from wastes/by-products rich in water. When the target compound has affinity for CO_2, pretreatments, such as dehydration need to be applied. However, CO_2 under supercritical conditions is a poor solvent for polar compounds. This problem can be overcome by using polar modifiers or cosolvents, such as water, ethanol, methanol, acetic acid, and ethylene glycol, to modify the polarity of the supercritical fluid and increase its solvating power toward substances of interest (Ghafoor et al., 2010; He et al., 2012; Seabra et al., 2010; Sovilj et al., 2011). Supercritical extraction using CO_2 presents great interest especially for the extraction of hydrophobic biomolecules, as a greener alternative of the extraction using classical organic solvents that, in general, are not ecofriendly.

Some of the research papers published in the last few years in which supercritical fluids were used to obtain compounds from by-products or food industry wastes are shown in Table 4.7. Many kinds of natural biomolecules were obtained thanks to SFE, such as lycopene from tomato peels (Machmudah et al., 2012), carotenoids from peach palm pulp (Espinosa-Pardo et al., 2014), or even phenolic compounds from different berries (Laroze et al., 2010).

Table 4.7: Examples of SFE of valuable compounds from agrifood by-products.

Target Compounds	Agrifood By-Product Sources	Recovery Yield	Treatment Conditions	References
Oil-fatty acids	*Passiflora* seeds	25.83 g/100 g oil	$S = CO_2$, $T = 56°C$, $P = 26$ MPa, $t = 240$ min	Liu et al. (2009)
Oil	Grape seeds	10 g/100 g	$S = CO_2$, $T = 40°C$, $P = 35$ MPa, $t = 240$ min	Prado et al. (2012)
Oil, phytosterols (β-sitosterol)	Peach *(Prunus Pérsica)* seeds	35.3 g oil/100 g seed; 122 mg β-sitosterol/100 g	$S = CO_2$, $T = 40°C$, $P = 20$ MPa, $t = 180$ min	Ekinc and Gürü (2014)
Carotenoids	Peach palm fruit (*Bactris gasipaes*)	200 mg carotenoid/100 g	$S = CO_2$, $T = 40–60°C$, $P = 30$ MPa	Espinosa-Pardo et al. (2014)
Phenolic compounds, anthocyanins	Grape peels	Total phenols: 21.56 mg GAE/L; anthocyanins: 1.176 mg/mL	S = 6–7% ethanol in CO_2, $T = 45–46°C$, $P = 15.7–16.2$ MPa	Ghafoor et al. (2010)
Flavonoids	Pomelo (*C. grandis L. Osbeck*) peel	2.37 g/100 g	S = 85% aqueous ethanol in CO_2, $T = 80°C$, $P = 39$ MPa, $t = 49$ min	He et al. (2012)
Anthocyanins	Elderberry (*Sambucus nigra L.*) pomace	15,000 mg cyanidin-3-glucoside Eq./100 g	$S = CO_2/ethanol/H_2O$, $T = 40°C$, $P = 21$ MPa	Seabra et al. (2010)
γ-oryzanol	Soapstock derived from rice bran oil	16,000 mg γ-oryzanol/100 g	$T = 60°C$, $P = 30$ MPa	Jesus et al. (2013)
Wax	Flax (*Linum usitatissimum L.*) straw	1.26 g/100 g	$T = 75°C$, $P = 37.8$ MPa	Athukorala and Mazza (2010)
Lycopene	Tomato peels	56,000 mg lycopene/100 g	$T = 90°C$, $P = 40$ MPa, $D = 1.05$ mm	Machmudah et al. (2012)
Oil and diterpenes	Spent coffee grounds	11.97 g oil/100 g; 10.290 g diterpene/100 g oil	S = 5% ethanol in CO_2, $T = 55°C$, $P = 19$ MPa; $S = CO_2$, $T = 40°C$, $P = 14$ MPa	Barbosa et al. (2014)
Phenolic compounds	Orange (*C. sinensis L. Osbeck*) pomace	3000 mg GAE/100 g	S = 8% ethanol in CO_2, $T = 50°C$, $P = 25$ MPa, $t = 300$ min	Benelli et al. (2010)
Phenolic compounds	Grape bagasse from Pisco residues	2000 mg GAE/100 g	S =10% ethanol in CO_2, $T = 40°C$, $P = 20–35$ MPa	Farias-Campomanes et al. (2013)
Phenolic compounds	Blueberry residue after pressing	9 g GAE/100 g	$T = 60°C$, $P = 8–30$ MPa, $t = 120$ min	Laroze et al. (2010)

Briefly, advantages of SFE as a green extraction process resides in the possibility to work with nontoxic alternative solvent (such as supercritical CO_2) and to recover the extracted molecules by simple pressure release detent, avoiding further supplementary separation steps.

9 Instant Controlled Pressure Drop

The limiting factor in conventional solid–solvent extraction processes is often the slow diffusion of both the solvent to penetrate into solid matrix and the solute from the core to the surface (Allaf et al., 2014). To avoid this difficulty, a relatively new pretreatment called instant controlled pressure drop (DIC), from French expression Détente Instantanée Contrôlée, has been proposed as a solution. DIC technology has been mentioned as an appropriate way to improve the internal structure of solid matter while preserving its chemical composition, even heat-sensitive molecules. Basically, DIC consists in submitting the matter to a steam pressure in a processing chamber and at high temperature over a relatively short time (few seconds to few minutes). This high-temperature–short-time stage is followed by an instant pressure drop toward pressure of about 5 kPa. The very rapid pressure drop, at a rate $\Delta P/\Delta t > 0.5$ MPa/s, leads to an autovaporization of part of the water in the product, and a simultaneous and instantaneous cooling of the products. These phenomena provoke stopping thermal degradation and a swelling or even rupture of the cell walls allowing valuable compounds to be more available and accessible (Allaf et al., 2013a; Berka-Zougali et al., 2010). Whatever the type of solvent is used in the extraction, the resulting porous structure can dramatically intensify the extraction kinetics by improving diffusivity of solvent within the solid and enhancing mass transfer. Powders issued from these technologies are distinguished by their specific expanded granule texture, which enables them to get very high functional behavior. Open pores increase the exchange surface, giving a more effective diffusivity and accessibility to the solvent (Mounir et al., 2012, 2014; Téllez-Pérez et al., 2012).

The main applications of DIC are pretreatment by texturing/drying of raw material and extraction of volatile compounds (Rombaut et al., 2014). Most authors conclude that DIC enabled enhancement of extraction yield, but it depended on the conditioning temperature and the contact time. By optimizing those process parameters, the duration required to reach maximum extraction yield is less than 2 min when using DIC process, whereas it takes more than 3 h to reach a similar yield with hydrodistillation (Berka-Zougali et al., 2010). As a pretreatment process, the texture of the residue obtained by DIC, with an increased specific surface area, further facilitates extraction of other compounds of interest. Rochová et al. (2008) have used this pretreatment prior to SFE, to increase the extraction rate of oil from soybean.

Table 4.8 presents several examples of DIC used as pretreatment to improve extraction processes from different agrifood by-products. DIC pretreatment has been used to improve the extraction of oligosaccharides from *Tephrosia purpurea* seeds (Amor et al., 2008) and essential oil from orange peels (Allaf et al., 2013b). These studies showed an improvement of solid-liquid extraction of the secondary metabolites, compared to extraction without sample DIC pretreatment, thus confirming a beneficial effect of the destructuration of the tissue organization and texture of the raw material on extraction.

Table 4.8: Examples of instant controlled pressure drop (DIC) technology pretreatment to improve the extraction of valuable compounds from agrifood by-products.

Target Compound	Agrifood By-Product Source	Recovery Yield	Treatment Conditions[a]	References
Phenolic compounds	Olive leaves	18,731 mg/100 g	M = batch extraction; S = ethanol (95%, v/v); T = 55°C; P = 0.58 MPa; t = 22 s	Mkaouar et al. (2016)
Phenolic compounds	Grape stalk powder	2,056.7 mg/L	M = batch extraction; S = aqueous ethanol mixture (50%, v/v); T = 30°C; P = 0.3 MPa; t = 50 s	Sánchez-Valdepeñas et al. (2015)
Essential oils	Orange peel	1,660 mg/100 g	M = UAE; T = 30°C; P = 1 MPa; t = 120 s; Pw = 150 W	Allaf et al. (2013b)
Essential oils	Myrtle leaves	560 mg/100 g	P = 0.6 MPa; t = 120 s	Berka-Zougali et al. (2010)
Oligosaccharides (ciceritol and stachyose)	Wild indigo seeds	1,536 mg ciceritol/100 g 1,841 mg stachyose/100 g	M = Batch extraction; S = water-ethanol mixture (30:70, v/v); T = 45°C; P = 0.52–0.2 MPa; t = 197 s = 30 s	Amor et al. (2008)
Flavonoids	Sea buckthorn	1,800 mg/100 g	S = methanol-water (80:20, v/v); 1.6% HCl; T = room temperature	Allaf et al. (2014)

[a]M, Method; Pw, power; P, pressure; S, solvent; T, temperature; t, time.

In summary, DIC can be considered a green process, since the obtained texture of the raw material with an increased surface area can enhance extraction kinetics and consequently can have a positive impact on process duration and energy consumption (third principle of Green Extraction of Natural Products).

10 Combined (Hybrid) Green Extraction Processes

To achieve the development of new and greener extraction processes, the full potential of conventional and emerging extraction processes has to be extended searching an optimal use of solvents, process conditions, driving forces, energy consumption, and extraction duration. Several processes combining different green extraction methods have recently been developed. These combined processes associate the advantages of at least two green methods (for example, SFE with ultrasound assistance) applied successively (offline) or simultaneously (online). Online applied combined processes are also called hybrid processes. Chemat and Cravotto (2011) have recently reviewed the existing works on hybrid green extraction processes, such as processes combining UAE and MAE, combining UAE and SFE, combining DIC process and MAE, combining enzyme and UAE, and combining enzyme and MAE. However, even if the interest to extract biomolecules by such hybrid processes increases (Chen et al., 2010; Cravotto et al., 2008; Seidi and Yamini, 2012), nowadays only very few studies are available concerning their applications on agrifood by-products.

Barrales et al. (2015) have applied ultrasound assistance during SFE of passion-fruit oil from a by-product of the pulp processing industry (seeds mixed with pulp). The authors have concluded that the application of ultrasound helped increasing the global yield up to 29% when compared to SFE without ultrasound. In online coupling of ultrasound and SFE, the use of high-intensity ultrasound represents an efficient manner of producing small-scale agitation, enhancing mass transfer in supercritical fluids extraction processes (Seidi and Yamini, 2012).

Konwarh et al. (2012) have reported great increase of extraction yield by ultrasound assistance (150%) and enzyme biocatalysis (225%) during lycopene recovery from tomato peels. However, combining online these two green methods has allowed obtain 662% better attainment results than conventional extraction (Konwarh et al., 2012). This great increase of extraction yield shows a clear synergic effect of enzyme and ultrasound assistances when applied simultaneously.

The extraction process combining UAE and MAE is already known in the literature as UMAE process. It has been applied for the recovery of various valuable compounds from natural sources (Chen et al., 2010; Cravotto et al., 2008; Seidi and Yamini, 2012; Zheng et al., 2015). Considering the use of UMAE process for the recovery of biomolecules from agrifood by-products specifically, in our knowledge one work only has currently been published. Bai et al. (2015) have extracted pectin from jujube waste using off-line UMAE (UAE followed by a MAE). On the basis of the results of an experimental design, the authors have concluded that both ultrasound and microwave assistances increased the yield of extracted pectin with higher impact of microwaves.

A typical example of offline combined processes is the combination of DIC pretreatment with a classical (Amor et al., 2008; Mkaouar et al., 2016) or a green extraction process (Allaf et al., 2013b; Rochová et al., 2008).

11 Integrated Green Extraction Processes

As mentioned earlier, if the content of target biomolecules in the extract is not satisfactory, the extraction step could be completed by an appropriate separation/purification process allowing enriched extracts with higher quality to be obtained. Moreover, the efficiency of biomolecules recovery from vegetal sources could be improved by integrating the extraction to a purification step in a single process (Dimitrov et al., 2005; Galván D'Alessandro et al., 2013; Zhang et al., 2011). Such integration of processes is widely used in bioindustries to extract in situ the produced molecules, providing improvement of productivity and yield,

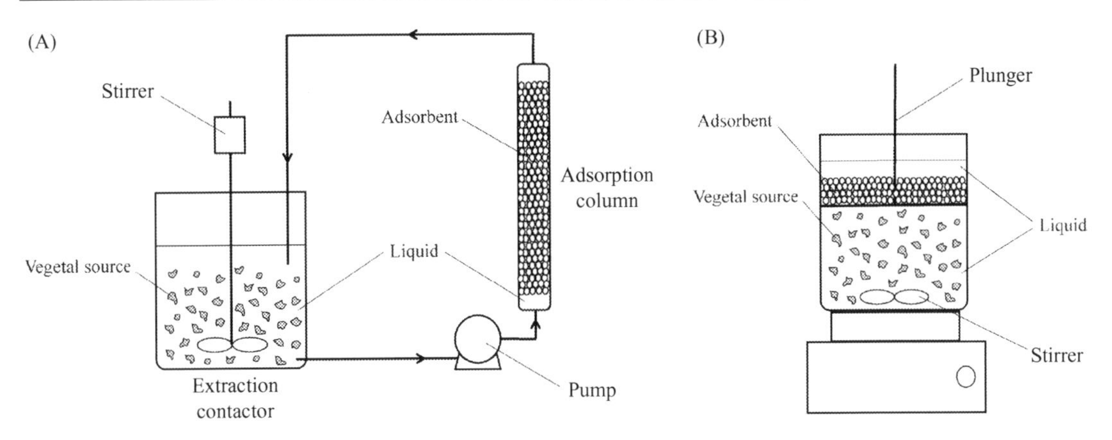

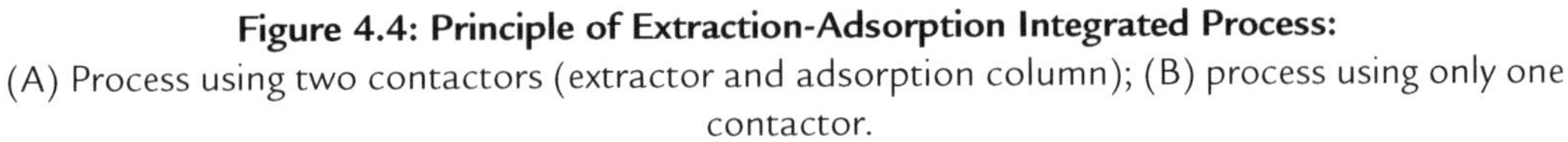
Figure 4.4: Principle of Extraction-Adsorption Integrated Process:
(A) Process using two contactors (extractor and adsorption column); (B) process using only one contactor.

and economy savings (Schugerl and Hubbuch, 2005). Recently, a new integrated extraction-adsorption process has been developed for selective recovery of antioxidant polyphenols from black chokeberry (Galván D'Alessandro et al., 2013). This green process combines the extraction of phenolic compounds and simultaneous enrichment of the extracts by adsorption. In this process, the liquid phase circulates through the extraction contactor and adsorption column in a closed loop (Fig. 4.4A). The phenolic compounds extracted from the vegetal source in the extractor are rapidly fixed to the adsorbent allowing maintenance of low concentrations in the liquid phase and maximal driving force in the extraction step. At the end of the process, the adsorbed phenolics are eluted by ethanol-water mixture allowing extracts that are concentrated and enriched in antioxidants to be obtained. This integrated process permits recovery of selectively antioxidant phenolics in a single and relatively simple green process (extraction and purification at room temperature, food grade solvents only used, reduced extraction duration, and reduced consumption of solvent and energy). Besides the high enrichment of the extracts in polyphenols due to selectivity of the adsorbent used (about 15 times), the process integration also clearly enhances the extraction yields (about 25%) compared to offline conducted extraction and adsorption steps (Galván D'Alessandro et al., 2013). Later, this process was also applied for the recovery of antioxidant phenolics from black chokeberry by-products (residues from fruit juice production) (Vauchel et al., 2015). The similar results obtained at laboratory and pilot scale (scale-up factor of 50) have shown that the integrated extraction-adsorption process could be used for production of large quantities of extracts highly rich in antioxidant phenolics (Vauchel et al., 2015). Jankowiak et al. (2015) also reported a similar integration of extraction and adsorption steps for the recovery of isoflavones from okara (a by-product from soymilk production). The authors have proposed another configuration of the process, putting the adsorbent directly in the extractor (Fig. 4.4B).

Many similar integrated processes could be developed since often purification or concentration steps are conducted after the extraction. For example, solid–liquid extraction of alkaloids has successfully been integrated to online liquid membrane (pertraction) purification of the obtained extracts allowing enriched extracts of atropine (Boyadzhiev et al., 2006; Dimitrov et al., 2005), glaucine (Lazarova and Dimitrov, 2009), and vincamine (Yordanov et al., 2009) from raw vegetal sources to be obtained. Peev et al. (2011) have combined extraction of rosmarinic acid from lemon balm to nanofiltration in a three-step extraction-nanofiltration procedure allowing nearly saturated retentate solutions and to be obtained and reuse of the permeates for rosmarinic acid extraction in place of pure solvent. Obviously, some integrated processes proposed for the extraction of biomolecules from raw sources could be adapted also for selective extraction of such molecules from agrifood by-products.

The integrated extraction processes, as well as the hybrid ones, respect the fifth principle of Green Extraction of Natural Products, since the number of unit operation is reduced. These processes also profit from the "green" impact of each of the used single processes. The obtained higher productivity and the energy savings observed in the case of integrated processes and some hybrid processes with clear synergic effect are in accord with the third principle (energy savings). The use of GRAS solvents correspond also to the second principle of Green Extraction of Natural Products (ecofriendly solvents).

For all presented green processes of recovery of valuable biomolecules from agrifood by-products, the fourth principle is clearly respected (valorization of wastes as by-products). Since, in general, in the extraction of valuable species ecofriendly solvents are used, the obtained extracts have no traces of toxic solvents and the products could be considered as nondenatured and without contaminants (the sixth principle). In most of the cited research works renewable plant resources are used as sources of the agrifood by-products, so the first principle of the Green Extraction of Natural Products is also considered.

12 Future Trends

Although the number of research publications concerning green extraction of biomolecules from agrifood by-products has greatly increased in the 21st century, their real and large application in industry has still not occurred. The aforementioned green extraction processes are very promising alternatives enabling savings of energy, extraction time, and unit operations, as well as the use of GRAS solvents only. Most of these processes, already applied for the recovery of valuable compounds from raw vegetal sources, have great potential for the extraction of such molecules from agrifood wastes. Then, some green extraction processes already used for the recovery of valuable species from specific agrifood by-products could also be applied to other kinds of agrifood wastes.

Different nature of valuable molecules in the same by-product suggests a combination of two or more processes to valorize maximally the source (biorefinery concept). Indeed, most of the studies on green extraction processes are related to recovery of phenolic compounds, especially because of their high antioxidant capacity and various biological activities, as well as their abundance in vegetal sources (and respectively, in agrifood by-products). Therefore, further studies dealing with the recovery of other valuable compounds, especially proteins and lipids, are still needed.

Some alternative solvents, such as agro- and biosolvents, fluorinated solvents, and ionic liquids, already used at laboratory scale for the extraction of biomolecules from raw vegetal sources, could be of interest in the case of by-products, also, and this path should be further investigated.

The large number of possible combinations of different extraction and separation processes represents a breeding ground for the development of new hybrid or integrated processes, with great potential for saving energy and time and higher extraction yields.

Even if many researchers point out low energy consumption during the used "green" processes, only a few have taken into account the real energy consumption of the process. Hence, it seems important to register and consider it while working on the development of green processes in which environmental impact is a key point. Of even greater importance is ensuring that the used processes are really "green," and in this perspective, the tools, such as Life Cycle Analysis seem to have high potential in the estimation of the environmental impact of all these proposed processes for the valorization of agrifood wastes.

Obviously, before selecting the more appropriate extraction process for a target biomolecule from a given agrifood by-product, it is important to optimize the experimental conditions and also to study different suitable extraction techniques. Although some authors have compared the extraction yields obtained using two or more conventional and green extraction processes, such studies are still quite rare; the experimental conditions are not optimized for each studied process and do not take into account the environmental impact of each process and operation conditions. Therefore, in extraction optimization, especially considering green processes, all these criteria should be considered.

Before real application of the aforementioned green extraction processes for the production of natural products from agrifood by-products, they need to be optimized and their efficiency should be proved at a pilot scale. The challenge of scaling up is great since, in some cases, the amounts of by-products are very important and, in high volumes, the extraction yields might not be so satisfactory (for example, the effect of sonication decreases with the distance from the transducer). Therefore, studies of these processes should be extended at least to a pilot scale.

References

Achat, S., Tomao, V., Madani, K., Chibane, M., Elmaataoui, M., Dangles, O., Chemat, F., 2012. Direct enrichment of olive oil in oleuropein by ultrasound-assisted maceration at laboratory and pilot plant scale. Ultrason. Sonochem. 19, 777–786.

Ahmad, J., Langrish, T.A.G., 2012. Optimisation of total phenolic acids extraction from mandarin peels using microwave energy: the importance of the Maillard reaction. J. Food Eng. 109, 162–174.

Aliakbarian, B., Fathi, A., Perego, P., Dehghani, F., 2012. Extraction of antioxidants from winery wastes using subcritical water. J. Supercrit. Fluids. 65, 18–24.

Allaf, T., Tomao, V., Ruiz, K., Bachari, K., ElMaataoui, M., Chemat, F., 2013a. Deodorization by instant controlled pressure drop autovaporization of rosemary leaves prior to solvent extraction of antioxidants. LWT - Food Sci. Tech. 51, 111–119.

Allaf, T., Tomao, V., Ruiz, K., Chemat, F., 2013b. Instant controlled pressure drop technology and ultrasound assisted extraction for sequential extraction of essential oil and antioxidants. Ultrason. Sonochem. 20, 239–246.

Allaf, T., Berka Zougali, B., Nguyen, C.V., Negm, M., Allaf, K., 2014. DIC texturing for solvent extraction. In: Allaf, T., Allaf, K. (Eds.), Instant Controlled Pressure Drop (D.I.C.) in Food Processing: From Fundamental to Industrial Applications. Springer-Verlag, New York, NY, pp. 127–150.

Alvarez, V.H., Cahyadi, J., Xu, D., Saldaña, M.D.A., 2014. Optimisation of phytochemicals production from potato peel using subcritical water: Experimental and dynamic modeling. J. Supercrit. Fluids 90, 8–17.

Amor, B.B., Lamy, C., Andre, P., Allaf, K., 2008. Effect of instant controlled pressure drop treatments on the oligosaccharides extractability and microstructure of Tephrosia purpea seeds. J. Chrom. A. 1213, 118–124.

Andrade, K.S., Goncalvez, R.T., Maraschin, M., Ribeiro-do-Valle, R.M., Martinez, J., Ferreira, S.R.S., 2012. Supercritical fluid extraction from spent coffee grounds and coffee husks: antioxidant activity and effect of operational variables on extract composition. Talanta 88, 544–552.

Angiolillo, L., Del Nobile, M.A., Conte, A., 2015. The extraction of bioactive compounds from food residues using microwaves. Curr. Opin. Food Sci. 5, 93–98.

Araujo, M., Pimentel, F.B., Alves, R.C., Oliveira, M.B.P.P., 2015. Phenolic compounds from olive mill wastes: health effects, analytical approach and application as food antioxidants. Trends Food Sci. Techl. 45, 200–211.

Arnous, A., Meyer, A.S., 2010. Discriminated release of phenolic substances from red wine grape skins (*Vitis vinifera* L.) by multicomponent enzymes treatment. Biochem. Eng. J. 49, 68–77.

Athukorala, Y., Mazza, G., 2010. Optimization of extraction of wax from flax straw by supercritical carbon dioxide. Separ. Sci. Technol. 46, 247–253.

Bagherian, H., Ashtiani, F.Z., Fouladitajar, A., Mohtashamy, M., 2011. Comparisons between conventional, microwave- and ultrasound-assisted methods for extraction of pectin from grapefruit. Chem. Eng. Process. 50, 1237–1243.

Bai, F., Wang, J., Guo, J., 2015. Optimization for ultrasound-microwave assisted extraction of pectin from jujube waste using response surface methodology. Adv. J. Food Sci. Tech. 7, 144–153.

Baiano, A., 2014. Recovery of biomolecules from food wastes: a review. Molecules. 19, 14821–14842.

Baiano, A., Bevilacqua, L., Terracone, C., Contò, F., Del Nobile, M.A., 2014. Single and interactive effects of process variables on microwave-assisted and conventional extractions of antioxidants from vegetable solid wastes. J. Food Eng. 120, 135–145.

Ballard, T.S., Mallikarjunan, P., Zhou, K., O'Keefe, S., 2010. Microwave-assisted extraction of phenolic antioxidant compounds from peanut skins. Food Chem. 120, 1185–1192.

Barba, F.J., Zhu, Z., Koubaa, M., Sant'Ana, A.S., Orlien, V., 2016. Green alternative methods for the extraction of antioxidant bioactive compounds from winery wastes and by-products: a review. Trends Food Sci. Tech. 49, 96–109.

Barbosa, H.M.A., de Melo, M.M.R., Coimbra, M.A., Passos, C.P., Silva, C.M., 2014. Optimization of the supercritical fluid coextraction of oil and diterpenes from spent coffee grounds using experimental design and response surface methodology. J. Supercrit. Fluids 85, 165–172.

Barrales, F.M., Rezende, C.A., Martinez, J., 2015. Supercritical CO2 extraction of passion fruit (*Passiflora edulis* sp.) seed oil assisted by ultrasound. J. Supercrit. Fluids 104, 183–192.

Ben Hamissa, A.M., Seffen, M., Aliakbarian, B., Casazza, A.A., Perego, P., Converti, A., 2012. Phenolics extraction from *Agave americana* (L.) leaves using high-temperature, high-pressure reactor. Food Bioprod. Proc. 90, 17–21.

Benelli, P., Riehl, C.S., Smania, Jr.A., Smania, E.F.A., Ferreira, S.R.S., 2010. Bioactive extracts of orange (*Citrus sinensis* L. Osbeck) pomace obtained by SFE and low pressure techniques: mathematical modeling and extract composition. J. Supercrit. Fluids 55, 132–141.

Berka-Zougali, B., Hassani, A., Besombes, C., Allaf, K., 2010. Extraction of essential oils from Algerian myrtle leaves using instant controlled drop technology. J. Chrom. A. 1217, 6134–6142.

Bobinaitė, R., Pataro, G., Raudonis, R., Vškelis, P., Bobinas, Č., Šatkauskas, S., Ferrari G., 2016. Improving the extraction of juice and anthocyanin compounds from blueberry fruits and their by-products by pulsed electric fields. 1st World Congress on Electroporation and Pulsed Electric Fields in Biology, Medicine and Food & Environmental Technologies. IFMBE Proceedings, vol. 53.

Boukroufa, M., Boutekedjiret, C., Petigny, L., Rakotomanomana, N., Chemat, F., 2015. Bio-refinery of orange peels waste: a new concept based on integrated green and solvent free extraction processes using ultrasound and microwave techniques to obtain essential oil, polyphenols and pectin. Ultrason. Sonochem. 24, 72–79.

Boussetta, N., Vorobiev, E., 2014. Extraction of valuable biocompounds assisted by high voltage electrical discharges: a review. Comptes Rendus Chim. 17, 197–203.

Boussetta, N., Vorobiev, E., Deloison, V., Pochez, F., Falcimaigne-Cordin, A., Lanoisellé, J.L., 2011. Valorisation of grape pomace by the extraction of phenolic antioxidants: application of high voltage electrical discharges. Food Chem. 128, 364–370.

Boussetta, N., Vorobiev, E., Le, L.H., Cordin-Falcimaigne, A., Lanoisellé, J.L., 2012a. Application of electrical treatments in alcoholic solvent for polyphenols extraction from grape seeds. LWT - Food Sci. Technol. 46, 127–134.

Boussetta, N., Vorobiev, E., Reess, T., De Ferron, A., Pecastaing, L., Ruscassie, R., Lanoisellé, J.L., 2012b. Scale-up of high voltage electrical discharges for polyphenols extraction from grape pomace: effect of the dynamic shock waves. Innov. Food Sci. Emerg. Technol. 16, 129–136.

Boussetta, N., Turk, M., De Taeye, C., Larondelle, Y., Lanoiselle, J.L., Vorobiev, E., 2013. Effect of high voltage electrical discharge, heating and ethanol concentration on the extraction of total polyphenols and lignans from flaxseed cake. Ind. Crop. Prod. 49, 690–696.

Boyadzhiev, L., Dimitrov, K., Metcheva, D., 2006. Integration of solvent extraction and liquid membrane separation: an efficient tool for recovery of bio-active substances from botanicals. Chem. Eng. Sci. 61, 4126–4128.

Brianceau, S., Turk, M., Vitrac, X., Vorobiev, E., 2016. High voltage electric discharges assisted extraction of phenolic compounds from grape stems: effect of processing parameters on flavan-3-ols, flavonols and stilbenes recovery. Innov. Food Sci. Emerg. Technol. 35, 67–74.

Bucic-Kojic, A., Planinic, M., Tomas, S., Bilic, M., Velic, D., 2007. Study of solid–liquid extraction kinetics of total polyphenols from grape seeds. J. Food Eng. 81, 236–242.

Çam, M., His¸il, Y., 2010. Pressurised water extraction of polyphenols from pomegranate peels. Food Chem. 123, 878–885.

Carabias-Martinez, R., Rodriguez-Gonzalo, E., Revilla- Ruiz, P., Hernandez-Mendez, J., 2005. Pressurized liquid extraction in the analysis of food and biological samples. J. Chrom. A. 1089, 1–17.

Carciochi, R.A., Manrique, G.D., Dimitrov, K., 2015. Optimization of antioxidant phenolic compounds extraction from quinoa (*Chenopodium quinoa*) seeds. J Food Sci. Tech. Mys. 52, 4396–4404.

Casazza, A.A., Aliakbarian, B., Mantegna, S., Cravotto, G., Perego, P., 2010. Extraction of phenolics from *Vitis vinifera* wastes using non-conventional techniques. J. Food Eng. 100, 50–55.

Cheigh, C.I., Chung, E.Y., Chung, M.S., 2012. Enhanced extraction of flavanones hesperidin and narirutin from *Citrus unshiu* peel using subcritical water. J. Food Eng. 110, 472–477.

Chemat, F., Cravotto, G., 2011. Combined extraction techniques. In: Lebovka, N., Vorobiev, E., Chemat, F. (Eds.), Enhancing Extraction Processes in the Food Industry. CRC Press, Boca Ratón, FL, pp. 173–193.

Chemat, F., Albert-Vian, M., Cravotto, G., 2012. Green extraction of natural products: concept and principles. Int. J. Mol. Sci. 13, 8615–8627.

Chemat, F., Huma, Z., Khan, M.K., 2011. Applications of ultrasound in food technology: processing, preservation and extraction. Ultrason. Sonochem. 18, 813–835.

Chen, Y., Gu, X., Huang, S., Li, J., Wang, X., Tang, J., 2010. Optimization of ultrasonic/microwave assisted extraction (UMAE) of polysaccharides from *Inonotus obliquus* and evaluation of its anti-tumor activities. Int. J. Biol. Macromolec. 46, 429–435.

Choudhari, S.M., Ananthanarayan, L., 2007. Enzyme aided extraction of lycopene from tomato tissues. Food Chem. 102, 77–81.

Chumnanpaisont, N., Niamnuyb, C., Devahastinc, S., 2014. Mathematical model for continuous and intermittent microwave-assisted extraction of bioactive compound from plant material: extraction of β-carotene from carrot peels. Chem. Eng. Sci. 116, 442–451.

Corrales, M., Toepfl, S., Butz, P., Knorr, D., Tauscher, B., 2008. Extraction of anthocyanins from grape by-products assisted by ultrasonics, high hydrostatic pressure or pulsed electric fields: a comparison. Innov. Food Sci. Emerg. Tech. 9, 85–91.

Cravotto, G., Boffa, L., Mantegna, S., Perego, P., Avogadro, M., Cintas, P., 2008. Improved extraction of vegetable oils under high-intensity ultrasound and/or microwaves. Ultrason. Sonochem. 15, 898–902.

Dahmoune, F., Boulekbache, L., Moussi, K., Aoun, O., Spigno, G., Madani, K., 2013. Valorization of *Citrus limon* residues for the recovery of antioxidants: evaluation and optimization of microwave and ultrasound application to solvent extraction. Ind. Crop. Prod. 50, 77–87.

Dahmoune, F., Nayak, B., Moussi, K., Remini, H., Madani, K., 2015. Optimization of microwave-assisted extraction of polyphenols from *Myrtus communis* L. leaves. Food Chem. 166, 585–595.

Dimitrov, K., Metcheva, D., Boyadzhiev, L., 2005. Integrated processes of extraction and liquid membrane isolation of atropine from *Atropa belladonna* roots. Sep. Purif. Tech. 46, 41–45.

Dong, X., Zhao, M., Shi, J., Yang, B., Li, J., Luo, D., Jiang, G., Jiang, Y., 2011. Effects of combined high-pressure homogenization and enzymatic treatment on extraction yield, hydrolysis and function properties of peanut proteins. Innov. Food Sci. Emerg. Technol. 12, 478–483.

Ekinc, M.S.I., Gürü, M., 2014. Extraction of oil and β-sitosterol from peach (*Prunus persica*) seeds using supercritical carbon dioxide. J. Supercrit. Fluids 92, 319–323.

Eller, F.J., Moser, J.K., Kenar, J.A., Taylor, S.L., 2010. Extraction and analysis of tomato seed oil. J. Am. Oil Chem. Soc. 87, 755–762.

Esclapez, M.D., Garcia-Perez, J.V., Mulet, A., Cárcel, J.A., 2011. Ultrasound-assisted extraction of natural products. Food Eng. Rev. 3, 108–120.

Espinosa-Pardo, F.A., Martinez, J., Martinez-Correa, H.A., 2014. Extraction of bioactive compounds from peach palm pulp (*Bactris gasipaes*) using supercritical CO2. J. Supercrit. Fluids 93, 2–6.

Farias-Campomanes, A.M., Rostagno, M.A., Meireles, M.A.A., 2013. Production of polyphenol extracts from grape bagasse using supercritical fluids: yield, extract composition and economic evaluation. J. Supercrit. Fluids 77, 70–78.

Fernández-Ponce, M.T., Casas, L., Mantell, C., Rodríguez, M., Martínez de la Ossa, E., 2012. Extraction of antioxidant compounds from different varieties of *Mangifera indica* leaves using green technologies. J. Supercrit. Fluids 72, 168–175.

Galanakis, C.M., 2012. Recovery of high added-value components from food wastes: conventional, emerging technologies and commercialized applications. Trends Food Sci. Tech. 26, 68–87.

Galanakis, C.M., 2015. Food Waste Recovery: Processing Technologies and Industrial Techniques. Elsevier, Amsterdam.

Galván D'Alessandro, L., Vauchel, P., Przybylski, R., Chataigné, G., Nikov, I., Dimitrov, K., 2013. Integrated process extraction-adsorption for selective recovery of antioxidant phenolics from *Aronia melanocarpa* berries. Sep. Pur. Tech. 120, 92–101.

Galván D'Alessandro, L., Dimitrov, K., Vauchel, P., Nikov, I., 2014. Kinetics of ultrasound assisted extraction of anthocyanins from *Aronia melanocarpa* (black chokeberry) wastes. Chem. Eng. Res. Des. 92, 1818–1826.

Ghafoor, K., Park, J., Choi, Y.-H., 2010. Optimization of supercritical fluid extraction of bioactive compounds from grape (*Vitis labrusca* B.) peel by using response surface methodology. Innov. Food Sci. Emerg. Technol. 11, 485–490.

Gironi, F., Piemonte, V., 2011. Temperature and solvent effects on polyphenol extraction process from chestnut tree wood. Chem. Eng. Res. Des. 89, 857–862.

Gonzales, G.B., Smagghe, G., Raes, K., Van Camp, J., 2014. Combined alkaline hydrolysis and ultrasound-assisted extraction for the release of nonextractable phenolics from cauliflower (*Brassica oleracea* var. botrytis) waste. J. Agric. Food Chem. 62, 3371–3376.

González-García, E., Marina, M.L., García, M.C., 2014. Plum (*Prunus Domestica L.*) by-product as a new and cheap source of bioactive peptides: extraction method and peptides characterization. J. Funct. Foods. 11, 428–437.

Grassino, A.N., Brnčićb, M., Vikić-Topić, D., Roca, S., Dent, M., Brnčićb, S.R., 2016. Ultrasound assisted extraction and characterization of pectin from tomato waste. Food Chem. 198, 93–100.

Grigoras, C.G., Destandau, E., Fougere, L., Elfakir, C., 2013. Evaluation of apple pomace extracts as a source of bioactive compounds. Ind. Crop. Prod. 49, 794–804.

Guo, X., Han, D., Xi, H., Rao, L., Liao, X., Hu, X., Wu, J., 2012. Extraction of pectin from navel orange peel assisted by ultra-high pressure, microwave or traditional heating: A comparison. Carbohyd. Polym. 88, 441–448.

Hawthorne, S.B., Miller, D.J., 1994. Direct comparison of Soxhlet and low temperature and high-temperature supercritical CO2 extraction efficiencies of organics from environmental solids. Anal. Chem. 66, 4005–4012.

Hayat, K., Hussain, S., Abbas, S., Farooq, U., Ding, B., Xia, S., Jia, C., Zhang, X., Xia, W., 2009. Optimized microwave-assisted extraction of phenolic acids from citrus mandarin peels and evaluation of antioxidant activity in vitro. Sep. Purif. Tech. 70, 63–70.

He, J., Shao, P., Liu, J., Ru, Q., 2012. Supercritical carbon dioxide extraction of flavonoids from pomelo (*Citrus grandis* (L.) Osbeck) peel and their antioxidant acivity. Int. J. Mol. Sci. 13, 13065–13078.

Ho, C.H.L., Cacace, J.E., Mazza, G., 2007. Extraction of lignans, proteins and carbohydrates from flaxseed meal with pressurized low polarity water. LWT- Food Sci. Tech. 40, 1637–1647.

Ho, K.K.H.Y., Ferruzzi, M.G., Liceaga, A.M., San Martín-González, M.F., 2015. Microwave-assisted extraction of lycopene in tomato peels: effect of extraction conditions on all-trans and cis-isomer yields. LWT- Food Sci. Tech. 62, 160–168.

Holtung, L., Grimmer, S., Aaby, K., 2011. Effect of processing of black currant press-residue on polyphenol composition and cell proliferation. J. Agric. Food Chem. 59, 3632–3640.

Hossain, M.B., Tiwari, B.K., Gangopadhyay, N., O'Donnell, C.P., Brunton, N.P., Rai, D.K., 2014. Ultrasonic extraction of steroidal alkaloids from potato peel waste. Ultrason. Sonochem. 21, 1470–1476.

Hossain, M.B., Aguiló-Aguayo, I., Lying, J.G., Brunton, N.P., Rai, D.K., 2015. Effect of pulsed electric field and pulsed light pre-treatment on the extraction of steroidal alkaloids from potato peels. Innov. Food Sci. Emerg. Technol. 29, 9–14.

Jankowiak, L., Sevillano, D.M., Boom, R.M., Ottens, M., Zondervan, E., van der Goot, A.J., 2015. A process synthesis approach for Isolation of isoflavones from okara. Ind. Eng. Chem. Res. 54, 691–699.

Jesus, S.P., Calheiros, M.N., Hense, H., Meireles, M.A.A., 2013. A simplified model to describe the kinetic behavior of supercritical fluid extraction from a rice bran oil byproduct. Food Public Health 3, 215–222.

Kaderides, K., Goula, A.M., Adamopoulos, K.G., 2015. A process for turning pomegranate peels into a valuable food ingredient using ultrasound-assisted extraction and encapsulation. Innov. Food Sci. Emerg. Tech. 31, 204–215.

Kalamara, E., Goula, A.M., Adamopoulos, K.G., 2015. An integrated process for utilization of pomegranate wastes: seeds. Innov. Food Sci. Emerg. Tech. 27, 144–153.

Kammerer, D., Claus, A., Schieber, A., Carle, R., 2005. A novel process for the recovery of polyphenols from grape (*Vitis vinifera*) pomace. J. Food Sci. 70, 157–163.

Kammerer, D.R., Kammerer, J., Valet, R., Carle, R., 2014. Recovery of polyphenols from the by-products of plant food processing and application as valuable food ingredients. Food Res. Int. 65, 2–12.

Karcheva, M., Kirova, E., Alexandrova, S., Georgieva, S., 2013. Comparison of citrus peels as a valuable components—polyphenols and antioxidants. J. Chem. Technol. Metall. 48, 475–478.

Konwarh, R., Pramanik, S., Kalita, D., Mahanta, C.L., Karak, N., 2012. Ultrasonication – A complementary "green chemistry" tool to biocatalysis: a laboratory-scale study of lycopene extraction. Ultrason. Sonochem. 19, 292–299.

Koubaa, M., Roselló-Soto, E., Šic Žlabur, J., Režek Jambrak, A., Brnčić, M., Grimi, N., Boussetta, N., Barba, F.J., 2015. Current and new insights in the sustainable and green recovery of nutritionally valuable compounds from *Stevia rebaudiana* bertoni. J. Agric. Food Chem. 63, 6835–6846.

Koubaa, M., Barba, F.J., Grimi, N., Mhemdi, H., Koubaa, W., Boussetta, N., Vorobiev, E., 2016. Recovery of colorants from red prickly pear peels and pulps enhanced by pulsed electric field and ultrasound. Innov. Food Sci. Emerg. Technol. 37, 336–344.

Lapornik, B., Prosek, M., Wondra, A.G., 2005. Comparison of extracts prepared from plant by-products using different solvents and extraction time. J. Food Eng. 71, 214–222.

Laroze, L.E., Díaz-Reinoso, B., Moure, A., Zúñiga, M.E., Domínguez, A., 2010. Extraction of antioxidants from several berries pressing wastes using conventional and supercritical solvents. Eur Food Res Technol. 231, 669–677.

Lazarova, M., Dimitrov, K., 2009. Selective recovery of alkaloids from *Glaucium Flavum Crantz* using integrated process extraction-pertraction. Sep. Sci. Technol. 44, 227–242.

Li, B.B., Smith, B., Hossain, Md.M., 2006. Extraction of phenolics from citrus peels II: enzyme-assisted extraction method. Sep. Purif. Technol. 48, 189–196.

Liu, C., Wang, L., Wang, J., Wu, B., Liu, W., Fan, P., Liang, Z., Li, S., 2013. Resveratrols in Vitis berry skins andleaves: their extraction and analysis by HPLC. Food Chem. 136, 643–649.

Liu, S., Yang, F., Zhang, C., Ji, H., Hong, P., Deng, C., 2009. Optimization of process parameters for supercritical carbon dioxide extraction of *Passiflora* seed oil by response surface methodology. J. Supercrit. Fluids. 48, 9–14.

Luengo, E., Alvarez, I., Raso, J., 2013. Improving the pressing extraction of polyphenols of orange peel by pulsed electric fields. Innov. Food Sci. Emerg. Technol. 17, 79–84.

Ma, Y.Q., Chen, J.C., Liu, D.H., Ye, X.Q., 2009. Simultaneous extraction of phenolic compounds of citrus peel extracts: effect of ultrasound. Ultrason. Sonochem. 16, 57–62.

Machmudah, S., Zakaria, Winardi, S., Sasaki, M., Goto, M., Kusumoto, N., Hayakawa, K., 2012. Lycopene extraction from tomato peels by-product containing tomato seed using supercritical carbon dioxide. J. Food Eng. 108, 290–296.

Madej, K., 2009. Microwave-assisted and cloud-point extraction in determination of drugs and other bioactive compounds. Trends Anal. Chem. 28 (4), 436–446.

Makris, D., Boskou, G., Chiou, A., Andrikopoulos, N.K., 2008. An investigation on factors affecting recovery of antioxidant phenolics and anthocyanins from red grape (*Vitis vinifera* L.) pomace employing water/ethanol-based solutions. Am. J. Food Tech. 3, 164–173.

Maldonado, A.F.S., Mudge, E., Ganzle, M.G., Schieber, A., 2014. Extraction and fractionation of phenolic acids and glycoalkaloids from potato peels using acidified water/ethanol-based solvents. Food Res. Int. 65, 27–34.

Marino, M.G., Gonzalo, J.C.R., Ibanez, E., Moreno, C.G., 2006. Recovery of catechins and proanthocyanidins from winery by-products using subcritical water extraction. Anal. Chim. Acta. 563, 44–50.

Meyer, A.S., 2010. Enzyme technology for precision functional food ingredients processes. Ann. N. Y. Acad. Sci. 1190, 126–132.

Meyer, A.S., Jepsen, S.M., Sørensen, N.S., 1998. Enzymatic release of antioxidants for human low-density lipoprotein from grape pomace. J. Agric. Food Chem. 60, 5571–5582.

Mkaouar, S., Gelicus, A., Bahloul, N., Allaf, K., Kechaou, N., 2016. Kinetic study of polyphenols extraction from olive (*Olea europea* L.) leaves using instant controlled pressure drop texturing. Sep. Purif. Tech. 161, 165–171.

Monteil-Rivera, F., HaiHuang, G., Paquet, L., Deschamps, S., Beaulieu, C., Hawari, J., 2012. Microwave-assisted extraction of lignin from triticale straw: optimization and microwave effects. Bioresour. Tech. 104, 775–782.

Morais, A.R.C., Bogel-Lukasik, R., 2013. Green chemistry and the biorefinery concept. Sust. Chem. Proc., 1–18.

Mounir, S., Allaf, T., Mujumdar, A.S., Allaf, K., 2012. Swell drying: Coupling instant controlled pressure drop DIC to standard convection drying processes to intensify transfer phenomena and improve quality—an overview. Dry. Tech. 30, 1508–1531.

Mounir, S., Allaf, T., Berka, B., Hassani, A., Allaf, K., 2014. Instant controlled pressure drop technology: from a new fundamental approach of instantaneous transitory thermodynamics to large industrial applications on high performance—high controlled quality unit operations. Comptes Rendus Chim. 17, 261–267.

Moure, A., Cruz, J.M., Franco, D., Domínguez, J.M., Sineiro, J., Domínguez, H., Núñez, M.J., Parajó, J.C., 2001. Natural antioxidants from residual sources. Food Chem. 72, 145–171.

Mustafa, A., Turner, C., 2011. Pressurized liquid extraction as a green approach in food and herbal plants extraction: a review. Anal. Chim. Acta. 703, 8–18.

Nayak, B., Dahmoune, F., Moussi, K., Remini, H., Dairi, S., Aoun, O., Khodir, M., 2015. Comparison of microwave, ultrasound and accelerated-assisted solvent extraction for recovery of polyphenols from *Citrus sinensis* peels. Food Chem. 187, 507–516.

Nieto, A., Borrull, F., Pocurull, E., Marcé, R.M., 2010. Pressurized liquid extraction: a useful technique to extract pharmaceuticals and personal-care products from sewage sludge. Trends Anal. Chem. 29, 752–764.

Paini, M., Casazza, A.A., Aliakbarian, B., Perego, P., Binello, A., Cravotto, G., 2016. Influence of ethanol/water ratio in ultrasound and high-pressure/high-temperature phenolic compound extraction from agri-food waste. Int. J. Food Sci. Tech. 51, 349–358.

Panouillé, M., Thibault, J.F., Bonnin, E., 2006. Cellulase and protease preparations can extract pectins from various plant byproducts. J. Agric. Food Chem. 54, 8926–8935.

Peev, G., Penchev, P., Peshev, D., Angelov, G., 2011. Solvent extraction of rosmarinic acid from lemon balm and concentration of extracts by nanofiltration: effect of plant pre-treatment by supercritical carbon dioxide. Chem. Eng. Res. Des. 89, 2236–2243.

Peschel, W., Sanchez-Rabaneda, F., Diekmann, W., Plescher, A., Gartzia, I., Jimenez, D., Lamuela-Raventos, R., Buxaderas, S., Codina, C., 2006. An industrial approach in the search of natural antioxidants from vegetable and fruit wastes. Food Chem. 97, 137–150.

Piasek, A., Kusznierewicz, B., Grzybowska, I., Malinowska-Panczyk, E., Piekarska, A., Azqueta, A., Collins, A.R., Namiesnik, J., Bartoszek, A., 2011. The influence of sterilization with EnbioJet® microwave flow pasteurizer on composition and bioactivity of aronia and blue-berried honeyskle juices. J. Food Comp. Anal. 24, 880–888.

Pinelo, M., Arnous, A., Meyer, A.S., 2006. Upgrading of grape skins: significance of plant cell-wall structural components and extraction techniques for phenol release. Trends Food Sci. Technol. 17, 579–590.

Pinelo, M., Zornoza, B., Meyer, A.S., 2008. Selective release of phenols from apple skin: mass transfer kinetics during solvent and enzyme-assisted extraction. Sep. Purif. Technol. 63, 620–627.

Pingret, D., Fabiano-Tixier, A.-S., Bourvellec, C.L., Renard, C.M.G.C., Chemat, F., 2012. Lab and pilot-scale ultrasound-assisted water extraction of polyphenols from apple pomace. J. Food Eng. 111, 73–81.

Pittia, P., Gharsallaoui, A., 2015. Conventional product formation. In: Galanakis, C.M. (Ed.), Food Waste Recovery: Processing Technologies and Industrial Techniques. Elsevier, Amsterdam, pp. 173–193.

Pourfarzad, A., Najafi, M.B.H., Khodaparast, M.H.H., Khayyat, M.H., 2014. Characterization of fructan extracted from *Eremurus spectabilis* tubers: a comparative study on different technical conditions. J Food Sci. Tech. Mys. 52, 2657–2667.

Pradal, D., Vauchel, P., Decossin, S., Dhulster, P., Dimitrov, K., 2016. Kinetics of ultrasound-assisted extraction of antioxidant polyphenols from food by-products: extraction and energy consumption optimization. Ultrason. Sonochem. 32, 137–146.

Prado, J.M., Dalmolin, I., Carareto, N.D.D., Basso, R.C., Meirelles, A.J.A., Oliveira, J.V., Batista, E.A.C., Meireles, M.A.A., 2012. Supercritical fluid extraction of grape seed: process scale-up, extract chemical composition and economic evaluation. J. Food Eng. 109, 249–257.

Prakash Maran, J., Sivakumar, V., Thirugnanasambandhama, K., Sridhar, R., 2013. Optimization of microwave assisted extraction of pectin from orange peel. Carbohyd. Polym. 97, 703–709.

Prakash Maran, J., Sivakumar, V., Thirugnanasambandhama, K., Sridhar, R., 2014. Microwave assisted extraction of pectin from waste *Citrullus lanatus* fruit rinds. Carbohyd. Polym. 101, 786–791.

Pronyk, C., Mazza, G., 2009. Design and scale-up of pressurized fluid extractors for food and bioproducts. J. Food Eng. 95, 215–226.

Puértolas, E., Luengo, E., Alvarez, I., Raso, J., 2012. Improving mass transfer to soften tissues by pulsed electric fields: fundamentals and applications. Annu. Rev. Food Sci. Technol. 3, 263–282.

Puri, M., Sharma, D., Barrow, C.J., 2012. Enzyme-assisted extraction of bioactives from plants. Trends Biotechnol. 30, 37–44.

Ramos, L., Kristenson, E.M., Brinkman, U.A.T., 2002. Current use of pressurised liquid extraction and subcritical water extraction in environmental analysis. J. Chrom. A. 975, 3–29.

Richter, B.E., Jones, B.A., Ezzell, J.L., Porter, N.L., Corporation, D., Way, T., Box, P.O., 1996. Accelerated solvent extraction: a technique for sample preparation. Anal. Chem. 68, 1033–1039.

Rochová, K., Sovová, H., Sobolík, V., Allaf, K., 2008. Impact of seed structure modification on the rate of supercritical CO_2 extraction. J. Supercrit. Fluids. 44, 211–218.

Rodrigues, S., Pinto, G.A.S., Fernandes, F.A.N., 2008. Optimization of ultrasound extraction of phenolic compounds from coconut (*Cocos nucifera*) shell powder by response surface methodology. Ultrason. Sonochem. 15, 95–100.

Rombaut, N., Tixier, S., Bily, A., Chemat, F., 2014. Green extraction processes of natural products as tools for biorefinery. Biofuel. Bioprod. Bior. 8, 530–544.

Roselló-Soto, E., Barba, F.J., Parniakov, O., Galanakis, C.M., Lebovka, N., Grimi, N., Vorobiev, E., 2015a. High voltage electrical discharges, pulsed electric field and ultrasounds assisted extraction of protein and phenolic compounds from olive kernel. Food Bioprocess Technol. 8, 885–894.

Roselló-Soto, E., Koubaa, M., Moubarik, A., Lopes, R.P., Saraiva, J.A., Boussetta, N., Grimi, N., Barba, F.J., 2015b. Emerging opportunities for the effective valorization of wastes and by-products generated during olive oil production process: nonconventional methods for the recovery of high-added value compounds. Trends Food Sci. Technol. 45, 296–310.

Sahin, S., Samli, R., 2013. Optimization of olive leaf extract obtained by ultrasound-assisted extraction with response surface methodology. Ultrason. Sonochem. 20, 595–602.

Sánchez-Valdepeñas, V., Barrajón, E., Vegara, S., Funes, L., Martí, N., Valero, M., Saura, D., 2015. Effect of instant controlled pressure drop (DIC) pre-treatment on conventional solvent extraction of phenolic compounds from grape stalk powder. Ind. Crop. Prod. 76, 545–549.

Sant'Anna, V., Brandelli, A., Marczak, L.D.F., Tessaro, I.C., 2012. Kinetic modeling of total polyphenol extraction from grape marc and characterization of the extracts. Sep. Purif. Technol. 100, 82–87.

Santana-Meridas, O., Gonzalez-Coloma, A., Sanchez-Vioque, R., 2012. Agricultural residues as a source of bioactive natural products. Phytochem. Rev. 11, 447–466.

Schieber, A., Stintzing, F.C., Carle, R., 2001. By-products of plant food processing as a source of functional compounds—recent developments. Trends Food Sci. Tech. 12, 401–413.

Schugerl, K., Hubbuch, J., 2005. Integrated bioprocesses. Curr. Opin. Microbiol. 8, 294–300.

Seabra, I.J., Braga, M.E.M., Batista, M.T., de Sousa, A.C., 2010. Effect of solvent (CO_2/ethanol/H_2O) on the fractionated enhanced solvent extraction of anthocyanins from elderberry pomace. J. Supercrit. Fluids. 54, 145–152.

Segovia, F.J., Luengo, E., Corral-Pérez, J.J., Raso, J., Almajano, M.P., 2015. Improvements in the aqueous extraction of polyphenols from borage (*Borago officinalis* L.) leaves by pulsed electric fields: pulsed electric fields (PEF) applications. Ind. Crop. Prod. 65, 390–396.

Seidi, S., Yamini, Y., 2012. Analytical sonochemistry; developments, applications, and hyphenations of ultrasound in sample preparation and analytical techniques. Cent. Eur. J. Chem. 10, 938–976.

Singh, P.P., Saldaña, M.D.A., 2011. Subcritical water extraction of phenolic compounds from potato peel. Food Res. Int. 44, 2452–2458.

Song, J., Li, D., Liu, L., Zhang, Y., 2011. Optimized microwave-assisted extraction of total phenolics (TP) from *Ipomoea batatas* leaves and its antioxidant activity. Innov. Food Sci. Emerg. Tech. 12, 282–287.

Sovilj, W.N., Nikolovski, B.G., Spasojevic´, M.D., 2011. Critical review of supercritical fluid extraction of selected spice plant materials. Maced. J. Chem. Chem. Eng. 30, 197–220.

Spigno, G., de Faveri, D.M., 2007. Antioxidants from grape stalks and marc: Influence of extraction procedure on yield, purity and antioxidant power of the extracts. J. Food Eng. 78, 793–801.

Stoll, T., Schweiggert, U., Schieber, A., Carle, R., 2003. Process for the recovery of a carotene-rich functional food ingredient from carrot pomace by enzymatic liquefaction. Innov. Food Sci. Emerg. Tech. 4, 415–423.

Strati, I.F., Oreopoulou, V., 2014. Recovery of carotenoids from tomato processing by-products—a review. Food Res. Int. 65, 311–321.

Tabaraki, R., Heidarizadi, E., Benvidi, A., 2012. Optimization of ultrasonic assisted extraction of pomegranate (*Punica granatum* L.) peel antioxidants by response surface methodology. Sep. Purif. Tech. 98, 16–23.

Teixeira, A., Baenas, N., Dominguez-Perles, R., Barros, A., Rosa, E., Moreno, D.A., Garcia-Viguera, C., 2014. Natural bioactive compounds from winery by-products as health promoters: a review. Int. J. Mol. Sci. 15, 15638–15678.

Téllez-Pérez, C., Sabah, M.M., Montejano-Gaitán, J.G., Sobolik, V., Martínez, C.A., Allaf, K., 2012. Impact of instant controlled pressure drop treatment on dehydration and rehydration kinetics of green Moroccan pepper (*Capsicum annuum*). Procedia Engineering. 42, 978–1003.

The Food and Agricultural Organization (FAO), 2013. Food Wastage Footprint: Impacts on Natural Resources. Summary Report. The Food and Agricultural Organization (FAO), Rome, Italy.

Toma, M., Vinatoru, M., Paniwnyk, L., Mason, T.J., 2001. Investigation of the effects of ultrasound on vegetal tissues during solvent extraction. Ultrason. Sonochem. 8, 137–142.

Tomaz, I., Maslov, L., Stupić, D., Preiner, D., Ašperger, D., Kontić, J.K., 2016. Recovery of flavonoids from grape skins by enzyme-assisted extraction. Separ. Sci. Technol. 51, 255–268.

Vauchel, P., Galván D'Alessandro, L., Dhulster, P., Nikov, I., Dimitrov, K., 2015. Pilot scale demonstration of integrated extraction–adsorption eco-process for selective recovery of antioxidants from berries wastes. J. Food Eng. 158, 1–7.

Viganó, J., Martinez, J., 2015. Trends for the application of passion fruit industrial by-products: a review on the chemical composition and extraction techniques of phytochemicals. Food and Public Health 5, 164–173.

Virot, M., Tomao, V., Le Bourvellec, C., Renard, C.M.G.C., Chemat, F., 2010. Towards the industrial production of antioxidants from food processing by-products with ultrasound-assisted extraction. Ultrason. Sonochem. 17, 1066–1074.

Vorobiev, E., Lebovka, N., 2010. Enhanced extraction from solid foods and biosuspensions by pulsed electrical energy. Food Eng. Rev. 2, 95–108.

Wang, H., Ding, J., Ren, N., 2016. Recent advances in microwave-assisted extraction of trace organic pollutants from food and environmental samples. Trends Analyt. Chem. 75, 197–208.

Wang, J., Sun, B., Cao, Y., Tian, Y., Li, X., 2008. Optimisation of ultrasounds-assisted extraction of phenolic compounds from wheat bran. Food Chem. 106, 804–810.

Wang, W., Ma, X., Xu, Y., Cao, Y., Jiang, Z., Ding, T., Ye, X., Liu, D., 2015. Ultrasound-assisted heating extraction of pectin from grapefruit peel: optimization and comparison with the conventional method. Food Chem. 178, 106–114.

Wijngaard, H., Ballay, M., Brunton, N., 2012a. The optimization of extraction of antioxidants from potato peel by pressurized liquids. Food Chem. 133, 1123–1130.

Wijngaard, H., Hossain, M.B., Rai, D.K., Brunton, N., 2012b. Techniques to extract bioactive compounds from food by-products of plant origin. Food Res. Int. 46, 505–513.

Wu, T., Yan, J., Liu, R., Marcone, M.F., AkberAisa, H., Tsao, R., 2012. Optimization of microwave-assisted extraction of phenolics from potato and its downstream waste using orthogonal array design. Food Chem. 133, 1292–1298.

Xynos, N., Papaefstathiou, G., Gikas, E., Argyropoulou, A., Aligiannis, N., Skaltsounis, A.-L., 2014. Design optimization study of the extraction of olive leaves performed with pressurized liquid extraction using response surface methodology. Sep. Purif. Technol. 122, 323–330.

Yemiş, O., Mazza, G., 2012. Optimization of furfural and 5-hydroxymethylfurfural production from wheatstrawby a microwave-assisted process. Bioresour. Technol. 109, 215–223.

Yilmaz, C., Gokmen, V., 2013. Compositional characteristics of sour cherry kernel and its oil as influenced by different extraction and roasting conditions. Ind. Crop. Prod. 49, 130–135.

Yordanov, B., Dimitrov, K., Boyadzhiev, L., 2009. Extraction and liquid-membrane preconcentration of vincamine from periwinkle (*Vinca minor L.*) leaves. Proc. Model. Chem. Biochem. Eng. Q. 23, 135–141.

Zhang, M., Yang, H., Chen, X., Zhou, Y., Zhang, H., Wang, Y., Hu, P., 2011. In-situ extraction and separation of salvianolic acid B from *Salvia miltiorrhiza* Bunge by integrated expanded bed adsorption. Sep. Purif. Technol. 80, 677–682.

Zheng, H.Z., Lee, H.R., Lee, S.H., Kim, S., Chung, K., 2008. Pectinase assisted extraction of polyphenols from apple pomace. Chinese J. Anal. Chem. 36, 306–310.

Zheng, S., Wu, H., Li, Z., Wang, J., Zhang, H., Qian, M., 2015. Ultrasound/microwave-assisted solid–liquid–solid dispersive extraction with high-performance liquid chromatography coupled to tandem mass spectrometry for the determination of neonicotinoid insecticides in *Dendrobium officinale*. J. Sep. Sci. 38, 121–127.

Zhu, Z., Bals, O., Grimi, N., Vorobiev, E., 2012. Pilot scale inulin extraction from chicory roots assisted by pulsed electric fields. Int. J. Food Sci. Tech. 447, 1361–1368.

Zibetti, A.W., Aydi, A., Livia, M.A., Bolzan, A., Barth, D., 2013. Solvent extraction and purification of rosmarinic acid from supercritical fluid extraction fractionation waste: economic evaluation and scale-up. J. Supercrit. Fluids. 83, 133–145.

Further Reading

Agustin-Salazar, S., Medina-Juárez, L.A., Soto-Valdez, H., Manzanares-López, F., Gámez-Meza, N., 2014. Influence of the solvent system on the composition of phenolic substances and antioxidant capacity of extracts of grape (*Vitis vinifera* L.) marc. Aust. J. Grape Wine R. 20, 208–213.

Allaf, T., Mounir, S., Tomao, V., Chemat, F., 2012. Instant controlled pressure drop combined to ultrasounds as innovative extraction process combination: fundamental aspects. Procedia Eng. 42, 1061–1078.

Maran, J.P., Prakash, K.A., 2015. Process variables influence on microwave assisted extraction of pectin from waste *Carcia papaya* L. peel. Int. J. Biol. Macromol. 73, 202–206.

Vigano, J., da Fonseca Machado, A.P., Martinez, J., 2015. Sub- and supercritical fluid technology applied to food waste processing. J. Supercrit. Fluids. 96, 272–286.

CHAPTER 5

Extraction of Bioactive Phenolic Compounds by Alternative Technologies

Jorge E. Wong-Paz*, Diana B. Muñiz-Márquez*, Pedro Aguilar-Zárate*, Juan A. Ascacio-Valdés, Karina Cruz**, Carlos Reyes-Luna*, Raúl Rodríguez**, Cristóbal N. Aguilar****

**Instituto Tecnológico de Ciudad Valles, Tecnológico Nacional de México, Ciudad Valles, San Luis Potosí, México; **Autonomous University of Coahuila, Saltillo, Coahuila, Mexico*

1 Introduction

Some natural sources, particularly the plants, fruit, and derived fruit products have been used in traditional medicine as an alternative to combat chronic degenerative diseases, such as cancer and cardiovascular problems (Rossi et al., 2010; Zhou et al., 2016). In recent years, the scientific community has focused on the study of bioactive compounds from plants, including phenolic compounds, for their high biological potential (Vermerris and Nicholson, 2006). Phenolic compounds have the ability to act as powerful antioxidants and antimicrobial compounds, which may be used in different areas, such as the pharmaceutical and food industries (Fig. 5.1) (Amarowicz and Weidner, 2009; Ballard et al., 2010; Lucchesi et al., 2004; Sultana et al., 2009). Currently, various conventional methods have been developed for the extraction and the recovery of phenolic compounds from natural products, such as hydrodistillation (HD), Soxhlet (SX), and infusion extraction (IE), steam distillation (SD), maceration (MA), and so on (Lucchesi et al., 2004; Sun et al., 2011; Wang et al., 2008); however, these methods usually are time consuming and therefore the bioactive compounds may be degraded by long exposure times. In addition, these classical methods of extraction require large amounts of solvent. On the other hand, the nature of compounds and the operating factors in the extraction process impact the yields and kind of molecules to be extracted (Routray and Orsat, 2013).

Emerging technologies are more efficient methods for the extraction of bioactive compounds. This is because they allow the use of the most suitable solvents in minor proportions, as well as reduced extraction times with respect to conventional techniques. The most notable of these emerging technologies are microwave-assisted extraction (MAE), supercritical fluid extraction (SFE), and ultrasound-assisted extraction (UAE) (Routray and Orsat, 2013). In MAE, the microwaves are combined with traditional solvent extraction (Hao et al., 2002).

Ingredients Extraction by Physicochemical Methods in Food
http://dx.doi.org/10.1016/B978-0-12-811521-3.00005-3

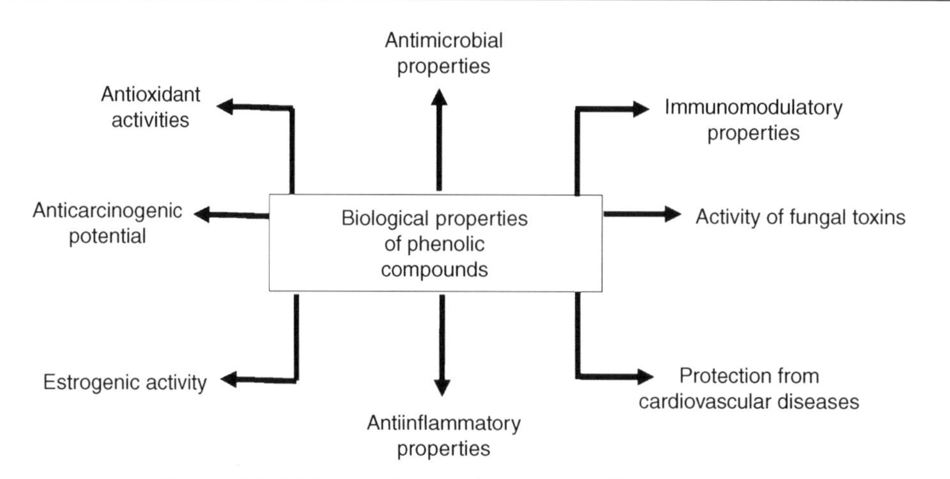

Figure 5.1: Biological Properties of Phenolic Compounds.

This methodology of extraction has been accepted as a potential alternative to conventional extraction methods for obtaining natural compounds (Yan et al., 2010). This is because it presents important advantages, such as short extraction time, less solvent, and a higher extraction rate thereby obtaining better products with lower cost (Ballard et al., 2010; Hao et al., 2002; Lucchesi et al., 2004). These reasons explain why some authors (Ballard et al., 2010; Hao et al., 2002; Martins et al., 2010b; Pan et al., 2012; Yan et al., 2010) have used microwave technology to extract natural antioxidants compounds, especially phenolic compounds.

The microwave process involves the interaction between solvent and plant material, where the absorbing properties are determined by the dielectric constant of the solvent. It is well known that solvents with high dielectric constant have the ability to absorb microwave energy and therefore produce a fast dissipation of energy into the solvent and solid plant matrix, which generates an efficient and homogeneous heating effect (Martins et al., 2010b). On the other hand, SFE is possible when high temperatures are combined with pressures that are higher than their critical values for a particular solvent, where the dissolvent is said to be in its supercritical moment. On the other hand, at high temperature and pressure, viscosity of the supercritical fluid is lower compared with the other liquids; for example, CO_2 is the most commonly used solvent in SFE because it is nontoxic, nonflammable, and less expensive, and it generally does not have any residue during the extraction process. Finally, UAE is an interesting alternative because the process is inexpensive, simple, and efficient. Here, the ultrasound uses ultrasonic waves, which generate mechanical vibrations in the system. In this extraction process, cavitation bubbles are formed in the liquid medium. After the bubbles form, they grow and collapse, thereby causing a strong impact on the solid surface (vegetal matrix) (Routray and Orsat, 2013). This chapter concentrates on the use of emerging technologies for the extraction of phenolic compounds including the critical factors that affect extraction efficiency. Also, some biological properties of phenolic compounds are discussed through in vitro studies.

2 Bioactive Phenolic Compounds and Human Health Benefits

Synthetic antioxidant compounds, such as tert-butylhydroquinne (TBHQ), butylated hydroxytoluene (BHT), propyl gallate (PG), butylated hydroxyanisole (BHA), and ascorbyl palmitate (PA) are widely used in the food industries to prevent the rancidity and deterioration in the food products. However, it is known that these compounds are promoters of chronic degenerative diseases, particularly cancerous tumors. Therefore, it has been of great interest to find natural sources of safe and inexpensive antioxidant phenolic compounds to use them in the pharmaceutical and food industries (Tusevski et al., 2014). Added to this, there is an increasing demand by consumers for functional foods to reduce disease and thereby increase longevity (Kim et al., 2016). Functional foods are those complemented with natural bioactive compounds, such as vitamins, minerals, prebiotics, probiotics, sterols, carotenoids, flavonoids, and phenolic compounds (Viera da Silva et al., 2016).

2.1 Phenolic Compounds

Phenolic compounds are phytochemicals that act in the secondary metabolism of a great variety of plants and are responsible for the plant's defense system. More than 800 types of phenolic compounds have been identified in fruits, vegetables, and plants. The World Health Organization (WHO) predicts that 80% of the world population uses traditional medicine as primary healthcare, principally through plant extracts and their bioactive components, due to the substantial increase of chronic health issues, such as diabetes, cardiovascular diseases, and obesity (Kim et al., 2016; Viera da Silva et al., 2016). Phenolic compounds have been reported as having antioxidant and antimicrobial properties as well.

2.1.1 Antioxidant activity

Oxidative stress (OS) has been correlated with molecular damage because it promotes oxygen-centered free radicals and other reactive oxygen molecules as by-products (ROS) in the biochemical reactions of the body (Thong and Nam, 2015). OS is induced when there is an imbalance between the generation and consumption of oxidative species that involves the mechanisms of a reaction between radicals and biological macromolecules, such as lipids, proteins, and DNA (Galano et al., 2016). Phenolic compounds have been documented to present antioxidant properties by different mechanisms of action. For example, phenolic compounds from green tea provide protection against oxidants, oxidative reactions, and reactive spices (Fig. 5.2). Existing preventive antioxidants demonstrate the ability to fight oxidative reactions by decreasing the oxygen concentration and avoiding chain reaction initiation by free radicals (HO, O_2), thus avoiding the generation of radicals and breaking lipid peroxides to peroxyl and alkoxyl molecules. On the other hand, primary antioxidants show effectiveness in the posterior events, such as decomposition of peroxides to nonradical molecules and the inhibiting of hydrogen removal from oxidable materials by intermediate

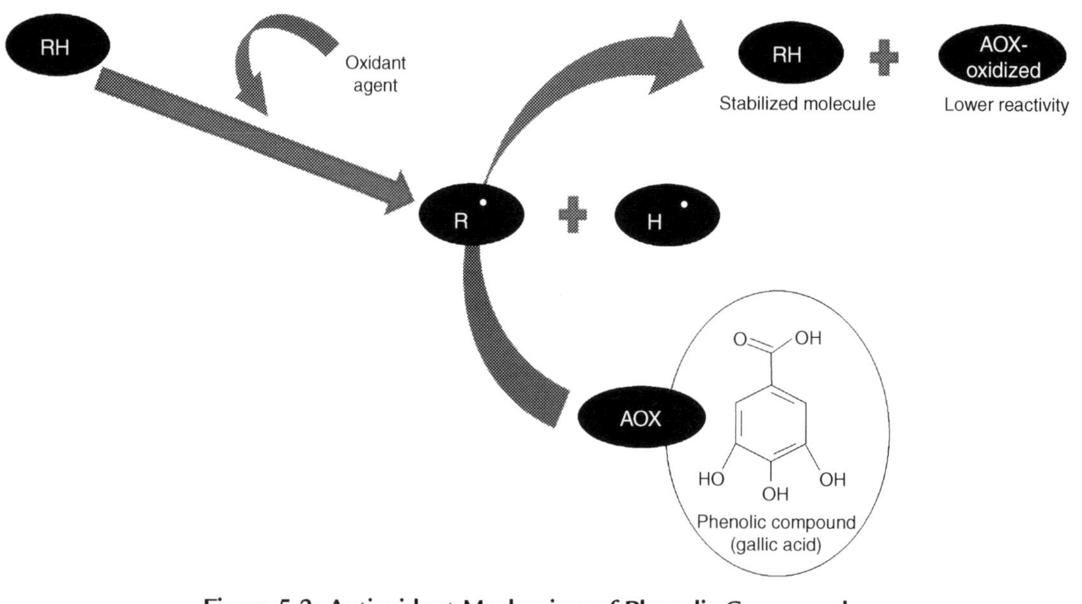

Figure 5.2: Antioxidant Mechanism of Phenolic Compounds.
AOX, antioxidant; *R*, free radical; *RH*, susceptible molecule to oxidation.

radicals. These compounds are part of the reactive oxygen species that are associated with oxidative damage of biological and food systems (Lorenzo and Sichetti, 2016; Miguel, 2010).

The study of the antioxidant properties of phenolic compounds obtained from plants is one of the hot topics among the scientific community through in vitro and in vivo studies (Martins et al., 2016). Tusevski et al. (2014) investigated the antioxidant potential of 27 plant species collected in the Republic of Macedonia. They used various antioxidant assays, which include reducing power assay (RP), phosphomolybdenum method (PM), cupric reducing antioxidant capacity assay (CUPRAC), ABTS radical scavenging activity, and DPPH radical scavenging activity. Based on the results, among the studied plants extracts, *Origanum vulgare* showed the strongest antioxidant activity and the highest total phenolic compounds. On the other hand, *Salvia ringens* and *Melissa officinalis* also proved to be a promising potential alternative source of natural antioxidants. In this work, a positive linear correlation between antioxidant capacity and total phenolic compounds, flavonoids and phenylpropanoids, was found. Luyen et al. (2014) studied the antioxidant activity of phenolic compounds present in the extracts from *Euphorbia maculate*. They prepared fractions and revealed that methanolic extracts, ethyl acetate, and aqueous fractions showed a high potential power antioxidant with the ORAC values of 27.24 ± 0.40, 28.47 ± 0.36, 27.07 ± 0.31; and CUPRAC values of 46.67 ± 0.34, 43.86 ± 0.26. This is 46.58 ± 0.58-fold higher than the protection provided by 1.0 μM of Trolox. De Fernández et al. (2014) tested the antioxidant properties of phenolic compounds from olive oil, as well as determined the phenolic profile, thus confirming the presence of 15 compounds: tyrosol, vinylphenol, oleuropein, hydroxytyrosol, rutin, catechin, naringenin, cinnamic acid, chlorogenic acid, syringic acid, luteolin, apigenin, vanillin acid, quercetin, and caffeic acid. This vegetal material showed the high antioxidant and free-radical scavenging capacities in all assays used.

2.1.2 Antimicrobial activity

Besides antioxidant activities, phenolic compounds also possess antimicrobial properties against certain pathogenic microorganisms. Recently, the growing occurrences of drug-resistant microorganisms, such as bacteria and fungi have been significant. Therefore, the scientific research is concerned with the discovery of novel natural antimicrobial compounds and the study of their therapeutic effects to combat microbial diseases, such as foodborne diseases (Patra, 2012). In addition, antimicrobial phenolic compounds could be used for extending the shelf life of foods, thereby acting as a protective barrier to preventing the microbial contamination. This is interesting because food poisoning is still a concern for consumers and the food industries, originating from foods contaminated by phatogenic microorganisms, especially, Gram-positive and Gram-negative bacteria (Kallel et al., 2014).

Mechanisms of action of the bioactive compounds of plants extracts involved with their antimicrobial action are the following: degradation or decomposition of the cell wall, damage to the cytoplasmic membrane, act against membrane proteins, release of the intracellular content, coagulation of cytoplasm, and depletion of the proton motive force (Fig. 5.3). Also, it

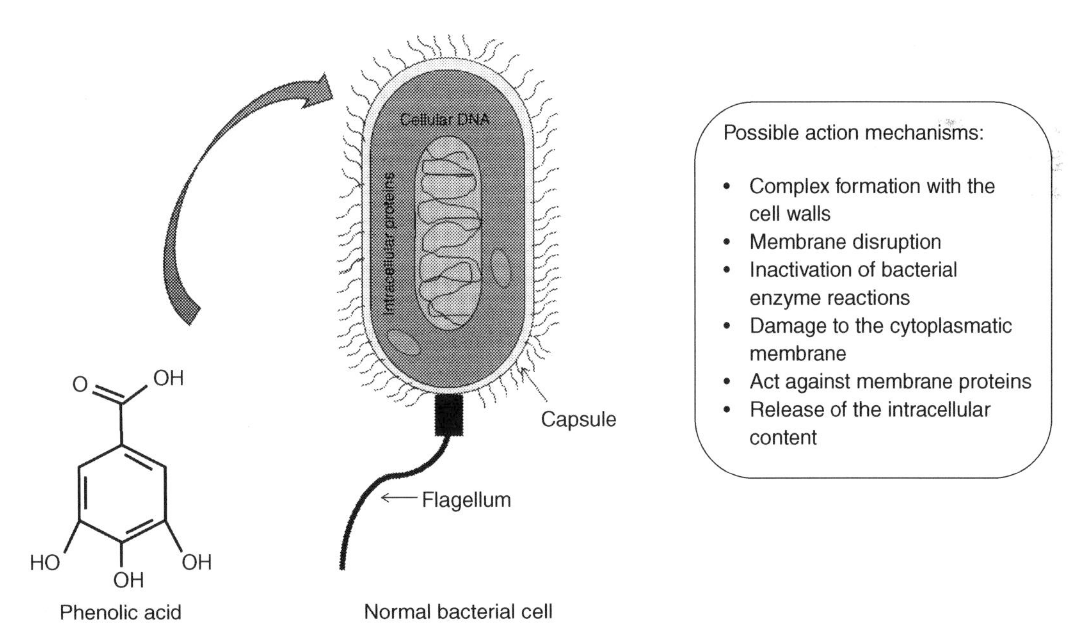

Figure 5.3: Antimicrobial Mechanism of Phenolic Compounds. *From Cetin-Karaca, H., Newman, M.C., 2015. Antimicrobial efficacy of plant phenolic compounds against Salmonella and Escherichia coli. Food Biosci. II, 8–16; Alkan, D., Yemenicioğlu, A., 2016. Potential application of natural phenolic antimicrobials and edible film technology against bacterial plant pathogens. Food Hydrocoll. 55, 1–10.*

is important to consider the cell surface structures between Gram-negative and Gram-positive microorganisms. For example, the Gram-positive bacteria are more susceptible to the phenolic acids than the Gram-negative microorganisms. On the other hand, the number, type, and position of substituents in the benzene ring of the phenolic compounds, as well as the saturated side-chain length influence the antimicrobial properties of the phenolic compounds against different microorganisms (Cetin-Karaca and Newman, 2015). Kallel et al. (2014) evaluated the antimicrobial potential of garlic husk waste extracts (GH) against *Escherichia coli*, *Salmonella typhimurium*, *Klebsiella pneumoniae*, *Pseudomonas aeruginosa*, *Bacillus thuringiensis*, *Bacillus subtilis*, and *Staphylococcus aureus*, and determined the minimal inhibitory concentration (MIC). According to their results, all tested extracts showed inhibitory response, but the growth inhibition was dependent on extract concentration; at the lowest concentrations (1 mg/mL), no inhibition zones were observed. Moreover, Gram-positive microorganisms were the most sensitive, being inhibited by all the extracts. Here, *S. aureus* was found to be the most sensitive bacteria, showing the MIC of 2 mg/mL. Others researchers have demonstrated the antimicrobial properties of the extract of plants rich in polyphenols. Gutiérrez-Larraínzar et al. (2012), studied the minimal inhibitory concentration (MIC) of seven pure phenolic compounds (thymol, carvacrol, eugenol, hydroquinone, *p*-hydroxybenzoicacid, protocatechuic acid, and gallic acid) against a Gram-negative (*E. coli* and *Pseudomonas fluorescens*) and Gram-positive (*S. aureus* and *Bacillus cereus*) bacteria and demonstrated that carvacrol and thymol were the phenolic compounds most effective for all bacteria used, except *S. aureus*, which was inhibited by hydroquinone compound. The findings of these authors are in accordance with other aforementioned studies because they also report that Gram-positive microorganisms were more sensitive than Gram-negative microorganisms. Another application of extracts rich in bioactive phenolic compounds is conversion into edible films to obtain antimicrobial packaging materials. Alkan and Yemenicioğlu (2016) evaluated the use of microbial phenolic compounds applied in edible films for the inhibition of major plant pathogens bacteria (*Erwinia amylovora*, *Erwinia carotovora*, *Xanthomonas vesicatoria*, and *Pseudomonas syringae*). The results demonstrated that the films containing phenolic acids between 1 and 4 mg/cm^2 inhibited all pathogens. The most potent films were obtained by using gallic acid (GA) against *E. amylovora* and *P. syringae*.

3 Extraction Processes

Recovery of BPCs from the vegetable matrix is a critical stage to the isolation, identification, and overall maintenance of biological activity. Unsuitable extraction conditions may result in a loss of the BPCs and even in the biological activity. Currently, two large groups can be classified in the processes of extraction, namely, conventional and alternative extraction technologies. The conventional extraction technologies of BPCs from plants, fruits, and their by-products is accomplished by means of agitation, heating, boiling, or refluxing extraction, which result in several disadvantages, such as high impurity contents, long extraction time, and loss of thermolabile compounds (Peng et al., 2010; Wong-Paz et al., 2015b). In contrast, alternative extraction technologies have found new applications in the field with advantages in terms of operational simplicity, safety (green solvents), improvements in the extraction yield, less thermolabile compound degradation, and increased rate of extraction in comparison to conventional extraction technologies (Izadiyan and Hemmateenejad, 2016; Muñiz-Márquez et al., 2013; Tiwari, 2015). The subsequent section examines the advances in the research done on BPC extraction using alternative extraction technologies. Modifications to improve the performance of alternative extraction technologies, such as UAE, MAE, fermentation-assisted extraction (FAE), and enzyme-assisted extraction (EAE) are also presented.

3.1 Conventional Extraction Technologies

Extraction of compounds, such as antioxidants and antimicrobial compounds (polyphenols) from plants is important research to obtain drugs/chemotherapeutics that have been manufactured for decades by conventional extraction methods, such as soxhlet and reflux, as well as infusion, decoction, digestion, maceration, and percolation (Lianfu and Zelong, 2007; Manish et al., 2012; Mueen, 2008).

The conventional extraction processes are time consuming, for example, maceration is done for 2–7 days, involving bulk amounts of solvents and resulting in damage or loss of antioxidant activity in the compounds extracted by the use of high temperatures for long periods of time (Manish et al., 2012; Muñiz-Márquez et al., 2013; Rodrigues et al., 2008). The disadvantage of using solvents (soxhlet and reflux) is in the increased operating costs and additional environmental problems (Tarke and Rajan, 2014). But these methodologies continue to be used because the laboratory equipment is widely available.

3.2 Alternative Extraction Technologies

By the disadvantages that conventional technologies have shown on the extraction of polyphenols (antioxidants). Several lines of research have emerged for the search and implementation of alternative methodologies, such as UAE, MAE, FAE, and EAE.

UAE and UAE have advantages, such as shortened extraction time, reduced organic solvent consumption, increased pollution prevention, and increased yields (Tarke and Rajan, 2014). FAE and EAE are ecological technologies, in that they use microorganisms, such as stain fungal and enzymes (catalytic activity proteins) to obtain polyphenols or antioxidant compounds. For example, ellagic acid has been obtained from submerged cultures using pomegranate husk powder and *Aspergillus niger* GH1 (Sepúlveda et al., 2014). Also, these techniques can be adapted on a small or large scale, they are cheaper than traditional methods, and they can be performed with fewer instrumental requirements (Muñiz-Márquez et al., 2013).

3.2.1 Ultrasound-assisted extraction

Extraction of bioactive compounds from plants, fruits, and derived products using UAE has been reported from the 1980s and 1990s (Routray and Orsat, 2013; Salisova et al., 1997). However, today the UAE remains a technology commonly studied to recovery BPCs due to its high efficiency showed (Tiwari, 2015). This statement is supported by several published articles focusing the use of ultrasonic system to improve the extraction of some class of bioactive compounds (Liu et al., 2015; Tekin et al., 2015). Therefore, it is really clear that in the last few years, ultrasound system has impacted as valuable tool in food engineering processes (Esclapez et al., 2011; Tiwari, 2015).

UAE has the advantages of penetrating force, less solvent consumption, short extraction time, and as a new kind of nonthermal extraction technology, it is especially suitable for extracting heat-sensitivity components and natural active constituents. UAE applications are more common in solid–liquid systems.

In the liquid–solid extraction, a mass transport phenomenon is observed. The solids contained in the matrix migrate into the solvent to achieve equilibrium (Ghafoor et al., 2009). In UAE this mass transfer phenomenon could be accelerate by the acoustic cavitation effect of the system. Cavitation is produced when ultrasonic waves pass through the solvent, then the generation, growth, and the subsequent collapse of millions of tiny vapor bubbles (voids) in the liquid or at liquid–solid interfaces (Roohinejad et al., 2016). The violent implosion of the vapor bubbles releases large amounts of energy with an increase in the temperature and pressure of the medium, yielding shock waves of several hundred atmospheres (Esclapez et al., 2011). Therefore, this phenomenon produces an enlargement in the pore walls or the disruption of the cell walls in a short period of time, reducing the particle size to allow greater penetration of the solvent into the sample and finally releases target ingredients, such as BPCs (Aspé and Fernández, 2011; Wong-Paz et al., 2015a). In food bioengineering, the recovery of active ingredients has received increasing interest due to the human health properties of some natural extracts. Besides, the extraction of active ingredients can increase the value of some food industry by-products. The UAE of BPCs, such as anthocyanins, flavonols, or phenolic acids has been specially conducted (Aspé and Fernández, 2011). The properties exhibited by BPCs are of interest in the production of functional foods.

UAE has been applied in a several vegetable materials to recovery bioactive compounds. Nowadays, it is a powerful tool in extraction processes and has been used in the extraction of BPCs from several plants, fruit, and derived by-products, such as red raspberries (Chen et al., 2007), coconut (*Cocos nucifera*) (Rodrigues et al., 2008), orange (*Citrus sinensis* L.) peel (Khan et al., 2010), black chokeberry (Galvan d'Alessandro et al., 2012), hawthorn seed (Pan et al., 2012), spices, such as *Origanum majorana* L. (Hossain et al., 2012) and *Laurus nobilis* leaves (Muñiz-Márquez et al., 2013), vegetal tissues (Wong-Paz et al., 2015a), xoconostle (*Opuntia oligacantha*) (Espinosa-Muñoz et al., 2017), pomegranate peel (Kazemi et al., 2016), blueberry (*Vaccinium ashei*) wine pomace (He et al., 2016), and spent filter coffee (Michail et al., 2016). Interestingly, all authors highlight the UAE system efficiency as compared to conventional extraction techniques.

Some minor disadvantages are found in the UAE system. An important factor to consider is wave distribution in the extractor because this is usually not uniform. Additionally, ultrasonic power decreases with the increase in distance from the radiating surface. To prevent this drawback in the extraction, shaking or agitation can be applied. Ultrasound also produces a slight heat increase, so the temperature level should be checked for prevention of excess heating (to prevent degradation of thermolabile compunds) (Routray and Orsat, 2013).

3.2.2 Microwave-assisted extraction

MAE is also a good alternative for the extraction of natural antioxidants in plants, fruits, and derived products. Several reports have shown the efficiency of this technology. The most important advantage is the increase in BPC yield extraction with reduction of the time from hours to minutes (Wong-Paz et al., 2014). The effect of the MAE and maceration on the recovery of bioactive compounds from eucalyptus (*Eucalyptus globulus*) wood industrial wastes was studied. Comparing both techniques, maceration led to the extract with the best antioxidant properties, but MAE allowed for significant reduction in the extraction time (Fernández-Agulló et al., 2015). It is important to note that two critical drawbacks are found in the MAE system. The first problem is the high cost of the equipment as compared to other extraction methodologies. And the second is the use of high temperatures to accelerate the diffusion of the BPCs. It is clear that moderated temperatures could be used. However, the use of the device would remain limited. At present, the use of MAE performed in low temperatures and in vacuo for the extraction of labile compounds in food samples has proven to be a good alternative in the use of microwave systems (Xiao et al., 2012). The report mentioned MAE-Vacuo was especially important to prevent the degradation of thermolabile components, as well as to have good potential for the extraction of compounds in foods, pharmaceuticals, and natural products. Other modifications to improve the microwave system have been successfully used. In this regard, Chan et al. (2011) reported the use of nitrogen-protected microwave-assisted extraction (NPMAE), ultrasonic microwave-assisted extraction (UMAE), dynamic microwave-assisted extraction (DMAE), and solvent-free microwave-assisted extraction (SFME).

Microwave systems have also been used to improve the extraction of BPCs from diverse materials, such as black tea (Spigno and De Faveri, 2009), plants, such as creosote bush (Martins et al., 2010b), wine lees (Pérez-Serradilla and Luque de Castro, 2011), Tunisian olive leaves (Taamalli et al., 2012), grape seed (Li et al., 2011), lettuce (Périno et al., 2016), and goji berries (Mendes et al., 2016). Despite good extraction of BPCs, it seems that UAE is preferred in this context. As we discussed earlier, the MAE apparatus is more expensive than ultrasonic equipment, and even more so in pilot or industrial scale. Some improvements in the equipment are needed to decrease the cost of the equipment in solving this problem.

3.2.3 Fermentation-assisted extraction

An alternative technology to obtain bioactive phenolic compounds is FAE. This technology has produced few reports and more research is necessary to elucidate and understand the FAE processes (Scalbert, 1992; Vivas et al., 2004). Few papers about FAE can be found in the scientific bibliography.

Novotanin O, novotanin P, and gallic acid have been extracted from Tiouchina multiflora using *A. niger* (Yoshida et al., 1999). Also, cranberry pomace polyphenols have been extracted by *Lentinusedodes* in solid state fermentation (Vattem and Shetty, 2002, 2003), and the reports reveal the extraction and recovery of ellagintannins and ellagic acid, the latter having potential due to its biological properties (Ascacio-Valdés et al., 2011).

Vaquero et al. (2004) obtained antocianins, galotannins, and ellagitannins by muscadine grape extract fermentation by *Leuconostoc*, *Lactobacillus*, *Oenococcus*, and *Pediococcus* strains. *A. niger* SHL 6 has been used in liquid fermentation with oak tree as substrate to the bioactive compounds (polyphenols) extraction as ellagitannins; comparing this method with chemical methods, the fermentation process is viable (Huang et al., 2005). According to this research, after a microorganism fermentation using a substrate rich in bioactive compounds, the latter are accumulated in the system and can be recovered by specific techniques. Shi et al. (2005) proved that valonia extracts fermented with *A. niger* and *Candida utilis* was accumulated and extracted antioxidant phenolic compounds, finally characterized by analytic techniques.

There is a lack of available information regarding the FAE of bioactive compounds (Saavedra et al., 2005). Aguilera-Carbó et al. (2007) reported that *A. niger* GH1 can ferment polyphenols using solid-state fermentation, with microbiological strains isolated from gobernadora (*Larrea tridentata*) (Cruz-Hernández et al., 2005). *A. niger* GH1 was used in solid-state fermentation to obtain bioactive compounds from gobernadora leaves (Aguilera-Carbó et al., 2009) and these compounds were identified and quantified.

Recently, Ascacio-Valdés et al. (2014) described the fermentation of pomegranate shell extracts using *A. niger* GH1in solid state fermentation, wherein the compounds of interest were recovered. Ascacio-Valdés et al. (2016) studied a second solid-state fermentation using the same mold strain and bioactive compounds source. The authors achieved the recovery of the compounds and reported the identification by analytic techniques as HPLC coupled to mass spectrometry. Information about FAE is scarce; however, in recent years, it has been demonstrated that the extraction of desired compounds is possible using biotechnological processes, such as fermentation. This is an attractive alternative to chemical extractions, as it diminishes the cost of the process, uses solvents in the recovery process, and supports environmental preservation.

3.2.4 Enzyme-assisted extraction

EAE (also referred to as enzyme-based extraction) of BPCs from different sources is an alternative, ecofriendly technology that allows extraction with less use of solvents. Conventional extraction methods, such as Soxhlet, are still being considered as reference methods for comparison with newly developed technologies. Enzyme-assisted methods have shown high extraction yields for many compounds, such as polysaccharides, oils, natural pigments, flavors, and medicinal compounds (Sowbhagya and Chitra, 2010; Yang et al., 2010).

Some phytochemicals are imbibed in plant matrices and dispersed in cell cytoplasm, tannosome (Brillouet et al., 2013), or interact with the polysaccharide-lignin network by ester, hydrogen, or hydrophobic bonding (Fig. 5.4)(Azmir et al., 2013). Sometimes these matrices are not accessible for solvents in extraction processes. Enzymatic pretreatment has

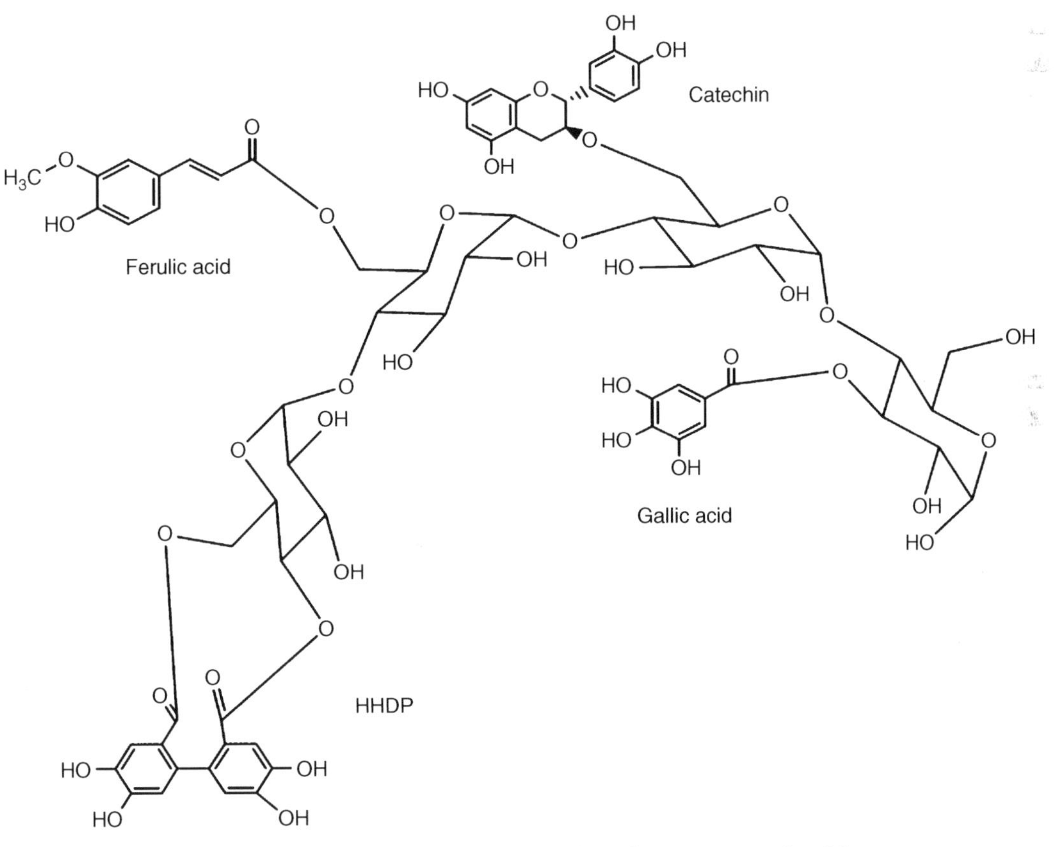

Figure 5.4: Representative Interaction of Bioactive Compounds with Hexose Chain Polysaccharide.
The compounds are interacting by ester bonds. HHDP represents the hexahydroxydiphenic acid prior to formation of ellagic acid.

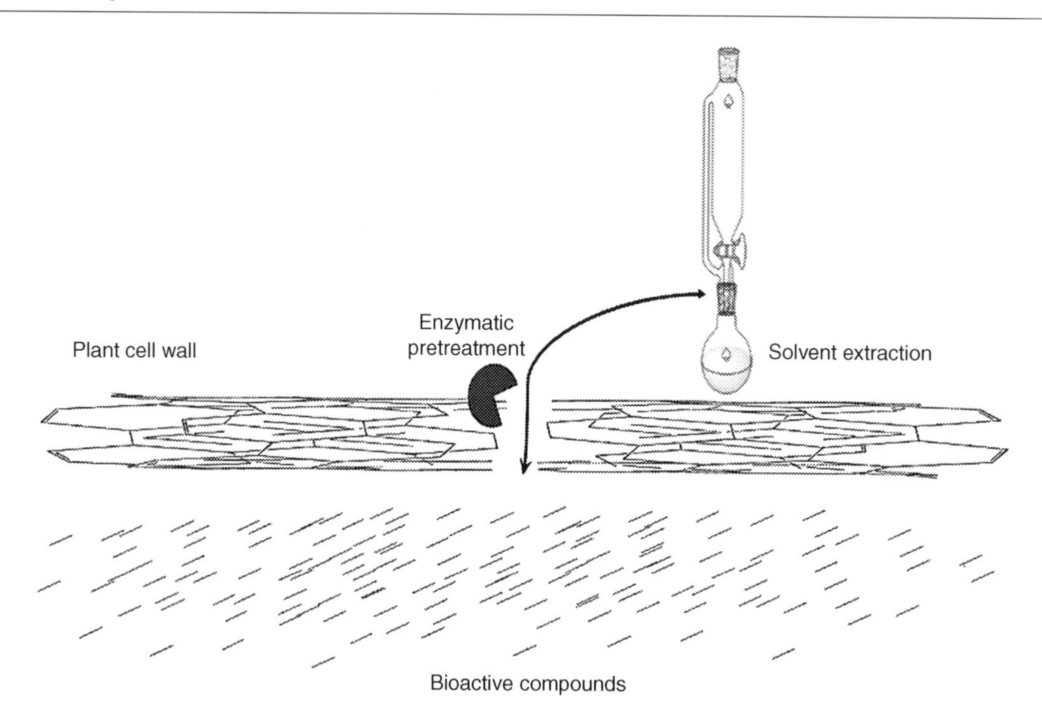

Figure 5.5: Schematic Representation of Enzymatic Pretreatment and Solvent Extraction Strategies for the Recovery of Bioactive Compounds from Plant Materials.

been used prior to conventional methods to enhance the releasing of polyphenols. Enzymes, such as cellulases, pectinases, hemicellulases, and α-amylases, among others, have been used to disrupt the plant cell wall, thereby allowing the entrance of solvents and increasing the recovery yield of polyphenols (Fig. 5.5)(Puri et al., 2012).

For example, Manasa et al. (2013) applied α-amylase, cellulose, protease, pectinase, and viscozyme (mixture of enzymes) for the pretreatment of ginger (*Zingiber officinale* Roscoe) to extract oleoresin and 6-gingerol by conventional methods using acetone and ethanol. Higher yields of 6-gingerol were obtained using viscozyme, amylase, pectinase, and cellulase followed by acetone extraction. Oleoresins and polyphenols were extracted using cellulase followed by extraction with ethanol.

Enzymatic clarification of pomegranate juice is another example of pectinolitic enzyme application as a means to increase the extraction of polyphenols. Rinaldi et al. (2013) used different pectolytic enzyme concentrations for the releasing of polyphenols from pomegranate juice mixed with arils. The results showed an increase in polyphenols content, as well as turbidity of the juice, thus recommending further clarification.

The extraction of phenolic compounds from grape waste using celluclast, pectinex, and novoferm has been reported by Gómez-García et al. (2012). They obtained better results using novoferm and proposed the enzyme-extraction technology as an alternative for the recovery of bioactive compounds from agroindustrial wastes.

Solid-state fermentation (SSF) has become an important tool for the recovery of bioactive compounds. The use of agricultural and agro-industrial materials as support and substrate for the growth of microorganisms has resulted in the increase in phenolic compounds content (Martins et al., 2011). This is due to the production of enzymes, such as cellulase, xylanase, β-glucosidase, tannase, ellagitannase, among others (Ascacio-Valdés et al., 2014). More recent studies were reported by Fernández et al. (2015). The authors reported an enzymatic extraction (pectinase, cellulose, and tannase) of proanthocyanidins from País grape seeds and skins. They mentioned that enzymes have a favorable effect on phenol release with a higher concentration of total phenols obtained from grape seeds than from skins.

Extraction of bioactive compounds by the action of enzymes has important disadvantages. The costs of this process increase due to the high costs of enzymes required for processing large amounts of raw material. The available enzymes cannot completely degrade the cell wall, thus limiting the solvent extraction. Laboratory scale conditions are difficult to apply in industrial scale (Puri et al., 2012). Despite the above, EAE remains a green technology for the enhancement of bioactive compounds recovery and for the use of low amounts of solvents.

4 Important Parameters in the Extraction Process

Several factors affect the extraction efficiency in the extraction process (Routray and Orsat, 2013). Among them we can mention the following: solvent nature, solid/solvent ratio, extraction temperature, extraction time and cycle, effect of stirring, and apparatus power.

The solvent nature has an impact in the extraction by the fact that some BPCs have different polarity. In the case of phenolic acids, hydroxycinnamic acids, and low molecular flavanols, water or alcohols, such as ethanol or methanol could be used (Ramirez-Coronel et al., 2004). In contrast, polymerized procyanidins, aqueous acetone, could be used with high yield of extraction. In this sense, alternative solvents have been sought out in the context of green chemistry that has the same level of efficiency (Mendez et al., 2012).

The ratio of vegetal material and solvent is also important to decrease the consumption of solvent used. Several ratios have been used, such as 1/4, 1/8, 1/12, 1/16, and even 1/40 (g of material/mL). However, the most appropriate ration depends on the vegetal material being used. Currently, the ratio of 1/12 seems to be the most commonly established in research.

The temperature is an important factor in the extraction of BPCs from plants, fruits, and derived products because many compounds are sensitive at high temperature extractions. In the same way as the nature of the solvent determines the ratio used, the temperature depends on the amount of BPCs to be extracted. Low temperatures could range from 40 to

60°C, medium temperatures from 61 to 80°C, and high temperatures are typically greater than 90° C. A recent study examined the effect of the MAE and maceration on the recovery of bioactive compounds from eucalyptus (*E. globulus*) wood industrial wastes. The authors reported that in maceration, the extraction yield increased with increasing temperature and with reducing particle size. However, extract properties decreased when the temperature was increased (Fernández-Agulló et al., 2015).

Extraction time is significantly linked to the temperature of the extraction of BPCs. In a case wherein low temperatures are used, a long time period is needed. In contrast, when moderate or high temperatures are used, shorter time periods can be applied. The cycles of extraction and stirring are other factors to be considered in BPC recovery. Stirring increases the diffusion of the component to the extracellular medium. Two cycles are recommended to extract the major percentage of BPCs. Finally, the apparatus power depends on the technique used. It is important to note that the high extraction yield of BPCs does not reflect the bioactivity efficacy.

5 Extraction Optimization

Many strategies have been reported for the optimization of extraction processes, the bulk of which can be classified as response surface methodology (RSM). This tool is effective for statistical optimization and modeling of processes and has been used for the optimization of processing parameters owing to more efficient and easier interpretation of experiments (Yan et al., 2011). It is an experimental design, that is, preferred to traditional designs and has the advantage of allowing researchers to evaluate different parameters and to determine the influence of variables over the process, as well as to reach high yields under the best possible economic conditions.

RSM has been applied as an important complement to conventional, ultrasound, and microwave extraction technologies for the aim of recovering phytochemicals from different sources. For example, a central composite design was applied for the extraction of polyphenols, pectin, and essential oils from orange peels by applying UEA and MAE (Boukroufa et al., 2015). Jeganathan et al. (2014) applied a Box-Behnken design for the optimization of solvent extraction of polyphenols from red grapes. MAE, UAE, and conventional technologies supported by RSM have been applied for the extraction of polyphenols from desert plants (*Jatropha dioica, Flourencia cernua, Turnera diffusa*, and *Eucalyptus camaldulensis*). The extraction time, solvent concentration, and liquid–solid ratio were optimized (Wong-Paz et al., 2014, 2015a,b). Also, the authors have used randomized complete block design with factorial arrangement for the extraction of coumaric acid and hydroxycinnamic acid from *L. nobilis*, thereby optimizing the extraction time, mass-solvent ratio, and solvent concentration by applying UAE (Muñiz-Márquez et al., 2013) and conventional (Muñiz-Márquez et al., 2014) extraction.

6 Advantages and Drawbacks of Alternative Extraction Technologies

Bioactive compound extractions using alternative technologies is an interesting option as compared to the conventional extraction methods (solvent extraction, reflush, Soxhlet, etc.), mainly due to the characteristics and advantages absent in the conventional methods. UAE offers short-time analysis with high reproducibility, reduces the solvent consumption, enhances the yield, and consumes little energy (Kellner et al., 2004; Khan et al., 2010; Routray and Orsat, 2011).

Also, UAE prevents heat-sensitive compounds degradation, thus volatile and semivolatile molecules can be extracted using UAE because the temperature and pressure used during the process are the environmental conditions (Pérez et al., 2008). One other advantage offered by UAE is the variability in the bioactive compounds extracted; there are reports of polyphenols, flavonols, antocianins, isoflavons, and so on, recovered from soy (Rostagno et al., 2003), citrus (Ma et al., 2008), pomegranate (Abbasi et al., 2008), and grape (Ghafoor et al., 2009), among others, in times significantly lower than the conventional extraction technologies. The main disadvantage of UAE is the initial cost with regard to the purchase of the ultrasonic equipment.

MAE is a rapid and efficient technique because of the heat transfer to the solvent, which penetrates the sample and instantaneously heats the trapped solvent in the sample pores (Shao et al., 2012). MAE uses 95% of the energy provided, thus reducing environmental pollution with short time analysis and lower CO_2 emission (Wang et al., 2007; Yemis and Mazza, 2012). MAE facilitates the desired compounds extraction due to the migration of dissolved ions, which enhances the solvent penetration and the release of the compounds from its matrix (Martins et al., 2010b). The main disadvantage of MAE is also the initial cost.

FAE offers several interesting advantages. Solid-state fermentation is one of them; the main characteristic of this is the use of capable microorganisms to release the bioactive compounds from a solid substrate, for example, an agroindustrial waste (Holker et al., 2004). FAE requires simple culture media; also, the low water activity prevents contamination by other microorganisms, specifically bacteria and yeasts (Viniegra-González et al., 2003); aeration is favored by the pores from the substrate/support; recovery and yield of the obtained compounds is high; and the wastes can be integrally used, for example, in animal feed (Doelle et al., 1992). Additionally, the bioactive compounds extracted from the specific sources of the produced enzymes (by the microorganisms) involved in the extraction process can be recovered, purified, and applied in the industry (Sabu et al., 2006; Singhania et al., 2007). Nevertheless, FAE has some disadvantages. Its application is limited to microorganisms capable of growth with low humidity; generated metabolic heat can be a problem if the process is not controlled; pH and temperature are difficult to determine due to the substrate solid nature; and scale and reactor design is poorly characterized (Doelle et al., 1992).

These technologies have emerged as potential options for obtaining bioactive products usable in the food, chemical, pharmaceutical industries, as well as for achieving integral exploitation of the resources used, thus reducing waste.

7 Industrial Approaches

Alternative extraction technologies have interesting industrial approaches. The main interest in UAE is based in short-time analysis and the enhanced yield of the obtained compounds (Robles-Ozuna and Ochoa-Martínez, 2012); these characteristics make UAE an important industrial application. Bioactive compounds of different characteristics and industrial interest can be extracted by UAE, for example, plant essential oils or vegetal foods, proteins, and bioactive compounds with high added value as gallic acid, ellagic acid, catechin, quercetin, and so on, from different vegetal sources. Reports about UAE applied to grape phenolic compounds and wine wastes mentioned an enhanced yield of 16%–23% in comparison with conventional extractions (Vila et al., 1999). Isoflavonoids have been extracted from soy with a yield increase of 15% with respect to the yield of conventional extractions (Rostagno et al., 2003). While the obtaining of bioactive compounds by UAE at the laboratory level has important benefits, scaling to industrial level remains the primary challenge (Vilkhu et al., 2008).

Scaling to industrial level is the main challenge associated with MAE and FAE, as well, due to the inversion cost (MAE) and the challenges regarding control of main parameters (FAE), such as heat transfer, biomass recovery, and moisture quantification.

The need for bioactive compounds has generated new technologies that guarantee high yields, low production cost, and environmental friendliness. The technologies described herein represent viable options to cover these requirements; however, more research is necessary regarding process scaling to achieve industrial-level objectives.

8 Future Prospects

Phenolics can be extracted from different plant tissues. But it has been reported that freeze-drying of the sample retains higher levels of phenolics content. Yet, drying processes, including freeze-drying, can cause undesirable effects on the constituent profiles (Dai and Mumper, 2010). Therefore, there is a necessity to test different sample processing conditions to find the best freeze-drying conditions of the sample. The development of modern sample-preparation techniques for the extraction and analysis of polyphenols is likely to play an important role to ensure availability of high-quality products to consumers worldwide (Gupta et al., 2016). Polyphenols are reported as an alternative to control propagation of bacteria resistant to many antibiotics, which are also called multidrug-resistant (MDR) bacteria (Djeussi et al., 2013).

One of the most important factors affecting the yield of chemical extraction is the solvent used for this purpose, which vary in polarity (Dai and Mumper, 2010). In most cases, to obtain bioactive plant extracts, solvents are used that may be unfit according to organic agriculture standards; these include methanol, hexane, chloroform, and diethyl ether (Guerrero et al., 2007). For this reason, it is very important to increase the use of more environment-friendly solvents (Castillo et al., 2010). One example of this philosophy is the paper by Mendez et al. (2012), who determined the antimicrobial activity of extracts of the plants of semidesert Mexican vegetation. These extracts were obtained with ethanol, lanolin, cocoa butter, and water. The extraction of phenolic compounds may also be influenced by solvent-to-solid ratio and the particle size of the sample. Equilibrium between the uses of high and low solvent-to-solid ratios need to be found to obtain an optimized value (Dai and Mumper, 2010).

The recovery of phenolic compounds is also influenced by the extraction time and temperature. An increase in the extraction temperature promotes higher analyte solubility by increasing both solubility and mass transfer rate. However, many phenolic compounds are easily hydrolyzed and oxidized under these extraction conditions (Dai and Mumper, 2010). The risk of degradation of thermolabile active compounds is of great interest (Routray and Orsat, 2011). Also because of increase in the cost of energy and momentum to reduce CO_2 emissions from the industry, the challenge has emerged to discover new technologies or processes to reduce energy consumption and meet the legal emissions requirements (Périno-Issartier et al., 2010). For these reasons, research using alternative extraction technologies is one of the topics of innovation that could contribute to sustainable growth in polyphenol extraction. In recent years, alternatives have been developed for plant extract technologies, which are also known as green techniques (Ballard et al., 2010) because they allow a greater recovery of bioactive compounds of interest, shorter periods of extraction time required (i.e., hours reduced to seconds), and an increase in the quality of the extracts. All these benefits are achieved with lower processing costs, or with the extraction of compounds that are difficult to obtain through conventional techniques (Martins et al., 2010a). Some of these techniques are UAE, ohmic heating, high pressure, supercritical fluid extraction, and MAE (Taamalli et al., 2012). Green techniques have the advantage of being more efficient and environment friendly because they are using fewer solvents and energy, less-time consuming, and simpler, as compared with conventional techniques. Furthermore, these techniques result in reduced degradation of thermolabile compounds, higher-quality products, and higher yields (Taamalli et al., 2012). Due to many advantages of green technologies, it is clear that they are going to play an increasingly important role in polyphenol extraction in the near future. In addition, lab equipment combining MAE and UAE green technologies has already been developed.

Recently, two important areas are being explored in the scientific medium. First, the hybrid or combined methods are being explored to further improve the extraction systems. Second, negative pressure cavitation extraction (NPCE) seems to be an alternative for better ultrasonic extraction (Roohinejad et al., 2016).

Finally, separation of polyphenols compounds has been performed using C18 cartridges, in most of the cases (Dai and Mumper, 2010). For this reason, it is important to develop alternative ways for polyphenol separation and analyses. NMR spectroscopy is a powerful technique for identifying and quantifying components in plant extracts because multiple compounds can be identified in the same spectrum.

9 Conclusions

The increasing interest in plant bioactive compounds and their wide array of applications has promoted the exploration of different alternatives for their extraction. Industrially, the recovery yield, the low cost of processes, and environmental protection are topics of concern. For that reason, the efforts have been focused on the development of alternative extraction methods. However, the success of every method depends on many factors, for example, the plant material composition, the pretreatment of materials, the interaction capacity of solvents with phytochemicals, and the mechanism applied for extraction, among others. In that sense, the generation of experimental data is still important because of the the ongoing need for knowledge and understanding of different plant compounds with or without biological activities and the effectiveness of methods for the extraction of such compounds to select the proper extraction method and plant material. Finally, the economic importance of bioactive compounds is leading the research efforts toward the discovery of new and effective extraction methods to improve the yield and quality of several compounds.

Acknowledgments

The authors are grateful to the Mexican Council for Science and Technology (CONACYT) for the financial support of students via graduate scholarships.

References

Abbasi, H., Rezaei, K., Emamdjomeh, Z., Mousavi, S.M.E., 2008. Effect of various extraction conditions on the phenolic contents of pomegranate seed oil. Eur. J. Lipid Sci. Technol. 110, 435–440.

Aguilera-Carbó, A., Hernandez-Rivera, J.S., Prado-Barragan, L.A., Augur, C., Favela-Torres, E., Aguilar, C.N., 2007. Ellagic acid production by solid state culture using a *Punica granatum* husk aqueous extract as culture broth. Proceedings of the 5th International Congress of Food Technology, Thessaloniki, Greece.

Aguilera-Carbó, A., Hernández, J., Augur, C., Prado-Barragán, A., Favela-Torres, E., Aguilar, C., 2009. Ellagic acid production from biodegradation of creosote bush ellagitannins by *Aspergillus niger* in solid state fermentation. Food Bioprocess Technol. 2, 208–212.

Alkan, D., Yemenicioğlu, A., 2016. Potential application of natural phenolic antimicrobials and edible film technology against bacterial plant pathogens. Food Hydrocoll. 55, 1–10.

Amarowicz, R., Weidner, S., 2009. Biological activity of grapevine phenolic compounds. In: Roubelakis-Angelakis, K.A. (Ed.), Grapevine Molecular Physiology & Biotechnology. Springer, Dordrecht, Netherlands, pp. 389–405.

Ascacio-Valdés, J., Buenrostro-Figueroa, J., Aguilera-Carbó, A., Prado-Barragán, A., Rodríguez-Herrera, R., Aguilar, C., 2011. Ellagitannins: biosynthesis, biodegradation and biological properties. J. Med. Plant Res. 19, 4696–4703.

Ascacio-Valdés, J.A., Buenrostro, J.J., De la Cruz, R., Sepulveda, L., Aguilera, A.F., Prado, A., Contreras, J.C., Rodriguez, R., Aguilar, C.N., 2014. Fungal biodegradation of pomegranate ellagitannins. J. Basic Microbiol. 54, 28–34.

Ascacio-Valdés, J., Aguilera-Carbó, A., Buenrostro, J., Prado-Barragán, A., Rodríguez-Herrera, R., Aguilar, C., 2016. The complete biodegradation pathway of ellagitannins by *Aspergillus niger* in solid-state fermentation. J. Basic Microbiol. 56, 229–236.

Aspé, E., Fernández, K., 2011. The effect of different extraction techniques on extraction yield, total phenolic, and anti-radical capacity of extracts from *Pinus radiata* bark. Ind. Crops Prod. 34, 838–844.

Azmir, J., Zaidul, I., Rahman, M., Sharif, K., Mohamed, A., Sahena, F., Jahurul, M., Ghafoor, K., Norulaini, N., Omar, A., 2013. Techniques for extraction of bioactive compounds from plant materials: a review. J. Food Eng. 117, 426–436.

Ballard, S.T., Mallikarjunan, P., Zhou, K., O'Keefe, S., 2010. Microwave-assisted extraction of phenolic antioxidant compounds from peanut skins. Food Chem. 120, 1185–1192.

Boukroufa, M., Boutekedjiret, C., Petigny, L., Rakotomanomana, N., Chemat, F., 2015. Bio-refinery of orange peels waste: a new concept based on integrated green and solvent free extraction processes using ultrasound and microwave techniques to obtain essential oil, polyphenols and pectin. Ultrason. Sonochem. 24, 72–79.

Brillouet, J.-M., Romieu, C., Schoefs, B., Solymosi, K., Cheynier, V., Fulcrand, H., Verdeil, J.-L., Conéjéro, G., 2013. The tannosome is an organelle forming condensed tannins in the chlorophyllous organs of Tracheophyta. Ann. Bot. 112 (6), 1003–1014.

Castillo, F., Hernández, D., Gallegos, G., Méndez, M., Rodríguez-Herrera, R., Reyes, A., Aguilar, C.N., 2010. In vitro antifungal activity of plant extracts obtained with alternative organic solvents against Rhizoctonia solani Kühn. Ind. Crops Prod. 32, 324–328.

Cetin-Karaca, H., Newman, M.C., 2015. Antimicrobial efficacy of plant phenolic compounds against *Salmonella* and *Escherichia coli*. Food Biosci. II, 8–16.

Chan, C.-H., Yusoff, R., Ngoh, G.-C., Kung, F.W.-L., 2011. Microwave-assisted extractions of active ingredients from plants. J. Chromatograph. A. 37, 6213–6225.

Chen, F., Sun, Y., Zhao, G., Liao, X., Hu, X., Wu, J., Wang, Z., 2007. Optimization of ultrasound-assisted extraction of anthocyanins in red raspberries and identification of anthocyanins in extract using high-performance liquid chromatographyâ€ mass spectrometry. Ultrason. Sonochem. 14, 767–778.

Cruz-Hernández, M.A., Contreras-Esquivel, J.C., Lara, F., Rodríguez-Herrera, R., Aguilar, C.N., 2005. Isolation and evaluation of tannin-degrading strains from the Mexican desert. Z. Naturfor. C. 60, 844–848.

Dai, J., Mumper, R.J., 2010. Plant Phenolics: extraction, analysis and their antioxidant and anticancer properties. Molecules 15, 7313–7352.

De Fernández, M.A., Soto, V.C., Silva, M.F., 2014. Phenolic compounds and antioxidant capacity of monovarietal olive oils produced in Argentina. J. Am. Oil Chem. Soc. 91, 2021–2033.

Djeussi, D.E., Noumedem, J.A.K., Seukep, J.A., Fankam, A.G., Voukeng, I.K., Tankeo, S.B., Nkuete, A.H.L., Kuete, V., 2013. Antibacterial activities of selected edible plants extracts against multidrug-resistant Gram-negative bacteria. BMC Complement. Altern. Med. 13, 164.

Doelle, H.W., Mitchell, D.A., Rolz, C.E., 1992. Solid Substrate Cultivation. Elsevier Applied Science, London, New York, Chapters 3, 35.

Esclapez, M.D., García-Pérez, J.V., Mulet, A., Cárcel, J.A., 2011. Ultrasound-assisted extraction of natural products. Food Eng. Rev. 3, 108–120.

Espinosa-Muñoz, V., Roldán-cruz, C.A., Hernández-Fuentes, A.D., Quintero-Lira, A., Almaraz-Buendía, I., Campos-Montiel, R.G., 2017. Ultrasonic-assisted extraction of phenols, flavonoids, and biocompounds with inhibitory effect against *Salmonella typhimurium* and *Staphylococcus aureus* from cactus pear. J. Food Process Eng. 40, e12358.

Fernández, K., Vega, M., Aspé, E., 2015. An enzymatic extraction of proanthocyanidins from País grape seeds and skins. Food Chem. 168, 7–13.

Fernández-Agulló, A., Freire, M.S., González-Álvarez, J., 2015. Effect of the extraction technique on the recovery of bioactive compounds from eucalyptus (*Eucalyptus globulus*) wood industrial wastes. Ind. Crops Prod. 64, 105–113.

Galano, A., Mazzone, G., Alvarez, R., Marino, T., Alvarez, J.R., Russo, N., 2016. Food antioxidants: chemical insights at the molecular level. Annu. Rev. Food Sci. Technol. 7, 335–352.

Galvan d'Alessandro, L., Kriaa, K., Nikov, I., Dimitrov, K., 2012. Ultrasound assisted extraction of polyphenols from black chokeberry. Sep. Purific. Technol. 93, 42–47.

Ghafoor, K., Choi, Y.H., Jeon, J.Y., Jo, I.H., 2009. Optimization of ultrasound-assisted extraction of phenolic compounds, antioxidants, and anthocyanins from grape (*Vitis vinifera*) seeds. J. Agric. Food Chem. 57, 4988–4994.

Gómez-García, R., Martínez-Ávila, G.C.G., Aguilar, C.N., 2012. Enzyme-assisted extraction of antioxidative phenolics from grape (*Vitis vinifera* L.) residues. J. Biotech. 2, 297–300.

Guerrero, R.E., Solís, G.S., Hernández, C.F.D., Flores, O.A., Sandoval, L.V., 2007. In vitro biological activity of Florencia cernua D.C. extracts on postharvest pathogens: Alternaria alternata (Fr.:Fr.) Keissl., Colletrichum gloeosporoides (Penz.) Penz and Sacc. and Penicillium digitatum (Pers.:Fr.) and Sacc. Rev. Mex. Fitopatol. 25, 48–53, In Spanish.

Gupta, A., Naraniwal, M., Kothari, V., 2016. Modern extraction methods for preparation of bioactive plant extracts. Int. J. Appl. Natur. Sci. 1, 8–26.

Gutiérrez-Larraínzar, M., Rúa, J., Caro, I., de Castro, C., de Arriaga, D., García-Armesto, M.R., del Valle, P., 2012. Evaluation of antimicrobial and antioxidant activities of natural phenolic compounds against foodborne pathogens and spoilage bacteria. Food Control. 26, 555–563.

Hao, J.-Y., Han, W., Huang, S.-D., Deng, X., 2002. Microwave-assisted extraction of artemisinin from Artemisia annua L. Sep. Purif. Technol. 28, 191–196.

He, B., Zhang, L.-L., Yue, X.-Y., Liang, J., Jiang, J., Gao, X.-L., Yue, P.-X., 2016. Optimization of ultrasound-assisted extraction of phenolic compounds and anthocyanins from blueberry (*Vaccinium ashei*) wine pomace. Food Chem. 204, 70–76.

Holker, U., Hofer, M., Lenz, J., 2004. Biotechnological advantages of laboratory-scale solid-state fermentation with fungi. Appl. Microbiol. Biotechnol. 64, 175–186.

Hossain, M.B., Brunton, N.P., Patras, A., Tiwari, B., O'Donnell, C.P., Martin-Diana, A.B., Barry-Ryan, C., 2012. Optimization of ultrasound assisted extraction of antioxidant compounds from marjoram (*Origanum majorana* L.) using response surface methodology. Ultrason. Sonochem. 19, 582–590.

Huang, W., Ni, J., Borthwick, A.J.L., 2005. Biosynthesis of valonia tanín hydrolase and hydrolysis of valonia tannin to ellagic acid by *Aspergillus niger* SHL 6. Proc. Biochem. 40, 1245–1249.

Izadiyan, P., Hemmateenejad, B., 2016. Multi-response optimization of factors affecting ultrasonic assisted extraction from Iranian basil using central composite design. Food Chem. 190, 864–870.

Jeganathan, P.M., Venkatachalam, S., Karichappan, T., Ramasamy, S., 2014. Model development and process optimization for solvent extraction of polyphenols from red grapes using Box–Behnken design. Prep. Biochem. Biotechnol. 44, 56–67.

Kallel, F., Driss, D., Chaari, F., Belghitha, L., Bouaziz, F., Ghorbel, R., Chaabouni, S.E., 2014. Garlic (*Allium sativum* L.) husk waste as a potential source of phenolic compounds: influence of extracting solvents on its antimicrobial and antioxidant properties. Ind. Crops Prod. 62, 34–41.

Kazemi, M., Karim, R., Mirhosseini, H., Abdul Hamid, A., 2016. Optimization of pulsed ultrasound-assisted technique for extraction of phenolics from pomegranate peel of Malas variety: punicalagin and hydroxybenzoic acids. Food Chem. 206, 156–166.

Kellner, P., Mermet, J., Otto, M., Valcarei, M., Widmer, M., 2004. Analytical Chemistry, third ed. Wiley-VCH, Germany.

Khan, M.K., Abert-Vian, M., Fabiano-Tixier, A.-S., Dangles, O., Chemat, F., 2010. Ultrasound-assisted extraction of polyphenols (flavanone glycosides) from orange (*Citrus sinensis* L.) peel. Food Chem. 119, 851–858.

Kim, M.J., Moon, Y., Tou, J.C., Mou, B., Waterland, N.L., 2016. Nutritional value, bioactive compounds and health benefits of lettuce (*Lactuca sativa* L.). J. Food Compost. Anal. 49, 19–34.

Li, Y., Skouroumounis, G.K., Elsey, G.M., Taylor, D.K., 2011. Microwave-assistance provides very rapid and efficient extraction of grape seed polyphenols. Food Chem. 129, 570–576.

Lianfu, Z., Zelong, L., 2007. Optimization and comparison ultrasound/microwave assisted extraction (UMAE) and ultrasonic assisted extraction (UAE) of lycopene from tomatoes. Ultrason. Sonochem. 15, 731–737.

Liu, J.-L., Zheng, S.-L., Fan, Q.-J., Yuan, J.-C., Yang, S.-M., Kong, F.-L., 2015. Optimisation of high-pressure ultrasonic-assisted extraction and antioxidant capacity of polysaccharides from the rhizome of Ligusticum chuanxiong. Int. J. Biol. Macromolec. 76, 80–85.

Lorenzo, J.M., Sichetti, P.E., 2016. Phenolic compounds of green tea: health benefits and technological application in food. Asian Pac. J. Trop. Biomed. 6 (8), 709–719.

Lucchesi, M.E., Chemat, F., Smadja, J., 2004. Solvent-free microwave extraction of essential oil from aromatic herbs: comparison with conventional hydro-distillation. J. Chromatograph. A. 1043, 323–327.

Luyen, B.T.T., Tai, B.H., Thao, N.P., Lee, S.H., Jang, H.D., Lee, Y.M., Kim, Y.H., 2014. Evaluation of the anti-osteoporosis and antioxidant activities of phenolic compounds from *Euphorbia maculata*. J. Korean Soc. Appl. Biol. Chem. 57, 573–579.

Ma, Y.Q., Ye, X.Q., Fang, Z.X., Chen, J.C., Xu, G.H., Liu, D.H., 2008. Phenoliccompounds and antioxidant activity of extracts from ultrasonic treatment of Satsuma mandarin (*Citrus unshiu* Marc.) peels. J. Agric. Food Chem. 56, 5682–5690.

Manasa, D., Srinivas, P., Sowbhagya, H., 2013. Enzyme-assisted extraction of bioactive compounds from ginger (*Zingiber officinale* Roscoe). Food Chem. 139, 509–514.

Manish, D., Arun, N., Shahid, A., 2012. Comparison of conventional and non-conventional methods of extraction of heartwood of *Pterocarpus marsupium* RoxB. *Polish Pharmaceutical Society*. Acta Pol. Pharm. 69 (3), 475–548.

Martins, S., Mercado, D., Mata-Gómez, M., Rodríguez, L., Aguilera-Carbó, A., Rodríguez, R., Aguilar, C., 2010a. Microbial production of potent phenolic-antioxidants through solid state fermentation. In: Singh, O.V., Harvey, S.P. (Eds.), Sustainable Biotechnology, Source of Renewable Energy. Springer, Dordrecht Heidelberg, London, New York, pp. 229–246.

Martins, S., Aguilar, C.N., De la Garza, I., Mussatto, S., Teixeira, J.A., 2010b. Kinetic study of nordihydroguaiaretic acid recovery from *Larrea tridentata* by microwave-assisted extraction. J. Chem. Technol. Biotechnol. 85, 1142–1147.

Martins, S., Mussatto, S.I., Martínez-Avila, G., Montañez-Saenz, J., Aguilar, C.N., Teixeira, J.A., 2011. Bioactive phenolic compounds: production and extraction by solid-state fermentation: a review. Biotechnol. Adv. 29, 365–373.

Martins, N., Barros, L., Ferreira, I., 2016. In vitro antioxidant activity of phenolic compounds: facts and gaps. Trends Food Sci. Tech. 48, 1–12.

Mendes, M., Carvalho, A.P., Magalhães, J.M.C.S., Moreira, M., Guido, L., Gomes, A.M., Delerue-Matos, C., 2016. Response surface evaluation of microwave-assisted extraction conditions for *Lycium barbarum* bioactive compounds. Innov. Food Sci. Emerg. Technol. 33, 319–326.

Mendez, M., Rodríguez, R., Ruiz, J., Morales-Adame, D., Castillo, F., Hernández-Castillo, F.D., Aguilar, C.N., 2012. Antibacterial activity of plant extracts obtained with alternative organic solvents against food-borne pathogen bacteria. Ind. Crops Prod. 37, 445–450.

Michail, A., Sigala, P., Grigorakis, S., Makris, D.P., 2016. Kinetics of ultrasound-assisted polyphenol extraction from spent filter coffee using aqueous glycerol. Chem. Eng. Commun. 203, 407–413.

Miguel, M.C., 2010. Antioxidant activity of medicinal and aromatic plants. Flavour Fragr. J. 25, 291–312.

Mueen, A.K.K., 2008. Introduction to isolation, identification and estimation of lead compounds from natural products. In: Hiremath, S.R.R. (Ed.), Textbook of Industrial Pharmacy: Drug Delivery Systems, and Cosmetic and Herbal Drug Technology. Orient Longman Private Ltd., Chennai, India, p. 345.

Muñiz-Márquez, D.B., Martínez-Ávila, G.C., Wong-Paz, J.E., Belmares-Cerda, R., Rodríguez-Herrera, R., Aguilar, C.N., 2013. Ultrasound-assisted extraction of phenolic compounds from *Laurus nobilis* L. and their antioxidant activity. Ultrason. Sonochem. 20, 1149–1154.

Muñiz-Márquez, D., Rodríguez, R., Balagurusamy, N., Carrillo, M., Belmares, R., Contreras, J., Nevárez, G., Aguilar, C., 2014. Phenolic content and antioxidant capacity of extracts of *Laurus nobilis* L., *Coriandrum sativum* L. and *Amaranthus hybridus* L. CyTA-J. Food 12, 271–276.

Pan, G., Yu, G., Zhu, C., Qiao, J., 2012. Optimization of ultrasound-assisted extraction (UAE) of flavonoids compounds (FC) from hawthorn seed (HS). Ultrason. Sonochem. 19, 486–490.

Patra, A.K., 2012. An overview of antimicrobial properties of different classes of phytochemicals. In: Patra, A.K. (Ed.), Dietary Phytochemicals and Microbes. Springer Science Business Media, Dordrecht, Holland.

Peng, L., Jia, X., Wang, Y., Zhu, H., Chen, Q., 2010. Ultrasonically assisted extraction of rutin from *Artemisia selengensis* Turcz: comparison with conventional extraction techniques. Food Anal. Methods 3, 261–268.

Pérez, C., Sáenz, A., Barajas, L., 2008. La química verde como herramienta indispensable en el aprovechamiento integral y sustentable de los recursos del semidesierto mexicano. Fitoquímicos Sobresalientes del Semidesierto Mexicano: de la Planta a los Químicos Naturales y a la Biotecnología. Editorial: Path Design S.A., México, 67-82.

Pérez-Serradilla, J.A., Luque de Castro, M.D., 2011. Microwave-assisted extraction of phenolic compounds from wine lees and spray-drying of the extract. Food Chem. 124, 1652–1659.

Périno, S., Pierson, J.T., Ruiz, K., Cravotto, G., Chemat, F., 2016. Laboratory to pilot scale: microwave extraction for polyphenols lettuce. Food Chem. 204, 108–114.

Périno-Issartier, S., Abert-Vian, M., Petitcolas, E., Chemat, F., 2010. Microwave turbo hydrodistillation for rapid extraction of the essential oil from Schinus terebinthifolius Raddi berries. Chromatographia 72, 347–350.

Puri, M., Sharma, D., Barrow, C.J., 2012. Enzyme-assisted extraction of bioactives from plants. Trends Biotechnol. 30, 37–44.

Ramirez-Coronel, M.A., Marnet, N., Kolli, V.S.K., Roussos, S., Guyot, S., Augur, C., 2004. Characterization and estimation of proanthocyanidins and other phenolics in coffee pulp (coffea arabica) by thiolysis-high-performance liquid chromatography. J. Agric. Food Chem. 52, 1344–1349.

Rinaldi, M., Caligiani, A., Borgese, R., Palla, G., Barbanti, D., Massini, R., 2013. The effect of fruit processing and enzymatic treatments on pomegranate juice composition, antioxidant activity and polyphenols content. LWT Food Sci. Technol. 53, 355–359.

Robles-Ozuna, L., Ochoa-Martínez, L., 2012. Ultrasonido y sus aplicaciones en el procesamiento de alimentos. Revista Iberoamericana de Tecnología Poscosecha 13, 109–122.

Rodrigues, S., Pinto, G.A.S., Fernandes, F.A.N., 2008. Optimization of ultrasound extraction of phenolic compounds from coconut (*Cocos nucifera*) shell powder by response surface methodology. Ultrason. Sonochem. 15, 95–100.

Roohinejad, S., Koubaa, M., Barba, F.J., Greiner, R., Orlien, V., Lebovka, N.I., 2016. Negative pressure cavitation extraction: a novel method for extraction of food bioactive compounds from plant materials. Trends Food Sci. Technol. 52, 98–108.

Rossi, M., Negri, E., Parpinel, M., Lagiou, P., Bosetti, C., Talamini, R., Montella, M., Giacosa, A., Franceschi, S., La Vecchia, C., 2010. Proanthocyanidins and the risk of colorectal cancer in Italy. Cancer Causes Control 21, 243–250.

Rostagno, D., Palma, M., Barroso, C., 2003. Ultrasound-assisted extraction of soy isoflavones. J. Chromatogr. A. 1012, 119–128.

Routray, W., Orsat, V., 2011. Blueberries and their anthocyanins: factors affecting biosynthesis and properties. Compr. Rev. Food Sci. Food Saf. 10, 303–320.

Routray, W., Orsat, V., 2013. Preparative extraction and separation of phenolic compounds. In: Ramawat, G.K., Mérillon, J.-M. (Eds.), Natural Products: Phytochemistry, Botany and Metabolism of Alkaloids, Phenolics and Terpenes. Springer, Berlin, Heidelberg, pp. 2013–2045.

Saavedra, P.G.A., Couri, S., Ferreira, L.S.G., de Brito, E.S., 2005. Tanase: conceitos, produção e aplicação. B. CEPPA, Curitiba 23, 435–462.

Sabu, A., Augur, C., Swati, C., Pandey, A., 2006. Tannase production by Lactobacillus sp. ASR-S1 under solid-state fermentation. Process Biochem. 41, 575–580.

Salisova, M., Toma, S., Mason, T.J., 1997. Comparison of conventional and ultrasonically assisted extractions of pharmaceutically active compounds from Salvia officinalis. Ultrason. Sonochem. 4, 131–134.

Scalbert, A., 1992. Quantitative methods for the estimation of tannins in plants tissues. In: Hemingway, R.W., Laks, P.E. (Eds.), Plant Polyphenols: Synthesis, Properties, Significance. Plenum, New York, NY, pp. 249–280.

Sepúlveda, L., Buenrostro-Figueroa, J.J., Ascacio-Valdés, J.A., Aguilera-Carbó, A.F., Rodríguez -Herrera, R., Contreras-Esquivel, J.C., Aguilar, C.N., 2014. Submerged culture for production of ellagic acid from pomegranate husk by *Aspergillus niger* GH1. Micol. Aplicada Int. 26 (2), 27–35.

Shao, P., He, J., Sun, P., Zhao, P., 2012. Analysis of conditions for microwave-assisted extraction of total water-soluble flavonoids from Perilla Frutescens leaves. J. Food Sci. Technol. 49, 66–73.

Shi, B., Qiang, H., Kai, Y., Huang, W., Quin, L., 2005. Production of ellagic acid from degradation of valonea tannins by *Aspergillus niger* and *Candida utilis*. J. Chem. Technol. Biotechnol. 80, 1154–1159.

Singhania, R., Sukumaran, R., Pandey, A., 2007. Improved cellulase production by *Trichoderma reesei* RUT C30 under SSF through process optimization. Appl. Biochem. Biotechnol. 142, 60–70.

Sowbhagya, H., Chitra, V., 2010. Enzyme-assisted extraction of flavorings and colorants from plant materials. Crit. Rev. Food Science Nutr. 50, 146–161.

Spigno, G., De Faveri, D.M., 2009. Microwave-assisted extraction of tea phenols: a phenomenological study. J. Food Eng. 93, 210–217.

Sultana, B., Anwar, F., Ashraf, M., 2009. Effect of extraction solvent/technique on the antioxidant activity of selected medicinal plant extracts. Molecules 14, 2167–2180.

Sun, Y., Liu, Z., Wang, J., 2011. Ultrasound-assisted extraction of five isoflavones from Iris tectorum Maxim. Sep. Purific. Technol. 78, 49–54.

Taamalli, A., Arráez-Román, D., Barrajón-Catalán, E., Ruiz-Torres, V., Pérez-Sánchez, A., Herrero, M., Ibañez, E., Micol, V., Zarrouk, M., Segura-Carretero, A., Fernández-Gutiérrez, A., 2012. Use of advanced techniques for the extraction of phenolic compounds from Tunisian olive leaves: phenolic composition and cytotoxicity against human breast cancer cells. Food Chem. Toxicol. 50, 1817–1825.

Tarke, P., Rajan, M., 2014. Comparison of conventional and novel extraction techniques for the extraction of Scopoletin from Convolvulus pluricaulis. Indian J. Pharm. Educ. Res. 48, 27–31.

Tekin, K., Akalın, M.K., Şeker, M.G., 2015. Ultrasound bath-assisted extraction of essential oils from clove using central composite design. Ind. Crops Prod. 77, 954–960.

Thong, N.M., Nam, P.C., 2015. Theoretical investigation on antioxidant activity of phenolic compounds extracted from *Artocarpus altilis*. In: Van Toi, V., Lien Phuong, T.H. (Eds.), 5th International Conference on Biomedical Engineering in Vietnam. IFMBE Proceedings, p. 46.

Tiwari, B.K., 2015. Ultrasound: a clean, green extraction technology. Trends Anal. Chem. 71, 100–109.

Tusevski, O., Kostovska, A., Iloska, A., Trajkovska, L., Simic, S.G., 2014. Phenolic production and antioxidant properties of some Macedonian medicinal plants. Cent. Eur. J. Biol. 9, 888–900.

Vaquero, I., Marcobal, A., Muñoz, R., 2004. Tannase activity by lactic bacteria isolated from grape must and wine. Int. J. Food Microbiol. 96, 199–2004.

Vattem, D.A., Shetty, K., 2002. Solid-state production of phenolic antioxidants from cranberry pomace by *Rhizopus oligosporum*. Food Biotechnol. 16, 189–210.

Vattem, D.A., Shetty, K., 2003. Ellagic acid production and phenolic antioxidant activity in cranberry pomace (*Vaccinium macrocarpon*) mediated by Lentinus edodes using a solid-state system. Process Biochem. 39, 367–379.

Vermerris, W., Nicholson, R., 2006. Chemical Properties of Phenolic Compounds Phenolic Compound Biochemistry. Springer, Dordrecht, Netherlands, pp. 35–62.

Viera da Silva, B., Barreira, J.C.M., Oliveira, M.B.P.P., 2016. Natural phytochemicals and probiotics as bioactive ingredients for functional foods: extraction, biochemistry and protected-delivery technologies. Trends Food Sci. Tech. 50, 144–158.

Vila, D., Mira, H., Lucena, R., Fernández, R., 1999. Optimization of an extraction method of aroma compounds in white wine using ultrasound. Talanta 50, 413–421.

Vilkhu, K., Mawson, R., Simmons, L., Bates, D., 2008. Applications and opportunities for ultrasound assisted extraction in the food industry: a review. Innov. Food Sci. Emerg. Technol. 9, 161–169.

Viniegra-González, G., Favela-Torrez, E., Aguilar, C., Romero-Gómez, S., Díaz-Godinez, G., Augur, C., 2003. Advantages of fungal enzyme production in solid-state over liquid fermentation systems. Biochem. Eng. J. 13, 157–167.

Vivas, N., Laguerre, M., Pianet de Biossel, I., Vivas, D.G.N., Nonier, M.F., 2004. Conformational interperation of vescalagin and castalagin physicochemical porperties. J. Agric. Food Chem. 52, 2073–2078.

Wang, S., Chen, F., Wu, J., Wang, Z., Liao, X., Hu, X., 2007. J. Food Eng. 78, 693–700.

Wang, J., Sun, B., Cao, Y., Tian, Y., Li, X., 2008. Optimization of ultrasound-assisted extraction of phenolic compounds from wheat bran. Food Chem. 106, 804–810.

Wong-Paz, J., Contreras-Esquivel, J., Muniz-Marquez, D., Belmares, R., Rodriguez, R., Flores, P., Aguilar, C., 2014. Microwave-assisted extraction of phenolic antioxidants from semiarid plants. Am. J. Agric. Biol. Sci. 9, 299–310.

Wong Paz, J.E., Muñiz Márquez, D.B., Martínez Ávila, G.C.G., Belmares Cerda, R.E., Aguilar, C.N., 2015a. Ultrasound-assisted extraction of polyphenols from native plants in the Mexican desert. Ultrason. Sonochem. 22, 474–481.

Wong-Paz, J.E., Contreras-Esquivel, J.C., Rodríguez-Herrera, R., Carrillo-Inungaray, M.L., López, L.I., Nevárez-Moorillón, G.V., Aguilar, C.N., 2015b. Total phenolic content, in vitro antioxidant activity and chemical composition of plant extracts from semiarid Mexican region. Asian Pac. J. Trop. Dis. 8, 104–111.

Xiao, X., Song, W., Wang, J., Li, G., 2012. Microwave-assisted extraction performed in low temperature and in vacuo for the extraction of labile compounds in food samples. Anal. Chim. Acta 712, 85–93.

Yan, M.M., Liu, W., Fu, Y.J., Zu, Y.G., Chen, C.Y., Luo, M., 2010. Optimization of the microwave-assisted extraction process for four main astragalosides in *Radix Astragali*. Food Chem. 119, 1663–1670.

Yan, Y.-l., Yu, C.-h., Chen, J., Li, X.-x., Wang, W., Li, S.-q., 2011. Ultrasonic-assisted extraction optimized by response surface methodology, chemical composition and antioxidant activity of polysaccharides from *Tremella mesenterica*. Carbohydr. Polym. 83, 217–224.

Yang, Y.-C., Li, J., Zu, Y.-G., Fu, Y.-J., Luo, M., Wu, N., Liu, X.-L., 2010. Optimisation of microwave-assisted enzymatic extraction of corilagin and geraniin from Geranium sibiricum Linne and evaluation of antioxidant activity. Food Chem. 122, 373–380.

Yemis, O., Mazza, G., 2012. Optimization of furfural and 5-hydroxymethylfurfural production from wheat straw by a microwave-assisted process. Bioresource Technology 109, 215–223.

Yoshida, T., Amakura, Y., Koyura, N., Ito, H., Isaza, J.H., Ramírez, S., Peláez, D.P., Renner, S.S., 1999. Oligomeric hydrolizable tannins from Tiouchina multiflora. Phytochem. 52, 1661–1666.

Zhou, Q., Bennett, L.L., Zhou, S., 2016. Multifaceted ability of naturally occurring polyphenols against metastatic cancer. Clin. Exp. Pharmacol. Physiol. 43, 394–409.

Further Reading

Riera, E., Golás, A., Blancom, J., Gallego, M., Blasco, M., Mulet, A., 2004. Mass transfer enhancement in supercritical fluids extraction by means of power ultrasound. Ultrason. Sonochem. 11, 241–244.

CHAPTER 6

The Extraction of Heavy Metals From Vegetable Samples

Amra Odobasic, Indira Sestan, Amra Bratovcic
University of Tuzla, Tuzla, Bosnia and Herzegovina

1 Introduction

The process of urbanization and rapid industrial development has led to ever-increasing amounts of environmental pollution, global warming, and climate change. Materials that can disrupt the natural ecosystem of land, water, and air are referred to as hazardous substances. The most common hazardous toxicants of the agroecosystem are heavy metals, radionuclides, synthetic organic matters, and pesticides (Milenković, 2014). Increased concentrations of heavy metals represent highly toxic pollutants that, in case they get into the food chain, are accumulated in living beings, which carry a high risk to human and animal health. Heavy metals usually enter the food chain by consuming contaminated plants (Kebert, 2014). Plants have an important role in the cycling of toxic elements ions in the nature. They mainly enter the food chain through the plants that grow on contaminated soil (Zeremski-Škorić et al., 2011).

The pollution of arable land with heavy metals is a major problem in developed countries, since the arable areas are reduced by contamination. The degree of soil pollution with heavy metals is not easy to determine and varies depending on the soil type. Thus the presence of a required amount of a compound in the soil does not cause disturbances in the plant production, while it does in the grown culture, which can be seen in decreased yield and quality. It is considered that the heavy metal toxicity is evident only if their concentration in plant tissue is increased above the average values (Marić, 2014). Agricultural lands are especially susceptible to degradation, due to the numerous, often-radical technical and technological procedures in plant growth, leading to the environmental pollution. Precisely, because they cannot degrade or be destroyed, heavy metals remain persistent in all parts of the environment, and their concentration in the soil increases cumulatively as a result of the accelerated industrial activity. Due to their toxicity, bioavailability, mobility, and persistence, heavy metals have now become the most studied environmental pollutants.

Ingredients Extraction by Physicochemical Methods in Food
http://dx.doi.org/10.1016/B978-0-12-811521-3.00006-5

The problem of heavy metal accumulation in the soil is so far solved by expensive, abrasive, chemical, and physical methods, which have not been sufficiently effective, easily acceptable, and cost-effective due to the lack of the universal chemicals that would be used for all metals. In recent years, more attention is given to the application of biological, less aggressive obstructive technologies in the domain of the new interdisciplinary scientific discipline called *phytoremediation*. One method of contaminated soil remediation is phytoextraction, that is, planting plants that accumulate heavy metals from the soil (Varga, 2010).

The total metal concentrations in the trace in the soil are very often used as risk indicators and the assessment of the metal availability. However, recent studies have shown that the total concentration is a poor availability indicator, while the concentrations in the soil solution show a good correlation with the concentrations in the plant. Metals are found in several forms in the soil (Marić, 2014):

1. in the soil solution, as metal ions and soluble metal complexes,
2. adsorbed as inorganic soil components for ion exchange,
3. linked to the organic soil matter,
4. combined, as oxides, hydroxides, carbonates, and
5. embedded in the structure of silicate minerals.

Metals are extracted from the soil by extraction and their quantity in various forms is determined. For the phytoextraction to occur, the pollutants must be bioavailable (ready to be absorbed in the root). Bioavailability depends on the metal solubility in the soil solution. Only metals in the forms given in 1 and 2 are immediately available for use by plants.

The purpose of the study was to apply a variety of newer extraction methods for heavy metals from tested plant species and soil, as well as to determine the correlation between them in order to find possible applications of the used procedures by subsequent research (development of existing and/or discovery of new technologies) contributing the environmental protection. Also, one of the aims was to determine the ability of used plant types (parsley, onion) and cereals (barley) for lead and copper accumulation from the soil by phytoremediation process.

Monitoring and determining copper and lead content in the plants widely represented in households, as well as in the soil where the growing was done under controlled conditions, is extremely important to assess the prevention of contamination of these elements.

2 Heavy Metals in Soil and Plants

The group of metals includes 80 elements of the periodic system, 17 elements that are nonmetals, while only 7 are metalloids (Hogan, 2010). The group of heavy metals includes those metals that have a density greater of 5 g/cm^3. Heavy metals are natural environmental constituents, varying in concentrations depending on the geographic area. According to the biological function of organism, the metals are divided into essential and nonessential. The essential metals include Cu, Fe, Zn, Mn, Mo, and Ni. They got their names due to the fact that they are necessary for life: reducing their introduction or their complete absence may lead to a serious disruption in organism functioning and may even lead to death (Grozdić, 2015). The group of nonessential metals include elements, which are not biogenic and operate exclusively toxic (Pb, Cd, Al, Hg, and As). Heavy metals have diverse roles in industry, as a raw material for the production, and in agriculture as a component of many fertilizers used to increase crop yield. Despite all positive effects, these metals mainly represent significant environmental pollutants (Grozdić, 2015). Heavy metals may enter the soil in several ways: the basic way is by leaching from the parent rock and the other is by anthropogenic means.

As the most common sources of the anthropogenic soil pollution, metallurgical, metal processing, and electronic industries stand out, as well as mining, plant for waste water treatment, waste disposal, agricultural fertilizers, pesticides, fossil fuels combustion (Lončarić et al., 2012). Heavy metals are very stable, so the entire amount of their emissions from nature and anthropogenic activities reaches the soil and water. Due to its stability, high toxicity, and tendency to accumulate in the ecosystem, heavy metals are dangerous for living organisms (Šarkanj et al., 2010). Exposure to heavy metals even in small traces represents risk for human health (Jamil et al., 2010; Khan et al., 2008; Peng et al., 2004; Singh et al., 2010).

A series of scientific discoveries and researches have led to the improvement of the original idea of using plants for environmental remediation and removal of various heavy metals from polluted media. The plants have an important role in heavy metal cycling in the nature, primarily because by consummation heavy metals enter the food chain and negatively impact human health. The soil type and pH value, as well as the content of organic matter in soil especially affect the accumulation and heavy metals availability to the plants.

In many scientific works, it was shown that some species might be used for phytoextraction of heavy metals from contaminated soil and water (Rai et al., 1995; Salt et al., 1998).

Great attention is given today to the study of the effects of different heavy metal concentration on plants, and not just from the standpoint of their metabolism, but also to create genotypes tolerant to their surplus, which would be distinguished by a high hyperaccumulation of heavy metals and thus find application in the decontamination of polluted soil (Kebert, 2014).

One of the biological methods of polluted soil remediation is phytoremediation, which involves the use of certain plant species that have phytoaccumulation potential toward the pollutants (Bajsić and Dobrotić, 2014). Heavy metals link in the soil to the adsorption complex or they are in ion form in the soil solution. They are accessible for the plants from an aqueous solution or bound on the adsorption soil complex (Bajsić and Dobrotić, 2014). The ability of a metal ion sorption mostly depends on the form in which it is in the soil, and depends the least on its quantity. The accumulation ability of some heavy metals is different with different plant species (Lasat, 2002). It has been shown that the plants from the soil can remove different types of pollutants, and are very successful in removing heavy metals (Ali et al., 2013; Rascio and Navari-Izzo, 2011).

In 1983 Chaney was one of the first to suggest the use of plants for remediation of the soil polluted with heavy metals (Prasad and Freitas, 2003). A large number of studies since then have been devoted to the examining the phytoremediation possibility using different plant species (Šyc et al., 2012). There are over 400 known plant species that can hyperaccumulate metals. When using plants for food, animals and people take heavy metals in their body and thus start biological processes: bioaccumulation and incorporation occur in the food chain (Stojanović, 2014). Vegetables show the greatest ability for heavy metals accumulation. High concentrations of zinc, boron, molybdenum, and cobalt can be found in plants, while to a lesser extent manganese, iron, aluminum, copper, lead, and chromium are only in the trace (Kerovec, 2010).

In phytoextraction, plants grow on polluted soil and from time to time they are cut so their biomass with accumulated inorganic pollutants could be used for different purposes, depending on the pollution type. For example, wood material can be burned for energy purposes. The main goal is to remove pollutants from the soil and their concentration in the biomass of above ground plants, where with the final burning of plant material there shall occur concentration of pollutants to the bioconcentration factor 10 and more with dry matter (Kebert, 2014).

2.1 Copper

Copper (Cu) is an essential redox-active transition metal, which is included in many physiological processes in plants and which exists in different oxidation states in vivo. Plants adopt it in the form of copper ion Cu^{2+}or in the form of chelates. For adoption of copper from soil energy is necessary, although it is considered that there is a specific receptor that has the role of a copper transferor. The copper in the soil originates from the primary and secondary minerals (chalcopyrite, malachite, azurite, cuprite, atacamite). With decomposition of these minerals, copper oxidizes and goes into the divalent state. The divalent copper ion (Cu^{2+}) creates different forms in soil, which are of different solubility, that is, accessibility for plants (Džamić and Stevanović, 2014).

The largest copper concentration is at the plant root, due to its relatively poor translocation through other parts of the plant. The mobility of copper through different plant parts is average. The upward transport and reutilization depend on the degree of plant provision with this element. The concentration of this element is approximately in the range of 5–30 mg/kg of dry matter. If the copper content is less than 4% of dry plant, then it is considered that the given plants lack in copper, while in cases when its concentration is from 20 to 100 mg/kg, it is considered that the given plants have great concentration of this element. The sensitivity and plant reaction on the lack of Cu^{2+} are very different. In the group of extremely sensitive plants are alfalfa, tobacco, spinach, oats, wheat, winter barley, and early barley. The copper availability in the soil depends on numerous factors. It can be said that between the total copper in the soil and its accessibility there is a close connection, so that the level of the soil provision with copper can be appreciated compared to their total reserves in the soil. The dynamics of the available forms of copper (Cu^{2+}: the soil solution and interchangeability Cu^{2+}) in soil is influenced by the following factors:

1. pH of the soil: availability of copper in the soil for plant nutrition depends on soil pH. The availability of copper is reduced at the pH value above 7. Below pH 6 the copper availability is increased (Džamić and Stevanović, 2014).
2. The amount of organic matter in soil: copper is tightly bound to organic matter in soil, whereby the soils with increased organic matter content reduce the copper availability (Džamić and Stevanović, 2014).

Toxic effect of this element occurs when its total content in the soil is from 25 to 40 mg/kg and if it is accompanied with the acid soil whose pH is about 5.5. Mainly the high copper concentrations occur in the acid soils. Copper, as an ecological factor, should be given special attention, considering that it is very toxic in high concentrations, and it has a very narrow, so-called "concentration window," that is, the line where from preferred becomes toxic.

2.2 Lead

Lead (Pb) is a heavy metal, as a pollutant of soil it is widely distributed in the environment, since the lead is present from natural sources, and may enter soil through underground fuel reservoirs or through the dust from old lead paint and from different industrial activities, but also from the lead pipelines buried in the ground. The major part of lead released from cars falls in a distance of about 100 m away from the road so the plants along the roads may contain up to 150 ppm of lead. In the surface layers of the soil where it mostly accumulates its value may reach even 3000 ppm. The lead bioavailability increases with increased soil acidity. The solubility of lead compounds in water does not reduce its toxicity, since due to the lead ion ability to build lipophilic complexes the bioavailability of this most common toxic element is increased. If it reaches the organism in high concentrations it may inhibit the activity of some enzymes, and it may cause paralysis and brain damage (Vukadinovic and Loncaric, 1998.). Plants take lead directly from the soil, but lead may also enter through leaves and stalk, and therefore into the food chain. Lead may enter plants during production, distribution, and cooking (Šarkanj et al., 2010).

Having in mind that roots have a great power to absorb lead, it can be predicted that this ability represents a certain protection of the above-ground parts of plants from the damaging lead effect. In high concentrations lead inhibits the growth of leaves and roots, inhibits the photosynthesis process, and impacts the morphological and anatomical structure of the plant. Wheat and soybeans have high tolerance to lead. Spinach is particularly sensitive to high lead concentration in soil. With these plant species at the concentration of 0 mg/kg of dry matter, a significant reduction of yield occurs (Simic et al., 2015).

2.3 Selection of Plant Material Samples

2.3.1 Barley

Barley (*Hordeum vulgare* L.) is one of the most important cereals in the world. According to the sown areas and its yield it ranks the fourth in the world, and third in the European Union, as well as in Croatia (http://faostat.fao.org, 2010). Barley is quite well adapted to different environmental conditions and it is more stress tolerant in comparison to wheat (Nevo et al., 2012). Barley products (grain, straw, whole green plant) are used for stock food production, starch industry, and alcohol. Barley varieties used for stock food are those with higher protein content (12%–14%), vitreous material, and varieties chosen for beer are those with powdered, soft, and finer grain. Barley has a very short growing vegetation period during which it has to have a sufficient amount of nutrients (Džamić and Stevanović, 2014). In recent years, the society of developed countries has recognized the high potential value of soluble fibers of barley grain (polysaccharides cell walls); therefore barley is increasingly used in the preparation of different dietary products with the aim of preventing illnesses of the cardiovascular system, diabetes, and colon cancer, which are suffered by people worldwide (Collins et al., 2010). Environmental stresses are limiting factors of agricultural production all over the world (Gaspar et al., 2002; Roy et al., 2011). Drought, as well as other abiotic factors have negative impacts on the quality (Wang and Frei, 2011) and quantity (Dolferus et al., 2011) of all produced cereals, including barley, which would create extremely great problems in the future concerning cereals, which are the foundatio of nutrition for most of the world population.

2.3.2 Parsley

Parsley (*Petroselinum crispum*) originates from the Mediterranean region; however, today it is grown throughout the world. It is a biennial plant growing up to 60–100 cm; the roots are thin, thick, and vertical. The fruit is rounded with a greenish gray color, and a length of 2.5 mm (Mozafarian, 2007). Root and leaves are used, and sometimes even seeds. It is used for purposes of traditional medicine, contains essential oils, flavonoids, sugar, and starch.

Leaves are rich in vitamin C, iron, iodine, and magnesium (Farzaei et al., 2013). Parsley is therefore, a medicinal herb with proven pharmacological properties, as well as antioxidant, analgesic, and antibacterial qualities (Farzaei et al., 2013).

2.3.3 Onion

Onion (*Allium cepa* L.) is herbaceous, biennial plant. It occupies a significant place in human nutrition. It is used fresh, dried, or prepared in different ways. The root of onion has weak suction and penetrating power. The optimum acidity of the soil for growing onion is pH 6.2–6.5. At lower and higher soil acidity, adoption of certain nutrients is uneven, which may directly result in poor bulb quality and reduced yield (Bernardoni et al., 2003). Today onion is very popular in agriculture worldwide. Asia is the largest producer, and along with China and Japan takes a share of 27% in total production. Due to the high capacity of adaptation, numerous populations, and varieties adapted to different environmental conditions have been developed. Onion is today known for its medicinal and biological properties, but the mechanisms of its characteristic components effect are still not completely explored, such as flavonoids, sulfur compounds, and selenium compounds. The beneficial effects of these specific components on consumers' health are reflected in a favorable impact in the battle against cardiovascular and carcinogenic diseases, for lowering blood pressure and cholesterol levels (Deak, 2008).

3 Extraction

Extraction is the technological operation of complete or partial separation of a mixture of substances having an unequal solubility in various solvents. The mixture is treated with a solvent to be separated from it much easily. A soluble component is used as a solution. By extraction we are getting the substance for which separation in pure form needed the resulting solution to evaporate or crystallize (Lianfu and Zelong, 2008).

For extraction the various conventional methods are used, such as:

1. The distillation: direct distillation of essential oils, steam distillation, or distillation with water and steam;
2. Extraction solvents: solvent extraction, maceration, extraction with oils;
3. Cold pressing;
4. Unconventional techniques: extraction with supercritical fluids, turbo-extraction, extraction with electricity, extraction assisted by ultrasound.

When extracting solids, there is a need to increase the surface area of interaction between the phases (grinding); in the middle phase, there is a need to increase the speed of movement; to increase the amount of sample, it is necessary to extend the duration of the extraction (Eskilsson and Bjorklund, 2000).

Extraction can be carried out as a continuous and discontinuous, single-stage, multistage, multistage countercurrent, or multistage cross-flow with one solvent. In our work, the sample that has been the subject of extraction is plant origin.

Namely, plant materials have a complex nature, and the extraction of the substances they contain is influenced by process conditions, such as temperature, mechanical action (such as pressure and shaking), extraction solvent type, and solubilization of the target compounds, which effectively depend on the solvent polarity and physical conditions.

In this chapter we will describe in detail the types of extraction that we used in the experimental section. These are:

1. Microwave extraction
2. Dry digestion
   a. Dry digestion with melts
   b. Dry digestion without melts
3. Wet digestion
   a. Wet digestion using ultrasonic radiation
   b. Wet digestion using microwave radiation
4. Ultrasonic extraction

3.1 Microwave Extraction

Microwaves are nonionizing electromagnetic energy with a frequency from 0.3 to 300 GHz. This energy is transmitted as waves, which can penetrate in biomaterials and interact with polar molecules inside the materials, such as water, to generate heat. *Microwave-assisted extraction* (MAE) is a process that uses the effect of microwaves to extract biological materials. MAE has been considered an important alternative to low-pressure extraction because of its advantages: lower extraction time, lower solvent usage, selectivity, and volumetric heating and controllable heating process.

The physical principle of this technique is based on the ability of polar chemical compounds to absorb microwave energy according to its nature, mainly the dielectric constant. This absorbed energy is proportional to the medium dielectric constant, resulting in dipole rotation in an electric field and migration of ionic species. The ionic migration generates heat as a result of the resistance of the medium to the ion flow, causing collisions between molecules because the direction of ions changes as many times as the field changes the sign. Rotation movements of the polar molecules occur while these molecules are trying to line up with the electric field, with consequent multiple collisions that generate energy and increase the medium temperature (Angela and Meireles, 2008; Luque de Castro et al., 1999; Romanik et al., 2007).

Therefore, a higher dielectric constant leads to a higher absorbed energy by the molecules, promoting a faster solvent heating, and extraction at higher temperatures. However, other solvents with low dielectric constants are also used, and in these cases the matrix is heated and the microwave heating leads to the rupture of cell walls by expansion, promoting the delivery of the target compounds into a cooler solvent; this technique is used for the extraction of thermally labile compounds of low polarity (Eskilsson and Bjorklund, 2000).

Microwave extraction is a procedure for extracting water insoluble or slightly water-soluble organic compounds from soils, clays, sediments, sludges, and solid wastes. The procedure uses microwave energy to produce elevated temperature and pressure conditions (i.e., 100–115°C and 50–175 psi; 1 atm = 14.6956 psi) in a closed vessel containing the sample and organic solvent(s) to achieve analyte recoveries equivalent to those from Soxhlet extraction, using less solvent and taking significantly less time than the Soxhlet procedure. This method is applicable to the extraction of semivolatile organic compounds, organophosphorus pesticides, organochlorine pesticides, chlorinated herbicides, phenoxyacid herbicides, substituted phenols, and on solid samples with small size particles. If practical, soil/sediment samples may be air-dried and ground to a fine powder prior to extraction. The total mass of material to be prepared depends on the specifications of the determinative method and the sensitivity required for the analysis, but 2–20 g of material are usually necessary and can be accommodated by this extraction procedure.

The use of dielectric heating in the laboratory using microwaves started in the late 1970s; however, this is the first use in the food industry. Dielectric heating depends on the material's ability to absorb the microwave energy and convert it into heat. Microwaves heat the total volume of the sample simultaneously and damage the hydrogen bonds by encouraging the rotation of the dipole. The movement of dissolved ions increases the penetration of the solvent into the matrix and thus promotes dissolution. There are two types of commercially available microwave extraction systems, and they are extraction in closed containers at a controlled pressure and temperature, and in microwave ovens at atmospheric pressure (Kaufmann and Christen, 2002).

The system of microwave extraction in closed containers is generally used for extraction under conditions of low- or high-temperature extraction. The pressure in the vessel is strongly dependent on the amount and boiling point of the solvent.

Microwave extraction can be used to extract the heat-sensitive compounds, such as essential oils (Brachet et al., 2002). However, it was found that the microwave extraction is inefficient if it is carried out from completely dry or fresh wet materials with hexane as extraction solvent (Molins et al., 1997). The particle size and size distribution usually have a significant impact on the efficiency of microwave extraction. Particle sizes of the extracted substances are typically in the range from 100 μm to 2 mm.

The choice of solvent in a microwave extraction is very important. The choice depends on the solubility of the desired extract, on the interaction between the solvent and the matrix and the properties of the solvent determined by specific dielectric constant. The selected solvent should have a high dielectric constant and the possibility of a good absorption of microwave energy (Bousbia et al., 2009). Solvents, such as ethanol, methanol, and water are polar enough to be able to be heated by microwave energy.

Generally, the temperature is another important factor for the microwave extraction, and increase in temperature results in better extraction effect. However, for the extraction of thermal-labile species, high temperatures can cause degradation of the extracts. The selected microwave power during microwave extraction must be set properly to avoid excessive temperature, which leads to degradation of the thermo-sensitive substances (Font et al., 1998).

Microwaves have limited energy potential, so that it does not cause changes in the structure of matter; their molecules vibrate and touch each other causing a rise in temperature. Microwave extraction is often applied for the analysis of organic compounds in solid samples. By microwave extraction or by extraction with combined ultrasound and microwave, it is possible to get a similar ratio of the extracted substances as by standard methods but with a much shorter time, if economically feasible and the energy is available (Abert et al., 2008).

However, it is necessary to pay attention to the negative effects of the application of microwaves and ultrasound, such as increased temperatures, which can negatively affect the bioactive compounds and quality of the extracted material.

3.1.1 Digestion

To analyze the heavy metals by atomic absorption spectrometry, it is required that all analytes from the samples are converted from the solid phase into solution, since all spectrometers are based on an analysis of the solution. Because of that, digestion is used. Digestion is the process of melting of sample mixed with some agent for melting (melting agent). The compounds or ions in the sample and melting agent chemically react by heating at a high temperature and forming new compounds, which are soluble in water or in acid. Reagents for digestion can be a base or acid. Most often Na_2CO_3 is used as a base agent and $KHSO_4$ as an acidic. In order to prepare the sample as the best possible, it is necessary to add a sufficient amount of reagents and bring enough energy to break bonds in the crystal lattices. For this purpose two basic methods used are: wet and dry digestion (Šoljić, 2005).

3.1.1.1 Dry digestion

Dry digestion is used for samples containing a large amount of organic matter and it is analyzed on the nonvolatile components. This type of digestion is usually used to analyze samples of food, plant, and biological material. The method involves burning the sample at high temperatures.

- *Dry digestion with melting agent* is used in the chemical decomposition of silicates, refractory materials, certain mineral oxides, and iron alloy. The sample is mixed with melting agents whereby upon melting leading to products that are easily dissolved in water or in diluted acid. A very high temperature (300–1000°C) is required, which is achieved by flame or microwave heating. The disadvantages of this method are impurities in the melts, the high content of electrolytes in the final solution, especially in the case of some analytical spectroscopic techniques, and the risk of contamination and losses in evaporation (Dizdar, 2012).
- *Dry digestion without melting agent* is used to remove organic materials; before determining the elements and minerals, it consists of the ignition of the organic compound in a stream of air or oxygen. To avoid losses of volatile elements, such as As, Cd, Hg, and Pb, various additives are used.

3.1.1.2 Wet digestion

Wet digestion is a method of converting the components from the complex matrix in the simple chemical form. The digestion is performed by applying energy/heat, by application of chemical reagents, or by a combination of these two methods.

The type of reagent used will depend on the type of sample, and various oxidizing agents that enable the determination of metal content or extract metals from inorganic materials are commonly used. Normally, a combination of concentrated acid and heating is used, and important factors to take into account are the strength of the acid, its oxidizing power, boiling point, solubility of obtained salt, and cleanliness.

The most commonly used acids are nitric, chloride, sulfuric, phosphoric, hydrofluoric acid, and hydrogen peroxide and mixtures of these. Samples can be metal alloys, minerals, soil, rock, clay, carbon, and silicates (Krug, 2000).

The advantage of the wet digestion is that it is equally effective for both organic and inorganic materials, which often leads to complete destruction of the sample matrix and enables the elimination of some interference. The wet methods may be performed using various forms of energy: thermal, ultrasonic energy, and radiation.

3.1.1.2.1 Wet digestion assisted by ultrasonic radiation Wet digestion assisted by ultrasonic radiation is mainly used for low contamination of water matrices. Mostly it is performed with a small addition of hydrogen peroxide or nitric acid. Generally, ultrasonic digestion of samples is carried out by probes or a water bath, but more often an ultrasonic bath is used, because they are cheaper, while probes require a shorter retention time than baths (Priego-Capote and Luque de Castro, 2007). The main advantages of ultrasound-assisted wet digestion are rapid digestion, high processing capacity of the sample treatment, and small amounts of reagents. Also, digestion can be performed in an ultrasonic bath using a plastic container with a screw (screw-top bottles) or polypropylene centrifuge tube, so the samples can be centrifuged rather than filtered (Vaisanen and Kiljunen, 2005; Vaisanen et al., 2002).

3.1.1.2.2 Wet digestion assisted by microwave radiation Microwave technology has significantly improved some traditional operations in chemistry and engineering and has introduced the acronym MEC, which means *microwave-enhanced chemistry* (Richter et al., 2001). The application of microwave has improved the efficiency and speed of digestion for samples that are difficult to dissolve. It is used for digestion of geological, biological, food samples, sludge, ash, and so on.

The main characteristics of commercially available focused-microwave technology are:

- Safety work at atmospheric pressure.
- Handling of large samples that can generate a large amount of gas, mainly when working with organic materials.
- The use of different types of materials for the construction of the reaction vessel (borosilicate glass, quartz, and PTFE).
- Programmed addition of reagents (or samples) at any time during digestion, allowing for a gradual attack of acid.
- Low-energy directional microwave field that can accelerate leaching of organometallic species without affecting the carbon-metal bond or can extract organic component. The consequence of the oriented nature of microwave energy is high efficiency and avoiding the application of high-power.

Focused microwave oven is operated at atmospheric pressure, can be used for samples weighing up to 10 g, and is suitable for samples with high concentrations of organic material. It allows the control of the temperature versus time and power versus time (with simultaneous measurement of the temperature). If using hydrofluoric acid, pipes made of glass, quartz, or PTFE are used, which are transparent to microwave radiation. Temperature control is programmed by the boiling point of the acid or acid mixture used (Kingston and Jassie, 1988).

In general, a mixture containing nitric acid and hydrogen peroxide is used for botanical and biological samples and food samples (cereals, algae, pasta). Sulfuric acid and hydrogen peroxide are used for lubricating oil, and for samples containing PVC, polypropylene, polyamide, polyester, and food (starch, spinach, peanut butter, etc.). Acid mixture (nitric and hydrochloric, aqua regia) is recommended for samples of inorganic materials, such as certain metals, alloys, ash, minerals, and metal extractions from soil (Kingston and Haswell, 1997).

The conventional microwave oven operates at high pressures, which depends on the type of vessel in which the digestion is conducted. For decomposition of samples, nitric acid and other mineral acids are used. For pressures up to 7 MPa, PTFE sealed containers are used, while the quartz tube is used for pressures up to 12 MPa (especially when using sulfuric acid). Pots and microwaves have different characteristics, depending on the manufacturer (Krug, 2000).

3.1.2 The advantages of microwave extraction

Conventional digestion processes, such as wet digestion and dry ashing digestion, are the most time consuming steps of the analysis, because they require long and tedious work and have a high potential of contamination. The time is reduced significantly by using microwave digestion analysis, and can be up to 20–60 times shorter (Julshamn et al., 1998; Siitonen and Thompson, 1998) than conventional methods and today is widely accepted by various scientists (Carlosena et al., 1996; Mester et al., 1999).

The contamination is reduced by using a microwave digestion, less reagents and samples are used, the loss of volatile components is reduced (As, Sb, Se, Sn) (Bettineli et al., 2000; Carlosena et al., 1996), the safety of users is improved, and microwave digestion gives more possibility to control the process and reproducible results.

Microwave-assisted sample preparation has broad applications, including the decomposition of organic and inorganic materials (Priego-Capote and Luque de Castro, 2007). Except for a short degradation time, the direct heating of the sample and reagent are the advantages of the method. Microwave techniques are widely applicable and have become a process of degradation of the matrix of various samples (food, silicate rocks, samples of water, vegetables, samples of ash, soil, and sediment) (Kubrakova, 1997; Smith and Arnesault, 1996; Takenaka et al., 1997). The destruction of the plant material with a mixture of HNO_3 and H_2O_2 provides the fastest, safest, and most precise analytical results with an accuracy of more than 5% for the determination of heavy metals in vegetable matrix.

By using *focused microwave technique,* the sample preparation for the analysis can be improved, because it allows better control of energy, that is, input into the sample. In addition, the use of closed containers enables achieving greater temperature by the increased pressure. Complete digestion of large amounts of samples or samples rich in organic components can be conducted in an open vessel at atmospheric pressure (Nóbreg et al., 2002).

3.2 Ultrasound Extraction

The development of modern techniques, such as extraction assisted by microwave or ultrasound, is intended to overcome these difficulties by increasing extraction efficiency, selectivity, and kinetics. In this context, the use of ultrasound or sonication to break the cell membranes has the advantage of reducing considerably the extraction time and increasing the extract yield. Ultrasonic extraction uses ultrasonic vibrations to extract samples with polar solvents in an ultrasonic bath. This is often used for chemical extraction from solid samples because it is simple. Ultrasound waves were employed to extract active compounds, such as saponins, steroids, and triterpenoids from *Chresta* spp. about three times faster than with the traditional extraction methods (Schinor et al., 2004).

The ultrasonic field enables generation, locally, of microcavitations in the liquid surrounding the plant material. The effects are twofold: mechanical disruption of the cell's wall, releasing its content and local heating of the liquid, increasing the extract diffusion. The kinetic energy is introduced in the whole volume following the collapse of cavitation bubbles at or near walls or interfaces, thus improving the mass transfer across the solid–liquid interface. The mechanical effects of ultrasounds induce a greater penetration of solvent into cellular membranes walls, facilitating the release of contents of the cells and improving mass transfer (Alupului et al., 2009; Keil, 2007).

4 Experimental Part

Within the experimental work, plant raw materials were selected that were used in large quantities for the food industry and households, which can be used in the process of detoxification and show tolerance to stress caused by heavy metals. Phytoaccumulation of heavy metals was the criteria for selection of plant species, to demonstrate the advantages of the use of certain methods of extraction for the wider area of concentration of metals in plant samples.

The following plant species were selected during the experimental work:

- Parsley
- Onion
- Barley

The concentration of lead and copper in the soil after 15 days planting of selected plant species was set to 100 mg/kg within the experiment. The samples were cultivated for a total of 45 days. After a laboratory phase of cultivation of plant species in the preset concentrations of heavy metals, plant material is dried to a constant weight on the paper towels, which are periodically changed. The soil is dried in Petri dishes. After drying, the plant material was ground to a size below 0.2 mm and stored in a plastic vial until the time of analysis.

4.1 Materials Methods

The following methods for preparation of plant materials for analysis of heavy metals were used:

- The method of dry burning
- Ultrasound method
- Method of microwave digestion

Chemicals used during the experiment:

- Nitric acid, 65% HNO_3 p.a
- Hydrogen peroxide, 30% H_2O_2
- Hydrochloric acid, 35% HCl, sp
- Solutions of Cu $(NO_3)_2$ and Pb $(NO_3)_2$
- Humus
- Redistilled water
- Blue tibbon, filter discs (Quant.), pore size of 45 μm

The methods for the analysis of heavy metals are used:

- Atomic absorption spectrophotometry, by using the atomic absorption spectrophotometer "Perkin Elmer"
- Optical emission spectrometer "Perkin Elmer" ICP Optima 2100 DV

Chemicals of high purity (Merck Co. and Kemika) are used in the process of preparation and analysis of samples for chemical analysis, as well as analysis of the heavy metals.

4.2 Preparation of Plant Material

4.2.1 Dry digestion

Samples were prepared to determine the Pb and Cu in the samples of plant material. The procedure consisted in a certain amount of plant material (3 g), which was burned by moderate heating for several hours while the carbon, hydrogen, nitrogen, and oxygen were partially translated into gases, while the nonvolatile oxides remained in a porcelain vessel. The process of combustion produced an ash that was completely free of organic matter, which is a basic precondition for further analytical testing. The ash that was obtained by the combustion process was dissolved in acid, filtered, diluted to 50 mL, and analyzed. The mixture of HCl and HNO_3 acid and H_2O_2 are used for dissolution of organic compounds.

4.2.2 Ultrasound extraction

The dried samples of plant material were weighed from 0.5 g, were transferred to a Teflon container, and then 7 mL of HNO_3 and 4 mL of H_2O_2 was added. After addition of reagents, the samples were transferred to an ultrasonic bath, where the conditions of ultrasonic digestion are set as follows: power of 1200 W, time of 183 min, and temperature of 80°C. After this process, the samples were cooled for 10–15 min, then filtered and diluted to a volume of 50 mL. Thus prepared samples were analyzed for the content of Pb and Cu.

4.2.3 Microwave digestion

The microwave method is one of the easiest and fastest, certainly. In addition to being performed, it requires the least time; it is suitable because it can be used for a variety of complex samples of plants or soil. A sample of plant origin that is being analyzed was placed in a closed vessel, which is a part of the microwave system—5 mL of HNO_3 and 3 mL of H_2O_2were added to sample weighing 0.5 g and was exposed to microwave digestion. The program of microwave digestion is: power 600 W, time 40 min, and temperature of 200°C. The pressure that has been made in a microwave vessel was around $2 \cdot 10^5$ bar. The clear samples were obtained by using this method. By this method, better and cleaner samples were obtained, compared to the samples prepared by other methods, where filtering of the sample was necessary. The time that it took to prepare the sample was also significantly shorter, as well as the quantity of reagents used. The possibility of contamination or errors due to washing by this method is reduced to a minimum. After completing the microwave digestion, the sample is quantitatively transferred into glass flasks and diluted to a volume of 50 mL. Then, the prepared sample was analyzed on the content of Pb and Cu.

4.2.4 The degradation of soil by aqua regia

The samples of soil (in which the plant material is grown) were prepared by extraction with aqua regia (mixture of hydrochloric and nitric acids in a ratio of 3:1) for the analysis of heavy metals as a control method. Deionized water was used to prepare the samples. In glass beakers of 50 ml were weighed 3 g of crushed and sieved air-dried soil. To each sample were added 15 mL of hydrochloric acid and 5 mL of nitric acid. The samples were then digested in a water bath at 50°C for 6 h. After cooling, the samples were filtered through a filter paper (a blue ribbon) and diluted by deionized water in a volumetric flask of 50 mL. In this way, prepared samples are used to determine the concentration of copper and lead that remained in the soil after cultivation of selected plants.

5 Results and Discussions

The experimental results showed that the concentration of the adoption of heavy metals of lead and copper was observed in samples of parsley, and the lowest was observed in the sample of onion. From the three methods used for the extraction of metals from parsley, barley, and onions, the best results were obtained from microwave digestion. Tables 6.1–6.4 show the results obtained after the extraction process of metals from plants and soil.

Figs. 6.1 and 6.2 show the correlation of three methods used for the extraction of metals of lead and copper from the studied plant species (parsley, barley, and onion).

By examination of alternative methods of application for extraction of lead and copper from the plants, it is shown that microwave and ultrasonic digestion methods give better results compared to the traditional method of dry-burning materials. Better process control of digestion with constant monitoring of process conditions, temperature, and pressure; short retention time; and sample preparation are the result of better experimental results of microwave and ultrasonic extraction. The biggest advantage of using microwave digestion

Table 6.1: Results of dry digestion for metals of Cu and Pb from plants.

Samples	Lead (mg/kg)	Copper (mg/kg)
Parsley	17	8.9
Barley	10.7	5.4
Onion	8.9	2.9

Table 6.2: Results of ultrasound digestion for metals of Cu and Pb from plants.

Samples	Lead (mg/kg)	Copper (mg/kg)
Parsley	17.5	9.7
Barley	9.8	6.5
Onion	10.8	3.2

Table 6.3: Results of microwave digestion for metals of Cu and Pb from plants.

Samples	Lead (mg/kg)	Copper (mg/kg)
Parsley	18.7	10.8
Barley	11.7	7.95
Onion	12.5	3.48

Table 6.4: Results of digestion with aqua regia for metals of Cu and Pb from soil.

Samples	Lead (mg/kg)	Copper (mg/kg)
Parsley	77.0	89
Barley	88.6	89.5
Onion	88.2	92.7

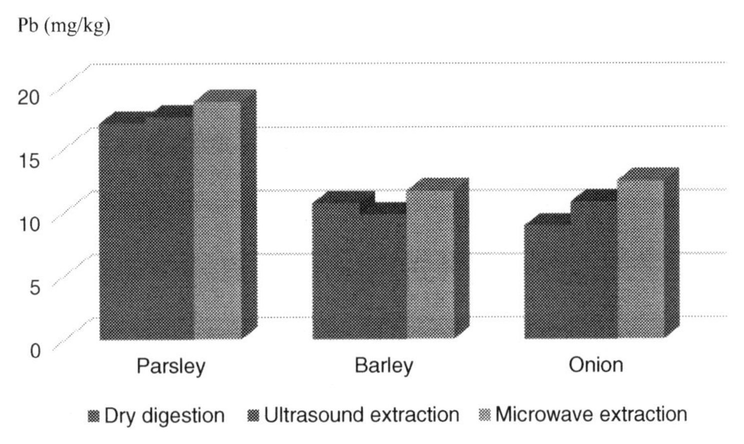

Figure 6.1: Results of Lead Extraction From Plant Material.

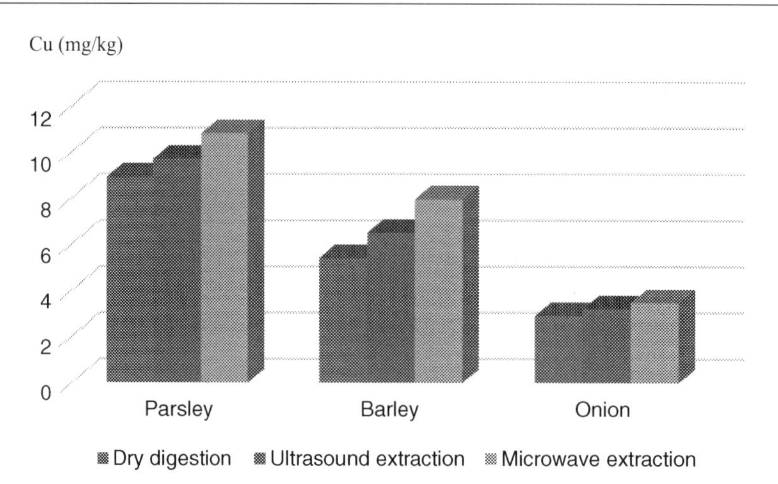

Figure 6.2: Results of Copper Extraction From Plant Material.

is reflected, because the duration of the process of sample preparation is more economical compared to other methods used, and a solutions obtained by this method can be immediately analyzed for metal content.

6 Conclusions

Based on the implemented experiments that are related to the determination of the degree of accumulation of heavy metals, such as lead and copper in parsley, barley and onion, it was concluded that the highest level of adoption of lead and copper occurred in parsley, and the lowest level was observed in onion. Researchers attempted to choose the best method of degradation for this type of sample by changing methods and parameters of degradation (temperature, power, retention time, different reagents).

From the results obtained it can be concluded that the method of ultrasonic and microwave digestion produces better results than traditional methods of preparing samples of plant material for analysis. Microwave method proved to be the most economical, with a short interval of sample preparation, less used reagents, and a sample obtained that does not need further purification. Also, microwave extraction has proved to be particularly suitable for plant material that shows lower affinity for the adoption of heavy metals from the soil and that is characterized by a lower concentration of heavy metals.

References

Abert, V.M., Fernandez, X., Visinoni, F., Chemat, F., 2008. Microwave hydrodiffusion and gravity: a new technique for extraction of essential oils. J. Chromatogr. A 1190, 14–17.

Ali, H., Khan, E., Sayad, M.A., 2013. Phytoremediation of heavy metals: concepts and applications. Chemosphere 91, 869–881.

Alupului, A., Calinescu, I., Lavric, V., 2009. Ultrasonic vs. microwave extraction intensification of active principles from medicinal plants. In: AIDIC conference series. 9.

Angela, M., Meireles, A. (Eds.), 2008. Extracting Bioactive Compounds for Food Products Theory and Applications. CRC Press Taylor & Francis Group, London, New York, p. 151.

Bajsić, Z., Dobrotić, I., 2014. Removal of heavy metals from the soil by phytomediation with the help of wild growing plants in the area of the town Varaždin. University in Zagreb, Geotechnical Faculty.

Bernardoni, P., et al., 2003. Black onion, *Allium cepa* L, Food and Agriculture Organization of the United Nations. Available from: http://www.elestra.rs/wp-content/uploads/2013/11/brosura-crni-luk.pdf

Bettineli, M., Beone, G.M., Spezia, S., Baffi, C., 2000. Determination of heavy metals in soils and sediments by microwave-assisted digestion and inductively coupled plasma optical emission spectrometry analysis. Anal. Chim. Acta 424, 289–296.

Bousbia, N., Vian, A.M., Ferhat, M.A., Petitcolas, E., Meklati, B.Y., Chemat, F., 2009. Comparison of two isolation methods for essential oil from rosemary leaves: hydrodistillation and microwave hydrodiffusion and gravity. Food Chem. 114, 355–362.

Brachet, A., Christen, P., Veuthey, J.L., 2002. Focused microwave-assisted extraction of cocaine and benzoylecgonine from coca leaves. Phytochem. J. Anal. 13, 162–169.

Carlosena, A., Prada, D., Andrade, J.M., Lopez, P., Muniategui, S., 1996. Cadmium determination in soil by microwave acid digestion and graphite furnace atomic absorption spectrometry. Fresenius' J. Anal. Chem. 355, 289–291.

Collins, H.M., Burton, R.A., Topping, D.L., Liao, M.L., Bacic, A., Fincher, G.B., 2010. Variability in fine structures of noncellulosic cell wall polysaccharides from cereal grains: potential importance in human health and nutrition. Cereal Chem. 87, 272–282.

Deak, 2008. Food Technology, Onion—growing, maturity, harvest and storage. Available from: http://www.tehnologijahrane.com/enciklopedija/crni-luk-uzgoj-tehnoloska-zrelost-berba-i-skladistenje

Dizdar, A.M., 2012. Microwave and ultrasound digestion of solid fuel ash. Undergraduate thesis. Zagreb, 2012.

Dolferus, R., Ji, X., Richards, R.A., 2011. Abiotic stresses and control of grain number in cereals. Plant Sci. 181, 331–341.

Džamić, R., Stevanović, D., 2014. Agrochemistry, third ed. Beograd.

Eskilsson, C.S., Bjorklund, E., 2000. Analytical-scale microwave-assisted extraction. J. Chromatogr. A 902, 227–250.

Farzaei, M.H., Abbasabadi, Z., Ardekani, M.R.S., Rahimi, R., Farzaei, F., 2013. Parsley: a review of ethnopharmacology, phytochemistry and biological activities. J. Tradit. Chin. Med. 33 (6), 815–826.

Font, N., Hernandez, F., Hogendoorn, E.A., Baumann, R.A., van Zoonen, P., 1998. Microwave-assisted solvent extraction and reversed-phase liquid chromatography: UV detection for screening soils for sulfonylurea herbicides. J. Chromatogr. A 798, 179–186.

Gaspar, T., Franck, T., Bisbis, B., Kevers, C., Jouve, L., Hausman, J.F., Dommes, J., 2002. Concepts in plant stress physiology: application to plant tissue cultures. Plant Growth Regul. 37, 263–285.

Gabrijela G., 2015. Master's thesis, Niš Determination of metal content in plant species *Seselirigidum* and *Seseli pallasii.*

Hogan, M., 2010. Heavy metal. In: Monosson, E., Cleveland, C. (Eds.), The Encyclopedia of Earth. National Council for Science and the Environment, Washington, D.C.

Jamil, M., Zia, M.S., Qasim, M., 2010. Contamination of agro-ecosystem and human health hazards from wastewater used for irrigation. J. Chem. Soc. Pak. 32, 370–378.

Julshamn, K., Maage, A., Wallin, H.C., 1998. Determination of magnesium and calcium in foods by atomic absorption spectrometry after microwave digestion: NKML collaborative study. J. AOAC Int. 81, 1202–1208.

Kaufmann, B., Christen, P., 2002. Recent extraction techniques for natural products: microwave-assisted extraction and pressurized solvent extraction. Phytochem. Anal. 13, 105–113.

Kebert, M., 2014. Biochemical and physiological characterization of the poplar (*Populus* spp.) clones in the phytoextraction process of copper, nickel and cadmium. Doctoral dissertation, Novi Sad.

Keil, F.J., 2007. Modeling of Process Intensification. Wiley–VCH Verlag GmbH & Co. KGaA, Weinheim.

Kerovec, D., Diploma paper, Osijek 2010. Determination of heavy metals concentration by AAS and ICP-OES in the soil and plant samples.

Khan, S., Cao, Q., Zheng, Y.M., Huang, Y.Z., Zhu, Y.G., 2008. Health risks of heavy metals in contaminated soils and food crops irrigated with wastewater in Beijing, China. Environ. Pollut. 152, 686–692.

Kingston, H.M., Haswell, S.J., 1997. Microwave-enhanced chemistry: fundamentals, sample preparation, and applications. Am. Chem. Soc. 119, 772.

Kingston, H.M., Jassie, L.B., 1988. Introduction to Microwave Sample Preparation: Theory and Practice. ACS Professional Reference Book, Washington, DC.

Krug, F.J., 2000. Métodos de Decomposição de Amostras, III Workshop sobre Preparo de Amostras, FAPESP, São Carlos.

Kubrakova, I., 1997. Microwave-assisted sample preparation and preconcentration for ETAAS. Spectrochim. Acta B 52, 1469–1481.

Lasat, M.M., 2002. Phytoextraction of toxic metals: a review of biological mechanisms. J. Environ. Qual. 31, 109–120.

Lianfu, Z., Zelong, L., 2008. Optimization and comparison of ultrasound/microwave assisted extraction (UMAE) and ultrasonic assisted extraction (UAE) of lycopene from tomatoes. Ultrason. Sonochem. 15, 731–737.

Lončarić, Z., Kadar, I., Jurković, Z., Kovačević, V., Popović, B., Karalić, K., 2012. Heavy metals from the land to the table. Proc. 47th Croatian International Symposium on Agriculture, Opatija, Croatia, pp. 14–23.

Luque de Castro, M.D., Jimenez-Carmona, M.M., Fernandez-Perez, V., 1999. Towards more rational techniques for the isolation of valuable essential oils from plants. Trends Anal. Chem. 18, 708–716.

Marić, M.J., 2014. The possibility of using some wild cultivated plants for soil remediation. Doctoral dissertation, Bor, Serbia.

Mester, Z., Angelone, M., Brunori, C., Cremisini, C., Muntau, H., Morabito, R., 1999. Digestion methods for analysis of fly ash samples by atomic absorption spectrometry. Anal. Chim. Acta 395, 157–163.

Milenković, M.Z., 2104. Study the effect of heavy metals on the growth of industrial hemp. Master's thesis, Niš, Serbia.

Molins, C., Hogendoorn, E.A., Heusinkveld, H.A.G., Van Zoonen, P., Baumann, R.A., 1997. Microwave assisted solvent extraction (MASE) of organochlorine pesticides from soil samples. Int. J. Environ. Anal. Chem. 68, 155–169.

Mozafarian, V., 2007. Flora of Iran. Forest & Rangelands Research Institute Press, Tehran, p. 54.

Nevo, E., Fu, J.B., Pavlicek, T., Khalifa, S., Tavasi, M., Beiles, A., 2012. Evolution of wild cereals during 28 years of global warming in Israel. Proc. Natl. Acad. Sci. USA 109, 3412–3415.

Nóbreg, J.A., Trevizana, L.C., Araújo, G.C.L., Nogueira, A.R.A., 2002. Focused-microwave-assisted strategies for sample preparation. Spectrochim. Acta B 57, 1855–1876.

Peng, S.H., Wang, W.X., Li, X.D., Yen, Y.F., 2004. Metal partitioning in river sediments measured by sequential extraction and biomimetic approaches. Chemosphere 57, 839–851.

Prasad, M.N.V., Freitas, H.M.O., 2003. Metal hyperaccumulation in plants: biodiversity prospecting for phytoremediation technology. Elect. J. Biotechnol. 6 (3), 225–321.

Priego-Capote, F., Luque de Castro, M.D., 2007. Ultrasound-assisted digestion: a useful alternative in sample preparation. J. Biochem. Biophys. 70, 299.

Rai, U.N., Sinha, S., Tripathi, R.D., Chandra, P., 1995. Wastewater treatability potential of some macrophytes: removal of heavy metals. Ecol. Eng. 5, 5–12.

Rascio, N., Navari-Izzo, F., 2011. Heavy metal hyperaccumulating plants: how and why do they do it? And what makes them so interesting? Plant Sci. 180, 169–181.

Richter, R.C., Link, D., Skip Kingston, H.M., 2001. Microwave-enhanced chemistry. Anal. Chem. 73, 31A–37A.

Romanik, G., Gilgenast, E., Przyjazny, A., et al., 2007. Techniques of preparing plant material for chromatographic separation and analysis. J. Biochem. Biophys. 70, 253–261.

Roy, S.J., Tucker, E.J., Tester, M., 2011. Genetic analysis of abiotic stress tolerance in crops. Curr. Opin. Plant Biol. 14, 232–239.

Salt, D.E., Smith, R.D., Raskin, I., 1998. Phytoremediation. Annu. Rev. Plant Physol. Plant Mol. Biol. 49, 643–668.

Šarkanj, B., Kipčić, D., Vasić-Rački, Đ., Delaš, F., Katalenić, K.G.M., Dimitrov, N., Klapec, T., 2010. Chemical and physical dangers in the food. Croatian Food Agency, Osijek, Croatia.

Schinor, E.C., Salvador, M.J., Turatti, I.C.C., 2004. Comparison of classical and ultrasound-assisted extractions of steroids and triterpenoids from three *Chresta* spp. Ultrason. Sonochem. 11, 415.

Siitonen, P.H., Thompson, Jr., H.C., 1998. Determination of calcium by inductively coupled plasma-atomic emission spectrometry, and lead by graphite furnace atomic absorption spectrometry, in calcium supplements after microwave dissolution or dry-ash digestion: method trial. J. AOAC Int. 81 (6), 1233–1239.

Simic, A.S., et al., 2015. Usability value and heavy metals accumulation in forage grasses grown on power station ash deposit. Hem. Ind. 69 (5), 459–467.

Singh, A., Sharma, R.K., Agrawal, M., Marshall, F.M., 2010. Health risk assessment of heavy metals via dietary intake of foodstuffs from the wastewater irrigated site of a dry tropical area of India. Food Chem. Toxicol. 48, 611–619.

Smith, F.E., Arnesault, E.A., 1996. Microwave-assisted sample preparation in analytical chemistry. Talanta 43, 1207–1268.

Šoljić, Z., 2005. Kvalitativna kemijska analiza anorganskih tvari, Fakultet kemijskog inženjerstva i tehnologije, Zagreb, 51, 52.

Stojanović, B.T., 2014. Chemical composition and antioxidant activity of methanol and acetone extracts of the pulp and the peel of selected species of fruit from the Southeast Serbia. Doctoral dissertation, Niš, Serbia.

Šyc, M., Pohořelý, M., Kameníková, P., Habart, J., Svoboda, K., Punčochář, M., 2012. Willow trees from heavy metals phytoextraction as energy crops. Biomass Bioenerg. 37, 106–113.

Takenaka, M., Kozuka, S., Hayashi, M., Endo, H., 1997. Determination of ultratrace amounts of metallic and chloride ion impurities in organic materials for microelectronics devices after a microwave digestion method. Analyst 122, 129–132.

Vaisanen, A., Kiljunen, A., 2005. Ultrasound-assisted sequential extraction method for the evaluation of mobility of toxic elements in contaminated soils. Int. J. Environ. Anal. Chem. 85, 1037–1049.

Vaisanen, A., Suontamo, R., Silvonen, J., Rintala, J., 2002. Ultrasound-assisted extraction in the determination of arsenic, cadmium, copper, lead, and silver in contaminated soil samples by inductively coupled plasma atomic emission spectrometry. Anal. Bioanal. Chem. 373, 93.

Varga, I., 2010. The effect of calcisation and phosphorus fertilization on the concentration of Zn and Cd in the leaf and soybean. Diploma paper, Osijek, Croatia.

Vukadinovic, V., Loncaric, Z., 1998. Plant nutrition. Agriculture Faculty in Osijek, Osijek, Croatia.

Wang, Y., Frei, M., 2011. Stressed food: the impact of abiotic environmental stresses on crop quality. Agr. Ecosyst. Environ. 141, 271–286.

Zeremski-Škorić, T., Ninkov, J., Sekulić, P., Milić, S., Vasin, J., Lazić, N., 2011. Quality of soils in kindergarten playgrounds in the city of Novi Sad. XV International Eco-conference, Novi Sad, September 21–24, 2011, pp. 185–192.

CHAPTER 7

Extraction and Use of Functional Plant Ingredients for the Development of Functional Foods

Rudi Radrigán, Pedro Aqueveque, Margarita Ocampo
Development of Agro industries Technology Center, University of Concepción, Chillán, Chile

1 Introduction

Some natural compounds are used as supplements in diets as a means to increase important nutrients in the diet of humans and animals; these compounds are known as nutraceuticals. The majority of bioactive compounds characterized as nutraceuticals are derived from plant material. Numerous bioactives isolated from mushrooms, legumes, cereals, grains, vegetables, and fruits have been shown to be efficacious in reducing lipid and cholesterol levels, as antioxidants, and in increasing bone calcium density or status, as well as possessing anticancer properties.

However, of the hundreds of plant-derived nutraceuticals identified, few have been incorporated into common food for regular consumption. The extraction from natural sources is a new paradigm, because the technology for extraction must not only be done with organic solvents for the selective separation of specific constituents from the organic material, but also must meet the legal requirements that ensure food quality and safety, which vary from country to country. These requirements are: a high degree of purity, chemical stability, inert, a low boiling point, and—a very important aspect—no toxic effects.

The functional food ingredients specific to food and nutraceuticals are at the center of the industries of the 21st century. They promise value-added opportunities in the food industry and new market opportunities for the pharmaceutical industry. They offer advances in public health, as marketing messages with health claims empower consumers to select healthier food choices.

Ingredients Extraction by Physicochemical Methods in Food
http://dx.doi.org/10.1016/B978-0-12-811521-3.00007-7

2 Conventional Extraction Techniques

There are several techniques for the extraction of bioactive compounds, based on plant materials. The classic technique is based on extraction by means of different pure solvents, or a mixture of solvents, coupled with heat. In order to obtain bioactive compounds, existing traditional techniques are as follows:

- *Soxhlet extraction* was proposed by the German chemist Franz Ritter von Soxhlet in 1879. It was designed for the extraction of lipids. Soxhlet extraction is used widely to extract many bioactive compounds coming from various plants. It is currently used as a comparative model for new extraction techniques. Extraction is performed by placing a minimal quantity of dry sample in a thimble, which is then placed in the distillation flask containing the solvent. After reaching a certain level of overflow, the solution in the thimble is sucked by a siphon down into the distillation flask. The solution has solutes, which are extracted into the liquid. The solutes are left in the distillation flask and the solvent passes toward the bedding solid of the material. The process is repeated several times until extraction is complete.
- *Maceration* has been used in the preparation of home tonics since antiquity. It was developed an economical and fast way to obtain compounds. Maceration is an extraction process in which the plant materials are in a solid-state and are put into an extraction liquid. To improve the process, it's necessary to grind the plant materials into small particles; this increases the uniformity of the surface area mixing with the solvent. When using maceration, it is also important to use the appropriate solvent and to keep it in a closed container. After a time, the liquid is filtered and the solid residue from the extraction process is pressed to retrieve large amounts of occluded solution. Filters or a press are used to get a liquid without impurities. Casual agitation or an increase in temperature facilitates the extraction of bioactive principles.
- *Hydrodistillation* is a traditional extraction method used for essential oils, which does not require the use of organic solvents and which can be carried out before plant material is dehydrated. Vankar (2004) mentions that there are three types of hydrodistillation: water distillation, water and steam distillation, and direct-steam distillation. The process consists of introducing the plant material into the flask with either water, which is brought to boiling point, or vapor, which is introduced directly, or a mixture of water and steam. The heat allows the compounds to be released from the material in the steam, which is then cooled and the compounds separated by density in a funnel. Silva et al. (2005) explains that three processes occur—creation, hydrolysis, and decomposition by heat—and that each compound has a particular temperature associated with extraction, which hinders your use and is not appropriate for all compounds.

Efficient extraction using a conventional method depends mainly on the choice of solvents (Cowan, 1999). The most important factor in the choice of solvent is its polarity.

Table 7.1: Examples of different solvents used to extract bioactive compounds.

Water	Ethanol	Methanol	Chloroform	Dichloromethanol	Ether	Acetone
Anthocyanins	Tannins	Anthocyanins	Terpenoids	Terpenoids	Alkaloids	Flavonoids
Tannins	Polyphenols	Terpenoids	Flavonoids	–	Terpenoids	–
Saponins	Flavonols	Saponins	–	–	–	–
Terpenoids	Terpenoids	Tannins	–	–	–	–
	Alkaloids	Flavones	–	–	–	–
		Polyphenols	–	–	–	–

Source: Adapted from Cowan, M., 1999. Plant products as antimicrobial agents. Clin. Microbiol. Rev. 12 (4), 564–582.

The molecular affinity between the solvent and solute, use of a cosolvent, transfer of mass, environmental safety, toxicity, and cost also must be considered in the selection of the solvent. Table 7.1 shows examples of solvents used to extract some active principles.

3 Nonconventional Extraction Techniques

The extraction of bioactive compounds by traditional methods presents challenges in terms of processing time, cost, and the limited selection of solvent temperatures that can be applied to thermolabile compounds without damaging them. But these limitations have allowed for the development of promising new extraction techniques, nonconventional ones among them, including ultrasound-assisted extraction (UAE), pulsed electric-field extraction (PEF), supercritical-fluid extraction (SFE), microwave-assisted extraction (MAE), pressurized-liquid extraction (PLE), and enzyme-assisted extraction (EAE).

3.1 Ultrasound-Assisted Extraction

Ultrasound is sound that has a frequency above sounds audible by human beings. This means ultrasound is considered to be sound vibrations that exceed 15,000 or 16,000 cycles per second (hertz), depending on the source; there are even some who consider it to be sound exceeding 20,000 hertz, although human beings can't hear such frequencies.

The chemical processes used normally between 20 kHz–100 MHz; these waves are capable of causing cross material biological resonance which produces the phenomenon of the cavitation, therefore the electric energy is transformed in kinetic power and subsequently the other half in heat energy.

According to Suslick and Doktycz (1990), the bubbles of cavitation can get to 5000 K, with pressure of 1000 atm in heating and 1010 K/s in cooling. This is the principle of UAE, which affects mostly liquids, facilitating the leaching of organic and inorganic compounds from the matrix. (Herrera and Luque de Castro, 2005).

UAE facilitates the transfer of mass and accelerates the extraction of the solvent and is made possible by two physical phenomena: (1) the broadcasting through the cell wall, and (2) the clarifying of the cell's contents after the walls are broken (Mason et al., 1996).

Other factors influencing the process are the unit-transfer surface (the size of the grinding and milling particle), the humidity quotient, and the solvent's polarity. There are also factors related to the process that have an effect, such as temperature, pressure, resonance frequency, and time of application.

UAE has advantages compared to other methods, since it reduces the time for extraction, improves leaching, makes the transfer of mass and energy more efficient, and does not require much energy, meaning the size of the team can be small (Chemat et al., 2008).

UAE is a good extraction method for bioactive compounds. Rostagno et al. (2003) showed removal efficiency of four derivatives of isoflavones, daidzin, genistin, glycitin and-malonyl genistin bean from soy, by using agitation with solvent method and gradients of extraction time.

These authors determined that UAE improves the performance and power of the extraction solvent, depending on its polarity.

Herrera and Luque de Castro (2005) developed a semiautomatic method to extract the phenolic compounds naringin, naringenin, ellagic acid, quercetin, and kaempferol from strawberries, using 0.8 s application with cycles of 30 s (Li et al., 2005).

3.2 Pulsed Electric-Field Extraction

During the last decade, PEF has been used to improve the processes of pressing, drying, extraction, and broadcasting. PEF works by destroying the structure of the membrane of the cell to facilitate extraction.

Electric power passes through the membrane of cell while the live cell is suspended in the electric field.

Depending on the composition of its dipoles, the molecules of the membrane act on the cell membranes, according to your load and electric power. After overcome a value critical of approximately 1 V of potential transmembrane, the pulse breaks the walls of the membrane cell; it necessarily associates a circuit simple of exponential decay which is used in the vegetable material.

The system has a camera feature that consists of two electrodes, which are placed into the plant material. Depending on the design, the treatment chamber can work in either continuous or batch mode (Puértolas et al., 2010). PEF depends strictly on the parameters used in the process, including the strength of the field, input of specific energy, the number of pulses, the temperature, and the properties of the materials to be treated. PEF increases the transfer of mass during the extraction because it destroys the structure of the cell membrane of vegetable material and decreases the time needed for extraction as it improves the release of intracellular compounds woven through the plant while helping to increase the permeability of the cell membrane. PEF involves a moderate electric field (500–1000 V/cm for 10^{-4}–10^{-2} s) that is responsible for damage to the cell membrane of the plant tissue using a low rise in temperature. For this reason, PEF can minimize the deterioration of heat-sensitive compounds (Ade-Omowaye et al., 2001). PEF can be applied to plant material as a pretreatment process before traditional extraction by solvent to lower effort involved in extraction: PEF treatment (using kV/1 cm with the low consumption of energy of 7 kJ/kg) using a solid–liquid extraction process for the extraction of betanin from beets resulted in more extracted material compared to mechanical freezing and crimping. Guderjan et al. (2005) showed that phytosterols in corn increased 32.4% and isoflavones (genistein and daidzein) from soybeans increased 20%–21% when PEF was used as pretreatment process. PEF treatment of grape skin before maceration can reduce the technique's duration and improve the stability of the bioactive compounds (polyphenols and anthocyanins) during winemaking (López et al., 2008).

3.3 Enzyme-Assisted Extraction

Some phytochemical compounds in the vegetal matrices are retained in the polysaccharide-lignin network by hydrogen or hydrophobic bonding, or are dispersed in cell cytoplasm, and some compounds are not accessible using a routine extraction process with a solvent. EAE is an enzyme pretreatment regarded as a novel and effective way to release limited compounds and increase the overall performance of the extraction process. Moreover, adding specific enzymes, such as cellulase, α-amylase, and pectinase during extraction facilitates rupturing the cell wall and improves recovery by hydrolyzing structural lipid and polysaccharide frameworks.

Latif and Anwar (2009) showed that there are two methods for extraction assisted by enzymes:

1. enzyme-assisted aqueous extraction (EAAE)
2. enzyme-assisted cold pressing (EACP)

In general, several seed-oil extraction methods involve EAAE. In EACP, enzymes are used to hydrolyze the cell walls of seeds, but because in this method colloidal polysaccharide proteins are not available (Concha et al., 2004).

Different factors key to the extraction, including the size of the plant particle, the proportion of solid material to water, the time for composition, the concentration, and the hydrolysis enzyme used. Domínguez et al. (1995) demonstrated that an important factor was the moisture content of the plant material, which is also important for enzymatic hydrolysis.

Bhattacharjee et al. (2006) described EACP as ideal alternative for the extraction of the bioactive components of seed oil, due to their nontoxic and nonflammable properties. The oil extracted by enzyme-assisted methods was found to contain more free fatty acids and phosphorus than the oil extracted using a traditional hexane method. EAE is recognized as an ecofriendly technology for the extraction of bioactive compounds and oil because it uses water as a solvent instead of organic chemical products (Puri et al., 2012).

The extraction of phenolic antioxidants from raspberry waste was increased in hydroalcoholic extraction by the use of enzymes, in comparison to the nonenzymatic control (Laroze et al., 2010). Several authors have depicted enzyme technology as an alternative for extracting bioactive compounds from agroindustrial by-products.

3.4 Microwave-Assisted Extraction

MAE is another new method for liquid-phase extraction; considering the ease of working with a large number of materials, this extraction method is a promising technique.

The frequencies between 300 MHz–300 GHz, are called microwave. These oscillating fields of the electric and magnetic fields which are perpendicular generate much heat because of the resistance of the medium to the passage of the ionic current, the same that keeps your address in all the fields but changes with a known frequency.

This frequent change of addresses causes collision between the molecules and thus generates heat. MAE involves three sequential steps, as described by Alupului et al. (2012): (1) the separation of solutes in active sites of the matrix with a low increase of temperature and pressure, (2) broadcasting of the solvent through the matrix of the sample, and (3) release of the solutes from the matrix of the sample to the solvent. Several advantages of MAE have been described by Cravotto et al. (2008); namely, faster warming for the extraction of bioactive substances from vegetable material, reduced thermal gradients, and the small size and higher performance of the extraction equipment. MAE can extract bioactive compounds faster, and better recovery is possible, than conventional extraction processes. It is a selective technique for extracting organic and organoleptic compounds without modifications. MAE also is recognized as an ecological technology that reduces the use of organic solvents (Alupului et al., 2012).

For example, in extracting polyphenols and caffeine from green tea leaves, MAE achieves a higher performance level of extraction in 4 min than any other method of extraction at room temperature in 20 h (Pan et al., 2003). Asghari et al. (2011) extracted certain bioactive compounds (*E*-guggulsterone, *Z*-guggulsterone, cinnamaldehyde, and tannin) from several plants in excellent condition and showed that MAE is a faster and easier method in comparison with conventional extraction processes.

3.5 Pressurized-Liquid Extraction

PLE is currently known by a few other names: accelerated solvent extraction (ASE), enhanced-solvent extraction (ESE), and high-pressure solvent extraction (HPSE) (Nieto et al., 2010). The concept behind PLE is the application of high pressure so that more liquid solvent remains beyond the normal point of boiling. High pressure facilitates the process of extraction. Automation techniques are the main reason for development of more techniques based on PLE, with their reduced requirements in terms of time and extraction solvents. PLE requires only small amounts of solvents, due to the combination of high pressure and temperature that allows for quicker extraction. The higher extraction temperature promotes greater solubility of the analyte; by increasing the solubility and rate of mass transfer and decreasing the viscosity and surface tension of solvents, the rate of extraction is improved.

Richter et al., 1996 compared PLE with traditional Soxhlet extraction and found that the consumption of solvent decreases considerably; they even think that PLE could compete with SFE for polar compounds. Wang and Weller (2006) studied the extraction of organic contaminants of them dies environmental that are stable at high temperatures. PLE has also been used for the extraction of bioactive compounds. Furthermore, due to the use of organic solvents, PLE has obtained wide recognition as a green extraction technique.

PLE has been applied with success to the extraction of active principles of different materials. Shen and Shao (2005) compared this technique to the extraction of terpenoids and sterols of tobacco using UAE and traditional Soxhlet extraction. Taking into account performance, reproducibility, extraction time, and solvent consumption, PLE was considered as an alternative to conventional methods, as the process was faster and used less solvent.

3.6 Supercritical-Fluid Extraction

SFE using carbon dioxide has been shown to be a viable alternative to conventional solvent extraction techniques to extract bioactive components from agricultural material. It offers the unique advantage of adding value to agricultural waste by extracting antioxidants and flavonoids, which are then used in the fortification of foods and other applications. Its drawbacks are the difficulties in extracting polar compounds and its susceptibility of extracting compounds from a complex matrix where the phase interaction with the intrinsic properties of the product inhibits its effectiveness. Many of these drawbacks can be ameliorated by using cosolvents.

However, much investigation is required to understand the solvation effect on the targeted bioactive components being extracted.

3.6.1 Process concept schemes and systems

SFE technology was conceptualized on the basis of obtaining pure extract without solvent residues, which could be detrimental to consumers of food and pharmaceutical products. Extraction is an analytical process used to separate and isolate a targeted component from other substances. The success of the process is dependent on the distribution of the analyte between two phases: the separation and stationary phases (King et al., 1993).

However, in most cases it was found that a single extraction step is not enough, and that further steps are required to improve the purity and composition of the extract from a natural product source. Hence, a number of sequential processing steps, as shown in Fig. 7.1, in which the extract resulting from the SFE step is enriched using supercritical-fluid fractionation (SFF) or supercritical-fluid reaction (SFR), becomes necessary. Studies related to the concept of sequential unit processing with multiple fluids have been discussed in detail by King and Srinivas (2009).

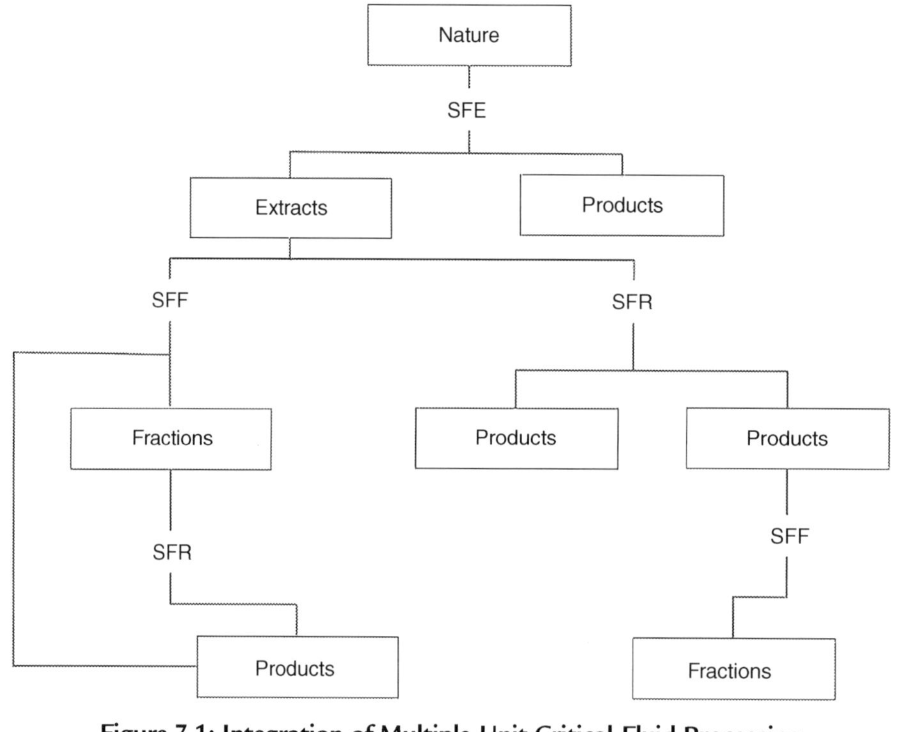

Figure 7.1: Integration of Multiple-Unit Critical-Fluid Processing for the Processing of Natural Products.
SFE, Supercritical fluid extraction; *SFF*, supercritical fluid fractionation; *SFR*, supercritical fluid refining.

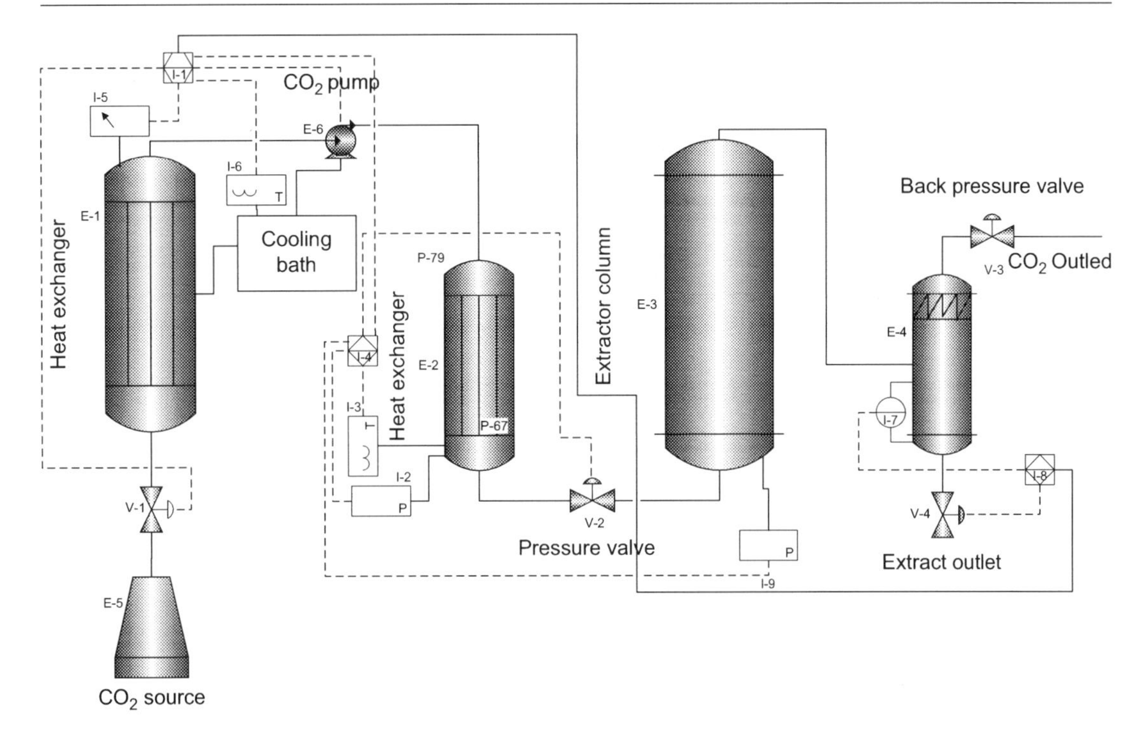

Figure 7.2: Schematic Diagram of Supercritical-Fluid Extraction System.

Extraction with supercritical fluids (SCFs) is comparable to liquid-liquid solvent extraction; however, compressed gas is used instead of organic solvents and the applied pressure is critical. The SFE process is governed by four key steps: extraction, expansion, separation, and solvent conditioning. These steps are accompanied by four generic primary components: an extractor column (high-pressure vessel), pressure-control valves, a separator column, and a pressure intensifier (pump) for the recyclable solvent (Fig. 7.2) (Brunner 2005).

The system has other built-in accessories, such as a heat exchanger for providing a heating source, a condenser for condensing SCFs into liquid, storage vessels, and a SCF source. The raw materials are usually ground and discharged into a temperature-controlled extractor column forming a fixed bed, which is usually the case for batch and single-stage mode.

3.6.1.1 Single-stage extraction process

SCF is fed at high pressure by means of a pump, which pressurizes the extraction tank and also circulates the supercritical medium throughout the system. Fig. 7.3 shows an example of a typical single-stage supercritical extraction system; once the SCF and the feed reach equilibrium in the extraction vessel through the manipulation of pressure and temperature to achieve the ideal operating conditions, the extraction process proceeds. The mobile phase

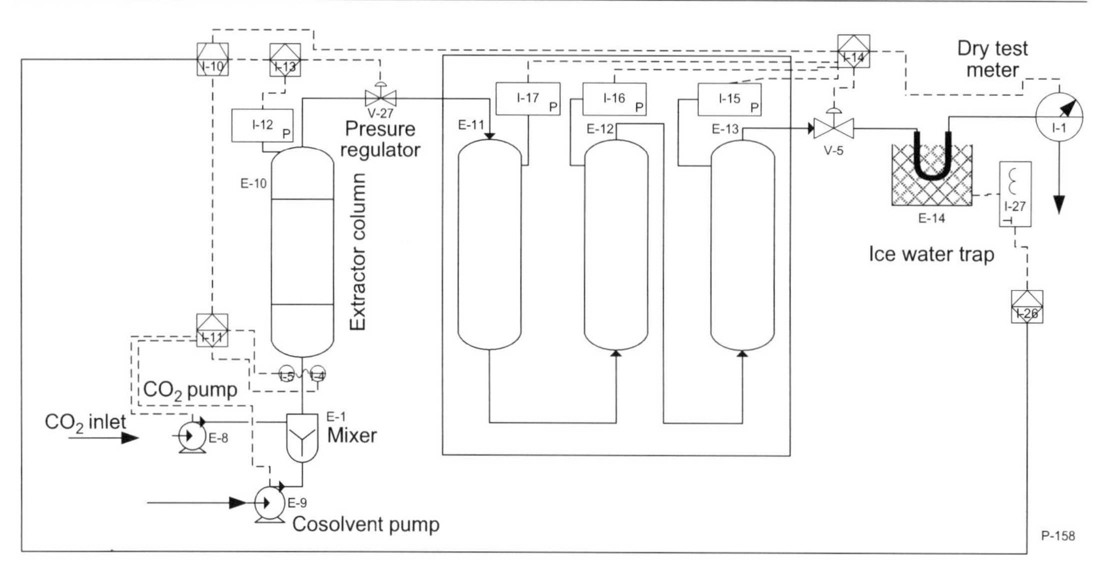

Figure 7.3: Schematic Diagram of Supercritical-Fluid Extraction System Used to Fractionate Bioactive Components From a Plant Mix Using Supercritical Carbon Dioxide.

consists of the supercritical CO_2 (SC-CO_2) fluid and the solubilized components being transferred to the separator, where the solvating power of the fluid is decreased by increasing the temperature or decreasing the pressure of the system. The extract precipitates in the separator while the SCF is either released or recycled back to the extractor. In cases, in which highly volatile components are being extracted, a multistage configuration may have to be employed, as shown in Fig. 7.3.

3.6.1.2 Multistage extraction process

A semicontinuous approach on a commercial scale uses a multiple-stage extraction process, which involves running the system concurrently by harnessing a series of extraction vessels in tandem (Fig. 7.4). In this system, the process is not interrupted at the end of each extraction period for each vessel because the process is switched by control valves to the next vessel prepared for extraction while unloading or loading the spent vessels; although imperfect, continuity is attained.

This is effective in cases in which more than one targeted component is to be extracted, providing the flexibility to vary the extraction parameters, such as pressure and temperature, to achieve different solubility for different components being extracted at each stage of the operation.

3.6.2 Physicochemical properties of supercritical fluids

The physicochemical properties of SCFs are crucial to understanding process-design calculations and for modeling of the extraction process. Therefore, the selection of the

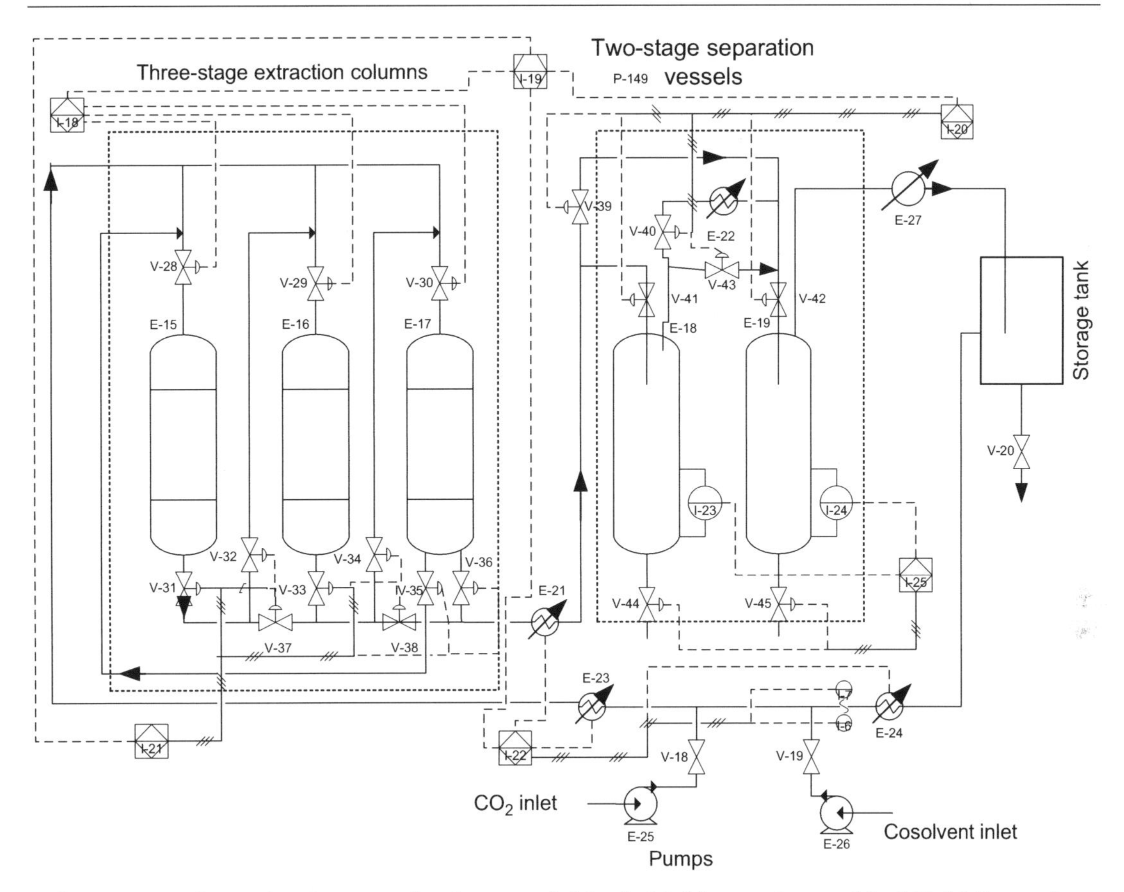

Figure 7.4: Schematic Diagram of a Commercial Scale Multistage Supercritical-Fluid Extraction System Used to Fractionate Bioactive Components.

solvent that will segregate the solutes is a key property for the process engineer. Physical characteristics, such as density and interfacial tension are important for separation to proceed; the density of the extract phase must be different from that of the raffinate phase, while the interfacial properties will influence coalescence, a step that must occur if the extract and raffinate phase are to be separate.

3.6.2.1 Phase diagram

The supercritical state of fluid is determined by temperature and pressure above the critical point. The critical point is at the end of the vapor–liquid coexistence curve, as shown on the pressure–temperature curve in Fig. 7.5, where a single gaseous phase is generated. When pressure and temperature are further increased beyond this critical point, the fluid enters a supercritical state. In this state no phase transition will occur regardless of any increase in pressure or temperature, nor will it transit to the liquid phase. Hence, diffusion and mass

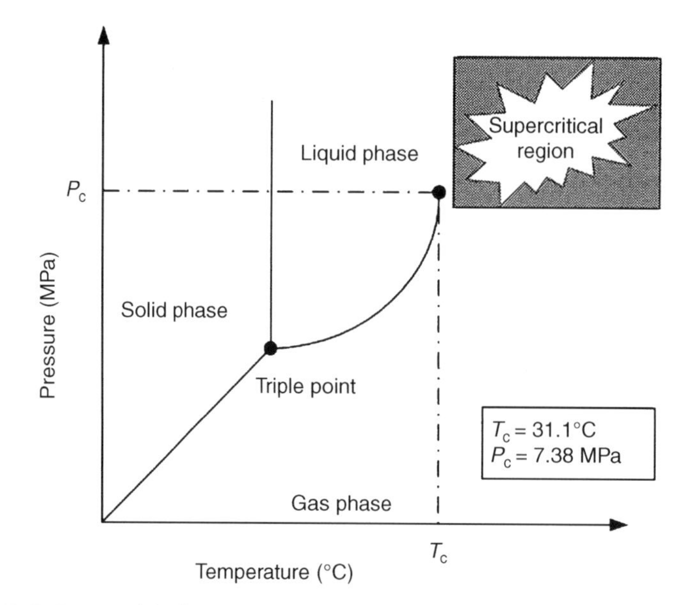

Figure 7.5: Supercritical Pressure–Temperature Diagram for Carbon Dioxide.

transfer during supercritical extraction are about two orders of magnitude greater than in the liquid state.

3.6.2.2 Physical properties

Substances that have similar polarities will be soluble in each other, but increasing deviation in polarity will make solubility increasingly difficult. Intermolecular polarities exist as a result of van der Waals forces, and although solubility behaviors depend on the degree of intermolecular attraction between molecules, the discrimination between different types of polarities is also important. Substances dissolve in each other if their intermolecular forces are similar, or if the composite forces are made up in the same way. Properties, such as the density, diffusivity, dielectrical constant, viscosity, and solubility are paramount to supercritical-extraction process design. The dissolving power of an SCF depends on its density and mass transfer characteristics, and is superior due to its high diffusivity, low viscosity, and interfacial tension of liquid solvents.

Although many different types of SCFs are in existence and have many industrial applications, CO_2 is the most desired for SFE of bioactive components. Table 7.2 shows some physical properties of compressed (20 MPa) SC-CO_2 at 55°C compared to condensed liquids commonly used as extraction solvents at 25°C. It should be noted that SC-CO_2 exhibited densities similar to those of the liquid solvents, although less viscous and highly diffusive. This fluid-like attribute of CO_2, coupled with its ideal transport properties and other attributes outlined before, make it a better choice than other solvents.

Table 7.2: Comparison of the physical properties of SC-CO_2 at 20 MPa and 55°C with selected liquid solvents at 25°C.

Properties	CO_2	*n*-Hexane	Methylene Chloride	Methanol
Density (g/mL)	0.75	0.66	1.33	0.79
Kinematic viscosity (m^2/s)	1.00	4.45	3.09	6.91
Diffusivity (m^2/s)	6.0×10^9	4.0×10^9	2.9×10^9	1.8×10^9
Cohesive energy density ($\delta\Pi$)	10.80	7.24	9.93	14.28

Source: Modified from King, J.W., Hill, H.H., Jr., Lee, M.L., 1993. Analytical supercritical fluid chromatografy and extraction. In: Supplement and Cumulative Index. John Wiley & Sons, New York, pp. 1–83.

3.6.3 Factors affecting supercritical-fluid extraction

To better optimize the extraction process for maximizing the yield of functional-food components from natural products, it is necessary to understand the thermodynamic and mass-transfer parameters that affect the activity of structurally complex biological solutes in SCFs. The basic principle outlining the extraction of bioactive compounds using SCFs is that the heated and pressurized solvent diffuses through the feed matrix where it dissolves the bioactive compounds and SFE of such value-added compounds from natural products are as follows:

- solubility of the bioactive compounds as a function of temperature and pressure;
- mass transfer or diffusion coefficients of the bioactive compounds in the SCFs;
- the chosen extraction or fractionation temperature, pressure, pH, and solvent flow rate;
- substrate or matrix particle size;
- moisture content of the matrix;
- morphology of the sample matrix.

3.6.4 Solubility of food components in SFE

Because solubility is important to the process design of SFE, a lot of experimental work has been done on measuring the solubilities of food components in SCFs. Typical solubility behavior of a solid solute in an SCF solvent is shown in Fig. 7.6.

Two convergence points are shown at P_L and P_U, and one minima (P_{min}). In the low-pressure region ($<P_{min}$), solubility decreases as pressure increases, while in the region of pressure ($P_{min} < P_L < P_U$), the solubility increases sharply with the increase in pressure. This region is usually observed in the near-critical and highly compressed region of an SCF solvent. It indicates that in SFE, solubility can be controlled by pressure. In other words, it is related to the state of the SCF. Hence, the prediction of solubility is usually based on the equation of state (EOS) of the solvent.

The crossover points P_L and P_U are attracting some interest as a method of separating components with a small difference in selectivity, such as isomers. For a multicomponent

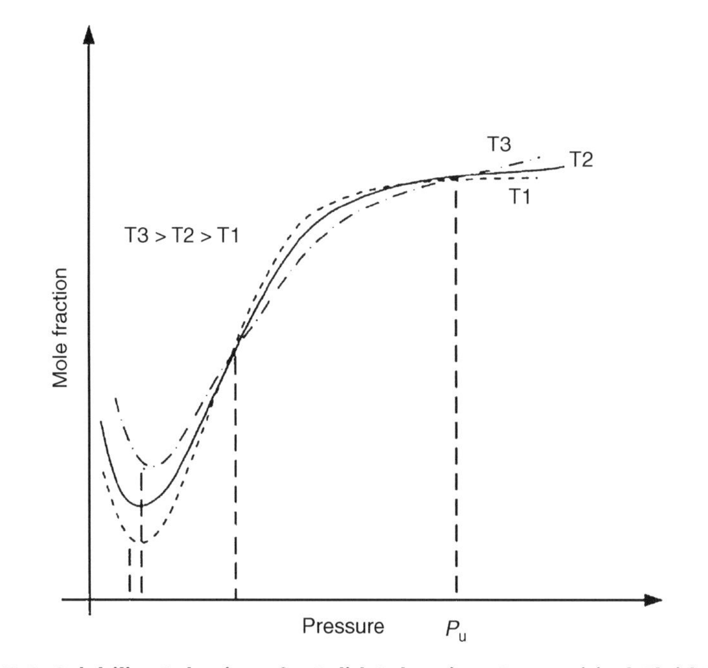

Figure 7.6: Solubility Behavior of a Solid Solute in a Supercritical-Fluid Solvent.

system, the crossover point of each component may not overlap, so there is a crossover region where most of the crossover points are located. At this point or in this region, the solubility of components are similar. Johnston et al. (1987) pointed out that the crossover point in fact is the turning point of the isotherm line of solubility, where y_2 is the fraction of solute in gas phase and T is the temperature Eq. (7.1):

$$\frac{d(\ln y_2)}{dt} = 0 \tag{7.1}$$

For one component, a slight increase in temperature at P_U will cause the solubility to increase above the crossover pressure and decrease below the crossover pressure. The results are reversed at P_L. Under these conditions a retrograde region is formed. In multicomponent systems solubility will not change with temperature at the crossover point of one component, while it will for others, so the selectivity of those components increases. Through several cyclings of retrograde crystallization or solvation, the components are separated. The operation becomes similar to distillation and requires a temperature gradient. Chimowitz and Pennisi (1986) and Foster et al. (1991) gave a detailed description of the operation in the crossover-pressure region. Because fewer data are available in the critical region, this "distillation" operation runs more often in the upper crossover-pressure region than in lower crossover-pressure region.

The other interesting observation is the sharp change of solubility with pressure. In view of the macroscopic thermodynamics, the influence of the pressure on the solubility can be explained by partial molar volume $\overline{V}_2$ directly, as shown in Eq. (7.2):

$$\ln \phi_2 = \frac{1}{RT} \int_0^P \overline{V}_2 \, dp \tag{7.2}$$

where V_2^s is the saturate volume, ϕ_2 is the fugacity coefficient, and R is the gas constant. The partial molar volume is a differential quantity that describes the solution behavior at a particular pressure. Here, the fugacity coefficient ϕ_2 is the pressure integral of the partial molar volume Eq. (7.3). Using Eq. (7.2), the solubility behavior versus the pressure can be easily explained. When the pressure is much lower that the critical point, the partial molar volume is not a function of composition, so the equation is simplified, as in Eq. (7.3):

$$\frac{\partial (lny_2)}{\partial P} = \frac{V_2^S - \overline{V}_2}{RT} \tag{7.3}$$

At low pressure $V_2^S \ll \overline{V}_2$, hence solubility decreases with pressure, and as the pressure increases V_2^s decreases more slowly than $\overline{V}_2$. When the partial molar volume equals the saturate volume, the solubility reaches its minimum at a certain pressure. At high pressures that are significantly above the critical pressure, $\overline{V}_2$ increases slowly; where $\overline{V}_2$ exceeds V_2^s because of the repulsive force, the solubility will reach a maximum and decrease slowly with the increase of pressure. In this region, the solubility does not change much. The quickest increase of solubility occurs at a pressure corresponding to the minimum $\overline{V}_2$. This theory explains solubility behavior in the supercritical region (Fig. 7.6) and is supported by some experimental data. Unfortunately, due to the lack of adequate experimental data for biosubstances, the practical application of the partial molar volume model is limited. So under these circumstances, the factors that influence solubility are usually analyzed phenomenologically.

Biocompounds are usually present as a mixture in natural tissues and are often not used in their pure form, thus they are regarded as a pseudocomponent in research. Some researchers will measure the pure substance, while others just measure the oil solubility. Staby and Mollerup (1993) published a list of fish oil-related components from the 1970s. Bartle et al. (1991) compiled solubility data from the 1980s, which included solubility in SC-CO_2 of not only bioactive components but also other chemical components. Güçlü-Üstündağ and Temelli (2000) collected data covering the period 1970–99 on the solubility of pure lipids in SC-CO_2. Foster et al. (1991) collected some solubility data for the years 1968–90 on the solubility of various chemical substances in various SCFs. Recent solubility information for bioactive components from different sources is listed in Table 7.3.

Table 7.3: Solubility of selected bioactive components in SC-CO_2.

Components	Substance	Temperature (°C)	Pressure (bar)	Sources
Sterols	Syringic acid	40–60	100–500	Murga et al. (2004)
Phenolic compounds	Vanillic acid	40–60	100–500	Murga et al. (2004)
	Echium, borrage, and lunaria	10–55	60–300	Gaspar et al. (2003)
	Hazelnut oil	40–60	150–600	Ozkal et al. (2005)
Fatty acids	Lauric acid	40	345–483	Nik Norulaini et al. (2004b)
	Lauric acid and oleic acid	80	276–483	Nik Norulaini et al. (2004a)
Fatty acid esters	2-Ethyl-1-hexanol, 2-ethylhexanoic acid	40–50	68–180	Ghaziaskar et al. (2003)
	2-Ethyl-1-hexanol, 2-ethylhexanoic acid	40–100	138	Ghaziaskar et al. (2003)

4 Defining Functional Foods

What are functional foods? The term is difficult to define, but if you look at the words "functional" and "foods" separately, it suggests the following definition, which is consistent with our understanding of the term: *A food can be regarded as functional if it is satisfactorily demonstrated to affect one or more target functions in the body, beyond adequate nutrition, in a way that improves health and well-being or reduces the risk of disease.*

This definition suggests that a product must remain a food to be included within this category. On this basis, a functional food can be:

- a natural food
- a food to which a positive component has been added, or from which a deleterious component has been removed
- a food in which the nature of one or more components has been modified

The idea of functionality reflects a major shift in attitude about the relationship between diet and health. Nutritionists have traditionally concentrated on defining a balanced diet; that is, one ensuring adequate intake of nutrients and avoiding certain dietary imbalances (e.g., excessive consumption of fat, cholesterol, and salt) that can contribute toward disease. That this lies behind all sound nutritional principles and guidelines is important. However, now the focus is on achieving optimized nutrition, maximizing life expectancy, and food quality by identifying ingredients, which when added to a balanced diet, improve the capacity to resist disease and enhance health. Functional foods are one of the outcomes of this focus.

4.1 Functional Food Science

Being foods, functional foods need to be safe according to all criteria defined in current food regulations. But in many cases, new concepts and new procedures will need to be developed and validated to assess functional food risks. In Europe, some, but certainly not all, functional foods will be classified as "novel foods" and consequently will require the decision-tree assessment regarding safety that is described in the EU novel food regulation (European Commission, 1997).

However, it must be emphasized that this regulation does not address the nutritional properties or the physiological effects of these novel foods. It is strictly a safety regulation. The requirement for safety is a prerequisite to the development of any functional food. Indeed, the risk-versus-benefit concept, which is familiar to pharmacologists developing new drugs, does not apply to functional foods except perhaps in very specific conditions for disease risk reduction when the scientific evidence is particularly strong. As described in the European consensus document (European Commission, 1999):

> The design and development of functional foods is a key issue, as well as a scientific challenge, which should rely on basic scientific knowledge relevant to target functions and their possible modulation by food components. Functional foods themselves are not universal, and a food-based approach would have to be influenced by local considerations. In contrast, a science-based approach to functional food is universal.

The approach to functional foods has a scientific basis as its foundation to gain a broader understanding of the interactions between diet and health. Emphasis is then put on the importance of the effect of food components on well-identified and well-characterized target functions in the body that are relevant to well-being and health issues, rather than solely on the reduction of disease risk.

Referencing the new concepts in nutrition just outlined, it is the role of functional food science to stimulate research into and development of functional foods (Fig. 7.7). By referencing basic knowledge of nutrition and related biological sciences, such development requires the identification of, and, at least partly, an understanding of, the mechanism(s) by which a potential functional food or functional food component can modulate the target

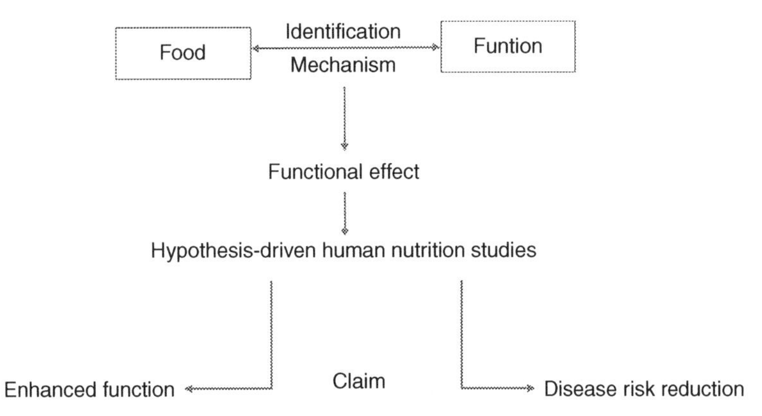

Figure 7.7: Strategies for Developing Functional Foods.

function(s) that is/are recognized or proven to be relevant to the state of well-being and health, and/or the reduction of a risk for disease. Epidemiological data demonstrating a statistically validated and biologically relevant relationship between the intake of specific food components and a particular health benefit will be, if available, very useful. The conclusion of that first step will be the demonstration of a functional effect that should serve to formulate hypotheses to be tested in a new generation of human nutrition studies aimed to show that relevant (in terms of dose, frequency, duration, etc.) intake of the specified food will be associated with improvements in one or more target functions, either directly or indirectly in terms of a valid marker of an improved state of well-being and health and/or reduced disease risk. If well supported by strong scientific evidence, the conclusion could be a recommendation for improved or new dietary guidelines. The new-generation human nutrition studies should be hypothesis driven, but in many cases they will differ quite substantially from what is classically referred to as clinical studies. The main differences are that the new nutrition studies will aim at testing the effect of a food as part of the ordinary diet; may concern the general population or generally large, at-risk target groups; will not be diagnostic or symptom based; and will not be planned to evaluate a risk versus benefit approach. Most of these studies will rely on change(s) in validated/relevant markers to demonstrate a positive modulation of target functions after (long-term) consumption of each potential functional food.

A (double) blind type of design based on parallel groups rather than a crossover study will generally be appropriate. Data from these studies should be collected and handled according to good standards for data management, and data analysis should prove statistical, as well as biological significance. Finally, the long-term consequences of interaction(s) between functional foods and body function(s) will have to be carefully monitored.

4.2 Adaptation of General Technologies Traditionally Used in Food Processing

4.2.1 Formulation and blending

Formulation and mixing has been widely used in food processing since it is the most simple and cheap technology to develop new functional foods. Their use in the development of functional foods for deficiencies of vitamins A and D, several B vitamins (thiamin, riboflavin, and niacin), iodine, and iron has a long history of successful control. Iodization of salt was introduced at the beginning of 1920 in Switzerland (Burgi et al., 1990) and the United States (Marine and Kimball, 1920), and its use has expanded progressively over the entire world, to the extent that iodized salt is now used in most countries.

Fortifying cereals with thiamin, riboflavin, and niacin has been common practice since the early 1940s. Margarine is fortified with vitamin A (FAO and WHO, 2006) and wheat grown in the United States is fortified with folic acid (Samaniego-Vaesken et al., 2010), a strategy that has been adopted by Canada, the United States, and about 20 countries in Latin America.

In recent years, the emergence of dietary compounds with health benefits offers an excellent opportunity to improve public health; therefore, this category of compounds has received a lot of attention from consumers, food manufacturers, and the scientific community. The list of active compounds for the diet is endless (vitamins, probiotics, bioactive peptides, antioxidants, etc.) and the type of these products that can be obtained is growing constantly. Among the items being developed are plain milk, yogurt, and milk formula enriched with prebiotics, probiotics, vitamins, and long-chain polyunsaturated fatty acids that can be provided to babies for optimal growth and development; juices enriched mainly with vitamins, flavonoids, and resveratrol; snacks and pasta rich in vegetables; and meat enriched with a great number of bioactive compounds.

4.2.2 Cultivation and animal breeding techniques

Agriculture and livestock are the primary sources of nutrients required by humans. There is a general consensus among nutritionists that the best way of addressing micronutrient deficiencies is through diversification of the diet, which should include vegetables, fruits, meat, and fish (FAO and WHO, 2001). However, this is not always possible. Agriculture and livestock have traditionally been presented as a way to obtain products with high nutrients. Biotechnology offers a useful alternative, in cases in which agricultural development and reproduction are unable to achieve a significant improvement in food (Zhao and Shewry, 2011).

Biotechnology has been practiced in crops and livestock for 12,000 years, led by the wish to get important features dictated by social, nutritional, and environmental needs but with no understanding of the molecular processes involved; all this was done by the evaluation and selection of different traits beginning with the domestication of animal and plant species.

With the use of the tools of molecular biology and the development of genetically modified seeds not so long ago, biotechnology became a modern technique that provides a complementary way to modify the composition of foods. The best-known enriched crop product is golden rice (Ye et al., 2000), which has 1.6 μg/g total carotenoids in the rice endosperm. The 2nd generation of golden rice (Golden Rice 2) (Paine et al., 2005) contains up to 37 μg/g of total carotenoids. A recent clinical trial shows that Golden Rice 2 is an effective source of vitamin A for humans, with a β-carotene to promote retinol conversion efficiency. Fortification with vitamin A has also been carried out in other food crops, such as potatoes.

Animal breeding also offers the possibility of obtaining improved food products. To this end, many studies have been done to examine the sources of nutrients available for inclusion in animal diets and their subsequent transfer into the products obtained. Matsushita et al. (2007) carried out a study in which they characterized the fatty acid profiles and physicochemical parameters of milk samples from Saanen goats fed diets enriched with 3% vegetable oil of three different kinds (soybean, canola, and sunflower). The milk obtained presented a different concentration of conjugated linoleic acid depending on the vegetable oil added to the animal feed. In addition, Laible (2009) has suggested several other modifications to improve the nutritional quality of milk and its related dairy products. Woods and Fearon (2009) examined in a review the sources of fatty acids available for inclusion in animal diets and their subsequent transfer into meat, eggs, and milk.

4.3 Specific Technologies for the Manufacture of Functional Foods That Prevent the Deterioration of Physiologically Active Compounds

4.3.1 Microencapsulation

Microencapsulation is the coating of small solid particles, liquid droplets, or gases in a layer (Thies, 1987).

Microencapsulation is based on the barrier effect of a polymer matrix, which creates a microenvironment in the capsule capable of controlling the interactions between the inner and the outer parts of the substance. Microencapsulation allows for the protection of a wide range of materials of biological interest, from small molecules and proteins (enzymes, hormones, and other compounds) to the cells of bacteria, yeast, and animals (Thies, 2005).

Therefore, this technology is widely studied and exploited in the high technology fields of biomedicine and biopharmaceuticals, for applications ranging from cell therapy to drug delivery. The characteristics of microencapsulation make it suitable for applications in the food industry, in particular for the production of high-value and nutraceutical foods.

Many encapsulation procedures have been proposed, but none of them can be considered as universally applicable procedure for bioactive food components. This is due to the fact that individual bioactive food components have their own characteristic molecular structures. However, compatibility with bioactives is not the only requirement an encapsulation procedure has to meet; it should also have specific characteristics to withstand influences from the environment (Augustin and Hemar, 2009).

An important requirement is that the encapsulation system has to protect the bioactive component from chemical degradation (e.g., oxidation or hydrolysis) to keep the bioactive component fully functional. A major obstacle in the efficacious delivery of bioactive food components is not only the hazardous events that occur during passage through the gastrointestinal tract but also the deleterious circumstances during storage in the product that serves as vehicle for the bioactive components. Many food components may interfere with the bioactivity of the added bioactive food component. It is therefore mandatory that the encapsulation procedure protects the bioactive component during the whole period of processing, storage, and transport. Another requirement is that the encapsulation system allows an efficient package load. How efficient this package load should be depends on the type of molecule that is desired as the bioactive component and the specific product that serves as the vehicle. Administration of large structures, such as probiotics will require a higher efficiency of package than molecular structures, such as vitamins. When choosing an encapsulation system with high-package efficiency, it is always essential to choose a system that can be easily incorporated into the food without interfering with the texture and taste of the food. And last, but certainly not least, it might be necessary to design the encapsulation system such that the bioactive component is released in a specific site of the gastrointestinal tract (de Vos et al., 2010).

The studies addresses a broad array of questions and challenges related to microencapsulation in four main research areas:

1. microencapsulating materials;
2. wall (matrix) materials for microencapsulation;
3. processes for microencapsulation;
4. properties and functionality of encapsulated systems.

Some studies have reported on the successes in encapsulating bioactive compounds. The most commonly used bioactive food molecules that have already been encapsulated in industrial applications are lipids, proteins, and carbohydrates. They cannot be easily dealt with in food products because of their extremely low solubility in water and polyunsaturated fatty acids, which are highly susceptible to oxidation, and are now widely applied in powdered products thanks to encapsulation processes that form an effective barrier to oxygen. Therefore, many different approaches to encapsulation have been proposed for the encapsulation of lipids to be able to apply them in a large variety of food products. Bioactive proteins also might require encapsulation.

Many food-derived peptides act as growth factors, immune regulator factors, antioxidant, or antiretinopathy agents. To exert a beneficial health effect, some of these proteins must reach the site of absorption in the small intestine in an intact configuration. Some peptides require hydrolysis in the stomach and the small intestine to release amino acids or specific bioactive peptides. Therefore, the intended type of protein to be encapsulated should be considered, including its effect on health and siel product that serves as a vehicle for the bioactive protein (de Vos et al., 2010). The carbohydrates that can benefit from microencapsulation are mainly the bioactive carbohydrates that are found in dietary fibers (Redgwell and Fischer, 2005). The fibers or their components that would benefit most from encapsulation are the soluble nondigestible polysaccharides. These fibers have been added for cholesterol reduction, reduction of glycemic fluctuations, prevention of constipation, prebiotic effects, and even for the cancer prevention (McClements et al., 2009). The main challenge in this area is not in targeting the fibers to specific parts of the gut but in increasing the amount of the fibers in food to achieve the aforementioned health benefits. The major encapsulation effort in this area is therefore improving the food load of fibers by packing enough fibers in capsules without interfering with product quality, such as changes in texture, mouth feel, or flavor.

4.3.2 Edible films and coatings

Edible films and coatings are any type of material used for enrobing (i.e., coating or wrapping) to extend shelf life and include various food products that may be eaten together with food (Pavlath and Orts, 2009). Edible films and coatings are applied to many products to control moisture transfer, gas exchange, or oxidation processes. For film-forming materials dispersed in aqueous solutions, solvent removal is required to achieve solid film formation and control of its properties. Edible films can be formed in two different ways: by immersion in biopolymers solubilized to form a film (solution casting) followed by evaporation of the solvent; and through a "dry process," which relies on the thermoplastic behavior exhibited by some proteins and polysaccharides at low moisture levels in compression molding and extrusion.

The important advantages of coating food with edible films are the ease of incorporating compounds that modify or favor nutritional or preservative properties, as well as those lengthen shelf life and decrease the growth or avoid the oxidation of pathogens on the surface of food, achieving the best functionality for the consumer.

Some studies report that the functionality of edible films is affected by the addition of bioactive compounds.

For example, Mei and Zhao (2003) evaluated edible films using milk protein with high concentrations of calcium (5% or 10% w/v) and vitamin E (0.1% or 0.2% w/v). Gómez-Estaca et al. (2007) studied the functional effect of edible film enriched with extracts of rosemary or oregano, a gelatin–chitosan film coating, and/or high-pressure processing, on the microbiological and oxidative stability of cold-smoked sardines. Park and Zhao (2004) reported that the water-barrier property of the chitosan-based films was improved by increasing the concentration of minerals (5%–20% w/v zinc lactate) or vitamin E in the film matrix.

Nevertheless, the tensile strength of the films was affected by the incorporation of high concentrations of calcium or vitamin E. Films enriched with oregano or rosemary extract were able to slow lipid oxidation, but they failed to slow microbial growth. Gómez-Guillén et al. (2007) obtained edible films with extracts of two ecotypes of murta leaves (*Ugni molinae* Turcz) using tuna-fish gelatin. The edible films of tuna-fish gelatin were transparent and demonstrated acceptable barrier properties for water vapor and UV light, as well as viscoelastic properties. In the case of films using Tintorera ecotypes, it was possible to significantly increase the antioxidant properties of the film when natural extracts with high content were added, producing only minor modifications of the film properties. When using an extract with a higher content of polyphenols, such as the blue-shark ecotype, the antioxidant capacity of the film increased, but the viscoelastic properties decreased due to the interaction between the proteins and polyphenols. Several authors have endeavored to improve the nutritional value of some fruits and vegetables through the incorporation of minerals, vitamins, and fatty acids in edible film, where these micronutrients are present in low quantities. Han (2002) indicates that coatings based on chitosan had the capacity to maintain high concentrations of calcium and of vitamin E, which significantly increased their content in fresh and frozen strawberries and red raspberries. Tapia et al. (2008) reported that the addition of ascorbic acid (1% w/v) to alginate- and gellan-based edible coatings helped keep the natural ascorbic acid content in papayas fresh, thus maintaining their nutritional quality during storage.

Tapia et al. (2008) developed the first edible films for probiotic coatings in fresh apples and papayas, noting that both fruits were successfully coated with alginate or gellan solutions, forming film that contained viable *Bifidobacterium* content of 106 ufc/g.

4.3.3 Vacuum impregnation

Vacuum impregnation (VI) has been useful to introduce solutes desirable in porous structures; a technology ideal for food, it modifies its original composition as a way to develop new products. Using this technique active compounds can be introduced into food without modifying its integrity. This technique differs from other methods of processing.

The use of VI is used to develop functional foods. On the one hand, several authors have used VI to modify the original nutritional composition of a porous food. Grass et al. (2003) assessed the fortification of eggplants and carrots with calcium using VI and oyster mushroom sucrose solutions, and noted that the levels of impregnation varied significantly with the variability of the raw material. Several authors studied the fortification of fresh apples with corn syrup and that of fruit and vegetables with calcium and zinc.

Fito et al. (2001) realized the feasibility of fortifying fruits and vegetables with vitamins and minerals using VI. They also evaluated the enrichment of apple products with vitamin E (100% fresh apple IDR/200 g).

Mathematical models have been developed to determine the concentration of various minerals in soaking solutions needed to achieve a strengthening of intake (per dietary reference intake, or DRI, recommendations) in the diet of 20%–25%, using 200 g of sample.

It is necessary that experimental validation confirms what models have predicted, that VI can be a good method for fruits and vegetables to be enriched with vitamins, minerals, and other compounds.

The impregnation of apples with calcium using VI and atmospheric pressure has been compared to determine the effect of these treatments on the behavior of the mechanical properties of the material. Others authors have compared the osmotic effect of dehydration in fruits fortified with minerals. Betoret et al. (2003) using the technique of VI to develop fruits enriched with probiotics, and also used slices of apple with *Lactobacillus casei* (*L. rhamnosus* spp.) in a concentration equivalent to dairy market products (106 cfu/g). Apple juice was fortified with *Saccharomyces cerevisiae* in a concentration of 107 or 108 cfu/mL.

More recently, some studies have focused on the protection that this technology can provide for bioactive compounds. Watanabe et al. (2011) studied the stability of anthocyanins in the preparation of strawberry jam made from strawberries impregnated with 0.29–1.46 mol/L of sucrose. The analysis showed that the final product had a greater amount of anthocyanins than the same product without impregnation.

4.3.4 Future trends

Referencing the conclusions of the European Commission Concerted Action on FUFOSE (European Commission, 1999), the future trends in development of food are as follows:

- Components in foods have the potential to modulate target functions in the body so as to enhance these functions and/or contribute toward reducing the risk of disease, and functional food science will contribute to human health in the future provided that evidence is supported by sound scientific, mostly human, data.
- Scientists, food engineers, and nutritionists have the possibility through the development of functional foods to offer beneficial opportunities related to health, welfare, and reduction in the risk of disease. The new nutrition depends strongly on the identification, characterization, measurement, and validation of markers that allow for a relevant design as defined earlier. The design of such studies still needs to be carefully analyzed and specifically developed by reference to, but differently from, classical clinical studies that have been elaborated to help in developing drugs, not food products.
- Major target functions in the body that are or can be modulated by specific food products will have to be identified or characterized. It is necessary to develop a science basis to understand these functions and how they relate to welfare and health, and in particular a methodical process must be developed to provide the necessary scientific basis to develop new functional-food products.
- Advances in food regulations, which are the means to guarantee the validity of claims, as well as food safety, will have to be made.

The challenge for nutritional science in the 21st century is development of functional foods that facilitate optimum nutrition. This challenge must be met by scientific evidence and not by marketing claims. The proper scientific validation of functional claims is critical to the success of functional foods, both for the benefit of human health and of the food industry.

5 Functional Food Product Design: Case Studies

5.1 Designing Dietetic Products by Using Aerated Foodstuffs

5.1.1 Stage 1: innovation and creativity

Bubbles are desirable elements in some foods, offering novel structures and textures, an attractive appearance, and improved volumes in culinary art. Aerated systems are thermodynamically unstable and in a fluid system will eventually oxidize. Controlling the size distribution of air bubbles and the spatial dispersion of the gaseous phase is crucial in controlling the quality of ice cream products. Mousses and soufflés are classic examples of aerated foods in which the incorporation and retention of bubbles are a critical factor in the success of the dish. An aerated structure can also facilitate mastication and enhance flavor delivery.

The original idea was to create a network of bubbles through a protein solution that would lower interfacial tension, promoting the formation of bubbles and creating a memory interface to stabilize the foam and retain as much air as possible inside the mixture. Meanwhile, the aerated product could potentially also achieve a reduction in caloric density and induce satiety through novel gastronomic structures. Common strategies for promoting healthy eating habits are reducing calories in the meal and controlling portion size, but satiety may not be reached through portion size alone.

Given the same volume of food, it may be advantageous to replace more energy-dense ingredients with equally satisfying flavor components. Different signals of satiety have been reported, including taste and texture (bouquet) as food moves through the gastric system (distention and clearing) and intestinal track (distention and absorption of nutrients).

5.1.2 Stage 2: product design

Protein-polysaccharide mixtures are widely used, as they play an essential role in the microstructure of many foodstuffs. Egg-white proteins and seaweed flour, for instance, are optimal ingredients for foam formation. By controlling the rheology of the aqueous phase, food polysaccharides, such as those in seaweed flour, reduce the thinning rate of aqueous films between bubbles and hence increase the stability of the foam. Formation of protein–polysaccharide complexes is sometimes related to an enhancement in functionality.

Our research was therefore focused on studying how seaweed-flour concentration and pH affected the surface properties of an egg-white protein solution (Fig. 7.8). The solution showed higher values in shear stress at pH 4.6 than at pH 7, suggesting that at pH 4.6 there was a substantial increase in the firmness of the foam. The egg-white protein and seaweed solutions at pH 4.6 gave more foam stability and produced smaller and more compact foams (Fig. 7.9A), whereas solutions at pH 7 made less stable foam and produce bigger bubbles

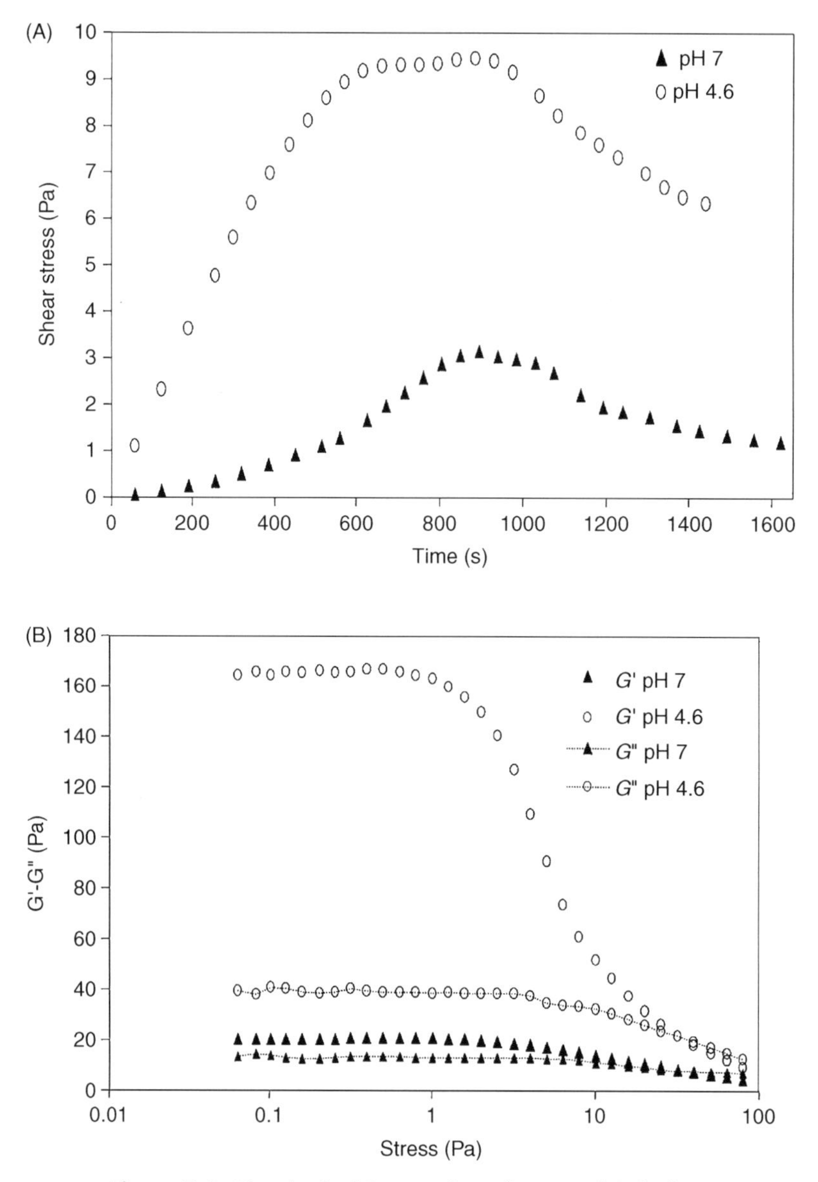

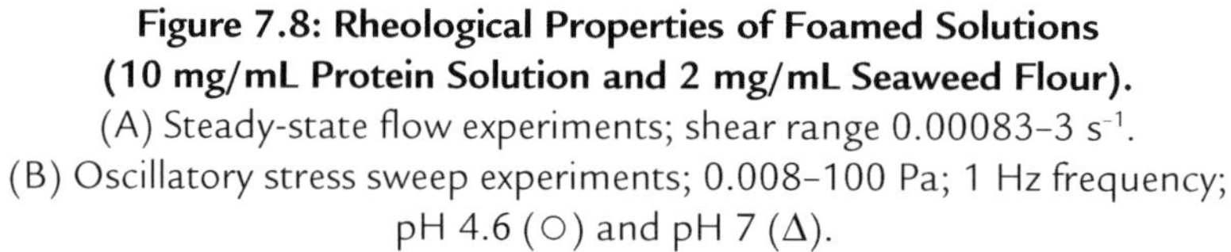

Figure 7.8: Rheological Properties of Foamed Solutions (10 mg/mL Protein Solution and 2 mg/mL Seaweed Flour).
(A) Steady-state flow experiments; shear range 0.00083–3 s^{-1}.
(B) Oscillatory stress sweep experiments; 0.008–100 Pa; 1 Hz frequency; pH 4.6 (○) and pH 7 (Δ).

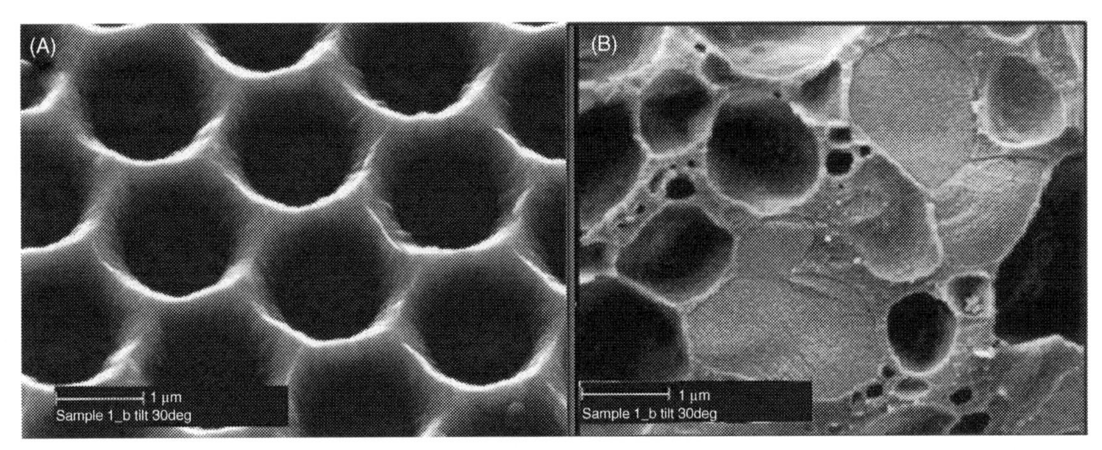

Figure 7.9: Microstructure of Ice Cream.
(A) Egg-white protein and seaweed solution at pH 4.6;
(B) egg-white protein and seaweed solution at pH 7.0.

(Fig. 7.9B). Using the knowledge gained from earlier studies, the next step in the design process is to create a product cell that induces a feeling of fullness, based on the principle that the satiation begins when first you see the product.

The recipe used ovalbumin, methylcellulose, maltodextrin, and amorphous silica, which helped with the reinforcement of the foam structure. Apart from the considerable interest in the effects of silica and selenium on human health, these partially hydrophobic particles are also known to stabilize interfaces, especially in the absence of any further surfactants, and increase stability in storage. Remarkably, stable foams are generated even from polymers that are liquid at room temperature and hence otherwise cannot be made into foams.

Regarding organoleptic requirements, the food aerated that was created seems to initially induce a feeling of fullness despite the reduction in product intake. There is some evidence that food intake is influenced by both the volume and weight of foods. It has been reported that increasing air content may reduce calorie intake from high-energy foods. Aroma is another key factor in the acceptance of foods by consumers.

The aromas stored in aerated products are rapidly released into the oral cavity during the mastication process. The addition of aromas to aerated food structures would provide new opportunities in terms of culinary ideas and product design. For instance, it has been suggested that the brain's response to a food odor sensed retronasally is related to satiation. The grade of the release of bouquet during consumption depends on the food's physical structure; solid foods produce a greater release of aroma than liquid foods. This fact implies that perceived satiation is increased by altering the extent of retronasal aroma release.

A step forward for this product might be the inclusion of different flavors by encapsulating aromas on each cavity.

The microstructure of these aerated dishes influences the rate of uptake into the body, and thus satiety. Reducing the rate of uptake of nutrients, such as sugars and glucids can also have a positive influence on the health of the consumer. In the case of ice cream, sugar was completely replaced by maltodextrin. By using this compound, we could obtain similar texture without having an intense taste of sweetness.

5.2 Industrial Approach

The same concepts that apply to the products just mentioned could easily be used for appropriate commercial health products in the food industry. Problems related to aerated food products are measuring their expansion, understanding their behavior, and translating this knowledge into commercial advantages. Designing aerated foodstuffs with customized textures, calorie reductions, and flavor properties will surely contribute toward developing new dietetic foods for the treatment of obesity. This knowledge could be a strategy for the design of food: keeping food energy dense with the organoleptic perception required while imperceptibly adding bubbles to the system to reduce the calories per portion.

Food microstructure has a significant influence on the feeling of fullness and various aspects of health when breakdown in the gastrointestinal tract is slowed. Some studies suggest that it is possible to increase the sense of satiety by altering the physical properties of meals, for instance, by increasing viscosity in the presence of some polysaccharides or fibers. Certain types of gels that are useful in controlling satiety and improving health, therefore, show important potential in terms of the development of functional foods. Combining these gels with air can create an edible-foam product that could be used in the treatment or prevention of obesity. The intake of the edible-foam product can produce a more apparent feeling of fullness than a similar product that does not have the in-mouth stability of the edible-foam product. In this sense, the use of the cells of gels could increase the feeling of fullness, but the effect of a cellular structure on the rate of assimilation in the gastrointestinal tract needs to be determined.

6 Conclusions

The first functional foods arose through the necessity of enriching the diet with vitamins or minerals. The first generation of functional foods involved the addition of trace minerals to appropriate foodstuffs, with cereals being a popular vehicle. Most current thinking and product development is centered specifically on gastrointestinal activity and microbial interactions.

Another key requirement for functional-food product design is a better understanding of the mechanisms underlying the functional behavior of a given food structure and how it affects consumer health. Including chefs in the food-product development process would have great benefits for the design of functional foods, with a focus on originality and health while retaining the cultural, social, and anthropological aspects that hugely affect consumer feelings of well being. This could encourage the food industry to seek new developments in functional foods. Chefs would be invaluable for the food industry, not only because they can be extremely useful in covering all the stages of the functional-food product design process, but also because of their ability to connect easily with the public. They could well be an interesting bridge for communicating the use of functional foods and a balanced diet in the pursuit of health to consumers. The food industry would do well to seek out such an integration that may be the key to unlock the next big revolutions in the food industry and in consumer market.

References

Ade-Omowaye, B., Angersbach, A., Taiwo, K., Knorr, D., 2001. Use of pulsed electric field pre-treatment to improve dehydration characteristics of plant based foods. Trends Food Sci. Technol. 12 (8), 285–295.

Alupului, A., Călinescu, I., Lavric, V., 2012. Microwave extraction of active principles from medicinal plants. UPB Sci. Bull. B 74 (2), 129–142.

Asghari, J., Ondruschka, B., Mazaheritehrani, M., 2011. Extraction of bioactive chemical compounds from the medicinal Asian plants by microware irradiation. J. Med. Plant Res. 5 (4), 495–506.

Augustin, M., Hemar, Y., 2009. Nano-and micro-structured assemblies for encapsulation of food ingredients. Chem. Soc. Rev. 38, 902–912.

Bartle, K.D., Cifford, A.A., Jafar, S.A., Shilstone, G.F., 1991. Solubilities of solids and liquids of low volatility in supercritical carbon dioxide. J. Phys. Chem. Ref. Data 20 (4), 756.

Betoret, N., Puente, L., Díaz, M., Pagán, M., García, M., Gras, M., 2003. Development of probiotic enriched dried fruits by vacuum impregnation. J. Food Eng. 56, 273–277.

Bhattacharjee, P., Singhal, R., Tiwari, S., 2006. Supercritical carbon dioxide extraction of cottonseed oil. J. Food Eng. 79 (3), 892–989.

Brunner, G., 2005. Supercritical fluids: technology and application to food processing. J. Food Eng. 67, 21–33.

Burgi, H., Supersaxo, Z., Selz, B., 1990. Iodine deficiency diseases in Switzeland one hundred years after Theodor Kocher's survery: a historical review with some new goitre prevalence data. Acta Endocrinol. 123, 577–590.

Chemat, F., Tomao, V., Virot, M., 2008. Ultrasound-assisted extraction in food analysis. In: Otles, S. (Ed.), Handbook of Food Analysis Instruments. CRC Press, Boca Raton, FL, pp. 85–94.

Chimowitz, E.H., Pennisi, K.J., 1986. Process synthesis concepts for supercritical gas extraction in the crossover region. AIChE J. 32 (10), 1665–1676.

Concha, J., Soto, C., Chamy, R., Zuñiga, M., 2004. Enzymatic pretreatment on rose-hip oil extraction: hydrolysis and pressing conditions. J. Am. Oil Chem. Soc. 81 (6), 549–552.

Cowan, M., 1999. Plant products as antimicrobial agents. Clin. Microbiol. Rev. 12 (4), 564–582.

Cravotto, G., Boffa, S., Mantegna, S., Perego, P., Avogadro, M., Cintas, P., 2008. Improved extraction of vegetables oils under high-intesity ultrasound and/or microwaves. Ultrason. Sonochem. 15 (5), 898–902.

de Vos, P., Faas, M., Spasojevic, M., Sikkema, J., 2010. Encapsulation for preservation of functionality and targeted delivery of bioactive food components. Int. Dairy J. 20, 292–302.

Domínguez, H., Núñez, M., Lema, M., 1995. Enzyme-assisted hexane extraction of soya bean oil. Food Chem. 54 (2), 223–231.

European Commission, 1997. Regulation (EC) No 258/97 of the European Parliament and of the Council of 27 January 1997 concerning novel foods and novel food ingredients. Regulation. Available from: http://eur-lex.europa.eu/legal-content/EN/ALL/?uri=CELEX:31997R0258

European Commission Concerted Action on Functional Food Science in Europe (FUFOSE), 1999. Scientific concepts of functional foods in Europe: consensus document. Report. Available from:

https://www.google.com/search?client=safari&rls=en&q=The+European+Commission+Concerted+Action+on+Functional+Food+Science+in+Europe&ie=UTF-8&oe=UTF-8

FAO and WHO, 2006. Guidelines on food fortification with micronutrients. Report. WHO Press, Geneva, Switzerland. Available from: http://www.who.int/nutrition/publications/micronutrients/GFF_Contents_en.pdf?ua=1

FAO and WHO, 2001. Human vitamin and mineral requirements. Report of a joint FAO/WHO expert consultation. FAO Food and Nutrition Division, Rome. Available from: http://www.fao.org/3/a-y2809e.pdf

Fito, P., Chiralt, A., Betoret, N., Gras, M., Chafer, M., Martinez-Monzó, J., 2001. Vacuum impregnation and osmotic dehydration in matrix engineering application in functional fresh food development. J. Food Eng. 49, 175–183.

Foster, N.R., Gurdial, G.S., Yun, J.S.L., Lons, K.K., Tilly, K.D., Ting, S.S.T., Singh, H., Lee, J.H., 1991. Significance of the crossover pressure in solid-supercritical fluid phase equilibra. Ind. Eng. Chem. Res. 30 (8), 1955–1964.

Gaspar, F., Lu, T., Marriott, R., Mellor, S., Watkinson, C., Al-Durin, B., Santos, R., Sevilla, J., 2003. Solubility of echium, borage, and lunaria seed oils in compressed CO_2. J. Chem. Eng. Data 48 (1), 107–109.

Ghaziaskar, H.S., Eskandari, H., Daneshfar, A., 2003. Solubility of 2-Ethyl-1-hexanol, 2-Ethylhexanoic acid, and their mixtures in supercritical carbon dioxide. J. Chem. Eng. Data 48 (2), 236–240.

Gómez-Estaca, J., Montero, P., Giménez, B., Gómez-Guillén, M., 2007. Effect of functional edible films and high pressure processing on microbial and oxidative spoilage in cold-smoked sardine (*Sardina pilchardus*). Food Chem. 105, 511–520.

Gómez-Guillén, M., Ihl, M., Bifani, V., Silva, A., Montero, P., 2007. Edible films made from tuna-fish gelatin with antioxidant extracts of two different murta ecotypes leaves (*Ugni molinae* Turcz). Food Hydrocoll. 21, 1133–1143.

Grass, M.L., Vidal, D., Betoret, N., Chiralt, A., Fito, P., 2003. Calcium fortification of vegetables by vacuum impregnation. J. Food Eng. 56, 279–284.

Güçlü-Üstündağ, Ö., Temelli, F., 2000. Correlating the solubility behaviour of fatty acids, mono-, di-, and triglycerides, and fatty acid esters in supercritical carbon dioxide. Ind. Eng. Chem. Res. 39 (12), 4756–4766.

Guderjan, M., Töpfl, S., Angersbach, A., Knorr, D., 2005. Impact of pulse electric field tratment on the recovery and quality of plant oils. J. Food Eng. 63 (3), 281–287.

Han, J.H., 2002. Protein-based edible films and coatings carrying antimicrobial agents. In: Gennadios, A. (Ed.), Protein-Based Films and Coatings. CRC Press, Boca Raton, FL, pp. 485–498.

Herrera, M.C., Luque de Castro, M.D., 2005. Ultrasound-assisted extraction of phenolic compounds from strawberries prior to liquid chromatographic separation and photodiode array ultraviolet detection. J. Chromatogr. 1, 1–7.

Johnston, K.P., Lemert, R., Kim, S., Wong, J., 1987. Multicomponent polar mixtures at supercritical fluid conditions. AIChE, New York.

King, J.W., Hill, Jr., H.H., Lee, M.L., 1993. Analytical supercritical fluid chromatography and extraction. Supplement and Cumulative Index. John Wiley & Sons, New York, pp. 1–83.

King, J.W., Srinivas, K., 2009. Multiple unit fluid processing using sub- and supercritical fluids. J. Supercrit. Fluid. 47, 598–610.

Laible, G., 2009. Enhancing livestock through genetic engineering—recent advances and future prospects. Comp. Immunol. Microbiol. Infect. Dis. 32, 123–137.

Laroze, L., Soto, C., Zuñiga, M., 2010. Phenolic antioxidants extraction from raspberry wastes assisted by enzymes. Electron. J. Biotechnol. 13, 6.

Latif, S., Anwar, F., 2009. Physicochemical studies of hemp (*Cannabis sativa*) seed oil using enzyme-assisted cold-pressing. Eur. J. Lipid Sci. Technol. 111 (10), 1042–1048.

Li, H., Chen, B., Yao, S., 2005. Application of ultrasonic technique for extracting chlorogenic acid from *Eucommia ulmodies* Oliv. (*E. ulmodies*). Ultrason. Sonochem. 12 (4), 295–300.

López, N., Puértolas, E., Condòn, S., Álvares, I., Raso, J., 2008. Effects of pulsed electric fields on the extraction of phenolic compounds during the fermentation of must of Tempranillo grapes. Innov. Food Sci. Emerg. Technol. 9 (4), 477–482.

Marine, D., Kimball, O.P., 1920. Prevention of simple goiter in man. Arch. Intern. Med. 25, 661–672.
Mason, T.J., Paniwnyk, L., Lorimer, J.P., 1996. The uses of ultrasound in food technology. Ultrason. Sonochem. 3, 253–260.
Matsushita, M., Tazinafo, N., Padre, R., Olivera, C., Souza, N., Visentrainer, J., 2007. Fatty acid profice of milk from Saanen goats fed a diet enriched with three vegetables oils. Small Ruminant Res. 72, 127–132.
McClements, D., Decker, E., Park, Y., 2009. Controlling lipid bioavailability through physicochemical and structural approaches. Crit. Rev. Food Sci. Nutr. 49, 48–67.
Mei, Y., Zhao, Y., 2003. Barrier and mechanical properties of milk protein-based edible films incorporated with nutraceuticals. J. Agric. Food Chem. 26, 1914–1918.
Murga, R., Sanz, M.T., Beltran, S., Cabezas, J.L., 2004. Solubility of syringic and vanillic acids in supercritical carbon dioxide. J. Chem. Eng. Data 49 (4), 779–782.
Nieto, A., Borull, F., Pocurull, E., Marcé, R.M., 2010. Pressurized liquid extraction: a useful technique to extract pharmaceuticals and personal-care product from sewage sludge. TrAC-Trend. Anal. Chem. 29 (7), 752–764.
Nik Norulaini, N.A., Md Zaidul, I.S., Anuar, O., Mohd Omar, A.K., 2004a. Supercritical enhancement for separation of lauric acid and oleic acid in palm kernel oil (PKO). Sep. Purif. Technol. 35 (1), 55–60.
Nik Norulaini, N.A., Md Zaidul, I.S., Anuar, O., Mohd Omar, A.K., 2004b. Supercritical reduction of lauric acid in palm kernel oil (PKO) to produce cocoa butter equivalent (CBE) fat. J. Chem. Eng. Jpn. 37 (2), 194–203.
Ozkal, S.G., Salgin, U., Yene, M.E., 2005. Supercritical carbon dioxide extraction of hazelnut oil. J. Food Eng. 69 (2), 217–223.
Paine, J., Shipton, C., Chaggar, S., Howells, R., Kennedy, M., Vernon, G., 2005. Improving the nutritional value of Golden Rice through increased pro-vitamin A content. Nature Biotechnol. 23, 482–487.
Pan, X., Niu, G., Liu, H., 2003. Microwave-assisted extraction of tea polyphenols and tea caffeine from green tea leaves. Chem. Eng. Process. 42, 129–133.
Park, S., Zhao, Y., 2004. Incorporation of a high concentration of mineral or vitamin into chitosan-based films. J. Agric. Food Chem. 52, 1933–1939.
Pavlath, A.E., Orts, W., 2009. Edible films and coating: why, what, and how? In: Huber, C., Embuscado, M. (Eds.), Edible Films and Coating for Food Applications. Springer, New York, pp. 1–23.
Puértolas, E., López, N., Sldaña, G., Álvarez, I., Raso, J., 2010. Evaluation of phenolic extraction during fermentation of red grapes treated by a continuous pulsed electric fields process at pilot-plant scale. J. Food Eng. 119 (3), 1063–1070.
Puri, M., Sharma, D., Barrow, C., 2012. Enzyme-assisted extraction of bioactives from plant. Trends Biotechnol. 30 (1), 37–44.
Redgwell, R., Fischer, M., 2005. Dietary fiber as a versatile food component: an industrial perspective. Mol. Nutr. Food Res. 49, 521–535.
Richter, B.E., Jone, B.A., Ezzell, J.L., Porter, N.L., Avdalovic, N., Pohl, C., 1996. Accelerated solvent extraction: a technology for sample preparation. Anal. Chem. 68 (6), 1033–1039.
Rostagno, M.A., Palma, M., Barroso, C.G., 2003. Ultrasound-assisted extraction of soy isoflavones. J. Chromatogr. 2, 119–128.
Samaniego-Vaesken, M.L., Alonso-Aperte, E., Valera-Moreira, G., 2010. Analysis and evaluation of voluntary folic acid fortification of breakfast cereal in the Spanish market. J. Food Comp. Anal. 23, 419–423.
Shen, J., Shao, X., 2005. A comparison of accelerated solvent extraction, Soxhlet extraction, and ultrasonic-assisted extraction for analysis of terpenoids and sterols in tobacco. Anal. Bioanal. Chem. 383 (6), 1003–1008.
Silva, L.V., Nelson, D.L., Drummond, M.F.B., Dufossé, L., Glória, M.B., 2005. Comparison of hydrodistillation methods for the deodorization of turmeric. Food Res. Int. 38 (8), 1087–1096, 9.
Staby, A., Mollerup, J., 1993. Separation of constituents of fish oil using supercritical fluids: a review of experimental solubility, extraction, and chromatographic data. Fluid Phase Equilibr. 91 (2), 349–386.
Suslick, K.S., Doktycz, S.J., 1990. The effects of ultrasound on solids. Mason, T.J. (Ed.), Advances in Sonochemistry, vol. 1, Jai Press, New York, pp. 197–230.
Tapia, M., Rojas-Grau, M., Carmona, A., Rodriguez, F., Soliva-Fortuny, R., Martin-Belloso, O., 2008. Use of alginate and gellan-based coatings for improving barrier, texture and nutritional properties of fresh-cut papaya. Food Hydrocoll. 22, 1493–1503.

Thies, C., 1987. Microencapsulation. In: Mark, H.F. (Ed.), Encyclopedia of Polymer Science and Engineering. John Wiley & Sons, New York, pp. 724–745.
Thies, C., 2005. A Survery of Microencapsulation Processes. In: Benita, S. (Ed.), Microencapsulation. Marcel Dekker, New York, pp. 1–20.
Vankar, P.S., 2004. Essential oils and fragrances from natural sources. Resonance 9 (4), 30–41.
Wang, L., Weller, C.L., 2006. Recent advances in extraction of nutraceuticals from plants. Trends Food Sci. Technol. 17 (6), 300–312.
Watanabe, Y., Yoshimoto, K., Okada, Y., Nomura, M., 2011. Effect of impregnation using sucrose solution on stability of anthocyanin in strawberry jam. LWT—Food Sci. Technol. 44 (4), 891–895.
Woods, V., Fearon, A., 2009. Dietary sources of unsatured fatty acids for animals and their transfer into meat, milk and eggs: a review. Livest. Sci. 126, 1–20.
Ye, X., Al-Babili, S., Kloti, A., Zhang, J., Lucca, P., Beyer, P., 2000. Engineering the provitamin A (beta-carotene) biosynthetic pathway into (carotenoid-free) rice endosperm. Science 287, 303–305.
Zhao, F.-J., Shewry, P.R., 2011. Recent developments in modifying crops and agronomic practice to improve human health. Food Policy 36, S94–S101.

Further Reading

Anino, S., Salvatori, D., Alzamora, S., 2006. Changes in calcium level and mechanical properties of apple tissue due to impregnation with calcium salts. Food Res. Int., 154–164.
Mukhopadhyay, M., 2000. Natural Extracts Using Supercritical Carbon Dioxide. CRC Press, Boca Raton, FL.
Rozzi, N.L., Singh, R.K., 2002. Supercritical fluids and the food industry. Compr. Rev. Food Sci. Food Safety, 33–34.

CHAPTER 8

Extracting Bioactive Compounds From Natural Sources Using Green High-Energy Approaches: Trends and Opportunities in Lab- and Large-Scale Applications

Thalia Tsiaka[*,,‡], Vassilia J. Sinanoglou[**], Panagiotis Zoumpoulakis[*,†,‡]**

[*]Institute of Biology, Medicinal Chemistry and Biotechnology, National Hellenic Research Foundation, Athens, Greece; [**]Technological Education Institution of Athens, Egaleo, Greece; [†]National Hellenic Research Foundation, Athens, Greece; [‡]University of Athens, Athens, Greece

1 Extraction: The Cornerstone for Recovery of Bioactive Compounds

Rephrasing the famous quote of Hippocrates, "Let food be thy medicine and medicine be thy food," modern scientists could state that natural products, either of plant, microbial, or animal origin, could not only be the food but also the medicine of present and future societies. Following the example of ancient world physicians and pharmacy practitioners, nowadays cosmetic and nutraceutical companies have focused their research to production and launching of active molecules from natural sources. Over the last 20 years, natural products and their extracts play a key role in almost every field of chemistry and biology, from microbiology and biochemistry to medicine and bioinformatics. Almost half of the products in the drug market are derived from bioactive natural compounds. According to projections, an increasing trend in the number of novel natural chemical substances, which could cover the needs of pharmaceuticals, food and cosmetic fields, is predicted due to the breakthroughs in in silico screening and drug discovery field (Azmir et al., 2013; Cragg and Newman, 2013; Harvey, 2008).

By the term "bioactive compounds," we refer mostly to secondary and some primary metabolites, which exhibit certain biological effects and act as functional ingredients, in low concentrations. A generalized classification, based on chemical structures, would assort them to a number of different groups, like phenolic acids and polyphenols, carotenoids, alkaloids, terpenes and terpenoids, tannins, anthocyanins, flavonoids, fatty

Ingredients Extraction by Physicochemical Methods in Food
http://dx.doi.org/10.1016/B978-0-12-811521-3.00008-9

acids (FA), and lipids, amino acids and proteins, polysaccharides and essential oils (EOs) (Azmir et al., 2013). Nature is an endless well of these molecules, which are present in different parts of vegetables, fruits, flowers, plants and herbs, animal products, wine and winery by-products, marine organisms, algae, bacteria, crustaceans, eggs, and plant oils (Fernández-Mar et al., 2012; Plaza et al., 2010). Furthermore, since the amount of waste produced by agroindustry and seafood sectors reaches excessive numbers, natural by-products consist of a really cheap source of biomolecules and also a sustainable, profitable, and ecocompatible idea for creating health-beneficial coproducts from residuals of natural origin (Chemat et al., 2012; Wijngaard et al., 2012).

But why are these molecules so important? Modern drug research is based on bioactive compounds from natural products since these components are known, among others, for their therapeutic effects including antimicrobial, antihypertensive, anticancer, cardioprotective, antioxidant, antidiabetic, neuroprotective, chemoprotective, antiaging, and immunoregulatory activity (Fernández-Mar et al., 2012; Ghasemzadeh and Ghasemzadeh, 2011). They also can act as functional food constituents, coloring, flavoring, and preserving food additives, fragrances, authenticity indices, and biomarkers in metabolomic pathways (Amorim-Carrilho et al., 2014). Recent studies show that their high bioavailability enhance their potential health benefits and this is the reason why countries that follow Mediterranean diet rich in bioactive content, present a lower percentage of morbidity and mortality caused by cancer, cardiovascular, and neurodegenerative disease (Cicerale et al., 2010).

Extraction process is the key point for successful recovery of bioactive compounds since ancient times. Its fundamental principal is based on the targeted retrieval of certain chemical groups from a solid or liquid matrix by a liquid solvent. Extraction efficiency is strongly related to variables, like extraction duration, pressure, temperature, solvent system, and substrate nature. Extraction techniques are of utmost significance in the workflow of a validated analytical technique as they are characterized as "sample preparation" methods. Research departments in pharmaceutical and cosmetic companies use extraction processes as the first step for the targeted isolation of compounds right after in silico studies and before biological activity tests that will allow the promotion of new formulations in the market (Azmir et al., 2013; Fig. 8.1).

A number of various extraction methods, from traditional procedures to contemporary nonconventional techniques, are recorded and used in modern chemistry applications. Classic extraction methods include classical diffusion through steering, maceration, distillation, and specific modifications of Soxhlet, Bligh-Dyer, and Folch procedures, while modern approaches enclose supercritical (SFE) and accelerated solvent extraction (ASE), also known as pressurized liquid extraction (PLE), ultrasound-assisted extraction (UAE), microwave-assisted extraction (MAE), subcritical water extraction (SWE), and pulsed electric field extraction (PEFE). Also liquid microextraction techniques, like solid phase microextraction (SPME), stir bar

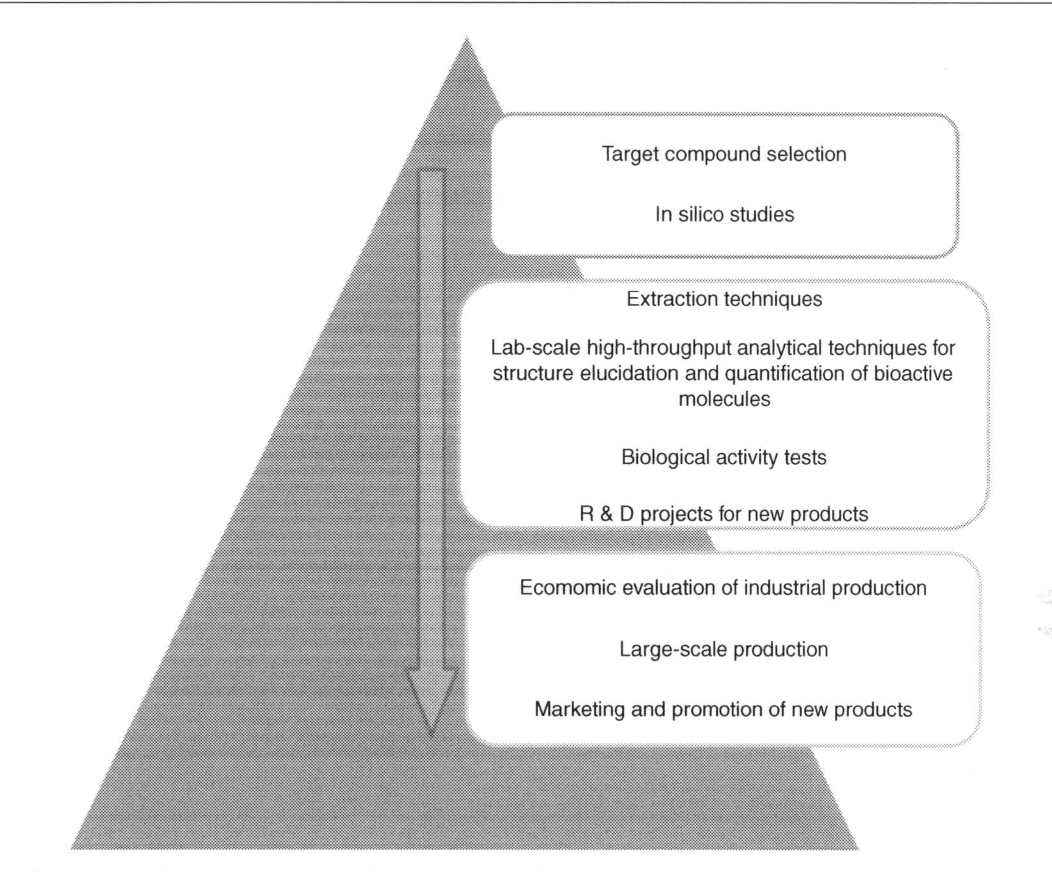

Figure 8.1: Bioactive Compounds From Natural Products: From Lab-Research to the Market.

sorptive extraction (SBSE), liquid phase microextraction (LPME), dispersive liquid–liquid microextraction (DLLME), extrusion, membrane-assisted extraction (hollow fiber renewal liquid membranes, HFRLMs/hollow fiber strip dispersion, PEHFSD), and enzymatic extraction belong to this group of extraction practices. The majority of these techniques refer to high-energy extraction, a fast and nonconventional way to give high amounts of energy to a chemical system. Some of these techniques, like SFE and PLE, require heating of the extraction mix in order to isolate target-molecules, while others, like UAE and MAE, are described as nonthermal extraction types. However, all these methods belong to what is called "green extraction practices," which are defined according to Chemat et al. (2012) as "extraction practices based on the discovery and design of extraction processes which will reduce energy consumption, allows use of alternative solvents and renewable natural products, and ensure a safe and high quality extract/product." Modern extraction techniques tend to sideline traditional methods as they are more selective, ecofriendly, fast, energy-saving processes with less solvent consumption and better extraction yields. As green practices, high-energy techniques comply also with other requirements of green chemistry, like instrumentation and industry extraction units diminution, automatization and monitoring of extraction processes, solvent recycling, exploitation of waste and residues by introducing innovative coproducts, and application of pioneering technologies. Nonetheless, contemporary research should focus on improving repeatability, accuracy, and reproducibility of modern techniques because classic extraction methods are still acknowledged and used in large-scale units as robust and reliable reference methods (Azmir et al., 2013; Chemat et al., 2012; Jeleń et al., 2012; Pabby and Sastre, 2013; Viñas et al., 2013).

Furthermore, there is an urgent need to validate and optimize every step of extraction techniques in order to standardize extraction conditions for each compound group and substrate and to deliver a cost-effective procedure and competitive final product. Experimental design (DOE), a multivariate set of mathematical algorithms and statistical analysis, is the most promising tool for achieving optimal extraction yields and maximum profits (Fig. 8.2). Through DOE models, simultaneous optimization of many factors in different value levels

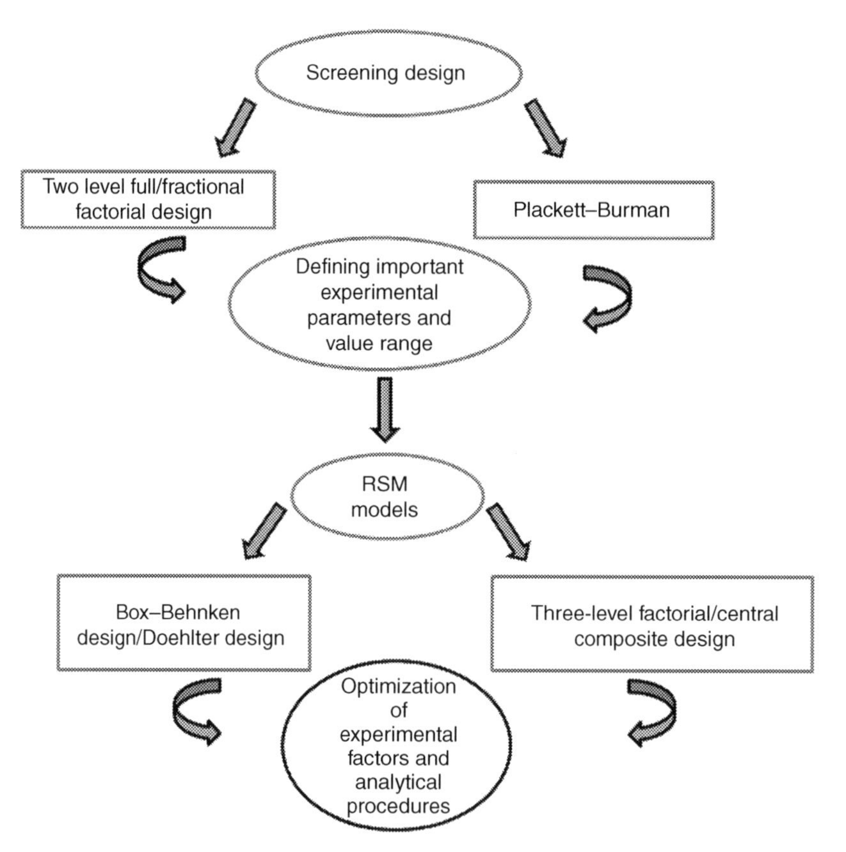

Figure 8.2: DOE Flowchart.

could take place by performing the minimum necessary number of experimental runs. The significance of the experimental factors and also of their interactions could be estimated by the proposed setup of the experiments. Therefore, a preliminary screening design, like Plackett–Burman or two-level full factorial design, could point out the important variables affecting crucially the measured experimental response and could define the experimental domain where optimal values of the investigated parameters lie. The next step after determining factors with the most critical effect and their value range is the optimization of the analytical procedure by response surface methodologies (RSM). Through RSM, where studied factors are usually set to three or more levels, the variable-value combination that results to the optimal experimental response could be extracted in less time and with higher accuracy compared to the one-variable-at-the-time (OVAT) optimization approach. Putting under the microscope DOE implementation in high-energy extraction field, extraction flow rate, solvent composition, extraction time, temperature and pressure of processes, amount of extraction energy, particle size, number of consecutively extractions, and solvent/material ratio could be considered as important independent variables under study, while extraction yield, extract purity, quantitative recovery of targeted compounds, total target-compound content and final extraction cost could be labeled as dependent variables or experimental responses (Dejaegher and Vander Heyden, 2011; Leardi, 2009).

Forming in line with the modern trade flow toward the flourishing use of health-beneficial natural products, this review aims to explore the potential of high-energy extraction methods for the recovery of bioactive molecules from natural sources and residues. The principals, instrumentation, advantages, and drawbacks of existing high-energy practices are presented by unfolding the array of their applications in different compound groups and natural matrices over the last 6 years. Also, the hyphenation of extraction processes with high throughput analytical techniques and powerful optimization statistical models for elucidation and quantification of bioactive targets in laboratory scale will be thoroughly investigated. In addition, future perspective of industrial-scale implementation will be scrutinized along with evaluation of environmental impact and brief economical assessment. To conclude, the aim of this review is to offer a detailed guide of high-energy extraction options and to propose new application fields in natural products chemistry for large-scale commercialization and promotion of food, drug, and cosmetic merchandize through the establishment of nonconventional extraction processes.

2 High-Energy Extraction: A Modern Approach for Recovering Bioactive Compounds

The 21st century inaugurated a new era for drug and food companies with the preface of products based on molecular structures of natural origin known for their reported functional and therapeutic effects. The group of nutraceuticals includes natural nonfood pharmaceutical formulations, like dietary supplements that contain high concentrations of bioactive molecules present initially in food substrates. Due to high amounts of bioactive compounds contained in these preparations, their health-promoting outcome is more intense compared to that of their original source. Additionally, they present minimum detrimental side effects set side-by-side with similar synthetic formulations. Foods or components of everyday diet, which demonstrate beneficial health effects besides their elevated nutritional value, are characterized as functional foods. Their advantageous properties are drastically correlated to bioactive compounds presence. Except for fruits and vegetables, the most consumed functional foods are meat and egg products, cereals, beverages and drinks, bakery products, and probiotic milk products. Polyphenols, such as anthocyanins, ellagitannins, coumarins, vegetable and marine oils, sterol esters, and carotenoids are nutraceutical functional constituents recognized for their antiinflammatory, antiviral, antihypertensive, antioxidative, and antineurodegenerative activity against chronic disease, like cancer, high cholesterol levels, Crohn's disease, cardiovascular disease, and lactose intolerance (Joana Gil-Chávez et al., 2013).

High-energy extraction methods are the building blocks of nutraceuticals and functional foods production since the amount of energy delivered to the system could cause matrix cell rupture and more quantitative release of the natural compounds. These techniques came as an environmentally friendly and sustainable remedy to the drawbacks of the old-fashioned techniques. Soxhlet extraction, maceration, hydrodistillation, and their modifications are used for more than a century as classic extraction approaches for the recovery of bioactive ingredients. All these methods rest on the molecular affinity between target compound and used solvent, therefore adequate solvent selection is of utmost importance. They are simple and inexpensive practices with no need of filtration, which results in quite high yields due to the continuous fresh solvent addition and the achieved high temperatures. Nevertheless, conventional extraction techniques are laborious as extraction process cannot be accelerated, environmentally unsafe, not fully automated, solvent-consuming methods that demand large amounts of expensive organic solvents and extraoperational units for solvent removal without providing protection to thermolabile compounds, like carotenoids, and high selectivity toward specific compound groups, along with active molecules, several impurities could be extracted. A comparative study of conventional and modern techniques is presented in Table 8.1 (Azmir et al., 2013; Wang and Weller, 2006; Wijngaard et al., 2013).

As high-energy extraction fulfills green chemistry concepts and practices, these techniques are emerging as the most suitable way for eliminating any traditional polluting extraction technology and for off-line waste management in industrial level (Armenta et al., 2008). The new extraction technology includes environmentally effective reduced-solvent methods. Among them SFE, ASE, or PLE and SWE, UAE, MAE, and solvent-free extraction are techniques of substantial significance because they have opened a new route for producing and commercializing natural-based products through a totally automated in-line procedure. Recent breakthroughs in extraction discipline promote the replacement of classic solvents,

Table 8.1: Comparison of conventional and nonconventional techniques.

Levels of Extraction Influential Factors	Conventional Techniques	Nonconventional Techniques
Extraction efficiency	High	High
Extraction time	Hours to days	μs to less than 1 h
Extraction parameters	Time, solvent system, solvent/material ratio, temperature, particle size, pH	Time, solvent system, solvent/material ratio, solvent flow, temperature, pressure, particle size, pH, energy power, extraction mode, sample moisture
Thermolabile compound protection	Low	High
Sample size	Medium to high	Medium to small
Operational cost	Low	Medium to high
Operational difficulty level	Low	Medium to high
Automatization	Low	High
Need for additional treatment steps	High	Low
Environmental impact	Energy- and solvent-intensive, negative environmental effect	Energy- and solvent-saving, ecofriendly

such as *n*-hexane and chlorinated solvents, with alternative green extractants. In silico predictive methods, like conductor-like screening Model for real solvents (COSMO-RS), quantitative structure–properties relationship (QSPR), and molecular modeling, constitute powerful tools because they can provide an accurate classification of industrial solvents and identify new extractants. Thus, according to computational characterization, the main categories of sustainable solvents are presented in Table 8.2 (Moity et al., 2014). CO_2, water, and ionic liquids are the ones commonly applied in high-energy extraction.

Table 8.2: Main groups of green solvents.

Green Solvents	Characteristics	Examples
Supercritical solvents	(+): Nonflammable, nontoxic, inert, recyclable	CO_2, water
Biosolvents	(+): Derived from renewable sources	Natural acids derivatives (i.e., lactic acid), bioethanol, vegetable methyl-esters (i.e., methyl soyate), terpenes
Ecofriendly solvents	(+): Solvents that fulfills three green principals: environment-human-safety (ESH)	Succinic and adipic acid, 3-methoxy-3-methyl-butan-1-ol (MMB), glutaric dimethyl, diethyl and dibutyl esters
Liquid polymers	(+): Nonhazardous, nonvolatile, biocompatible; (−): Biodegradable	Polyethylene glycol (PEG), polydimethylsiloxane (PDMS)
Fluorinated solvents	(+): Nonhazardous, nonflammable; (−): Biodegradable	Perfluorooctanate (C_8F_{18}), perfluorotributylamine ($C_{12}F_{27}N$)
Ionic liquids/Eutectic mixtures	(+): Thermostable, nonvolatile; (−): Questionable toxicity	Imidazolium salts, choline acetate

However, in order to proceed with a complete replacement of decade-old extraction methods, a fruitful combination of analytical chemistry skills, updated validation protocols, sophisticated optimization statistical analysis (DOE), and thorough surveys for economic feasibility are required.

2.1 High-Energy Extraction Techniques: Concepts and Instrumentation

2.1.1 Supercritical fluid extraction (SFE)

The innovative technology of supercritical and near-critical fluids eased significantly extraction processes in industrial level since supercritical solvents could provide residue-free high-added value products. The chemistry behind supercritical fluids (SF) is based on density-dependent properties of a multicomponent system induced at a state above their critical temperature and pressure. Carbon dioxide (CO_2), which is considered SF above 31.1°C and 7.38 MPa, could be an ideal solvent for supercritical applications related to thermosensitive substances. SFE extraction time varies from a few hours even to 20 min depending on target compound nature (Sahena et al., 2009).

Supercritical fluid extraction (SFE) (Fig. 8.3) is an attractive substitute due to SFs physical features, like thermal conductivity and heat capacity, which correspond to a phase between gas and liquid. Their ability to adapt to temperature, pressure, and composition alterations near critical points makes them adequate solvents for complex substrate cell rupture and selective extraction of thermolabile molecules since operating temperatures varies between 40 and 80°C. In addition, low performing temperatures of SFE could impede undesirable reactions of analytes, like hydrolysis, oxidation, rearrangement, and degradation, in extractor cells. When classic methods need up to 100 g of sample for quantitative recovery of biomolecules, SFE only requires 0.5–1.5 g. Their low surface tensions and viscosity allow penetration in solid matrices with low friction flow and therefore more sufficient

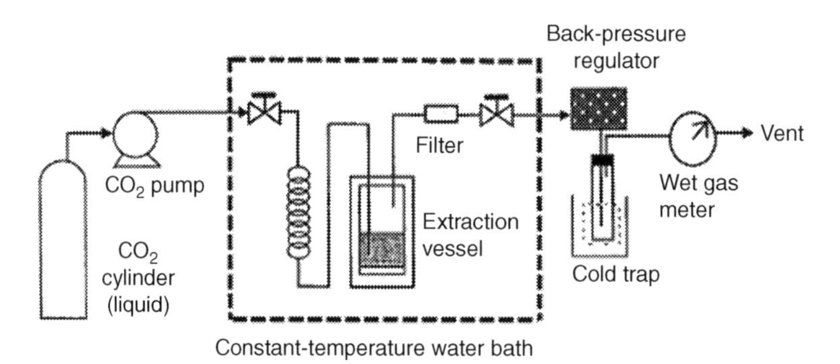

Figure 8.3: SFE Instrumentation (Markom et al., 2001).

and rapid mass transfer. High-density and diffusion coefficients increase target solubility. CO_2 is relatively miscible with hydrocarbons and other organic molecules, therefore CO_2-SFE shows high selectivity to certain groups of bioactive compounds, like low-weight molecules and lipids, and high-extraction yields. However, extractability of analytes with polar functional groups, amino acids, and sugars is almost negligible. It is worth mentioning that in the case of nonpolar solvents, process solubility and selectivity could be enhanced by using an appropriate cosolvent, like methanol or acetone. Compared to the rest of high-energy techniques, SFE identifies with a key strength of liquid–liquid extraction, constant solvent reload. Separation of solute and solvent mixture is less difficult and time-consuming in comparison with other methods because target compounds could be isolated by depressurization, a procedure where more than one fraction could be recovered sequentially by a high-density SF. Also, final extracts are heavy metal and inorganic salts free because they are not extracted even though they may be present in raw material. One of SFE greatest leverages is that fractions of different bioactive compounds could be obtained when different pressures and temperatures are applied. Also, SFE could be directly hyphenated with high-throughput analytical techniques, such as GC-MS and LC-MS. These facts make SFE a very tempting alternative for natural-based pharmaceutical and food companies. In terms of economic figures, start-up capital and operational costs are factors that should be taken under consideration in selecting SFE as an extraction method in production chain (Pereda et al., 2007; Sahena et al., 2009).

Two major steps of SFE procedure is extraction and separation of compounds of interest. SFE plants consist of extractors, where bioactive compounds from solid matrices are recovered when supercritical solvents flow uniformly in compressed form into the platform through pumps. Before SF enter extraction cell, it is preheated by passing through a heat exchanger. Normally, extraction cells are immersed in water baths to stabilize their temperature. Solvent-extract mixture is transferred from extraction cell to flash tanks, where pressure decreased quickly by insenthalpic expansion, which causes fluid density minimization and solvent power reduction. The recovery of analytes from SF could also take place by adjusting SFE temperature. In the case that compound solubility increases with temperature reduction, a decrease in temperature if the pressure is kept constant, will remove soluble molecules from SF. Another way to filtrate extract from solvent is the use of an absorbent as an auxiliary agent, which is a less expensive but sometimes more complicated procedure. Heavier and lighter fractions of valuable compounds could be obtained simultaneously by applying high pressures, from 40 to 60 MPa. Through depressurization of high-pressure solvents more than one fraction could be diluted in SF. Pure extract is acquired after precipitation into the flash tanks, while SF follows a recycling-reuse process after cooling, recompression, and storage in special tanks. Drastic decrease of waste production due to solvent recycling increases SFE application range in industrial level (Jesus and Meireles, 2014; Mendes et al., 2003).

2.1.2 Accelerated solvent extraction (ASE)-subcritical water extraction (SWE)

ASE, which is also popular with the synonym denominations of PLE, pressurized solvent extraction (PSE), and enhanced solvent extraction (ESE), was first developed by Dionex company in the 1990s for environmental analysis purposes. The theory behind ASE comprises a liquid solvent extraction performed in relatively high pressure (3.5–20 MPa) and temperature (25–200°C) with 5–15 min extraction time for every extraction cycle. When the extractant is water, ASE is defined as SWE or pressurized hot water extraction (PHWE) (Fig. 8.4). This method is an intriguing option because it is based on water's unique physical properties. SWE refers to water extraction when this fluid is at a liquid state at temperature over its boiling point (100°C) and up to its critical temperature (374°C) under pressures from 0.1 to 22.1 MPa for avoiding water vaporization. This practice could also be described by the terms of superheated water extraction, pressurized low polarity water extraction, high-temperature water extraction, and hot liquid water extraction (Mustafa and Turner, 2011).

ASE operational units function in two different modes, the static mode and the dynamic mode. Static extraction, where solvent is replaced in every extraction cycle, is commonly used in industrial scale. Extraction yield is influenced by time, temperature, which is adjusted by an oven, a heating tape or a heating jacket, analyte solubility, and particle size. Extraction cycles improve extraction efficiency as one-cycle extraction may not be quantitative due to

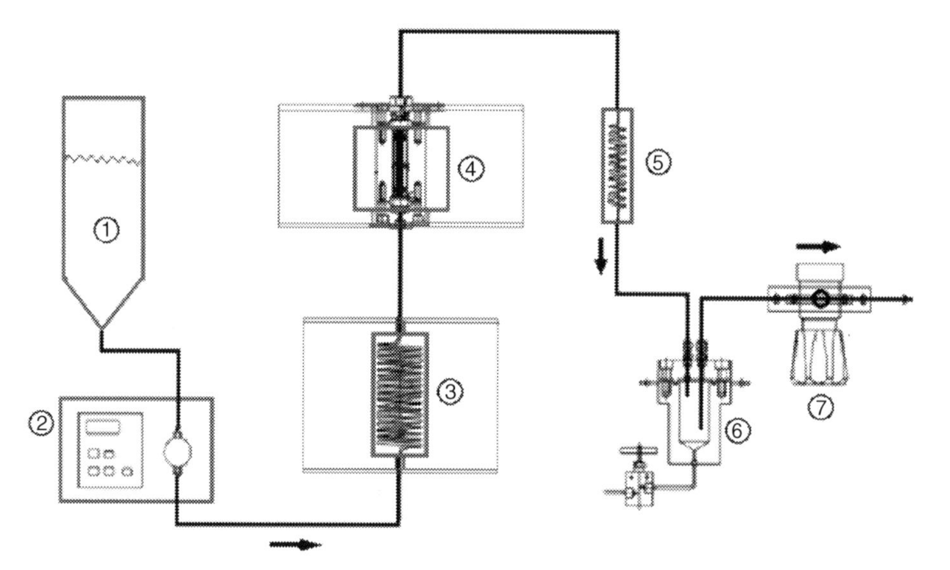

Figure 8.4: SWE Instrumentation (1) water tank; (2) high-pressure pump; (3) pre-heater; (4) main heater with a reactor; (5) chiller; (6) solid/water separator; (7) back press regulator. (Islam et al., 2012).

the low solvent volumes used in each cycle. Extraction cells are usually autoclave vessels, where a stirrer accelerates mass transfer and extraction rate. In higher temperature and smaller particle size, extraction yield is normally higher since long diffusion into matrix paths and firm solute–matrix interactions are reduced. Replenishment of extraction solvent is also critical because it can favor analyte partitioning to extraction solvent instead of the matrix. In dynamic mode, the solvent is continuously inserted into extraction cells, manufactured from stainless steel. A pump channels high pressure to the system and preserves its flow constant by using a pressure restrictor. An HPLC column can operate as an extraction chamber when the target compounds are thermolabile molecules and moderate conditions are required. Although a plain dynamic mode platform is not commercially available, a static-dynamic combination setup, where fresh solvent is pumped into the extraction vessel, is currently in the market. After equilibrating pressure and temperature at set values, static mode is on for a fixed time period and predetermined extraction cycles number. The ultimate step of each batch is extract introduction in inert gas, normally nitrogen, and cleaning of solvent remains before the extract's final collection in order to prevent analyte losses and carryover effects. In many cases, extraction outcome is enhanced by using auxiliary agents. Dehydration agents, like hydromatrix, an inert diatomaceous earth, can provide dry moisture-free solid substrates for increasing extraction yield. Dispersion agents, like glass beads, diminish solvent consumption in extraction vessels by filling up and decreasing the extra space. The latest innovations in material science promote the substitution of stainless steel cells, which are not compatible with extreme pH values, with dionium vessels, which is pH hardy. The operational difficulty of ASE is quite comparable to SFE because its layout is analogous to the platforms used in this technique. A future application of this technology could permit in-vessel reactions for substrate treatment, such as enzymatic reactions or saponification of lipids (Mustafa and Turner, 2011; Plaza and Turner, 2015).

Concerning SWE, this method draws attention from traditional techniques due to water's physical properties. The most essential of them is dielectric constant (ε) variability caused by temperature changes. Due to dielectric constant changeability, water polarity, density, and viscosity vary affecting bioactive molecules desorption from substrates and, as a result, extraction speed, selectivity, and efficiency. Thus, water presents tunable solvent power and it can extract polar molecules when it behaves as a polar solvent at room temperatures but also hydrophobic and lipophilic organic compounds at temperatures more than 250°C, when dielectric constant value is really close to that of volatile alcohols, like methanol or ethanol ($\varepsilon = 33$). In the case of SWE, normally, lower flow rates prolong extraction time and result to higher yields of relatively moderate thermolabile compounds, while higher flow rates lead to less time-consuming and possibly more cost-effective procedures. Nonpolar compounds are less soluble with the increase of their carbon chain, the side chain oxygenated groups and the decrease of conjugated bonds. According to recent studies, SWE is more efficient method when raw moisture-containing samples are processed compared to dry lyophilized matrix. Over the last years, SWE is in the first line of analytical extraction techniques as water is the most ecocompatible, nontoxic, and food-grade solvent with inexpensive removal costs compared to apolar organic solvents. Food and pharmaceutical industry have made a turn to SWE because it is an organic solvent-free technique adequate for high-quality nutraceutical merchandise production with high-extraction yields of health-beneficial compounds. At the same time, SWE requires the smallest initial quantities of solid matrices compared to techniques, such as maceration or micellar extraction. Nonetheless, SWE is not a very suitable technique for thermosensitive compounds and for extractable molecules that require high pressures. In addition, SWE free-solvent nature could be compromised in the case of lipid constituents' extraction because after extract cooling, these ingredients are not miscible with water and therefore an amount of organic solvent is necessary for their retrieval (Herrero et al., 2006; Petigny et al., 2015; Zakaria and Kamal, 2016).

2.1.3 Ultrasound-assisted extraction (UAE)

Over the last decades, ultrasounds (US) stand out as a powerful tool in food-processing industry, which shows more and more concern to protect nutritional value, quality, and biofunctionality of nutraceutical products. US are mechanical waves of frequencies over 16 kHz, boundary for audible frequencies. Their application spectrum includes emulsification, homogenization, crystallization, low-temperature pasteurization, defoaming, dewatering, particle-size reduction, and changing viscosity, degassing, activation and inactivation of enzymes. However, UAE, first studied in the 1950s, has managed to intensify the extraction of a large number of bioactive ingredients by speeding up extraction rate and by augmenting extraction quality (Soria and Villamiel, 2010).

UAE theoretical background relies on acoustic cavitation phenomenon, which is based on contraction and rarefaction of ultrasonic gas bubbles incited on matrix molecules during the propagation of an ultrasound wave that penetrates it. During ultrasonication cycles, bubbles reach to a critical size, where they are no longer stable due to gas diffusion and they break down by releasing high amounts of energy and turbulence at extreme temperatures (~5000°C) and pressures (100 MPa). US intensity and energy plus vapor pressure, surface tension, temperature, and solvent viscosity influence cavitation magnitude. The shear forces of microstreaming and microjetting effects produced by cavitation could facilitate solvent penetration for cell rupture and release of active compounds from matrix to extraction solvent. Categorized by frequency range, US technology is divided in two groups. High-frequency (100 kHz–1 MHz) low-power ($<1\ W\ cm^{-2}$) low-intensity US are normally taking part in quality assessment, such as physicochemical properties evaluation (acidity, firmness, sugar content, ripeness), as nondestructive technique. Low-frequency (16–100 kHz) high-power ($10–1000\ W\ cm^{-2}$) high-intensity is the type used in UAE applications (Picó, 2013). UAE could reach to an extraction yield comparable to that of dated techniques within 10–90 min and could support higher sample throughput. Furthermore, this procedure is performed with generally recognized-as-safe (GRAS) solvents in relatively low processing temperatures, which cannot cause structural or functional alterations to analytes of interest. Therefore, UAE is an ideal extraction type for an environmental safe and economically remunerative recovery of heat-sensitive molecules. Operational simplicity, high reproducibility, energy effectiveness, and high automatization level count up to the UAE advantages (Chandrapala et al., 2013).

UAE instrumentation is quite simple as it only consists of the extraction medium, an energy generator and a transducer for energy conversion to its acoustic form. An indispensable apparatus in every laboratory is the ultrasonic water bath, which most of the times operates only at a two-frequency mode (40 and 56 kHz). In this case, sonication is indirect as acoustic waves need to be transmitted from the water to the sample through sample vessels. Present technological improvements deliver ultrasonic baths with multifrequency setup and also sonoreactors, cup, and microplate horns, which supply energies 50 times higher than baths in sealed matrix tubes. Nevertheless, a complete extraction of bioactive constituents from complex natural matrices requires more advanced and energy-boosting platforms. Ultrasonic probe (Fig. 8.5) is a state-of-art equipment in this field because it can provide high extraction yields within a few minutes due to the great amounts of energy (100-fold higher than water bath energy) that emancipate directly to extractant/matrix solution. Nowadays, companies' engineering departments manufacture different types of probes, such as titanium alloy probes, spiral probes, silica-glass probes, dual and multiprobes. In both instrumentations, temperature could be stabilized in controlled values. With ultrasonic probe, extraction intensity could be regulated in different intended levels (Picó, 2013).

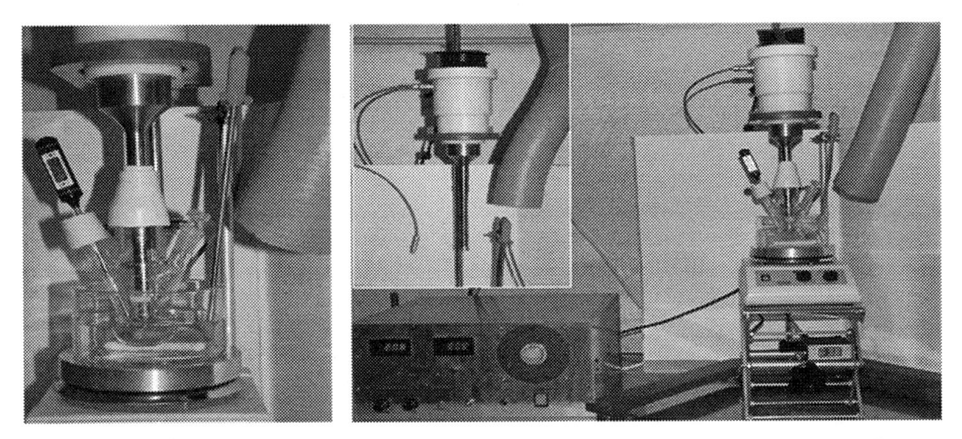

Figure 8.5: UAE Probe Instrumentation (Sonics and Materials INC., Vibra-Cell VCX 750 (20 kHz, 750 W), ultrasonics processor, equipped with piezoelectric converter and 13 mm diameter probe fabricated from titanium alloy Ti-6Al-4V.

2.1.4 Microwave-assisted extraction (MAE)

Electromagnetic waves with frequencies between 300 MHz and 300 GHz are accountable for MAE. MAE mechanism is based on microwave (MW) heating, which is produced by the synergistic combination of ionic conduction and dipole rotation. The physical principals behind these two phenomena are responsible for converting electromagnetic to thermal energy, which is transferred directly to the extractable material via molecular interactions of MW and polar solvents. Dielectric solvents or solvents with permanent dipole properties interact strongly with MW. Dielectric constant (ε), dielectric loss (ε'') and dissipation factor (tan δ) are estimators of solvents MW heating capacity. The aforementioned indices represent the ability of a solvent to absorb MW, to transform MW to thermal energy and to warm under MW, respectively. Higher dielectric constant and dissipation factor values indicate more sufficient MW absorption and higher extraction rates. Polar solvents, like water, which have high dielectric constant at lower dissipation value, could absorb amounts of energy higher than that it can deplete. The extreme induced temperatures provoke superheating effect inside the matrix that results to cell collapse, maximum diffusivity, and solute release into the extractant. As a typical high-energy extraction example MAE gathers all the advantages of these techniques (short extraction times, low solvent consumption, high-extraction yields). Although it can be used for thermolabile compounds extraction, MAE could be performed in temperatures near solvent boiling point and for that reason there is a possibility for compounds biofunctionality loss. Nonetheless, MAE could favor bioactive compounds quantitative isolation (Jain et al., 2009; Veggi et al., 2013).

Four major components constitute MAE devices, a magnetron or MW generator, a wave guide for MW diffusion, an applicator for samples, and a circulator for forward MW movement. Homogenous MW distribution is achieved by beam reflectors. Two different MAE types are developed for laboratory and large-scale applications. Closed vessel MAE is accomplished in controlled temperature and pressure, while open vessel or focused MAE is based on the solvent reflux and on sample-focused MW irradiation. Both MAE modes are available as single- and multimode devices. Closed vessel MAE is a pressurized system that requires less solvent volumes and operates in higher temperatures resulting in efficient recovery of volatile compounds. Despite higher yields, high pressures increase operational risk. Additionally, closed mode vials do not permit large amounts of material processing. Solvent addition during extraction and atmospheric pressure, make focused mode (Fig. 8.6) a more secure option. Instrumentation setup is simple and can be programmed to elaborate larger matrix quantities. Precision issues, lack of simultaneous sample processing, and longer extraction times are the main pitfalls of focused MAE (Tatke and Jaiswal, 2011). Despite disadvantages, like extra filtration or centrifugation step, poorer recoveries of nonpolar or volatile compounds, and a risk of thermosensitive molecules deterioration, MAE is one of the most widely applicable high-energy techniques due to its competitive advantages, such as more than 60% energy saving, compact systems, and high recoveries, compared to other methods (Veggi et al., 2013).

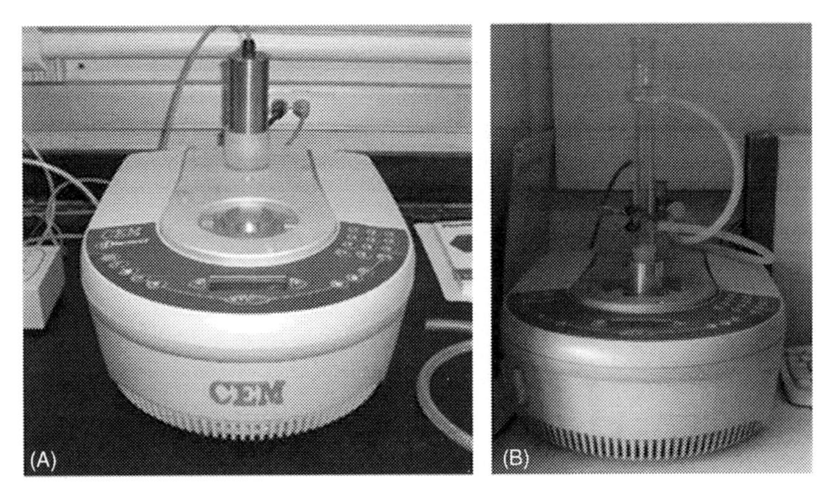

Figure 8.6: Closed- (A) and Open- (B)Vessel MAE Instrumentation (Discover LabMate, CEM).

2.1.5 Solvent-free extraction

Health and environment are liable to being harmed from industrial disposal of volatile organic extraction solvents, like *n*-hexane. A greener economically feasible answer is solvent-free extraction. Solvent-free microwave extraction (SFME), instant controlled pressure drop (DIC) methods, and PEFE are the most known solvent-free techniques (Abert Vian et al., 2014). Under the umbrella of SFME fall techniques, such as pressurized SFME (PSFME), vacuum microwave hydro distillation (VMHD), improved SFME (ISFME), microwave steam diffusion (MSDf), and also versions of microwave hydrodiffusion and gravity (MHG), and microwave dry-diffusion and gravity (MDG) (Li et al., 2013b).

ISFME is an advanced SFME modification, which includes the addition of a MW-absorbing agent, that is, graphite and activated carbon powders. In VMHD, MW processing is combined with sequential vacuum implementation. MSD provides EOs condensed extracts, while MSDf finds application in EOs extraction from natural by-products. PSFME is a SFME at pressures higher than atmospheric. MHG is based on MW principals only in this method, cell rupture and release of analytes takes place without any solvent under thermal energy generated by MW at atmospheric pressure, a phenomenon called hydrodiffusion. Compounds' diffusion outside matrix is followed by gravity dropping out of MW reactor. Although MHG implementation is yet at an early stage, it catches scientific world attention as a nonsolvent technique. A more sophisticated version of MHG, called vacuum MHG (VMHG) is already evolving. As matrix moisture is an essential component of MHG, dry samples could be processed by MDG (Chemat et al., 2010; Li et al., 2013b; Zill-E-Huma et al., 2011). DIC, the most newly developed technique, provokes rapid cell disruption by injecting high-temperature and high-pressure steam into the matrix. Sudden pressure decrease causes autovaporization of matrix-containing moisture, which lead to substrate texture and porosity changes. Short durations of extreme conditions employment does not destroy bioactive molecules providing pure valuable extracts. High-quality products are DIC's prime advantage. Besides volatile compounds extraction, DIC has been used in bacterial decontamination, treatment of water-logged-archaeological wood and swell-drying (Abert Vian et al., 2014; Besombes et al., 2010). PEFE technology adopted electroporation concepts. PEFE driving force is based on formation and charging of cell membranes pores due to moderate and high intensity (10–40 kV cm^{-1}) electric field application during really short period pulses (μs–ms). As biological membranes become electrically perforated and more permeable, intracellular target compounds migrate out of tissues (Grimi et al., 2007; Puértolas et al., 2013).

In general, SFME device includes a multimode MW reactor and an IR temperature sensor. Supplementary equipment is added according to SFME modifications. DIC setup consists of a processing vessel, which is a heated autoclave where sample is introduced, a vacuum tank-vacuum pump system, which is cooled for extract condensation and permits saturated steam influx under regulated pressure, and an instant valve between autoclave and vacuum parts, which is used for pressure vent. PEFE equipment contains two electrodes, a pulse generator and a nonconductor that adjusts pulse sequences of any type, which can be batch or continuous. Solvent-free high-energy extraction exhibits tremendous possibilities for a wider nonpilot application as free-solvent procedures assist nonresidue and safer scale-up of these techniques by eliminating the risk of explosion and overpressure accidents (Abert Vian et al., 2014; Fig. 8.7).

2.1.6 Coupled high-energy techniques

As the demand for natural-based products marks an exponential ascent, an even more fruitful scale-up of the existing extraction methods is top priority for massive production. A promising scheme toward this direction could be the synergistic combination of the already developed techniques in order to enhance extraction outcome by overcoming their implicit flaws.

Sustainability and ecoefficiency of large-scale processes is strongly supported by the replacement of hazardous organic solvents from ionic liquids (ILs) state-of-art technology, commonly known as molten salts. A vast number of salts that remain in their liquid phase at temperatures under 100°C due to their quiet large size and high-molecular asymmetry constitute the group of ILs. Their intriguing characteristics arise from their pure ionic nature, which is formed from organic cations (i.e., ammonium, pyridinium, imidazolium phosphonium, pyrrolidinium) and both inorganic (Cl^-, PF_6^-, BF_4^-) and organic anions (trifluoromethylsulfonate $[CF_3SO_3]^-$, *bis*[(trifluoromethyl)sulfonyl]imide $[(CF_3SO_2)_2N]^-$, trifluoroethanoate $[CF_3CO_2]^-$). Their ionic character is responsible for their advantageous properties, for example, ecoviability, thermal stability, almost zero vapor pressure and low volatility, electrolytic conductivity, flexibility in viscosity and miscibility values and liquid types, ability for fit-to-purpose IL synthesis, recycling and reuse, inflammability, distinct

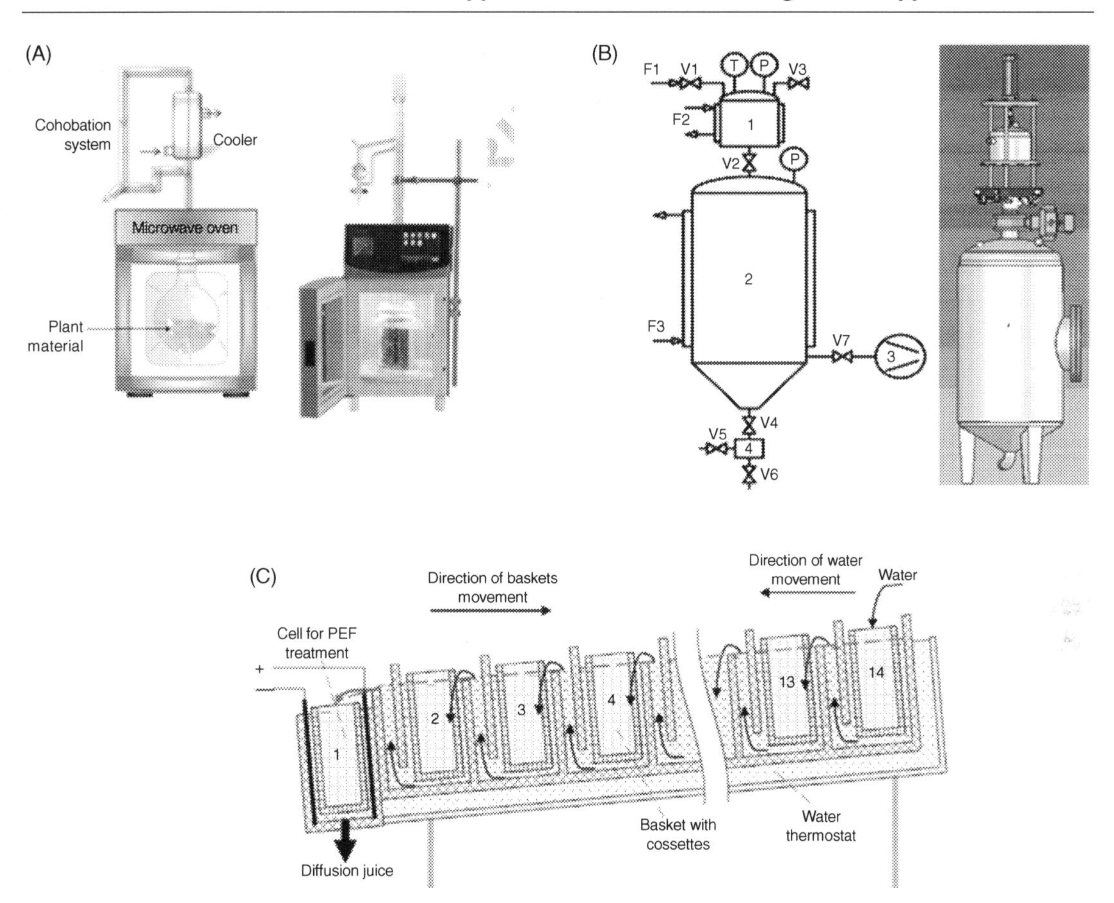

Figure 8.7: (A) SFME instrumentation; (B) DIC instrumentation (1) autoclave with heating jacket, (V2) rapid valve, (2) vacuum tank with cooling water jacket, (3) vacuum pump, (4) extract container, F1 and F2 steam flow, F3 cooling water flow; (C) PEFE instrumentation (Loginova et al., 2011).

physicochemical properties based on cation–anion combination, and coupling with almost every known analytical technique (Sun and Armstrong, 2010). Deep eutectic solvents (DESs), which are naturally composed multimixtures between hydrogen bond donors and hydrogen bond acceptors with lower melting points than their forming components, are also a breakthrough in the green substitution of classic volatile solvents because they are biodegradable, cheap, easily made mediums. DES examples are urea mixtures (choline chloride-urea, 2:5 mixture of L-proline and glycerol), quaternary ammonium halide salts and phosphonium halide salts (Nam et al., 2015; Pena-Pereira and Namieśnik, 2014). Despite the fact that both technologies are still in their early stages, they are considered among the most auspicious suggestions for viable extraction scale-up of bioactive components as they have incorporated in classic liquid extraction and microextraction (Ho et al., 2011; Trujillo-Rodríguez et al., 2013). They are also effectively hyphenated with

ASE (Wu et al., 2012), UAE (Bi et al., 2010; Cvjetko Bubalo et al., 2016; Ma et al., 2011; Nam et al., 2015; Yilmaz and Soylak, 2015; Zhang et al., 2014a; Zu et al., 2012), MAE (Cui et al., 2015; Cvjetko Bubalo et al., 2016; Liu et al., 2011; Ma et al., 2010; Wei et al., 2013, 2015a) or platforms of multiple high energy techniques, that is, UAE-MAE-ILs (Lou et al., 2012; Lu et al., 2012).

Since Soxhlet extraction still stands as a reliable approach for in-factory extraction, there is a serious attempt for modern modifications in order to cut down extraction time and surpass recurring disadvantages. High-pressure Soxhlet extraction simulates to SFE. Although solvents do not reach their critical conditions, they preserve their liquid character in high pressures (~7–10 MPa). Automated Soxhlet devices combine a dual-mode extractor with Soxhlet heating and a reflux system. UAE benefits could be integrated to Soxhlet extraction, when a US probe offers energy through a thermostated water bath, where Soxhlet flask is immersed. Anyhow, Soxhlet-MAE coupling is the most commercially feasible platform. In Soxhlet-MAE setup, extraction is ensued at atmospheric pressures and focused MW mode, while extraction basis is established on Soxhlet principals (Zhou et al., 2012). Although these combinations have not yet caught the eye of industrial community, faster process, enriched final extracts, lower solvent consumption, automation, extraction cycles reduction, and quantitative recoveries are the highlights of Soxhlet-high energy methods hyphenation (Luque de Castro and Priego-Capote, 2010).

Apart from novel extractants and Soxhlet-based modernized techniques, combination of high-energy methods in a single-step process or in a sequential continuous procedure empowers the positive effects of individual techniques. Lab studies worldwide have deployed equipment and principals of US and MW (UMAE) (Chen et al., 2010; Li et al., 2014; Lianfu and Zelong, 2008; You et al., 2014), US and SF (UAE-SFE (UASFE) (Dias et al., 2016; Klejdus et al., 2010; Santos et al., 2015; Tang et al., 2010; Yang et al., 2014, 2013), and MAE-SFE (Lu et al., 2014b), UAE-DLLME (Sereshti et al., 2014, 2013, 2011), and solid-phase (SPE)-high energy extraction (UAE-SPE, UAE-SPE-ASE, SFE-SPE, ASE-SPE, MAE-SPE) for natural products recovery. UMAE results to faster heating due to MW energy and more efficient analyte release due to US effect. UASFE higher yields are attributed to improved mass transfer due to US cavitation and diffusion impact on SF properties. DLLME is coupled with MAE and UAE usually for extracting and determining metal, amines types, and hydrocarbons. SPE conjoining in extraction layout contribute to sample cleanup and preconcentration and therefore it enhances procedure selectivity by reducing at the same time total analysis steps. SPE is more adaptable to ASE since the use of organic pressurized solvent activates better the solid phase material. Howbeit, one of the greatest achievements of combined high-energy methods is the acceleration or initiation of targeted chemical reactions, that is, hydrolysis concurrently with the extraction process. Notwithstanding, further studies concerning the increased number of critical extraction factors and system complexity is a vital action for commercialization of coupled techniques (Rostagno et al., 2010). Typical examples of combinatory/hyphenated extraction are exhibited in Table 8.3.

Table 8.3: Combinatory/hyphenated extraction techniques.

Technique	Matrix	Target Compounds	Reference
MAE-Soxhlet	Ginseng	Pesticides	Zhou et al. (2012)
ILs-DLLME	Water, biological fluids, cosmetics, wine, agricultural products (i.e., honey), pharmaceutical formulations, milk, and infant formulas	Pesticides, parabens, phenols, metals, aflatoxins, antimicrobial compounds	Trujillo-Rodríguez et al. (2013)
ILs-ASE	Medicinal plants	Rutin, quercetin	Wu et al. (2012)
ILs-UAE	Fruits	Lignans	Ma et al. (2011)
	Rosemary	Carnosic acid, rosmarinic acid	Zu et al. (2012)
	Shrimp waste	Astaxanthin	Bi et al. (2010)
ILs-MAE	Lotus leaves	Alkaloids	Ma et al. (2010)
	Rosemary	Carnosic acid, rosmarinic acid, EOs	Liu et al. (2011) Wei et al. (2013)
	Pigeon pea leaves	Flavanones Stilbenois	
ILs-UAE-MAE	Burdock leaves	Phenolic compounds	Lou et al. (2012)
	Galla chinensis herb	Tannins	Lu et al. (2012)
DES-UAE	Sheep, bovine, chicken liver	Metals	Yilmaz and Soylak (2015)
	Shrimp by-products	Astaxanthin	Zhang et al. (2014a)
	Flos sophorae herb	Quercetin, kaempferol, isorhamnetin glycosides	Nam et al. (2015)
	Grape skin	Phenolics	Cvjetko Bubalo et al. (2016)
DES-MAE	Pigeon pea roots	Genistin-genistein apigenin	Cui et al. (2015)
	Radix scutellariae plant	Flavonoids	Wei et al. (2015b)
	Grape skin	Phenolics	Cvjetko Bubalo et al. (2016)
UMAE	Tomatoes	Lycopene	Lianfu and Zelong (2008)
	Inonotus obliquus fungus, *Tricholoma mongolicum Imai* mushroom	Polysaccharides	Chen et al. (2010); You et al. (2014)
	Gelsemium elegans plant	Indole alkaloids	Lu et al. (2014b)
UASFE	*Capsicum* peppers	Capsaicinoids, phenolics	Santos et al. (2015) Dias et al. (2016)
	Algae	Isoflavones	Klejdus et al. (2010)
	Clove buds	Oils	Yang et al. (2014)
	Tea	Caffeine	Tang et al. (2010)
	Scutellaria barbata D. Don-Hedyotis diffusa plants	Oleanolic and ursolic acid	Yang et al. (2013); Wei et al. (2015a)
	Blackberry bagasse	Antioxidant compounds	Pasquel Reátegui et al. (2014)
MAE-SFE	*Gynura segetum* plant	Alkaloids	Lu et al. (2014a)
UAE-DLLME	*Oliveria decumbens* Vent.	EOs	Sereshti et al. (2011)
	Saffron-tea plants	Volatiles	Sereshti et al. (2013, 2014)
	Edgeworthia chrysantha Lindl flower	EOs	Wen et al. (2014)
UAE-SPE	Pharmaceutical samples	Thymol and carvacrol	Roosta et al. (2015)

3 Experimental Design (DOE) in the Spotlight of Optimization Strategies

The alpha and omega of industrial online production is the optimization of each procedure step in order to support a viable remunerative outcome. Classic optimization approaches, like one-variable-at-time (OVAT), a univariate method where each factor is optimized individually by keeping the rest of the parameters at a predefined constant value, are an obsolete experimental planning. OVAT unsuitability is ascribed to the large number of experimental runs needed to the lack of estimating the interactions between the experimental factors and the deficiency of determining the global optimum since the acquired data only provide information about the points where the experiments were conducted (Ebrahimi-Najafabadi et al., 2014). Instead of applying this underachieving approach, laboratories, and corporations use multivariate optimization strategies, known as experimental design (DOE) in order to improve effectiveness. These statistical tools allow the simultaneous optimization of more than one experimental parameter or independent factor at several predefined value levels and estimation of their interactions effects on more than one experimental response or dependent factor by performing the minimum necessary number of experimental runs. Built on mathematical algorithms and transformations, DOE supplies accurate, reproducible, and validated models through the interpretation of statistical indices (R^2, R^2_{adjusted}, MS, ANOVA-test, P-value, lack of fit), charts and plots (Pareto chart, predicted vs. experimental value charts, contour plots, 3D surface plots). Usually, a DOE setup consists of two sequential designs. Screening models, for example, two-level full or fractional factorial (2^{k-n}, where 2 applies to the number of value levels, k to the number of experimental independent factors and n to the size of fraction used from full factorial matrix) and Plackett–Burman designs, are applied for a preliminary screening of the experimental variables and their values aiming at finding the parameters with crucial effects (P-value ≤ 0.05) on the response and confining the range of near-to-optimum experimental values. Both full and fractional factorial designs could identify major parameters and their interactions effect, while Plackett–Burman could estimate only the important main effects of numerous factors by performing really few experimental runs. Therefore, this model is basically used for LC/GC-MS optimization purposes. On the other hand, factorial designs can be applied when the number of experimental variables is less than 16 because multiplying factors numbers increase significantly total experimental runs. Concerning extraction methods, full factorial designs are highly recommended as by reckoning in all interactions terms, a more reliable model could be produced (Sharif et al., 2014). Screening designs conclusions point out the right direction of optimization models, otherwise noted as response surface models (RSM) in terms of important factors and value range. RSM are three level-minimum models, adequate for determining the exact optimal conditions of a process. The most popular RSM designs are symmetrical three-level full factorial, central composite (CCD), Doehlert and Box–Behnken (BBD) designs, asymmetrical D-optimal and Kennan-Stone algorithm designs and mixture designs, like the three-component simplex lattice design used for solvent mixture composition optimization (Dejaegher and Vander Heyden, 2011). CCD and BBD are the most ubiquitous models in extraction applications. Each step of a DOE planning is illustrated in Fig. 8.2. In high-energy extraction fields, DOE is applied for optimization of various critical parameters illustrated in Table 8.4.

Table 8.4: Common extraction parameters optimized by DOE.

Experimental Parameters	SFE	ASE-SWE	UAE	MAE
Pressure	+	+	−	±*
Temperature	+	+	+	+
Extraction solvent system	+	+	+	+
Extraction time	+	+	+	+
Energy power	−	−	+	+
Solvent/material ratio	+	+	+	+
Particle size	+	+	+	+
Moisture content	+	+	−	+
Solvent flow rate	+	+	−	−
pH	−	+	−	−
Modifiers addition	−	+	−	−
Extraction mode	+	+	−	+

+, Closed-vessel mode; −, open-vessel mode.

More explicitly, higher pressures are applied for keeping solvent in the liquid form and for facilitating the traversal of the analytes through matrix pores. Elevated temperatures increase mass transfer and compound solubility, but they may affect negatively extraction selectivity. Especially for bioactive molecules extraction, temperatures less than 100°C should be preferred due to compounds thermosensitivity (Chandrapala et al., 2013; Sun et al., 2012).

Extraction solvent systems depend crucially not only on the group of extracted compounds and its affinity to the solvent, on the extractant properties, like polarity, density, boiling point, and toxicity but also on the principals of each extraction method (Carr et al., 2011). In UAE, acoustic cavitation is intensified when extraction solvents are liquids with low viscosity, which enable cell penetration, and low vapor pressure. Notwithstanding, low vapor pressure solvents could develop locally really extreme pressure and temperature causing molecule degradation. Thus, medium vapor pressure solvents with high-molecular affinity to the target molecules are selected (Dey and Rathod, 2013; Ruen-ngam et al., 2011). MAE extraction solvent systems include solvents that absorb MW energy, mixtures of solvents with high- and low-dissipation factors and MW transparent solvents with matrices of high dielectric loss. Among them, mixtures of low and high absorbing ability, like hexane and ethanol, give better yields because their heating is milder and varies according to their exact composition. Facing the possible deterioration of heat-sensitive components, use of very absorbing MW solvents, that is, water, leads to shorter exposure times, but in general, longer duration renders better results (Tatke and Jaiswal, 2011).

Extraction time effect is almost similar to that of temperature. Although, efficient yields are obtained in very short exposure times, higher extraction periods provide higher extraction yields. For instance, in UAE with ultrasonic probe optimal extraction time is less than 30 min. However, optimal extraction time should be a compromise between extraction yields and compounds susceptibility to thermodegradation (Lianfu and Zelong, 2008; Tatke and Jaiswal, 2011). Maximum mass transfer, solvent diffusion, and matrix disruption is achieved normally at higher (UAE) or with a succession of low and moderate power (MAE) and at higher solvent/material ratio. While higher solvent/material ratios increase extraction yields, they could extend analysis time by adding up extra steps, like solvent evaporation (Tatke and Jaiswal, 2011; Wang et al., 2013; Xu and Pan, 2013). Smaller matrix particle size maximizes contact surface between substrate and solvent and thus leads to higher extraction recoveries. Higher flow rates reduce target compounds dwelling time in high temperatures and therefore side reactions. Since flow rates affect significantly analyte kinetics, their optimal value should be established in order to improve extraction yield. On the other hand, an optimal flow rate controls overdilution of compounds of interest that could take place in excessively high flow rates (Plaza and Turner, 2015). Modifiers addition is commonly recommended in ASE cases. Strong acids and bases along with fluids with autoignition temperature of 40–200°C (carbon disulfide, diethyl ether, and 1,4-dioxane) cannot be used in ASE. Addition of inert modifiers, such as hydromatrix, sodium sulfate, quartz sands, and basic alumina, could be essential to extraction efficiency improvement due to better solute dispersion. Uniform dispersion could be also achieved with agitation that blocks agglomerates formation. Organic and inorganic agents improve solubility by enhancing analyte–solvent interactions (Sun et al., 2012).

Parameters with minor effects that also need to be set to predetermined values for higher extraction yields, especially in the case of more automated techniques like SFE and ASE, are preheating time (~5 min), flush volume for moving away all analytes before new extraction cycle (~60%), and purge time. Optimization strategies should also be applied in order to amend extraction kinetics (i.e., analyte desorption from substrate, analyte diffusion into the matrix, solute dissolution into the solvent, and target compound elution) (Plaza and Turner, 2015; Sun et al., 2012). Normally, a pulsed sonication in UAE is preferred from a continuous mode for energy-saving reasons. A zero-ramping time value in MAE reduces extraction time without any significant losses of target compounds (Tsiaka et al., 2015). Extraction mode is pretty much a deciding factor when we refer to ASE and MAE. Regarding ASE, even though extraction conditions in static mode are stable, high-concentrated samples, or low solubility analytes could be recovered completely if extraction process is repeated several times. A faster and better extraction recovery could be ensued in dynamic mode due to the variations in extraction process but higher solvent volumes are an obstacle for its wide application. A static-dynamic extraction mode could be an optimal solution in terms of profit since it combines lower solvent volumes of static extraction and shorter extraction times of dynamic mode. By succeeding more intense extract suspension in shorter periods, the dual model eludes solute degradation and results to higher extraction yields. In MAE, open-vessel mode is the more applicable version because it is compatible with a wide range of extracted molecules with the exception of volatile components, which are generally extracted by closed-vessel systems to avoid possible losses (Sun et al., 2012; Tatke and Jaiswal, 2011). Overall, the same group of bioactive molecules could necessitate totally different extraction conditions depending on substrate nature and optimal conditions could differ according to the aim of the process. In some cases, extraction objective is higher yields and in others, maximal bioactivity of extracted compounds. Table 8.5 presents some of the most recent indicative examples of DOE models used in high-energy extraction optimization.

Table 8.5: DOE models examples in high-energy extraction applications.

DOE Models	Extraction Technique	Substrate—Target Compounds	References
Two-level full factorial designs	SFE	Rosemary leaves—carnosol and carnosic acid	Ramandi et al. (2011)
		Borage flower—FAs	Caldera et al. (2012)
	ASE	Kaki, peaches, apricots—carotenoids	Zaghdoudi et al. (2015)
	PSFME	Sea buckthorn—antioxidants	Michel et al. (2011)
	MAE	*Cicerbita alpina* plant—phenolics/flavonoids	Alexandru et al. (2013)
	UAE	Citrus peel—EOs/antioxidants	Omar et al. (2013)
	UAE-DLLME	*O. decumbens* Vent. plant—EOs	Sereshti et al. (2011)
	SFMAE	Ginger—EOs	Shah and Garg (2014)
Two-level fractional factorial designs	SFE	Rapeseed cake—oil and lipids	Uquiche et al. (2012)
		Summer savory—EOs	Khajeh (2011)
	ASE	Rice grains—phenolics	Setyaningsih et al. (2016)
Plackett-Burman	ASE	Olive leaves—oleuropein	Xynos et al. (2014)
		Brown algae *Eisenia bicyclis*—fucoxanthin	Shang et al. (2011)
		White grape marc—polyphenols	Álvarez-Casas et al. (2014)
	UAE	Haskap berries—anthocyanins	Celli et al. (2015)
Three-level factorial designs	ASE	Drumstick tree leaves—antioxidants	Rodriguez-Perez et al. (2016)
	UAE	Olive leaf—antioxidants	Şahin and Şamlı (2013)
CCD	SFE	Sunflower oil—FAs	Rai et al. (2016)
		Propolis—polyphenols	de Zordi et al. (2014)
		Flixweed—EOs/FAs	Ara et al. (2015)
	ASE-SWE	Paprika leaves—lutein	Kang et al. (2016)
		Amaranthus seeds—lipids	Kraujalis et al. (2013)
		Grape pomace—phenolics	Rajha et al. (2014)
		Chili peppers—capsacinoids	Bajer et al. (2015)
	UAE	White button mushrooms—polysaccharides	Tian et al. (2012)
		Basil—antioxidants	Izadiyan and Hemmateenejad (2016)
		Clove—EOs	Tekin et al. (2015)
		Sunflower oil—carotenoids	Li et al. (2013a)
		Grapefruit—flavonoids	Garcia-Castello et al. (2015)
	MAE	Sour cherry—anthocyanins/phenolic acids	Garofulić et al. (2013)
		Rice grains—phenolics	Setyaningsih et al. (2015)
		Spent espresso coffee grounds-antioxidants	Ranic et al. (2014)
		Agaricus bisporus mushrooms—ergosterol	Heleno et al. (2016)
		Gac fruit aril—lipids/FAs	Kha et al. (2013)
	ILs-MAE	Ge gen plant—isoflavones	Zhang et al. (2014b)
	UAE-DLLME	Saffron—volatile compounds	Sereshti et al. (2014)
BBD	SFE	Tea leaves—flavonoids	Prakash Maran et al. (2013)
		H. pluvialis algae—astaxanthin	Reyes et al. (2014)
		Spent coffee grounds—oil/diterpenes	Barbosa et al. (2014)
		Muskmelon seeds—oil	Maran and Priya (2015)
	SFE-PHW	Lingzhi mushroom—β-glucan	Benito-Román et al. (2016)

(*Continued*)

Table 8.5: DOE models examples in high-energy extraction applications. (*cont.*)

DOE Models	Extraction Technique	Substrate–Target Compounds	References
	ASE-SWE	Fennel–EOs/estragole	Rodríguez-Solana et al. (2014)
		Hullless barley bran–β-glucan	Du et al., 2014
		Saffron petals–phenolics	Ahmadian-Kouchaksaraie et al. (2016)
		Artichoke external bracts–carbohydrates	Ruiz-Aceituno et al. (2016)
		Blackcurrant–polysaccharides	Xu et al. (2016)
	UAE	Onion solid waste–polyphenols	Katsampa et al. (2015)
		Amino acids–grape	Carrera et al. (2015)
		Shrimp waste–carotenoids	Tsiaka et al. (2015)
		Pomegranate peel–pectin	Moorthy et al. (2015)
		Blueberry wine pomace-phenolics/ anthocyanins	He et al. (2016)
		Spinach–lutein/zeaxanthin	Altemimi et al. (2015)
		Sugar beet molasses–phenolics/ anthocyanins	Chen et al. (2015b)
	MAE	Mulberry leaves/waste mango peel–polysaccharides	Thirugnanasambandham et al. (2015)/Maran et al. (2015)
		Ashoka tree barklignans	Mishra and Aeri (2015)
		Almond skin byproducts–phenolics	Valdés et al. (2015)
		Tomato–phenolics/flavonoids	Pinela et al. (2016)
		Sour orange peel–pectin	Hosseini et al. (2016)
		Flax seed cake–polyphenols/ antioxidants	Teh et al. (2015a)
	UAE-ILs	Amur cork tree-alkaloids	Wang et al. (2015)
		Turmeric–antioxidants	Xu et al. (2015)
	VMAE-ILs	*Sorbus tianschanica* plant–flavonoids	Gu et al. (2016)
	UAE-DLLME-DES	Olive/almond/sesame/cinammon oil–ferulic/caffeic/cinammic acid	Khezeli et al. (2016)
Doehlert design	ASE-SWE	Carrob kibbles–phenolic/ carbohydrates	Almanasrah et al. (2015)
		Stevia leaves–steviol glycosides	Jentzer et al. (2015)
		Soybean flour-isoflavones	Benjamin et al. (2017)
	UAE	Morning glory flowers–alkaloids	Nowak et al. (2016)
	MAE	Sheanut kernels–oil	Nde et al. (2016)
		Neem–oil	Nde et al. (2015)
D-optimal design	UAE	Virgin olive oil waste–phenolics	Icyer et al. (2016)
	MAE	Lime bagasse–pectin	Thirugnanasambandham and Sivakumar (2015)

4 Bioactive Compounds Recovery: A High-Energy-Oriented Approach for Extracting Them From Natural Sources

Since high-energy extraction techniques are a fast ecocompatible solution, which processes' layout require lower solvent quantities and provide quantitative extraction yields without any special pretreatment, they are gaining momentum in biofunctional compounds' extraction field. Their application range broadens as these methods shield molecules susceptible to light and oxygen degradation. They could also be a money- and time-saving alternative for extraction of fats, polyunsaturated and monounsaturated FAs, lipids, triglycerides, vegetable and fish oils, EOs, phenolic compounds and polyphenols, tannins, anthocyanins and flavonoids, carotenoids and other natural pigments, vitamins and terpenoids, sterols and polysaccharides, capsaicinoids, alkaloids and curcuminoids from almost every type of natural sources, from spices and plants to marine organisms and algae (macro- and microalgae, cyanobacteria, invertebrates, crustaceans) (Chemat and Cravotto, 2013; Clodoveo et al., 2013; Ibañez et al., 2012).

Among the four main techniques described in this review, SFE could also provide information for the nutritional value and labeling of fat-containing products, could generate high-quality perfumes in aroma industry and could be used as a cleanup step for obtaining cholesterol-free lipid fractions. Pharmaceutical corporations use SFE as a tool for microencapsulation, bacterial and polymeric material sterilization, particle, tissue and crystal engineering, biotechnology, foaming and solvent removal (Martinez and Vance, 2007; Sahena et al., 2009). ASE is preferred for extracting lignans from grain, cereals, and seeds, carotenoids from high-lipid foods, isoflavones and xanthones from plants, fruits, and vegetables, FAs and lipids from oils, EOs and tannins from plants and medicinal herbs. ASE is considered a really suitable procedure for food safety applications based on determination of veterinary drugs (sulfonamides, tetracyclines, and antibiotics, benzimidazoles, and barbiturate molecules, heterocycle amines) all types of pesticides, organochlorines, alkylphenols, and bisphenols, polycyclic aromatic hydrocarbons (PAHs), and organometallic complexes (Mustafa and Turner, 2011; Sun et al., 2012). As far as it concerns SWE practices, although this technique has not found yet a mass appeal in industrial scale, it can be applied for extraction of EOs from herbs (thyme, cilantro, fennel), phenolic compound glycosides and flavonoids from vegetables and grains (potato, rice, onions), fruits (pomegranate, citrus fruits, mango), herbs and spices (sea buckthorn, cinnamon, tea, ginseng, oregano, rosemary, clove), poly-/monosaccharides and carbohydrates, β-glucan from mushrooms FAs, carotenoids from vegetable oils (palm, corn, soybean, coconut, sunflower oil), pectin and sugars from fruits (apple, citrus fruits, coconut, grape), volatile aroma compounds from herbs (savory, rosemary, peppermint), steviosides and diterpenes, capsaicin from peppers, mannitol from olive leaves, terpene derivatives from ginkgo biloba and PAHs. Algae (*Haematococcus pluvialis*, *Chlorella vulgaris*, *Cystoseira abiesmarina*, *Spirulina platensis*), a sustainable natural source for antioxidants structures, have also been used as SWE matrix (Petigny et al., 2015; Zakaria and Kamal, 2016). On the other hand, UAE and MAE are gaining popularity for extraction of thermosensitive components of natural origin as two of the most viable techniques in commercial-scale applications because they require less invested capital and they present an operational simplicity compared to ASE and SFE (Zhang et al., 2011). Between others they have been used from extracting proteins, lipids, pigments, antioxidants, metals, polysaccharides, volatile compounds, foam stabilizers, preservatives, chemical contaminants, and pollutants. US could be applied for enzyme activation and deactivation, for molecular weight reduction of biopolymers, like chitin, for antioxidant capacity enhancement, and for lipophilic molecules encapsulation, when MAE is widely applied also for pasteurization and sterilization purposes (Chan et al., 2011; Vilkhu et al., 2011). While SFE, ASE, UAE, and MAE entrench their position in large-scale extraction procedures, PEF, DLLME, ILs, and DES are still in the early stages of their extraction-based implementation.

It will only take a few minutes for someone to find out that the number of peer-reviewed reports related to high-energy extraction of biofunctional molecules has been soared especially the last 5 years. To have a small taste from the enormous variety of high-energy targeted extraction, some of its principal applications are presented in Tables 8.6 and 8.7.

Alkaloids are nature's most potent drugs. They are nitrogen-containing phytochemicals, used traditionally in Asian folklore medicine and mostly found in medicinal plants, which exhibit strong pharmacological activities, such as hyperlipidemia and cholesterol reduction, cytotoxicity, antimicrobial, antiplasmodial, antitumor, high antiinflammatory, anti-HIV, skin protective, and karyokinesis-resisting activity (Ma et al., 2010). Gañán et al. (2016) applied SFE with isopropanol-diethylamine mixture as cosolvent for the recovery of major isoquinolones from *Chelidonium majus* and other researchers studied the effect of SFE factors in galanthamine recovery from floricultural crop waste of *Narcissus pseudonarcissus*. Following the same principals, Hossain et al. (2014, 2015a) used potato peels for glycoalkaloids ASE and UAE, while SWE was implemented by Mokgadi et al. (2013) for extracting hydrastine and berberine, the two major alkaloids of goldenseal. In Teng and Choi's (2014) study, a CCD was employed for optimizing alkaloids UAE from rhizoma coptidis. RSM was also used by Petigny et al. (2013) for simulating UAE of boldine from boldo leaves. Jiao et al. (2015) managed to extract six different alkaloids from *Isatis tinctoria* hairy root cultures by coupling MAE and high-speed homogenization. An aqueous two-phase MAE with ethanol/ammonium sulfate system was used by Zhang et al. (2015) for extraction and cleanup of *Sophora flavescens* alkaloids. Using PEFE, Bai et al. (2013) acquired alkaloid Guanfu base A in less than a minute.

Carotenoids, which are the most important natural pigments, consist partly of the lipophilic fraction of animal matrices, like egg yolk and crustaceans, and horticultural substrates, such as peaches, spinach, melon, and pumpkins. Carotenes (lycopene, α-, β-, and γ-carotene) are

Table 8.6: High energy extraction examples of alkaloids and lipophilic molecules from 2013 to 2016.

Target Compounds	High-Energy Extraction Method	Extracted Molecules—Substrate	References
Alkaloids	SFE	Isoquinolones—*C. majus* plant	Gañán et al. (2016)
		Galanthamine—*N. pseudonarcissus* floricultural crop waste	Rachmaniah et al. (2014)
		Lycopodine—*Lycopodium clavatum* clubmoss	Silva et al. (2015)
		Sesquiterpene pyridine alkaloids—*Tripterygium wilfordii* (thunder god vine)	Fu et al. (2015)
		Protopine, tetrahydropalmatine, bicuculline, and egenine—*Corydalis decumbens* rhizomes	Wu et al. (2012)
	ASE-SWE	Yunaconitine and 8-deacetylyunaconitine—*Aconitum vilmorinianum* flower	Shu et al. (2013)
		Steroidal alkaloids—potato peel waste	Hossain et al. (2015a)
		Hydrastine and berberine—goldenseal	Mokgadi et al. (2013)
	UAE	Steroidal alkaloids—potato peel waste	Hossain et al. (2014)
		Berberine—rhizoma coptidis	Teng and Choi (2014)
		Boldine—boldo leaves	Petigny et al. (2013)
		Cepharanthine—*Stephania rotunda*	Desgrouas et al. (2014)
		Sparteine, lupanine, and 17-oxosparteine—*Lupinus mirabilis* leaves	Castañeda-González et al. (2014)
	MAE	Sinoacutine, palmatine, isocorydine, and *l*-tetrahydropalmatine—*Stephania sinica* herb	Xie et al. (2014)
		Six alkaloids—*I. tinctoria* hairy root cultures	Jiao et al. (2015)
		Oxymatrine and matrine—*S. flavescens* plant	Zhang et al. (2015)
		Oxoisoaporphine alkaloids—*Menispermum dauricum* plant	Wei et al. (2016)
		Vasicine, harmaline and harmine—*Peganum harmala* plant	Shang et al. (2016)
	PEFE	Steroidal alkaloids—potato peel waste	Hossain et al. (2015b)
		Guanfu base A—*Aconitum coreanum* herbs	Bai et al. (2013)
	ILs-UAE	Berberine, jatrorrhizine, and palmatine—*Phellodendron amurense* plant	Wang et al. (2015)
Carotenoids	SFE	Total carotenoid content—pink shrimp	Mezzomo et al. (2013)
		Total carotenoid yield/astaxanthin—*N. oleoabundans*/*H. pluvialis* microalgae paste	Reyes et al. (2016a,b)
		β-carotene and lycopene—Gac oil	Kha et al. (2014)

(*Continued*)

Table 8.6: High energy extraction examples of alkaloids and lipophilic molecules from 2013 to 2016. (*cont.*)

Target Compounds	High-Energy Extraction Method	Extracted Molecules—Substrate	References
		α-carotene, β-carotene, β-cryptoxanthin, lutein, and zeaxanthin—pumpkin	Durante et al. (2014)
		Xanthophylls—rice bran	Sookwong et al. (2016)
	ASE-SWE	Twelve carotenoids/lutein—paprika	Kim et al. (2016)/Kang et al. (2016)
		Lutein, zeaxanthin, β-cryptoxanthin and β-carotene—Tunisian kaki, peaches, and apricots	Zaghdoudi et al. (2015)
		Lutein—green tea and its by-products	Heo et al. (2014)
		α-carotene and β-carotene—pressed palm fiber	Cardenas-Toro et al. (2014)
		Fucoxanthin, lutein, zeaxanthin, and β-carotene—*Laminaria japonica* Aresch seaweed	Lu et al. (2014a)
	UAE	Lycopene and β-carotene—tomato waste	Luengo et al. (2014a)/Kumcuoglu et al. (2014)
		Total carotenoids—peach palm fruit	Ordóñez-Santos et al. (2015)
		β-carotene—carrot waste	Purohit and Gogate (2015)
		Total carotenoids yield—red shrimp body and head	Tsiaka et al. (2015)
		Lutein and β-carotene—spinach	Altemimi et al. (2015)
		Lycopene—red grapefruit	Xu and Pan (2013)
		Astaxanthin——*H. pluvialis* microalgae	Zou et al. (2013)
		Bixin—annatto seeds	Yolmeh et al. (2014)
	MAE	Crocetin, crocin, and picocrocin—saffron	Sobolev et al. (2014)
		β-carotene—carrot peel	Hiranvarachat and Devahastin (2014)
		cis- and *trans*-lycopene—tomato peels	Ho et al. (2015)
		β-carotene—gac aril	Kha et al. (2013)
	UMAE	Lycopene—tomato paste	Lianfu and Zelong (2008)
	PEFE	Carotenoid content/β-carotene—carrot puree/carrot pomace	Roohinejad et al. (2014)
		Carotenoid content/lutein—*C. vulgaris* microalgae	Luengo et al. (2014b)/Luengo et al. (2015)
	ILs-UAE	Lycopene—tomato	Martins and de Rosso (2016)
Lipids and FAs	SFE	α-Linolenic acid—microalgae *Scenedesmus obliquus, C. protothecoides*, and *Nannochloropsis salina*	Solana et al. (2014)

Table 8.6: High energy extraction examples of alkaloids and lipophilic molecules from 2013 to 2016. (*cont.*)

Target Compounds	High-Energy Extraction Method	Extracted Molecules—Substrate	References
		Fatty acid methyl esters—black sesame seeds	Botelho et al. (2014)
		PUFA—longtail tuna head	Sahena et al. (2014)
		Oleic acid and triglycerides—carrot seed oil	Gao et al. (2016)
		Palmitic and linoleic acid—roasted coffee beans	Hurtado-Benavides et al. (2016)
	ASE-SWE	Lipids—marine microorganisms	Golmakani et al. (2014)
		FAs, sterols, waxes, sterols, and steroids—spruce bark	Jablonský et al. (2015)
		Polar and neutral lipids—cow, goat, and ewe milk	Castro-Gómez et al. (2014)
		Stearic, linoleic, and oleic acids—safflower oil	Conte et al. (2016)
	UAE	Higher FAs—grape berries	Duan et al. (2016)
		Oleic and linoleic acids—pumpkin	Hernández-Santos et al. (2016)
		Lipids and FAs—*Dunaliella tertiolecta* microalgae	Qv et al. (2014)
		Linoleic and linolenic acid—chia seed	de Mello et al. (2015)
		Phospholipids, glycolipids, high unsaturated FAs—*Laetiporus sulphureus* mushroom	Sinanoglou et al. (2014)
	MAE	Palmitic, stearic, oleic, and linoleic acids—cottonseed	Taghvaei et al. (2014)
		Lipids—meat	Medina et al. (2015)
		PUFAs—pumpkin seeds	Jiao et al. (2014)
		FAs—milk	González-Arrojo et al. (2015)
		PUFAs—purslane	Stroescu et al. (2013)
	PEFE	FAMEs—*Scenedesmus* microalgae	Lai et al. (2014)
	ILs-MAE	Lipids—microalgae	Pan et al. (2016)

C_{40} isoprenoid asymmetrical tetraterpenic hydrocarbons, while xanthophylls (i.e., lutein, zeaxanthin, astaxanthin, canthaxanthin) are their oxygenated derivatives. Epidemiological surveys reveal a rich antioxidant profile responsible for chronic degenerative, eye [age-related macular degeneration (AMD) and cataracts] and cardiovascular disease, cancer, and aging prevention. In addition, due to their antioxidant properties and their ability to absorb light, they could terminate food degradation caused by light and oxygen. As they are implicated in many physiological mechanisms (e.g., β-carotene is considered a precursor of vitamin A), they could constitute possible biomarkers in metabolomic studies (Amorim-Carrilho

Table 8.7: High energy extraction examples of volatile compounds and carbohydrate-containing molecules from 2014 to 2016.

Target Compounds	High-Energy Extraction Method	Extracted Molecules—Substrate	References
VOCs-EOs	SFE	Linalool, eugenol, sabinene hydrate, and terpineol—marjoram and sweet basil	Arranz et al. (2015)
		Germacrene D and 1-octadecene—juniper needles	Larkeche et al. (2015)
		Total EOs—Algerian myrtle	Zermane et al. (2014)
		p-Cymene, thymol, and γ-terpinene—Tunisian oregano	Mechergui et al. (2014)
		1-Docosene, neophytadiene, methyl abietate and cedrane-8,13-diol-pine hay needles	Cheng et al. (2015)
	ASE-SWE	Menthol/menthone, eucalyptol/camphor, carvacrol/thymol—peppermint, rosemary, oregano and thyme	Rodríguez-Solana et al. (2014)
		Oxygenated monoterpenes—coriander	Pavlić et al. (2015)
		Thymol—thyme plants	Villanueva Bermejo et al. (2015)
		Thymol—ajowan caraway	Khajenoori et al. (2015)
	UAE	Estragole—tarragon leaves	Gholivand et al. (2014)
		Mono- and sesquiterpenes—cannabis	Da Porto et al. (2014)
		Eugenol, α-caryophyllene, and 2-methoxy-4-(2-propenyl) phenol acetate—clove	Tekin et al. (2015)
		trans-cinnamaldehyde—cinnamon bark	Li et al. (2015)
		p-anisaldehyde and anethole—star anise fruit	Lee et al. (2014)
	MAE	Total EOs—lemongrass leaves	Desai and Parikh (2015)
		Total EOs—lemon peel	Golmakani and Moayyedi (2015)
		Total EOs—orange peel	Franco-Vega et al. (2016)
		Total EOs—peppermint	Gavahian et al. (2015)
	UAE-SWE	Ethyl *trans-p*-methoxycinnamate—aromatic ginger	Ma et al. (2015)
	UAE-DLLME	VOCs—saffron	Sereshti et al. (2014)
	UAE-MAE	EOs—*Angelica dahurica* dry roots	Feng et al. (2014)
	MAE-hydrostillation	Santalol, α-bergamotol, nuciferol, γ-elemene, *cis*-lanceol, and α-cedrol—sandalwood	Kusuma and Mahfud (2016)
		Total EOs—lavender, sage, rosemary, fennel, and clove	González-Rivera et al. (2016)
		β-elemene and β-phellandrene—magnolia	Chen et al. (2015a)
	ILs-MAE	Total EOs—Tasmanian blue-gum leaves	Liu et al. (2016b)
	MAE-DLLME	Total EOs—cardamom	Ye et al. (2014)

Table 8.7: High energy extraction examples of volatile compounds and carbohydrate-containing molecules from 2014 to 2016. (*cont.*)

Target Compounds	High-Energy Extraction Method	Extracted Molecules—Substrate	References
Polysaccharides-Carbohydrates	SFE	Mannose, rhamnose, galactose, glucose, arabinose, xylose, and fucose-containing polysaccharides—*Artemisia sphaerocephala* seeds	Chen et al. (2014)
		Total polysaccharides—ginseng	Yu et al. (2015)
	ASE-SWE	β-Glucan—hullless barley brans	Du et al. (2014)
		Rhamnose, arabinose, xylose, mannose, galactose, and glucose-containing polysaccharides—blackcurrant	Xu et al. (2016)
		β-glucan—edible mushrooms	Palanisamy et al. (2014)
		Inositol and inulin—artichoke external bracts	Ruiz-Aceituno et al. (2016)
		Lignin—ground spruce wood	Sumerskiy et al. (2015)
	UAE	Polysaccharide yield—pomegranate peel	Zhu et al. (2015)
		Polysaccharide yield—mulberry leaves	Zhang et al. (2016a)
		Hemicelluloses, xyloglycans, mannans, and xylans—grape pomace	Minjares-Fuentes et al. (2016)
		Pectin—tomato waste	Grassino et al. (2016)
		Water soluble polysaccharides—hazelnut skin	Yılmaz and Tavman (2016)
	MAE	Total polysaccharides—bamboo leaves	Zhang et al. (2016b)
		Rhamnogalacturonan I—potato pulp	Khodaei et al. (2016)
		Pectin—pumpkin biomass	Košťálová et al. (2016)
		Pectic polysaccharides—tangerine peels	Chen et al. (2016)
		Inulin—chicory roots	Tewari et al. (2015)
	UMAE	Total polysaccharides-mulberry leaves	Liu (2015)
		Total polysaccharides—kumquat swingle	Zeng et al. (2015)
		Total polysaccharides—*T. mongolicum Imai* mushroom	You et al. (2014)
		Crude polysaccharides—*Pericarpium granati*	Zhou et al. (2014)
	PEFE	Total polysaccharides—cabernet sauvignon grape berry skins	Cholet et al. (2014)
		Endopolysaccharide—*Morchella esculenta* mushroom	Liu et al. (2016a)
		Polysaccharide yield—garlic	Yang et al. (2016)
		Pectin—passion fruit peel	de Oliveira et al. (2015)
	PEFE-ASE	Total polysaccharides—mushroom *A. bisporus*	Parniakov et al. (2014)
	ILs-UAE	Secoisolariciresinol diglucoside—flaxseed	Tan et al. (2015)

et al., 2014; Arathi et al., 2015; Singh et al., 2015). High-energy alternative is nowadays the number one option for extracting carotenoids (Strati and Oreopoulou, 2014). Contrary to animal and plant organisms, algae can produce high amounts of carotenoids, therefore Reyes et al. (2016a,b) considered *Neochloris oleoabundans* and *H. pluvialis* microalgae as suitable matrices for SFE of carotenoids, especially astaxanthin. Nevertheless, SFE is also a handy solution for carotenoids recovery from pink shrimp (Mezzomo et al., 2013), pumpkin (Durante et al., 2014), and pigmented rice bran (Sookwong et al., 2016). Through ASE, Kim et al. (2016) delivered carotenoid profile of 27 types of red, yellow, and orange paprika based on their shapes and cultivation type. An LC-MS profiling took place in ASE carotenoid extracts from Tunisian kaki, apricots, and peaches (Zaghdoudi et al., 2015). Heo et al. (2014) succeeded the separation of the isomers lutein and zeaxanthin by ASE and UPLC. Carrot and tomato waste are two of the most opportune substrates for extraction high yields of carotenoids. Luengo et al. (2014a) recovered carotenoids from tomato pomace by combing moderate pressure with UAE and Purohit and Gogate (2015) investigated the use of vegetable oils as UAE solvents for β-carotene from carrot residues. An NMR assay was performed by Sobolev et al. (2014) in MAE extracts of saffron for structural elucidation of crocin, crocetin, and picocrocin. As shown in the work of Hiranvarachat and Devahastin (2014), a noncontinuous MW irradiation enhances the antioxidant capacity of carrot peels carotenoids extracts. In order to improve β-carotene extractability Hiranvarachat et al. (2012) tested the effect of different pretreatment procedures, like soaking in citric acid and blanching in water and citric acid. Ho et al. (2015) isolated *cis*- and *trans*-lycopene in tomato peels MAE extracts. Tomato paste was also used as substrate in a Lianfu and Zelong (2008) project for a comparative study between UAE and UMAE, where UMAE resulted in higher yields. Roohinejad et al. (2014) and Luengo et al. (2014b, 2015) teams become active in carotenoids PEFE from carrot by-products and microalgae *C. vulgaris*, respectively (Fig. 8.8).

Recovery and fractionation of lipids and FAs is a current trend issue in generating high-value nutraceutical commodities due to their benign properties. Lipids, water-insoluble bioactive compounds, which include steroids, waxes, prostaglandins, are assorted in two general classes, unsaturated and saturated FAs and glycerides. It is known that these molecules play an important role on regulatory and neuroprotective mechanisms, heart attack, stroke, coronary disease, blood clots and pressure, cholesterol levels, rheumatoid arthritis, atherosclerosis, diabetes, depression, and cancer (Sahena et al., 2009). An optimization of SF-CO_2 extraction for characterizing FAs of Colombian roasted coffee bean through GC-FID was realized in Hurtado-Benavides et al. (2016) work. Sahena et al. (2014) examined storage stability and SFE polyunsaturated FA (PUFA) quality in tuna head. Also, Botelho et al. (2014) delivered SF extracts of black sesame seeds rich in palmitic and linolenic acid with possible biofunctionality against stroke. Pressurized ethanol was selected by Conte et al. (2016) for chemical profile of lipid fraction in safflower oil. Jablonský et al. (2015) project was oriented to exploitation of biomass waste through recovery of lipid substances from spruce bark. Castro-Gómez et al. (2014) combined ASE and chromatography for quantification of

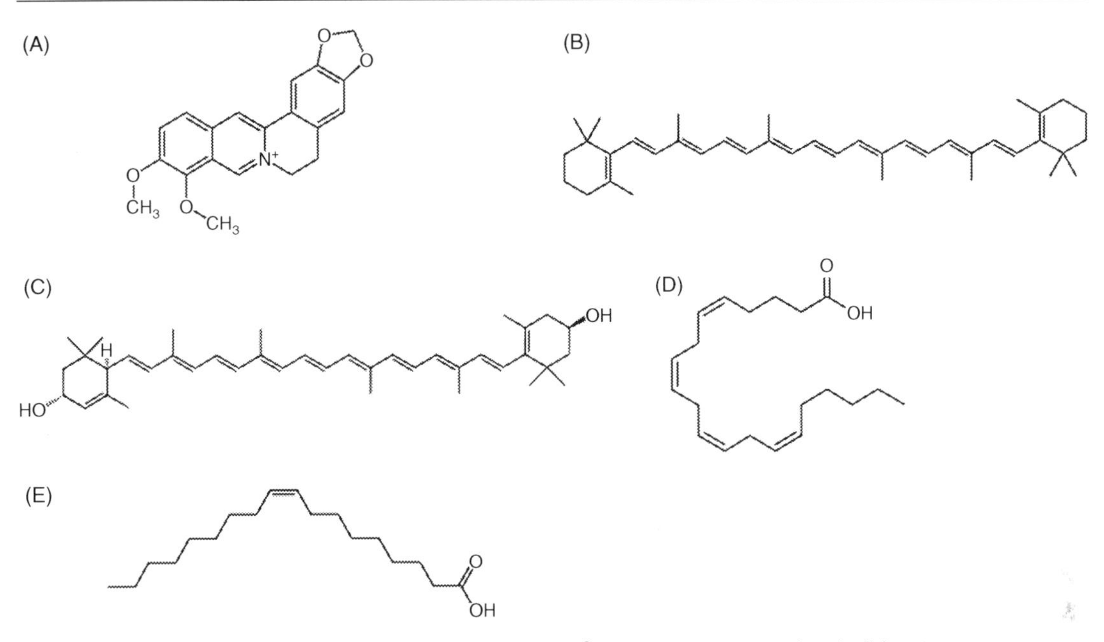

Figure 8.8: (A) Alkaloid berberine, (B) β-carotene, (C) xanthophyll lutein, (D) arachidonic acid, (E) oleic acid.

neutral and polar lipids in different types of animal milk. UAE hexane samples of mushroom *Laetiporus sulphureus* presented significant antifungal and antibacterial activity in the report of Sinanoglou et al. (2014). According to Qv et al. (2014), UAE was a promising alternative for extraction microalgae lipids for biodiesel production. Duan et al. (2016) noticed that α-linolenic acid deterioration is lower in UAE samples of grape berries compared to Soxhlet extracts. Medina et al. (2015) extracted meat lipids without affecting their oxidative stability by MAE, while González-Arrojo et al. (2015) developed a one-step MAE for recovering and transmethylating milk FAs. Jiao et al. (2014) proposed a combination of MAE and an aqueous enzymatic mix of cellulase, pectinase, and proteinase for acquiring pumpkin seeds PUFAs. PEFE proved to enhance FAMEs extractability from microalgae *Scenedesmus* in the work of Lai et al. (2014). Pan et al. (2016) extracted successfully lipids from three different microalgae using MAE and IL [BMIM][HSO_4] as extraction solvent.

Volatile organic compounds (VOCs) and EOs hide behind the chemistry of "aroma" in every food, beverage, cosmetic, pharmaceutical, and perfumery product. The branch of fragrances and aromatherapy is based on their antimicrobial and sedative activity. EOs, mainly derived from plants and herbs, dispose an hydrocarbon skeleton with different functional groups, which segregate them in classes, like terpenoids, aldehydes, ketones, alcohols, esters, and phenols. The majority of their biological properties refer to antiviral, antioxidant, antiphlogistic, antinociceptive, anticancer, and psychoactive effects (Baser and Buchbauer, 2015). Cheng et al. (2015) discovered the novel EO compound, cedrane-8,13-diol in SFE fraction of pine needles.

RSM enhanced SFE yields for EOs in juniper (Larkeche et al., 2015) and myrtle (Zermane et al., 2014). Arranz et al. (2015) explored EOs antiinflammatory and antiatherogenic properties from SFE samples of marjoram and sweet basil. Thorough studies have been realized for the application of ASE-SWE in herbal extraction (peppermint, coriander, thyme, oregano, etc.) EOs (Pavlić et al., 2015; Rodríguez-Solana et al., 2014). SWE of aromatic ginger EOs was promoted by in tandem UAE as reported by Ma et al. (2015). Compared to classic techniques, like hydrostillation, the UAE method resulted in much faster ($\leq$10 min) and equal or increased efficiency for EOs from herbs (Da Porto et al., 2014; Gholivand et al., 2014). MAE merits were useful for enhancing classic EOs hydrostillation, therefore MAE-hydrostillation is the most promising technique for EOs extraction as proved by Kusuma and Mahfud (2016), González-Rivera et al. (2016), and Chen et al. (2015a) (Fig. 8.9).

Polysaccharides are long-chain carbohydrate polymers, whose building units are glycoside-bound monosaccharides, like glucose, mannose, and xylose. One of the most valuable and

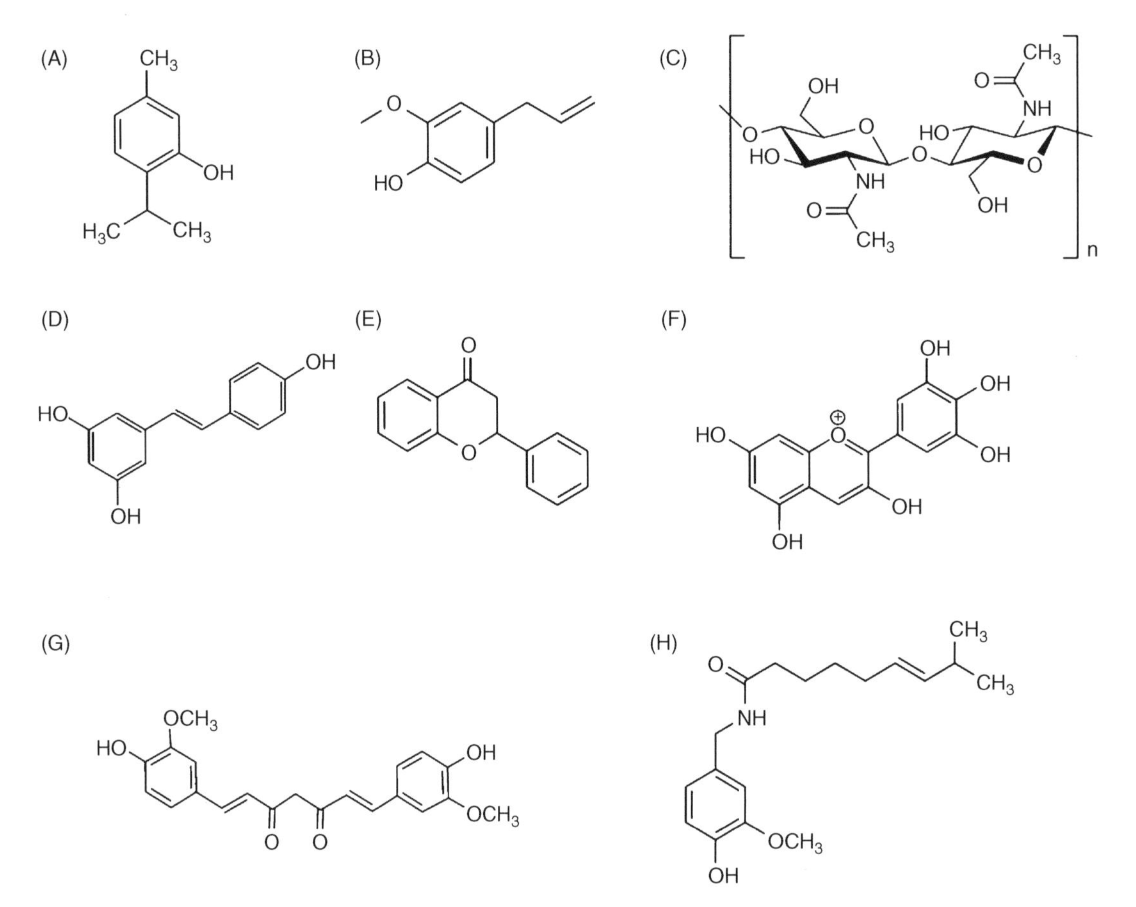

Figure 8.9: (A) EO thymol, (B) EO eugenol, (C) building unit of chitin, (D) *trans*-resveratrol, (E) lavanone, (F) delphinidin, (G) curcumin, (H) capsaicin

health-beneficial polysaccharides in modern market is β-glucan. In the cell of plant and marine organisms are also present the form of aggregate blocks, like pectin, lignin, and chitin. Their bioactivity is related to hematopoietic, anticancer, antimutagenic, antioxidant, and immunostimulatory therapeutic properties. Their role as dietary fibers is very important as they can reduce cholesterol level and help digestion. Due to their complexity, pectin and chitin are major components of active packaging technologies, fat-replacers to low-calories foods owing to their organoleptic aspects, thickeners, stabilizers, and water binders (Prakash Maran et al., 2014; You et al., 2014). In contrast to carotenoids and lipid fractions extraction from natural sources, SFE is not the most common choice for polysaccharide acquisition. On the other hand, PHWE was selected for recovering priceless molecules, like β-glucan (Du et al., 2014; Palanisamy et al., 2014), lignin (Sumerskiy et al., 2015), and inulin (Ruiz-Aceituno et al., 2016). However, the number of papers referring to polysaccharides UAE, MAE or their synergistic effect through DOE models from plant waste and by-products (i.e., tomato waste, hazelnut skin, grape pomace, pumkin biomass, potato pulp, tangerine peel) is enormous even by taking under consideration 2016 projects (Chen et al., 2016; Grassino et al., 2016; Khodaei et al., 2016; Košťálová et al., 2016; Minjares-Fuentes et al., 2016; Yılmaz and Tavman, 2016). Polyethylene glycol was tested successfully as an innovative ecocompatible solvent for polysaccharides (Zhou et al., 2014). Although PEFE do not count numerous applications for extraction of other groups of biomolecules yet, over the last 2 years it is used more and more frequently for polysaccharides extraction especially from fungus and mushrooms (Liu et al., 2016b; Parniakov et al., 2014) and for enhancing the extraction of quality biomarkers of wines and beverages (Cholet et al., 2014).

In recent years, antioxidant health-promoting effects of natural products have caught the eye of marketing departments in nutraceutical industry. Thus, a lot of effort, money, and workforce have been bought into the production of drugs and foods that reveal activity against oxidative stress. The antioxidant capacity of these products is ascribed principally to phenolic acids and polyphenols, which are aromatic compounds with various functional groups. Flavonoids, stilbenes, anthocyanins, and tannins, which are also considered pH-depended natural dyes, curcuminoids [i.e., curcumin and *bis*-(demethoxy) curcumin], also known as the bioactive phytoconstituents of turmeric and capsacinoids (i.e., capsaicin, dihydrocapsaicin, norcapsaicin, homodihydrocapsaicin, etc.), which are *N*-vanillylamides derivatives of chili peppers, are the most distinguishing exponents of this category. Clinical trials gave prominence to their antimicrobial, spasmolytic, cardioprotective, antitumor, antimutagenic, antiviral, antibacterial, antifungal, antiparasitic, antiinflammatory, hypolipidemic, anticoagulative, analgesic, and coloring characteristics as antioxidant agents (Bajer et al., 2015; Casazza et al., 2010; Dandekar and Gaikar, 2002). The infinite types and bioactivities of polyphenols and antioxidants make these compounds the most studied group of active molecules. High energy methods examples for extracting phenolic compounds and polyphenols are presented in Table 8.8. SFE and PHWE provided high yields of phenolic acids, flavanols, and anthocyanins, particularly when water and ethanol are

Table 8.8: High energy extraction examples of phenolic compounds and polyphenols from 2014 to 2016.

Target Compounds	High-Energy Extraction Method	Extracted Molecules—Substrate	References
Phenolic compounds-Flavonoids-Anthocyanins	SFE	Polyphenols, epicatechin/catechin—Guaraná	Marques et al. (2016)
		Vanilic, ferulic, syringic acid—pomegranate peel	Mushtaq et al. (2015)
		Total phenolics—pepper-rosmarin	Garmus et al. (2015)
		Mangiferin, isomangiferin, quercetin 3-*o*-galactoside, quercetin 3-*o*-glucoside, quercetin 3-*o*-xyloside, quercetin 3-*o*-arabinoside, quercetin and kaempferol—mango by-products	Meneses et al. (2015)
		Gallic acid, cyanidin-3-glycoside, and catechin-purple corn cobs	Monroy et al. (2016)
		Luteolin and 3-desoxyanthocyanidins—Arrabidaea chica leaves	Paula et al. (2014)
		Anthocyanins—bilberry press cake	Kerbstadt et al. (2015)
	ASE-SWE	Sixteen anthocyanins—blueberry	Paes et al. (2014)
		(+)-Catechin, (−)-epicatechin, phloridzin, chlorogenic acid, hyperoside, isoquercitrin, quercitrin, ideain—apples	Franquin-Trinquier et al. (2014)
		Phenolic acids and 3-*o*-feruloylquinic acid—asparagus	Solana et al. (2015)
		Oleuropein—olive leaves	Xynos et al. (2014)
		Total phenolics—saffron petals	Ahmadian-Kouchaksaraie et al. (2016)
		Flavonoids—*Momordica foetida* plant	Khoza et al. (2016)
		Total phenolics—larch waste wood	Ravber et al. (2015)
		Polyphenols/catechin, epicatechin, anthocyanins—grape pomace	Monrad et al. (2014); Vergara-Salinas et al. (2015)
		Anthocyanins and quercetin—red onions	Liu et al. (2014)
		Anthocyanins—black carrot	Aşkin Uzel (2016)
		Anthocyanins and biflavonoids—Brazilian peppers	Feuereisen et al. (2017)
	UAE	Naringin—grapefruit waste	Garcia-Castello et al. (2015)

Table 8.8: High energy extraction examples of phenolic compounds and polyphenols from 2014 to 2016. (*cont.*)

Target Compounds	High-Energy Extraction Method	Extracted Molecules—Substrate	References
		Hesperidin and gallic acid—mandarin and lime peel	Singanusong et al. (2015)
		Total phenolics and total anthocyanins—wine lees	Tao et al. (2014)
		Chlorogenic acid—artichoke leaves	Saleh et al. (2016)
		Cyanidin 3,5-diglucoside, cyanidin 3-glucoside, cyanidin 3-rutinoside, pelargonidin 3-glucoside, and peonidin 3-glucoside—Haskap berries	Celli et al. (2015)
		Catechin, myricetin, and quercetin—curry leaves	Ghasemzadeh et al. (2014)
		Anthocyanins and flavonoids—purple sweet potatoes	Cai et al. (2016)
		Nonextractable phenolic—cauliflower waste	Gonzales et al. (2014)
		Total phenolics—pomegranate arils, juice, and seeds	Lantzouraki et al. (2015)
		Gallic, chlorogenic, and ferulic acids—sugarcane rinds	Feng et al. (2015)
		Stilbenes—grape cane	Piñeiro et al. (2016)
		Total phenolics—spent coffee grounds	Al-Dhabi et al. (2017)
		Total monomeric anthocyanins—eggplant peel	Dranca and Oroian (2016)
		Polyphenols and anthocyanins—black chokeberry waste	D'Alessandro et al. (2014)
	MAE	Polyphenol content—almond skin by-products	Valdés et al. (2015)
		Phenolics and flavonoids—tomato	Pinela et al. (2016)
		Total phenolic, total flavonoids, and tannins—myrtle leaves	Dahmoune et al. (2015)
		Gallic, ellagic, caffeic acid, rutin, and magniferin—*Physalis angulata* plant	Carniel et al. (2016)
		Flavonoid yield—*Terminalia bellerica* plant	Krishnan and Rajan (2016)
		Total phenolics, total flavonoids, and proanthocyanidins—eucalyptus leaves	Bhuyan et al. (2015)

(Continued)

Table 8.8: High energy extraction examples of phenolic compounds and polyphenols from 2014 to 2016. (*cont.*)

Target Compounds	High-Energy Extraction Method	Extracted Molecules—Substrate	References
		Total phenolics, flavonoid, and proanthocyanidin—grape seeds	Dang et al. (2014)
		Rosmarinic and carnosic acid—rosemary leaves	Bellumori et al. (2016)
		Total phenolic and total flavonoid content—green and black tea	Shah et al. (2015)
		Total anthocyanins—blackberry	Wen et al. (2015)
		Anthocyanin composition—blue wheat, purple corn, and black rice	Abdel-Aal et al. (2014)
		Anthocyanins content—black corn hull	Jiang et al. (2013)
		Betalains—red beet	Cardoso-Ugarte et al. (2014)
	PEFE	Polyphenols—flaxseed hulls	Boussetta et al. (2014)
		Anthocyanins and flavanols—plum and grape peels	Medina-Meza and Barbosa-Cánovas (2015)
		Polyphenols—defatted canola seed cake	Teh et al. (2015b)
		Total phenolics—spearmint	Fincan (2015)
		Quercetin and ellagic acid—*Emblica officinalis* fruit	Bansal et al. (2014)
		Phenolic yield—graciano, tempranillo, and grenache grape varieties	López-Giral et al. (2015)
	ILs-ASE	Total phenolics—barley hull	Sarkar et al. (2014)
	ILs-UAE	Gallic acid, chlorogenic acid, rutin, psoralen, and bergapten—figs	Qin et al. (2015)
		Flavonoids—*Apocynum venetum* leaves	Tan et al. (2016)
	ILs-MAE	Gallic acid and ellagic acid—eucalyptus leaves	Li et al. (2016)
	ILs-VMAE	Rutin, hyperoside and hesperidin—*S. tianschanica* leaves	Gu et al. (2016)
Capsacinoids	SFE	Capsaicin, dihydrocapsaicin, nordihydrocapsaicin and homodihydrocapsaicin—malagueta peppers	Santos et al. (2015)
		Capsaicinoids and capsinoids—biquinho pepper	de Aguiar et al. (2014)
	ASE-SWE	Capsaicin, dihydrocapsaicin, nordihydrocapsaicin, and nonivamide—chili peppers	Bajer et al. (2015)

Table 8.8: High energy extraction examples of phenolic compounds and polyphenols from 2014 to 2016. (*cont.*)

Target Compounds	High-Energy Extraction Method	Extracted Molecules—Substrate	References
Curcuminoids	SFE	Curcuminoids—ethanoli turmeric extract	Osorio-Tobón et al. (2016)
		Turmeric oil and ar-turmerone—turmeric	Carvalho et al. (2015)
	ASE-SWE	Curcumin—turmeric rhizome	Kiamahalleh et al. (2016)
		Curcumin, demethoxycurcumin, and bisdemethoxycurcumin—turmeric	Kwon and Chung (2015)
	UAE	Methyl β-cyclodextrin-complexed curcumin—turmeric rhizome oleoresin	Hadi et al. (2015a)
	MAE	Curcumin	Bener et al. (2016)
		Methyl β-cyclodextrin-complexed curcumin—turmeric rhizome oleoresin	Hadi et al. (2015b)
		Curcumin, demethoxycurcumin, and bisdemethoxycurcumin—turmeric	Laokuldilok et al. (2015)
	UAE and MAE	Curcuminoids—turmeric	Li et al. (2014)
	ILs-UAE	Curcuminoids—turmeric rhizome	Xu et al. (2015)

used as cosolvents (Garmus et al., 2015). These two techniques are also methods-of-choice for capsacinoids extraction from peppers (Bajer et al., 2015; de Aguiar et al., 2014; Santos et al., 2015). In the work of Ahmadian-Kouchaksaraie et al. (2016) and Khoza et al. (2016), distribution patterns of solvent/material ratio, temperature, and extraction time in phenolic extracts were evaluated by applying principal component analysis (PCA). The optimum values of these parameters are strongly correlated to the chemical structure of phenolic compounds, that is, number of double bonds, type of sugar units, polarity of side chains (Ko et al., 2014). State-of-art analytical techniques as UHPLC-qTOF-MS contribute in the accurate identification and quantification of phenolic molecules (Khoza et al., 2016). UAE extracts of curry leaves showed anticancer activity due to the presence of catechin, myricetin, and quercetin (Ghasemzadeh et al., 2014). As stated by Cai et al. (2016), UAE presented higher anthocyanins yields than classic extraction but lower recoveries from ASE, while the classic method is better from both high-energy techniques in the case of total phenolics and flavonoids. Gonzales et al. (2014) achieved nonextractable phenolics recovery from cauliflower after alkaline hydrolysis and sonication. Piñeiro et al. (2016) affirmed that sugar cane by-product is a suitable matrix for stilbenes UAE. Krishnan and Rajan (2016) provided useful observations related to the kinetics of flavonoid MAE. An aqueous two-phase MAE system developed by Dang et al. (2014) was applied for obtaining phenolic constituents from wine making process residues. Ethanol and acetone increase dramatically MAE and UAE rosmarinic and carnosic acid yields of rosemary leaves (Bellumori et al., 2016). MAE green tea samples exhibited higher antioxidant and antiproliferative activity as shown in the work of Shah et al. (2015). Based on the remarks of Abdel-Aal et al. (2014), MAE and ASE were equally efficient methods to classic techniques for the determination of anthocyanin fraction of colored grains. MAE in red beets produced extracts enriched in betalanines and betaxanthins (Cardoso-Ugarte et al., 2014). In the case of phenolic extracts, PEFE is used as a preliminary step for intensifying in tandem high-energy techniques (Medina-Meza and Barbosa-Cánovas, 2015). In López-Giral et al. (2015) project, PEFE was implemented for ameliorating the phenolic profile of spanish wine varieties during two vintages. VMAE of rutin, hesperidin and hyperoside showed an efficient recovery and reproducibility when [C_6mim][BF_4] was used as extractant (Gu et al., 2016). As it is shown in Table 8.8, ASE and MAE are relatively selective techniques for the recovery of curcuminoids.

The combination of high-energy methods with high-throughput analytical techniques, more generic or specialized in each type of bioactive compounds, as for example GC-MS for EOs and FAs, UPLC-MS-TOF, and LC-MS for phenolic compounds and polyphenols, NMR for polysaccharides, and FTIR for sugars and carbohydrates, make these extraction approaches a quite competitive answer for developing and validating large-scale production line operations.

5 Large-Scale High Energy Extraction: Is It an Economically Feasible Solution?

The winning bet for industrial scale high-energy extraction is its intensification in the production of health-promoting bioactive extracts and in waste management processes for residues of food and drug companies in local and international level. A plethora of economic studies are available in the literature including business plans and feasibility studies for the installation of medium or large scale extraction units. High-energy extraction processes are mostly complex thermodynamic systems with higher capital costs, which are mainly attributed to installation. The engineering design of these systems requires good knowledge of the thermodynamic constraints of solubility and selectivity, and kinetic constraints of mass transfer rate. Modeling of the extraction processes can provide a better understanding of the mechanisms behind extraction and be used to quickly optimize the operating conditions and scale-up any design (Wang and Weller, 2006). Although the economic feasibility analysis for each technique is different, several common general assumptions have to be made including (1) unit working period (hours per day and per year), (2) number of workers per unit extractor and necessary level of expertise, (3) scale-up criteria for each technique like solid feed, solvent flow rate, extraction cell dimensions (i.e., extraction bed for SFE), (4) minimum time to load and unload in each extraction cycle, (5) market price of the extract, (6) matrix initial pretreatment if necessary. Another commonly made assumptions that although in many cases the scale-up from the laboratory to the industrial scale under the same conditions increases the extraction yield, in many studies the yield is kept identical at both scales. Finally, it is necessary to establish the optimal equilibrium between extraction efficiency, extraction duration, physicochemical, and biological properties of the extract and extract quality. Besides the high initial equipment capital required for their industrial-scale implementation, high-energy techniques could result to net profit for the companies because operational expenses and raw material cost, especially when we refer to natural products waste, are, in many cases, lower compared to conventional methods, like steam distillation. Economic evaluation studies showed that scale-up of extraction units is favored as manufacturing cost is decreasing proportionally (over 50%) when operational units' capacity is enlarging. Increase of extractors' size is translating into more utilities charges (energy consumption, staff payments, maintenance cost), therefore, extraction procedures efficiency in corporate groups normally outcompete those of small-scale businesses (Pereira and Meireles, 2010; Santos et al., 2010). Economic assessment reports of high-energy extraction processes have been increasingly found in literature during the last 5 years. A previous economic evaluation indicated the feasibility of an industrial SFE plant with a capacity of 0.5 m^3 for producing an extract with an approximate concentration of phenolic compounds 23 g kg^{-1} of extract, at an estimated cost of manufacturing of US $133.16 kg^{-1} (Farías-Campomanes et al., 2013). Several factors can attenuate uncertainty in an economic estimation of large-scale high-energy extraction processes. One of the most important sources of uncertainty is the price of the produced extracts (final product), which frequently hinders the correct economic evaluation causing unrealistic estimations.

6 Conclusions and Future Perspectives

The use of emerging technologies in the field of high-energy extraction is spreading widely in many industrial sectors and mainly food, pharmaceuticals, chemicals, and cosmetics. Safety, sustainability, environmental, and economic factors are all forcing laboratories and industry to turn to nonconventional technologies and greener protocols. The high-energy techniques discussed in the present review have certain advantages compared to conventional ones because, in principle, they manage to shorten the extraction time, increase the yield and quality of extracts, and decrease the solvent consumption. However, most of these high-energy extraction techniques are still performing successfully mainly at the laboratory or bench-scale although several industrial applications (medium- or large-scale) can be found as for the case of SFE. A relatively recent review on SFE (Souza Machado et al., 2013) based on patents concluded that the research applied to SFE shows the food and agricultural sector as the main application and development field for this technique. ASE under elevated temperature and high pressure can be supplementary to SFE for extraction of polar compounds but still more research is needed in order to reduce the capital and operating costs of SFE and ASE.

Presently, the use of new extraction techniques can boost the production and development of functional food, food additives, and ingredients for food and pharmaceutical products. The number of potential applications continues to grow globally, which is reflected to the increase of research articles and patents deposited during the last 5 years. The need for adaptation of these techniques is further supported by the increasing demand of consumers in developed mainly countries toward high-value natural products.

Currently, there is cumulative knowledge in the area, which may provide substantial progress on analytical, engineering, scale-up, and economic issues and may help in its implementation at industrial level. The combination of more comprehensive scale-up works together with the corresponding economic assessment studies would foster a clearer perception of the technoeconomic requirements. Overall, optimized strategies and technological advances in the field of high-energy extraction can provide an economically viable and competitive solution for the preparation and marketing of high-added value natural products for a variety of substrates.

References

Abdel-Aal, E.-S.M., Akhtar, H., Rabalski, I., Bryan, M., 2014. Accelerated, microwave-assisted, and conventional solvent extraction methods affect anthocyanin composition from colored grains. J. Food Sci. 79, C138–C146.

Abert Vian, M., Allaf, T., Vorobiev, E., Chemat, F., 2014. Solvent-free extraction: myth or reality? In: Chemat, F., Abert Vian, M. (Eds.), Alternative Solvents for Natural Products Extraction. Springer, New York, NY, pp. 25–38.

Ahmadian-Kouchaksaraie, Z., Niazmand, R., Najafi, M.N., 2016. Optimization of the subcritical water extraction of phenolic antioxidants from *Crocus sativus* petals of saffron industry residues: Box-Behnken design and principal component analysis. Innov. Food Sci. Emerg. Technol. 36, 234–244.

Al-Dhabi, N.A., Ponmurugan, K., Maran Jeganathan, P., 2017. Development and validation of ultrasound-assisted solid-liquid extraction of phenolic compounds from waste spent coffee grounds. Ultrason. Sonochem. 34, 206–213.

Alexandru, L., Pizzale, L., Conte, L., Barge, A., Cravotto, G., 2013. Microwave-assisted extraction of edible *Cicerbita alpina* shoots and its LC-MS phenolic profile. J. Sci. Food Agric. 93, 2676–2682.

Almanasrah, M., Roseiro, L.B., Bogel-Lukasik, R., Carvalheiro, F., Brazinha, C., Crespo, J., Kallioinen, M., Mänttäri, M., Duarte, L.C., 2015. Selective recovery of phenolic compounds and carbohydrates from carob kibbles using water-based extraction. Ind. Crops Prod. 70, 443–450.

Altemimi, A., Lightfoot, D.A., Kinsel, M., Watson, D.G., 2015. Employing response surface methodology for the optimization of ultrasound assisted extraction of lutein and β-carotene from spinach. Molecules 20, 6611–6625.

Álvarez-Casas, M., García-Jares, C., Llompart, M., Lores, M., 2014. Effect of experimental parameters in the pressurized solvent extraction of polyphenolic compounds from white grape marc. Food Chem. 157, 524–532.

Amorim-Carrilho, K.T., Cepeda, A., Fente, C., Regal, P., 2014. Review of methods for analysis of carotenoids. TrAC Trends Anal. Chem. 56, 49–73.

Ara, K.M., Jowkarderis, M., Raofie, F., 2015. Optimization of supercritical fluid extraction of essential oils and fatty acids from flixweed (*Descurainia Sophia* L.) seed using response surface methodology and central composite design. J. Food Sci. Technol. 52, 4450–4458.

Arathi, B.P., Sowmya, P.R.R., Vijay, K., Baskaran, V., Lakshminarayana, R., 2015. Metabolomics of carotenoids: the challenges and prospects: a review. Trends Food Sci. Technol. 45, 105–117.

Armenta, S., Garrigues, S., de la Guardia, M., 2008. Green Anal Chemistry. Trends Anal. Chem. 27 (6), 497–511.

Arranz, E., Jaime, L., López de las Hazas, M.C., Reglero, G., Santoyo, S., 2015. Supercritical fluid extraction as an alternative process to obtain essential oils with anti-inflammatory properties from marjoram and sweet basil. Ind. Crops Prod. 67, 121–129.

Aşkin Uzel, R., 2016. A practical method for isolation of phenolic compounds from black carrot utilizing pressurized water extraction with in-site particle generation in hot air assistance. J. Supercrit. Fluids 120, 320–327.

Azmir, J., Zaidul, I.S.M., Rahman, M.M., Sharif, K.M., Mohamed, A., Sahena, F., Jahurul, M.H.A., Ghafoor, K., Norulaini, N.A.N., Omar, A.K.M., 2013. Techniques for extraction of bioactive compounds from plant materials: a review. J. Food Eng. 117 (4), 426–436.

Bai, Y., Li, C., Zhao, J., Zheng, P., Li, Y., Pan, Y., Wang, Y., 2013. A high yield method of extracting alkaloid from *Aconitum coreanum* by pulsed electric field. Chromatographia 76, 635–642.

Bajer, T., Bajerová, P., Kremr, D., Eisner, A., Ventura, K., 2015. Central composite design of pressurised hot water extraction process for extracting capsaicinoids from chili peppers. J. Food Compost. Anal. 40, 32–38.

Bansal, V., Sharma, A., Ghanshyam, C., Singla, M.L., 2014. Optimization and characterization of pulsed electric field parameters for extraction of quercetin and ellagic acid in *Emblica officinalis* juice. J. Food Meas. Charact. 8, 225–233.

Barbosa, H.M.A., De Melo, M.M.R., Coimbra, M.A., Passos, C.P., Silva, C.M., 2014. Optimization of the supercritical fluid coextraction of oil and diterpenes from spent coffee grounds using experimental design and response surface methodology. J. Supercrit. Fluids 85, 165–172.

Baser, K.H.C., Buchbauer, G., 2015. Handbook of Essential Oils: Science, Technology and Applications. CRC Press, Taylor & Francis Group, Boca Raton, Florida, FL.

Bellumori, M., Innocenti, M., Binello, A., Boffa, L., Mulinacci, N., Cravotto, G., 2016. Selective recovery of rosmarinic and carnosic acids from rosemary leaves under ultrasound- and microwave-assisted extraction procedures. Comptes Rendus Chim. 19, 699–706.

Bener, M., Özyürek, M., Güçlü, K., Apak, R., 2016. Optimization of microwave-assisted extraction of curcumin from *Curcuma longa* L. (turmeric) and evaluation of antioxidant activity in multi-test systems. Rec. Nat. Prod. 10, 542–554.

Benito-Román, Ó., Alonso, E., Cocero, M.J., Goto, M., 2016. β-Glucan recovery from *Ganoderma lucidum* by means of pressurized hot water and supercritical CO_2. Food Bioprod. Process 98, 21–28.

Benjamin, M., Stéphane, R., Gérard, V., Pierre-Yves, P., 2017. Pressurized water extraction of isoflavones by experimental design from soybean flour and soybean protein isolate. Food Chem. 214, 9–15.

Besombes, C., Berka-Zougali, B., Allaf, K., 2010. Instant controlled pressure drop extraction of lavandin essential oils: fundamentals and experimental studies. J. Chromatogr. A 1217, 6807–6815.

Bhuyan, D.J., Van Vuong, Q., Chalmers, A.C., van Altena, I.A., Bowyer, M.C., Scarlett, C.J., 2015. Microwave-assisted extraction of *Eucalyptus robusta* leaf for the optimal yield of total phenolic compounds. Ind. Crops Prod. 69, 290–299.

Bi, W., Tian, M., Zhou, J., Row, K.H., 2010. Task-specific ionic liquid-assisted extraction and separation of astaxanthin from shrimp waste. J. Chromatogr. B 878, 2243–2248.

Botelho, J.R.S., Medeiros, N.G., Rodrigues, A.M.C., Araújo, M.E., Machado, N.T., Guimarães Santos, A., Santos, I.R., Gomes-Leal, W., Carvalho, R.N., 2014. Black sesame (*Sesamum indicum* L.) seeds extracts by CO_2 supercritical fluid extraction: isotherms of global yield, kinetics data, total fatty acids, phytosterols and neuroprotective effects. J. Supercrit. Fluids 93, 49–55.

Boussetta, N., Soichi, E., Lanoisellé, J.-L., Vorobiev, E., 2014. Valorization of oilseed residues: extraction of polyphenols from flaxseed hulls by pulsed electric fields. Ind. Crops Prod. 52, 347–353.

Cai, Z., Qu, Z., Lan, Y., Zhao, S., Ma, X., Wan, Q., Jing, P., Li, P., 2016. Conventional, ultrasound-assisted, and accelerated-solvent extractions of anthocyanins from purple sweet potatoes. Food Chem. 197, 266–272.

Caldera, G., Figueroa, Y., Vargas, M., Santos, D.T., Marquina-Chidsey, G., 2012. Optimization of supercritical fluid extraction of antioxidant compounds from Venezuelan rosemary leaves. Int. J. Food Eng. 8 (4), 1–14.

Cardenas-Toro, F.P., Forster-Carneiro, T., Rostagno, M.A., Petenate, A.J., Maugeri Filho, F., Meireles, M.A.A., 2014. Integrated supercritical fluid extraction and subcritical water hydrolysis for the recovery of bioactive compounds from pressed palm fiber. J. Supercrit. Fluids 93, 42–48.

Cardoso-Ugarte, G.A., Sosa-Morales, M.E., Ballard, T., Liceaga, A., San Martín-González, M.F., 2014. Microwave-assisted extraction of betalains from red beet (*Beta vulgaris*). LWT Food Sci. Technol. 59, 276–282.

Carniel, N., Dallago, R.M., Dariva, C., Bender, J.P., Nunes, A.L., Zanella, O., Bilibio, D., Luiz Priamo, W., 2016. Microwave-assisted extraction of phenolic acids and flavonoids from *Physalis angulata*. J. Food Process Eng. 40 (3), 40(3), 1–11.

Carr, A.G., Mammucari, R., Foster, N.R., 2011. A review of subcritical water as a solvent and its utilisation for the processing of hydrophobic organic compounds. Chem. Eng. J. 172, 1–17.

Carrera, C., Ruiz-Rodríguez, A., Palma, M., Barroso, C.G., 2015. Ultrasound-assisted extraction of amino acids from grapes. Ultrason. Sonochem. 22, 499–505.

Carvalho, P.I.N., Osorio-Tobón, J.F., Rostagno, M.A., Petenate, A.J., Meireles, M.A.A., 2015. Techno-economic evaluation of the extraction of turmeric (*Curcuma longa* L.) oil and ar-turmerone using supercritical carbon dioxide. J. Supercrit. Fluids 105, 44–54.

Casazza, A.A., Aliakbarian, B., Mantegna, S., Cravotto, G., Perego, P., 2010. Extraction of phenolics from *Vitis vinifera* wastes using non-conventional techniques. J. Food Eng. 100, 50–55.

Castañeda-González, L., Angarita-Pabón, S., Bernal, F.A., Coy-Barrera, E., 2014. Comparison of conventional and ultrasound-assisted extraction of alkaloids from *Lupinus mirabilis* leaves. Planta Med. 80, PD74.

Castro-Gómez, M.P., Rodriguez-Alcalá, L.M., Calvo, M.V., Romero, J., Mendiola, J.A., Ibañez, E., Fontecha, J., 2014. Total milk fat extraction and quantification of polar and neutral lipids of cow, goat, and ewe milk by using a pressurized liquid system and chromatographic techniques. J. Dairy Sci. 97, 6719–6728.

Celli, G.B., Ghanem, A., Brooks, M.S.-L., 2015. Optimization of ultrasound-assisted extraction of anthocyanins from haskap berries (*Lonicera caerulea* L.) using response surface methodology. Ultrason. Sonochem. 27, 449–455.

Chan, C.H., Yusoff, R., Ngoh, G.C., Kung, F.W.L., 2011. Microwave-assisted extractions of active ingredients from plants. J. Chromatogr. A 1218, 6213–6225.

Chandrapala, J., Oliver, C.M., Kentish, S., Ashokkumar, M., 2013. Use of power ultrasound to improve extraction and modify phase transitions in food processing. Food Rev. Int. 29, 67–91.

Chemat, F., Cravotto, G., 2013. Microwave-Assisted Extraction for Bioactive Compounds. Springer, Verlag Berlin Heidelberg.

Chemat, F., Abert Vian, M., Visinoni F., 2010. Microwave hydro-diffusion for isolation of natural products. United States Patent, US 0,062,121.

Chemat, F., Abert Vian, M., Cravotto, G., 2012. Green extraction of natural products: concept and principles. Int. J. Mol. Sci. 13, 8615–8627.

Chen, F., Zu, Y., Yang, L., 2015a. A novel approach for isolation of essential oil from fresh leaves of *Magnolia sieboldii* using microwave-assisted simultaneous distillation and extraction. Sep. Purif. Technol. 154, 271–280.

Chen, J., Li, J., Sun, A., Zhang, B., Qin, S., Zhang, Y., 2014. Supercritical CO_2 extraction and pre-column derivatization of polysaccharides from *Artemisia sphaerocephala* Krasch. seeds via gas chromatography. Ind. Crops Prod. 60, 138–143.

Chen, M., Zhao, Y., Yu, S., 2015b. Optimisation of ultrasonic-assisted extraction of phenolic compounds, antioxidants, and anthocyanins from sugar beet molasses. Food Chem. 172, 543–550.

Chen, R., Jin, C., Tong, Z., Lu, J., Tan, L., Tian, L., Chang, Q., 2016. Optimization extraction, characterization and antioxidant activities of pectic polysaccharide from tangerine peels. Carbohydr. Polym. 136, 187–197.

Chen, Y., Gu, X., Huang, S., Li, J., Wang, X., Tang, J., 2010. Optimization of ultrasonic/microwave assisted extraction (UMAE) of polysaccharides from *Inonotus obliquus* and evaluation of its anti-tumor activities. Int. J. Biol. Macromol. 46, 429–435.

Cheng, M.C., Chang, W.-H., Chen, C.-W., Li, W.-W., Tseng, C.-Y., Song, T.-Y., 2015. Antioxidant properties of essential oil extracted from *Pinus morrisonicola* hay needles by supercritical fluid and identification of possible active compounds by GC/MS. Molecules 20, 19051–19065.

Cholet, C., Delsart, C., Petrel, M., Gontier, E., Grimi, N., L'Hyvernay, A., Ghidossi, R., Vorobiev, E., Mietton-Peuchot, M., Gény, L., 2014. Structural and biochemical changes induced by pulsed electric field treatments on cabernet sauvignon grape berry skins: impact on cell wall total tannins and polysaccharides. J. Agric. Food Chem. 62 (13), 2925–2934.

Cicerale, S., Lucas, L., Keast, R., 2010. Biological activities of phenolic compounds present in virgin olive oil. Int. J. Mol. Sci. 11, 458–479.

Clodoveo, M.L., Durante, V., La Notte, D., 2013. Working towards the development of innovative ultrasound equipment for the extraction of virgin olive oil. Ultrason. Sonochem. 20, 1261–1270.

Conte, R., Gullich, L.M.D., Bilibio, D., Zanella, O., Bender, J.P., Carniel, N., Priamo, W.L., 2016. Pressurized liquid extraction and chemical characterization of safflower oil: a comparison between methods. Food Chem. 213, 425–430.

Cragg, G.M., Newman, D.J., 2013. Natural products: a continuing source of novel drug leads. Biochim. Biophys. Acta 1830, 3670–3695.

Cui, Q., Peng, X., Yao, X.H., Wei, Z.F., Luo, M., Wang, W., Zhao, C.J., Fu, Y.J., Zu, Y.G., 2015. Deep eutectic solvent-based microwave-assisted extraction of genistin, genistein and apigenin from pigeon pea roots. Sep. Purif. Technol. 150, 63–72.

Cvjetko Bubalo, M., Ćurko, N., Tomašević, M., Kovačević Ganić, K., Radojcic Redovnikovic, I., 2016. Green extraction of grape skin phenolics by using deep eutectic solvents. Food Chem. 200, 159–166.

D'Alessandro, G.L., Dimitrov, K., Vauchel, P., Nikov, I., 2014. Kinetics of ultrasound assisted extraction of anthocyanins from *Aronia melanocarpa* (black chokeberry) wastes. Chem. Eng. Res. Des. 92, 1818–1826.

Da Porto, C., Da Porto, C., Decorti, D., Natolino, A., 2014. Ultrasound-assisted extraction of volatile compounds from industrial *Cannabis sativa* L. inflorescences. Int. J. Appl. Res. Nat. Prod. 7, 8–14.

Dahmoune, F., Nayak, B., Moussi, K., Remini, H., Madani, K., 2015. Optimization of microwave-assisted extraction of polyphenols from *Myrtus communis* L. leaves. Food Chem. 166, 585–595.

Dandekar, D.V., Gaikar, V.G., 2002. Microwave assisted extraction of curcuminoids from *Curcuma longa*. Sep. Sci. Technol. 37, 2669–2690.

Dang, Y.-Y., Zhang, H., Xiu, Z.-L., 2014. Microwave-assisted aqueous two-phase extraction of phenolics from grape (*Vitis vinifera*) seed. J. Chem. Technol. Biotechnol. 89, 1576–1581.

de Aguiar, A.C., dos Santos, P., Coutinho, J.P., Barbero, G.F., Godoy, H.T., Martínez, J., 2014. Supercritical fluid extraction and low pressure extraction of Biquinho pepper (*Capsicum chinense*). LWT Food Sci. Technol. 59, 1239–1246.

de Mello, B.T.F., dos Santos Garcia, V.A., da Silva, C., 2015. Ultrasound-assisted extraction of oil from chia (*Salvia hispânica* L.) seeds: optimization extraction and fatty acid profile. J. Food Process Eng. 40 (1), 1–8.

de Oliveira, C.F., Giordani, D., Gurak, P.D., Cladera-Olivera, F., Marczak, L.D.F., 2015. Extraction of pectin from passion fruit peel using moderate electric field and conventional heating extraction methods. Innov. Food Sci. Emerg. Technol. 29, 201–208.

de Zordi, N., Cortesi, A., Kikic, I., Moneghini, M., Solinas, D., Innocenti, G., Portolan, A., Baratto, G., Dallácqua, S., 2014. The supercritical carbon dioxide extraction of polyphenols from *Propolis*: a central composite design approach. J. Supercrit. Fluids 95, 491–498.

Dejaegher, B., Vander Heyden, Y., 2011. Experimental designs and their recent advances in set-up, data interpretation, and analytical applications. J. Pharm. Biomed. Anal. 56, 141–158.

Desai, M.A., Parikh, J., 2015. Extraction of essential oil from leaves of lemongrass using microwave radiation: optimization, comparative, kinetic, and biological studies. ACS Sustain. Chem. Eng. 3 (3), 421–431.

Desgrouas, C., Baghdikian, B., Mabrouki, F., Bory, S., Taudon, N., Parzy, D., Ollivier, E., 2014. Rapid and green extraction, assisted by microwave and ultrasound of cepharanthine from *Stephania rotunda* Lour. Sep. Purif. Technol. 123, 9–14.

Dey, S., Rathod, V.K., 2013. Ultrasound assisted extraction of β-carotene from *Spirulina platensis*. Ultrason. Sonochem. 20, 271–276.

Dias, A.L.B., Arroio Sergio, C.S., Santos, P., Barbero, G.F., Rezende, C.A., Martínez, J., 2016. Effect of ultrasound on the supercritical CO_2 extraction of bioactive compounds from dedo de moça pepper (*Capsicum baccatum* L. var. *pendulum*). Ultrason. Sonochem. 31, 284–294.

Dranca, F., Oroian, M., 2016. Optimization of ultrasound-assisted extraction of total monomeric anthocyanin (TMA) and total phenolic content (TPC) from eggplant (*Solanum melongena* L.) peel. Ultrason. Sonochem. 31, 637–646.

Du, B., Zhu, F., Xu, B., 2014. β-Glucan extraction from bran of hull-less barley by accelerated solvent extraction combined with response surface methodology. J. Cereal Sci. 59, 95–100.

Duan, L., Jiang, R., Shi, Y., Duan, C., Wu, G., 2016. Optimization of ultrasonic-assisted extraction of higher fatty acids in grape berries (seed-free fruit sections). Anal. Methods. 8, 6208–6215.

Durante, M., Lenucci, M.S., Mita, G., 2014. Supercritical carbon dioxide extraction of carotenoids from pumpkin (*Cucurbita* spp.): a review. Int. J. Mol. Sci. 15, 6725–6740.

Ebrahimi-Najafabadi, H., Leardi, R., Jalali-Heravi, M., 2014. Experimental design in analytical chemistry. Part I: theory. J. AOAC Int. 97, 3–11.

Farías-Campomanes, A.M., Rostagno, M.A., Angela, M., Meireles, A., 2013. Production of polyphenol extracts from grape bagasse using supercritical fluids: yield, extract composition and economic evaluation. J. Supercrit. Fluids 77, 70–78.

Feng, S., Luo, Z., Tao, B., Chen, C., 2015. Ultrasonic-assisted extraction and purification of phenolic compounds from sugarcane (*Saccharum officinarum* L.) rinds. LWT Food Sci. Technol. 60, 970–976.

Feng, X.F., Jing, N., Li, Z.-G., Wei, D., Lee, M.-R., 2014. Ultrasound-microwave hybrid-assisted extraction coupled to headspace solid-phase microextraction for fast analysis of essential oil in dry traditional Chinese medicine by GC–MS. Chromatographia 77, 619–628.

Fernández-Mar, M.I., Mateos, R., García-Parrilla, M.C., Puertas, B., Cantos-Villar, E., 2012. Bioactive compounds in wine: resveratrol, hydroxytyrosol and melatonin: a review. Food Chem. 130, 797–813.

Feuereisen, M.M., Gamero Barraza, M., Zimmermann, B.F., Schieber, A., Schulze-Kaysers, N., 2017. Pressurized liquid extraction of anthocyanins and biflavonoids from *Schinus terebinthifolius* Raddi: a multivariate optimization. Food Chem. 214, 564–571.

Fincan, M., 2015. Extractability of phenolics from spearmint treated with pulsed electric field. J. Food Eng. 162, 31–37.

Franco-Vega, A., Ramírez-Corona, N., Palou, E., López-Malo, A., 2016. Estimation of mass transfer coefficients of the extraction process of essential oil from orange peel using microwave assisted extraction. J. Food Eng. 170, 136–143.

Franquin-Trinquier, S., Maury, C., Baron, A., Le Meurlay, D., Mehinagic, E., 2014. Optimization of the extraction of apple monomeric phenolics based on response surface methodology: comparison of pressurized liquid–solid extraction and manual-liquid extraction. J. Food Compos. Anal. 34, 56–67.

Fu, Q., Li, Z., Sun, C., Xin, H., Ke, Y., Jin, Y., Liang, X., 2015. Rapid and simultaneous analysis of sesquiterpene pyridine alkaloids from *Tripterygium wilfordii* hook. f. Using supercritical fluid chromatography-diode array detector-tandem mass spectrometry. J. Supercrit. Fluids 104, 85–93.

Gañán, N.A., Dias, A.M.A., Bombaldi, F., Zygadlo, J.A., Brignole, E.A., De Sousa, H.C., Braga, M.E.M., 2016. Alkaloids from *Chelidonium majus* L.: fractionated supercritical CO_2 extraction with co-solvents. Sep. Purif. Technol. 165, 199–207.

Gao, F., Yang, S., Birch, J., 2016. Physicochemical characteristics, fatty acid positional distribution and triglyceride composition in oil extracted from carrot seeds using supercritical CO_2. J. Food Compos. Anal. 45, 26–33.

Garcia-Castello, E.M., Rodriguez-Lopez, A.D., Mayor, L., Ballesteros, R., Conidi, C., Cassano, A., 2015. Optimization of conventional and ultrasound assisted extraction of flavonoids from grapefruit (*Citrus paradisi* L.) solid wastes. LWT Food Sci. Technol. 64, 1114–1122.

Garmus, T.T., Paviani, L.C., Queiroga, C.L., Cabral, F.A., 2015. Extraction of phenolic compounds from pepper-rosmarin (*Lippia sidoides* Cham.) leaves by sequential extraction in fixed bed extractor using supercritical CO_2, ethanol and water as solvents. J. Supercrit. Fluids 99, 68–75.

Garofulić, E.I., Dragović-Uzelac, V., Režek Jambrak, A., Jukić, M., 2013. The effect of microwave assisted extraction on the isolation of anthocyanins and phenolic acids from sour cherry Marasca (*Prunus cerasus* var. Marasca). J. Food Eng. 117, 437–442.

Gavahian, M., Farahnaky, A., Farhoosh, R., Javidnia, K., Shahidi, F., 2015. Extraction of essential oils from *Mentha piperita* using advanced techniques: microwave versus ohmic assisted hydrodistillation. Food Bioprod. Process 94, 50–58.

Ghasemzadeh, A., Ghasemzadeh, N., 2011. Flavonoids and phenolic acids: role and biochemical activity in plants and human. J. Med. Plants Res. 5 (31), 6697–6703.

Ghasemzadeh, A., Jaafar, H.Z., Karimi, E., et al., 2014. Optimization of ultrasound-assisted extraction of flavonoid compounds and their pharmaceutical activity from curry leaf (*Murraya koenigii* L.) using response surface methodology. BMC Complement. Altern. Med. 14, 318.

Gholivand, M.B., Yamini, Y., Dayeni, M., 2014. Optimization and comparison of ultrasound-assisted extraction of estragole from tarragon leaves with hydro-distillation method. Anal. Bioanal. Chem. Res. 1, 99–107.

Golmakani, M.T., Moayyedi, M., 2015. Comparison of heat and mass transfer of different microwave-assisted extraction methods of essential oil from *Citrus limon* (Lisbon variety) peel. Food Sci. Nutr. 3, 506–518.

Golmakani, M.T., Mendiola, J.A., Rezaei, K., Ibáñez, E., 2014. Pressurized limonene as an alternative bio-solvent for the extraction of lipids from marine microorganisms. J. Supercrit. Fluids 92, 1–7.

Gonzales, G.B., Smagghe, G., Raes, K., Camp, J., Van, 2014. Combined alkaline hydrolysis and ultrasound-assisted extraction for the release of nonextractable phenolics from cauliflower (*Brassica oleracea* var. botrytis) waste. J. Agric. Food Chem. 62 (15), 3371–3376.

González-Arrojo, A., Soldado, A., Vicente, F., de la Roza-Delgado, B., 2015. Microwave-assisted methodology feasibility for one-step extraction and transmethylation of fatty acids in milk for GC-mass spectrometry. Food Anal. Methods 8, 2250–2260.

González-Rivera, J., Duce, C., Falconieri, D., Ferrari, C., Ghezzi, L., Piras, A., Tine, M.R., 2016. Coaxial microwave assisted hydrodistillation of essential oils from five different herbs (lavender, rosemary, sage, fennel seeds and clove buds): chemical composition and thermal analysis. Innov. Food Sci. Emerg. Technol. 33, 308–318.

Grassino, A.N., Brnčić, M., Vikić-Topić, D., Roca, S., Dent, M., Brnčić, S.R., 2016. Ultrasound assisted extraction and characterization of pectin from tomato waste. Food Chem. 198, 93–100.

Grimi, N., Praporscic, I., Lebovka, N., Vorobiev, E., 2007. Selective extraction from carrot slices by pressing and washing enhanced by pulsed electric fields. Sep. Purif. Technol. 58 (2), 267–273.

Gu, H., Chen, F., Zhang, Q., Zang, J., 2016. Application of ionic liquids in vacuum microwave-assisted extraction followed by macroporous resin isolation of three flavonoids rutin, hyperoside and hesperidin from *Sorbus tianschanica* leaves. J. Chromatogr. B 1014, 45–55.

Hadi, B., Sanagi, M.M., Wan Ibrahim, W.A., Jamil, S., Mu'azu, M.A., Aboul-Enein, H.Y., 2015a. Ultrasonic-assisted extraction of curcumin complexed with methyl-β-cyclodextrin. Food Anal. Methods 8, 1373–1381.

Hadi, B.J., Sanagi, M.M., Aboul-Enein, H.Y., Ibrahim, W.A.W., Jamil, S., Mu'azu, M.A., 2015b. Microwave-assisted extraction of methyl β-cyclodextrin-complexed curcumin from turmeric rhizome oleoresin. Food Anal. Methods 8, 2447–2456.

Harvey, A.L., 2008. Natural products in drug discovery. Drug Discov. Today 13, 894–901.

He, B., Zhang, L.-L., Yue, X.-Y., Liang, J., Jiang, J., Gao, X.-L., Yue, P.-X., 2016. Optimization of ultrasound-assisted extraction of phenolic compounds and anthocyanins from blueberry (*Vaccinium ashei*) wine pomace. Food Chem. 204, 70–76.

Heleno, S.A., Diz, P., Prieto, M.A., Barros, L., Rodrigues, A., Barreiro, M.F., Ferreira, I.C.F.R., 2016. Optimization of ultrasound-assisted extraction to obtain mycosterols from *Agaricus bisporus* L. by response surface methodology and comparison with conventional Soxhlet extraction. Food Chem. 197, 1054–1063.

Heo, J.Y., Kim, S., Kang, J.H., Moon, B., 2014. Determination of lutein from green tea and green tea by-products using accelerated solvent extraction and UPLC. J. Food Sci. 79, 816–821.

Hernández-Santos, B., Rodríguez-Miranda, J., Herman-Lara, E., Torruco-Uco, J.G., Carmona-García, R., Juárez-Barrientos, J.M., Chávez-Zamudio, R., Martínez-Sánchez, C.E., 2016. Effect of oil extraction assisted by ultrasound on the physicochemical properties and fatty acid profile of pumpkin seed oil (*Cucurbita pepo*). Ultrason. Sonochem. 31, 429–436.

Herrero, M., Cifuentes, A., Ibañez, E., 2006. Sub- and supercritical fluid extraction of functional ingredients from different natural sources: plants, food-by-products, algae and microalgae: a review. Food Chem. 98, 136–148.

Hiranvarachat, B., Devahastin, S., 2014. Enhancement of microwave-assisted extraction via intermittent radiation: extraction of carotenoids from carrot peels. J. Food Eng. 126, 17–26.

Hiranvarachat, B., Devahastin, S., Chiewchan, N., 2012. In vitro bioaccessibility of β-carotene in dried carrots pretreated by different methods. Int. J. Food Sci. Technol. 47, 535–541.

Ho, T.D., Canestraro, A.J., Anderson, J.L., 2011. Analytica Chimica Acta ionic liquids in solid-phase microextraction: a review. Anal. Chim. Acta 695, 18–43.

Ho, K.K.H.Y., Ferruzzi, M.G., Liceaga, A.M., San Martín-González, M.F., 2015. Microwave-assisted extraction of lycopene in tomato peels: effect of extraction conditions on all-*trans* and *cis*-isomer yields. LWT Food Sci. Technol. 62, 160–168.

Hossain, M.B., Tiwari, B.K., Gangopadhyay, N., O'Donnell, C.P., Brunton, N.P., Rai, D.K., 2014. Ultrasonic extraction of steroidal alkaloids from potato peel waste. Ultrason. Sonochem. 21, 1470–1476.

Hossain, M., Rawson, A., Aguiló-Aguayo, I., Brunton, N., Rai, D., 2015a. Recovery of steroidal alkaloids from potato peels using pressurized liquid extraction. Molecules 20, 8560–8573.

Hossain, M.B., Aguiló-Aguayo, I., Lyng, J.G., Brunton, N.P., Rai, D.K., 2015b. Effect of pulsed electric field and pulsed light pre-treatment on the extraction of steroidal alkaloids from potato peels. Innov. Food Sci. Emerg. Technol. 29, 9–14.

Hosseini, S.S., Khodaiyan, F., Yarmand, M.S., 2016. Optimization of microwave assisted extraction of pectin from sour orange peel and its physicochemical properties. Carbohydr. Polym. 140, 59–65.

Hurtado-Benavides, A., Dorado A., D., Sánchez-Camargo, P. del A., 2016. Study of the fatty acid profile and the aroma composition of oil obtained from roasted Colombian coffee beans by supercritical fluid extraction. J. Supercrit. Fluids 113, 44–52.

Ibañez, E., Herrero, M., Mendiola, J.A., Castro-Puyana, M., 2012. Extraction and characterization of bioactive compounds with health benefits from marine resources: macro and micro algae, cyanobacteria, and invertebrate. In: Hayes, M. (Ed.), Marine Bioactive Compounds. Springer, Verlag Berlin Heidelberg, pp. 55–98.

Icyer, N.C., Toker, O.S., Karasu, S., Tornuk, F., Bozkurt, F., Arici, M., Sagdic, O., 2016. Combined design as a useful statistical approach to extract maximum amount of phenolic compounds from virgin olive oil waste. LWT Food Sci. Technol. 70, 24–32.

Islam, M.N., Jo, Y.T., Park, J.H., 2012. Remediation of PAHs contaminated soil by extraction using subcritical water. J. Ind. Eng. Chem. 18, 1689–1693.

Izadiyan, P., Hemmateenejad, B., 2016. Multi-response optimization of factors affecting ultrasonic assisted extraction from Iranian basil using central composite design. Food Chem. 190, 864–870.

Jablonský, M., Vernarecová, M., Ház, A., Dubinyová, L., Škulcová, A., Sladková, A., Šurina, I., 2015. Extraction of phenolic and lipophilic compounds from spruce (*Picea abies*) bark using accelerated solvent extraction by ethanol. Wood Res. 6 (4), 583–590.

Jain, T., Jain, V., Pandey, R., 2009. Microwave assisted extraction for phytoconstituents—an overview. Asian J. Res. Chem. 2, 19–25.

Jeleń, H.H., Majcher, M., Dziadas, M., 2012. Microextraction techniques in the analysis of food flavor compounds: a review. Anal. Chim. Acta 738, 13–26.

Jentzer, J.-B., Alignan, M., Vaca-Garcia, C., Rigal, L., Vilarem, G., 2015. Response surface methodology to optimise accelerated solvent extraction of steviol glycosides from *Stevia rebaudiana Bertoni* leaves. Food Chem. 166, 561–567.

Jesus, P.S., Meireles, M.A.M., 2014. Supercritical fluid extraction: a global perspective of the fundamental concepts of this eco-friendly extraction technique. In: Chemat, F., Abert Vian, M. (Eds.), Alternative Solvents for Natural Product Extraction. Springer, Verlag Berlin Heidelberg, pp. 39–72.

Jiang, Y.H., Jiang, X.L., Ma, Q.H., 2013. Optimization of microwave-assisted extraction conditions of anthocyanins from black corn hull using response surface methodology. Adv. Mater. Res. 864-867, 536–540.

Jiao, J., Li, Z.-G., Gai, Q.-Y., Li, X.-J., Wei, F.-Y., Fu, Y.-J., Ma, W., 2014. Microwave-assisted aqueous enzymatic extraction of oil from pumpkin seeds and evaluation of its physicochemical properties, fatty acid compositions and antioxidant activities. Food Chem. 147, 17–24.

Jiao, J., Gai, Q.-Y., Zhang, L., Wang, W., Luo, M., Zu, Y.-G., Fu, Y.-J., 2015. High-speed homogenization coupled with microwave-assisted extraction followed by liquid chromatography–tandem mass spectrometry for the direct determination of alkaloids and flavonoids in fresh *Isatis tinctoria* L. hairy root cultures. Anal. Bioanal. Chem. 407, 4841–4848.

Joana Gil-Chávez, G., Villa, J.A., Fernando Ayala-Zavala, J., Basilio Heredia, J., Sepulveda, D., Yahia, E.M., González-Aguilar, G.A., 2013. Technologies for extraction and production of bioactive compounds to be used as nutraceuticals and food ingredients: an overview. Compr. Rev. Food Sci. Food Saf. 12, 5–23.

Kang, J.-H., Kim, S., Moon, B., 2016. Optimization by response surface methodology of lutein recovery from paprika leaves using accelerated solvent extraction. Food Chem. 205, 140–145.

Katsampa, P., Valsamedou, E., Grigorakis, S., Makris, D.P., 2015. A green ultrasound-assisted extraction process for the recovery of antioxidant polyphenols and pigments from onion solid wastes using Box–Behnken experimental design and kinetics. Ind. Crops Prod. 77, 535–543.

Kerbstadt, S., Eliasson, L., Mustafa, A., Ahrné, L., 2015. Effect of novel drying techniques on the extraction of anthocyanins from bilberry press cake using supercritical carbon dioxide. Innov. Food Sci. Emerg. Technol. 29, 209–214.

Kha, T.C., Nguyen, M.H., Phan, D.T., Roach, P.D., Stathopoulos, C.E., 2013. Optimisation of microwave-assisted extraction of Gac oil at different hydraulic pressure, microwave and steaming conditions. Int. J. Food Sci. Technol. 48, 1436–1444.

Kha, T.C., Phan-Tai, H., Nguyen, M.H., 2014. Effects of pre-treatments on the yield and carotenoid content of Gac oil using supercritical carbon dioxide extraction. J. Food Eng. 120, 44–49.

Khajeh, M., 2011. Optimization of process variables for essential oil components from *Satureja hortensis* by supercritical fluid extraction using Box-Behnken experimental design. J. Supercrit. Fluids 55, 944–948.

Khajenoori, M., Asl, A.H., Eikani, M.H., 2015. Subcritical water extraction of essential oils from *Trachyspermum ammi* Seeds. J. Essent. Oil Bear Pl. 18 (5), 1165–1173.

Khezeli, T., Daneshfar, A., Sahraei, R., 2016. A green ultrasonic-assisted liquid–liquid microextraction based on deep eutectic solvent for the HPLC-UV determination of ferulic, caffeic and cinnamic acid from olive, almond, sesame, and cinnamon oil. Talanta 150, 577–585.

Khodaei, N., Karboune, S., Orsat, V., 2016. Microwave-assisted alkaline extraction of galactan-rich rhamnogalacturonan I from potato cell wall by-product. Food Chem. 190, 495–505.

Khoza, B.S., Dubery, I.A., Byth-Illing, H.-A., Steenkamp, P.A., Chimuka, L., Madala, N.E., 2016. Optimization of pressurized hot water extraction of flavonoids from *Momordica foetida* using UHPLC-qTOF-MS and multivariate chemometric approaches. Food Anal. Methods 9, 1480–1489.

Kiamahalleh, V.M., Najafpour-Darzi, G., Rahimnejad, M., Moghadamnia, A.A., Valizadeh Kiamahalleh, M., 2016. High performance curcumin subcritical water extraction from turmeric (*Curcuma longa* L.). J. Chromatogr. B 1022, 191–198.

Kim, J.S., An, C.G., Park, J.S., Lim, Y.P., Kim, S., 2016. Carotenoid profiling from 27 types of paprika (*Capsicum annuum* L.) with different colors, shapes, and cultivation methods. Food Chem. 201, 64–71.

Klejdus, B., Lojková, L., Plaza, M., Šnóblová, M., Štěrbová, D., 2010. Hyphenated technique for the extraction and determination of isoflavones in algae: ultrasound-assisted supercritical fluid extraction followed by fast chromatography with tandem mass spectrometry. J. Chromatogr. A 1217, 7956–7965.

Ko, M.J., Cheigh, C.I., Chung, M.S., 2014. Relationship analysis between flavonoids structure and subcritical water extraction (SWE). Food Chem. 143, 147–155.

Košťálová, Z., Aguedo, M., Hromádková, Z., 2016. Microwave-assisted extraction of pectin from unutilized pumpkin biomass. Chem. Eng. Process. Process Intensif. 102, 9–15.

Kraujalis, P., Venskutonis, P.R., Pukalskas, A., Kazernavičiute, R., 2013. Accelerated solvent extraction of lipids from *Amaranthus* spp. seeds and characterization of their composition. LWT Food Sci. Technol. 54, 528–534.

Krishnan, Y.R., Rajan, K.S., 2016. Microwave assisted extraction of flavonoids from *Terminalia bellerica*: study of kinetics and thermodynamics. Sep. Purif. Technol. 157, 169–178.

Kumcuoglu, S., Yilmaz, T., Tavman, S., 2014. Ultrasound assisted extraction of lycopene from tomato processing wastes. J. Food Sci. Technol. 51, 4102–4107.

Kusuma, H.S., Mahfud, M., 2016. Chemical composition of essential oil of Indonesia sandalwood extracted by microwave-assisted hydrodistillation. AIP Conf. Proc. 1755, 050001.

Kwon, H.-L., Chung, M.-S., 2015. Pilot-scale subcritical solvent extraction of curcuminoids from *Curcuma longa* L. Food Chem. 185, 58–64.

Lai, Y.S., Parameswaran, P., Li, A., Baez, M., Rittmann, B.E., 2014. Effects of pulsed electric field treatment on enhancing lipid recovery from the microalga, *Scenedesmus*. Biores. Technol. 173, 457–461.

Lantzouraki, D.Z., Sinanoglou, V.J., Zoumpoulakis, P., Proestos, C., 2015. Comparison of the antioxidant and antiradical activity of pomegranate (*Punica granatum* L.) by ultrasound-assisted and classical extraction. Anal. Lett. 49 (7), 969–978.

Laokuldilok, N., Kopermsub, P., Thakeow, P., Utama-Ang, 2015. Microwave assisted extraction of bioactive compounds from turmeric (*Curcuma longa*). J. Agric. Technol. 11, 1185–1196.

Larkeche, O., Zermane, A., Meniai, A.-H., Crampon, C., Badens, E., 2015. Supercritical extraction of essential oil from *Juniperus communis* L. needles: application of response surface methodology. J. Supercrit. Fluids 99, 8–14.

Leardi, R., 2009. Experimental design in chemistry: a tutorial. Anal. Chim. Acta 652, 161–172.

Lee, A.Y., Kim, H.S., Choi, G., Moon, B.C., Chun, J.M., Kim, H.K., 2014. Optimization of ultrasonic-assisted extraction of active compounds from the fruit of star anise by using response surface methodology. Food Anal. Methods 7, 1661–1670.

Li, Y., Fabiano-Tixier, A.S., Tomao, V., Cravotto, G., Chemat, F., 2013a. Green ultrasound-assisted extraction of carotenoids based on the bio-refinery concept using sunflower oil as an alternative solvent. Ultrason. Sonochem. 20, 12–18.

Li, Y., Fabiano-Tixier, A.S., Vian, M.A., Chemat, F., 2013b. Solvent-free microwave extraction of bioactive compounds provides a tool for green analytical chemistry. TrAC Trends Anal. Chem. 47, 1–11.

Li, M., Ngadi, M.O., Ma, Y., 2014. Optimisation of pulsed ultrasonic and microwave-assisted extraction for curcuminoids by response surface methodology and kinetic study. Food Chem. 165, 29–34.

Li, P., Tian, L., Li, T., 2015. Study on Ultrasonic-Assisted Extraction of Essential Oil from Cinnamon Bark and Preliminary Investigation of Its Antibacterial Activity. Springer, Berlin Heidelberg, pp. 349–360.

Li, S., Chen, F., Jia, J., Liu, Z., Gu, H., Yang, L., Wang, F., Yang, F., 2016. Ionic liquid-mediated microwave-assisted simultaneous extraction and distillation of gallic acid, ellagic acid and essential oil from the leaves of *Eucalyptus camaldulensis*. Sep. Purif. Technol. 168, 8–18.

Lianfu, Z., Zelong, L., 2008. Optimization and comparison of ultrasound/microwave assisted extraction (UMAE) and ultrasonic assisted extraction (UAE) of lycopene from tomatoes. Ultrason. Sonochem. 15, 731–737.

Liu, F., 2015. Optimisation of ultrasonic-microwave-assisted extraction conditions for polysaccharides from mulberry (*Morus atropurpurea* Roxb) leaves and evaluation of antioxidant activities in vitro. Med. Chem. 5 (2), 90–95.

Liu, T., Sui, X., Zhang, R., Yang, L., Zu, Y., Zhang, L., 2011. Application of ionic liquids based microwave-assisted simultaneous extraction of carnosic acid, rosmarinic acid and essential oil from *Rosmarinus officinalis*. J. Chromatogr. A 1218, 8480–8489.

Liu, J., Sandahl, M., Sjöberg, P.J.R., Turner, C., 2014. Pressurised hot water extraction in continuous flow mode for thermolabile compounds: extraction of polyphenols in red onions. Anal. Bioanal. Chem. 406, 441–445.

Liu, C., Sun, Y., Mao, Q., Guo, X., Li, P., Liu, Y., Xu, N., 2016a. Characteristics and antitumor activity of *Morchella esculenta* polysaccharide extracted by pulsed electric field. Int. J. Mol. Sci. 17, 986.

Liu, Z., Chen, Z., Han, F., Kang, X., Gu, H., Yang, L., 2016b. Microwave-assisted method for simultaneous hydrolysis and extraction in obtaining ellagic acid, gallic acid and essential oil from *Eucalyptus globulus* leaves using Brönsted acidic ionic liquid $[HO_3S(CH_2)4mim]HSO_4$. Ind. Crops Prod. 81, 152–161.

Loginova, K.V., Vorobiev, E., Bals, O., Lebovka, N.I., 2011. Pilot study of countercurrent cold and mild heat extraction of sugar from sugar beets, assisted by pulsed electric fields. J. Food Eng. 102, 340–347.

López-Giral, N., González-Arenzana, L., González-Ferrero, C., López, R., Santamaría, P., López-Alfaro, I., Garde-Cerdán, T., 2015. Pulsed electric field treatment to improve the phenolic compound extraction from Graciano, Tempranillo and Grenache grape varieties during two vintages. Innov. Food Sci. Emerg. Technol. 28, 31–39.

Lou, Z., Wang, H., Zhu, S., Chen, S., Zhang, M., Wang, Z., 2012. Analytica chimica acta ionic liquids based simultaneous ultrasonic and microwave assisted extraction of phenolic compounds from burdock leaves. Anal. Chim. Acta 716, 28–33.

Lu, C., Wang, H., Lv, W., Ma, C., Xu, P., Zhu, J., Xie, J., Liu, B., 2012. Ionic liquid-based ultrasonic/microwave-assisted extraction combined with UPLC for the determination of tannins in *Galla chinensis*. Nat. Prod. Res. 26 (19), 1842–1847.

Lu, J., Feng, X., Han, Y., Xue, C., 2014a. Optimization of subcritical fluid extraction of carotenoids and chlorophyll a from *Laminaria japonica* Aresch by response surface methodology. J. Sci. Food Agric. 94, 139–145.

Lu, Q.F., Pan, L.F., Chen, M., Qiu, Y., Xie, B.H., 2014b. Research on the microwave-assisted supercritical CO_2 extraction of alkaloids from *Gynura segetum (Lour.) Merr.* Adv. Mater. Res. 988, 390–396.

Luengo, E., Condón-Abanto, S., Condón, S., Àlvarez, I., Raso, J., 2014a. Improving the extraction of carotenoids from tomato waste by application of ultrasound under pressure. Sep. Purif. Technol. 136, 130–136.

Luengo, E., Condón-Abanto, S., Álvarez, I., Raso, J., 2014b. Effect of pulsed electric field treatments on permeabilization and extraction of pigments from *Chlorella vulgaris*. J. Membr. Biol. 247, 1269–1277.

Luengo, E., Martínez, J.M., Bordetas, A., Álvarez, I., Raso, J., 2015. Influence of the treatment medium temperature on lutein extraction assisted by pulsed electric fields from *Chlorella vulgaris*. Innov. Food Sci. Emerg. Technol. 29, 15–22.

Luque de Castro, M.D., Priego-Capote, F., 2010. Soxhlet extraction: past and present panacea. J. Chromatogr. A 1217, 2383–2389.

Ma, W., Lu, Y., Hu, R., Chen, J., Zhang, Z., Pan, Y., 2010. Application of ionic liquids based microwave-assisted extraction of three alkaloids *N*-nornuciferine, O-nornuciferine, and nuciferine from lotus leaf. Talanta 80, 1292–1297.

Ma, C., Liu, T., Yang, L., Zu, Y., Wang, S., Zhang, R., 2011. Study on ionic liquid-based ultrasonic-assisted extraction of biphenyl cyclooctene lignans from the fruit of *Schisandra chinensis Baill.* Anal. Chim. Acta 689, 110–116.

Ma, Q., Fan, X.-D., Liu, X.-C., Qiu, T.-Q., Jiang, J.-G., 2015. Ultrasound-enhanced subcritical water extraction of essential oils from *Kaempferia galangal* L. and their comparative antioxidant activities. Sep. Purif. Technol. 150, 73–79.

Maran, J.P., Priya, B., 2015. Supercritical fluid extraction of oil from muskmelon (*Cucumis melo*) seeds. J. Taiwan Inst. Chem. Eng 47, 71–78.

Maran, J.P., Manikandan, S., Priya, B., Gurumoorthi, P., 2015. Box-Behnken design based multi-response analysis and optimization of supercritical carbon dioxide extraction of bioactive flavonoid compounds from tea (*Camellia sinensis* L.) leaves. J. Food Sci. Technol. 52, 92–104.

Markom, M., Singh, H., Hasan, M., 2001. Supercritical CO_2 fractionation of crude palm oil. J. Supercrit. Fluids 20, 45–53.

Marques, L.L.M., Panizzon, G.P., Aguiar, B.A.A., Simionato, A.S., Cardozo-Filho, L., Andrade, G., de Oliveira, A.G., Guedes, T.A., Mello, J.C.P. de, 2016. Guaraná (*Paullinia cupana*) seeds: selective supercritical extraction of phenolic compounds. Food Chem. 212, 703–711.

Martinez, J.L., Vance, S.W., 2007. Supercritical extraction plants, equipment, process and costs. In: Martinez, J.L. (Ed.), Supercritical Fluid Extraction of Nutraceuticals and Bioactive Compounds. Taylor & Francis Group, New York, NY, pp. 25–49.

Martins, P.L.G., de Rosso, V.V., 2016. Thermal and light stabilities and antioxidant activity of carotenoids from tomatoes extracted using an ultrasound-assisted completely solvent-free method. Food Res. Int. 82, 156–164.

Mechergui, K., Jaouadi, W., Coelho, J.P., Serra, M.C., Marques, A.V., Palavra, A.M.F., Khouja, M.L., Boukhchina, S., 2014. Chemical composition and antioxidant activity of Tunisian *Origanum glandulosum* Desf. essential oil and volatile oil obtain by supercritical CO_2 extraction. Int. J. Adv. Res. 2, 337–343.

Medina, A.L., da Silva, M.A.O., de Sousa Barbosa, H., Arruda, M.A.Z., Marsaioli, A., Bragagnolo, N., 2015. Rapid microwave assisted extraction of meat lipids. Food Res. Int. 78, 124–130.

Medina-Meza, I.G., Barbosa-Cánovas, G.V., 2015. Assisted extraction of bioactive compounds from plum and grape peels by ultrasonics and pulsed electric fields. J. Food Eng. 166, 268–275.

Mendes, R.L., Nobre, B.P., Cardoso, M.T., Pereira, A.P., Palavra, A.F., 2003. Supercritical carbon dioxide extraction of compounds with pharmaceutical importance from microalgae. Inorganica Chim. Acta 356, 328–334.

Meneses, M.A., Caputo, G., Scognamiglio, M., Reverchon, E., Adami, R., 2015. Antioxidant phenolic compounds recovery from *Mangifera indica* L. by-products by supercritical antisolvent extraction. J. Food Eng. 163, 45–53.

Mezzomo, N., Martínez, J., Maraschin, M., Ferreira, S.R.S., 2013. Pink shrimp (*P. brasiliensis* and *P. paulensis*) residue: supercritical fluid extraction of carotenoid fraction. J. Supercrit. Fluids 74, 22–33.

Michel, T., Destandau, E., Elfakir, C., 2011. Evaluation of a simple and promising method for extraction of antioxidants from sea buckthorn (*Hippophae rhamnoides* L.) berries: pressurised solvent-free microwave assisted extraction. Food Chem. 126, 1380–1386.

Minjares-Fuentes, R., Femenia, A., Garau, M.C., Candelas-Cadillo, M.G., Simal, S., Rosselló, C., 2016. Ultrasound-assisted extraction of hemicelluloses from grape pomace using response surface methodology. Carbohydr. Polym. 138, 180–191.

Mishra, S., Aeri, V., 2015. Optimization of microwave-assisted extraction conditions for preparing lignan-rich extract from *Saraca asoca* bark using Box–Behnken design. Pharm. Biol. 54 (7), 1255–1262.

Moity, L., Durand, M., Benazzouz, A., Molinier, V., Aubry, J.M., 2014. In silico search for alternative green solvents. In: Chemat, F., Abert Vian, M. (Eds.), Alternative Solvents for Natural Products Extraction. Springer, New York, NY, pp. 1–24.

Mokgadi, J., Turner, C., Torto, N., 2013. Pressurized hot water extraction of alkaloids in goldenseal. Am. J. Anal. Chem. 04, 398–403.

Monrad, J.K., Suárez, M., Motilva, M.J., King, J.W., Srinivas, K., Howard, L.R., 2014. Extraction of anthocyanins and flavan-3-ols from red grape pomace continuously by coupling hot water extraction with a modified expeller. Food Res. Int. 65, 77–87.

Monroy, Y.M., Rodrigues, R.A.F., Sartoratto, A., Cabral, F.A., 2016. Influence of ethanol, water, and their mixtures as co-solvents of the supercritical carbon dioxide in the extraction of phenolics from purple corn cob (*Zea mays* L.). J. Supercrit. Fluids. 118, 11–18.

Moorthy, I.G., Maran, J.P., Surya, S.M., Naganyashree, S., Shivamathi, C.S., 2015. Response surface optimization of ultrasound assisted extraction of pectin from pomegranate peel. Int. J. Biol. Macromol. 72, 1323–1328.

Mushtaq, M., Sultana, B., Anwar, F., Adnan, A., Rizvi, S.S.H., 2015. Enzyme-assisted supercritical fluid extraction of phenolic antioxidants from pomegranate peel. J. Supercrit. Fluids 104, 122–131.

Mustafa, A., Turner, C., 2011. Pressurized liquid extraction as a green approach in food and herbal plants extraction: a review. Anal. Chim. Acta 703, 8–18.

Nam, M.W., Zhao, J., Lee, M.S., Jeong, J.H., Lee, J., et al., 2015. Enhanced extraction of bioactive natural products using tailor-made deep eutectic solvents: application to flavonoid extraction from *Flos sophorae*. Green Chem. 17, 1718–1727.

Nde, D.B., Boldor, D., Astete, C., 2015. Optimization of microwave assisted extraction parameters of neem (*Azadirachta indica* A. Juss) oil using the Doehlert's experimental design. Ind. Crops Prod. 65, 233–240.

Nde, D.B., Boldor, D., Astete, C., Muley, P., Xu, Z., 2016. Oil extraction from sheanut (*Vitellaria paradoxa Gaertn* C.F.) kernels assisted by microwaves. J. Food Sci. Technol. 53, 1424–1434.

Nowak, J., Woźniakiewicz, M., Klepacki, P., Sowa, A., Kościelniak, P., 2016. Identification and determination of ergot alkaloids in Morning Glory cultivars. Anal. Bioanal. Chem. 408, 3093–3102.

Omar, J., Alonso, I., Garaikoetxea, A., Etxebarria, N., 2013. Optimization of focused ultrasound extraction (FUSE) and supercritical fluid extraction (SFE) of citrus peel volatile oils and antioxidants. Food Anal. Methods 6, 1244–1252.

Ordóñez-Santos, L.E., Pinzón-Zarate, L.X., González-Salcedo, L.O., 2015. Optimization of ultrasonic-assisted extraction of total carotenoids from peach palm fruit (*Bactris gasipaes*) by-products with sunflower oil using response surface methodology. Ultrason. Sonochem. 27, 560–566.

Osorio-Tobón, J.F., Carvalho, P.I.N., Rostagno, M.A., Petenate, A.J., Meireles, M.A.A., 2016. Precipitation of curcuminoids from an ethanolic turmeric extract using a supercritical antisolvent process. J. Supercrit. Fluids 108, 26–34.

Pabby, A.K., Sastre, A.M., 2013. State-of-the-art review on hollow fibre contactor technology and membrane-based extraction processes. J. Memb. Sci. 430, 263–303.

Paes, J., Dotta, R., Barbero, G.F., Martínez, J., 2014. Extraction of phenolic compounds and anthocyanins from blueberry (*Vaccinium myrtillus* L.) residues using supercritical CO_2 and pressurized liquids. J. Supercrit. Fluids 95, 8–16.

Palanisamy, M., Aldars-García, L., Gil-Ramírez, A., Ruiz-Rodríguez, A., Marín, F.R., Reglero, G., Soler-Rivas, C., 2014. Pressurized water extraction of β-glucan enriched fractions with bile acids-binding capacities obtained from edible mushrooms. Biotechnol. Prog. 30, 391–400.

Pan, J., Muppaneni, T., Sun, Y., Reddy, H.K., Fu, J., Lu, X., Deng, S., 2016. Microwave-assisted extraction of lipids from microalgae using an ionic liquid solvent [BMIM][HSO_4]. Fuel 178, 49–55.

Parniakov, O., Lebovka, N.I., Van Hecke, E., Vorobiev, E., 2014. Pulsed electric field assisted pressure extraction and solvent extraction from mushroom (*Agaricus Bisporus*). Food Bioproc. Technol. 7, 174–183.

Pasquel Reátegui, J.L., Machado, A.P.D.F., Barbero, G.F., Rezende, C.A., Martínez, J., 2014. Extraction of antioxidant compounds from blackberry (*Rubus* sp.) bagasse using supercritical CO_2 assisted by ultrasound. J. Supercrit. Fluids 94, 223–233.

Paula, J.T., Paviani, L.C., Foglio, M.A., Sousa, I.M.O., Duarte, G.H.B., Jorge, M.P., Eberlin, M.N., Cabral, F.A., 2014. Extraction of anthocyanins and luteolin from *Arrabidaea chica* by sequential extraction in fixed bed using supercritical CO_2, ethanol and water as solvents. J. Supercrit. Fluids 86, 100–107.

Pavlić, B., Vidović, S., Vladić, J., Radosavljević, R., Zeković, Z., 2015. Isolation of coriander (*Coriandrum sativum* L.) essential oil by green extractions versus traditional techniques. J. Supercrit. Fluids 99, 23–28.

Pena-Pereira, F., Namieśnik, J., 2014. Ionic liquids and deep eutectic mixtures: sustainable solvents for extraction processes. ChemSusChem 7, 1784–1800.

Pereda, S., Bottini, S.B., Brignole, E.A., 2007. Fundamentals of supercritical fluid technology. In: Martinez, J.L. (Ed.), Supercritical Fluid Extraction of Nutraceuticals and Bioactive Compounds. Taylor & Francis Group, New York, NY, pp. 1–24.

Pereira, C.G., Meireles, M.A.A., 2010. Supercritical fluid extraction of bioactive compounds: fundamentals, applications and economic perspectives. Food Bioproc. Technol. 3, 340–372.

Petigny, L., Périno-Issartier, S., Wajsman, J., Chemat, F., 2013. Batch and continuous ultrasound assisted extraction of boldo leaves (*Peumus boldus* Mol.). Int. J. Mol. Sci. 14, 5750–5764.

Petigny, L., Özel, M.Z., Périno, S., Wajsman, J., Chemat, F., 2015. Water as green solvent for extraction of natural products. In: Chemat, F., Strube, J. (Eds.), Green Extraction of Natural Products: Theory and Practice. Wiley-VCH Verlag GmbH & Co. KGaA, Weinheim, pp. 237–263.

Picó, Y., 2013. Ultrasound-assisted extraction for food and environmental samples. TrAC Trends Anal. Chem. 43, 84–99.

Piñeiro, Z., Marrufo-Curtido, A., Serrano, M., Palma, M., 2016. Ultrasound-assisted extraction of stilbenes from grape canes. Molecules 21, 784.

Pinela, J., Prieto, M.A., Carvalho, A.M., Barreiro, M.F., Oliveira, M.B.P.P., Barros, L., Ferreira, I.C.F.R., 2016. Microwave-assisted extraction of phenolic acids and flavonoids and production of antioxidant ingredients from tomato: a nutraceutical-oriented optimization study. Sep. Purif. Technol. 164, 114–124.

Plaza, M., Turner, C., 2015. Pressurized hot water extraction of bioactives. TrAC Trends Anal. Chem. 71, 39–54.

Plaza, M., Santoyo, S., Jaime, L., García-Blairsy Reina, G., Herrero, M., Señoráns, F.J., Ibañez, E., 2010. Screening for bioactive compounds from algae. J. Pharm. Biomed. Anal. 51, 450–455.

Prakash Maran, J., Manikandan, S., Thirugnanasambandham, K., Vigna Nivetha, C., Dinesh, R., 2013. Box-Behnken design based statistical modeling for ultrasound-assisted extraction of corn silk polysaccharide. Carbohydr. Polym. 92, 604–611.

Prakash Maran, J., Sivakumar, V., Thirugnanasambandham, K., Sridhar, R., 2014. Microwave assisted extraction of pectin from waste *Citrullus lanatus* fruit rinds. Carbohydr. Polym. 101, 786–791.

Puértolas, E., Cregenzán, O., Luengo, E., Álvarez, I., Raso, J., 2013. Pulsed-electric-field-assisted extraction of anthocyanins from purple-fleshed potato. Food Chem. 136, 1330–1336.

Purohit, A.J., Gogate, P.R., 2015. Ultrasound-assisted extraction of β-carotene from waste carrot residue: effect of operating parameters and type of ultrasonic irradiation. Sep. Sci. Technol. 50, 1507–1517.

Qin, H., Zhou, G., Peng, G., Li, J., Chen, J., 2015. Application of ionic liquid-based ultrasound-assisted extraction of five phenolic compounds from fig (*Ficus carica* L.) for HPLC-UV. Food Anal. Methods 8, 1673–1681.

Qv, X.-Y., Zhou, Q.-F., Jiang, J.-G., 2014. Ultrasound-enhanced and microwave-assisted extraction of lipid from *Dunaliella tertiolecta* and fatty acid profile analysis. J. Sep. Sci. 37, 2991–2999.

Rachmaniah, O., Choi, Y.H., Arruabarrena, I., Vermeulen, B., Van Spronsen, J., Verpoorte, R., Witkamp, G.J., 2014. Environmentally benign supercritical CO_2 extraction of galanthamine from floricultural crop waste of *Narcissus pseudonarcissus*. J. Supercrit. Fluids 93, 7–19.

Rai, A., Mohanty, B., Bhargava, R., 2016. Supercritical extraction of sunflower oil: a central composite design for extraction variables. Food Chem. 192, 647–659.

Rajha, H.N., Ziegler, W., Louka, N., Hobaika, Z., Vorobiev, E., Boechzelt, H.G., Maroun, R.G., 2014. Effect of the drying process on the intensification of phenolic compounds recovery from grape pomace using accelerated solvent extraction. Int. J. Mol. Sci. 15, 18640–18658.

Ramandi, N.F., Najafi, N.M., Raofie, F., Ghasemi, E., 2011. Central composite design for the optimization of supercritical carbon dioxide fluid extraction of fatty acids from *Borago Officinalis* L. Flower. J. Food Sci 76, C1262–C1266.

Ranic, M., Nikolic, M., Pavlovic, M., Buntic, A., Siler-Marinkovic, S., Dimitrijevic-Brankovic, S., 2014. Optimization of microwave-assisted extraction of natural antioxidants from spent espresso coffee grounds by response surface methodology. J. Clean. Prod. 80, 69–79.

Ravber, M., Knez, Ž., Škerget, M., 2015. Isolation of phenolic compounds from larch wood waste using pressurized hot water: extraction, analysis and economic evaluation. Cellulose 22, 3359–3375.

Reyes, F.A., Mendiola, J.A., Ibañez, E., Del Valle, J.M., 2014. Astaxanthin extraction from *Haematococcus pluvialis* using CO_2-expanded ethanol. J. Supercrit. Fluids 92, 75–83.

Reyes, F.A., Mendiola, J.A., Suárez-Alvarez, S., Ibañez, E., Del Valle, J.M., 2016a. Adsorbent-assisted supercritical CO_2 extraction of carotenoids from *Neochloris oleoabundans* paste. J. Supercrit. Fluids 112, 7–13.

Reyes, F.A., Sielfeld, C.S., del Valle, J.M., 2016b. Effect of high-pressure compaction on supercritical CO_2 extraction of astaxanthin from *Haematococcus pluvialis*. J. Food Eng. 189, 123–134.

Rodriguez-Perez, C., Gilbert-Lopez, B., Mendiola, J.A., Quirantes-Pine, R., Segura-Carretero, A., Ibanez, E., 2016. Optimization of microwave-assisted extraction and pressurized liquid extraction of phenolic compounds from *Moringa oleifera* leaves by multi-response surface methodology. Electrophoresis 37, 1938–1946.

Rodríguez-Solana, R., Salgado, J.M., Domínguez, J.M., Cortés-Diéguez, S., 2014. Characterization of fennel extracts and quantification of estragole: optimization and comparison of accelerated solvent extraction and Soxhlet techniques. Ind. Crops Prod. 52, 528–536.

Roohinejad, S., Everett, D.W., Oey, I., 2014. Effect of pulsed electric field processing on carotenoid extractability of carrot purée. Int. J. Food Sci. Technol. 49, 2120–2127.

Roosta, M., Ghaedi, M., Daneshfar, A., Sahraei, R., 2015. Ultrasound assisted microextraction-nano material solid phase dispersion for extraction and determination of thymol and carvacrol in pharmaceutical samples: experimental design methodology. J. Chromatogr. B Anal. Technol. Biomed. Life Sci. 975, 34–39.

Rostagno, M.A., D'Arrigo, M., Martínez, J.A., Martínez, J.A., 2010. Combinatory and hyphenated sample preparation for the determination of bioactive compounds in foods. TrAC Trends Anal. Chem. 29, 553–561.

Ruen-ngam, D., Shotipruk, A., Pavasant, P., 2011. Comparison of extraction methods for recovery of astaxanthin from *Haematococcus pluvialis*. Sep. Sci. Technol. 46, 64–70.

Ruiz-Aceituno, L., García-Sarrió, M.J., Alonso-Rodriguez, B., Ramos, L., Sanz, M.L., 2016. Extraction of bioactive carbohydrates from artichoke (*Cynara scolymus* L.) external bracts using microwave assisted extraction and pressurized liquid extraction. Food Chem. 196, 1156–1162.

Sahena, F., Zaidul, I.S.M., Jinap, S., Karim, A.A., Abbas, K.A., Norulaini, N.A.N., Omar, A.K.M., 2009. Application of supercritical CO_2 in lipid extraction—a review. J. Food Eng. 95, 240–253.

Sahena, F., Zaidul, I.S.M., Norulaini, N.N.A., Jinap, S., Jahurul, M.H.A., Omar, M.A.K., 2014. Storage stability and quality of polyunsaturated fatty acid rich oil fraction from Longtail tuna (Thunnus tonggol) head using supercritical extraction. Available from: http://mc.manuscriptcentral.com/tcyt

Şahin, S., Şamlı, R., 2013. Optimization of olive leaf extract obtained by ultrasound-assisted extraction with response surface methodology. Ultrason. Sonochem. 20, 595–602.

Saleh, I.A., Vinatoru, M., Mason, T.J., Abdel-Azim, N.S., Aboutabl, E.A., Hammouda, F.M., 2016. A possible general mechanism for ultrasound-assisted extraction (UAE) suggested from the results of UAE of chlorogenic acid from *Cynara scolymus* L. (artichoke) leaves. Ultrason. Sonochem. 31, 330–336.

Santos, D.T., Veggi, P.C., Meireles, M.A.A., 2010. Extraction of antioxidant compounds from Jabuticaba (*Myrciaria cauliflora*) skins: yield, composition and economical evaluation. J. Food Eng. 101, 23–31.

Santos, P., Aguiar, A.C., Barbero, G.F., Rezende, C.A., Martínez, J., 2015. Supercritical carbon dioxide extraction of capsaicinoids from malagueta pepper (*Capsicum frutescens* L.) assisted by ultrasound. Ultrason. Sonochem. 22, 78–88.

Sarkar, S., Alvarez, V.H., Saldaña, M.D.A., 2014. Relevance of ions in pressurized fluid extraction of carbohydrates and phenolics from barley hull. J. Supercrit. Fluids 93, 27–37.

Sereshti, H., Izadmanesh, Y., Samadi, S., 2011. Optimized ultrasonic assisted extraction-dispersive liquid-liquid microextraction coupled with gas chromatography for determination of essential oil of *Oliveria decumbens* vent. J. Chromatogr. A 1218, 4593–4598.

Sereshti, H., Samadi, S., Jalali-Heravi, M., 2013. Determination of volatile components of green, black, oolong and white tea by optimized ultrasound-assisted extraction-dispersive liquid-liquid microextraction coupled with gas chromatography. J. Chromatogr. A 1280, 1–8.

Sereshti, H., Heidari, R., Samadi, S., 2014. Determination of volatile components of saffron by optimised ultrasound-assisted extraction in tandem with dispersive liquid-liquid microextraction followed by gas chromatography-mass spectrometry. Food Chem. 143, 499–505.

Setyaningsih, W., Saputro, I.E., Palma, M., Barroso, C.G., 2015. Optimisation and validation of the microwave-assisted extraction of phenolic compounds from rice grains. Food Chem. 169, 141–149.

Setyaningsih, W., Saputro, I.E., Palma, M., Barroso, C.G., 2016. Pressurized liquid extraction of phenolic compounds from rice (*Oryza sativa*) grains. Food Chem. 192, 452–459.

Shah, M., Garg, S.K., 2014. Application of 2^k full factorial design in optimization of solvent-free microwave extraction of ginger essential oil. J. Eng. 2015, 1–5.

Shah, S., Gani, A., Ahmad, M., Shah, A., Gani, A., Masoodi, F.A., 2015. In vitro antioxidant and antiproliferative activity of microwave-extracted green tea and black tea (*Camellia sinensis*): a comparative study. Nutrafoods 14, 207–215.

Shang, Y.F., Kim, S.M., Lee, W.J., Um, B.H., 2011. Pressurized liquid method for fucoxanthin extraction from *Eisenia bicyclis (Kjellman) Setchell*. J. Biosci. Bioeng. 111, 237–241.

Shang, X., Guo, X., Pan, H., Zhang, J., Zhang, Y., Miao, X., 2016. Microwave-assisted extraction of three bioactive alkaloids from *Peganum harmala* L. and their acaricidal activity against *Psoroptes cuniculi in vitro*. J. Ethnopharmacol. 192, 350–361.

Sharif, K.M., Rahman, M.M., Azmir, J., Mohamed, A., Jahurul, M.H.A., Sahena, F., Zaidul, I.S.M., 2014. Experimental design of supercritical fluid extraction—a review. J. Food Eng. 124, 105–116.

Shu, X.K., Li, J., Liu, F., Lin, X.J., Wang, X., Song, C.X., 2013. Accelerated solvent extraction and pH-zone-refining counter-current chromatographic purification of yunaconitine and 8-deacetylyunaconitine from *Aconitum vilmorinianum* Kom. J. Sep. Sci. 36, 2680–2685.

Silva, G.F. da, Gandolfi, P.H.K., Almeida, R.N., Lucas, A.M., Cassel, E., Vargas, R.M.F., 2015. Analysis of supercritical fluid extraction of lycopodine using response surface methodology and process mathematical modeling. Chem. Eng. Res. Des. 100, 353–361.

Sinanoglou, V.J., Zoumpoulakis, P., Heropoulos, G., Proestos, C., Ćirić, A., Petrovic, J., Glamoclija, J., Sokovic, M., 2014. Lipid and fatty acid profile of the edible fungus *Laetiporus sulphurous*. Antifungal and antibacterial properties. J. Food Sci. Technol. 52, 3264–3272.

Singanusong, R., Nipornram, S., Tochampa, W., Rattanatraiwong, P., 2015. Low power ultrasound-assisted extraction of phenolic compounds from Mandarin (*Citrus reticulata Blanco* cv. Sainampueng) and lime (*Citrus aurantifolia*) peels and the antioxidant. Food Anal. Methods 8, 1112–1123.

Singh, A., et al., 2015. Green extraction methods and environmental applications of carotenoids: a review. RSC Adv. 5, 62358–62393.

Sobolev, A., Carradori, S., Capitani, D., Vista, S., Trella, A., Marini, F., Mannina, L., 2014. Saffron samples of different origin: an NMR study of microwave-assisted extracts. Foods 3, 403–419.

Solana, M., Rizza, C.S., Bertucco, A., 2014. Exploiting microalgae as a source of essential fatty acids by supercritical fluid extraction of lipids: comparison between *Scenedesmus obliquus, Chlorella protothecoides* and *Nannochloropsis salina*. J. Supercrit. Fluids 92, 311–318.

Solana, M., Boschiero, I., Dall'Acqua, S., Bertucco, A., 2015. A comparison between supercritical fluid and pressurized liquid extraction methods for obtaining phenolic compounds from *Asparagus officinalis* L. J. Supercrit. Fluids 100, 201–208.

Sookwong, P., Suttiarporn, P., Boontakham, P., Seekhow, P., Wangtueai, S., Mahatheeranont, S., 2016. Simultaneous quantification of vitamin E, γ-oryzanols and xanthophylls from rice bran essences extracted by supercritical CO_2. Food Chem. 211, 140–147.

Soria, A.C., Villamiel, M., 2010. Effect of ultrasound on the technological properties and bioactivity of food: a review. Trends Food Sci. Technol. 21, 323–331.

Souza Machado, B.A., Gambini Pereira, C., Baptista Nunes, S., Ferreira Padilha, F., Umsza-Guez, M.A., 2013. Supercritical fluid extraction using CO_2: main applications and future perspectives. Sep. Sci. Technol. 48 (18), 2741–2760.

Strati, I.F., Oreopoulou, V., 2014. Recovery of carotenoids from tomato processing by-products: a review. Food Res. Int. 65, 311–321.

Stroescu, M., Stoica-Guzun, A., Ghergu, S., Chira, N., Jipa, I., 2013. Optimization of fatty acids extraction from *Portulaca oleracea* seed using response surface methodology. Ind. Crops Prod. 43, 405–411.

Sumerskiy, I., Pranovich, A., Holmbom, B., Willför, S., 2015. Lignin and other aromatic substances released from spruce wood during pressurized hot-water extraction. Part 1: extraction, fractionation and physico-chemical characterization. J. Wood Chem. Technol. 35 (6), 387–397.

Sun, P., Armstrong, D.W., 2010. Ionic liquids in analytical chemistry. Anal. Chim. Acta 661, 1–16.

Sun, H., Ge, X., Lv, Y., Wang, A., 2012. Application of accelerated solvent extraction in the analysis of organic contaminants, bioactive and nutritional compounds in food and feed. J. Chromatogr. A 1237, 1–23.

Taghvaei, M., Jafari, S.M., Assadpoor, E., Nowrouzieh, S., Alishah, O., 2014. Optimization of microwave-assisted extraction of cottonseed oil and evaluation of its oxidative stability and physicochemical properties. Food Chem. 160, 90–97.

Tan, Z.J., Wang, C.-Y., Yang, Z.-Z., Yi, Y.-J., Wang, H.-Y., Zhou, W.-L., Li, F.-F., 2015. Ionic liquid-based ultrasonic-assisted extraction of secoisolariciresinol diglucoside from flaxseed (*Linum usitatissimum* L.) with further purification by an aqueous two-phase system. Molecules 20, 17929–17943.

Tan, Z., Yi, Y., Wang, H., Zhou, W., Wang, C., 2016. Extraction, preconcentration and isolation of flavonoids from *Apocynum venetum* l. leaves using ionic liquid-based ultrasonic-assisted extraction coupled with an aqueous biphasic system. Molecules 21, 262.

Tang, W.Q., Li, D.C., Lv, Y.X., Jiang, J.G., 2010. Extraction and removal of caffeine from green tea by ultrasonic-enhanced supercritical fluid. J. Food Sci. 75, C363–C368.

Tao, Y., Wu, D., Zhang, Q.-A., Sun, D.-W., 2014. Ultrasound-assisted extraction of phenolics from wine lees: modeling, optimization and stability of extracts during storage. Ultrason. Sonochem. 21, 706–715.

Tatke, P., Jaiswal, Y., 2011. An overview of MAE and its applications in herbal drug research. Res. J. Med. Plant 5 (1), 21–31.

Teh, S.-S., Niven, B.E., Bekhit, A.E.-D.A., Carne, A., Birch, J., 2015a. Optimization of polyphenol extraction and antioxidant activities of extracts from defatted flax seed cake (*Linum usitatissimum* L.) using microwave-assisted and pulsed electric field (PEF) technologies with response surface methodology. Food Sci. Biotechnol. 24, 1649–1659.

Teh, S.-S., Niven, B.E., Bekhit, A.E.-D.A., Carne, A., Birch, E.J., 2015b. Microwave and pulsed electric field assisted extractions of polyphenols from defatted canola seed cake. Int. J. Food Sci. Technol. 50, 1109–1115.

Tekin, K., Akalın, M.K., Şeker, M.G., 2015. Ultrasound bath-assisted extraction of essential oils from clove using central composite design. Ind. Crops Prod. 77, 954–960.

Teng, H., Choi, Y.H., 2014. Optimization of ultrasonic-assisted extraction of bioactive alkaloid compounds from rhizoma coptidis (*Coptis chinensis* Franch.) using response surface methodology. Food Chem. 142, 299–305.

Tewari, S., Ramalakshmi, K., Methre, L., Mohan Rao, L.J., 2015. Microwave-assisted extraction of inulin from chicory roots using response surface methodology. J. Nutr. Food Sci. 5 (1), 342.

Thirugnanasambandham, K., Sivakumar, V., 2015. Application of D-optimal design to extract the pectin from lime bagasse using microwave green irradiation. Int. J. Biol. Macromol. 72, 1351–1357.

Thirugnanasambandham, K., Sivakumar, V., Maran, J.P., 2015. Microwave-assisted extraction of polysaccharides from mulberry leaves. Int. J. Biol. Macromol. 72, 1–5.

Tian, Y., Zeng, H., Xu, Z., Zheng, B., Lin, Y., Gan, C., Lo, Y.M., 2012. Ultrasonic-assisted extraction and antioxidant activity of polysaccharides recovered from white button mushroom (*Agaricus bisporus*). Carbohydr. Polym. 88, 522–529.

Trujillo-Rodríguez, M.J., Rocío-Bautista, P., Pino, V., Afonso, A.M., 2013. Ionic liquids in dispersive liquid-liquid microextraction. TrAC Trends Anal. Chem. 51, 87–106.

Tsiaka, T., Zoumpoulakis, P., Sinanoglou, V.J., Makris, C., Heropoulos, G.A., Calokerinos, A.C., 2015. Response surface methodology toward the optimization of high-energy carotenoid extraction from *Aristeus antennatus* shrimp. Anal. Chim. Acta 877, 100–110.

Uquiche, E., Romero, V., Ortíz, J., Del Valle, J.M., 2012. Extraction of oil and minor lipids from cold-press rapeseed cake with supercritical CO_2. Braz. J. Chem. Eng. 29, 585–597.

Valdés, A., Vidal, L., Beltrán, A., Canals, A., Garrigós, M.C., 2015. Microwave-assisted extraction of phenolic compounds from almond skin byproducts (*Prunus amygdalus*): a multivariate analysis approach. J. Agric. Food Chem. 63, 5395–5402.

Veggi, P.C., Martinez, J., Meireles, M.A.A., 2013. Fundamentals of microwave extraction. In: Chemat, F., Cravotto, G. (Eds.), Microwave-Assisted Extraction for Bioactive Compounds: Theory and Practice. Springer, New York, NY, pp. 15–52.

Vergara-Salinas, J.R., Vergara, M., Altamirano, C., Gonzalez, Á., Pérez-Correa, J.R., 2015. Characterization of pressurized hot water extracts of grape pomace: chemical and biological antioxidant activity. Food Chem. 171, 62–69.

Vilkhu, K., Manasseh, R., Mawson, R., Ashokkumar, M., 2011. Ultrasonic recovery and modification of food ingredients. In: Feng, H., Barbosa-Cánovas, G.V., Weiss, J. (Eds.), Ultrasound Technologies for Food and Bioprocessing. Springer, New York, NY.

Villanueva Bermejo, D., Angelov, Vincente, G., Stateva, R.P., García-Risco, M.R., Reglero, G., Ibáñez, E., Fornari, T., 2015. Extraction of thymol from different varieties of thyme plants using green solvents. J. Sci. Food Agric. 95, 2901–2907.

Viñas, P., Campillo, N., López-García, I., Hernández-Córdoba, M., 2013. Dispersive liquid-liquid microextraction in food analysis: a critical review. Anal. Bioanal. Chem. 406 (8), 1–33.

Wang, L., Weller, C.L., 2006. Recent advances in extraction of nutraceuticals from plants. Trends Food Sci. Technol. 17, 300–312.

Wang, X., Wu, Y., Chen, G., Yue, W., Liang, Q., Wu, Q., 2013. Optimisation of ultrasound assisted extraction of phenolic compounds from *Sparganii rhizoma* with response surface methodology. Ultrason. Sonochem. 20, 846–854.

Wang, W., Li, Q., Liu, Y., Chen, B., 2015. Ionic liquid-aqueous solution ultrasonic-assisted extraction of three kinds of alkaloids from *Phellodendron amurense* Rupr and optimize conditions use response surface. Ultrason. Sonochem. 24, 13–18.

Wei, Z., Zu, Y., Fu, Y., Wang, W., Luo, M., Zhao, C., Pan, Y., 2013. Ionic liquids-based microwave-assisted extraction of active components from pigeon pea leaves for quantitative analysis. Sep. Purif. Technol. 102, 75–81.

Wei, M.C., Yang, Y.C., Hong, S.J., 2015a. Determination of oleanolic and ursolic acids in hedyotis diffusa using hyphenated ultrasound-assisted supercritical carbon dioxide extraction and chromatography. Evid. Based. Complement. Alternat. Med., 2015.

Wei, Z.F., Wang, X.Q., Peng, X., Wang, W., Zhao, C.J., Zu, Y.G., Fu, Y.J., 2015b. Fast and green extraction and separation of main bioactive flavonoids from *Radix Scutellariae*. Ind. Crops Prod. 63, 175–181.

Wei, J., Chen, J., Liang, X., Guo, X., 2016. Microwave-assisted extraction in combination with HPLC-UV for quantitative analysis of six bioactive oxoisoaporphine alkaloids in *Menispermum dauricum* DC. Biomed. Chromatogr. 30, 241–248.

Wen, Y., Nie, J., Li, Z.-G., Xu, X.-Y., Wei, D., Lee, M.-R., 2014. The development of ultrasound-assisted extraction/dispersive liquid-liquid microextraction coupled with DSI-GC-IT/MS for analysis of essential oil from fresh flowers of *Edgeworthia chrysantha* Lindl. Anal. Methods 6, 3345–3352.

Wen, Y., Chen, H., Zhou, X., et al., 2015. Optimization of the microwave-assisted extraction and antioxidant activities of anthocyanins from blackberry using a response surface methodology. RSC Adv. 5, 19686–19695.

Wijngaard, H., Hossain, M.B., Rai, D.K., Brunton, N., 2012. Techniques to extract bioactive compounds from food by-products of plant origin. Food Res. Int. 46, 505–513.

Wijngaard, H.H., Trifunovic, O., Bongers, P., 2013. Novel extraction techniques for phytochemicals. Wiley, New Jersey, USA, pp.412–433.

Wu, H., Chen, M., Fan, Y., Elsebaei, F., Zhu, Y., 2012. Determination of rutin and quercetin in Chinese herbal medicine by ionic liquid-based pressurized liquid extraction-liquid chromatography—chemiluminescence detection. Talanta 88, 222–229.

Xie, D.-T., Wang, Y.-Q., Kang, Y., Hu, Q.-F., Su, N.-Y., Huang, J.-M., Che, C.-T., Guo, J.-X., 2014. Microwave-assisted extraction of bioactive alkaloids from *Stephania sinica*. Sep. Purif. Technol. 130, 173–181.

Xu, Y., Pan, S., 2013. Effects of various factors of ultrasonic treatment on the extraction yield of all-trans-lycopene from red grapefruit (*Citrus paradise* Macf.). Ultrason. Sonochem. 20, 1026–1032.

Xu, J., Wang, W., Liang, H., Zhang, Q., Li, Q., 2015. Optimization of ionic liquid based ultrasonic assisted extraction of antioxidant compounds from *Curcuma longa* L. using response surface methodology. Ind. Crops Prod. 76, 487–493.

Xu, Y., Cai, F., Yu, Z., Zhang, L., Li, X., Yang, Y., Liu, G., 2016. Optimisation of pressurised water extraction of polysaccharides from blackcurrant and its antioxidant activity. Food Chem. 194, 650–658.

Xynos, N., Papaefstathiou, G., Gikas, E., Argyropoulou, A., Aligiannis, N., Skaltsounis, A.L., 2014. Design optimization study of the extraction of olive leaves performed with pressurized liquid extraction using response surface methodology. Sep. Purif. Technol. 122, 323–330.

Yang, Y.C., Wei, M.C., Hong, S.J., Huang, T.C., Lee, S.Z., 2013. Development/optimization of a green procedure with ultrasound-assisted improved supercritical carbon dioxide to produce extracts enriched in oleanolic acid and ursolic acid from *Scutellaria barbata* D. Don. Ind. Crops Prod. 49, 542–553.

Yang, Y.C., Wei, M.C., Hong, S.J., 2014. Ultrasound-assisted extraction and quantitation of oils from *Syzygium aromaticum* flower bud (clove) with supercritical carbon dioxide. J. Chromatogr. A 1323, 18–27.

Yang, N., Jin, Y., Jin, Z., Xu, X., 2016. Electric-field-assisted extraction of garlic polysaccharides via experimental transformer device. Food Bioproc. Technol. 9 (9), 1–11.

Ye, Z.-Y., Li, Z.-G., Wei, D., Lee, M.-R., 2014. Microwave-assisted extraction/dispersive liquid–liquid microextraction coupled with DSI-GC-IT/MS for analysis of essential oil from three species of cardamom. Chromatographia 77, 347–358.

Yilmaz, E., Soylak, M., 2015. Ultrasound assisted-deep eutectic solvent extraction of iron from sheep, bovine and chicken liver samples. Talanta 136, 170–173.

Yılmaz, T., Tavman, S., 2016. Modeling and optimization of ultrasound assisted extraction parameters using response surface methodology for water soluble polysaccharide extraction from hazelnut skin. J. Food Proc. 41 (2), 1–13.

Yolmeh, M., Habibi Najafi, M.B., Farhoosh, R., 2014. Optimisation of ultrasound-assisted extraction of natural pigment from annatto seeds by response surface methodology (RSM). Food Chem. 155, 319–324.

You, Q., Yin, X., Zhang, S., Jiang, Z., 2014. Extraction, purification, and antioxidant activities of polysaccharides from *Tricholoma mongolicum Imai*. Carbohydr. Polym. 99, 1–10.

Yu, I.L., Yu, Z.-R., Koo, M., Wang, B.-J., 2015. A continuous fractionation of ginsenosides and polysaccharides from *Panax ginseng* using supercritical carbon dioxide technology. J. Food Proc. 4, 743–748.

Zaghdoudi, K., Pontvianne, S., Framboisier, X., Achard, M., Kudaibergenova, R., Ayadi-Trabelsi, M., Kalthoum-Cherif, J., Vanderesse, R., Frochot, C., Guiavarc'h, Y., 2015. Accelerated solvent extraction of carotenoids from: Tunisian Kaki (*Diospyros kaki* L.), peach (*Prunus persica* L.) and apricot (*Prunus armeniaca* L.). Food Chem. 184, 131–139.

Zakaria, S.M., Kamal, S.M.M., 2016. Subcritical water extraction of bioactive compounds from plants and algae: applications in pharmaceutical and food ingredients. Food Eng. Rev. 8, 23–34.

Zeng, H., Zhang, Y., Lin, S., Jian, Y., Miao, S., Zheng, B., 2015. Ultrasonic–microwave synergistic extraction (UMSE) and molecular weight distribution of polysaccharides from *Fortunella margarita* (Lour.) Swingle. Sep. Purif. Technol. 144, 97–106.

Zermane, A., Larkeche, O., Meniai, A.-H., Crampon, C., Badens, E., 2014. Optimization of essential oil supercritical extraction from Algerian *Myrtus communis* L. leaves using response surface methodology. J. Supercrit. Fluids 85, 89–94.

Zhang, D.Y., Wan, Y., Xu, J.Y., Wu, G.H., Li, L., Yao, X.H., 2016a. Ultrasound extraction of polysaccharides from mulberry leaves and their effect on enhancing antioxidant activity. Carbohydr. Polym. 137, 473–479.

Zhang, H.F., Yang, X.H., Wang, Y., 2011. Microwave assisted extraction of secondary metabolites from plants: current status and future directions. Trends Food Sci. Technol. 22, 672–688.

Zhang, H., Tang, B., Row, K.H., 2014a. A green deep eutectic solvent-based ultrasound-assisted method to extract astaxanthin from shrimp byproducts. Anal. Lett. 47, 742–749.

Zhang, W., Zhu, D., Fan, H., Liu, X., Wan, Q., Wu, X., Liu, P., Tang, J.Z., 2015. Simultaneous extraction and purification of alkaloids from *Sophora flavescens* Ait. by microwave-assisted aqueous two-phase extraction with ethanol/ammonia sulfate system. Sep. Purif. Technol. 141, 113–123.

Zhang, X., Li, M., Zhong, L., Peng, X., Sun, R., 2016b. Microwave-assisted extraction of polysaccharides from bamboo (*Phyllostachys acuta*) leaves and their antioxidant activity. BioResources 11, 5100–5112.

Zhang, Y., Liu, Z., Li, Y., Chi, R., 2014b. Optimization of ionic liquid-based microwave-assisted extraction of isoflavones from *Radix puerariae* by response surface methodology. Sep. Purif. Technol. 129, 71–79.

Zhou, T., Xiao, X., Li, G., 2012. Microwave accelerated selective Soxhlet extraction for the determination of organophosphorus and carbamate pesticides in ginseng with gas chromatography/mass spectrometry. Anal. Chem. 84, 5816–5822.

Zhou, X.-Y., Liu, R.-L., Ma, X., Zhang, Z.-Q., 2014. Polyethylene glycol as a novel solvent for extraction of crude polysaccharides from *pericarpium granati*. Carbohydr. Polym. 101, 886–889.

Zhu, C., Zhai, X., Li, L., Wu, X., Li, B., 2015. Response surface optimization of ultrasound-assisted polysaccharides extraction from pomegranate peel. Food Chem. 177, 139–146.

Zill-E-Huma, Abert-Vian, M., Elmaataoui, M., Chemat, F., 2011. A novel idea in food extraction field: study of vacuum microwave hydrodiffusion technique for by-products extraction. J. Food Eng. 105, 351–360.

Zou, T.-B., Jia, Q., Li, H.-W., Wang, C.-X., Wu, H.-F., 2013. Response surface methodology for ultrasound-assisted extraction of astaxanthin from *Haematococcus pluvialis*. Mar. Drugs 11, 1644–1655.

Zu, G., Zhang, R., Yang, L., Ma, C., Zu, Y., Wang, W., Zhao, C., 2012. Ultrasound-assisted extraction of carnosic acid and rosmarinic acid using ionic liquid solution from *Rosmarinus officinalis*. Int. J. Mol. Sci. 13, 11027–11043.

Further Reading

Lu, Q., Zeng, R., Wu, S., Chen, J., 2014c. Ultrasound/microwave-assisted extraction and comparative analysis of bioactive/toxic indole alkaloids in different medicinal parts of *Gelsemium elegans Benth* by ultra-high performance liquid chromatography with MS/MS. J. Sep. Sci. 37, 308–313.

Rodríguez-Solana, R., Salgado, J.M., Domínguez, J.M., Cortés-Diéguez, S., 2015. Comparison of Soxhlet, accelerated solvent and supercritical fluid extraction techniques for volatile (GC-MS and GC/FID) and phenolic compounds (HPLC-ESI/MS/MS) from Lamiaceae species. Phytochem. Anal. 26, 61–71.

Roohinejad, S., 2014. Extraction of β-carotene from carrot pomace using microemulsions and pulsed electric fields. Available from: http://hdl.handle.net/10523/5098

CHAPTER 9

Assessment of the State-of the-Art Developments in the Extraction of Antioxidants From Marine Algal Species

Mayyada El-Sayed*,, Daisy Fleita*, Dalia Rifaat*, Hanaa Essa*,†**

*American University in Cairo, New Cairo, Egypt; **National Research Centre, Giza, Egypt;
†Agriculture Research Centre, Giza, Egypt

1 Introduction

Antioxidants have potential positive impact on human health due to their ability to attack reactive oxygen species (ROS) that are harmful to the body, such as membrane lipids, proteins, and DNA. The excessive presence of these species may lead to health problems, such as cancer, diabetes mellitus, inflammatory diseases, neurological, and immunological disorders. In addition, food deterioration in many instances has been attributed to lipid oxidation and generation of undesirable products. Various synthetic antioxidants have been commercially utilized in food and pharmaceutical industries to retard the oxidation processes. The use of synthetic antioxidants causes potential health hazards due to its toxicity. Thus, in recent years, interest in natural antioxidants particularly those of plant origin, has increased (Goyal et al., 2013).

Marine organisms are rich in a variety of bioactive compounds that vary in their structure and hence could have promising potential in different industrial fields (Barrow and Shahidi, 2008). Examples of these compounds, include polyunsaturated fatty acids, polysaccharides that are mostly sulfated, proteins, bioactive peptides, natural pigments, essential minerals, and vitamins (Chandini et al., 2008; Ortiza et al., 2006). Recently, sulfated polysaccharides (SPs) extracted from marine algal species have been of much interest due to their various applications in functional foods, nutraceuticals, pharmaceuticals, and cosmetics industries.

SPs constitute complex macromolecules with myriad biological activities, such as antiviral, anticoagulant, antiinflammatory, antitumor, and antioxidant activities.

The chemical structure of SPs varies according to the type of algal species from which it is extracted. For example, fucoidans and laminarans are found in brown algae, carrageenans in red species, and ulvans in green species.

Ingredients Extraction by Physicochemical Methods in Food
http://dx.doi.org/10.1016/B978-0-12-811521-3.00009-0

It has been reported that the antioxidant activity of SPs is dependent on their composition and structural characteristics, such as degree of sulfation, molecular weight, phenolic content, type of major sugars, content and sequence of monosaccharides, and glyscosidic branching (Qi et al., 2005a,b; Zhang et al., 2003). These features vary from one algal species to the other according to its type, composition, as well as the environment and habitat in which it existed. Seasonal variations, for example, were found to affect the antioxidant activity. In a study on three different Egyptian algal species, the green algae *Ulva lactuca*, and the red algal species *Jania rubens* and *Pterocladia capillacea*, β-carotene content of the latter was the highest in summer, whereas the total phenolic content significantly increased in the same season. This, in turn, was reflected on the antioxidant activity of these species (Khairy and El-Sheikh, 2015).

Different techniques have been employed in the extraction of SPs from marine algal cell walls. Some of these techniques are conventional, such as soaking and/or shaking in cold or hot water, in solvent, or in a mixture of both. This is in addition to Soxhlet and hydrodistillation. Most conventional extraction methods consume time and substantial amounts of possibly polluting and hazardous organic solvents for the precipitation of polysaccharides. Because the cell wall of algae consists of complex polymers, extraction of active polysaccharides using conventional solvent extraction processes is not always efficient. Thus, modern techniques have been introduced in an attempt to overcome issues pertaining to conventional techniques and to enhance the efficiency of extracting bioactive compounds from seaweeds. Examples of modern techniques are ultrasonic-, microwave-, and enzyme-assisted extraction; in addition to supercritical fluid extraction (SFE), subcritical water extraction, and high-pressure liquid extraction (Sheng et al., 2007; Ye et al., 2006). More recently, it has been proven that the extraction method, as well as its relevant operating conditions may affect the yield of the extracted SPs, as well as their antioxidant activity.

2 Algae

Algae are organisms of the plant type, which convert carbon dioxide and water into glucose molecules via a process called photosynthesis. This process takes place with the aid of sunlight energy that is absorbed by chlorophyll and other pigments present in the algae. In most algal species, energy is stored in the form of carbohydrates (complex sugars).

Algae exhibit diverse forms ranging from unicellular organisms to giant multicellular ones, which lack stems, leaves, and roots. Algae can be found in both fresh and salt water. It can also be seen floating on the surfaces of moist soil and rocks. Seaweeds or macroalgae are those that grow in oceans, rivers, and lakes.

Algal cell walls are mainly composed of cellulose, along with pectin in some cases. The chemical composition of seaweeds is known to constitute primarily carbohydrates, along with minerals and proteins, in addition to bioactive functional compounds, such as polyphenols, terpenoids, carotenoids, and tocophenols.

2.1 Types of Algae

The botanical classification of seaweeds is based primarily on their morphology; however, the division into classes is based on the nature of the pigments present. Marine algae are characterized by the presence of chlorophyll and other colored pigments that sometimes mask the green color of chlorophyll. Accordingly, algal species are divided into three main classes; chlorophyta (green algae), phaeophyta (brown algae), and rhodophyta (red algae).

Green algae are more predominant in freshwater than in seawater. They range from unicellular organisms to more complex multicellular organisms, that can either form spherical colonies or straight or branched filaments. Red algae live mainly in seawater; very few exist in freshwater. They vary in their morphological nature from single-celled to branched filaments. Brown seaweeds are usually found along rocky coasts, however, some are found floating on the surface of the ocean. They have large sizes and could be gigantic as in case of *Macro cystispyrifera*. Common rockweeds are *Fucus* and *Sargassum*. Chrysophyta or golden-brown algae and diatoms are named after their yellow pigments. They both live in seawater and freshwater bodies. Their cell walls do not contain cellulose but constitute mostly pectin, and are often silica-filled. Diatoms have two glass shells composed mainly of silica (Medlin et al., 1997).

2.2 Marine Algal Polysaccharides

A general characteristic of marine algae is probably the presence of at least one polysaccharide linked with sulfate ester groups. These substances are not present in terrestial plants. Polysaccharides, in general, are different from proteins and nucleic acids in that they are made up of a high condensation of polymers of monosaccharides, which are formed by the elimination of a molecule of water between each pair of monosaccharide units, to form what is known as a glycosidic linkage. The structural features of a polysaccharide determine its biological function(s). In the determination of the fine structure of a polysaccharide, it is important to study the following: constituent monosaccharides, which make up the macromolecule, hydroxyl groups, the type and nature of the glycosidic linkage (α or β), location of the ester sulfate groups if present, average chain length, and in branched molecules the position of branches, and finally the molecular weight of the molecule.

These variations in structural features allow for a possibly very high degree of diversity of polysaccharides and consequently variety in their structure-activity relationships (SAR) (Percival and McDowell, 1990).

2.2.1 Polysaccharides of chlorophyta

The majority of polysaccharides in all green algal species are heteropolysaccharides. The hydroxyl groups of their sugar residues, or some of them, are substituted by half ester sulfate groups. This is in addition to some homopolysaccharides (Costa et al., 2010). The sugar proportions in these polysaccharides differ from genus to genus and within species of the same genus. All species that have been examined contained a high proportion of complex water-soluble SPs. In *U. lactuca*, *Enteromorpha*, and *Acrosiphonia*, the major monosaccharides are L-rhamnose, D-xylose, and D-glucuronic acid. While those from *Cladophora*, *Chaetomorpha*, *Caulerpa*, and *Codium* encompass other major sugars, such as D-galactose, L-arabinose, and D-xylose but in rather different proportions (Jiao et al., 2011; Lahaye, 1998).

SPs extracted from *U. lactuca* in aqueous medium were characterized by the presence of L-rhamnose, D-xylose, and D-glucose in the molar proportions of about 4.2: 1.3: 1.0, with some traces of D-mannose, 24% of D-glucuronic acid, and 19% of ester sulfate (Haq and Percival, 1966; Percival and Wold, 1963).

In their biochemical studies on the Egyptian marine algae *U. lactuca*, workers reported the isolation of uronic acid-containing polysaccharides by extraction with H_2O, dilute HCl, and ammonium oxalate solution. The latter solution extracted the highest amount of polysaccharides, which comprised (after hydrolysis) 11.4% glucuronic acid, 3.64% glucose, 5.01% arabinose, 4.55% xylose, 27.33% rhamnose, and 9.00% protein (Abdel-Fattah and Edress, 1972).

Additionally, under slight acidic conditions (pH 5), the isolation of two fractions of H_2O-soluble polysaccharide materials from the Egyptian *U. lactuca* species was reported. The first type was a neutral glucan (starch) while the second was an acidic macromolecule complex that comprised residues of glucose, rhamnose, xylose, glucuronic acid, arabinose, half-ester sulfate, and a protein moiety (Hussien, 1977).

Water-soluble SPs of *Codium fragile* were also investigated and they comprised D-galactose, L-arabinose, and D-xylose in molar ratios of 2:2:1, respectively together with traces of uronic acid and L-rhamnose (Love and Percival, 1964).

2.2.2 Polysaccharides of phaeophyta

All the species of brown algae examined contain SPs typified by the presence of the sugar L-fucose. These SPs may exist in different forms as fucoidan, ascophyllan, saragassan, and glucorono xylofucan. They include a group of polydisperse heteromolecules comprising different proportions of galactose, xylose, mannose, and glucuronic acid, along with fucose (Berteau and Mulloy, 2003; Li et al., 2008; Percival, 1979).

Sargassan extracted from *Saragassum linifolium* contained D-glucuronic acid, D-galactose, D-mannose, D-xylose, L-fucose in the molar ratios of 4.57:8.40:1.00:2.48:2.50, respectively, as well as a protein moiety (Abdel-Fattah et al., 1974). These sugars were found to be commonly present in all Sargassan extracted from the *Saragassum* genus located in different geographical areas (Dore et al., 2013).

Similarly, SPs containing 96% fucose as the major component and 4% of glucuronic acid were extracted from the brown algae, *S. horneri* (Turner) C.Agardh via hot water extraction and the extract was identified as fucan (Hoshino et al., 1998).

Fucoidan is one kind of complex SPs that exist mainly in the cell walls of brown algal species. It constitutes L-fucose and sulfate ester groups. In recent years, many researchers were interested in studying fucoidan due to its various biological activities especially antioxidation (Wijesinghe and Jeon, 2012).

2.2.3 Polysaccharides of rhodophyta

The main polysaccharides of most examined red seaweeds are galactans. The sugar units present in them are D- and L-galactose (some carrying half ester sulfate), 3,6-anhydro-D- and L-galactose and 6-O-methyl-D-galactose. A chain of alternating 1,3- and 1,4-glycosidic links allows a number of different types of polymers to be built up. These included three main groups; agar-, porphoran-, and carrageenan-type polysaccharides. The main difference between agars and carrageenans is the presence of D- and L-galactose units in agars, while carrageenans are only composed of the D-sugar (Kadam et al., 2015; Percival, 1979).

Carrageenans exist as a mixture of SPs with various types, such as λ, κ, Ł, ε, μ. *Chondruscrispus* is the primary source of Ł carrageenan, while *Eucheuma cottonii* and *E. spinosum* are sources of Ḳ and į carrageenans, respectively (McLellan and Jurd, 1992).

The effect of seasonal variations on the chemical composition of the red algae *P. capillacea* was investigated. Chromatographic examination of acid hydrolysates of this algal species confirmed the presence of mainly galactose along with glucose, glucuronic acid, and xylose residues. Changes in the sugar content were realized as the season was varied (Abdel-Fattah et al., 1973b).

3 Algae as Antioxidant Sources

3.1 Antioxidants as Free Radical Scavengers

Antioxidants are molecules that stabilize free radicals by either donating or accepting an electron and hence suppress their high reactivity that is caused by their unpaired electrons. Antioxidants act as either inhibitors to the formation of free radicals or retardants to the propagation step of the peroxidation process as is discussed in detail in a later section. The oxidative damage of free radicals is caused by their ability to undergo chain reactions where the unpaired electrons are passed on to the recipient molecule converting it into a free radical, and thus neutralizing the donor molecule.

Free radicals in biological systems are mostly derivatives of oxygen known as "ROS" (reactive oxygen species), however, nitrogen derivatives also exist ("reactive nitrogen species," *RNS*). Molecules with loosely bound hydrogen atoms can donate these atoms, in a similar manner as electrons, to free radicals to reduce or neutralize them. Examples of some free radicals are hydroxyl (OH), peroxyl (LOO), glutathione (GSH), tocopherol (Toc), ascorbate (Asc), and Fe^{3+}-EDTA (Duan et al., 2006; Yang et al., 2011).

3.2 Antioxidants From Algae

Algal species possess reactive antioxidant compounds, such as carotenoids (α- and β-carotene, fucoxanthin, astaxanthin), mycosporine-like amino acids (mycosporine-glycine) and catechins (catechin, epigallocatechin), gallate, phlorotannins (phloroglucinol), eckol, and tocopherols (α-, χ-, δ-tocopherols) (Vadlapudi, 2012).

3.3 Mechanisms of Antioxidation

The formation of ROS in food occurs as a result of the oxidation of unsaturated fatty acids. Lipid oxidation takes place via a free-radical chain mechanism, which entails three main steps. The first step is the initiation of free peroxyl radicals. This is followed by a propagation step where lipid hydroperoxides are formed through the reaction of these free radicals with unsaturated fatty acids. In the last step, each of the two produced free radicals combine to form a nonreactive nonradical species and hence the reaction is terminated.

Antioxidants retard the propagation of free radicals by either of two mechanisms; chelating with metal ions thus preventing them from decomposing peroxides, or quenching oxygen radicals that take part in formation of peroxyl radicals (Ueda et al., 1996).

In seaweeds, compounds with potential antioxidant activity include polysaccharides that are mostly sulfated, polyphenols, such as phlorotannins, bromophenols, flavonoids, phospholipids, pigments, such as fucoxanthin and astaxanthin, as well as chlorophyll-related compounds. In polyphenolic compounds, the phenolic rings act as electron acceptors thus scavenging free radicals, such as hydroxyl, peroxyl, and superoxide anions. This scavenging ability is the reason behind their antioxidative nature. The ability of phenolic compounds to chelate metals was found to depend on their phenolic structure, as well as number of hydroxyl groups and their location. Algal phlorotannins are better antioxidants than plant polyphenols owing to their possession of up to eight interconnected rings. In one study, workers owed the hydrogen peroxide scavenging activity of *Grateloupia filicina* extracts to the presence of phenolic compounds as electron donors. In another study, it was suggested that the bioactive compounds extracted from the green seaweeds of *Enteromorpha* sp. may as well function as reductones thus donating electrons to free radicals and suppressing their reactivity (Gupta and Abu Ghannam, 2011).

Compounds with phenolic rings can thus undergo antioxidation by two main mechanisms: hydrogen transfer or single electron transfer. In the former mechanism, the phenolic compounds donate hydrogen atoms to the free radicals. In the latter mechanism, a single

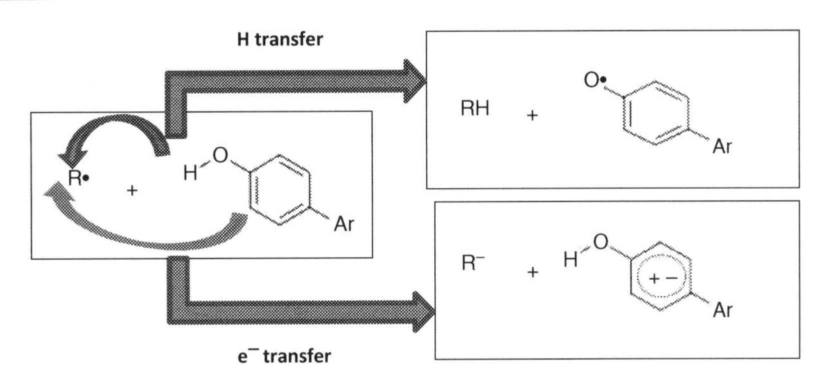

Figure 9.1: Free Radical Scavenging Mechanisms.

electron is either transferred from the free radical and stabilizes the phenolic ring, or transferred from the ring to the free radical and stabilizes the free radical by converting it into an anion (Choe and Min, 2009). The two mechanisms are illustrated in Fig. 9.1.

3.4 Determination of Antioxidant Activity

There are several methods for measuring antioxidant activity. These include 2,2- diphenyl-1-picryl-hydrazyl (DPPH); hydrogen peroxide (H_2O_2); 2-thiobarbituric acid (TBA); 2,2′-azino-bis(3-ethylbenzothiazoline-6-sulphonic acid) ABTS; and oxygen radical absorbance capacity (ORAC) methods.

DPPH is a free radical that has a purple color in solution and turns yellow when it is scavenged by the antioxidant through a hydrogen transfer mechanism. The amount of color change corresponds to a change in the absorbance, which is measured at a wavelength of 517 nm. The lower absorbance indicates higher antioxidant activity (Fleita et al., 2015).

In the hydrogen peroxide method, the antioxidant either prevents the formation of new hydroxyl radicals or scavenges the already existing ones. The scavenging mechanism takes place via metal chelation to prevent the reaction between the metal and H_2O_2 to form hydroxyl radicals (Ueda et al., 1996).

In the TBA assay, the degree of lipid peroxidation is measured by the residual amount of malondialdehyde (MDA) that was degraded as a result of lipid peroxidation. The quantification of MDA is based on measuring the absorbance of a red MDA-TBA complex formed as a result of the reaction of two TBA molecules with one MDA molecule.

The ABTS assay is based on oxidizing ABTS to its radical cation using sodium persulfate, such that it becomes blue in color and absorbs light at 734 nm. Antioxidants inhibit this reaction because they act as hydrogen donors reducing ABTS radical cation and converting it back to its colorless form. In this decolorization assay, the antioxidant activity is measured spectrophotometrically (Re et al., 1999).

The ORAC method measures the degree of oxidative degeneration caused by the reaction of free radicals with a fluorescent molecule where the latter molecule loses its fluorescence. Hence, the measured fluorescence intensity decreases indicating the proceeding of the oxidative degeneration. Antioxidants hinder this reaction and prevent the degeneration. Compounds, such as azo-initiators, are used to generate free radicals (Garrett et al., 2010; Huang et al., 2005).

4 Extraction Methods

The process of producing antioxidants from algae involves pretreatment, extraction, and postextraction steps as illustrated in Fig. 9.2. Pretreatment prepares the substrate for the subsequent extraction step. For example, grinding of the raw algal material facilitates the release of bioactive compounds in the subsequent extraction step. Decreasing the particle size of the algal material increases the exposed surface area, which allows for better dissolution of the active ingredient in the solvent. It enriches the substrate with the targeted antioxidant by increasing the concentration of the antioxidant as it helps eliminate or reduce the amount of other interfering substances. Pretreatment techniques, include drying, grinding, hydrolysis, and freeze-drying. Extraction is the main crucial step that can be undertaken by either conventional or modern techniques. Modern techniques have many advantages over conventional ones, such as reduced extraction time, solvent amount, and sample degradation, in addition to improved extraction efficiency and selectivity. The details of the different extraction methods are explained in the following subsection. As for postextraction, it comes as a polishing step to further purify the targeted antioxidants from other unwanted compounds. For example, cellulose anion-exchanges, such as DEAE, are utilized in column chromatography to fractionate water-soluble high molecular weight polysaccharides. Acidic polysaccharides are adsorbed onto the cellulose at acidic pH and depending on their

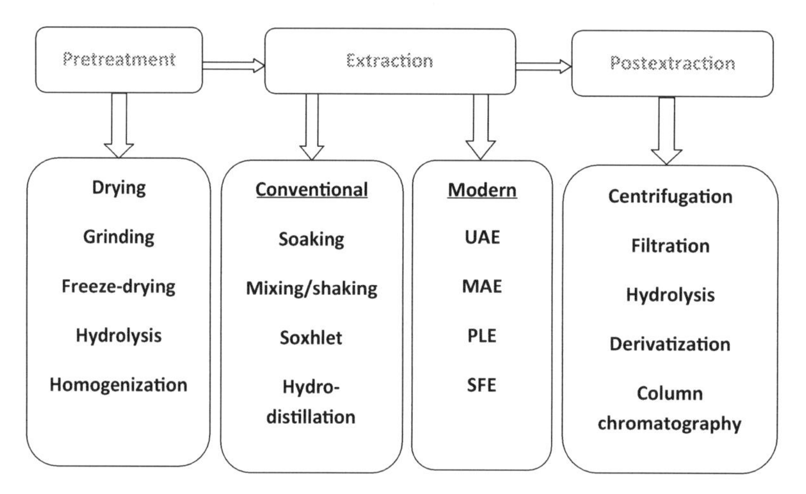

Figure 9.2: Steps of Pretreatment, Extraction and Postextraction of SPs From Algae.
MAE, Microwave-assisted extraction; *SFE*, supercritical fluid extraction; *UAE*, ultrasound-assisted extraction; *PLE*, high pressure liquid extraction.

content of the acidic groups, elution is achieved by increasing concentrations of either salts, alkaline, or acidic solutions. Enzymatic hydrolysis could also be used as a postextraction step. It is possible in some cases to use enzymes to degrade contaminating components in a polysaccharide mixture. For example, papain can remove protein, while α-amylase can remove starch-type polysaccharides (Hahn et al., 2012; Percival and McDowell, 1967; Percival and McDowell, 1990).

4.1 Conventional Methods

Different methods have been employed for the extraction of bioactive compounds from algae, including Soxhlet, hydrodistillation, and maceration with alcohol. Selection of the appropriate method depends on the nature of the targeted bioactive compound, as well as the amount of yield and purity required. The type and structure of these compounds may change with altering pH, temperature, and pressure-working conditions. This is in addition to the solvent type and solvent-to-algae ratio.

In general, the conventional method is relatively simple and cheap compared to other methods; however, it suffers from low selectivity and yield, as well as the need to use large amounts of solvents that might in some cases be organic solvents that are toxic and harmful to the environment (Herrero et al., 2006).

4.1.1 Affecting parameters

4.1.1.1 Solvent type

One of the factors affecting conventional extraction is the solvent type. Different solvents were employed in that regard, such as water, acetone, methanol, ethanol, and hexane.

The use of water as a solvent has its attractive environmental and economic merits; however, it might suffer from low-yield issues. The dried powder of the brown seaweed (*Sargassum tenerrimum*) was extracted in water for 16 h at 90–95°C. The filtrate was precipitated with ethanol after being concentrated to a quarter of its original volume. Produced precipitate was centrifuged then dehydrated with diethyl ether. Free radical scavenging activity was the highest when measured as ABTS (70.33 ± 2.33)% followed by DPPH (64.66 ± 2.08)% and H_2O_2 (61.56 ± 2.05)%, while the total antioxidant capacity (TAC) was found to be (62.55 ± 1.40)% (Vijayabaskar and Vaseela, 2012). In a similar study by the same authors, water extracts of the brown seaweed (*S. swartzii*) were obtained using the same method outlined earlier. The antioxidant activity was found to be (55.00 ± 3.61)%, (47.23 ± 2.81)%, and (25.33 ± 2.52)% as measured by ABTS, H_2O_2, and DPPH assays, respectively (Vijayabaskar et al., 2012). In a study on red seaweed *Gracilaria birdiae*, SPs were extracted using1.5% (w/v) water stirred for 15 h at 25°C. The algal residue obtained after filtering off the supernatant was extracted again in water for 45 min at 90°C. The mixture was then centrifuged and the supernatant was precipitated using a 1:3 volume ratio of supernatant to ethanol. The antioxidant properties of *G. birdiae* SPs were tested by measuring DPPH free radical scavenging effect and results showed that the extracted polysaccharides have moderately inhibited the formation of DPPH radicals (Souza et al., 2012).

The *Laminaria japonica* dry powdered material was first pretreated to remove fats using anhydrous ethanol at 60°C for 3 h, and then it was soaked in distilled water for 3 h at 100°C. After concentrating the supernatant, it was precipitated with 20% ethanol containing 0.2% $CaCl_2$, then filtered. A 75% ethanolic solution was consequently produced after ethanol was added to the supernatant and left overnight at 40°C. Three fractions (WPS-1, WPS-2, and WPS-3) were obtained at different retention times after purification of the extract onto an anion-exchange chromatography column. The WPS-1 fraction exhibited the highest superoxide radical scavenging activity while WPS-2 revealed potent scavenging ability for hydroxyl radicals. Based on cytotoxicity tests, WPS-2 fraction was recommended as a safe antioxidant candidate (Peng et al., 2012).

Hot water was also utilized to facilitate the extraction. Five different types of dry algal material were cut and autoclaved in water at 115–125°C for 3–4 h. The obtained solution was successively filtrated with gauze and siliceous earth. Filtration was followed by dialysis against tap and distilled water for 48 h, concentration of the filtrate under reduced pressure, and finally precipitation of the polysaccharides with 75% (by volume) ethanol. The results confirmed that all extracts possessed antioxidant activities in certain assays. It was suggested that the extracts could be used as possible supplements in food and pharmaceuticals. They also maintained a stable radical scavenging activity at high temperatures and could therefore be potentially utilized in medicine (Zhang et al., 2010). In another study, the dried seaweed powder of the brown algae *Turbinaria ornate* was depigmented in acetone and extracted by soaking in hot water for 3–4 h at 90–95°C. After concentrating, cooling and precipitating the filtrate overnight at 4°C with three volumes of ethanol, the precipitate was obtained at a % yield of 10% of dried brown crude SPs. In vitro and in vivo results showed its potential antioxidant and antiinflammatory activities. The study concluded that future work on total phenolic content (TPC) could be beneficial in managing oxidative stress and inflammatory diseases (Ananthi et al., 2010).

Other solvents were used for extraction, such as alcohols (methanol, ethanol, butanol) or ketones (acetone), or a mixture of both. The extracts of brown and red algae were obtained after 8 h using absolute methanol in a Soxhlet extractor. The extract suspension was then concentrated under reduced pressure at 600°C using rotary evaporation then filtered, washed with distilled water and stored in the dark at 40°C. Antioxidant activities of algal extracts were determined by their inhibitory effect on lipoxygenase activity or their ability to be oxidized and decolorize the DPPH. Aqueous and methanolic extracts of Chlorophyta *U. lactuca* (U), Rhodophyta *Digenea simplex* (D), and Phaeophyta *S. crassifolia* (S) were tested for their antioxidant activities. The two aqueous (U and D) extracts and the methanol (S) extract inhibited lipoxygenase activity by less than 50% of the control. The most potent methanolic extract was that of (S), which caused 76% inhibition. In all cases, the methanolic extract had higher inhibitory activity than the aqueous extract. The sequence of antioxidant activity for the methanol extracts of the three algal species as assayed by lipoxygenase inhibition was as follows: (S) > (D) > (U). The algal species (S) showed the highest DPPH antioxidant activity of 69% in aqueous extract. The antioxidant activity of brown algae was much higher than that of its red or green counterparts. SPs from the brown algae exhibited a higher antioxidant activity than that of the green, as well as that of the agar-like sulfated galactans obtained from the red algae (Al-Amoudi et al., 2009). Algal extracts of different species were obtained using 70% methanol at room temperature. The combined extracts were evaporated under vacuum. Then, the concentrated extract was partitioned between 9:1 (by volume) for methanol-H_2O and *n*-hexane, 8:2 for methanol-H_2O and $CHCl_3$, and 1:1 for methanol-H_2O and ethanol-acetone. Methanol-H_2O and chloroform fractions of *S. swartzii* showed the highest antioxidant activity of 73.92 ± 12.30 and 55.32 ± 4.80 mmol Fe (II) per 100 g dried plant, as well as total phenolic contents of 12.0 ± 0.5 and 11.05 ± 0.64 mg gallic acid equivalents per 100 g dried plant, respectively (Sadati et al., 2011). Methanol in diethyl ether was utilized for the extraction of three Indian seaweeds namely; *Halimeda tuna*, *T. conoides*, and *G. foliifera*. Repetitive extraction was carried out to obtain a substantial amount of the extract which was then concentrated at 40°C. *T. conoides* showed better antioxidant properties as was evident from its higher total phenolic content (1.675± 0.361 mg GAE/g) and total antioxidant activity (1.231 ± 0.173 mg GAE/g) relative to other tested algal species (Devi et al., 2011). In a similar study, the antioxidant activities of methanol extract, as well as successive fractions of water, petroleum ether, ethyl acetate, dichloromethane, and butanol extracts of Indian brown seaweeds of *S. marginatum*, *Padina tetrastomatica*, and *T. conoides* were investigated. Activities were determined according to DPPH radical scavenging and deoxyribose assays. The latter assay recorded scavenging activities that were concentration dependent and ranged from 17.79 to 23.16% at 1000 μg/mL extract concentration (Chandini et al., 2008). A mixture of methanol and chloroform in the ratio 2:1 was successful in extracting antioxidants from the dried red algae *Polysiphonia urceolata* with about 12.1% yield. The extraction was conducted twice for 6 h in a Soxhlet apparatus. After precipitation and drying of the extract, the solution was partitioned in petroleum ether to yield 1.4%, then in water. This was followed by successive extractions in ethyl acetate and butanol to produce fractions with about 1.2 and 1.6% yields, respectively. Among these fractions, the highest antioxidant activity as measured by DPPH scavenging activity and β-carotene linoleate assays, was exhibited by the ethyl acetate fraction. This fraction was further fractionated onto a silica gel column into seven fractions with different polarities. All fractions showed potent antioxidant activities with the first fraction having the highest potency (Duan et al., 2006).

One class of bioactive compounds that was successfully extracted from algae was the class of bromophenols. They were extracted from the red algae *Rhodomela confervoides* using ethanol. The dried algal material was suspended in ethanol, and then the mixture was filtered to collect the residue, which was first suspended in water before being successfully partitioned in petroleum ether, ethyl acetate, and *n*-butanol. The ethyl acetate extract was further fractionated onto a silica gel column to isolate 37 fractions, 9 were known bromophenol derivatives and another 5 were new nitrogen-containing bromophenols. These compounds showed potent DPPH antioxidant activity and ABTS radical scavenging activity that ranged from 2.11 to 3.58 mM (Li et al., 2012a).

In some studies, the antioxidant activities of alcohol and water algal extracts were compared. SPs were extracted from the blue-green algae *S. platensis*, using two different extraction methods. First, extraction with hot water was conducted wherein algae were mixed with distilled water, and heated in a water bath at 100°C for 2 h, and the liquid was subsequently filtered then SPs were precipitated using absolute ethanol. The mixture was centrifuged then residue was collected after solvent was evaporated under vacuum at 60°C using rotary evaporation. Second, extraction with 85% ethanol was undertaken following the same steps of water extraction but with ethanol. DPPH scavenging activities of the extracts were evaluated relative to those of the commercial synthetic antioxidants: BHA (butylated hydroxyanisole) and BHT (butylated hydroxytoluene). Potent DPPH radical scavenging activities were obtained in hot water extracts of *S. platensis* grown on media containing different concentrations of nitrogen. However, all extracts exhibited activities that are lower than the synthetic antioxidants at the same concentration. Scavenging activities of extracts with different concentrations were also evaluated according to ABTS radical cation assay. Significant ABTS radical cation scavenging activities were also reported (Abd El Baky et al., 2013). In another study, extracts of 16 seaweed species were obtained using both ethanol and water as solvents. Ethanol extraction was performed by soaking in ethanol for 1 week. Extracts were filtered and concentrated then dried out by rotary vacuum evaporation at 40°C. Water extraction was conducted by homogenizing the seaweed in 100 mL of water. Extracts were then filtered and stored in a refrigerator at 4°C. The extracts were tested for their ascorbic acid (vitamin C) content. Out of all extracts, 14 contained appreciable amounts of ascorbic acid, which among its other merits in boosting the activities of the immune system and improving absorption of iron in intestine, acts as a strong antioxidant (Ambreen et al., 2012).

Acetone and methanolic extracts of various algal species were compared in terms of their antioxidant activity, phenolic, and flavonoid contents. Dried algal powders were left shaking overnight and the obtained extracts were filtered. The filtrate was concentrated at 50°C under reduced pressure and it subsequently yielded 12% and 11% acetone and methanolic concentrated extracts, respectively. The methanolic extracts of *Desmococcus olivaceous*, and *Chlorococcum humicola* had higher phenolic and flavonoid contents than the acetone extract. Furthermore, DPPH scavenging activity was 95.8% for *Desmococcus* and 93% for *Chlorococcum*. As for the hydrogen peroxide scavenging assay, 2.5 mg/0.5 mL *D. olivaceous* methanolic extract exhibited 39% scavenging activity whereas the *C. humicola* acetone extract showed scavenging activity of 15%. Furthermore, *Desmococcus* acetone extract and *Chlorococcum* methanolic extract showed low H_2O_2 scavenging activities. According to 2-thiobarbituric acid reactive substances (TBARS) assay, it was also deduced that *D. olivaceous* and *C. humicola* methanolic extracts had more realized inhibitory effects on lipid peroxidation than their acetone counterparts. Based on the ferric reducing antioxidant power (FRAP) assay, both *Desmococcus* methanolic extracts and *Chlorococcum* acetone extracts showed the highest antioxidant capacities (Uma et al., 2011).

Antioxidants were also extracted from brown and red seaweeds using alcohol/water mixtures. The dried algae were shaken at room temperature and pH of 2.0 in each of methanol/water (1:1, by volume) and acetone/water (70:30, by volume) mixtures. The solution was then centrifuged and the supernatant was recovered. DPPH antioxidant activity, polyphenol, and sulfated polysaccharide contents were determined. The results showed a correlation between radical scavenging activities, reduction power and total polyphenol content. It also suggested that the phenolic compounds are involved in the antioxidant mechanisms in *Phaeophyceae*, whereas sulfate-containing polysaccharides are involved in the reduction power mechanism in case of *Florideophyceae* (Jiménez-Escrig et al., 2012).

Acetone, butanol, and ethanol extracts of *U. lactuca* were obtained using Soxhlet apparatus. After 6 h of extraction, solvents were evaporated to yield 3.15, 4.14, and 5.10%, respectively. Nine free phenolic compounds were identified by the reversed-phase high performance liquid chromatography (HPLC). Compound identification was based on comparison of their relative retention times compared to those obtained with the different standard compounds. Vanillin and *p*-coumaric acid were the major components found at about 27.6 and 21.0%, respectively. On the other hand, ferulic and salicylic acids were quantified as 2.06 and 3.00%, respectively (Hassan and Ghareib, 2009).

In addition to alcohols, alkanes, such as hexane and chloromethane, were also employed for extraction. Extraction of antioxidants from 12 microalgal strains isolated from the soil was conducted twice. In each extraction, the algal material was left in hexane for 30 min at room temperature, and then the suspension was centrifuged. Similarly, ethyl acetate and water extracts were also obtained at 80°C. Supernatants from each extraction were collected. Hexane and ethyl acetate extracts were dried with nitrogen. Antioxidant activities of both cells and extracellular substances were determined according to FRAP and DPPH-HPLC assays. The FRAP assay measurements revealed that the algal cells possess reasonable amounts of antioxidants of 0.56 ± 0.06 and 31.06 ± 4.00 μmol Trolox g^{-1} for the *Microchaete tenera* hexane extract and the *Chlorella vulgaris* water extract, respectively. In water extracts of extracellular substances, the amount of antioxidants ranged from 1.30 ± 0.15 μmol Trolox g^{-1} for *Fischerella musicola* to 73.20 ± 0.16 μmol Trolox g^{-1} for *F. ambigua*. In addition,

DPPH-HPLC assay revealed high antioxidant potential for the water fractions. Relevant radical scavenging activities of the tested microalgae ranged from a minimum value of 0.15 ± 0.02 in *Nostoc ellipsosporum* cell mass to a maximum value of 109.02 ± 8.25 in *C. vulgaris* extracellular substance. Furthermore, total phenolic contents were also determined by Folin–Ciocalteu method. Their amounts varied from one algal strain to the other and varied from zero in hexane extract to 19.15 ± 0.04 mg GAE g^{-1} in *C. vulgaris* extracellular water fraction (Hajimahmoodi et al., 2010). In another study, the freeze-dried powders of each of *Chaetoceros* sp., a diatom, and *Nannochloropsis* sp., a unicellular green microalga, were soaked in hexane, dichloromethane, chloroform, or methanol, and were left to shake for 24 h. The suspension was then filtered and the solvent was evaporated at 35°C. Extracts were dried and stored at −20°C. It was observed that the highest antioxidant power belonged to the nonpolar solvent extracts from the diatom. On the other hand, potent antioxidant activity was shown for both polar and nonpolar solvent extracts of green microalgae. It was concluded that *Chaetoceros* sp. had generally better antioxidant capacities than *Nannochloropsis* sp. (Goh et al., 2010).

A mixture of alkanes with alcohols was utilized for the extraction of antioxidants from *Colpomenia sinuosa*, *J. rubens*, and *U. lactuca*. Extraction was carried out three successive times with CH_2Cl_2: MeOH (1:1) for 1 week at 25°C. Yields of the obtained residues were 0.34, 0.48, and 0.65% from *C. sinuosa*, *J. rubens*, and *U. lactuca*, respectively. DPPH scavenging activity of the crude algal extracts was estimated to be 50%. The results showed a clear correlation between phenolic content and antioxidant activity of the algal extracts (Kabbash and Shoeib, 2012).

Extraction conditions differ from one algae to another due to variation in composition. SPs were extracted from five algal species and their antioxidant activities were measured. The tested algal species were *U. pertusa*, *L. japonica*, *Enteromorpha linza*, *Porphyra hai-Tanensis*, and *Bryopsis plumose*. All five species required the same amount of water as a solvent, except *L. japonica*, which needed 25-fold less water than the others. As for the extraction temperature, it varied from 115 to 125°C and *U. pertusa* required the highest temperature. The extraction time also varied from 3 to 4 h according to the type of algae, however, both *U. pertusa* and *P. hai-Tanensis* required 1 h extra than the others (Zhang et al., 2010).

4.1.1.2 Temperature

One of the factors that could influence extraction is the temperature. Few studies were conducted to show the effect of temperature on the extraction of plant organisms (Conde et al., 2010). In general, increasing the temperature enhances the diffusion and dissolution of the bioactive compounds in solvent and hence improves the antioxidant activity. However, high temperatures in some cases may denature these compounds (Delgado-Andrade and Morales, 2005). As for algae, conventional water extraction was applied for the extraction of SPs from the red algae *G. fisheri* at two different temperatures: 25 and 55°C. The purified SPs fractions showed high galactose and sulfate contents at lower temperatures; in addition, it exhibited the highest content of phenolic compounds in comparison with fractions at higher temperatures. It was found that extraction temperature affects SPs yield, as well as both sugar and sulfate contents. It was suggested that there is a positive correlation between high antioxidant activity and content of phenolic compounds in these SPs fractions (Imjongjairaka et al., 2015).

4.1.1.3 Sulfation

The influence of sulfation on the antioxidant activity of SPs was also investigated in a study performed on the red seaweed *Corallina officinalis*. The dried algal tissue was suspended in sodium acetate buffer with pH 5.0 and molarity of 0.1 M, and to which 5 mM EDTA and 5 mM cysteine were added. The suspension was incubated along with papain for 24 h at 60°C. More acetate buffer was added and left for 6 h. The mixture was filtered and centrifuged and the supernatant was collected then dialyzed for 48 h and finally concentrated. The concentrate was precipitated with ethanol and the obtained precipitate was washed and dialyzed. The resulting solution was collected and further concentrated then lyophilized. This crude SPs extract was fractionated onto a DEAE anion exchange column and the fractions with the highest SPs contents (F1 and F2) were collected and their yields were 28.56 and 6.60%, respectively. In vitro antioxidation measurements showed that the higher molecular weight fraction (F2) had stronger scavenging effect on superoxide radical, hydroxyl radical, DPPH radical, and higher reducing power than the lower molecular weight one (F1). On desulfating the two fractions, it was revealed that the sulfated fractions had relatively higher antioxidant activities than their desulfated counterparts (Yang et al., 2011). Using a similar extraction procedure that does not involve column chromatography fractionation, red algal crude SPs or what is known as sulfated galactans were isolated from the Lebanese red algae of the genus *Pterocladia*. In addition, carrageenans were extracted after pretreating the dried algae with acetone to extract the hydrophobic pigments and with ethanol to separate the hydrophilic ones. The pretreated algae were then heated in water at 90°C for 3 h at pH (8–9). Filtration, alcohol precipitation, and drying procedures then followed. About 2.7% and 11.5% yields of sulfated galactans and carrageenans were obtained, respectively. As tested by the electrolysis method, galactans had better antioxidation activity than carrageenans (Sebaaly et al., 2012).

Table 9.1 presents a summary of the conventional extraction methods reported in literature, their operating conditions, % yields, and maximum antioxidant activities of the produced extracts.

4.2 Modern Methods

Modern extraction technologies have been recently introduced, which include EAE, MAE, and UAE techniques. These methods enhance extraction by special driving forces, such as ultrasound waves, microwaves, high hydrostatic pressure, and enzymatic action. These effects help the solvent reach different parts of the algal cell and break the cell walls.

Table 9.1: Summary of the reported conventional extraction methods and their operating conditions.

Types of Algae	Methods	Conditions				Antioxidant Activity[a]	References
		t	T, °C	Solvent:Algae (ratio, %)	Yield (%)		
S. tenerrimum	Water	16 h	90–95	3:1		TAC (62.55 ± 1.40)%	Vijayabaskar and Vaseela (2012)
S.swartzii	Water	16 h	90–95	3:1	10.43	ABTS (55.00 ± 3.61)%, H_2O_2 (47.23 ± 2.81)%, DPPH (25.33 ± 2.52)%	Vijayabaskar et al. (2012)
G. birdiae	Water Stirring	15 h 45 min	25 90	Water (1.5% w/v)		DPPH 80% Hydroxyl 75%	Souza et al. (2012)
L. japonica	Water	3 h	100	50:1 (v/w)	2.5	Superoxide 95% Hydroxyl 100%	Peng et al. (2012)
5 Seaweeds	Hot water	3–4 h	115–125			Superoxide 90% Hydroxyl 70%	Zhang et al. (2010)
Turbinara ornata	Hot water	3–4 h	90–95		10	ABTS 89.69%	Ananthi et al. (2010)
Ulva lactuca, D. simplex *S. crassifolia*	Absolute methanol Soxhlet	8 h	Methanol b.p.		21.79 23.52 25.11	Lipoxygenase inhibition 76% DPPH 69%	Al-Amoudi et al. (2009)
S. swartzii	Methanol		r.t.	70% (w/v) Methanol		TPC 73.92 mmol Fe II/100 g Gallic acid 12 mg equiv/100 g	Sadati et al. (2011)
T.conoides	Methanol		40			TAC 1.231 ± 0.173 mg GAE/g	Devi et al. (2011)
S. marginatum,P. tetrastomatica,T. conoides	Methanol					DPPH 17.79%–23.16%	Chandini et al. (2008)
P. urceolata	Methanol: chloroform (2:1); Soxhlet apparatus	6 h	r.t.	5:1 (v/w)	12.1	96%	Duan et al. (2006)
R.confervoides	Ethanol	(3 × 72 h)	r.t.			ABTS 2.11–3.58 mM	Li et al. (2012a)
S. platensis	Hot water 85% ethanol	2 h	100			ABTS (71.12 ± 0.02%) DPPH (68.29 ± 0.02%)	Abd El Baky et al. (2013)
U. lactua	Ethanol Water	1 week		4:1 100:1 (v/w)			Ambreen et al. (2012)
D. olivaceous, C. humicola	Acetone Shaking Methanol Shaking	Overnight	r.t.		12 11	H_2O_2 15% DPPH 95.8%, H_2O_2 39%	Uma et al. (2011)

Brownand red seaweeds	Methanol/water (50:50 v/v), shaking Acetone/water (70:30 v/v) Shaking	1 h	r.t.	40:1 (v/w)	64.71	ABTS 89.79 ± 2.3	Jiménez-Escrig et al. (2012)
U. lactua	Acetone Butanol Ethanol Soxhlet	6 h	b.p. Solvent		3.15 4.14 5.1		Hassan and Ghareib (2009)
12 soil-isolated strains of microalgae	Hexane Water and ethyl acetate	30 min	r.t.80			FRAP 73.20 ± 0.16 μmol Trolox g^{-1} DPPH 109.02 ± 8.25	Hajimahmoodi et al. (2010)
Chaetoceros and *Nannochloropsis*	Hexane Dichloromethane Chloroform Methanol Soaking Shaking	24 h	r.t.			DPPH 30.60 ± 4.6% μmol TE/g superoxide anion 1029.11 ± 0.7 μmol TE/g FRAP 470.72 ± 0.59 μmol TE/g	Goh et al. (2010)
C. sinuosa *J. rubens* *U. lactuca*	CH_2Cl_2: methanol (1:1)	1 week (3 times)	25		0.34 0.48 0.65	DPPH 50%	Kabbash and Shoeib (2012)
U. pertusa, L. japonica, E. linza, P. hai-Tanensis, B. plumose	Water	3–5 h	115–125			Superoxide radical 90% Hydroxyl radical 70%	Zhang et al. (2010)
G. fisheri	Water		25, 55				Imjongjairaka et al. (2015)
C. officinalis	0.1 M acetate buffer (pH 5) + 5 mM EDTA + 5 mM cysteine + papain	24 h + 6 h	60	50:1 (v/w)	28.56 6.6	Superoxide 80% DPPH 30%	Yang et al. (2011)
Pterocladia	0.1 M acetate buffer (pH 5) + 5 mM EDTA + 5 mM cysteine + papain	3 h	90	10:1 (v/w)	2.7 Sulfated galactans 11.5 Carrageenans	ROS inhibition 71% 47%	Sebaaly et al. (2012)

b.p., Boiling point; r.t., room temperature.
[a]Maximum attained values are given.

4.2.1 Enzyme-assisted extraction (EAE)

Marine algal cell walls are composed of heterogeneous complex biomolecules known to be sulfated and branched polysaccharides. These molecules are associated with proteins, as well as different bound ions, such as calcium and potassium. In order to extract these bioactive compounds, the cell wall matrix should be ruptured and broken down. Degrading enzymes, such as carbohydrases, have the ability to break down these cell walls at optimal conditions of temperature and pH, and hence release the desired intracellular bioactive compounds, as well as remove unnecessary components from cell walls. In addition, they can break down the high molecular weight proteins and polysaccharides and this could possibly improve the antioxidant properties of the extracts. Enzyme-assisted extraction (EAE) technique can therefore be utilized to enhance the extraction efficiency of bioactive compounds from seaweeds.

EAE offers an environmental friendly and safe alternative for extraction of algal bioactives, since it alleviates consumption of toxic organic solvents as enzymes convert water insoluble materials into water soluble ones. It also produces compounds with high yields and remarkable biological activities as compared to water and organic extracts produced via conventional techniques. Generally, food grade enzymes, such as cellulase, α-amylase, and pepsin, are used. These are relatively noncostly and therefore the process may turn out to be cost effective if it provides reasonable yields of highly bioactive compounds. However, specific expensive enzymes could be required for some extractions. Furthermore, enzymes are known for their high catalytic efficiency, which speeds up the extraction process. The enzymatic extraction also preserves, to a high extent, the original efficacy, and bioactive properties of the extracts (Athukorala et al., 2006; Heo et al., 2005; Siriwardhana et al., 2008; Wang et al., 2010).

Seven species of brown seaweeds were hydrolyzed using 10 different enzymes; 5 of them are carbohydrases: viscozyme, celluclast, AMG, termamyl, and ultraflo and the other 5 are proteases: protamex, kojizyme, neutrase, flavourzyme, and alcalase. The extracts obtained showed a high hydrogen peroxide scavenging activity even when compared to commercial antioxidants, while DPPH activities were remarkable for both ultraflo and alcalase extracts (Heo et al., 2005). In an earlier study by the same author, the optimum enzymatic hydrolysis was conducted for 12 h and the hydrolysates were tested for both their DPPH antioxidant activity, as well as their ability to inhibit lipid peroxidation. Promising antioxidant activities were obtained (Heo et al., 2003).

A number of protease and carbohydrase treatments were performed to extract antioxidant compounds from the red algae *Palmaria palmate*. The enzyme-assisted extraction was conducted for 24 h under the optimum conditions pertaining to each enzyme. The enzymatic action was terminated by boiling and then cooling the solution. Conventional water extraction was also conducted for the same time period and at room temperature. Protease treatments yielded better antioxidant activities than carbohydrases, as determined by DPPH and ORAC assays. Umamizyme extract recorded the highest total phenolic content (TPC) and hence the highest antioxidant activities among the tested enzymes. It was concluded that this method have effectively enhanced the recovery of polyphenols and other antioxidant compounds from *P. palmate* (Wang et al., 2010).

Eleven seaweed species were collected along the Brazilian coast of Natal. Dried algae were suspended in 0.25M NaCl and the pH was maintained at 8.0 with NaOH. Enzymes of maxataze, an alkaline protease from *Esporo bacillus* were added to the mixture to undergo proteolytic digestion. After agitated incubation for 24 h at 60°C and at controlled pH, the suspension was filtered and precipitated with two volumes of ice-cold acetone while being agitated at 48°C. After centrifugation, the precipitate was collected and dried under vacuum, then resuspended in distilled water. All tested species showed potential antioxidant activities, as determined by total antioxidant capacity, reducing power and ferrous chelating ability. In particular, four algal species, *Caulerpa sertularioide*, *Dictyota cervicornis*, *S. filipendula*, and *Dictyo pterisdelicatula*, exhibited the highest antioxidation activity (Costa et al., 2010). The same procedure was pursued by the same research team to isolate fucans SPs from the brown algae *S. vulgare*. The crude extract showed DPPH scavenging activity of about 22% (Dore et al., 2013).

One theoretical study was undertaken to predict the optimal conditions for the enzymatic extraction of SPs from the red algae *Gelidium amansii* and the brown algae *Japonica aresch*. Response surface methodology using Box-Behnken design was used and optimal conditions were found to be 778.01and 997.03 mg/g concentration of cellulase, 86.80 and 98.76 water to raw algae ratio, and 129.93 and 117.69 min extraction time for the red and green algae, respectively. The corresponding predicted yields were 48.36 and 32.47%, respectively. Experimental results were in good agreement with theoretical predictions (Li et al., 2012b).

In all of the earlier studies, enzymes were applied during the extraction process. In addition to that, enzymatic hydrolysis could be utilized as a postextraction step. For example, hydrolysis of the commercial agarose produced novel agarose-oligosaccharides. These short-chain saccharides that were obtained through chemical cleavage of SPs exhibited DPPH and ferric-reducing antioxidant activities (Kang et al., 2014).

We recently studied the extraction of SPs from the red seaweed *P. capillacea*. Water extraction was first conducted, where the dried and ground seaweeds were shaken in distilled water at pH 4, for 48 h at room temperature. After filtration then neutralization and dialysis of the filtrate for 48 h, the solution was centrifuged and the supernatant was then treated with four volumes of absolute ethanol. Both the high molecular weight precipitate S1(M), and the low molecular weight supernatant S2(M) were enzymatically hydrolyzed using different enzymes. Hydrolyzed fractions were then purified onto an anion exchange column. Hydrolysate of viscozyme and its purified fraction showed remarkable antioxidant activities of higher than 90% (Fleita et al., 2015).

4.2.2 Microwave-assisted extraction (MAE)

Microwaves are a nonionizing form of electromagnetic radiation with frequencies ranging from 300 MHz to 300 GHz. During extraction, energy of this radiation is transferred to molecules of the algal intracellular liquids and induces vibration therein. Homogeneous heating of the medium takes place via a twin dipole rotation and ionic conduction heating mechanism. The vibration is in the form of rotational motion of molecules accompanied by an increase in the temperature of the algal intracellular liquids. This, in turn, disrupts hydrogen bonds and leads to evaporation of these liquids into a vapor that exerts pressure on the cell walls. Thus, cell wall matrix breaks down releasing its contents. This enables the extraction of target compounds. There are two main modes of operation for MAE systems, closed and open vessel modes. In the former, extraction of target compounds is conducted at higher temperature and pressure conditions, while extraction in the latter mode extraction is performed under atmospheric pressure conditions. Nowadays, there is a growing interest in MAE as it offers a viable option for the extraction of bioactive compounds from plant material. This is attributed to its potential promising advantages over conventional extraction techniques. Advantages may, include enhanced extraction rate, reduced utilization of solvents, and improved extraction yield (Hahn et al., 2012; Kadam et al., 2013).

Studies showed that MAE was able to reduce solvent consumption and extraction times, while providing yields that are equivalent to or even higher than those obtained with conventional methods.

Microwave-assisted extraction was utilized at 120 psi for 1 min to extract fucoidan from *Fucus vesiculosus*. A high yield of 18.22% was achieved and this was comparable to the yield obtained from successive conventional extractions at 70°C. The method was thus recommended as a timesaving and environmental friendly method that requires no solvents (Rodriguez-Jassoa et al., 2011). In another study conducted on brown seaweed *Ascophyllum nodosum* and applying microwave-assisted extraction, SPs (fucoidan) were extracted under different extraction conditions, such as, temperature (90–150°C) and extraction time intervals (5–30 min). The optimal fucoidan yield was 16.08%, obtained at 120°C temperature for 15 min extraction time. All fractions were evaluated for their antioxidant activities by DPPH scavenging, as well as reducing power. The highest antioxidant activity belonged to fucoidan extracted at 90°C. This study proved that MAE could provide an efficient method for extracting SPs from seaweeds that are potential resources for natural antioxidants (Yuan and Macquarrie, 2015).

In another study, six SPs with different molecular weights were extracted from the green algae *Enteromerpha prolifera* by the microwave-assisted acid hydrolysis method and the relationship between molecular weight (Mw) and antioxidant activities of polysaccharides was studied. In comparison to conventional heating, the reaction time required to produce polysaccharides with the same Mw was greatly reduced. This proved the efficacy of the microwaves in accelerating rate of reaction. All fractions possessed antioxidant activities and lower Mw fractions showed greater inhibitory effects as compared to vitamin C. The study revealed that the effect of Mw becomes significant when the sulfate content and monosaccharide composition of the fractions are similar (Li et al., 2013).

Microwave could also be used as a postextraction step. In a study on the extracellular polysaccharides of the unicellular marine microalga, hermetical microwave degradation of the polysaccharides and subsequent purification of the degradation fractions on a size exclusion column were performed. The fractions with lower molecular weights showed better DPPH, hydroxyl, superoxide anion radical scavenging activities than the higher molecular weight fractions, in addition to an inhibitory effect on lipid peroxidation (Sun et al., 2009).

4.2.3 Ultrasound-assisted extraction (UAE)

Ultrasound waves are sound waves of high frequency that exceeds that of the human hearing capacity of 20 kHz. When these waves propagate through the solvent in the form of rarefactions and compressions, they generate a negative pressure in the medium. When that pressure exceeds the tensile strength of the medium, then vapor bubbles form. These bubbles are compressed then they collapse under the effect of the high-intensity ultrasound field, resulting in a phenomenon known as cavitation. The compression of bubbles elevates the solvent pressure and temperature leading to more bubble collapse, which, in turn, creates shock waves that pass through the solvent. These shock waves improve the mixing between sample and solvent. In addition, they convert sound waves into mechanical waves causing macroturbulence and rapid collisions between molecules, thus facilitating the penetration of the solvent into the sample and, in turn, increasing the contact surface area between them. The two commonly used pieces of equipment for UAE are the ultrasonic water bath and the ultrasonic probe system fitted to horn transducers (Hahn et al., 2012; Kadam et al., 2013).

UAE has the advantages of being rather simple, cost-effective, and efficient. Additionally, it is not as destructive as conventional extraction and can be used to extract thermolabile compounds. It can also provide high yields with faster rate than the conventional extraction. Compared to other modern extraction methods, it utilizes lower cost equipment with versatile aqueous and organic solvents. The technology is suitable for both laboratory-scale, as well as large-scale applications (Kadam et al., 2013). However, there are many parameters that should be adjusted to improve the extraction process using ultrasound assisted techniques; some are similar to those of the conventional methods, such as the amount and type of solvent, time of extraction, sample to solvent ratio and ultrasound bath temperature, and others are related to the ultrasound source, such as frequency and intensity of pulses, which affect the extraction dynamics.

A study was conducted for the purpose of extracting and purifying polysaccharides from algae, where the ultrasound-assisted extraction of polysaccharides from the green algae *C. pyrenoidosa* was investigated. The yield was equivalent to 44.8 g/kg when using optimal conditions of 400 W of ultrasound power, with incubation in water at 100°C for 4 h and precipitation of polysaccharides with 80% ethanol (Shi et al., 2007).

In other studies, the antioxidant activities of the UAE extracts were determined. We recently conducted UAE studies on green and red algal species. Extraction was carried out using water for different time intervals at 50 and 60°C for *U. lactuca* and at room temperature for *J. rubens*. After filtration, dialysis of filtrate and centrifugation, ethanol was added to the supernatant in the ratio of 3.5:1 (by volume). The supernatant was then collected and lyophilized. For the green algae, optimum yields of SPs were obtained under UAE conditions of 60°C temperature and 4 h extraction time. As for the red algae, conventional extraction and UAE gave comparable yields. For the purpose of saving time and energy, it was suggested that UAE would be favored (El-Sayed et al., 2015).

4.2.4 Supercritical fluid extraction

This method is conducted using carbon dioxide at supercritical conditions, where mild pressure and temperature are applied such that the material used reaches its supercritical point. The commonly used gas is carbon dioxide, which is abundant and cheap, however, the cost of the supercritical equipment is high. In addition, sometimes a cosolvent is required to facilitate dissolving the extracted materials since the nonpolarity of carbon dioxide may hinder the dissolution of the extracts (Conde et al., 2010). Factors affecting the extraction are temperature, pressure, and type of cosolvent.

SPs were extracted from the brown seaweed *Saragasum pallidum* in three successive steps; supercritical CO_2 extraction for degreasing the polysaccharides at 55°C and 45 MPa for 4 h, ultrasonic-assisted extraction at 90°C for 5 h, and membrane ultrafiltration technology. The extracts were then purified using diethylaminoethyl 52 (DEAE 52) column chromatography. Crude fractions and fractions received from column were investigated for their antioxidant activity using DPPH assay. The antioxidant activities of the three crude extracts before purification were low ranging from 10% to 18%. Even lower activities were obtained for the purified fractions. It was suggested that the activity is related to the degree of sulfation and molecular weight (Ye et al., 2008).

4.2.5 Subcritical water extraction

In this method, temperature above the water boiling point (100°C) is employed along with high pressure to ensure water is in the liquid state (Plaza et al., 2010). This method has been applied for species other than algae (Yang et al., 2013).

4.2.6 High-pressure liquid extraction

Extraction via this method takes place under high pressure and temperature. Hence, it requires shorter time and consumes less volume of solvent relative to the conventional method. However, it utilizes specific expensive equipment to apply high pressure (Conde et al., 2010). This method has been reported for the extraction of plant material other than algae (Prasad et al., 2009).

Table 9.2 presents a summary of the modern extraction methods reported in literature, their operating conditions, % yields, and maximum antioxidant activities of the produced extracts.

Table 9.2: Summary of the reported modern extraction methods and their operating conditions.

Types of Algae	Methods	Conditions				Antioxidant Activity[a]	References
		t	*T*, °C	Solvent:Algae (ratio, %)	Yield (%)		
Seven species of brown seaweeds	EAE Carbohydrase/protease	12 h	Optimum T for enzyme	100:1 (v/w)		DPPH 72.46%/68.16% 80%	Heo et al. (2003) Heo et al. (2005)
P. palmate	EAE Carbohydrase/protease	24 h	Optimum T for enzyme	25:1 (v/w)		ORAC 629.5 ± 15.2	Wang et al. (2010)
11 species of Brazilian coast of Natal	EAE Maxataze, alkaline protease	24 h	60			Hydroxyl 11.8 ± 0.5% superoxide 32.5 ± 2.4%	Costa et al. (2010)
S. vulgare	EAE Maxataze, alkaline protease	24 h	60	5:1 (v/w)		DPPH 22%	Dore et al. (2013)
G. amansii *J. aresch*	EAE Cellulase	129.93 and 117.69 min	37	86.80 98.76	48.36 32.47		Li et al. (2012b)
Pterocldia capillacea	EAE (viscozyme)	48 h	r.t.	20:1 (v/w)		DPPH (>90%)	Fleita et al. (2015)
F. vesiculosus	MAE (120 psi)	1 min		No solvent	18.22		Rodriguez-Jassoa et al. (2011)
A. nodosum	MAE	5–30 min	90–150		16.08		Yuan and Macquarrie (2015)
E. prolifera	MAE	5 min	80 90	75:1 (v/w)	20.3	Superoxide 85% Hydroxyl 85%	Li et al. (2013)
C. pyrenoidosa	UAE (400 W, water)	4 h	100		44.8 g/kg		Shi et al. (2007)
U. lactua *J. rubens*	UAE (400 W, water)	4 h 2 h	50, 60 r.t.	20:1 (v/w)	0.25–8	DPPH 55%–60% 55%–68%	El-Sayed et al. (2015)
S. pallidumiin	SFE	4 h	55	2:1 (v/w)		DPPH 6.6%	Ye et al. (2008)

r.t., Room temperature.
[a]Maximum attained values are given.

5 Conclusions

The yield and antioxidant activity of the algal extracts are influenced by substrate-related, process-related, and ecorelated factors as illustrated in Fig. 9.3.

The substrate-related factors, include algae type being red, green, or brown; algal composition in terms of carbohydrate, sulfate, phenolic contents, as well as content of sugars and their structure, arrangement, and type of glycosidic linkages connecting them. The algal composition and structural properties are also function of the ecological conditions in which the algae exist and the surrounding environment that the algae strives on. Examples of these are location, water depth of the algal habitat, seasonal variations of climate, wind speed, water pH, and water salinity, and the microbial species that are hosted by the algal species and that are believed to affect their productivity for bioactive compounds.

As for the process-related factors, these comprise the extraction method; being conventional or modern and its relevant techniques and operating conditions of temperature, pressure, time, pH, solvent type, algae to solvent ratio, and so on. This is in addition to pretreatment and postextraction procedures.

Conventional methods, although relatively simple and nonexpensive, produce generally low yields with low selectivity and require long extraction durations along with large amounts of solvents. The use of excessive amounts of organic solvents poses health and risk hazards,

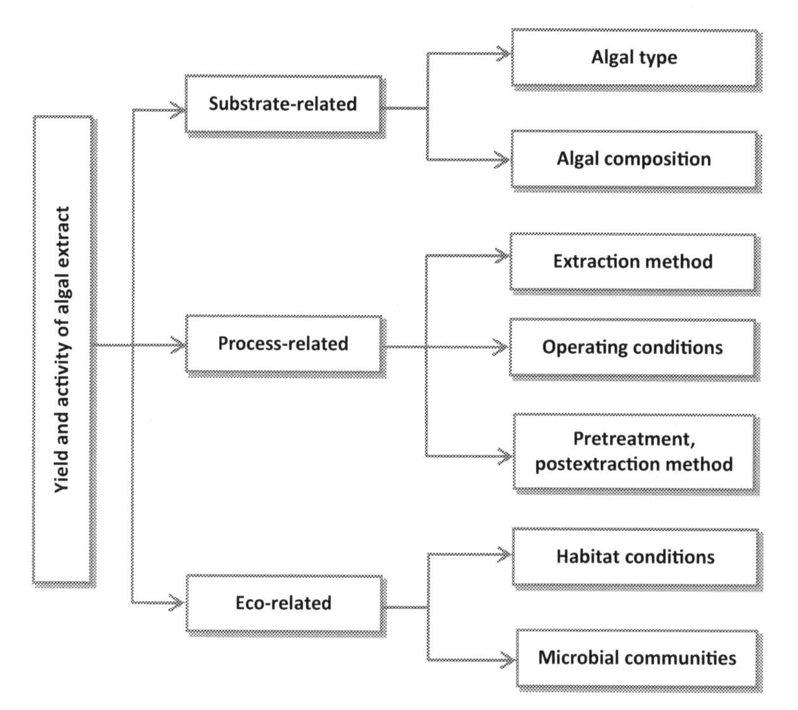

Figure 9.3: Factors Affecting Yields and Antioxidant Activity of Algal Extracts.

and calls for stringent safety measures. Attractive features of modern methods, on the other hand, include their high productivity and efficiency, selectivity, faster operation, and relatively low or no requirements for solvents. However, they usually demand equipment that is more expensive and sophisticated than the conventional one. EAE, in particular, could be performed under mild conditions, which is advantageous in preserving the bioactivity of the extract. It could therefore be potentially suitable for scaled-up industrial processes. However, cost and availability of specific enzymes is still an issue.

As for the operating conditions, reported studies have shown that extraction is favored by increasing temperature, unless the compound is at risk of being denatured. Temperature, in general, facilitates breaking down of algal cell walls and this leads to dissolving of the bioactive compound in the solvent. Previously, studies showed that the yields could be increased with prolonging time of extraction. The polarity of solvents also plays a role in extraction because it affects the dissolution process. Operating conditions could also affect the type of bioactive compounds that are being released.

The pretreatment procedure is important in getting rid of undesired compounds and enriching the extract with the targeted bioactive compounds. The postextraction procedures, on the other hand, are employed to either further purify the product or to fractionate it into different products with different molecular weights, as well as different structures and properties. Fractionation of SPs is usually performed on an anion exchange column onto which the polysaccharides ($pK_a > 12$) are retained under alkaline conditions provided by the mobile phase. This takes place via coulombic attraction forces by virtue of their negative charge that is opposite to that of the column packing. Then, monosaccharides and oligosaccharides are successively separated using gradient elution by gradually increasing the ionic strength of the alkaline mobile phase. In some studies, this fractionation was successful in producing fractions with higher antioxidation potency than the crude extracts. This was ascribed to the separation of lower molecular weight compounds that are more bioactive than the crude polysaccharide extracts, or the isolation of compounds with distinct chemical and structural properties. However, some previous reports concluded that the fractionation did not improve the antioxidant activity, and this could be due to alteration of the bioactive chemical structure during fractionation. Although it was reported that anion exchange chromatography does not affect the sulfate ester groups, yet the changes in ionic strength during elution may alter the special arrangement, and hence the molecular geometry of the target compounds which may, in turn, reduce their bioactivity (Hahn et al., 2012).

The antioxidant activity of the extracted product is also dependent on its chemical and structural properties, such as degree of sulfation, sugar, and phenolic contents, molecular weight, type of glycosidic linkages, type of monosaccharides, and their arrangement. To date, structure-activity correlations have not yet been fully elucidated. It was reported that there is a positive direct correlation between phenolic content and antioxidant activity. The same applies for degree of sulfation. It was also found that lower molecular weight compounds often have more potent antioxidant activities than their higher molecular weight counterparts.

The design of the appropriate extraction method should involve technical, economic, and environmental assessment of the extraction method under investigation. Technical assessment entails conducting a thorough optimization study to determine the conditions at which optimal yields and antioxidant activities are obtained. In the economic assessment, running (operating) costs and capital costs should be estimated to calculate the profit and return on investment. Finally, environmental and safety aspects should be evaluated.

6 Future Outlook

The chemical structure of the algal extract greatly affects its antioxidant activity. However, to date, SARs have not yet been fully understood and hence should be subject to future investigation.

Optimization of operating parameters is a crucial step in designing any extraction process. Very few studies utilized optimization tools, such as Box-Behnken and d-optimal designs. Future research should make better use of these tools and explore other tools, such as artificial neural networks (ANN). Maximizing the yields and antioxidant activities of the algal extracts is the goal of any extraction process. Most of the reported extraction yields did not exceed 30%. Therefore, further research should look into introducing new efficient and cost-effective techniques that produce bioactive compounds with high yields without compromising their antioxidant activity.

The production of bioactive compounds from algal species is a largely growing field, however, only lab-scale extraction processes have been reported so far. The potential for scaling up should be of interest in future studies. Furthermore, there is also a lack in studies that deal with economics of the extraction process and its environmental assessment. Reported studies mostly focus on the technical aspects only.

References

Abd El Baky, H., El Baz, K.F.H., EL-Latife, S.A., 2013. Induction of sulfated polysaccharides in *Spirulina platensis* as response to nitrogen concentration and its biological evaluation. J. Aquac. Res. Dev. 5 (1), 206–213.

Abdel-Fattah, A.F., Edress, M., 1972. Polysaccharides content of *Ulva lactuca*, seasonal changes in the constituents of *Ulva lactuca*. Qual. Plant. Mater. Veg. 22 (1), 15–18.

Abdel-Fattah, A.F., Hussien, M.M., Salem, M.M., 1974. Studies of the purification and some properties of sargassan, asulphated heteropolysaccharides from *Sargassum linifolium*. Carbohydr. Res. 33, 9–17.

Abdel-Fattah, A.F., Abed, N.M., Edrees, M., 1973b. Seasonal variation in the chemical composition of the agarophyte *Pterocladia capillacea*. Aust. J. Mar. Freshwat. Res. 24 (2), 177–181.

Al-Amoudi, O.A., Mutawie, H.H., Patel, A.V., Blunden, G., 2009. Chemical composition and antioxidant activities of Jeddah corniche algae. Saudi J. Biol. Sci. 16 (1), 23–29.

Ambreen, H., Hira, K., Tariq, A., Ruqqia, R., Sultana, V., Ara, J., 2012. Evaluation of biochemical component and antimicrobial activity of some seaweeds occurring at Karachi Coast. Pak. J. Bot. 445, 1799–1803.

Ananthi, S., BalajiRaghavendran, H.R., Sunil, A., Gayathri, V., Ramakrishnan, G., Vasanthi, H.R., 2010. In vitro antioxidant and in vivo anti-inflammatory potential of crude polysaccharide from *Turbinaria ornata* (Marine Brown Alga). Food Chem. Toxicol. 48, 187–192.

Athukorala, Y., et al., 2006. An anticoagulative polysaccharide from an enzymatic hydrolysate of *Ecklonia cava*. Carbohydr. Polym. 66, 184–191.

Barrow, C., Shahidi, F., 2008. Marine Nutraceuticals and Functional Foods. CRC Press, New York, USA.

Berteau, O., Mulloy, B., 2003. Sulfated fucans, fresh perspectives: structures, functions, and biological properties of sulfated fucans and an overview of enzymes active toward this class of polysaccharide. Glycobiology 13, 29R–40R.

Chandini, S.K., Ganesan, P., Bhaskar, N., 2008. In vitro antioxidant activities of three selected brown seaweeds of Indian. Food Chem. 107, 707–713.

Choe, E., Min, D.B., 2009. Mechanisms of antioxidants in the oxidation of foods. Compr. Rev. Food Sci. Food Saf. 8 (4), 345–358.

Conde, E., Moure, A., Domínguez, H., Parajó, J.C., 2010. Extraction of Natural Antioxidants From Plant Foods. In: Rizvi, S.S.H. (Ed.), Separation, extraction and concentration processes in the food, beverage and nutraceutical industries. Woodhead Publishing Series in Food Science, Technology and Nutrition. Woodhead Publishing, Sawston, Cambridge, United Kingdom, pp. 506–594.

Costa, L.S., Fidelis, G.P., Cordeiro, S.L., Oliveira, R.M., Sabry, D.A., Câmara, R.B.G., Nobre, L.T.D.B., Costa, M.S.S.P., Almeida-Lima, J., Farias, E.H.C., Leite, E.L., Rocha, H.A.O., 2010. Biological activities of sulfated polysaccharides from tropical seaweeds. Biomed. Pharmacother. 64, 21–28.

Delgado-Andrade, C., Morales, F., 2005. Unravelling the contribution of melanoidins to the antioxidant activity of coffee brew. J. Agric. Food Chem. 53, 1403–1407.

Devi, G.K., Manivannan, K., Thirumaran, G., Rajathi, F.A.A., Anantharaman, P., 2011. In vitro antioxidant activities of selected seaweeds from Southeast coast of India. Asian Pac. J. Trop. Med., 205–211.

Dore, C.M., das, C., Faustino Alves, M.G., Will, L.S., Costa, T.G., Sabry, D.A., de Souza Rêgo, L.A., Accardo, C.M., Rocha, H.A., Filgueira, L.G., Leite, E.L., 2013. A sulfated polysaccharide, fucans, isolated from brown algae *Sargassum vulgare* with anticoagulant, antithrombotic, antioxidant and anti-inflammatory effects. Carbohydr. Polym. 91 (1), 467–475, 2.

Duan, X.-J., Zhang, W.-W., Li, X.-M., Wang, B.-G., 2006. Evaluation of antioxidant property of extract and fractions obtained from a red alga, *Polysiphonia urceolata*. Food Chem. 95, 31–43.

El-Sayed, M.M.H., Fleita, D., Rifaat, D., Essa, H., Samy,S., 2015. Towards Optimizing the Conventional and Ultrasonic-assisted Extraction of Sulfated Polysaccharides from Marine Algae, EURECA Conference 2015, American University in Cairo, Egypt.

Fleita, D., El-Sayed, M., Rifaat, D., 2015. Evaluation of the antioxidant activity of enzymatically-hydrolyzed sulfated polysaccharides extracted from red algae; *Pterocladia capillacea*. LWT—Food Sci. Technol., 1–9.

Garrett, A.R., Murray, B.K., Robison, R.A., O'Neill, K.L., 2010. Measuring antioxidant capacity using the ORAC and TOSC assays. Methods Mol. Biol. 594, 251–262.

Goh, S.H., MdYusoff, F., Loh, S.P., 2010. A comparison of the antioxidant properties and total phenolic content in a diatom, *Chaetoceros*sp. and a Green Microalga, *Nannochloropsis* sp. J. Agric. Sci. 2, 3.

Goyal, A.K., Mishra, T., Bhattacharya, M., Kar, P., Sen, A., 2013. Evaluation of phytochemical constituents and antioxidant activity of selected actinorhizal fruits growing in the forests of Northeast India. J. Biosci. 38 (4), 1–7.

Gupta, S., Abu Ghannam, N., 2011. Recent developments in the application of seaweeds or seaweed extracts as a means for enhancing the safety and quality attributes of foods. Innov. Food Sci. Emerg. Technol. 12, 600–609.

Hahn, T., Lang, S., Ulber, R., Muffler, K., 2012. Novel procedures for the extraction of fucoidan from brown algae. Process Biochem. 47, 1691–1698.

Hajimahmoodi, M., Faramarzi, M.A., Mohammadi, N., Soltani, N., Oveisi, M.R., Nafissi-Varcheh, N., 2010. Evaluation of antioxidant properties and total phenolic contents of some strains of microalgae. J. Appl. Phycol. 22 (1), 43–50.

Haq, Q.N., Percival, E., 1966. Structural studies on the water: soluble polysaccharide from the green seaweed, *Ulva lactuca*. In: Barnes, H. (Ed.), Some Contemporary Studies in Marine Science. Allen and Unwin, London, p. 355.

Hassan, S.M., Ghareib, H.R., 2009. Bioactivity of *Ulva lactuca* L. Acetone extract on germination and growth of lettuce and tomato plants. Afr. J. Biotechnol. 8, 3832–3838.

Heo, S.-J., Lee, K.-W., Song, C.B., Jeon, Y.-J., 2003. Antioxidant activity of enzymatic from brown seaweeds. Algae 18, 71–81.

Heo, S., Park, F., Lee, K., Jeon, Y., 2005. Antioxidant activities of enzymatic extracts from brown seaweeds. Biores. Technol. 96, 1613–1623.

Herrero, M., Cifuentes, A., Ibanez, E., 2006. Sub- and supercritical fluid extraction of functional ingredients from different natural sources: plants, food-by-products, algae and microalgae. Food Chem. 98, 136–148.

Hoshino, T., Hayashi, T., Hayashi, K., Hamada, J., Lee, J.B., Sankawa, U., 1998. An antivirally active sulfated polysaccharide from *Sargassumhorneri* (TURNER) CAGARDH. Biol. Pharm. Bull. 21, 730–734.

Huang, D., Ou, B., Prior, R.L., 2005. The chemistry behind antioxidant capacity assays. J. Agric. Food Chem. 53 (6), 1841–1856.

Hussien, M.M., 1977. Biochemical studies on the Egyptian marine algae: the water soluble polysaccharides of *Ulva lactuca*. Pak. J. Biochem. 10 (1), 19–24.

Imjongjairaka, S., Ratanakhanokchaia, K., Laohakunjita, N., Tachaapaikoonb, C., PatthraPasonb, P., Waeonukulb, R., 2015. Biochemical characteristics and antioxidant activity of crude and purified sulfated polysaccharides from *Gracilariafisheri*. Food Nutr. Sci. 80, 524–532.

Jiao, G., Yu, G., Zhang, J., Ew art, H., 2011. Chemical structures and bioactivities of sulfated polysaccharides from marine algae. Mar. Drugs 9, 196–223.

Jiménez-Escrig, A., Gómez-Ordóñez, E., Rupérez, P., 2012. Brown and red seaweeds as potential sources of antioxidantnutraceuticals. J. Appl Phycol. 24, 1123–1132.

Kabbash, A., Shoeib, N., 2012. Bio-screening of some marine algae from the coasts of Egypt. J. Sci. Res. Pharm. 1, 17–21.

Kadam, S.U., Tiwari, B.K., O'Donnell, C.P., 2015. Extraction, structure and biofunctional activities of laminarin from brown algae. Int. J. Food Sci. Technol. 50, 24–31.

Kadam, S.U., Tiwari, B.K., O'Donnell, C.P., 2013. Application of novel extraction technologies for bioactives from marine algae. J. Agr. Food Chem. 61, 4667–4675.

Kang, O.L., Ghani, M., Hassan, O., Rahmati, S., Ramli, N., 2014. Novalagaro-oligosaccharide production through enzymatic hydrolysis: physicochemical properties and antioxidant activities. Food Hydrocoll., 1–5.

Khairy, H., El-Sheikh, M., 2015. Antioxidant activity and mineral composition of three Mediterranean common seaweeds from Abu-Qir Bay, Egypt. Saudi J. Biol. Sci. 22 (5), 623–630.

Lahaye, M., 1998. NMR spectroscopic characterization of oligosaccharides from two *Ulva rigidaulvan* samples (Ulvales, Chlorophyta) degraded by a lyase. Carbohydr. Res. 314, 1–12.

Li, B., Wei, X., Zhao, R., 2008. Fucoidan: structure and bioactivity. Molecules 13, 1671–1695.

Li, B., Liu, S., Xing, R., Li, K., Li, R., Qina, Y., Wanga, X., Wei, Z., Li, P., 2013. Degradation of sulfated polysaccharides from *Enteromorphaprolifera* and their antioxidant activities. Carbohydr. Polym. 92, 1991–1996.

Li, K., Li, X.M., Gloer, J.B., Wang, B.G., 2012a. New nitrogen-containing bromophenols from the marine red alga *Rhodomela confervoides* and their radical scavenging activity. Food Chem. 135, 868–872.

Li, S., Han, D., Row, K.H., 2012b. Optimization of enzymatic extraction of polysaccharides from some marine algae by response surface methodology. Kor. J. Chem. Eng. 29 (5), 650–656.

Love, J., Percival, E., 1964. The polysaccharides of the green seaweed *Codium fragile*. Part III. A β-1, 4-linked mannan. J. Chem. Soc., 3345–3350.

McLellan, D.S., Jurd, K.M., 1992. Anticoagulants from marine algae. Blood Coag. Fibrinol. 3, 69–80.

Medlin, L.K., Kooistra, W.H.C.F., Potter, D., Saunders, G.W., Anderson, R.A., 1997. Phylogenetic relationships of the "golden algae" (haptophytes, heterokontchromophytes) and their plastids. Plant Syst. Evol. 11, 187–219.

Ortiza, J., Romeroa, N., Roberta, P., Arayab, J., Lopez-Hernándezc, J., Bozzoa, C., Navarretea, E., Osorioa, A., Riosa, A., 2006. Dietary fiber, amino acid, fatty acid and tocopherol contents of the edible seaweeds *Ulva lactuca* and *Durvillaeaantarctica*. Food Chem. 99, 98–104.

Peng, Z., Liu, M., Fang, Z., Zhang, Q., 2012. In vitro antioxidant effects and cytotoxicity of polysaccharides extracted from *Laminaria japonica*. Int. J. Biol. Macromol. 50 (5), 1254–1259.

Percival, E., 1979. The polysaccharides of green, red and brown seaweeds: their basic structure, biosynthesis and function. Br. Phycol. J. 14, 103–117.

Percival, E., McDowell, R.H., 1990. Algal polysaccharides. In: Dey, P.M. (Ed.), Carbohydrates, vol. 2 of Methods in Plant Biochemistry. Academic Press, London, pp. 523–547.

Percival, E., McDowell, R.H., 1967. Chemistry and Enzymology of Marine Algal Polysaccharides. Academic Press, London, New York.

Percival, E., Wold, J.K., 1963. The acid polysaccharide from the green seaweed, *Ulva lactuca*. Part II. The site of the ester sulphate. J. Chem. Soc., 5459–5468.

Plaza, M., Amigo-Benavent, M., Del Castillo, M.D., Ibanez, E., Herrero, M., 2010. Facts about the formation of new antioxidants in natural samples after subcritical water extraction. Food Res. Int. 43, 2341–2348.

Prasad, K., Yang, E., Yi, C., Zhao, M., Jiang, Y., 2009. Effects of high pressure extraction on the extraction yield, total phenolic content and antioxidant activity of longan fruit pericarp. Innov. Food Sci. Emerg. Technol. 10, 155–159.

Qi, H., Zhang, Q., Zhao, T., Chen, R., Zhang, H., Niu, X., et al., 2005a. Antioxidant activity of different sulfate content derivatives of polysaccharide extracted from *Ulva pertusa* (Chlorophyta) in vitro. Int. J. Biol. Macromol. 37, 195–199.

Qi, H., Zhao, T., Zhang, Q., Li, Z., Zhao, Z., Xing, R., 2005b. Antioxidant activity of different molecular weight sulphated polyssacharides from *Ulva pertusa Kjellum* (Chlorophyta). J. Appl. Phycol. 17, 527–534.

Re, R., Nicoletta, P., Anna, P., Ananth, P., Rice-Evans, C., 1999. Antioxidant activity applying an improved ABTS radical cation decolorization assay. Free Rad. Biol. Med. 26 (9–10), 1231–1237.

Rodriguez- Jassoa, R., Mussatoa, S., Pastrana, L., Aguilar, C., 2011. Microwave-assisted extraction of sulfated polysaccharides (fucoidan) from brown seaweed. Carbohydr. Polym. 86, 1137–1144.

Sadati, N., Khanavi, M., Mahrokh, A., Nabavi, S.M.B., Sohrabipour, J., Hadjiakhoondi, A., 2011. Comparison of antioxidant activity and total phenolic contents of some Persian Gulf marine algae. J. Med. Plants 10, 37.

Sebaaly, C., Karaki, N., Chahine, N., Evidente, A., Yassine, A., Habib, J., Kanaan, H., 2012. Polysaccharides of the red algaePterocladia growing on the Lebanese coast: isolation, structural features with antioxidant and anticoagulant activities. J. Appl. Pharm. Sci. 2 (10), 1–10.

Sheng, J.C., Yu, F., Xin, Z.H., Zhao, L.Y., Zhu, X.J., Hu, Q.H., 2007. Preparation, identification and their antitumor activities in vitro of polysaccharides from *Chlorella pyrenoidosa*. Food Chem. 105, 533–539.

Shi, Y., Sheng, J., Yang, F., Hu, Q., 2007. Purification and identification of polysaccharides derived from *Chlorella pyrenoidosa*. Food Chem. 103, 101–105.

Siriwardhana, N., Kim, K.N., Lee, K.W., Kim, S.H., Ha, J.H., Song, C.B., Bak, L.J., Jeon, Y., 2008. Optimisation of hydrophilic antioxidant extraction from *Hizikia fusiformis* by integrating treatments of enzymes, heat and pH control. Int. J. Food Sci. Technol. 43, 587–596.

Souza, B.W.S., Cerqueira, M.A., Bourbon, A.I., Pinheiro, A.C., Martins, J.T., Teixeira, J.A., Coimbra, M.A., Vicente, A.A., 2012. Chemical characterization and antioxidant activity of sulfated polysaccharide from the red seaweed *Gracilaria birdiae*. Food Hydrocoll. 27, 287–292.

Sun, L., Wang, C., Shi, Q., Ma, C., 2009b. Preparation of different molecular weight polysaccharides from *Porphyridium cruentum* and their antioxidant activities. Int. J. Biol. Macromol. 45 (1), 42–47.

Ueda, J.-I., Saito, N., Shimazu, Y., Ozawa, T., 1996. Oxidative DNA strand scission induced by copper(II) complexes and ascorbic acid. Arch. Biochem. Biophys. 333, 377–384.

Uma, R., Sivasubramanian, V., Devaraj, S.N., 2011. Evaluation of in vitro antioxidant activities and antiproliferative activity of green microalgae, *Desmococcus olivaceous* and *Chlorococcum humicola*. J. Algal Biomass Utiln. 2, 82–93.

Vadlapudi, V., 2012. Medicinal plants as antioxidant agents: understanding their mechanism of action and therapeutic efficacy. Antiox. Act. Marine Algae, 189–203.

Vijayabaskar, P., Vaseela, N., 2012. In vitro antioxidant properties of sulfated polysaccharide from brown marine algae *Sargassum tenerrimum*. Asian Pac. J. Trop. Dis. 2, S890–S896.

Vijayabaskar, P., Vaseela, N., Thirumaran, G., 2012. Potential antibacterial and antioxidant properties of a sulfated polysaccharide from the brown marine algae *Sargassum swartzii*. Chin. J. Nat. Med. 10 (6), 0421–0428.

Wang, T., Jónsdóttir, R., Kristinsson, H.G., Hreggvidsson, G.O., Jónsson, J.Ó., Thorkelsson, G., Ólafsdóttir, G., 2010. Enzyme-enhanced extraction of antioxidant ingredients from red algae *Palmaria Palmata*. LWT—Food Sci. Technol. 43, 1387–1393.

Wijesinghe, W.A., Jeon, Y.J., 2012. Enzyme-assistant extraction (EAE) of bioactive components: a useful approach for recovery of industrially important metabolites from seaweeds: a review. Fitoterapia 83, 6–12.

Yang, L., Qu, H., Mao, G., Zhao, T., Li, F., Zhu, B., Zhang, B., Wu, X., 2013. Optimization of subcritical water extraction of polysaccharides from Grifolafrondosausing response surface methodology. Pharmacognosy 9, 120–129.

Yang, Y., Liu, D., Wu, J., Chen, Y., Wang, S., 2011. In vitro antioxidant activities of sulfated polysaccharide fractions extracted from *Corallina officinalis*. Int. J. Biol. Macromol. 49, 1031–1037.

Ye, H., Wang, K., Zhou, C., Liu, J., Zeng, X., 2008. Purification, antitumor and antioxidant activities in vitro of polysaccharides from the brown seaweed *Sargassum pallidum*. Food Chem. 11, 428–432.

Ye, H., Wu, T., Zhou, C.H., 2006. Optimization of the extraction of *Sargassum* sp. polysaccharides. Food Res. Dev. 27, 22–24.

Yuan, Y., Macquarrie, D., 2015. Microwave assisted extraction of sulfated polysaccharides (fucoidan) from *Ascophyllum nodosum* and its antioxidant activity. Carbohydr. Polym. 129, 101–107.

Zhang, Q., Li, N., Zhou, G., Lu, X., Xu, Z., Li, Z., 2003. In vivo antioxidant activity of polysaccharide fraction from *Porphyrahaitanensis* (Rhodephyta) in aging mice. Pharmacol. Res. 48, 151–155.

Zhang, Z., Wang, F., Wang, X., Liu, X., Hou, Y., Zhang, Q., 2010. Extraction of the polysaccharides from five algae and their potential antioxidant activity in vitro. Carbohydr. Polym. 82, 118–121.

Further Reading

Alang, G., Kaur, R., Singh, A., Budlakoti, P., Singla, P., 2009. Antimicrobial activity of *Ulva lactuca* extracts and its fractions. Pharmacologyonline 3, 107–117.

Aruna, P., Mansuya, P., Sridhar, S., Kumar, J., Babu, S., 2010. Pharmacognostical and antifungal activity of selected seaweeds from Gulf of Mannarregion. Rec. Res. Sci. Technol. 2 (1), 115–119.

Babu, S., Johnson, M., Raja, D.P., Arockia, A., 2014. Chemical constituents and their biological activity of *Ulva lactuca* Linn. Int. J. Pharm. Drug Anal. 2 (7), 595–600.

Blois, M.S., 1985. Antioxidant determination by the use of a stable free radical. Nature 181, 1199–2000.

Dubois, M., Gillis, K.A., Hamilton, J.K., Rebers, P.A., Smith, F., 1956. Colorimetric methods for determination of sugars and related substances. Anal. Chem. 28, 350–356.

Elnabris, K.J., Elmanama, A.A., Chihadeh, W.N., 2013. Antibacterial activity of four marine seaweeds collected from the coast of Gaza Strip. Palestine Mesopot. J. Mar. Sci. 28, 81–92.

El-Rafie, H.M., El-Rafie, M.H., Zahran, M.K., 2013. Green synthesis of silver nanoparticles using polysaccharides extracted from marine macro algae. Carbohydr. Polym. 96 (2), 403–410.

Garrido, M.L., 1964. Determination of sulphur in "plant material". Analyst 89, 61–66.

Gorban, E.N., Kuprash, L.P., Gorban, N.E., 2003. Spirulina: perspectives of the application in medicine. LikSprava 7, 100–110.

Govindan, S.M., Thomas, J., Pratheesh, P.T., Kurup, G.M., 2012. *Ex vivo* anticoagulant activity of the polysaccharide isolated from *Ulva fasciata*. Int. J. Life Sci. Pharm. Res. 1 (3), 194–197.

Hemalatha, K.G., Parthiban, C., Saranya, C., Anantharaman, P., 2013. Antioxidant properties and total phenolic content of a marine diatom, *Naviculaclavata* and green microalgae, *Chlorella marina* and *Dunaliellasalina*. Adv. Appl. Sci. Res. 4, 151–157.

Hernández-Garibay, E., Zertuche-González, J., Pacheco-Ruíz, I., 2011. Isolation and chemical characterization of algal polysaccharides from the green seaweed *Ulva clathrata* (Roth) C. Agardh. J. Appl. Phycol. 23 (3), 537–542.

Hettiarachchy, N.S., Glenn, K.C., Ghanasambandan, R., Johnson, M.G., 1996. Natural antioxidant extract from fenugreek (*Trigonellafoenumgraecum*) for ground beef patties. J. Food Sci. 61, 516–519.

Kandhasamy, M., Arunachalam, K.D., 2008. Evaluation of *in vitro* antibacterial property of seaweeds of southeast coast of India. Afr. J. Biotechnol. 7, 1958–1961.

Koracevic, D., Koracevic, G., Djordjevic, V., Andrejevic, S., Cosic, V., 2001. Methods for the measurement of antioxidant activity in human fluids. J. Clin. Pathol. 54, 356–361.

Ktari, L., Alain Blond, M., 2000. 16β-Hydroxy-5α-cholestane-3,6-dione, a novel cytotoxic oxysterol from the Red Alga *Jania rubens*. Bioorg. Med. Chem. Lett. 10 (22), 2563–2565.

Larsen, B., Haug, A., Painter, J.T.J., 1966. Sulfated polysaccharides in brown algae I. Isolation and preliminary characterization of three sulfated polysaccharides from *Ascophyllumnodosum*. Acta Chem. Scand. 20, 219–230.

Leopoldini, M., Russo, N., Toscano, M., 2011. The molecular basis of working mechanism of natural polyphenolic antioxidants. Food Chem. 125, 288–306.

Nair, R., Chabhadiya, R., Chanda, S., 2007. Marine algae: screening for a potent antibacterial agent. J. Herb. Pharmacother., 73–86.

Osman, M.E.H., Abushady, A.M., Elshobary, M.E., 2010. In vitro screening of antimicrobial activity of extracts of some macroalgae collected from Abu-Qir bay Alexandria, Egypt. Afr. J. Biotechnol. 9, 7203–7208.

Park, P.J., Jung, W.K., Nam, K.S., Shahidi, F., Kim, S.K., 2001. Purification and characterization of antioxidative peptides from protein hydrolysate of lecithin-free egg yolk. J. Am. Oil Chem. Soc. 78, 651–656.

Percival, E., 1963. Polysaccharides of the green seaweeds, *Ulva lactuca* and *Enteromorpha compressa*. Proceedings of the 4th International Seaweed Symposium. Pergamon Press, Oxford, United Kingdom, pp. 360–365.

CHAPTER 10

The Use of Ultrasound as an Enhancement Aid to Food Extraction

Larysa Paniwnyk*, Alma Alarcon-Rojo, José C. Rodriguez-Figueroa**, Mihai Toma†**

**Coventry University, Coventry, United Kingdom; **Autonomous University of Chihuahua, Chihuahua, Mexico; †Costin D. Nenițescu–Institute of Organic Chemistry of the Romanian Academy, Bucharest, Romania*

1 Introduction

The use of ultrasound as a nonthermal processing technique is well known and it has been utilized in many industries to date. Ultrasound (sound at frequencies above human hearing generally 16 kHz and above) employs the formation of cavitation bubbles to induce many mechanical and chemicals effects within the processed media under study. When ultrasound is transmitted through a medium, the molecular structure is compressed and then pulled apart as the cycles of ultrasound pass through it. Voids are created in the rarefaction phase of the cycle if there is enough negative pressure present, and as a result cavitation bubbles are formed (Lauterborn et al., 2007). These cavitation bubbles are the basis of effects observed as a result of the application of ultrasound (Fig. 10.1).

Cavitation bubbles are short lived and eventually collapse resulting in physical and chemical effects within the bulk medium, which are often used to great advantage. In extraction processes, where a solid plant material is extracted into a liquid medium, the formation of cavitation bubbles near a solid surface will result in bubbles that collapse unevenly, sending a microjet of liquid toward that solid surface (Lauterborn and Ohl, 1997). In food materials the formation of cavitation bubbles near solid surfaces, such as those of a plant, root or seed, result in an effective penetration of the liquid medium into that material. Benefits are enhanced cleaning of the solids' surface and enhanced extraction of bioactive principles from within the food material itself. Acoustic factors will affect the amount and intensity of cavitation. As frequency is increased, the rarefaction phase shortens and cavitational bubble collapse occurs more frequently and more readily. As a result mechanical effects, such as enhanced mixing and stirring, become muted and radical effects become more prominent. Comprehensive details of radical production, determination and interactions are fully detailed in the text by Fang et al. (2014). External temperatures will also affect cavitation. Higher bulk temperatures result in the generation of

Ingredients Extraction by Physicochemical Methods in Food
http://dx.doi.org/10.1016/B978-0-12-811521-3.00010-7

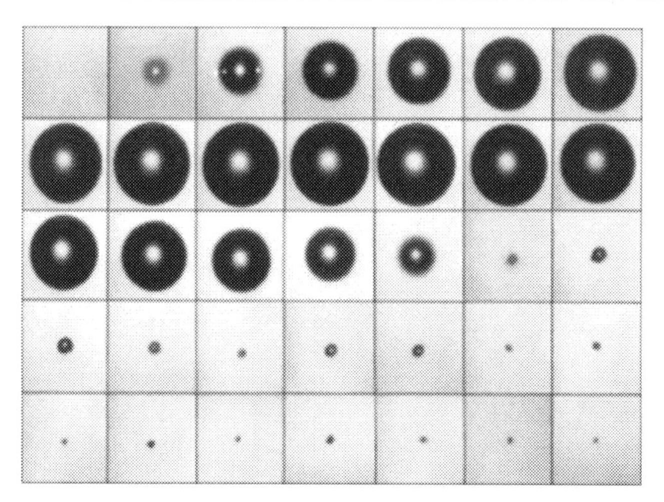

Figure 10.1: Cavitation Bubble Formation and Collapse (Lauterborn et al., 2007).

extra dissolved vapor, which saturates the cavitation bubbles formed and as a result prevents and cushions their eventual cavitational collapse, reducing the overall benefits of the ultrasound itself.

The mechanical and chemical effects of the collapsing cavitation bubble are thought to occur in three distinct regions, namely inside the bubble itself, at the bubble/solvent interface, and in the bulk solvent medium itself. Mechanical effects are due to cavitational bubble collapse with extreme conditions of several thousand Kelvin and 1000–4000 atm, being generated within these short-lived cavities (Mason et al., 2014). Volatile chemicals or solvent vapors that enter these cavities are therefore subjected to these extreme conditions and often result in the formation of radicals. At low ultrasonic frequencies (20–40 kHz) mechanical effects are dominant with enhanced mixing, stirring, cleaning, particle size reduction, and solvent penetration. These mechanical effects are thought to affect the texture of food materials and are the processes behind the enhanced extraction capabilities of ultrasound through the formation of microjets as a result of uneven cavitational collapse near a solid surface.

At higher frequencies (200 kHz and above) the shorter cavitation bubble life spans result in the release of radicals contained within them into the bulk media and chemical effects dominate, particularly at frequencies approximating 850 kHz. Here radicals affect food materials and may enhance oxidative processes, which can be either advantageous or detrimental to the flavors and antioxidant properties of the food materials themselves.

2 Ultrasonic Equipment

There is nowadays a wide range of ultrasonic equipment available with the choice of equipment being dependent upon its final purpose. Extraction relies upon the forcing of liquid into a plant material to increase the yield of desirable products. This is more favored by the mechanical effects of ultrasound and as a result frequencies approximating 20–40 kHz are the most utilized for these processes. The most readily available type of ultrasonic equipment is in the form of an ultrasonic bath.

2.1 Ultrasonic Bath

This type of ultrasonic device is preferred for applications involving low ultrasonic power using ultrasound frequencies between 40–100 kHz (Mason and Peters, 2002). The frequency and power of an ultrasonic bath depends upon the type and number of transducers, which are often in the base of the bath or clamped to its side, employed in its design. Bath sizes vary from small 1 L baths to larger industrial sized 1000 L versions (Chemat et al., 2011). Submersible transducers are also available to be used with existing vessels. These are simply inserted into the vessel and are an easy method of introducing ultrasound to existing equipment. These come in a range of frequencies with standard generators operating at approximately 38 kHz, but some are also available with variable or split frequencies between 25 and 120 kHz, and a range of surface areas (Fig. 10.2).

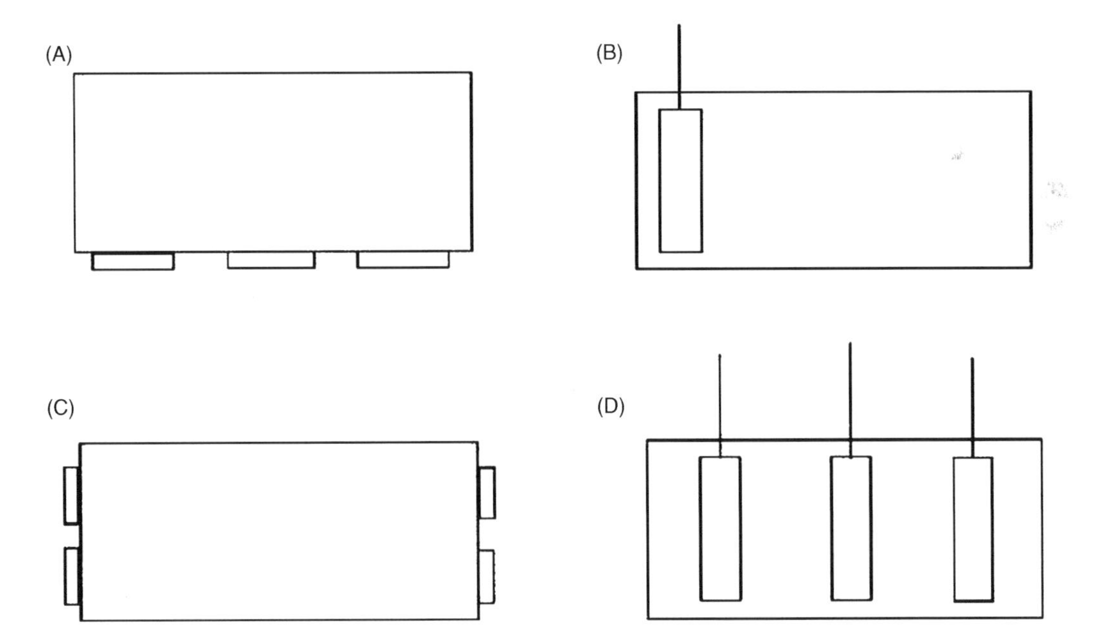

Figure 10.2: Ultrasonic Bath Configurations.
(A) "Normal" bath configuration with a number of transducers on the base, (B) submersible transducer can be introduced into existing vessels, (C) Some tanks have transducers clamped on the outside of the vessel, (D) several submersible transducers maybe used simultaneously.

2.2 Alternative Ultrasonic Equipment

A range of ultrasonic equipment with various designs are available. The 20 and 40 kHz ultrasonic probe/horn systems are very popular for processes that require strong mechanical effects; various stepped horns are now available (Gaete-Garretón et al., 2011) that prove an alternative to the traditional horn system. Recently multifrequency ultrasonic baths have been designed with ultrasonic frequencies between 100 and 1176 kHz. While this allows for a wider processing ability these baths are more suited to the processes where the chemical effects of ultrasound are likely to prove interesting to observe. Baths utilizing this range of ultrasound are of limited volumes (1–2 L) due to the significant attenuation levels of ultrasound at these frequencies within liquid media. Larger, industrial-sized ultrasonic systems are also currently being developed and are discussed at the end of this chapter.

3 Extraction

It is well known that the cavitation effects of ultrasound are particularly beneficial for low temperature extraction, which is particularly useful for natural products that can often degrade or denature on heating. These benefits also include the efficient mixing of solvent and solid materials and the effects of various factors, including the extraction time, solvent/material ratio, the ultrasonic intensity and duty cycle (ultrasound off vs. ultrasound on) of ultrasonic irradiation also have an impact on the final extraction yield. Ultrasound has been used in the extraction of a wide range of natural compounds ranging from natural antioxidants, oils, color, and flavor compounds to lycopenes, carotenoids, and polyphenols. Materials extracted include citrus peel, olive oil, palm oil, grapes, coffee, and tea, to name but a few of the many examined. All parts of plant materials have been extracted at some point ranging from flowers and leaves to bark and roots. Shirsath et al. (2012) produced a review paper on how ultrasound could benefit the extraction process and included examples from pharmaceutical and medicinal agents to natural antioxidants, algal oil, food additives, and flavorings. Other authors have produced book chapters describing the many advantages to using ultrasound either as a preparatory technique prior to extraction or during the extraction process itself (Asbahani et al., 2015; Barba et al., 2015; Jambrak and Herceg, 2014; Kha and Nguyen, 2014; Mason et al., 2014; Roselló-Soto et al., 2016; Tejada-Ortigoza et al., 2015; Udaya Sankar, 2014). Talmaciu et al. (2015) has produced a review article examining the extraction of polyphenols from food materials under a variety of extraction conditions.

There are many examples in the literature of the potential for use of ultrasound as an aid to the extraction process, with most papers extolling the virtues of enhanced yields, greater levels of antioxidants, and target materials, all often extracted at lower temperatures, and in shorter processing times than conventional processing techniques. Rather than being a novel extraction technique, ultrasound is now being viewed as an enhancement method to existing processes while maintaining the standard and quality of the extracted products themselves.

3.1 Fruit

Fruit often contains many active ingredients; however, supply is limited due to the often short times that fruit is mature enough to be worth processing. It is therefore important that the extraction of active ingredients from fruit is maximized to achieve the greatest yields possible in the shortest timescale possible. Often materials are sensitive to temperatures; thus, use of ultrasound at lower temperatures, used to prevent degradation of many antioxidants, for example, appears very promising. Levels of anthocyanins, flavonoids, and phenolics are often enhanced, coupled with significant antioxidant capabilities as measured by assays, such as ferric-reducing antioxidant power (FRAP) and 2,2-diphenyl-1-picrylhydrazyl (DPPH) radical scavenging capacity assays.

Ramli et al. (2014) examined the effects of several extraction methods on the antioxidant capacities of red dragon fruit peel and flesh. Results indicated that ultrasonic extraction increased the levels of flavonoids extracted and produced the highest amount of antioxidants, as determined via their radical scavenging activity, for dragon fruit peel; however, ultrasonic extraction reduced the antioxidant activity for dragon fruit flesh, thus indicating the different preferences of peel and flesh for extraction procedures.

Portu et al. (2014) viewed levels of anthocyanins extracted from grapes when using various extraction procedures, such as cold maceration, enzymes, ultrasound, and pulsed electric fields. Alternate extraction procedures offer significant advantages in the extraction of anthocyanins and other polyphenols with the much-enhanced speeds of extraction reducing the maceration of red wine by up to 3 days.

Nguyen et al. (2014) focussed on the levels of total phenolics and anthocyanins of mulberry fruit. The extraction rate constant for ultrasonic extraction increased by approximately 16.9 (phenolics) and 21.5 (anthocyanins), with yields of 11.3 (phenolics) and 15.9% (anthocyanins), respectively, a significant improvement when compared to enzyme-assisted extraction.

Ivanovic et al. (2014) investigated anthocyanins from the blackberry cultivar known as Čačanska Bestrna. Ultrasound-assisted extraction with acidified ethanol at room temperature, and also at 40°C, enabled rapid isolation (15–30 min) of anthocyanin extracts with reasonably high yields (5.3%–6.3%), while increasing sonication time and temperature had a positive effect. Subsequent HPLC analysis revealed that the cyanidin content in the blackberry extracts (0.7%–1.0%) was up to 20 times higher than when compared to the blackberry juice with the extracts exhibiting strong antioxidant effects as observed by FRAP and DPPH assays. This suggests that ultrasonically produced blackberry extracts retain high levels of target materials that would be of benefit for consumer health.

Golmohamadi et al. (2013) examined the total phenolic and anthocyanin content of red raspberry puree. The pureed fruit was subjected to high intensity 20 kHz sonication and higher frequency low-intensity 490 and 986 kHz ultrasound. They noted that 20 kHz ultrasound treatment produced the best results for the extraction of bioactive compounds from red raspberry when treatment times of 10 min were employed. However, levels of the compounds began to decrease after 30 min of sonication. Pineapple mash was treated by (Nguyen and Le, 2012) during the juice processing stage. Sonication increased extraction yield by 10.8% in comparison with the nonsonicated control. In addition, use of ultrasound in the treatment of pineapple mash highly improved the level of sugars, total acids, phenolics, and vitamin C in the final pineapple juice, thus enhancing its quality parameters.

El-Sharnouby et al. (2014) used a combination of enzymes, pectinase/cellulase, in combination with sonication to extract the maximum amount of date syrup from date fruits (*Phoenix dactylifera* L.). Conventional extraction often employs temperatures greater than 70°C with extended extraction periods that may denature many active ingredients. Various variables were examined, such as fruit/water ratio, ratio of the enzyme mixture used, and ultrasonic power. Results showed that using a sonication power of 25% with a date: water ratio of 1:3, and 0.1% enzyme mixture recovered 74.30% of solids in a shorter extraction time and resulted in a better physicochemical quality of the final syrup in comparison to other techniques (use of 1% enzyme mixture produced 66.70% solids extracted and traditional methods produced 58.45% yield) (Fig. 10.3).

Xu and Pan (2013) looked at the effects of various factors of ultrasonic treatment on the extraction yield of *trans*-lycopene from red grapefruit. In comparison with the conventional solvent extraction process ultrasonic extraction showed a greater extraction yield and reduced extraction time. Many other authors have also observed similar trends.

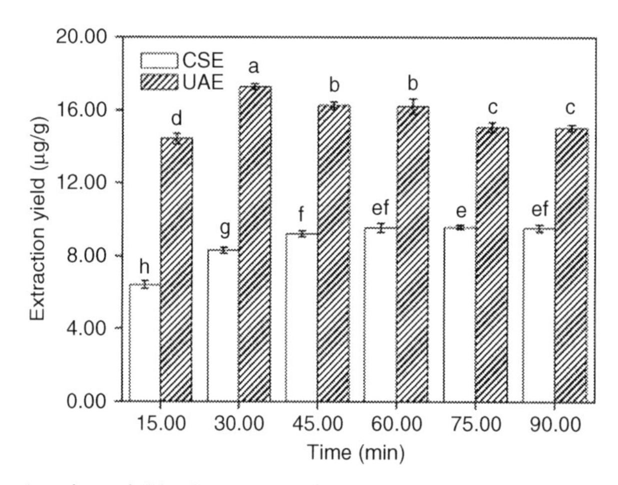

Figure 10.3: Extraction Yield of Lycopene from Red Grapefruit (Xu and Pan, 2013).
A time versus yield comparison of CSE versus UAE methods. *CSE*, Conventional stirred extraction; *UAE*, ultrasonic assisted extraction.

Table 10.1: Effect of sonication on fruit extraction.

Fruit Extracted	Target Materials	Enhancements	References
Dragon fruit	Flavonoids	Increased yields	Ramli et al. (2014)
Mulberry fruit	Phenolics and anthocyanins	Increased yields	Nguyen et al. (2014)
Pineapple	Phenolics and vitamin C	Increased yield by 10%	Nguyen and Le (2012)
Blackberries	Anthocyanins	Increased extraction and cyanidin content by up to 20 times	Ivanovic et al. (2014)
Blueberries	Anthocyanins	Increased yields	Dibazar et al. (2015)
Grapes	Phenolics and anthocyanins	Reduced extraction time by up to 3 days	Portu et al. (2014)
Raspberries	Phenolics and anthocyanins	Increased yields	Golmohamadi et al. (2013)
Dates	Sugar syrup	Enhanced extraction rate and quality of syrup	El-Sharnouby et al. (2014)
Grapefruit	Lycopene	Increased yield and rate of extraction	Xu and Pan (2013)
Dill fruit	Carvones	Enhanced selectivity for carvone	Chemat and Esveld (2013)

Fu et al. (2014); Zhang et al. (2013) optimized the extraction of oleanolic and ursolic acids from pomegranate (*Punica granatum* L.) flowers. They determined that the optimum extraction conditions were 90% ethanol solvent, ultrasound power 150 W, ratio of 20:1 (v/w), and extraction for 50 min at 40°C. In addition to enhanced yields of the target extracts, obtained after 30 min of sonication, higher antioxidant activity was exhibited than those of other conventional methods. Chemat and Esveld (2013) examined the recovery of carvone and limonene from flaked dill fruits (*Anethum graveolens* L.). They determined that use of ultrasound enhances extraction selectivity for carvone but has no great effect on limonene recovery. Dibazar et al. (2015) extracted anthocyanins from low-bush blueberries. Results obtained employing a mathematical model suggested that greatest yields of anthocyanins require 11.5 mins extraction at conditions of 60% acidified ethanol, solvent to solid ratio of 50 mL/g, and 65°C (Table 10.1).

Most literature indicates enhanced extraction of anthocyanins and phenolics when using ultrasound. Preferred frequencies are 20–40 kHz sonication with some work employing higher frequencies (up to 1000 kHz) also being undertaken. In addition to enhanced yields the sonicated extracts show little alteration of the target materials, thus indicating a retention of quality after processing.

3.2 Seeds, Bark, and Roots

The roots, barks, and seeds of plants can also be treated with ultrasound and target components extracted. The extraction of the bioactive compounds from plants, and their constituents, is a potential way to increase the value of these products; however, it often requires efficient extraction techniques to reduce processing costs and improve the productivity of what can be normally a slow and complex/difficult process. Ultrasound assisted extraction, being a nonthermal method, is considered a means of intensifying a slow process without being detrimental to the final product. Using ultrasound to extract oils from seeds increases the yield without affecting the final oil content; in addition, when used as a pretreatment method, extraction times are much reduced. Often ultrasound is combined with solvents or supercritical carbon dioxide to further enhance the extraction of valuable oils.

Da Porto et al. (2015) employed ultrasound as a pretreatment method prior to extraction. They sonicated dry intact hemp seeds for up to 40 min prior to supercritical carbon dioxide extraction at 40°C, 300 bar, and 45 kg CO_2/kg feed. A yield of 24% (w/w) was achieved with only 10 min sonication with little deterioration of target acids. Said et al. (2015) also combined ultrasound with supercritical carbon dioxide extraction to extract antioxidants and polyphenolics from Ginger oleoresin. The highest oleoresin yield of 8.15 ± 0.06% was obtained with the two combined techniques at 35°C and 25 MPa pressure. The ultrasound-assisted solvent extraction was also 1.75 times faster than simple Soxhlet extraction, indicating potential cost savings to manufacturers as a result of the shorter processing times and temperatures.

Assami et al. (2012) employed ultrasound to achieve extraction of essential oil from *Carum carvi* L. seeds. Results indicated that ultrasound treatment caused a rapid release of essential oil with 80% oils recovered after only 30 min of treatment compared to a normal 90 min. Similar yields of volatile fractions were obtained with a higher carvone/limonene concentration ratio for ultrasonically treated seeds being obtained in a shorter processing time. Bimakr et al. (2012) extracted oil from winter melon (*Benincasa hispida*) seeds using ultrasound and evaluated its antioxidant activity, total phenolic content, and fatty acid composition. They found that although the crude extract yield was lower when using ultrasound, the antioxidant activity and total phenolic content of the extract was clearly higher than those of the Soxhlet extracts with both types of extracts being rich in unsaturated fatty acids.

Conventional extraction of β-glucans from barley normally takes 3 h at 55°C and 1000 rpm stirring with an energy requirement of 1460 kJ/L. However, using sonication not only enhanced yields but significantly reduced time to only 3 min and energy requirements to 170 kJ/L (Benito-Román et al., 2013).

Fahmi et al. (2011) used ultrasound to determine its effect on the protein content and rheological properties of soymilk extracted from soybeans. Soybean slurry was exposed to 35 kHz ultrasound treatment at different treatment temperatures (20 and 40°C) and times (20, 40, and 60 min) then filtered. Results indicated that ultrasound treatment could significantly increase the protein content of the extracted soymilk by nearly 6.3%. They continued this work by examining levels of daidzin, genistin, and their respective aglycones, in the resultant soymilk (Fahmi et al., 2014) and determined that sonication also increased the isoflavone content, levels of glycosides, and levels of aglycones in the extract.

Gaete-Garretón et al. (2011) used ultrasound to extract bioactive principles, such as saponnins from the chips of branches and the bark of the *Quillaja saponaria* Molina tree. Using a 20 kHz stepped horn system, ultrasound was shown to enhance the extraction ratio alongside a reduction in extraction time at lower extraction temperatures with smaller grain sizes were treated. Wang et al. (2012a) examined the root bark of *Morus alba*. Experimental results showed that an ultrasound extraction can be applied to extract the resveratrol, myricetin, and flavonoids successfully.

Several authors have employed response surface methodology to determine optimum extraction conditions. Using these mathematical statistical modelling systems, the optimum reaction conditions are predicted and are often verified by confirming the results within laboratory or industrial environmental settings.

De Mello et al. (2015) employed Box-Behnken design (BBD), one of the most common response surface methodology (RSM) methods, to optimize the extraction of oil from chia seeds (*Salvia hispânica* L.). Process variables, such as temperature, solvent to seed ratio, and time were examined. The solvent to seed ratio and temperature had the greatest influence on oil yield, followed by the extraction time. An extraction temperature of 50°C, a solvent to seed ratio of 12mL/g and 40 min of extraction were identified as the optimal conditions, with final 27.24% oil yield. Ultrasound also resulted in a higher oil yield when compared with a nonsonicated process with similar fatty acid content resulting with both extraction techniques.

Jia et al. (2015) also used Box-Behnken design (BBD), to optimize the experimental conditions for ultrasound-assisted extraction of polysaccharides from *Rhynchosia minima* root (PRM). They determined that the optimal extraction conditions were 21 min sonication, 46 mL/g ratio of water: solids with an extraction temperature of 63°C. Extracts exhibited significant antioxidant activity and reducing power (Table 10.2).

Table 10.2: Effect of sonication on seeds, bark or roots.

Source Materials	Target Materials	Enhancements	References
Hemp seeds	Fatty acids	Reduced extraction time	Da Porto et al. (2015)
Ginger	Antioxidants polyphenolics and oleoresin	Increased yields	Said et al. (2015)
Carum seeds	Essential oil	Increased yields, higher carvone content, faster rate of extraction	Assami et al. (2012)
Winter melon seeds	Antioxidants, fatty acids/ phenolics	Greater antioxidant activity, increased total phenolic yield	Bimakr et al. (2012)
Barley grains	β-glucans	Reduced processing time	Benito-Román et al. (2013)
Soybeans	Soymilk proteins	Increased yield by 6.3%	Fahmi et al. (2011)
Molina tree bark	Saponnins	Enhanced extraction rate at lower temperatures	Gaete-Garretón et al. (2011)
Morus alba root	Flavonoids, resveratrol	Increased yields	Wang et al. (2012a)
Chia seeds	Oil	Increased oil yield	De Mello et al. (2015)

Most authors show that enhanced yields and shorter processing times are possible by using ultrasound. When using it as a pretreatment, or with additional solvents, the prospect of reducing of grain size and hence increasing surface areas of these materials enhances extraction rates and reduced extraction times. This offers potential economic savings.

3.3 Plant Body and Leaves

Efficient ultrasonic extraction of plant stems and leaves is dependent upon their structural characteristics. Ultrasound is known to perform well under heterogeneous systems with solids contained within solvents. However, the pliability of the solid is of importance. A harder, less flexible material that is more brittle responds better to cavitational collapse and hence particle size reduction as compared to a softer, more sponge-like mass. This means that leaves that are softer and more flexible are less likely to respond well to sonication when compared to more rigid leaf structures (e.g., sage versus rosemary). If the target material is close to the leaf surface within, say, trichrome glands, then it is more easily targeted by sonication and cavitational collapse. If the material of interest is embedded deep within the plant stem or leaf, it will therefore be more difficult to extract and the benefits of sonication will be lowered as compared to normal conventional extraction methods. Combining sonication with the correct solvent/leaf ratio is a powerful tool often exhibiting significant reductions in extraction times coupled with significant yields of active materials. Target materials, such as polyphenolics, anthocyanins, flavonoids, and tannins, have been investigated to give some examples. All retain their activity levels postultrasonic extraction with yields often significantly higher when compared to thermal conventional extraction methods.

Petigny et al. (2013) examined both batch and continuous ultrasound-assisted extraction of boldo leaves (*Peumus boldus* Mol.). The optimized ultrasonic extraction parameters provided a better yield of flavonoids and alkaloids compared to conventional maceration and reduced process times considerably (30 min instead of 120 min). Zhang et al. (2015) studied the enzyme enhanced (pectinase/cellulase) ultrasonic extraction of *Ginkgo biloba* leaves (GBLs). Results showed that the optimum conditions were an enzyme quantity of 0.5 g, pH of 4.5, and temperature of 45°C. Under these conditions, the yield of polyphenols was 0.80% ± 0.22%, which was 69.70% higher than that obtained by direct petroleum ether extraction.

Kostic et al. (2015) investigated Salvia verbena extracts with respect to their antioxidant and antimicrobial activities. Sonication of samples in 80% ethanol resulted in the highest levels of total polyphenols, catechins, gallic acid, tannins, and hydroxycinnamic derivatives. These samples also exhibited the highest antimicrobial activity. Rodríguez-Pérez et al. (2015) examined the extraction of the *Moringa oleifera* L. tree by calculating the total phenolic content using the Folin-Ciocalteu assay. A 50% water/ethanol solvent was most effective, with sonication producing highest yields.

Bernatoniene et al. (2016) studied the leaves of rosemary (*Rosmarinus officinalis* L.) and analyzed the extract for their triterpenoid and phenolic acid content. The highest yield of ursolic acid (15.8 ± 0.2 mg/g), rosmarinic acid (15.4 ± 0.1 mg/g), and oleanic acid (12.2 ± 0.1 mg/g) from rosemary leaves was obtained by ultrasound-assisted extraction with 90% ethanol, with 70% ethanol or water (at pH 9).

Ahmad-Qasem et al. (2013) studied the ultrasonic kinetics and composition of phenolic extracts from olive leaves (var. *Serrana*). Compared with conventional extraction, ultrasound assisted extraction reduced the extraction time from 24 h to 15 min and did not modify the final extract composition.

Yu et al. (2015) optimized the ultrasound-assisted extraction conditions of capsaicins from Yugan pepper using ultrasound. The optimal extraction conditions for capsaicins were obtained at 55% methanol, 108W ultrasonic power, 25-min extraction time at 60°C with a final yield of 1.07%.

Wiktor et al. (2016) analyzed the influence of ultrasound on electrical conductivity, color, total polyphenols, and antioxidant activity of apple tissue extracts. The plant tissue samples were either ultrasonically treated via an immersion method (using 21 or 40 kHz) or by direct contact ultrasound (24 kHz). The sonication lasted for up to 30 min. Ultrasound generally did not change the electrical conductivity of the sample; however, the application of ultrasound increased the total polyphenolic concentration up to 145.3% and the antioxidant activity by 64.5%, in comparison to the intact material. Apples sonicated by the immersion method at 40 kHz produced the highest total polyphenolic content regardless of sonication time.

Bi et al. (2014) examined the extraction of chlorogenic acid from *Chrysanthenum* using ionic liquid aqueous solutions and sonication. The optimum conditions for extraction were 30°C, extraction time of 30 min, ultrasonic power of 320 W with the ratio of liquid to solid of 30:1 under these conditions 2.92% of chlorogenic acid was obtained. Sivakumar et al. (2011) have also used ultrasound to extract natural dyes from different plant materials. By sonicating plant material at 80 W, 45°C for 3 h they indicated a significant improvement of dye extracted up to 100% enhancement for marigold flowers, 25% for pomegranate, and 12.5% for green wattle. Enhancement depended upon the penetrability of the plant material itself with reductions observed for materials with rubbery or hard outer layers. Pan et al. (2012) used continuous and pulsed ultrasound (interval sonication) to extract antioxidants from pomegranate peel. The use of pulsed ultrasound (at an intensity level of 59.2 W/cm^2 and pulse duration of 5 s with a resting interval of 5 s) increased the antioxidant yield by 22%, and reduced the extraction time by 87%. Continuous ultrasonic extraction at the same intensity level increased the antioxidant yield by 24% and reduced the extraction time by 90%. However, employing pulsed ultrasound encouraged an energy saving of 50% for the process. Zhang et al. (2013) examined citrus pectin and determined that sonication degrades the pectin chains, affects the methylation of pectin slightly with no effect on the monosaccharides present. The average molecular weight of pectin decreased from 464 to 296 kDa after 30 min of sonication. Allaf et al. (2013) employed controlled pressure drop technology and ultrasound assisted extraction for sequential extraction of essential oil and antioxidants from orange peel and Wang et al. (2012b) used ultrasound to aid the extraction of geniposide from *Gardenia jasminoides*. All results showed enhanced benefits when using sonication for extraction.

Once again several authors have employed response surface methodology and Box Behnken designs to determine optimum ultrasonic extraction conditions; namely, Ravanfar et al. (2015) used Taguchi L9 orthogonal design to optimize the process parameters for ultrasound-assisted extraction of anthocyanins from red cabbage. Zhang et al. (2016) employed Box-Behnken design (BBD) in the extraction of polysaccharides from mulberry leaves, Prakash Maran and Priya (2015) studied extraction of natural pigments (betacyanin and betaxanthin) from *Amaranthus tricolor* L. leaves, Luo et al. (2015) used extraction of *Crataegus pinnatifida* leaves, Sharmila et al. (2016) used extraction of total phenolics from *Cassia auriculata* leaves, Wang et al. (2016) used the ultrasonic extraction of *Artemisia selengensis* Turcz, and Liu et al. (2015) evaluated *Psidium guajava* leaf extraction. In all cases, optimum ultrasonic reaction conditions for maximum product yields were determined by evaluating factors, such as ultrasonic power, extraction time, and temperature. Results were confirmed by determining total phenolics content, antioxidant yields using FRAP (Ferric reducing antioxidant power), and DPPH (1,1′-diphenyl-2-2′-picrylhydrazyl) radical scavenging activities as responses. Often ultrasonic extraction produced higher levels of total phenolics and greater antioxidant, anticancer, and antimicrobial abilities of the extracts when compared to other techniques examined (Table 10.3).

Table 10.3: Effect of sonication on plant body and leaves.

Materials Extracted	Target Materials	Enhancements	References
Boldo leaves	Flavonoids and alkaloids	Reduced extraction time by 66%	Petigny et al. (2013)
Ginko balboa leaves	Polyphenols	Increased yields by 70%	Zhang et al. (2015)
Salvia verbena Leaves	Antioxidants, antimicrobials	Increased yields of polyphenols and catechins	Kostic et al. (2015)
Rosemary leaves	Antioxidants	Greater antioxidant yields	Bernatoniene et al. (2016)
Olive leaves	Phenolics	Reduced processing time from 24 h to 15 min	Ahmad-Qasem et al. (2013)
Yugan peppers	Capsaicins	Increased yield	Yu et al. (2015)
Apple tissue	Polyphenols/antioxidants	Increased polyphenol yields by 143%	Wiktor et al. (2016)
Marigold flowers	Natural dye extraction	Increased yield 100%	Sivakumar et al. (2011)
Pomegranate	Natural dye extraction	Increased yield 25%	Sivakumar et al. (2011)
Green wattle	Natural dye extraction	Increased yield 12.5%	Sivakumar et al. (2011)
Pomegranate peel	Antioxidants	Yield increased 22% in shorter extraction time	Pan et al. (2012)

Clear benefits from the use of ultrasound include significant enhancements in reducing processing times. Often this occurs at lower temperatures thus economic savings are made. Yields of target materials are also often enhanced with no detrimental effects observed with respect to their antioxidant or antimicrobial abilities.

3.4 Conclusions

Use of ultrasound appears to be very promising; however, the extent of enhancement is dictated by the form of the material to be extracted in the first instance. Extraction of the plant body and leaves, in particular for hard/brittle materials, shows very significant improvement on yield and times of extraction when utilizing some form of sonication during the extraction process. Extraction of seeds, bark or roots indicates enhancements in terms of yield as a result of ultrasonically reduced grain sizes and hence increasing surface areas. Some degree of enhanced extraction rates and reduced extraction times are often observed with these source materials. With regard to the extraction of fruit most literature indicates enhanced extraction of anthocyanins and phenolics, which are retained in the final extract thus potentially increasing the quality of the final juice itself.

In all cases ultrasonic frequencies from the power ultrasound region of 20–100 kHz are preferred, with 20–25 kHz being the most quoted. Many sources also state that the attributes of the final extracts are often enhanced with quality parameters being retained after processing, indicating the potential of using ultrasound as a method of extracting target materials from a wide range of plant types and sources.

4 Alteration of Functional Properties

4.1 Meat

With regard to meat products ultrasound is employed in several ways. First, to inhibit the action of microbes to preserve the meat and, second, to alter the meat structure as an aid to tenderization, sometimes also employing ultrasound to aid the marination process itself. Here the meat muscles are softened and relevant enzymatic reactions are enhanced during the processing stage. Extraction is not employed in the strictest terms, however, similar mechanisms which are present in extraction processes are beneficial here, also, such as ultrasonically enhanced solvent mass transfer, enhanced diffusion, and formation of microchannels for effective penetration into meat muscles.

4.1.1 Microbial inhibition

High-intensity ultrasound offers an alternative to the traditional method of food preservation (Feng et al., 2013; Turantas et al., 2015) and is regarded as a green, versatile, popular, and promising technology (Majid et al., 2015). This emerging technology can help to control microbial growth in foods, such as chicken meat (Haughton et al., 2010) and milk

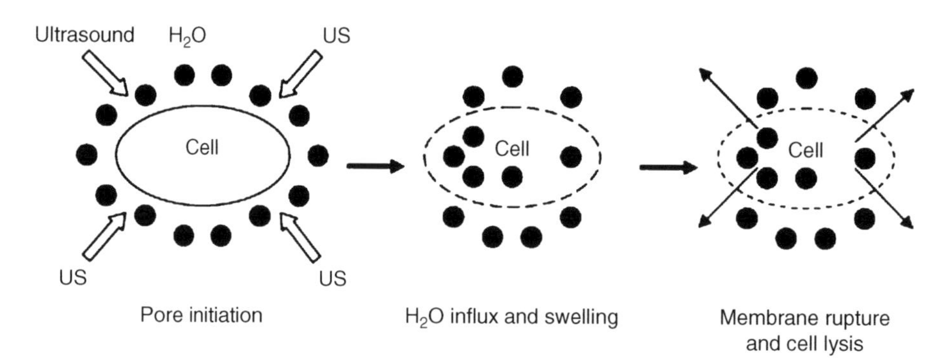

Figure 10.4: Mechanism of Ultrasonic Induced Cell Lysis (Chemat et al., 2011).

(Cameron, 2007). The antibacterial action of high-intensity ultrasound is attributable to the phenomenon of cavitation (Cameron, 2007). By contrast, low-intensity ultrasound does not cause sufficient cellular damage to eliminate microorganisms because it produces too few bubbles by cavitation (Jørgensen et al., 2008). It has been reported that high-intensity ultrasound generates gradients of intense pressure that can alter the structure of bacteria in food. The effect of ultrasound on microorganisms is complex, but the alteration of cell membranes and DNA chains is believed to be the main cause of the lethal or deactivation effect (Fig. 10.4).

The effect of high-intensity ultrasound on bacteria depends on many factors, the most important of which include the type of microorganism, ultrasonic frequency, ultrasonic intensity, and treatment times, as well as the characteristics of the material or treated food (Joyce et al., 2011). It has been reported that Gram-positive bacteria are more resistant to ultrasound treatments than Gram-negative bacteria (Feng et al., 2008; Piyasena et al., 2003). Studies on the microbiological effect of high-intensity ultrasound have focused on the deactivation of *Listeria monocytogenes* (Birk and Knøchel, 2009; Cameron, 2007), *Escherichia* (Cameron, 2007; Hunter, 2008; Patil et al., 2009; Smith, 2011; Zhou et al., 2009), a series of *Salmonella spp.* (Mukhopadhyay and Ramaswamy, 2012; Smith, 2011), *Staphylococcus aureus,* (Kalantar et al., 2010) and some other microorganisms (Char et al., 2010), which are among the most common food-poisoning bacteria highlighting the importance of ultrasound in the microbiological quality of food (Sango et al., 2014). High-intensity ultrasound helps to control the growth of mesophilic and psychrophilic bacteria and total coliforms in beef stored at 4°C (Caraveo et al., 2015). A relatively new concept in antimicrobial treatment has been proposed involving the combined effect of ultrasound and pressure or/and heat (Pagan et al., 1999). When ultrasound is used in combination with moderate heat, ultrasound can accelerate the decontamination rate of food, and reduce both the duration and intensity of heat treatments and their resulting damage (Piyasena et al., 2003). Morild et al. (2011) observed a reduction of pathogens levels in the pig skin by the application of pressurized steam simultaneously combined with high-power ultrasound.

The reduction of microorganism levels in the pig skin was significantly greater than the reduction found in the meat. Also ultrasound treatment in combination with lactic acid has been demonstrated to be a suitable method for decontamination of the skin of poultry (Kordowska-Wiater and Stasiak, 2011). Steam treatment and ultrasound applied to chicken carcasses in a processing line can significantly reduce by approximately three logs the number of *Campylobacter* on contaminated birds (Hanieh et al., 2014); and also ultrasound treatment in combination with high temperature and amplitude increased inactivation of pathogens after longer periods of treatment (Herceg et al., 2013). Piñon et al. (2015) reported that high-intensity ultrasound alone was not effective for controlling the growth of anaerobic bacteria, LAB, and mesophilic bacteria during the refrigerated storage of marinated chicken meat; however, when ultrasound treatment is combined with 0.3% oregano essential oil ,the growth of those bacteria was reduced. In contrast with these results, Smith (2011) reported no effect on *Salmonella* or *Escherichia coli* in chicken meat marinated with the help of ultrasound. Therefore, in some cases ultrasound alone might not be fully effective in inhibiting bacterial growth.

4.1.2 Technological improvement

Ultrasound treatment has been widely used experimentally for the tenderization of meat (Chemat et al., 2011; Siró et al., 2009; Stadnik et al., 2008). It is used for its ability to cause cell membrane destruction resulting from cavitation. Ultrasound can act in two ways in the meat tissue: by breaking the integrity of the muscle cells and by promoting enzymatic reactions (Boistier-Marquis et al., 1999). The effectiveness of ultrasound in softening the skeletal muscle structure have been mainly observed in beef (Jayasooriya et al., 2004; Pohlman et al., 1997; Smith et al., 1991; Stadnik and Dolatowski, 2011) and some results are shown in Table 10.4. Other effects in meat properties are changes in pH, particularly during aging (Jayasooriya et al., 2007; Stadnik et al., 2008), improvement in color (Pohlman et al., 1997) improvement in the water-holding capacity (Jayasooriya et al., 2007; Stadnik et al., 2008). However, others have failed to confirm the tenderization effect of ultrasound and the variation in meat quality as a result of the application of ultrasound (Got et al., 1999; Lyng et al., 1998). Ultrasound applied for 60 or 90 min to bovine semitendinosus muscle increases meat luminosity and lowers pH without affecting the redness or yellowness, or the water holding or drip loss properties (Caraveo et al., 2015). Jayasooriya et al. (2007) sonicated (24 kHz, 12 Wcm^{-2}) bovine muscles for a maximum of 4 min and observed that ultrasound increased meat tenderness and pH without significant interaction between ultrasound and maturation time. Ultrasound treatment did not affect the color or drip loss, but cooking losses and total losses decreased. The hypothesis that ultrasound causes mechanical disruption and muscle tenderizing has also been confirmed in poultry (Xiong et al., 2012) who found an improvement in muscle degradation, tenderness, and water-holding capacity with no change in cooking loss of hens by a combination of ultrasound (24 kHz for 4 min) and exogenous proteolytic enzyme

Table 10.4: Effects of ultrasound in dairy and meat processing.

Samples	Applications (Intensity/Freq/Time)	Effects of Ultrasound	References
Beef (longissimus thoracis and lumborum, and semimembranosus)	62 W cm^{-2}, 20 kHz, 15 s	No effect on mastication force, sensory traits, solubility of collagen or myofibrillar proteolysis	Lyng et al. (1998)
Semimembranosus pre- and postrigor	10 W cm^{-2}, 2.6 MHz, 2 × 15 s	Larger sarcomeres, Z-line disruption, increased calcium. No effect on collagen	Got et al. (1999)
Beef (semimembranosus)	2 W cm^{-2}, 25 kHz, 1 or 2 min	Lower loss of water after cooling, thawing and heating. No effect on pH. Higher water holding capacity	Dolatowski et al. (2000)
Beef (semimembranosus) matured for 24, 48, 72, or 96 h at 2°C	2 W cm^{-2}, 45 kHz, 2 min	No effect on meat color. Increased free calcium. Changes in protein structure. Improved WHC at 4 days postmortem	Dolatowski and Stadnik (2007)
Beef (semimembranosus) 24 h postmortem and matured for 24, 48, 72, or 96 h at 2°C	2 W cm^{-2}, 45 kHz, 2 min	No effect on pH or color. Reduced hardness	Stadnik and Dolatowski (2011)
Beef (semimembranosus) 24 h postmortem and matured for 24, 48, 72, or 96 h at 2°C	2 W cm^{-2}, 45 kHz	Acceleration of aging process. Fragmentation of protein structures. Increase WHC	Stadnik et al. (2008)
Beef Longissimus lumborum et thoracis and Semitendinosus aged up to 8.5 days	12 W cm^{-2}, 24 kHz, for up to 240 s	Reduced WBS force and hardness. Increased pH. No interaction between ultrasound and aging. No changes in meat color and drip loss. Ultrasound reduced cook and total loss	Jayasooriya et al. (2007)
Hen breast meat 0, 1, 3, or 7 days at 4°C	12 W cm^{-2}, 24 kHz, 15 s period	Reduced shear force. No change in cooking loss	Xiong et al. (2012)
Beef (semitendinosus)	1500 W; 40 kHz; 10, 20, 30, 40, 50, and 60 min	No effect on brightness and red color. Decreased the tendency to yellow. Decreased the muscle fiber diameter. No effect on heat-insoluble collagen. Weaken collagen stability	Chang et al. (2012)
Pork biceps femoris 24 h postmortem	150 W, 1 MHz, and 500 W, 25 kHz; 40 min plus kiwi protease (actinidin)	Ultrasound did not change in shear force. Ultrasound combined with actinidin decreased shear force more than actinidin alone	Jørgensen et al. (2008)
Beef (semimembranosus) 24 h postmortem and matured for 24, 48, 72, or 96 h at 2°C	2 W cm^{-2}, 45 kHz, 2 min	Slightly less stable color. No change in oxidative stability at 4 days storage	Stadnik et al. (2008)
Beef longissimus thoracic and deep pectoralis Matured 14 days at 2°C, cooked at 62 or 70°C	1000 W, 20 kHz	Faster cooking, higher water retention, decreased cooking loss, shear force, and soluble collagen. Higher sensory tenderness	Pohlman et al. (1997)

Chicken breast and soybean gels 4 to 8°C	450 W; 20 kHz; 0, 3, 6, 9, and 12 min (4 or 2 s pulses)	More viscoelastic gel improved WFB and textural properties. Homogeneous fine network microstructures.	Zhao et al. (2014b)
Chicken breast	22 W cm^{-2}; 40 kHz; 15 or 30 min	Increased mass transfer and higher meat weight	Leal-Ramos et al. (2011)
Pork loin in NaCl saturated solution	100 W and 2 kHz	Increased salt gain and water loss. Higher mass transfer at higher ultrasound intensity	Carcel et al. (2007)
Pork Longissimus dorsi	2–4 W cm^{-2}, 20 kHz	Higher salt diffusion. Diffusion coefficient increases with ultrasound intensity	Siró et al. (2009)
Longissimus dorsi cerdo	40 kHz, 37.5 W dm^{-3}	Higher salt and water diffusion	Ozuna et al. (2013)
Pork Longissimus thoracis and lumborum	4.2, 11, or 19 W cm^{-2}; 20 kHz; 10, 25, or 40 min	No effect on water holding capacity and structure of meat. Higher mass transfer and protein extraction. Myosin denaturation at higher intensities	McDonnell et al. (2014)
Pork meat and skin surface	High-intensity ultrasound, 0.5 at 2 s	Less skin and surface bacteria	Morild et al. (2011)
Chicken breast	Ultrasonic bath, 20 min	No effect on water retention capacity, shear force and cooking loss. No changes in *Salmonella* and *E. coli*	Smith (2011)
Chicken wing surface	2.5 W cm^{-2}, 40 kHz, 3 or 6 min	Microorganism reduction. Higher reduction with higher time. *E. coli* more sensible to ultrasound	Kordowska-Wiater and Stasiak (2011)
Pure culture suspensions	20 kHz; 3, 6, and 9 min; 20, 40, and 60°C	Bacteria inactivation is higher at higher time and temperature	Herceg et al. (2013)
Chicken carcasses	Campylobacter and total count reduction	Campylobacter and viable total count reduction	Hanieh et al. (2014)
Milk	400 W; 22 ± 1.7 kHz; 1, 3, and 5 min	Removal of lead (~50%), mercury (~14%), and arsenic (~7%), heavy metals contamination	Porova et al. (2014)
Milk	20 kHz, 700 W, 2.5–10 min	*Pseudomonas fluorescens* and *E. coli* viable cell counts were reduced by 100% and *Listeria monocytogenes* was reduced by 99%	Cameron et al. (2009)
Dairy proteins	~34 W cm^{-2}, 20 kHz, 2 min	Sodium caseinate, whey protein isolate, and milk protein isolate reduction in the micelle size and hydrodynamic volume	O'Sullivan et al. (2014)
Raw milk	24 kHz, 240 W, 15 min, and UV light (13.2 W/cm^{-2}) 65°C for 30 min ultrasound alone	Reduction for total and coliform bacteria (log cfu/mL^{-1}); 4.79 and 5.31, 3.29 and 5.31, 1.31 and 4.01	S¸engül et al. (2011)
Flax seed oil emulsions in dairy systems	176 W, 20 kHz, 1–8 min	Stable emulsions of 7% of flax eed oil and pasteurized homogenized skim milk for a minimum of 9 days at 4 ± 2°C	Shanmugam and Ashokkumar (2014)

(Continued)

Table 10.4: Effects of ultrasound in dairy and meat processing. (*cont.*)

Samples	Applications (Intensity/Freq/Time)	Effects of Ultrasound	References
Milk protein concentrate	12.5 ± 0.3 W, 20 kHz, 0–5 min, pretreatment	Improvement of functional properties, such as solubility, emulsification and gelation	Yanjun et al. (2014)
Whey proteins	43–48 W/cm^{-2}, 20 kHz; 1 W/cm^{-2}, 40 kHz; 15 and 30 min	Both treatments showed changes in particle size, molecular weight, and structure of whey protein isolate and whey protein concentrate	Jambrak et al. (2014)
Skim milk	101 kWm^{-2}, 20 kHz, 15 min; 86 kW/m^{-2}, 400 kHz, 17 min; 1600 kHz, ll kW/m^{-2}, 34 min	Disruption of casein micelles that formed aggregates of smaller size but with surface charge similar to the casein micelles in the original milk. Low frequency treatment presented higher disruption than high frequency treatments.	Liu et al. (2013)
Skim milk	20 W or 41 W; 20 kHz; 15, 30, 45, 60 min	Reduction in the turbidity and no effect in the viscosity. Whey proteins were denatured and formed soluble whey-whey/whey-casein aggregates, which further interacted with casein micelles to form micellar aggregates during the initial 30 min of sonication. Prolonged sonication resulted in the partial disruption of some whey proteins	Shanmugam et al. (2012)
Cheddar cheese whey	89 ± 2 W, 20 kHz; 223 ± 3 W, 400 kHz; 345 ± 3 W, 1000 kHz; 219 ± 3 W, 2000 kHz; 10 and 30 min	Lipid oxidation was not promoted at any tested frequency or specific energy. Treatments did not affect free fatty acid concentration	Torkamani et al. (2014)
Milk	200 W, 24 kHz, 0–16 min	Longer sonicated time affected the acceptability of pasteurized milk	Chouliara et al. (2010)
Skim milk	101 kW m^{-2}, 20 kHz, 15 min	The rennet gelation time, curd firming rate, curd firmness, and the connectivity of the rennet gel network were improved significantly in rennet gels made from milk ultrasonicated at pH 8.0 and readjusted back to pH 6.7 compared to those made from milk sonicated at the natural pH 6.7	Liu et al. (2014)
Milk	280 W, 2 MHz; 348 W, 1 MHz in a decreasing power level to 0	The lipid oxidation derived volatiles were below the human sensory detection level and it was not found oxidative changes in milk after both treatments	Johansson et al. (2016)

inhibitors. The results suggest that both ultrasound and endogenous proteases contributed to muscle degradation. Chang et al. (2012) found no difference to beef color or levels of heat insoluble collagen, however, but observed a decrease in the muscle fiber diameter. The combination of actinidin with ultrasound resulted in a further reduction of the toughness of the meat and the results suggest that the treatments weakened both the myofibrillar and the connective tissue components of the meat. Stadnik and Dolatowski (2011) highlighted the potential of using low-frequency and low-intensity ultrasound to tenderize meat. They found reduced meat toughness at 48 and 72 h postmortem. Ultrasound effects are also present during thawing of frozen meat. Dolatowski et al. (2000) reported that ultrasound treatment could help change the textural properties of meat and increase the WHC after thawing and thermal processing without effect on the pH of the treated meat. It was reported that the drip loss and shear force values in pale, soft, and exudative (PSE) meat can be decreased with ultrasound application (Dolatowski et al., 2001; Twarda and Dolatowski, 2006). The hypothesis that the application of ultrasound treatment may cause an acceleration of the maturation process has been repeatedly confirmed. Dolatowski and Stadnik (2007) and Stadnik and Dolatowski (2011) sonicated calf semimembranosus muscle at 24 h post mortem for 2 min and stored it for 24, 48, 72, or 96 h at 2°C. No changes in pH or color were observed, but there was an increase in the WHC in the sonicated samples, similar to that of the matured meat. Thus, the authors suggested that treatment with ultrasound accelerated rigor mortis, since they also observed fragmentation in the structures of cellular proteins (Stadnik et al., 2008). Improved tenderness and higher water holding capacity in the semimembranosus muscle during aging were also confirmed by Dolatowski et al. (2001) and Dolatowski and Twarda (2004). Ultrasound exerts changes in macromolecules, such as enzymes that remain active in the postmortem stage and are capable of inducing changes in meat structure. This could cause changes in physical properties, such as tenderness, cohesiveness, adhesiveness, and fibrosity, among others; and those changes could be positive, as well as negative. Mason (1998) reported that ultrasound with salt treatment improved binding strength, water-holding capacity, product color, and yield. The authors also indicated that ultrasound treatment resulted in an increase in the strength of the reformed meat product. Similarly Zhao et al. (2014b) have observed that ultrasound treatment has the potential for producing emulsified meat products with excellent functional properties and improved the fatty acid composition at high yields. Overall, it has been assumed that high power ultrasound technology is a promising process that can improve the gelation properties and thereby allowing for a partial reduction in the salt levels in chicken meat gels (Li et al., 2015b). It is believed that the mentioned effects are a consequence of the acoustic cavitation, where bubble formation, growth, and eventual collapse have thermal, chemical, and mechanical effects produced by ultrasound (Yusaf and Al-Juboori, 2014). Although ultrasound has been shown to accelerate tenderizaton during lab-scale trials, no studies have assessed the feasibility of an ultrasound system on meat tenderization at pilot-scale.

4.1.3 Mass transfer acceleration

The ultrasound-assisted marination technology is considered an alternative to the traditional marinating process. It is well known that ultrasound can reduce brining time without affecting meat quality (McDonnell et al., 2014). Acoustic waves applied to solid-liquid systems increases the permeability of the muscle tissue and this can be used to estimate the effect of ultrasound in the brining of meat or the rate of mass transfer (Carcel et al., 2007; Leal-Ramos et al., 2011; Siró et al., 2009). The mechanical effects are attributed to physically breaking down tissues, which creates microchannels, and the diffusion of salt increases with the intensity of ultrasound without significant changes in other characteristics of the meat (Carcel et al., 2007; Chemat et al., 2011). Improved distribution of solutes (Ozuna et al., 2013) and changes in water retention capacity result in less water loss (Siró et al., 2009), while conserving other properties. Also ultrasound can aid in the extraction, gelation, and restructuring of meat proteins (McDonnell et al., 2014; Zhao et al., 2014b). These results demonstrated that when ultrasound was applied, the rate of gain of NaCl increased compared with curing under static conditions, suggesting that ultrasound improved the transfer of both external and internal mass.

4.1.4 Possible negative effects

The impact of power ultrasound in meat processing has been rarely attributed to producing negative effects. However, ultrasonic treatment conditions, such as intensity, frequency, and application time may influence meat quality. Acoustic energy can be absorbed, giving rise to elevated temperatures due to cavitation and this can cause thermal damage (Kasaai, 2013), particularly when sonication is used in combination with heat application. The mode of action of sonication is attributed to several mechanical and chemical actions caused by cavitation with a possible impact on food quality attributes (Feng et al., 2008). It has been found that water-binding capacity was negatively affected by higher intensities and longer time treatments, causing denaturation of proteins (McDonnell et al., 2014; Siró et al., 2009). Studies on milk reported some perceived alterations in quality parameters post ultrasonic treatment. Effects,such as off-flavors, change of color, lipid oxidation, modifications of minor compounds, and decrease of sugar content have been the most remarked upon (Pingret et al., 2013). Also, high-intensity ultrasound (HIU) induces modifications on functional properties related to molecular modifications, and possibly improves production processes, but its success relies on both the knowledge of the protein response and the expertise of its use (Arzeni et al., 2012). Treatments should be performed at lower temperatures to avoid negative side effects on treated materials (Jambrak et al., 2008). High intensity ultrasound induces breakage of water molecules, leading to the generation of hydroxyl radicals, which could react with foods that are easily oxidized (Kasaai, 2013). The application of ultrasound at industrial level should consider the control of sonochemical reactions, such as the hydroxylation of food chemicals. Although in some cases the hydroxylation of phenolic compounds seems to be a viable process for enhancing the antioxidant properties of foods (Ashokkumar et al., 2008). Finally, the results obtained from studies using ultrasound in food systems are difficult to compare due to different ultrasound parameters used. As a consequence the interpretation of the results between studies varies a lot and makes harmonization of ultrasound effects very difficult.

4.2 Dairy

In the last two decades, several studies have evaluated the effects of ultrasound on dairy products (Benedito et al., 2000; Johansson et al., 2016; Mason et al., 1996; Torkamani et al., 2014; Zhao et al., 2014a). The mechanical and physical effects of ultrasound have been explored in dairy foods (Table 10.4) and the application of ultrasound-induced modifications on particle size, molecular weight, and structure of whey proteins have been studied (Jambrak et al., 2014). Ultrasound has demonstrated changes in the functional properties of dairy products, such as reduction of yogurt viscosity, improvement of solubility, emulsification, and gelation of milk protein concentrate, as well as inactivation of spoilage and pathogenic microorganisms (Cameron et al., 2009; Riener et al., 2010; Yanjun et al., 2014).

The food industry commonly uses conventional thermal pasteurization and sterilization techniques to inactivate microorganisms and enzymes. These techniques have shown the ability to kill vegetative microorganisms and some spores (Chandrapala et al., 2012; Piyasena et al., 2003). In fact, thermoduric aerobic sporeformers microbes are able to survive milk pasteurization and cause spoilages of dairy products (Gopal et al., 2015; Khanal et al., 2014a). Moreover, these thermal processes required high levels of energy, so they impact on the nutritional content, sensory properties, and final products quality (Chandrapala et al., 2012; Piyasena et al., 2003). However, alternative technologies of food processing and preservation, such as ultrasound have been explored in the last years (Higuera-Barraza et al., 2016; Khanal et al., 2014a; Yanjun et al., 2014). High-intensity, low-frequency, or power ultrasound (10–1000 W cm^2 or 20–100 kHz) generates acoustic cavitation in a liquid medium. The physical forces caused by this phenomenon are considered the main mechanism responsible for the ultrasonic microbial deactivation effect (Chandrapala et al., 2012). Cavitation is also associated with hot spots (up to 5000 K) of short lifetimes (10^{-6} s), pressure (up to 1200 bar) free-radical formation and shear disruption (Hosseini et al., 2013). Microscopic bubbles increase in size, reaching a critical diameter and collapse asymmetrically, which leads to a liquid jet rushing through the centre of the collapsing bubble. This microjet has the ability to achieve a velocity of a few hundred meters per second (Chandrapala et al., 2012). Therefore, mechanical forces result in the breaking and shearing of microorganisms cell walls (Cameron et al., 2009). Tabatabaie and Mortazavi (2008) identified microcracks, microvoids, and ruptures as three of the most usual kinds of microdamages produced by ultrasound. All of them promote the cell wall permeability bringing about the bactericidal effect. Several studies have shown the ability of ultrasound to inactivate spoilage and pathogenic microorganisms and enzymes in dairy products (Gao et al., 2013, 2014; Mohammadi et al., 2014; Pingret et al., 2013; Şengül et al., 2011). According to Khanal et al. (2014a), traditional pasteurization was ineffective in inactivating the vegetative cells of microbes associated with spoilage in milk, such as *Bacillus coagulans* and *Anoxybacillus*.

flavithermus, compared with ultrasound treatment. Cameron et al. (2009) evaluated the ultrasound effect on pathogens. They inoculated UHT milk with 1×10^6 colony forming units per mL (UFC mL^{-1}) of *Escherichia coli*, *Listeria monocytogenes*, and *Pseudomonas fluorescens*; this amount represents 5× greater levels than those permitted by the British milk legislation. Milk was ultrasonicated (750 W, 20 kHz) for durations of 2.5–10 min. Results showed that *P. fluorescens* and *E. coli* viable cell counts were reduced by 100% after 6 and 10 min, respectively; meanwhile *L. monocytogenes* was reduced by 99% after 10 min. This nonthermal process may be an alternative to conventional pasteurization. On the other hand, ultrasound has been ineffective in deactivating alkaline phosphatase and lactoperoxidase activities; both enzymes are used as indicators of effective thermal process by the dairy industry (Cameron et al., 2009; Khanal et al., 2014a). Therefore, further studies will be necessary to establish new parameters to specify the efficacy of this emergent technology during production processes.

The use of ultrasound coupled with other treatments has shown increased efficacy (Khanal et al., 2014a; Khanal et al., 2014b; Ganesan et al., 2015; Şengül et al., 2011) in preservation of dairy products. Ultrasound processing in combination with pressure (manosonication), heat (thermosonication), both pressure and heat (manothermosonication), ultraviolet irradiation (photosonication), and antimicrobial solutions have demonstrated more potential for microbes inactivation than alone (O'Donnell et al., 2010; Piyasena et al., 2003; Sango et al., 2014; Şengül et al., 2011). These technological combinations have been explored to minimize aerobic spore-forming bacteria associated with industrial dairy processing environments and products spoilage. Khanal et al. (2014b) evaluated the effect of ultrasonication (500 W, 20 KHz) and ultrasonication combined with heat on inactivation of vegetative cells of thermoduric *Bacillus* spp. in nonfat milk. Results showed that ultrasonication in combination with pasteurization (63°C for 30 min) reduced significantly ($P < 0.05$) more bacilli spores than ultrasound. The highest inactivation was observed on *Geobacillus stearothermophilus* endospores treated with ultrasound (10 min) and heat (80°C for 1 min) 75.3 ± 3.2%, comparing with 49.0 ± 3.2% obtained by ultrasound alone. This study suggested that the use of ultrasound combined with heat may inactivate bacterial endospores, improving the shelf life of skim milk.

The microbial lethality efficacy of photosonication also has been evaluated. The effect of this emergent technology on the population reduction of total and coliform bacteria in raw milk was investigated. The simultaneously application of ultrasound (240 W, 24 kHz) in combination with ultraviolet irradiation (UV-C, 13.2 W/cm^2) reduced 4.79 and 5.31 log cfu/mL^{-1} for total and coliform bacteria, respectively. On the other side, the application of ultrasound (240 W, 24 kHz) alone was able to reduce 1.31 and 4.01 log cfu/mL^{-1} for total and coliform bacteria, respectively. Moreover, they reported that traditional pasteurization (63°C for 30 min) reduced 3.29 and 5.31 log cfu/mL^{-1} for total and coliform bacteria, respectively. These results showed the potential of photosonication to inactivate relevant microorganisms in raw milk (Şengül et al., 2011).

In recent years, innumerous advantages of ultrasound technique have been reported in dairy foods (Adjonu et al., 2014; Gajendragadkar and Gogate, 2016; Ojha et al., 2016; O'Sullivan et al., 2014; Shanmugam and Ashokkumar 2014; Yanjun et al., 2014). In fact, several benefits associated with ultrasound treatments are shown in Table 10.4. However, changes in the components of dairy products have also been found during the processing or after treatment, thus affecting their quality (Pingret et al., 2013; Shanmugam et al., 2012). For example, the effect of both heat treatment (92°C, 102°C, 111°C) and manothermosonication (200 kPa, 20 kHz) was evaluated on several compounds produced in nonenzymatic browning in model systems. Higher amounts of free 5-(hydroxymethil)-2-furfural (HMF) were detected in milk that had been manothermosonicated compared to that treated by heat during the first 5 min of treatment (Vercet et al., 2001). The presence of high amounts of HMF in milk may impact on the final color of the product. Riener et al. (2009) investigated the effect of thermosonication (24 kHz, 400 W, 10 min) of milk on yogurt gels during fermentation. They found higher gelation pH (6.3) in the cultured thermosonicated milks compared to their conventionally processed (90°C for 10 min) counterparts pH (5.6). Moreover, Vercet et al. (2002) reported that yogurt made with manothermosonicated (2kg cm^{-2}, 40°C, 20 kHz) milk had a longer fermentation time (10%–20%) than the control milk. On the other hand, Jambrak et al. (2014) found significant changes in the particle size, molecular weight, and structure of whey proteins associated with partial cleavage of intermolecular hydrophobic interactions, rather than the peptide of disulphide bonds. Therefore, the application of ultrasound may modify native chemical structures of milk components.

The application of ultrasound may also be associated with the production of undesirable effects in dairy products (Jambrak et al., 2008; Pingret et al., 2013; Vercet et al., 2002). Differences in conductivity were observed in 10% whey protein model suspensions treated with ultrasound. The 40 and 500 kHz ultrasound treatments reduced conductivity; meanwhile the 20 kHz treatment increased it. This increment is related to the formation of hydroxyl radicals and other compounds able to induce oxidant species (Jambrak et al., 2008). Chouliara et al. (2010) also found that the concentration of volatile compounds, mainly products of lipid oxidation, was increased in sonicated milk (24 kHz, 200W, 0–16 min). Therefore, the application of ultrasound processing in the dairy industry should be continuously researched and validated to determine the effects on the physical, chemical and functional properties of food products, to make it adaptable to a factory environment (Shanmugam et al., 2012).

4.3 Conclusions

Novel techniques, such as thermosonication, manosonication, and manothermosonication may be a more relevant energy-efficient processing alternative for the food industry. Ultrasound has multiple applications in the food industry, most being focused on improving the product quality and extending the shelf life of fresh and processed products. In meat processing, power ultrasound can modify cell membranes that can help in curing, marinating, and tenderizing the tissue, and controlling the bacteria growth in meat. However, these processes need to be developed further before they can be implemented at a full industrial level. Some ultrasonic innovations are already close to being used on a large scale whereas the potential for many other applications exists in other areas. The research to date has not been sufficiently consistent to establish the effect of the ultrasound on the structure and properties of meat. Little research has included the effects of ultrasound on sensory attributes of meat and on consumer acceptance of the treated meat. On the other hand, the application of ultrasound to dairy products combined with other treatments has demonstrated a similar pathogenic and spoilage microorganism reduction to conventional thermal techniques. Moreover, ultrasound treatments have been able to modify the functional properties of these products, such as improvement of solubility, reduction of yogurt viscosity, emulsification and gelation of milk protein concentrate. However, changes in the components of dairy products have also been reported during ultrasound processing, thus affecting their quality. Therefore, further research is still required until ultrasound technology is broadly applied in the meat and dairy industries.

5 Recycling of Food Waste

The end products of a food process often offer the potential for valorization of the final product. In a lot of cases, once the production process has ended, the waste material is used for animal feedstock or used as fertilizer on fields. However, these remains often contain bioactive ingredients and flavorings that have not been extracted to their full potential. Vegetal extracts are widely used as primary ingredients for various products from creams to perfumes in the pharmaceutical, nutraceutical, and cosmetic industries. The primary extraction process must therefore extract as much soluble material as possible in a minimum time, using the least possible volume of solvent as the final product requires that a concentrated and active extract is essential. This is often not the case and sometimes quite substantial amounts of these "added value" materials are simply thrown away due to lack of recycling and second step extraction processes. As a result an emerging area of food processing is beginning to concentrate on these waste materials, from use in procedures as wide apart as wine and tomato juice production, to determine if extra bioactive ingredients may be extracted from these wastes. This is to increase the economic value of the original production, as well as offering the possibility of increased and new levels of bioactive materials.

Since the development of industrialized food manufacture and processing, materials that are not for consumption are removed at various stages of the food supply chain. Food waste at the manufacturer, such as residues from the processing of raw materials and by-products with no immediate interest for companies, can cause huge disposal problems. For example, citrus processing waste can account for 50% of the fresh biomass produced (Mamma and Christakopoulos, 2014). Pomaces, press cakes, bagasse, flakes, peels, grinds, hulls, bran, shells, skins, and wastewater from canning, olive mills, and so on accumulate at the end of different production lines and are traditionally reused on livestock feed formulation or for composting. However, large volumes of the food waste decompose on landfills to form greenhouse gases and leachates. As a result increasing efforts are now being focussed on procedures for the reuse and recycling of food waste in addition to energy recovery, as well as clean production strategies.

Back in the late 1970s, Toma et al. (1978) qualified potato and tomato peels as concentrated dietary fibre sources and Helen et al. (1979) identified a technique for protein recovery from beef bones, making waste recovery a subject of much interest and promise. However, major advances have been made in the last decade to the point that food waste is now acknowledged as a material containing a high concentration of valuable compounds. Vegetal waste contains structural macromolecules, such as pectin (Ninčević Grassino et al., 2016a), lignans, cellulose, hemicellulose (Elleuch et al., 2011), polysaccharides, crude protein (Zhou et al., 2013), dietary fibres (Nawirska and Kwasniewska, 2005), and valuable secondary metabolites, such as oils and fats, polyphenols, flavonoids, anthocyanins, carotenoids, gums, and tannins, to name but a few. Food waste from animals contains proteins, amino acids, collagen, chitin, chitosan, vitamins, fats, bioactive peptides, and omega fatty acids. All can potentially be extracted to form valuable sources of materials that can be used for consumer products with additional health benefits.

A wide range of technologies can be employed for the pretreatment and recovery of food waste, from the conventional: steam distillation, solvent and enzymatic extraction, membrane filtration, isoelectric separation to the advanced: microwave assisted, sub and supercritical fluid extraction, pulsed electric field, ultrasonically assisted techniques (Galanakis, 2012).

5.1 Ultrasonic Techniques

Ultrasound can be employed for waste processing as a single technique, such as ultrasonic assisted extraction (UAE), as a pretreatment step added to an existing process (Cheok et al., 2013; Zhou et al., 2013), or in combination with other modern methods (Chen et al., 2015; Rabelo et al., 2016). Often bespoke assessment is required for nonstandard methods of extraction to obtain optimized yields with no detrimental effect to the extracted material itself.

The influence of process variables on the yields and quality of the final product was studied for many food waste materials using response surface methods (RSM). Parameters with the most influence on the process optimization were: time, with respect to phenols extraction from wheat bran (Wang et al., 2008); ultrasonic power, temperature, time, solid/liquid ratio, for oil from pomegranate seeds (Tian et al., 2013); solvent (ethanol) concentration, for flavonoids from grapefruit waste (Garcia-Castello et al., 2015) or polyphenols, from rice bran (Tabaraki and Nateghi, 2011); solvent concentration, temperature, for lycopene from papaya waste (Li et al., 2015a); amplitude, time, on alkaloids from potato peels (Hossain et al., 2014). Ultrasonic devices employed for food waste processing are similar with those for raw material extraction and varies from ordinary low frequency, low power cleaning bath (Eh and Teoh, 2012) to the more dedicated, such as ultrasonic extraction reactors (Pingret et al., 2012; Virot et al., 2010). As expected, direct sonication with ultrasonic probe systems proved to be more efficient than indirect sonication (Cheok et al., 2013).

Recently, UAE have been extensively applied to food waste from fruits and vegetable, as can be seen in Table 10.5. By using a broad range of ultrasonic processing conditions, it

Table 10.5: Applications of UAE to fruits and vegetable food waste.

Waste materials	Recovered Products	Ultrasonic Techniques/Devices and Experimental Designs/ RSM Optimized Parameters	Process Benefits	References
Grapefruit solid waste	Flavonoids	40 ± 2 kHz, 100 W ultrasonic bath; 0%–100% aqueous EtOH; 15–60 min; RSM optimized extraction: 25°C, 40% EtOH, 55 min	UAE yield 80 mg GAE/g DW TCP with 36 mg/g DW naringin, exceed by 50% the yield with CE; UAE reduced the time and lowered the temperature	Garcia-Castello et al. (2015)
Satsuma peels	Phenolic acids	-60 kHz ultrasonic bath, 3.2–56 W; 15–40°C for 10–60 min; temperatures below 40°C, 20 min to avoid degradation effects	UAE increased significantly the yields of seven phenolic acids and two flavanone glycosides: narirutin and hesperidin, as compared with M	Ma et al. (2008)
Mandarin and lime peels	Flavonoids, flavanone glycosides	-38.5 kHz, 50.93 W ultrasonic bath; 40°C; 30 min; methanol, ethanol, acetone (20%–80% w/v); UAE in acetone 80% is more efficient and protect the antioxidant potential of phenols	UAE mandarin extract have more hesperidin (528.77 mg/100 g), TPC (2.45 mg GAE/100 g), total flavonoids (2.14 mg quercetin eq/100 g) than lime, which is richer on naringin (53.39 mg/100 g)	Singanusong et al. (2014)
Banana peels, Cinnamor bark	Polyphenols	Ultrasonic bath; 40–60°C; 30–90 min; 95% (v/v) ethanol; RSM: 40°C, 60 min for cinnamon bark and 60°C, 30 min for banana peels	UAE show significantly higher TPC yield than VME: 427.9 mg/g flavonoid from cinnamon and 196.1 mg/g flavonoid from banana peels	Anal et al. (2014)
Grape pomace	Phenolic acids, flavonoids	-47 kHz ultrasonic bath, 0%–90% aqueous EtOH; 0–20 min, S/L ratio 1:20; RSM: 30% EtOH, 6 min ultrasounds + 12 min agitation at 40°C	47.2 mg GAE/g DW TPC obtained from grape pomace by optimized UAE	Őzcan (2006)
	Phenols, flavonols	-40–120 kHz ultrasonic reactor, 50–120 W/L, 5–25 min for both us frequency and power optimization; RSM best parameter values: 40kHz, 150W/L, 25 min	UAE achieved 32.31 mg GAE/100 g FW TPC and 2.04 mg quercetin eq./100 g	Gonzales-Centeno et al. (2014)
Apple pomace	Polyphenols	-25 kHz, 150 W us extraction reactor, 0–100% EtOH-water, 0–50 min., S/L ratio 0%–35% w/v; scale-up approach: 30 L ultrasonic tank 25 kHz and 4 × 200 W; optimal settings by multivariate study for TPC: 40.1°C, 45 min., S/L ratio <15% w/v, 50% EtOH/water	UAE increased flavan-3-ols and procyanidins yields by 25% and TPC by 20% as compared with CE; flavonols yield rise to 105.3, phenolic acids to 49.3 and dihydrochalcones to 182.55 mg/100 g; UAE at large scale is cost suitable	Virot et al. (2010); Pingret et al. (2012)
Mangosteen hull	Phenolic compounds, anthocyanins	-20 kHz, 400 W ultrasonic probe and 38 kHz, 75 W ultrasonic bath; acidified methanol, ethanol; employed as a pretreatment step; RSM optimization for TPC: ethanol, 25 min, 80% amplitude while for anthocyanins were 15 min, 20% amplitude, 69.7% methanol	UAE improved TPC yield was 140.66 mg GAE/g DW, 8.8% higher that SE; anthocyanins obtained by optimized UAE were 45% higher than SE	Cheok et al. (2013)

Papaya waste	Lycopene	-40 kHz, 600 W ultrasonic bath; 15–40 min; 40–70°C; EtOH in ethyl acetate 40%; S/L ratio 1:5; 42.28% EtOH in ethyl acetate as a solvent, 25 min at 50.12°C	UAE yield of lycopene was 189.8 ± 4.5 μg/g FW versus 68.3 ± 41 μg/g by SE	Li et al. (2015a)
Grapefruit peels	Pectin	-20 kHz, 800 W ultrasonic probe, intensity: 10.18–14.26 W/cm^2; 20–40 min, 60–80°C, ultrasound pulse set at 50%, water; RSM best conditions: power intensity 12.5 W/cm^2; 28 min 66°C	UAE increased the yield by 16.34% to 27.34%, pectin, lowered the temperature with 13.3°C, and reduced the time by 37.78% as compared with CE	Wang et al. (2015)
Pomegranate peel	Pectin	-20 kHz, 130 W ultrasonic probe; distilled water; 50–70°C; 12–35 min; RSM best condition: S/L ratio 1:17.52 g/mL, 28.31 min, 61.9°C	UAE optimized yield: 23.87%; RMS predicted yield: 24.05%	Ganesh Moorthy et al. (2015)
Bendizao mandarin	All-*trans-b*-carotene	-21–25 kHz, 950 W us. probe; 60.51–1028.86 W/cm^2; hexane, ethanol, tetrahydrofuran, ethyl acetate, dichloromethane; 20–120 min; best extraction values: 25°C, 544.59 W/cm^2, ethanol, pulse mode	UEA increased the yield and reduced electrical energy consumption; UEA effectiveness rose on particle size > 0.2 mm	Sun et al. (2011)
Red prickly pear peels	Red pigments	24 kHz, 400 W ultrasonic probe; 5–15 min; water (pH 7)	UAE yield of up to 70% betanin and isobetanin in the case of the whole fruit as compared to a SE	Koubaa et al. (2016)
Tomato pomace	Pectin	-37 kHz ultrasonic bath; ammonium oxalate/oxalic acid; 60°C, 80°C; 15–90 min; a two-step UAE	UAE decreases the time from 24 h to 15 min; best yield: 35.7% at 60°C, and 90 min	Ninčević Grassino et al. (2016b)
	Lycopene	37 kHz, 180 W ultrasonic bath; 23–56 min; 31–48°C; 53–86 L/S ratio	UEA increase the yield with 26% over CE	Eh and Teoh (2012)
Artichoke bracts	Phenolic compounds	-20 kHz ultrasonic probe; 20–720 W; 0%–75% v/v aqueous EtOH; 5–60 min; S/L ratio 1:10 g/mL; optimum parameters: 20 kHz, 50% aqueous EtOH, 240 W, 10 min	UAE provided 16.4 mg of chlorogenic ac./g DW; most influential parameter on UAE was found to be the solvent concentration	Rabelo et al. (2016)
Red beet stalks waste	Betacyanin, betaxanthin	20 kHz 400 W ultrasonic probe, 60–120 W, 40–60°C; 15–45 min; optimum parameters: 89 W, 35 min, 53°C, S/L ratio 1:19 g/mL	UAE optimum yields are: 1.29 mg/g betacyanin, 5.32 mg/g betaxanthin	Prakash Maran and Priya (2016)
Orange peel	Polyphenols (TPC)	Biorefinery attempt: UAE after MWH extraction of essential oil; -25 kHz, 150W ultrasonic reactor; 20–60°C; 0.19–0.95 W/cm^2; RSM optimized parameters 0.956 W/cm^2; 59.8°C	UAE yield increase by 30% over SE; 50.02 mg GAE/100 g DW TPC obtained after optimization	Boukroufa et al. (2015)
	Flavanone glycosides	-25 kHz, ultrasonic reactor; 20%–80% aqueous EtOH; particle size: 0.5 2 cm^2, 30 min, 50–150 W; RSM optimized parameters: 150 W, 40°C, ethanol-water 4:l (v/v)	-UAE yields 275.8 mg GAE/lOOg FW TPC, 0.3 mg/100 g FW naringin and 205.2 mg/100 g FW hesperidin	Khan et al. (2010)

CE, Classical extraction; DW, dry weight; FW, fresh weigh; GAE, gallic acid equivalent; M, maceration; RSM, surface response method; SE, silent extraction, stirring instead of sonication; S/L, solid/liquid; TPC, total phenolic compounds; UAE, ultrasonically assisted extraction; VME, vacuum microwave extraction.

is possible to extract compounds with valuable chemical and functional properties. The ultrasonic applications employed for the extraction of saccharide type food waste can be seen in Table 10.6.

Low-frequency high power ultrasounds have been used to extract and reuse some of the food waste with protein and lipid compositions, as shown in Table 10.7. The relatively low occurrence of ultrasonic applications on treatment of food waste with lipids as the main composition can be explained by the competition with microwave techniques that are more fitted for lipid extraction.

5.2 Relevance for Industry

Several food waste materials have been shown to be important sources of compounds with technological functionalities (gelling, binding, surface activity, foaming, and emulsifying agents, fermentation substrate). Pectin is a structural polysaccharide and also a versatile product used in the food industry as a stabilizer (Chen et al., 2015), thickener, or functional food ingredient. Pectin extracted by ultrasonically assisted extraction from tomato peels acts as a good tin corrosion inhibitor (Ninčević Grassino et al., 2016a,b). Olive pomace pretreated with low-frequency high power ultrasound proved to be a good substrate for the production of lignocellulosic enzyme (Leite et al., 2016). More research reports and reviews were dedicated to the ultrasonic extraction of plants secondary metabolites. Based on their structural diversity and biological activity, many other commercial applications are developing year by year for the food industry, pharmaceutical industry, cosmetics, crops, and livestock.

Phenolic compounds are important radical scavengers with industrial value that prevent oxidative degradation of food and protects against reactive oxygen species damage to the human body. The impact of ultrasound power on their extraction and antioxidant activity was studied for a plethora of fruits and vegetable peels and pomaces (Table 10.5). Tabaraki and Nateghi (2011) extracted rice bran polyphenols, suitable for fat stability in food and cosmetics. γ-Oryzanol is another valuable antioxidant, antimicrobial, antiulcer, antiallergenic material extracted after superior valorization of rice bran, a saccharide food waste (Table 10.6). The advantages reported for ultrasonic processing are of significant economic importance and comprise increased yields, reduced extraction time, reduced energy consumption (Sun et al., 2011), low processing temperature (Wang et al., 2015), and solvent reduction (Kim et al., 2013). It should be emphasized that the antioxidant activity of ultrasonically extracted biomolecules is also higher than those extracted by a more classical procedure (Tang et al., 2015). It is clear that ultrasonically assisted techniques are especially important for the recovery of bioactive compounds with an initial low concentration in the source material but with high commercial value, such as conjugated linoleic acid (CLA) in pomegranate seed oil (Tian et al., 2013), peptides with inhibitory effect on angiotensin-converting enzyme (ACE) from bran (Zhou et al., 2013) as some examples.

Table 10.6: Applications of UAE on waste from food with saccharides.

Food Wastes	Recovered Products	Ultrasonic Techniques—Devices and Experimental Designs—RSM Optimized Parameters	Process Benefits	References
Rice bran	Polyphenols, antioxidants	-35 kHz, 140 W ultrasonic bath, 50–90% v/v ethanol; 40–60°C; 15–45 min; 10:1–80:1 v/w S/L ratio; RSM optimum conditions were as follows: 65–67% ethanol, 51–54°C, 40–15 min	UAE yield 6.05 mg GAE/g DW; UAE extract antioxidant activity is 54.14 mmol Fe^{2+}/g DW and antiradical activity is 52.83% inhibition; UAE reduced the time, solvent, temperature and water needed	Tabaraki and Nateghi (2011)
	γ-oryzanol, crude oil	-24 kHz, 120 W ultrasonic bath; petroleum ether, hexane, methanol; 38°C for 60 min; S/L ratio 1/3; best extraction solvent: methanol	γ-oryzanol yield is 82.0 ppm and UAE is more efficient than CE (73.51 ppm) and slightly less effective as microwaves (85.0 ppm); UAE reduced the processing solvent end time	Pramod et al. (2016)
Wheat bran	Phenolic compounds	-40 kHz ultrasonic bath EtOH-water mixtures 20%–95%; 50°C; 20 min; RSM optimized conditions: 64% ethanol; 60°C; 25 min	UAE yield 3.12 mg GAE/g DW; UAE reduced the time from 24 h as in the case of SE to 25 min	Wang et al. (2008)
Sugar beet pulp	Pectin	-25 kHz ultrasonic probe; l W/cm^2 power intensity; 10 min; UAE used as pretreatment step	Ultrasound used as pretreatment step; maximum pectin yield was 24.63% (with 59.12% galacturonic acid, 21.66% arabinose)	Chen et al. (2015)
Potato peels	Steroidal alkaloids	-20 kHz, 1500 W, ultrasonic probe; 24.40–61 μm amplitude; 9.24–22.79 W/cm^2; 3–17 min; pulse mode (5 s on, 5 s off) methanol; RSM-optimized parameters: 61 μm amplitude, 17 min, methanol	UAE recovered 1102 μg alkaloid/g DW vs. 710.5 μg alkaloid/g DW for CE	Hossain et al. (2014)
Defatted wheat germ	Peptides with inhibitory effect on angiotensin-converting enzyme (ACE)	-20 kHz, us probe system; 0–1800 W for 10 min followed by 600 W (pulsed mode) for 0–60 min; 63.70–573.25 W/cm^2 power intensities; us bath 24–68 kHz; at fixed and sweep freq.; 1.15 W/cm2 power intensity; best UAE conditions: ultrasonic probe, power intensity: 191.1 W/cm^2, 10 min	Ultrasound is used as a pretreatment of defatted wheat germ hydrolysis; ultrasonic probe is more efficient as compared with ultrasonic bath; combined 24/68 kHz frequencies increased the ACE activity as compared with singe freq. irradiation at 33 kHz	Zhou et al. (2013)
Rambutan	Polysaccharide	-20 kHz, 400 W ultrasonic probe; 40–60°C; 80–120 W; S/L ratio 1/20–1/40 g/mL; best extraction parameters: 110 W, 40 min, solid/liquid ratio 1/31 g/mL	UAE polysaccharide yields 8.31%	Prakash Maran and Priya (2014)

Table 10.7: Applications of UAE to food waste with protein or lipids as main nutritional component.

Food Waste Materials	Recovered Productd	Ultrasonic Techniqued—Deviced and Experimental Designd—RSM Optimized Parameters	Process Benefits	References
Food waste with protein composition				
Sea bass skins	Collagen	24 kHz, 750 W; ultrasonic probe sonicator; 20%–80%; 0–24 h, 0.01–0.1% acetic acid, S/L ratio 1:200 (w/v); 4°C; best extraction conditions: 80% amplitude, 0.1% acetic acid, 24 h	UAE resulted in 4.4-fold increase of collagen compared to yield extracted by SE	Kim et al. (2013)
Porcine placenta	Proteins/ peptide	-20 kHz, ultrasonic probe system; 120–360 W; 3–9 min; pulse mode; 40–60°C; RSM parameters were: 257 W, 49°C, 7 min	UEA increased the yield of water-soluble proteins from 17.7% (classical extraction) to 32.7% and reduced the time	Tang et al. (2015)
Defatted soy flakes	Protein, sugar	-20 kHz, 2.2 KW bench scale ultrasonic horn unit; 21–84 μm_{pp} amplitute; 15–120 s.; pulse mode (30 s on, 30 s off); 0.30–2.56 W/mL ultrasonic density; best extraction conditions: 1280 W, 120 s, 2.56 W/mL	UAE yields: 78% protein (46% increase) and 13.57 g/100g sugar (50% improve vs. CE); low power setting is more efficient as high power	Karki et al. (2010)
Food waste with lipids composition				
Pomegranate seeds	Pomegranate seed oil (PSO)	-40 kHz, 180 W ultrasonic cleaning bath; 140–180 W; petroleum ether, *n*-hexane, diethyl ether, isopropanol, acetone, and ethyl acetate; RSA: 140 W, petroleum ether; solvent/solid ratio 10 mL/g, 40°C, 36 min	UAE get 25.11% PSO, a yield higher than soxhlet (20.50%) or supercritical fluid extraction (15.72%); the fatty acids in POS were dominated by punicic acid (>65%)	Tian et al. (2013)
Exhausted olive pomace	Enzymes by solid-state fermentation	-20 KHz, 750 W ultrasonic processor; cellulase, xylanase; 5–15 min; 3–11 L/S ratio; ultrasounds used as a pretreatment for solid state fermentation	Ultrasonic pretreatment increased by threefold the xylanase activity	Leite et al. (2016)

It was reported that by reducing the ethanol concentration to 0 and the sonication time to 3 min the total phenolic content was only reduced from 80 to 75.3 mg gallic acid equivalents/g. A discovery that qualifies ultrasonic assisted extraction as a green technology (Garcia-Castello et al., 2015).

A recently introduced concept of a dry biorefinery promises a total valorization on the site of fruit and vegetable waste. Thus, by employing green processes and only the fluids from biomass structure as extraction solvent for ginger press cake, a total valorization of ginger rhizome that left a solid residue rich in fiber and phenolic acids can be achieved. In addition to the squeezed juice (the main product), essential oil was obtained by microwave extraction, and gingerols and 6-shogaol by ultrasonically assisted technique. Ultrasonic intensity of 16.7 W/cm^2 enhances the mass extraction yield by 125% as compared with conventional extraction (Jacotet-Navarro et al., 2016). There is no need for a new type of ultrasonic devices for food waste ultrasonic processing, and food waste is employed as the raw material and thus, cost effective scale-up is possible in this field (Pingret et al., 2012).

5.3 Conclusions

Given the current and further expectation for waste management and clean food technologies, ultrasonic processing is an efficient, modern, green technique for obtaining bioactive compounds from unusual sources. New commercial applications can be developed for extracting food waste by means of ultrasound for food and nonfood sectors, such as cosmetics, bio-based products, and pharmaceuticals. For the food industry, food waste reprocessing by means of ultrasound offers the prospect of economic benefits and sustainability offered by valuable materials obtained from saved resources, so as to lower the environmental impact.

6 Potential for Larger-Scale Applications

As stated earlier in this chapter, ultrasound has been used quite effectively during the extraction process to target materials from a wide variety of substrates.

The main benefits from using ultrasound and sonication are reduced processing time coupled with a higher yield of the target compounds while retaining their product quality and benefits within the extract material itself. Some examples of larger-scale equipment for use in industrial applications are as follows.

Several ultrasound reactor designs have been described by Vinatoru (2001) for industrial extraction of plant tissue. One of the first examples of large-scale extraction was the development of a batch extractor built to enhance extraction of plants. The equipment was a stainless steel tank with a capacity of 1 m^3 but with a working volume of 700–850 cm^3. Ultrasonic transducers were bonded to the outside of the tank and supplied ultrasound through the tank walls, a similar process to that in conventional smaller scale ultrasonic bath systems as described earlier. These systems can be used with or without an additional stirring mechanism. Results obtained using this simple design produced shorter extraction times with increased yields for a range of plant materials studied. Chemat et al. (2011) in a recent publication have examined the use of ultrasound for a variety of batch processing techniques, such as extraction, mixing, emulsification preservation, drying, and so on. In doing so, the possibilities for scale up are also discussed and batch reactors of volumes up to 1000 L are being described.

Several types of equipment are also now available for scale up of ultrasonic extraction processes using flow through systems. All have different designs but all show an increased benefit in that ultrasound provides enhanced yields, faster extraction rates and these can all be achieved at low temperatures. Designs include use of an ultrasonic horn placed directly into a stirred bath, a stirred reactor with ultrasound coupled to the vessel wall or a recycling system where the product from the stirred reactor then passes through an external ultrasonic flow cell where it is sonicated to encourage further extraction. These configurations may provide both intermittent/pulsed sonication or continuous application. Ultrasonic power intensities range from low intensity 0.01 to 0.1 W/cm^3, for a large reactor volume bath, for example, to higher intensity of 1–10 W/cm^3 for an external looped flow cell. Presently, 16 kW is the largest available single ultrasound flow cell, which can be configured either in series or in parallel module. Industrial ultrasound manufacturers have also recently promoted the use of ultrasound for food processing applications (Patist and Bates, 2008).

Some pilot scale work has also been done with (Clodoveo et al., 2013) developing a novel type of ultrasound equipment for the extraction of virgin olive oil. The sonication treatment was applied on olives submerged in a water bath (before crushing) and also on olive paste (after crushing). The ultrasound technology provided a reduction of the malaxing time and improved extract yields. Better extractibility and higher minor compounds content was obtained by sonicating the olives submerged in a water bath when compared to olive paste. Achat et al. (2012) used ultrasound to enrich olive oil with oleuropein both on a laboratory and pilot plant scale. They were able to enhance extraction of the target phenolic compounds above those obtained using conventional methods. By adding olive leaves to the olive oil extract and sonicating it they produced an "added value" olive oil that contained increased levels of beneficial phenolic compounds.

Cintas et al. (2010) employed a new pilot scale ultrasonic flow reactor to transesterify soybean oil with methanol for biodiesel production. Abramov et al. (2009) employed a flow reactor chamber to purify oil contaminated with a large amount of residual fuel oil. The reactor had an inside diameter of 150 mm attached to two magnetostrictive probe systems. The magnetostrictive transducers operate in the frequency range 25–40 kHz with variable power from 1 to 10 kW with the liquid being continuously treated as it passed through the reactor.

Montalbo-Lomboy et al. (2010) used a doughnut-shaped ultrasonic reactor to observe the effects of ultrasound on corn slurry saccharification, yield, and particle size distribution, at a frequency of 20 kHz. Continuous flow experiments were conducted by pumping corn slurry at various flow rates (10–28 L/min) through the ultrasonic horn. The sonicated samples were found to yield 2–3 times more reducing sugars than unsonicated controls after the saccharization process with continuous sonication being recommended for large throughput volumes.

Finally, Gallego-Juárez et al. (2010) examined the benefits of high-power ultrasonics with respect to a range of industrial processes. In these papers, he describes a range of novel ultrasonic transducers specifically adapted to the requirements of each individual process but which still operate at high efficiencies and powers. These comprise a variety of transducer types designed with the radiators adapted to different specific uses in fluids and multiphase media. A number of processes in the food and beverage industry, in the environment, and in manufacturing are described and the paper deals with the basic structure and main characteristics of novel transducers and their performance on a semiindustrial and industrial scale.

7 Conclusions

It is clear from the wealth of material available that the use of ultrasound in extraction provides positive benefits in terms of enhanced yields, reduced processing times, and in many cases also reduced processing temperatures. Shelf life is increased via the reduction of microbes and quality of products is enhanced. Many food materials have the potential for the use of ultrasound in their extraction process. These benefits would clearly be to the advantage of many food producers in terms of cost savings and increases in productivity. Extracted products are often of similar quality and maintain positive characteristics when compared to conventionally processed food materials. Advantages in extraction can also be translated to the reprocessing and recycling of food waste, thus adding extra value to the overall production process, itself an additional benefit to the food manufacturing industry, with potential for exploitation and valorization of previously unwanted waste materials. As a result of the development of new and more easily accessible larger-scale ultrasonic equipment, ultrasound is poised to make an even greater impact in future years in the food extraction arena, with additional benefits to the food industry and consumer alike.

References

Abramov, O.V., Abramov, V.O., Myasnikov, S.K., Mullakaev, M.S., 2009. Extraction of bitumen, crude oil, and its products from tar sand and contaminated sandy soil under effect of ultrasound. Ultrason. Sonochem. 16 (3), 408–416.

Achat, S., Tomao, V., Madani, K., Chibane, M., Elmaataoui, M., Dangles, O., Chemat, F., 2012. Direct enrichment of olive oil in oleuropein by ultrasound-assisted maceration at laboratory and pilot plant scale. Ultrason. Sonochem. 19, 777–786.

Adjonu, R., Doran, G., Torley, P., Agboola, S., 2014. Whey protein peptides as components of nanoemulsions: a review of emulsifying and biological functionalities. J. Food Eng. 122, 15–27.

Ahmad-Qasem, M.H., Cánovas, J., Barrajón-Catalán, E., Micol, V., Cárcel, J.A., García-Pérez, J.V., 2013. Kinetic and compositional study of phenolic extraction from olive leaves (var. Serrana) by using power ultrasound. Innov. Food Sci. Emerg. Technol. 17, 120–129.

Allaf, T., Tomao, V., Ruiz, K., Chemat, F., 2013. Instant controlled pressure drop technology and ultrasound assisted extraction for sequential extraction of essential oil and antioxidants. Ultrason. Sonochem. 20 (1), 239–246.

Anal, A.K., Jaisanti, S., Noomhorm, A., 2014. Enhance yield of phenolic extracts from banana peels (*Musa acuminata* Colla AAA) and cinnamon barks (*Cinnamomum varum*) and their antioxidative potentials in fish oil. Food Sci. Technol. 51, 2632–2639.

Arzeni, C., Martínez, K., Zema, P., Arias, A., Pérez, O.E., Pilosof, A.M.R., 2012. Comparative study of high intensity ultrasound effects on food proteins functionality. J. Food Eng. 108, 463–472.

Asbahani, A.E., Miladi, K., Badri, W., Sala, M., Addi, E.H.A., Casabianca, H., Mousadik, A.E., Hartmann, D., Jilale, A., Renaud, F.N.R., Elaissari, A., 2015. Essential oils: from extraction to encapsulation. Int. J. Pharma. 483 (1–2), 220–243.

Ashokkumar, M., Sunartio, D., Kentish, S., Mawson, R., Simons, L., Vilkhu, K., Versteeg, C., 2008. Modification of food ingredients by ultrasound to improve functionality: a preliminary study on a model system. Innov. Food Sci. Emerg. Technol. 9, 155–160.

Assami, K., Pingret, D., Chemat, S., Meklati, B.Y., Chemat, F., 2012. Ultrasound induced intensification and selective extraction of essential oil from *Carum carvi* L. seeds. Chem. Eng. Proc. 62, 99–105.

Barba, F.J., Puértolas, E., Brncic, M., Panchev, I.N., Dimitrov, D.A., Athès-Dutour, V., Moussa, M., Souchon, I., 2015. Emerging extraction, first ed. Food Waste Recovery, 249–272.

Benedito, J., Carcel, J., Clemente, G., Mulet, A., 2000. Cheese maturity assessment using ultrasonics. J. Dairy Sci. 83 (2), 248–254.

Benito-Román, Ó., Alonso, E., Cocero, M.J., 2013. Ultrasound-assisted extraction of β-glucans from barley. LWT – Food Science and Technology 50 (1), 57–63.

Bernatoniene, J., Cizauskaite, U., Ivanauskas, L., Jakstas, V., Kalveniene, Z., Kopustinskiene, D.M., 2016. Novel approaches to optimize extraction processes of ursolic, oleanolic and rosmarinic acids from *Rosmarinus officinalis* leaves. Ind. Crop. Prod. 84, 72–79.

Bi, Y., Su, J., Yang, D., 2014. Ultrasound assisted extraction of chrysanthemum chlorogenic acid with ionic liquid. J. Chin. Instit. Food Sci. Technol. 14 (10), 164–170.

Bimakr, M., Rahman, R.A., Taip, F.S., Adzahan, N.M., Islam Sarker, M.Z., Ganjloo, A., 2012. Optimization of ultrasound-assisted extraction of crude oil from winter melon (*Benincasa hispida*) seed using response surface methodology and evaluation of its antioxidant activity, total phenolic content and fatty acid composition. Molecules 17 (10), 11748–11762.

Birk, T., Knøchel, S., 2009. Fate of food-associated bacteria in pork as affected by marinade, temperature, and ultrasound. J. Food Protect. 72 (3), 549–555.

Boistier-Marquis, E., Lagsir-Oulahal, N., Callard, M., 1999. Applications des ultrasons de puissances en industries alimentaires. Industries Alimentaires et Agricoles 116, 23–31.

Boukroufa, M., Boutekedjiret, C., Petigny, L., Rakotomanomana, N., Chemat, F., 2015. Biorefinery of orange peels waste: A new concept based on integrate green and solvent free extraction processes using ultrasound and microwave techniques to obtain essential oil, polyphenols and pectin. Ultrason. Sonochem. 24, 72–79.

Cameron M., 2007. Impact of low-frequency high-power ultrasound on spoilage and potentially pathogenic dairy microbes. Doctor of Philosophy in Food Science dissertation. Faculty of AgriSciences, University of Stellenbosch. Stellenbosch, Sudafrica. Available from: http://scholar.sun.ac.za/handle/10019.1/1163

Cameron, M., Mcmaster, L.D., Britz, T.J., 2009. Impact of ultrasound on dairy spoilage microbes and milk components. Dairy Sci. Technol. 89 (1), 83–98.

Caraveo, O., Alarcon-Rojo, A.D., Renteria, A., Santellano, E., Paniwnyk, L., 2015. Physicochemical and microbiological characteristics of beef treated with high-intensity ultrasound and stored at 4°C. J. Sci. Food Agr. 95 (12), 2487–2493.

Carcel, J.A., Benedito, J., Bon, J., Mulet, A., 2007. High intensity ultrasound effects on meat brining. Meat Sci. 76, 611–619.

Chandrapala, J., Oliver, C., Kentish, S., Ashokkumar, M., 2012. Ultrasonics in food processing: food quality assurance and food safety. Trends Food Sci. Technol. 26 (2), 88–98.

Chang, H.-J., Xu, X.-L., Zhou, G.-H., Li, C.H.-B., Huang, M., 2012. Effects of characteristics changes of collagen on meat physicochemical properties of beef semitendinosus muscle during ultrasonic processing. Food Bioproc. Technol. 5, 285–297.

Char, C.D., Mitilinaki, E., Guerrero, S.N., Alzamora, S.M., 2010. Use of high-intensity ultrasound and UV-C light to inactivate some microorganisms in fruit juices. Food Bioproc. Technol. 3, 797–803.

Chemat, S., Esveld, E.D.C., 2013. Contribution of microwaves or ultrasonics on carvone and limonene recovery from dill fruits (*Anethum graveolens* L.). Innov. Food Sci. Emerg. Technol. 17, 114–119.

Chemat, F., Zill-E-Huma, Khan, M.K., 2011. Applications of ultrasound in food technology: processing, preservation, and extraction. Ultrason. Sonochem. 18 (4), 813–835.

Chen, H., Fu, X., Luo, Z., 2015. Properties and extraction of pectin-enriched materials from sugar beet pulp by ultrasonic-assisted treatment combined with subcritical water. Food Chem. 168, 302–310.

Cheok, C.Y., Chin, N.L., Yusof, Y.A., Talib, R.A., Law, C.L., 2013. Optimization of total monomeric anthocyanin (TMA) and total phenolic content (TPC) extractions from mangosteen (*Garcinia mangostana* Linn.) hull using ultrasonic treatments. Ind. Crop. Prod. 50, 1–7.

Chouliara, E., Georgogianni, K.G., Kanellopoulou, N., Kontominas, M.G., 2010. Effect of ultrasonication on microbiological, chemical, and sensory properties of raw, thermized, and pasteurized milk. Int. Dairy J. 20 (5), 307–313.

Cintas, P., Mantegna, S., Gaudino, E.C., Cravotto, G., 2010. A new pilot flow reactor for high-intensity ultrasound irradiation: application to the synthesis of biodiesel. Ultrason. Sonochem. 17 (6), 985–989.

Clodoveo, M.L., Durante, V., La Notte, D., 2013. Working towards the development of innovative ultrasound equipment for the extraction of virgin olive oil. Ultrason. Sonochem. 20 (5), 1261–1270.

Da Porto, C., Natolino, A., Decorti, D., 2015. Effect of ultrasound pre-treatment of hemp (*Cannabis sativa* L.) seed on supercritical CO_2 extraction of oil. J. Food Sci. Technol. 52 (3), 1748–1753.

De Mello, B.T.F., Dos Santos Garcia, V.A., Da Silva, C., 2015. Ultrasound-assisted extraction of oil from chia (*Salvia hispânica* L.) seeds: optimization extraction and fatty acid profile. J. Food Proc. Eng. 40 (1), DOI: 10.1111/jfpe.12298.

Dibazar, R., Bonat Celli, G., Brooks, M.S.-., Ghanem, A., 2015. Optimization of ultrasound-assisted extraction of anthocyanins from lowbush blueberries (*Vaccinium angustifolium* Aiton). J. Berry Res. 5 (3), 173–181.

Dolatowski, Z.J., Stadnik, J., 2007. Effect of sonication on technological properties of beef. Biosyst. Divers. 15 (1), 220–223.

Dolatowski, Z.J., Twarda, J., 2004. Einfluss von Ultraschall auf das Wasserbindungsvermögen von Rindfleisch. Fleischwirtschaft 12, 95–99.

Dolatowski, Z., Stasiak, D., Latoch, A., 2000. Effect of ultrasound processing of meat before freezing on its texture after thawing. EJPAU 3 (2), Available from: http://wwwejpaumediapl/volume3/issue2/engineering/art-02html.

Dolatowski, Z.J., Stasiak, D.M., Giemza, S., 2001. Effects of sonication on properties of reduced pH meat. Pol. J. Food Nutr. Sci. 10/51 (3(S)), 192–196.

Eh, A.L.-S., Teoh, S.-G., 2012. Novel modified ultrasonication technique for the extraction of lycopene from tomatoes. Ultrason. Sonochem. 19, 151–159.

Elleuch, M., Bedigian, D., Roiseux, O., Besbes, S., Blecker, C., Attia, H., 2011. Dietary fibre and fiber-rich by-products of food processing: characterization, technological functionality, and commercial applications: a review. Food Chem. 124, 411–421.

El-Sharnouby, G.A., Aleid, S.M., Al-Otaibi, M.M., 2014. Production of liquid sugar from date palm (*Phoenix dactylifera* L.) fruits. Adv. Environ. Biol. 8 (10), 93–100.

Fahmi, R., Khodaiyan, F., Pourahmad, R., Emam-Djomeh, Z., 2011. Effect of ultrasound assisted extraction upon the protein content and rheological properties of the resultant soymilk. Advance J. Food Sci. Technol. 3 (4), 245–249.

Fahmi, R., Khodaiyan, F., Pourahmad, R., Emam-Djomeh, Z., 2014. Effect of ultrasound assisted extraction upon the Genistin and Daidzin contents of resultant soymilk. J. Food Sci. Technol. 51 (10), 2857–2861.

Fang, Z., Smith, R., Qi, X., 2014. Production of biofuels and chemicals using with ultrasound. Springer, Dordrecht, Netherlands.

Feng, H., Yang, W., Hielscher, T., 2008. Power ultrasound. Food Sci. Technol. Int. 14, 433–436.

Feng, H., Zhou, B., Lee, H., Li, Y., Park, H.K., Palma, S., Pearlstein, A., 2013. Microbial inactivation by ultrasound for enhanced food safety. J. Acoust. Soc. Am. 133, 3596.

Fu, Q., Zhang, L., Cheng, N., Jia, M., Zhang, Y., 2014. Extraction optimization of oleanolic and ursolic acids from pomegranate (*Punica granatum* L.) flowers. Food Bioprod. Proc. 92 (3), 321–327.

Gaete-Garretón, L., Vargas-Hernández, Y., Cares-Pacheco, M.G., Sainz, J., Alarcón, J., 2011. Ultrasonically enhanced extraction of bioactive principles from *Quillaja saponaria* Molina. Ultrasonics 51 (5), 581–585.

Gajendragadkar, C.N., Gogate, P.R., 2016. Intensified recovery of valuable products from whey by use of ultrasound in processing steps: a review. Ultrason. Sonochem. 32, 102–118.

Galanakis, C.M., 2012. Recovery of high added-value components from food wastes: conventional, emerging technologies and commercialized applications. Trends Food Sci. Technol. 26 (2), 68–87.

Gallego-Juárez, J.A., Rodriguez, G., Acosta, V., Riera, E., 2010. Power ultrasonic transducers with extensive radiators for industrial processing. Ultrason. Sonochem. 17 (6), 953–964.

Ganesan, B., Martini, S., Solorio, J., Walsh, M.K., 2015..Determining the effects of high intensity ultrasound on the reduction of microbes in milk and orange juice using response surface methodology. Int. J. Food Sci. 2015 (1), 7.

Ganesh Moorthy, I., Prakash Maran, J., Muneeswari Surya, S., Naganyashree, S., Shivamathi, C.S., 2015. Response surface optimization of ultrasound assisted extraction of pectin from pomegaranate peel. Int. J. Biol. Macromol. 71, 1323–1328.

Gao, S., Hemar, Y., Lewis, G.D., Ashokkumar, M., 2013. Inactivation of *Enterobacter aerogenes* in reconstituted skim milk by high- and low-frequency ultrasound. Ultrason. Sonochem. 21, 2099–2109.

Gao, S., Lewis, G.D., Ashokkumar, M., Hemar, Y., 2014. Inactivation of microorganisms by low-frequency high-power ultrasound: 1. Effect of growth phase and capsule properties of the bacteria. Ultrason. Sonochem. 21 (1), 446–453.

Garcia-Castello, E.M., Rodriguez-Lopez, A.D., Major, L., Ballesteros, Conodi, C., Cassano, A., 2015. Optimization of conventional and ultrasound assisted extraction of flavonoids from grapefruit (*Citrus paradisi* L.) solid wastes. LWT – Food Sci. Technol. 64, 1114–1122.

Golmohamadi, A., Möller, G., Powers, J., Nindo, C., 2013. Effect of ultrasound frequency on antioxidant activity, total phenolic and anthocyanin content of red raspberry puree. Ultrason. Sonochem. 20 (5), 1316–1323.

Gonzales-Centeno, M.R., Knoerzer, K., Sabarez, H., Simal, S., Rosselló, C., Femenia, A., 2014. Effect of acoustic frequency and power density on the aqueous ultrasonic-assisted extraction of grape pomace (*Vitis vinifera* L.): a response surface approach. Ultrason. Sonochem. 21 (6), 2176–2184.

Gopal, N., Hill, C., Ross, P.R., Beresford, T.P., Fenelon, M.A., Cotter, P.D., 2015. The prevalence and control of *Bacillus* and related spore-forming bacteria in the dairy industry. Front. Microbiol. 6, 1–18.

Got, F., Culioli, J., Berge, P., Vignon, X., Astruc, T., Quideau, J.M., Lethiecq, M., 1999. Effects of high- intensity high-frequency ultrasound on ageing rate, ultrastructure and some physico-chemical properties of beef. Meat Sci. 51, 35–42.

Hanieh, S., Niels, H., Nonboe, K.U., Corry, J.E.L., Purnell, G., 2014. Combined steam and ultrasound treatment of broilers at slaughter: a promising intervention to significantly reduce numbers of naturally occurring *Campylobacters* on carcasses. Int. J. Food Macromol. 176, 23–28.

Haughton, P., Lyng, J., Morgan, D., Cronin, D., Noci, F., Fanning, S., Whyte, P., 2010. An evaluation of the potential of high-intensity ultrasound for improving the microbial safety of poultry. Food Bioproc. Technol. 5, 992–998.

Helen, P., Earle, M., Edwardson, W., 1979. Recovery of meat protein from alkaline extracts of beef bones. J. Food Sci. 44, 327–331.

Herceg, Z., Markov, K., Šalamon, B.S., Jambrak, A.R., Vukušić, T., Kaliterna, J., 2013. Effect of high-intensity ultrasound treatment on the growth of food spoilage bacteria. Food Technol. Biotechnol. 51, 352–359.

Higuera-Barraza, O.A., Del Toro-Sanchez, C.L., Ruiz-Cruz, S., Márquez-Ríos, E., 2016. Effects of high-energy ultrasound on the functional properties of proteins. Ultrason. Sonochem. 31, 558–562.

Hossain, M.B., Tiwari, B.K., Gangopadhyay, N., O'donnell, C.P., Brunton, N.P., Rai, D.K., 2014. Ultrasonic extraction of steroidal alkaloids from potato peel waste. Ultrason. Sonochem. 21 (4), 1470–1476.

Hosseini, S.M.H., Emam-Djomeh, Z., Razavi, S.H., Moosavi-Movahedi, A.A., Saboury, A.A., Atri, M.S., Van Der Meeren, P., 2013. β-Lactoglobulin–sodium alginate interaction as affected by polysaccharide depolymerization using high intensity ultrasound. Food Hydrocoll. 32 (2), 235–244.

Hunter, G., 2008. Bacterial inactivation using radial mode ultrasonic devices. Doctor of Philosophy dissertation. Department of Mechanical Engineering. University of Glasgow, Glasgow, Scotland.

Ivanovic, J., Tadic, V., Dimitrijevic, S., Stamenic, M., Petrovic, S., Zizovic, I., 2014. Antioxidant properties of the anthocyanin-containing ultrasonic extract from blackberry cultivar "Čačanska Bestrna". Ind. Crop. Prod. 53, 274–281.

Jacotet-Navarro, M., Rombaut, N., Deslis, S., Fabiano-Tixier, A.-S., Pierre, F.-X., Bily, A., Chemat, F., 2016. Towards a "dry" biorefinery without solvents or added water using microwaves and ultrasound for total valorization of fruit and vegetable byproducts. Green Chem. 18 (10), 3106–3115.

Jambrak, A.R., Herceg, Z., 2014. Application of ultrasonics in food preservation and processing. In: Bhattacharya, S. (Ed.), Conventional and Advanced Food Processing Technologies. John Wiley & Sons, Ltd, Chichester, pp. 515–536.

Jambrak, A.R., Mason, T.J., Lelas, V., Herceg, Z., Herceg, I.L., 2008. Effect of ultrasound treatment on solubility and foaming properties of whey protein suspensions. J. Food Eng. 86 (2), 281–287.

Jambrak, A.R., Mason, T.J., Lelas, V., Paniwnyk, L., Herceg, Z., 2014. Effect of ultrasound treatment on particle size and molecular weight of whey proteins. J. Food Eng. 121, 15–23.

Jayasooriya, S.D., Bhandari, B.R., Torley, P., D'Arey, B.R., 2004. Effect of high power ultrasound waves on properties of meat: a review. Int. J. Food Prop. 7, 301–319.

Jayasooriya, S.D., Torley, P.J., D'arcy, B.R., Bhandari, B.R., 2007. Effect of high-power ultrasound and ageing on the physical properties of bovine Semitendinosus and Longissimus muscles. Meat Sci. 75, 628–639.

Jia, X., Zhang, C., Hu, J., He, M., Bao, J., Wang, K., Li, P., Chen, M., Wan, J., Su, H., Zhang, Q., He, C., 2015. Ultrasound-assisted extraction, antioxidant and anticancer activities of the polysaccharides from *Rhynchosia minima* root. Molecules 20 (11), 20901–20911.

Johansson, L., Singh, T., Leong, T., Mawson, R., Mcarthur, S., Manasseh, R., Juliano, P., 2016. Cavitation and non-cavitation regime for large-scale ultrasonic standing wave particle separation systems: in situ gentle cavitation threshold determination and free radical related oxidation. Ultrason. Sonochem. 28, 346–356.

Jørgensen, A.S., Christensen, M., Ertbjerg, P., 2008. Marination with kiwifruit powder followed by power ultrasound tenderizes porcine m. Biceps femoris. In: International Conference of Meat Science and Technology, Cape Town, Sudafrica.

Joyce, E., Al-Hashimi, A., Mason, T.J., 2011. Assessing the effect of different ultrasonic frequencies on bacterial viability using flow cytometry. J. Appl. Microbiol. 110, 862–870.

Kalantar, E., Maleki, A., Khosravi, M., Mahmodi, S., 2010. Evaluation of ultrasound waves effect on antibiotic resistance seudomonas aeruginosa and *Staphylococcus aureus* isolated from hospital and their comparison with standard species. Iran J. Health Environ. 3, 319–326.

Karki, B., Lamsal, B.P., Jung, S., Van Leeuwen, J., Pometo Iii, A.L., Grewell, D., Khanal, S.K., 2010. Enhancing protein and sugar release from defatted soy flakes using ultrasound technology. J. Food Eng. 96 (2), 270–278.

Kasaai, M.R., 2013. Input power-mechanism relationship for ultrasonic irradiation: food and polymer applications. Nat. Sci. 5, 14–22.

Kha, T.C., Nguyen, M.H., 2014. Extraction and isolation of plant bioactives. Plant Bioact. Comp. Pan. Canc. Prev., 117–144.

Khan, M.K.K., Abert-Vian, M., Fabiano-Tixier, A.-S., Dangles, O., Chemat, F., 2010. Ultrasound assisted extraction of polyphenols (flavanone glycosides) from orange (*Citrus sinensis* L.) peel. Food Chem. 119, 851–858.

Khanal, S., Anand, S., Muthukumarappan, K., Huegli, M., 2014a. Inactivation of thermoduric aerobic sporeformers in milk by ultrasonication. Food Cont. 37, 232–239.

Khanal, S.N., Anand, S., Muthukumarappan, K., 2014b. Evaluation of high-intensity ultrasonication for the inactivation of endospores of 3 bacilli species in nonfat milk. J. Dairy Sci. 97 (10), 5952–5963.

Kim, H.K., Kim, Y.H., Park, H.J., Lee, N.H., 2013. Application of ultrasonic treatment to extraction of collagen from the skins of sea bass *Lateolabrax japonicas*. Fish Sci. 79, 849–856.

Kordowska-Wiater, M., Stasiak, D.M., 2011. Effect of ultrasound on survival of gram-negative bacteria on chicken skin surface. Bull. Veterin. Instit. Pulawy 55, 207–210.

Kostic, M., Zlatkovic, B., Miladinovic, B., Zivanovic, S., Mihaijlov-Krstev, T., Pavlovic, D., Kitic, D., 2015. Rosmarinic acid levels, phenolic contents, antioxidant and antimicrobial activities of the extracts from salvia verbenacal. Obtained with different solvents and procedures. J. Food Biochem. 39 (2), 199–208.

Koubaa, M., Barba, F.J., Grimi, N., Mhemdi, H., Koubaa, W., Boussetta, N., Vorobiev, E., 2016. Recovery of colorants from red prickly pear peels and pulps enhanced by pulsed electric field and ultrasound. Innov. Food Sci. Emerg. Technol. 37, 336–344.

Lauterborn, W., Ohl, C.D., 1997. Cavitation bubble dynamics. Ultrason. Sonochem. 4 (2), 65–75.

Lauterborn, W., Kurz, T., Geisler, R., Schanz, D., Lindau, O., 2007. Acoustic cavitation, bubble dynamics and sonoluminescence. Ultrason. Sonochem. 14 (4), 484–491.

Leal-Ramos, M.Y., Alarcón-Rojo, A.D., Mason, T.J., Paniwnyk, L., Alarjah, M., 2011. Ultrasound-enhanced mass transfer in Halal compared with non-Halal chicken. J. Sci. Food Agr. 91 (1), 130–133.

Leite, P., Saldago, J.M., Venancio, A., Dominguez, J.M., Belo, I., 2016. Ultrasound pretreatment of olive pomace to improve xylanase and cellulose production by solid-state fermentation. Biores. Technol. 214, 737–746.

Li, A.-N., Li, S., Xu, D.-P., Xu, X.-R., Chen, Y.-M., Ling, W.-H., Chen, F., Li, H-B., 2015a. Optimization of ultrasound-assisted extraction of lycopene from papaya processing waste by response surface methodology. Food Anal. Meth. 8, 1207–1214.

Li, K., Kang, Z., Zou, Y., Xu, X., Zhou, G., 2015b. Effect of ultrasound treatment on functional properties of reduced-salt chicken breast meat batter. J. Food Sci. Technol. 52 (5), 2622–2633.

Liu, Z., Juliano, P., Williams, R., Niere, J., Augustin, M., 2013. Ultrasound effects on the assembly of casein micelles in reconstituted skim milk. J. Dairy Res. 81, 1455.

Liu, Z., Juliano, P., Williams, R.P.W., Niere, J., Augustin, M.A., 2014. Ultrasound improves the renneting properties of milk. Ultrason. Sonochem. 21 (6), 2131–2137.

Liu, C.-., Wang, Y.-., Huang, C.-., Lu, H.-., Chiang, W.-., 2015. Optimization extraction conditions with ultrasound for anti-hyperglycemic activities from *Psidium guajava* leaf. Food Sci. Technol. Res. 21 (4), 615–621.

Luo, M., Hu, J.-., Song, Z.-., Jiao, J., Mu, F.-., Ruan, X., Gai, Q.-., Qiao, Q., Zu, Y.-., Fu, Y.-., 2015. Optimization of ultrasound-assisted extraction (UAE) of phenolic compounds from *Crataegus pinnatifida* leaves and evaluation of antioxidant activities of extracts. RSC Adv. 5 (83), 67532–67540.

Lyng, J.G., Allen, P., Mckenna, B., 1998. The effects of pre- and post-rigor high-intensity ultrasound treatment on aspects of lamb tenderness. LWT-Food Science Technol. 31, 334–338.

Ma, Y.Q., Ye, X.Q., Fang, Z.X., Chen, J.C., Xu, G.H., Liu, D.H., 2008. Phenolic compounds and antioxidant activity of extracts from ultrasonic treatment of Satsuma mandarin (*Citrus unshiu* Marc.) peels. J. Agr. Food Chem. 56 (14), 5682–5690.

Majid, I., Nayik, G.A., Nanda, V., 2015. Ultrasonication and food technology: a review. Cogent Food Agr. 1, 1071022.

Mamma, D., Christakopoulos, P., 2014. Biotransformation of citrus by-products into value-added products. Waste Biomass Valor. 5, 529–549.

Mason, T.J., 1998. Power ultrasound in food processing: the way forward. In: Povey, M.J.V., Mason, T.J. (Eds.), Ultrasound in Food Processing. Blackie Academic and Professional. Thomson Publishing, UK, pp. 105–127.

Mason, T.J., Chemat, F., Ashokkumar, M., 2014. Power ultrasonics for food processing. In: Gallego-Juarez, J.A., Graff, K.F. (Eds.), Power Ultrasonics: Ist edition, Applications of High-Intensity Ultrasound. Woodhead Publishing, pp. 815–843.

Mason, T.J., Paniwnyk, L., Lorimer, J.P., 1996. The uses of ultrasound in food technology. Ultrason. Sonochem. 3 (3), S253–S260.

Mason, T.J., Peters, D., 2002. Practical Sonochemistry: Power Ultrasound Uses and Applications, second ed. Horwood Publishing, Chichester.

McDonnell, C.K., Allen, P., Morin, C., Lyng, J.G., 2014. The effect of ultrasonic salting on protein and water–protein interactions in meat. Food Chem. 147, 245–251.

Mohammadi, V., Ghasemi-Varnamkhasti, M., Ebrahimi, R., Abbasvali, M., 2014. Ultrasonic techniques for the milk production industry. Measurement 58, 93–102.

Montalbo-Lomboy, M., Khanal, S.K., Van Leeuwen, J.H., Raj Raman, D, Dunn, Jr., L., Grewell, D., 2010. Ultrasonic pretreatment of corn slurry for saccharification: a comparison of batch and continuous systems. Ultrason. Sonochem. 17 (5), 939–946.

Morild, R.K., Christiansen, P.S., Anders, H., Nonboe, U., Aabo, S., 2011. Inactivation of pathogens on pork by steam-ultrasound treatment. J. Food Protect. 74, 769–775.

Mukhopadhyay, S., Ramaswamy, R., 2012. Application of emerging technologies to control salmonella in foods: a review. Food Res. Int. 45, 666–677.

Nawirska, A., Kwasniewska, M., 2005. Dietary fibre fractions from fruit and vegetable processing waste. Food Chem. 9, 221–225.

Nguyen, T.P., Le, V.V.M., 2012. Application of ultrasound to pineapple mash treatment in juice processing. Int. Food Res. J. 19 (2), 547–552.

Nguyen, T.N.T., Phan, L.H.N., Le, V.V.M., 2014. Enzyme-assisted and ultrasound-assisted extraction of phenolics from mulberry (*Morus alba*) fruit: comparison of kinetic parameters and antioxidant level. Int. Food Res. J. 21 (5), 1937–1940.

Ninčević Grassino, A., Halembek, J., Djaković, S., Rimac Brnčić, S., Dent, M., Grabarić, Z., 2016a. Utilization of tomato peel waste from canning factory as a potential source for pectin production and application as tin corrosion inhibitor. Food Hydrocoll. 52, 265–274.

Ninčević Grassino, A., Brnčić, M., Vikić-Topić, D., Roca, S., Dent, M., Rimac Brnčić, S., 2016b. Ultrasonically assisted extraction and characterization of pectin from tomato waste. Food Chem. 198, 93–100.

O'Donnell, C.P., Tiwari, B.K., Bourke, P., Cullen, P.J., 2010. Effect of ultrasonic processing on food enzymes of industrial importance. Trends Food Sci. Technol. 21 (7), 358–367.

Ojha, K.S., Mason, T.J., O'donnell, C.P., Kerry, J.P., Tiwari, B.K., 2016. Ultrasound technology for food fermentation applications. Ultrason. Sonochem. 17, 410–417.

O'Sullivan, J., Arellano, M., Pichot, R., Norton, I., 2014. The effect of ultrasound treatment on the structural, physical and emulsifying properties of dairy proteins. Food Hydrocoll. 42, 386–396.

Őzcan, E., 2006. Ultrasound assisted extraction of phenolics from grape pomace. Master of Science thesis. Middle East Technical University, pp. 26–42. Available from: http://etd.lib.metu.edu.tr

Ozuna, C., Puig, A., García-Pérez, J.V., Mulet, A., Cárcel, J.A., 2013. Influence of high intensity ultrasound application on mass transport, microstructure and textural properties of pork meat (*Longissimus dorsi*) brined at different NaCl concentrations. J. Food Eng. 119, 84–93.

Pagan, R., Manas, P., Alvarez, I., Condon, S., 1999. Resistance of *Listeria monocytogenes* to ultrasonic waves under pressure at sublethal (manosonication) and lethal (manothermosonication) temperatures. Food Microbiol. 16, 139–148.

Pan, Z., Qu, W., Ma, H., Atungulu, G.G., Mchugh, T.H., 2012. Continuous and pulsed ultrasound-assisted extractions of antioxidants from pomegranate peel. Ultrason. Sonochem. 19 (2), 365–372.

Patil, S., Bourke, P., Cullen, B., Frias, J.M., Cullen, P.J., 2009. The effects of acid adaptation on *Escherichia coli* inactivation using power ultrasound. Innov. Food Sci. Emerg. Technol. 10, 486–490.

Patist, A., Bates, D., 2008. Ultrasonic innovations in the food industry: from the laboratory to commercial production. Innov. Food Sci. Emerg. Technol. 9 (2), 147–154.

Petigny, L., Périno-Issartier, S., Wajsman, J., Chemat, F., 2013. Batch and continuous ultrasound assisted extraction of boldo leaves (*Peumus boldus* Mol.). Int. J. Mol. Sci. 14 (3), 5750–5764.

Pingret, D., Fabiano-Tixier, A.-S., Bourvellec, C.L., Renard, C.M.G.C., Chemat, F., 2012. Lab and pilot-scale ultrasound-assisted water extraction of polyphenols from apple pomace. J. Food Eng. 111 (1), 73–81.

Pingret, D., Fabiano-Tixier, A.-S., Chemat, F., 2013. Degradation during application of ultrasound in food processing: a review. Food Contr. 31 (2), 593–606.

Piñon, M.I., Alarcon-Rojo, A.D., Renteria, A.L., Mendez, G., Janacua-Vidales, H., 2015. Reduction of microorganisms in marinated poultry breast using oregano essential oil and power ultrasound. Acta Alimen. 44 (4), 527–533.

Piyasena, P., Mohareb, E., Mckellar, R., 2003. Inactivation of microbes using ultrasound: a review. Int. J. Food Macromol. 87 (3), 207–216.

Pohlman, F.W., Dikeman, M.E., Zayas, J.F., 1997. The effect of low-intensity ultrasound treatment on shear properties, color, stability and shelf-life of vacuum-packaged beef semitendinosus and biceps femoris muscles. Meat Sci. 45, 329–337.

Porova, N., Botvinnikova, V., Krasulya, O., Cherepanov, P., Potoroko, I., 2014. Effect of ultrasonic treatment treatment on heavy metal decontamination in milk. Ultrason. Sonochem. 21 (6), 2107–2111.

Portu, J., López-Giral, N., López, R., González-Arenzana, L., González-Ferrero, C., López-Alfaro, I., Santamaría, P., Garde-Cerdán, T., 2014. Different tools to enhance grape and wine anthocyanin content. In: Warner, L.M. (Ed.), Handbook of Anthocyanins: Food Sources, Chemical Applications and Health Benefits. Nova Science Publishers, pp. 51–88.

Prakash Maran, J., Priya, B., 2014. Ultrasound-assisted extraction of polysaccharide from *Nephelium lappaceum* L. fruit peel. Int. J. Biol. Macromol. 70, 530–536.

Prakash Maran, J., Priya, B., 2015. Optimization of ultrasound-assisted extraction of natural pigments from *Amaranthus tricolor* L. leaves. Journal of food processing and preservation, J. Food Proc. Preserv. 39 (6), 2314–2321.

Prakash Maran, J., Priya, B., 2016. Multivariate statistical analysis and optimization of ultrasound-assisted extraction of natural pigments from waste red beet stalks. J. Food Sci. Technol. 53 (1), p.792–799.

Pramod, K., Yadav, D., Kumar, P., Panesar, S.P., Bunkar, S.D.S., Mishra, D., Chopra, H.K., 2016. Comparative study on conventional, ultrasonication and microwave assisted extraction of γ-oryzanol from rice bran. J. Food Sci. Technol. 53 (4), 2047–2053.

Rabelo, R.S., Machado, M.T.C., Martinez, J., Hubinger, M.D., 2016. Ultrasound assisted-extraction and nanofiltration of phenolic compounds from artichoke solid vastes. J. Food Eng. 178, 170–180.

Ramli, N.S., Ismail, P., Rahmat, A., 2014. Influence of conventional and ultrasonic-assisted extraction on phenolic contents, betacyanin contents, and antioxidant capacity of red dragon fruit (*Hylocereus polyrhizus*). Sci. World J. 2014 (7), 964731.

Ravanfar, R., Tamadon, A.M., Niakousari, M., 2015. Optimization of ultrasound assisted extraction of anthocyanins from red cabbage using Taguchi design method. J. Food Sci. Technol. 52 (12), 8140–8147.

Riener, J., Noci, F., Cronin, D.A., Morgan, D.J., Lyng, J.G., 2009. The effect of thermosonication of milk on selected physicochemical and microstructural properties of yogurt gels during fermentation. Food Chem. 114 (3), 905–911.

Riener, J., Noci, F., Cronin, D.A., Morgan, D.J., Lyng, J.G., 2010. A comparison of selected quality characteristics of yogurts prepared from thermosonicated and conventionally heated milks. Food Chem. 119 (3), 1108–1113.

Rodríguez-Pérez, C., Quirantes-Piné, R., Fernández-Gutiérrez, A., Segura-Carretero, A., 2015. Optimization of extraction method to obtain a phenolic compounds-rich extract from *Moringa oleifera* Lam leaves. Ind. Crop. Prod. 66, 246–254.

Roselló-Soto, E., Parniakov, O., Deng, Q., Patras, A., Koubaa, M., Grimi, N., Boussetta, N., Tiwari, B.K., Vorobiev, E., Lebovka, N., Barba, F.J., 2016. Application of non-conventional extraction methods: toward a sustainable and green production of valuable compounds from mushrooms. Food Eng. Rev. 8 (2), 214–234.

Said, P.P., Arya, O.P., Pradhan, R.C., Singh, R.S., Rai, B.N., 2015. Separation of oleoresin from ginger rhizome powder using green processing technologies. J. Food Proc. Eng. 38 (2), 107–114.

Sango, M., Abela, D., Mcelhatton, A., Valdramidis, V.P., 2014. Assisted ultrasound applications for the production of safe foods. J. Appl. Microbiol. 116 (5), 1067–1083.

Şengül, M., Erkaya, T., Başlar, M., Ertugay, M.F., 2011. Effect of photosonication treatment on inactivation of total and coliform bacteria in milk. Food Contr. 22 (11), 1803–1806.

Shanmugam, A., Ashokkumar, M., 2014. Ultrasonic preparation of stable flaxseed oil emulsions in dairy systems: physicochemical characterization. Food Hydrocoll. 39, 151–162.

Shanmugam, A., Chandrapala, J., Ashokkumar, M., 2012. The effect of ultrasound on the physical and functional properties of skim milk. Innov. Food Sci. Emerg. Technol. 16, 251–258.

Sharmila, G., Nikitha, V.S., Ilaiyarasi, S., Dhivya, K., Rajasekar, V., Kumar, N., Muthukumaran, K., Muthukumaran, C., 2016. Ultrasound assisted extraction of total phenolics from *Cassia auriculata* leaves and evaluation of its antioxidant activities. Ind. Crop. Prod. 84, 13–21.

Shirsath, S.R., Sonawane, S.H., Gogate, P.R., 2012. Intensification of extraction of natural products using ultrasonic irradiations: a review of current status. Chem. Eng. Proc. 53, 10–23.
Singanusong, R., Nipornram, S., Tochampa, W., Rattanatraiwong, P., 2014. Low-power ultrasound-assisted extraction of phenolic compounds from mandarin (*Citrus reticulate* Blanco cv. Sainamueng) and Lime (*Citrus aurantifolia*) peels and the antioxidant. Food Anal. Meth. 8, 1112–1123.
Siró, I., Vén, C.S., Balla, C.S., Jónás, G., Zeke, I., Friedrich, L., 2009. Application of an ultrasonic assisted curing technique for improving the diffusion of sodium chloride in porcine meat. J. Food Eng. 91, 353–362.
Sivakumar, V., Vijaeeswarri, J., Anna, J.L., 2011. Effective natural dye extraction from different plant materials using ultrasound. Ind. Crop. Prod. 33 (1), 116–122.
Smith, D.P., 2011. Effect of ultrasonic marination on broiler breast meat quality and salmonella contamination. Int. J. Poult. Sci. 10, 757–759.
Smith, N.B., Cannon, J.E., Novakofski, J.E., Mckeith, F.K., O'brien, W.D., 1991. Tenderization of Semitendinosus muscle using high-intensity ultrasound. Ultrasonics Symposium, Proceedings IEEE 2, 1371–1374.
Stadnik, J., Dolatowski, Z.J., 2011. Influence of sonication on Warner–Bratzler shear force, colour and myoglobin of beef (m. semimembranosus). Eur. Food Res. Technol. 233, 553–559..
Stadnik, J., Dolatowski, Z.J., Baranowska, H.M., 2008. Effect of ultrasound treatment on water holding properties and microstructure of beef (m semimembranosus) during ageing. LWT – Food Sci. Technol. 41, 2151–2158.
Sun, Y., Liu, D., Chen, J., Ye, X., Yu, D., 2011. Effects of different factors of ultrasound treatment on the extraction yield of the all-*trans-b*-carotene from citrus peel. Ultrason. Sonochem. 18, 243–249.
Tabaraki, R., Nateghi, A., 2011. Optimization of ultrasonic-assisted extraction of natural antioxidants from rice bran using response surface methodology. Ultrason. Sonochem. 18 (6), 1279–1286.
Tabatabaie, F., Mortazavi, A., 2008. Studying the effects of ultrasound shock on cell wall permeability and survival of some LAB in milk. World Appl. Sci. J. 3 (1), 119–131.
Talmaciu, A.I., Volf, I., Popa, V.I., 2015. A comparative analysis of the "green" techniques applied for polyphenols extraction from bioresources. Chem. Biodivers. 12 (11), 1635–1651.
Tang, W., Zhang, M., Fang, Z., 2015. Optimization of ultrasound-assisted-extraction of porcine placenta water-soluble proteins and evaluation of the antioxidant activity. J. Food Sci. Technol. 52 (7), 4041–4053.
Tejada-Ortigoza, V., Garcia-Amezquita, L.E., Serna-Saldívar, S.O., Welti-Chanes, J., 2015. Advances in the functional characterization and extraction processes of dietary fiber. Food Eng. Rev., 1–21.
Tian, Y., Xu, Z., Zeng, X., Zeng, Y., Lo, M., 2013. Optimization of ultrasonic-assisted extraction of pomegranate (*Punica granatum* L.) seed oil. Ultrason. Sonochem. 20, 202–208.
Toma, R.B., Orr, P.H., Appolonia, B.D., Dintzis, F.R., Tabekhia, M.M., 1978. Physical and chemical properties of potato peel as a source of dietary fibre in bread. J. Food Sci. 44, 1403–1407.
Torkamani, A.E., Juliano, P., Ajlouni, S., Singh, T.K., 2014. Impact of ultrasound treatment on lipid oxidation of Cheddar cheese whey. Ultrason. Sonochem. 21 (3), 951–957.
Turantas, F., KıLıÇ, G.B., KıLıÇ, B., 2015. Ultrasound in the meat industry: general applications and decontamination efficiency. Int. J. Food Microbiol. 198, 59–69.
Twarda, J., Dolatowski, Z.J., 2006. The effect of sonication on the colour and WHC of normal and PSE pork. Anim. Sci. 1, 184–185.
Udaya Sankar, K., 2014. Chapter 7 Extraction processes. In: Bhattacharya, S. (Ed.), Conventional and Advanced Food Processing Technologies. John Wiley & Sons, Ltd, pp. 129–158.
Vercet, A., Burgos, J., López-Buesa, P., 2001. Manothermosonication of foods and food-resembling systems: effect on nutrient content and nonenzymatic browning manothermosonication of foods and food-resembling systems. J. Agr. Food Chem. 49, 483–489.
Vercet, A., Oria, R., Marquina, P., Crelier, S., Lopez-Buesa, P., 2002. Rheological properties of yogurt made with milk submitted to manothermosonication. J. Agr. Food Chem. 50 (21), 6165–6171.
Vinatoru, M., 2001. An overview of the ultrasonically assisted extraction of bioactive principles from herbs. Ultrason. Sonochem. 8 (3), 303–313.
Virot, M., Tomao, V., Lebourvellec, C., Renard, C.M.C.G., Chemat, F., 2010. Toward the industrial aproduction of antioxidants from food processing by-products with ultrasound-assisted extraction. Ultrason. Sonochem. 17 (6), 1066–1074.

Wang, J., Sun, B., Cao, Y., Yian, Y., Li, X., 2008. Optimisation of ultrasound-assisted extraction of phenolic compounds from wheat bran. Food Chem. 106, 804–810.

Wang, C.-., Jang, M.-., Sheu, S.-., 2012a. Analysis of ultrasound extraction techniques for root bark of *Morus alba*. Adv. Sci. Lett. 9, 518–522.

Wang, X.-., Wu, Y.-., Dai, S.-., Chen, R., Shao, Y., 2012b. Ultrasound-assisted extraction of geniposide from *Gardenia jasminoides*. Ultrason. Sonochem. 19 (6), 1155–1159.

Wang, W., Ma, X., Xu, Y., Cao, Y., Jiang, Z., Ding, T., Ye, X., Liu, D., 2015. Ultrasound-assisted extraction of pectin from grapefruit peel: optimization and comparison with the conventional method. Food Chem. 178, 106–114.

Wang, J., Lu, H.D., Muhammad, U., Han, J.Z., Wei, Z.H., Lu, Z.X., Bie, X.M., Lu, F.X., 2016. Ultrasound-assisted extraction of polysaccharides from *Artemisia selengensis* Turcz and its antioxidant and anticancer activities. J. Food Sci. Technol. 53 (2), 1025–1034.

Wiktor, A., Sledz, M., Nowacka, M., Rybak, K., Witrowa-Rajchert, D., 2016. The influence of immersion and contact ultrasound treatment on selected properties of the apple tissue. Appl. Acoust. 103, 135–142.

Xiong, G.-Y., Zhang, L.-L., Zhang, W., Wu, J., 2012. Influence of ultrasound and proteolytic enzyme inhibitors on muscle degradation, tenderness, and cooking loss of hens during aging. Czech J. Food Sci. 30, 195–205.

Xu, Y., Pan, S., 2013. Effects of various factors of ultrasonic treatment on the extraction yield of all-*trans*-lycopene from red grapefruit (*Citrus paradise* Macf.). Ultrason. Sonochem. 20 (4), 1026–1032.

Yanjun, S., Jianhang, C., Shuwen, Z., Hongjuan, L., Jing, L., Lu, L., Jiaping, L., 2014. Effect of power ultrasound pre-treatment on the physical and functional properties of reconstituted milk protein concentrate. J. Food Eng. 124, 11–18.

Yu, L., Shen, Y., Shangguan, X., Li, W., Li, Z., Tu, N., 2015. Optimization of ultrasound-assited extraction condition of capsaicins from Yugan pepper. J. Chin. Instit. Food Sci. Technol. 15 (1), 157–163.

Yusaf, T., Al-Juboori, R.A., 2014. Alternative methods of microorganism disruption for agricultural applications. Appl. Energ. 114, 909–923.

Zhang, D-., Wan, Y., Xu, J-., Wu, G-., Li, L., Yao, X-., 2016. Ultrasound extraction of polysaccharides from mulberry leaves and their effect on enhancing antioxidant activity. Carbohydr. Polym. 137, 473–479.

Zhang, C.-W., Wang, C.-Z., Tao, R., Fang, C., 2015. Enzymolysis-based ultrasound extraction and antioxidant activities of polyprenol lipids from *Ginkgo biloba* leaves. Proc. Biochem. 51 (3), 444–451.

Zhang, L., Ye, X., Xue, S.J., Zhang, X., Liu, D., Meng, R., Chen, S., 2013. Effect of high-intensity ultrasound on the physicochemical properties and nanostructure of citrus pectin. J. Sci. Food Agr. 93 (8), 2028–2036.

Zhao, Y., Wang, P., Zou, Y., Li, K., Kang, Z., Xu, X., Zhou, G., 2014b. Effect of pre-emulsification of plant lipid treated by pulsed ultrasound on the functional properties of chicken breast myofibrillar protein composite gel. Food Res. Int. 58, 98–104.

Zhao, L., Zhang, S., Uluko, H., Liu, L., Lu, J., Xue, H., Lv, J., 2014a. Effect of ultrasound pretreatment on rennet-induced coagulation properties of goat's milk. Food Chem. 165, 167–174.

Zhou, B., Feng, H., Luo., Y., 2009. Ultrasound enhanced sanitizer efficacy in reduction of *Escherichia coli* o157:H7 population on spinach leaves. J. Food Sci. 74 (6), 308–313.

Zhou, C., Ma, H., Yu, X., Liu, B., Abu El-Gasim, A.Y., Pan, Z., 2013. Pretreatment of defatted wheat germ proteins (byproducts of flour mill industry) using ultrasonic horn and bath reactors: effect on structure and preparation of ACE-inhibitory peptides. Ultrason. Sonochem. 20 (6), 1390–1400.

Further Reading

Cacciola, V., Batllò, I.F., Ferraretto, P., Vincenzi, S., Celotti, E., 2013. Study of the ultrasound effects on yeast lees lysis in winemaking. Eur. Food Res. Technol. 236 (2), 311–317.

Goula, A.M., Lazarides, H.N., 2015. Integrated processes can turn industrial food waste into valuable food by-products and/or ingredients: the cases of olive mill and pomegranate wastes. J. Food Eng. 167, 45–50.

Li, X., He, X., Lv, Y., He, Q., 2014. Extraction and functional properties of water-soluble dietary fiber from apple pomace. J. Food Proc. Eng. 37 (3), 293–298.

CHAPTER 11

Extraction of Bioactive Compounds From Olive Leaves Using Emerging Technologies

Rui M.S. Cruz*, Romilson Brito*, Petros Smirniotis, Zoe Nikolaidou**, Margarida C. Vieira***

**MeditBio and University of Algarve, Faro, Portugal; **Alexander Technological Educational Institute of Thessaloniki (ATEITh), Thessaloniki, Greece*

1 Introduction

The usual diet provides, in addition to macro and micronutrients, some chemical compounds, present mostly in fruits and vegetables, which exert a potent biological activity, as proven by several studies. These compounds are called bioactive compounds and can play many roles in human health (Carratu and Sanzini, 2005).

According to Silva et al. (2006), bioactive compounds are considered secondary metabolites of plants, which play an important role in resistance to disease, pests, and protection against spread of species. Their interest is growing every year and is related to its antioxidant activity and promotion of health benefits.

One of the most important fruit trees in Mediterranean countries is the olive tree (*Olea europaea* L.), and olive oil, which is obtained from olive fruits by mechanical processes, is valued throughout the world by consumers, due to its healthy and nutritional aspects. Health-promoting effects can be attributed to the antioxidant effect of phenolic compounds, and their pharmacological actions have been reported in the literature (Bilgin and Şahin, 2013).

The byproducts originated from the olive oil industry are a promising source of phenolic compounds, due to their strong antioxidant activity (Blasa et al., 2006; Javanmardi et al., 2002). In recent years, there has been a growing interest in adding value to these products, not only for environmental reasons but also due to dietary and nutritional characteristics. In this context, several studies have been developed to try to understand the role of the large number of natural compounds present in these products.

Phenolics can be classified into simple phenols and polyphenols, based exclusively on the number of phenol subunits present. Thus, the term *plant phenolics* includes simple phenols, phenolic acids, coumarins, flavonoids, and stilbenes, as well as hydrolysable and condensed

Ingredients Extraction by Physicochemical Methods in Food
http://dx.doi.org/10.1016/B978-0-12-811521-3.00011-9

tannins, lignans, and lignins. Phenolics have been observed to have biological in vitro effects, such as free radical scavenging, modulation of enzymatic activity, and inhibition of cellular proliferation. They also show antibiotic activities and can be used as antiallergic, antidiarrheal, antiulcer, and antiinflammatory agents (Cherng et al., 2007).

The extensive quantitative and qualitative changes in phenolic compounds are also dependent on the biological cycle of the olive tree (Brahmi et al., 2012). Phenolic compounds are found in a significant amount in fruits, as well as in leaves and are transferred during the fruit's processing to the olive oil, being responsible for its flavor and antioxidant activity, and thus playing an important role in the quality of olive oils (Tsimidou et al., 1992). Approximately 8000 naturally occurring compounds belong to the category of phenolics, all sharing a common structure: an aromatic ring bearing at least one hydroxyl substituent, that is, a phenol (Croteau et al., 2000).

Phenolic acids are widely distributed in the plant kingdom. The term *phenolic acids*, in general, designates phenols that have one carboxylic acid functionality (Clifford, 1999). Flavonoids are included in the group of polyphenols, possessing at least two phenol subunits. Flavonoids encompass around 6000 naturally occurring compounds that are widely distributed and occur in plants as glycosylated derivatives (Chebil et al., 2007). Both flavonoid and phenolic compounds from olive leaf are known to have diverse biological activities and may also be responsible for the pharmacological actions of olive leaf or, at least, for synergistically reinforcing those actions (Abaza et al., 2011).

The primary olive leaf constituents are secoiridoids (oleuropein and its derivatives); hydroxytyrosol; other polyphenols; and triterpenes, including oleanolic acid and flavonoids (rutin and diosmin). These constituents provide the tree and its fruits and leaves resistance to damage from pathogens and insects (Furneri et al., 2004).

Oleuropein was first isolated in 1908, and it is believed to be responsible for many of the therapeutic properties of olive leaf extracts. Oleuropein, the ester of 2-(3,4-dihydroxyphenyl) ethanol (hydroxytyrosol), produces a bitter taste and is present in the olive fruit, oil, and leaf (Omar, 2010). Oleuropein has been reported to possess strong antioxidant and free radical scavenger activity, as well as antimicrobial, hypoglycemic, antitoxoplasmosis, antiviral, antimycoplasmal, platelet antiaggregant, and hypolipidimic activities (Al-Azzawie and Alhamdani, 2006; Furneri et al., 2002; Jemai et al., 2008, 2009; Jiang et al., 2008; Kruk et al., 2005; Lee-Huang et al., 2007a,b; Zbidi et al., 2009; Zhao et al., 2009).

Hydroxytyrosol, is another major phenolic component of olive products, showing similar activities as oleuropein. Oleuropein and hydroxytyrosol have also been reported to act as potent selective anticancer compounds in both cancer cell lines and mouse tumor models. Although present in the olive fruit and oil, the content in the leaf is significantly higher than in other parts of the tree (Odiatou et al., 2013).

The fruit and leaves of the olive tree contain a series of compounds that represent multichemical mechanisms of defense against microbe and insect attacks, as previously mentioned. There is clear evidence regarding the antimicrobial activity of compounds contained in olives, olive oil, and olive leaves (Furneri et al., 2004). Several gram-negative and -positive bacteria, yeast, and parasites, including malaria, are also inhibited by the olive leaf extract. Its antibacterial activity is thought to be via either inactivation of cellular enzymes crucial for bacterial replication or direct attack on the cell membrane resulting in leakage of intracellular components, such as glutamate, potassium, and phosphorus (Alternative Medicine Review, 2009).

Olive variety, climatic conditions, tree age, wood proportion, agricultural practices, genetics, temperature, and extraction procedures are factors responsible for the chemical composition of olive leaves (Fares et al., 2011). It is fundamental to develop optimal extraction conditions and to characterize the extract in terms of antioxidant activity and composition, since each plant material presents different composition in terms of phenolic content. With regard to the type of extraction, the use of a solvent (liquid matrix), known as solvent extraction, allows the separation of soluble phenolic compounds by diffusion from olive leaves (solid matrix). The efficiency of this process can be affected by several factors, such as solvent type, temperature, pH, number of extraction steps, solvent/solid ratio, and particle size (Abaza et al., 2011; Chirinos et al., 2007; Lafka et al., 2007). Traditional leaching techniques use heat and/or agitation to increase the rate of mass transfer in the extraction of plant materials (Sánchez-Ávila et al., 2009). Nowadays, different extraction techniques are used to reduce extraction times and sample preparation costs (Ansari et al., 2011). Moreover, new extraction techniques have been studied to reduce the volume of solvents required to extract the important compounds from plants and enhance the precision of analyte recoveries. This chapter will focus on these extraction techniques, including microwave, supercritical fluid, superheated liquid, and ultrasound, which are being used to extract bioactive compounds from olive leaves, as well as their antioxidant and antimicrobial properties.

2 Extraction Methods

The extraction of bioactive compounds from plant materials has been studied for several years. Basically, the extraction begins with drying operations, size reduction, and homogenization of the samples under study. After the initial treatment, the samples are conducted to the extraction of bioactive compounds (Stalikas, 2007). However, the extractions are dependent on specific characteristics, such as the chemical nature of the plant, extraction method (extraction time, temperature, solvent/sample ratio, solvent used, number of extractions carried out), particle size, storage time and conditions, and presence of interfering substances. In addition to these factors, the type of compounds to be extracted (soluble or insoluble fraction) must also be taken into account, since the extraction of the insoluble fractions requires pretreatments to extraction, such as acid hydrolysis, alkali extraction, or enzymatic treatment, to break up the bonds between the phenolic conjugates (Khoddami et al., 2013; Naczk and Shahidi, 2004).

The behavior of extraction is not always linear as there is a wide variability in the plant material. The samples may exhibit different concentrations of several compounds, leading to extracts with different characteristics. Sometimes, within the same type of products, it is possible to observe compositional variations as a result of variability among species, environmental factors, and extent of processing feedstock, which may impact their antioxidant activity and biological properties (Ferreira et al., 2015).

The choice of extraction solvent is always an important point, since the yield of the processes depends on the used matrix and combinations. The liquid-liquid and solid–liquid extractions are techniques that use polar solvents. These solvents include water, acetone, and alcohols, such as methanol and ethanol, as well as mixtures of solvents (Khoddami et al., 2013).

Usually, the solubility of the samples is promoted by increasing the time and temperature. The traditional process for extraction of high value compounds using a solvent is constituted, first, by the dissolution of soluble compounds present in the sample and, second, by the mass transfer to the solution by diffusion and osmotic pressure of the soluble compounds in the plant material (called slow extraction). The last phase is much slower than the first stage and is primarily responsible for limiting the extraction rate. In addition, slow extractions require large equipment to increase yield, and long exposure to solvent may cause degradation of the components of interest (Alupului et al., 2012).

The extraction of high value compounds has gained interest every year. Several studies have been conducted and they all focus on the same goal: valorization of plant by-products. However, interest is not confined only to develop a new technique for extraction. The search for less polluting alternatives has been a constant concern, and researchers seek to provide alternative methods of physical nature to replace the conventional processes (Artajo et al., 2007; de Leonardis et al., 2008).

2.1 Microwave-Assisted Extraction

Microwaves are nonionizing electromagnetic radiation having a frequency in the range 300–300,000 MHz, which corresponds to wavelength (λ) of 1 mm–1 m. The microwave region lies between the infrared and radio waves in the electromagnetic spectrum (Sanseverino, 2002).

In the microwave, the heating mechanism occurs differently (the molecules of any object are heated by dual mechanism of ionic conduction and dipole rotation) from conventional heating (conduction, convection, and radiation). In conventional heating, the container and any material contained therein heats up, resulting in a temperature gradient from the warmer medium to the less heated (Tsukui and Rezende, 2014). The factors involved in microwave heating are: temperature, chemical bonding, molecular structure, dipole moment, polarization, heat capacity, and dielectric constant. In microwave-assisted extraction (MAE), the heat is dissipated volumetrically and each molecule exposed to the microwave field is affected directly (Alupului et al., 2012).

The microwave extraction process occurs in three different steps. First, the substrate of the equilibrium phase is removed, at an approximately constant velocity, from the outer surface of the particle. Second, the resistance to mass transfer starts to appear in the solid–liquid interface. In the last step, the interactions that bind the solute to the matrix must be overcome and the solute diffuses into the extracting solvent (Raynie, 2000). During the process, the microwave energy is converted into kinetic energy, thus allowing selective heating of the material under study. The heating leads to an increase in volume and consequent explosion of the cells, thereby releasing its contents into the liquid phase. When the liquid phase absorbs the microwaves, the kinetic energy of the molecules increases, as does the speed of diffusion, thereby resulting in a rapid mass transfer (Alupului et al., 2012).

In general, polar substances absorb microwave radiation (such as water, methanol, and ethanol), whereas the less polar substances (aliphatic or aromatic hydrocarbons) are not recommended for extraction in the microwave (Sanseverino, 2002).

Nevertheless, the mixture polarity increases with high water concentration to a degree where it is no longer favorable for extraction, thus reducing the extraction yield. The solvent–solid (feed) ratio (S/F) is also an important parameter to be optimized. The solvent volume must be enough to warrant that the entire sample is immersed in the solvent throughout the entire irradiation process (Eskilsson and Björklund, 2000).

The temperature also appears as an important factor in yield. On one hand, the yield of the extraction increases with a correct application; on the other hand, compounds present in the extracts may be degraded using high temperatures. Particle size of the samples also contributes to the efficiency of extraction. Typically, the particle size in the range of 0.01–2 mm is suitable for the extraction process (Di Khanh, 2015).

The heating duration is another important factor in MAE, in which the power and the temperature are important factors. Processing times are very short in MAE extraction and may vary from a few minutes to a half-hour compared to conventional techniques. Thus, it is possible to avoid thermal degradation and oxidation in compounds sensitive to overheating (Chan et al., 2011).

Microwave power at high levels allows the increasing of the temperature of the system, resulting in the increase of the extraction yield until it declines or becomes insignificant. Therefore, it is of extreme importance, when selecting the MAE power, to minimize the time required to reach the desired temperature and avoid a *bumping* phenomenon in temperature during the extraction (Hu et al., 2008).

Moreover, overexposure to microwave radiation may result in a decrease of the extraction yield, even at low temperature or low operating power, due to the loss of chemical structure of the active compounds (Eskilsson and Björklund, 2000).

In terms of equipment, there are two types of systems commercially available. The first type is the extraction vessels that are closed and perform extraction under controlled pressure and temperature. In the other systems, known as focused microwave ovens, the microwaves are charged only in the section of the extraction vessel containing the sample. Both are available as multimode and single mode or focused systems. In the multimode system, random scattering of microwave radiation within the microwave cavity occurs, guaranteeing that every part in the cavity and sample receives irradiation. In the single mode or focused systems, the focused microwave radiation occurs on a particular area where the sample is exposed to a much stronger electric field than in the case previously referred (Mandal et al., 2015).

According to Alupului et al. (2012), microwaves produce rapid heating of the matrix, thus facilitating the extraction of bioactive compounds. Extractions assisted by microwave allow a significant reduction in the consumption of solvents, as well as the possibility of multiple extractions (Costa, 2004). Di Khanh (2015) also reported that the main advantages of the process are: short time of processing; small amount of solvents used; simple process; and low cost compared to processes in which supercritical fluids (SCFs) are used (an extraction process that will be discussed in the next section). Di Khanh (2015) further cites disadvantages of this process, such as the need for additional process steps (centrifuging or filtration) and the correct choice of extraction solvent.

MAE system is a promising technique that presents different physical and chemical phenomena as compared to those in conventional extractions (Veggi et al., 2013).

2.2 Supercritical Fluid Extraction

Supercritical fluid extraction (SFE) is the process of separating one component (the extractant) from another (the matrix) using SCFs as the extracting solvent. A supercritical fluid is produced by heating a gas above its critical temperature or compressing a liquid above its critical pressure. Above its own critical temperature and pressure, the substance becomes a supercritical fluid and possesses the properties of gas, as well as that of a liquid. Thus, the gas-like mass transfer and the liquid-like solvating power of a supercritical fluid give them an edge over other extracting solvents (Mandal et al., 2015).

In the search for environmental friendly solvents, increasing attention is being paid to SCFs due to their great potential especially in food, cosmetic, and pharmaceutical industries. For instance, supercritical solvents, such as water, carbon dioxide, and ethanol under pressure are used in extractions, micronization, material processing, chemical reactions, cleaning, and drying, among several other applications. The density of SCFs can be easily adjusted to the process needs, with changes in temperature, pressure, and/or composition. SCFs also show other important properties, such as very low surface tensions, low viscosities, and moderately high diffusion coefficients (Pereda et al., 2008).

Several bioactive compounds can be obtained using SFE without any solvent residue and safety issues. SCF solvents can be classified as: (1) low-critical temperature for some condensable gases, such as carbon dioxide (CO_2), ethane, and propane, and (2) high-critical temperature solvents that include higher alkanes, methanol, and water. Solvent power and selectivity differences characterize the low and high critical temperature solvents (Pereda et al., 2008).

Among several possible fluids, carbon dioxide (CO_2) has been the most studied solvent for supercritical processes, due to its critical point (304°K, 7.4 MPa) that allows low temperature extraction conditions, consequently avoiding thermal degradation of the product. Its price and availability are also advantages (Sovilj et al., 2011). However, it exhibits strong liquid-liquid and gas–liquid immiscibility for hydrocarbons with carbon numbers greater than 13. Moreover, for reactions carried out at moderate or high temperatures, CO_2 presents, as previously stated, a rather low critical temperature to be used as a solvent (Pereda et al., 2008).

Depending on the objective of the extraction process, two different methods can be used: (1) carrier material separation, in which the feed material constitutes the final product after undesirable compounds are removed, for example, dealcoholization of alcohol beverages or removal of off-flavors, and (2) extract material separation, in which the extracted compounds from the feed material constitute the final product, for example, essential oil or antioxidant extraction (Mandal et al., 2015).

The separation of soluble compounds from the supercritical fluid can be achieved by modifying the thermodynamic properties of the supercritical solvent, or by an external agent. In the first situation, the solvent power is altered by changing the operating pressure or temperature, while in the second one, known as solid trapping, the separation can be achieved by absorption or adsorption (Mandal et al., 2015; Pereda et al., 2008).

For the extraction of phenolic compounds, CO_2-SFE is one of the best methods, presenting higher phenol recoveries as compared to liquid methanol SFE (only 45%) (Le Floch et al., 1998). Nevertheless, more recently, the use of CO_2-SFE alone, during oleuropein extraction from olive leaves, was not satisfactory, requiring a polar modifier to improve yield and selectivity of the process (Şahin and Bilgin, 2012).

2.3 Superheated Liquid Extraction

Superheated liquid extraction (SHLE) is the extraction using aqueous or organic solvents at a high pressure and temperature without reaching the critical point. Three different modes can be considered: static (volume of extractant is fixed), dynamic (the extractant flows continuously through the sample), and static–dynamic (a combination of the two modes) (Morales-Muñoz et al., 2003). The most significant drawback of the static mode is partition of the analytes between the solid and the fixed extractant volume, while in the dynamic mode, dilution of the analytes in the extract is one of the negative features. Thus, a combination of static and dynamic modes was found to reduce the extraction time and provide better extraction efficiency (Fernández-Pérez et al., 2000; Luque-García and Luque de Castro, 2003a).

The higher extraction efficiency obtained by SHLE as compared with a conventional method of extraction can be attributed to the higher capacity to disrupt matrix-analyte interactions of the former due to the high temperature and pressure. Moreover, at high temperatures, the surface tension and the viscosity of the extractant are reduced (maintaining high pressures to keep the solvent in the liquid state), leading to a better sample wetting and matrix diffusion, and thus, increasing mass transfer and enhancing extraction (Morales-Muñoz et al., 2003). Reduction of solvents and higher selectivity are additional advantages of SHLE. This technique has been used to extract phenols from different food matrices including apple, olive leaves, and carob pod (Alonso-Salces et al., 2001; Fernández-Pérez et al., 2000; Papagiannopoulos et al., 2004).

The mechanism of the extraction process can be divided into six sequential steps. (1) A fast entry of fluid, (2) the desorption of solutes from matrix active sites, (3) the diffusion of solutes through organic materials, (4) the diffusion of solutes through static fluid in porous materials, (5) the diffusion of solutes through a layer of stagnant fluid outside particles, and (6) the elution of solutes by the flowing bulk of fluid (Haghighi and Khajenoori, 2013).

Superheated water (water under pressure and above 100°C but below its critical temperature of 374°C; and pressure high enough to maintain the liquid state) has been used for the extraction process. Under these conditions, water is less polar, and consequently, the organic compounds are more soluble than at room temperature (Plaza and Turner, 2015).

One of the most important parameters to control efficiencies is the extraction temperature. As the temperature rises, there is a marked and systematic decrease in permittivity, an increase in the diffusion rate, and a decrease in the viscosity and surface tension. In consequence, more polar target materials with high solubility in water at ambient conditions are extracted most efficiently at lower temperatures, whereas moderately polar and nonpolar targets require a less polar medium induced by elevated temperature (Smith, 2006).

Superheated water has certain advantages, such as shorter extraction time, higher extraction ability for polar compounds, higher quality of the extract, and lower costs; however, it is not so suitable for thermally labile compounds (Golmohammad et al., 2012; Mandal et al., 2015). In addition, superheated water is an environmental friendly technique (Herrero et al., 2006).

2.4 Ultrasound-Assisted Extraction

Ultrasound-assisted extraction (UAE) uses the energy of sound waves (mechanical). The sound waves propagate in the matter in frequencies from 20 to 100 kHz (greater than the limit of human hearing), creating expansion and compression cycles. In a liquid, they produce negative pressure cycles, and may cause bubbles or cavitation, causing permanent physical and chemical changes, such as breakage and instability of solid at the interface liquid-liquid systems and liquid–gas (Luque-García and Luque de Castro, 2003b).

The changes are a consequence of the large quantity of energy generated by the cavitation phenomenon, which is defined as the phenomenon formation and subsequent disruption of the microbubbles in the center of a liquid. The importance of cavitation is not related to how bubbles are formed but what occurs when they collapse. At some point, the bubbles do not absorb more energy of the ultrasounds and implode, thus facilitating the diffusion of the solvent extractor into the matrix (Suslick et al., 1999). The energy associated with high temperatures and pressures destroys the walls of the sample cells, easily releasing the compounds (Di Khanh, 2015).

Another effect that occurs during cavitation is the formation of radicals, which can optionally react with the compounds of interest present in the sample, causing oxidation (Soria and Villamiel, 2010). These radicals are formed due to dissociation of the water molecule or other gases that may migrate to the interior of the bubble caused by heat and high pressure produced during the implosion of the cavitation bubbles (Luque de Castro and Capote, 2007).

Apart from being a viable, rapid, and efficient alternative as compared to conventional extraction, UAE is also characterized as being one of the most economical techniques and easy to apply on a large industrial scale (Di Khanh, 2015).

The extraction mechanism of the UAE process has two main stages. First, dissolution of soluble components on surfaces of the matrix occurs, which is also referred to as *washing*. Second, a *slow extraction*, in which mass transfer of the solute from the matrix into the solvent occurs by diffusion and osmotic processes (Toma et al., 2001; Veggi et al., 2013; Veličković et al., 2006, 2008; Vinatoru, 2001).

There are two common types of ultrasound equipment: (1) the ultrasonic bath (direct or indirect) and (2) ultrasonic probe (Suslick et al., 1999).

The ultrasonic bath is a relatively simple device and commonly used in laboratories for processing small amounts of sample. However, over time, the ultrasonic energy tends to lose intensity and be distributed nonuniformly, thus interfering in the results. Moreover, the position at which the sample is placed inside the container and its size contribute to the results variability. The ultrasonic probe has the advantage of applying power at any point of the mixture, favoring extraction processes (Luque-García and Luque de Castro, 2003b).

Ultrasound baths use frequencies between 20 and 40 kHz (Luque de Castro and Capote, 2007). Low frequencies, such as 20 kHz are effective for the extraction of compounds from plant sources, with the physical effects generated by cavitation being predominant. The bubbles formed in the low frequencies are larger than those formed at high frequencies, and they implode violently; thus, they are more effective in the extraction processes. Cavitation may also be influenced by factors, such as intensity of sonication, gases, particle size, applied external pressure, viscosity, surface tension, and vapor pressure of the solvent, among others. Therefore, it becomes necessary to optimize the aforementioned extraction conditions to obtain high yields at the end of the process (Luque de Castro and Capote, 2007).

UAE has as its advantages significantly short extraction times, low extraction temperature, and high extraction yields (Yang et al., 2013). UAE uses small amounts of fossil energy and contributes for reducing solvent consumption, thus resulting in higher yields and a more pure product (Gaete-Garretón et al., 2011). Moreover, the combination of sonication and microwaves in the extraction of lipids from vegetables and microalgae sources has shown excellent extraction efficiencies in terms of yield and time, presenting a 10-fold reduction in the time required with conventional methods and an increase of yields from 50% to 500% (Cravotto et al., 2008).

Di Khanh (2015) describes this technique as efficient, simple, cheap, and a good alternative to conventional extraction processes. Moreover, the equipment is easy to handle and less expensive compared to other methods, such as microwave or supercritical fluid. Di Khanh (2015) also found that the kinetics and yield of the process depends largely on the type and nature of the plant material. Moreover, Luque-García and Luque de Castro (2003b) reported, as disadvantages of UAE, the lack of uniformity in the distribution of ultrasonic energy and the loss of power during the process. The same authors also refer to the advantages of UAE being the possibility of use for a wide sample size range, speed in sample processing, and low cost.

3 Extraction of Bioactive Compounds From Olive Leaves

This section presents several studies reporting the effect of different techniques to extract bioactive compounds from olive leaves and their antioxidant and antimicrobial properties.

Fig. 11.1A presents the effect of the different treatments applied on the total phenolics content extracted from olive leaves, Maçanilha Algarvia [5 g olive leaves and 75 mL ethanol:water (70:30)]. The results showed that when combining grinding, vacuum impregnation, homogenization, and microwave (800 W and 2450 MHz during 30 s), a significant increase in extraction is obtained [total phenolic content of 21.82 ± 1.68 mg GAE/g dry leaf ($P < 0.05$)]. Moreover, adding ultrasound (20 kHz and 125 W during 10 min.) as an extra processing treatment, the combination presented the highest extraction value, 27.52 ± 2.91 mg GAE/g dry leaf ($P < 0.05$).

Figure 11.1B presents the effect of different applied treatments on the total flavonoid content extracted from the same olive leaves. The process combining all technologies also showed

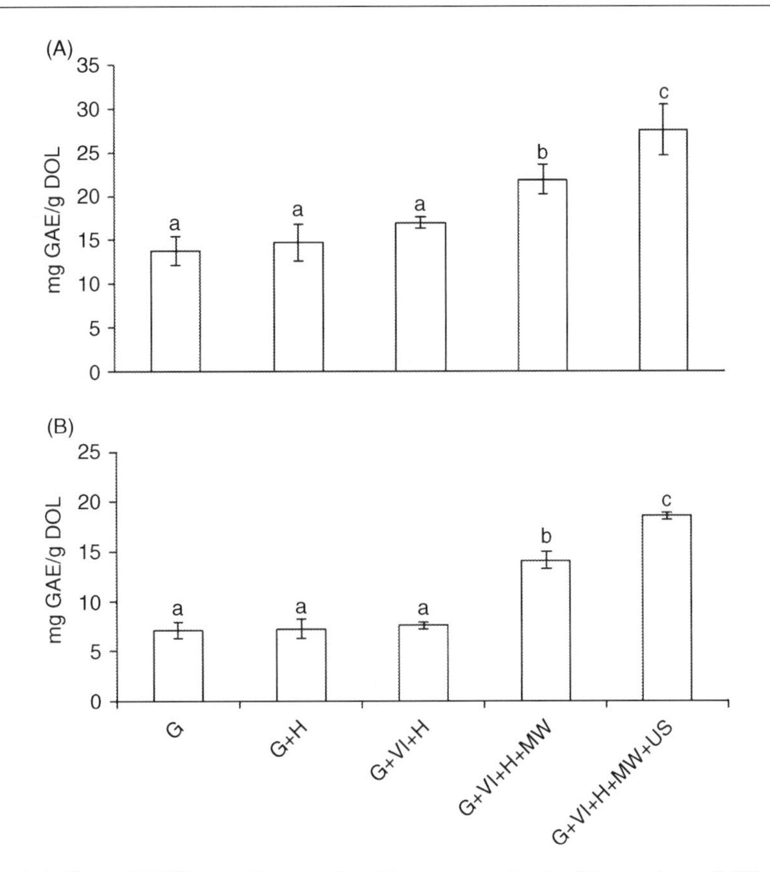

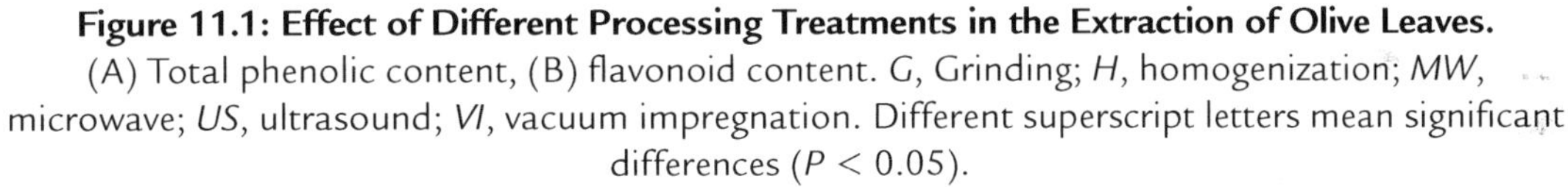

Figure 11.1: Effect of Different Processing Treatments in the Extraction of Olive Leaves. (A) Total phenolic content, (B) flavonoid content. *G*, Grinding; *H*, homogenization; *MW*, microwave; *US*, ultrasound; *VI*, vacuum impregnation. Different superscript letters mean significant differences ($P < 0.05$).

the highest value (18.55 ± 0.32 mg CE/g dry leaf). No significant differences ($P > 0.05$) were observed among the treatments applied without microwave and ultrasound processing. Abaza et al. (2011) reported values of 15.83 ± 1.26 mg CE/g dry leaf extracted with ethanol at 70% in Chétoui olive leaves. On the other hand, Salah et al. (2012) presented values of 94.03 ± 4.19, 125.64 ± 3.36, 120.88 ± 0.78, 82.74 ± 10.73, 56.75 ± 6.72, 97.74 ± 12.67, 76.01 ± 8.72, and 91.32 ± 1.13 mg CE/g dry leaf for the varieties Chétoui, Gerboua, Limouni, Chemlali, Sevillane, Lucques, Rosicola, and Meski, respectively, extracted with ethanol at 70%, for at least 1 week at room temperature in the dark. Differences in results, besides the equipment and processing treatments applied, are also dependent, as previously referred, on the olive leaves' intrinsic factors (Al-Rimawi et al., 2014a; Brahmi et al., 2012).

Table 11.1 presents several studies in which the total phenolic content in different varieties of olive leaves and extraction methods was determined.

Table 11.1: Total phenolic content in different varieties of olive leaves and methods of extraction.

References	Extraction Methods	Olive Leaves Variety or Geographical Sites	Total Phenolic Content
Abaza et al. (2011)	70% ethanol, 24 h, 1 g/10 mL	Chétoui[a]	24.36 ± 0.85 mg GAE/g dry leaf
Ahmad-Qasem et al. (2013)	Ultrasound (51.47 W), 15 min, 80% ethanol, 6.25 g/200 mL	Serrana[a]	66 ± 8 mg GAE/g dry leaf
Apostolakis et al. (2014)	9.3% glycerol, 80°C, 241 min, 0.5 g/30 mL	Koroneiki[a]	51.91 mg GAE/g dry leaf
Bilgin and Şahin (2013)	Ultrasound (50 Hz), 60 min, methanol, 500 mg/10 mL	Bursa[b]	38.66 mg GAE/g dry leaf
		Mardin[b]	~32.00 mg GAE/g dry leaf
		Ayvalik[b]	~26.00 mg GAE/g dry leaf
		Kas[b]	~24.00 mg GAE/g dry leaf
		Tekirdag[b]	~20.00 mg GAE/g dry leaf
		Canakkale[b]	7.35 mg GAE/g dry leaf
Bouallagui et al. (2011)	80% methanol during night, 60 g/300 mL	Chemlali[a]	15.55 mg GAE/g dry leaf
Mkaouar et al. (2016)	Instant controlled pressure drop, 2 h, 96% ethanol, 55°C, 5 g/250 mL	Chemlali[a]	187.31 mg GAE/g dry leaf
Şahin et al. (2015)	Ultrasound (40 kHz), 60 min, 43.61% ethanol, 34.18°C, 0.5 g/10 mL	Ayvalık[b]	43.82 mg GAE/g dry leaf
Şahin and Samli (2013)	Ultrasound, 60 min, 50% ethanol, 0.5 g/10 mL	Tavşan Yüreği[a]	25.06 mg GAE/g dry leaf
Salah et al. (2012)	70% ethanol at least 1 week, 1 g/20 mL	Chemlali[a]	99.71 ± 11.45 mg GAE/g dry leaf
		Chétoui[a]	102.32 ± 1.18 mg GAE/g dry leaf
		Gerboua[a]	142.21 ± 3.53 mg GAE/g dry leaf
		Limouni[a]	144.19 ± 10.27 mg GAE/g dry leaf
		Lucques[a]	106.80 ± 9.24 mg GAE/g dry leaf
		Meski[a]	110.03 ± 20.04 mg GAE/g dry leaf
		Rosicola[a]	91.90 ± 11.04 mg GAE/g dry leaf
		Sevillane[a]	73.05 ± 15.52 mg GAE/g dry leaf
Yancheva et al. (2016)	70% methanol, 15 min, 70°C, 1 g/1 mL	Chondrolia Halkidiki[a]	15.6 ± 0.5 mg GAE/g dry leaf
		Kalamon[a]	14.7 ± 0.5 mg GAE/g dry leaf
		Koroneiki[a]	9.2 ± 0.5 mg GAE/g dry leaf

[a]Olive leaves variety.
[b]Geographical sites.

Table 11.2: Antioxidant activity of "Maçanilha Algarvia" olive leaf extract.

DPPH radical scavenging activity (μmol TE/g of dried leaf)	266.67 ± 35.14
ABTS radical scavenging activity (μmol TE/g of dried leaf)	123.25 ± 17.19
Radical scavenging capacity for DPPH (%)	91.57 ± 1.52
Radical scavenging capacity for ABTS (%)	99.04 ± 0.55

Table 11.2 presents the results of radical scavenging activity of Maçanilha Algarvia olive leaf extract obtained by 1,1-diphenyl-2-picrylhydrazyl free radical (DPPH) and 2,2′-azino-bis(3-ethylbenzthiazoline-6-sulphonic acid) (ABTS). Briante et al. (2003) reported that phenolic compounds could be active as antioxidants by a number of potential pathways. The free radical scavenging in which the phenol can break the free radical chain reaction is considered the most important. The existence of different substituents within the backbone structure of phenols controls their antioxidant properties, in particular, their hydrogen-donating capacity. The DPPH assay allows for the measuring of the antioxidant activity and correlates with the phenolic content of olive leaf extract presented previously. In this study, the value of the DPPH radical scavenging activity of this extract was 266.37 ± 35.14 μmol TE/g of dry leaf, indicating that phenolic compounds may contribute directly to the antioxidant effect of the extract. Syrpas et al. (2010) reported lower values ranging from 48.21 to 140.90 μmol TE/g dry leaf in a study with olive leaves from Greece (e.g., Koroneiki, Kalamon), Italy (Frantoio), and Spain (e.g., Picual, Cordal). Maçanilha Algarvia olive leaf extract showed similar levels of antioxidant activity, when compared with Ginkgo biloba, presenting values of 312 and 274 μmol TE/g dry weight for aqueous and methanolic extracts, respectively (Tawaha et al., 2007). The ABTS radical scavenging activity was 123.25 ± 17.19 μmol TE/g of dry leaf. Sevim and Tuncay (2012) reported values of 8.95 ± 0.01 μmol TE/g of leaf in ABTS radical scavenging activity in a study with Memecik olive leaves. The relatively high levels of phenolics and flavonoids are probably related to the strong activity observed in both methods. The correlation between polyphenolics and antioxidant activities from plant materials has been reported in several studies (Leong and Shui, 2002; Skerget et al., 2005; Turkoglu et al., 2007).

Several studies, besides showing the total phenolic content, determined the amount of specific compounds, such as oleuropein, from olive leaves with different extraction methods. Yateem et al. (2014) studied the effect of extraction solvent and temperature on oleuropein content using a fixed amount of olive leaves in different solvents and different composition (volume fraction). Seven extraction procedures were performed (10 g of olive leaves) using 100 mL of seven extraction solvents (water, 80% methanol, 100% methanol, 50% ethanol, 80% ethanol, 100% ethanol, and 20% acetonitrile) for 4 h. The results showed that the highest oleuropein content was obtained when the olive leaves were extracted with 80% ethanol (13 mg/g) followed by 20% acetonitrile (10.0 mg/g), 80% methanol (5.31 mg/g), and 50% ethanol (2.57 mg/g). Moreover, the amount of oleuropein extracted using pure water, pure methanol, and pure ethanol was low (0.16, 0.10, and 0.02 mg/g for water, methanol, and ethanol, respectively). These results showed that a mixture of an organic solvent with water is required to effectively extract oleuropein from olive leaves as very low amounts of oleuropein were extracted using pure solvents as compared to solvent mixtures. With regard to the effect of temperature in the extraction solvent (25, 40, and 60°C), the results showed that oleuropein content increased with increasing temperature of extraction solvent. There was an 18-fold increase in oleuropein content as temperature increased from 25 to 40°C, and 43-fold increase as temperature increased from 25 to 60°C. This can be attributed to an increase in solubility of oleuropein with increasing temperature.

Altıok et al. (2008) also used different extractant solvents, ethanol, methanol, acetone, and their aqueous form (10%–90%, v/v) during 24 h in the extraction of polyphenols from olive leaves. The results showed high content of phenolics and antioxidant capacity (oleuropein: 13.4%, rutin: 0.18%) when 70% ethanol was used as an extractant solvent.

Talhaoui et al. (2014) compared the phenolic compounds of "Sikitita" olive leaves determined by HPLC-DAD-TOF-MS with "Arbequina" and "Picual" olive leaves. In this study, 0.5 g of dry leaves were mixed with 10 mL of MeOH/H_2O (80/20) and extracted with an ultrasonic bath during 10 min. The extraction was repeated twice and 30 phenolic compounds were identified. The total phenolic compounds ranged from 52.12 to 60.64 mg/kg.

Japón-Luján et al. (2006b) performed a multivariate optimisation of MAE of oleuropein and related biophenols from olive leaves. The experimental conditions were: MAE power 100–200 W, irradiation time 5–15 min and ethanol 80%–100%. The conditions of 200 W for 8 min and 80% ethanol allowed the extraction of oleuropein (2.32%), verbacoside (631 mg/kg), apigenin-7-glucoside (1076 mg/kg), and luteolin-7-glucoside (1016 mg/kg). However, simple phenols were not found in the extracts obtained by MAE.

In another study reported by the same authors (Japón-Luján et al., 2006a), UAE was used for extracting oleuropein and related biophenols from olive leaves. Several conditions were tested: ultrasound (20 kHz, 450 W), probe position: 0–4 cm, radiation amplitude: 10%–50%, duty cycle: 30%–70%, irradiation time: 6–30 min, extractant flow-rate: 4–6 mL/min, ethanol: 50%–90%, and water bath: temperature 25–40°C. A multivariate methodology optimization was performed and the best method was the following combination: 1 g of milled leaves in a 59:41 ethanol–water mixture, bath temperature 40°C, extraction time 25 min, ultrasonic irradiation (duty cycle 0.7 s, output amplitude 30% of the converter, and applied power of 450 W). Oleuropein, verbacoside, apigenin-7-glucoside, and luteolin-7-glucoside presented values of 22610 ± 632, 488 ± 21, 1072 ± 38, and 970 ± 43 mg/kg, respectively.

Xynos et al. (2012) developed a green extraction method with super/subcritical fluids and olive leaves to obtain extracts enriched in oleuropein. The combination of SC–CO_2 modified by 5% ethanol at 30 MPa, 50°C, and subcritical water showed high extract yield (44.1%), high recovery of oleuropein (4.6%), and good antioxidant activity.

In another study, Xynos et al. (2014) tested an ethanolic extraction at 190°C for three consecutive cycles. A mixture of H_2O/EtOH 43:57 at 190°C using only 1 extraction cycle provided the optimal results regarding the oleuropein content of the extract.

Olive leaves were also extracted using ethanol and water as extraction solvents at 150 and 200°C for 20 min. There were 31 different phenolic compounds found in the extracts, including secoiridoids, simple phenols, flavonoids, cinnamic-acid derivatives, and benzoic acids. The results showed that the ethanolic extract proved to be especially rich in flavonoids, while the aqueous extract was richer in hydroxytyrosol (Quirantes-Piné et al., 2013).

In a study reported by Japón-Luján and Luque de Castro (2006), a static–dynamic superheated liquid extraction (140°C and 6 bar) was used to extract biophenols from olive leaves. Only 13 min were necessary to extract up to 23,000 mg/kg of oleuropein. The authors pointed out as main advantages of this process, the efficacy of superheated liquids to extract biophenols (compounds highly demanded for their nutraceutical properties), the use of low-toxic ethanol–water mixtures as extractant, and the low costs of the raw material.

Lafka et al. (2013) used several techniques/solvents, including methanol, ethanol, ethanol:water 1:1, and SFE/CO_2, to extract phenolic compounds from wild olive leaves. The total phenolics were affected significantly ($P < 0.05$) due to differences in solvent polarities, which probably influence the solubility of several constituents present in olive leaves. Ethanol was the most effective solvent for phenol extraction, showing the following optimum conditions: 180 min, solvent to sample ratio 5:1 v/w, and pH 2. Ethanol extract also showed the highest antiradical activity among solvent and SFE extracts.

Luo (2011) studied the effect of aqueous ethanol, boiling water, and UAE in olive leaves. The results showed that higher total phenolic content and oleuropein concentration were obtained by increasing temperature and solvent:solid ratio. The 80% ethanol at 40°C and a solvent:solid ratio of 30 showed a good recovery of phenolic compounds, presenting an extraction efficiency of 80%. On the other hand, boiling water showed a poor recovery of phenolic compounds, presenting an extraction efficiency of 40%. UAE accelerated the extraction rate and reduced extraction time.

Besides the total phenolic content and the antioxidant properties in olive leaves, several studies also tested the effect of olive leaves extract in a wide range of microorganisms. Al-Rimawi et al. (2014b) reported antibacterial activities around 80%–90% of olive leaves extract against *Staphylococcus aureus* and *S. epidermidis*, as compared to neomycin.

Sudjana et al. (2009) studied the antibacterial activity of a commercial olive leaf extract in different bacteria. The results showed that the olive leaf extract did not present broad-spectrum antibacterial activity, but had considerable activity on *Helicobacter pylori* (0.6%–1.2% v/v) and *Campylobacter jejuni* (0.3%–2.5% v/v).

Nora et al. (2012) studied the antibacterial activity and phytochemical screening of olive leaves from Algeria: 5 g of olive leaves were boiled with 50 mL of distilled water for 30 min. The aqueous extract reacted positively on all bacterial strains tested, *Escherichia coli* ATCC 25922, *S. aureus* ATCC 6538, *S. aureus* ATCC 25923, *Klebsiella pneumoniae*, *Enterobacter cloacae* ATCC 13047, *E. coli*, *Pseudomonas aeruginosa* ATCC 10145, and *Bacillus stearothermophilus* ATCC 11778. The extract of olive leaves showed the best inhibition against *E. coli* with a minimal inhibitory concentration of 150 μl/mL and an inhibition area of 15.3 mm in diameter. Markin et al. (2003) investigated the antimicrobial effect of olive leaves against bacteria and fungi. The olive leaves were extracted in water at a 20% (w/v) concentration (200 g dried leaves per liter water) and autoclaved at 121°C for 20 min. The microorganisms tested were inoculated in various concentrations of olive leaf water extract. Olive leaf 0.6% (w/v) water extract killed almost all bacteria tested within 3 h. Dermatophytes were inhibited by 1.25% (w/v) of the extract following a 3-day exposure, and *Candida albicans* was killed following a 24 h incubation in the presence of 15% (w/v) of the extract. Scanning electron microscopic observations of *C. albicans*, exposed to 40% (w/v) olive leaf extract, showed invaginated and amorphous cells. *E. coli* cells, exposed to a similar treatment, showed complete destruction with 0.6% (w/v) of olive leaf extract.

In another study, 10 g of olive leaves were extracted for 2 h with 200 mL of 70% (v/v) aqueous ethanol at 38°C and a thermo-shaker at 180 rpm. Relative antioxidant capacity and total phenol content of the extract were determined and found as 966 μg ascorbic acid eq./mg and 197.42 mg GAE/g sample, respectively. The most susceptible bacteria to the effect of the olive leaf extract were *E. coli*, *Listeria innocua*, and *S. carnosus* (Aytul, 2010).

Olive leaves from Gilan province (Northern Iran) were extracted by mechanical shaking (1 L of water and 50 g of powder obtained from leaves). Olive leaf aqueous extracts were screened for their antimicrobial activity against pathogenic bacteria (*S. aureus* PTCC 1431, *Salmonella typhimurium* PTCC 1639, *E. coli* PTCC 1399, *K. pneumonia* PTCC 1053, and *B. cereus* PTCC 1274). Olive leaf extract was found to be most active against *S. typhimurium* PTCC 1639 with an inhibition area of 11.5 mm (Aliabadi et al., 2012). Pereira et al. (2007) tested the antimicrobial activity of aqueous olive leaf extract against *S. aureus*, *B. subtilis*, *P. aeruginosa*, *E. coli*, *K. pneumoniae*, *B. cereus*, *C. albicans*, and *C. neoformans*. The results revealed that the growth rates of *S. aureus* and *E. coli* decreased while olive leaf extract concentration increased. The olive leaf extract showed an IC25 (25% inhibitory concentration) value of 2.68 and 1.81 mg/mL for *S. aureus* and *E. coli*, respectively. In another study, Korukluoglu et al. (2010) investigated the effect of the extraction solvent on the antimicrobial efficiency of *S. aureus*, *E. coli*, *S. enteritidis*, and *S. thypimurium*. They reported that solvent type affected the phenolic distribution and concentration in extracts, and the antimicrobial activity against tested bacteria. The highest antimicrobial efficiency against

E. coli and *S. enteritidis* was shown by the ethanol olive leave extract while the acetone extract showed the highest antimicrobial efficiency against *S. thypimurium*.

4 Conclusions

Olive leaves are a by-product of the olive oil production and are rich in phenolic compounds, which provide antioxidant and antimicrobial properties. Several studies have contributed to the extraction of these bioactive compounds by emerging technologies with higher efficiency, in terms of yield and energy. These promising technologies allow, on one hand, an opportunity window for the valorization of olive leaf bioactive compounds in different industries as valuable ingredients; and on the other hand, they offer new ways to reduce the environmental impact of conventional extraction techniques.

References

Abaza, L., Youssef, N.B., Manai, H., Haddada, F.M., Methenni, K., Zarrouk, M., 2011. Chétoui olive leaf extracts: influence of the solvent type on phenolics and antioxidant activities. Grasas Aceites 62, 96–104.

Ahmad-Qasem, M., Cánovas, J., Barrajón-Catalán, E., Micol, V., Cárcel, A.J., García-Pérez, J.V., 2013. Kinetic and compositional study of phenolic extraction from olive leaves (var. Serrana) by using power ultrasound. Innov. Food Sci. Emerg. Technol. 17, 120–129.

Al-Azzawie, H.F., Alhamdani, M.S., 2006. Hypoglycemic and antioxidant effect of oleuropein in alloxan-diabetic rabbits. Life Sci. 78, 1371–1377.

Aliabadi, M.A., Darsanaki, R.K., Rokhi, M.L., Nourbakhsh, M., Raeisi, G., 2012. Antimicrobial activity of olive leaf aqueous extract. Ann. Biol. Res. 3, 4189–4191.

Alonso-Salces, R.M., Korta, E., Barranco, A., Berruela, L.A., Gallo, B., Vicente, F.J., 2001. Pressurized liquid extraction for the determination of polyphenols in apple. J. Chromatogr. A 933, 37–43.

Al-Rimawi, F., Odeh, I., Bisher, A., Abbadi, J., Qabbajeh, M., 2014a. Effect of geographical region and harvesting date on antioxidant activity, phenolic and flavonoid content of olive leaves. J. Food Nutr. Res. 12, 925–930.

Al-Rimawi, F., Odeh, I., Bisher, A., Yateem, H., Taraweh, M., 2014b. Natural antioxidants, antibacterials from olive leaf extracts used in cosmetics, pharmaceutical, and food industries. Health and Biomedical. Qatar Foundation Annual Research Conference Proceedings. ARC '14 (1), HBPP0116.

Alternative Medicine Review, 2009. Olive leaf. Monogr. Altern. Med. Rev. 14 (1), 62–66.

Altıok, E., Bayçın, D., Bayraktar, O., Ülkü, S., 2008. Isolation of polyphenols from the extracts of olive leaves (*Olea europaea* L.) by adsorption on silk fibroin. Sep. Purif. Technol. 62, 342–348.

Alupului, A., Călinescu, I., Lavric, V., 2012. Microwave extraction of active principles from medicinal plants. UPB Sci. Bull. 74, 129–142.

Ansari, M., Kazemipour, M., Fathi, S., 2011. Development of a simple green extraction procedure and HPLC method for determination of oleuropein in olive leaf extract applied to a multi-source comparative study. J. Iran. Chem. Soc. 8, 38–47.

Apostolakis, A., Grigorakis, S., Makris, D.P., 2014. Optimisation and comparative kinetics study of polyphenol extraction from olive leaves (*Olea europaea*) using heated water/glycerol mixtures. Sep. Purif. Technol. 128, 89–95.

Artajo, L.S., Romero, M.P., Suárez, M., Motilva, M.J., 2007. Partition of phenolic compounds during the virgin olive oil industrial extraction process. Eur. Food Res. Technol. 225, 617–625.

Aytul, K.K., 2010. Antimicrobial and antioxidant activities of olive leaf extract and its food applications. Master thesis, İzmir Institute of Technology, Turkey.

Bilgin, M., Şahin, S., 2013. Effects of geographical origin and extraction methods on total phenolic yield of olive tree (*Olea europaea*) leaves. J. Taiwan Inst. Chem. Eng. 44, 8–12.

Blasa, M., Candiracci, M., Accorsi, A., Piacentini, M.P., Albertini, M.C., Piatti, E., 2006. Raw Millefiori honey is packed full of antioxidants. Food Chem. 97, 217–222.

Bouallagui, Z., Han, J., Isoda, H., Sayadi, S., 2011. Hydroxytyrosol rich extract from olive leaves modulates cell cycle progression in MCF-7 human breast cancer cells. Food Chem. Toxicol. 49, 179–184.

Brahmi, F., Mechri, B., Dabbou, S., Dhibi, M., Hammami, M., 2012. The efficacy of phenolics compounds with different polarities as antioxidants from olive leaves depending on seasonal variations. Ind. Crops Prod. 38, 146–152.

Briante, R., Febbraio, F., Nucci, R., 2003. Antioxidant properties of low molecular weight phenols present in the Mediterranean diet. J. Agric. Food Chem. 51, 6975–6981.

Carratu, E., Sanzini, E., 2005. Sostanze biologicamente attive presenti negli alimenti di origine vegetable. Ann. Ist. Super. Sanita 41, 7–16.

Chan, C.-H., Yusoff, R., Ngoh, G.-C., Kung, F.W.-L., 2011. Microwave-assisted extractions of active ingredients from plants. J. Chromatogr. A. 1218, 6213–6225.

Chebil, L., Anthoni, J., Humeau, C., Gerardin, C., Engasser, J.M., Ghoul, M., 2007. Enzymatic acylation of flavonoids: effect of the nature of the substrate, origin of lipase, and operating conditions on conversion yield and regioselectivity. J. Agric. Food Chem. 55, 9496–9502.

Cherng, J.M., Shieh, D.E., Chiang, W., Chang, M.Y., Chiang, L.C., 2007. Chemopreventive effects of minor dietary constituents in common foods on human cancer cells. Biosci. Biotechnol. Biochem. 71, 1500–1504.

Chirinos, R., Rogez, H., Campos, D., Pedreschi, R., Larondelle, Y., 2007. Optimization of extraction conditions of antioxidant phenolic compounds from mashua (*Tropaeolum tuberosum* Ruíz and Pavón) tubers. Sep. Purif. Technol. 55, 217–225.

Clifford, M., 1999. Chlorogenic acids and other cinnamates-nature, occurrence and dietary burden. J. Sci. Food Agric. 79, 362–372.

Costa, A., 2004. Desenvolvimento de métodos de extração por micro-ondas para pesticidas em solos, Estágio-PRODEP. Instituto Superior de Engenharia do Porto, Porto.

Cravotto, G., Boffa, L., Mantegna, S., Perego, P., Avogadro, M., Cint, P., 2008. Improved extraction of vegetable oils under high-intensity ultrasound and/or microwaves. Ultrason. Sonochem. 15, 898–902.

Croteau, R., Kutchan, T.M., Lewis, N.G., 2000. Natural products (secondary metabolites). In: Buchanan, B., Gruissem, W., Jones, R. (Eds.), Biochemistry and Molecular Biology of Plants. American Society of Plant Physiologists, Rockville, MD, pp. 1250–1318.

de Leonardis, A., Aretini, A., Alfano, G., Macciola, V., Ranalli, G., 2008. Isolation of a hydroxytyrosol-rich extract from olive leaves (*Olea europaea* L.) and evaluation of its antioxidant properties and bioactivity. Eur. Food Res. Technol. 226, 653–659.

Di Khanh, N., 2015. Advances in the extraction of anthocyanin from vegetables. J. Food Nutr. Sci. 3, 126–134.

Eskilsson, C.S., Björklund, E., 2000. Analytical-scale microwave-assisted extraction. J. Chromatogr. A 902, 227–250.

Fares, R., Bazzi, S., Baydoun, S.E., Abdel-Massih, R.M., 2011. The antioxidant and anti-proliferative activity of the Lebanese *Olea europaea* extract. Plant Foods Hum. Nutr. 66, 58–63.

Fernández-Pérez, V., Jiménez-Carmona, M.M., Luque de Castro, M.D., 2000. An approach to the static-dynamic subcritical water extraction of laurel essential oil: comparison with conventional techniques. Analyst 125, 481–485.

Ferreira, M.S.L., Santos, M.C.P., Moro, T.M.A., Basto, G.J., Andrade, R.M.S., Gonçalves, É.C.B.A., 2015. Formulation and characterization of functional foods based on fruit and vegetable residue flour. J. Food Sci. Technol. 52, 822–830.

Furneri, P.M., Marino, A., Saija, A., Uccella, N., Bisignano, G., 2002. In vitro antimycoplasmal activity of oleuropein. Int. J. Antimicrob. Agents 20, 293–296.

Furneri, P.M., Piperno, A., Sajia, A., Bisignano, G., 2004. Antimycoplasmal activity of hydroxytyrosol. Antimicrob. Agents Chemother. 48, 4892–4894.

Gaete-Garretón, L., Vargas-Hernández, Y., Cares-Pacheco, M.G., Sainz, J., Alarcón, J., 2011. Ultrasonically enhanced extraction of bioactive principles from *Quillaja saponaria* Molina. Ultrasonics 51, 581–585.

Golmohammad, F., Eikani, M.H., Maymandi, H.M., 2012. Cinnamon bark volatile oils separation and determination using solid-phase extraction and gas chromatography. Procedia Eng. 42, 247–260.

Haghighi, A.A., Khajenoori, M., 2013. Subcritical water extraction. In: Nakajima, H. (Ed.), Mass Transfer: Advances in Sustainable Energy and Environment-Oriented Numerical Modeling. InTech, Rijeka, Croatia.

Herrero, M., Cifuentes, A., Ibañez, E., 2006. Sub- and supercritical fluid extraction of functional ingredients from different natural sources: plants, food-by-products, algae and microalgae: a review. Food Chem. 98, 136–148.

Hu, Z., Cai, M., Liang, H.H., 2008. Desirability function approach for the optimization of microwave-assisted extraction of saikosaponins from *Radix bupleuri*. Sep. Purif. Technol. 61, 266–275.

Japón-Luján, R., Luque de Castro, M.D., 2006. Superheated liquid extraction of oleuropein and related biophenols from olive leaves. J. Chromatogr. A 1136, 185–191.

Japón-Luján, R., Luque-Rodríguez, J.M., Luque de Castro, M.D., 2006a. Dynamic ultrasound-assisted extraction of oleuropein and related biophenols from olive leaves. J. Chromatogr. A 1108, 76–82.

Japón-Luján, R., Luque-Rodríguez, J.M., Luque de Castro, M.D., 2006b. Multivariate optimisation of the microwave-assisted extraction of oleuropein and related biophenols from olive leaves. Anal. Bioanal. Chem. 385, 753–759.

Javanmardi, J., Khalighi, A., Kashi, A., Bais, H.P., Vivanco, J.M., 2002. Chemical characterization of basil (*Ocimum basilicum* L.) found in local accessions and used in traditional medicines in Iran. J. Agric. Food Chem. 50, 5878–5883.

Jemai, H., Bouaziz, M., Fki, I., El Feki, A., Sayadi, S., 2008. Hypolipidimic and antioxidant activities of oleuropein and its hydrolysis derivative-rich extracts from Chemlali olive leaves. Chem. Biol. Interact. 176, 88–98.

Jemai, H., El Feki, A., Sayadi, S., 2009. Antidiabetic and antioxidant effects of hydroxytyrosol and oleuropein from olive leaves in alloxan-diabetic rats. J. Agric. Food Chem. 57, 8798–8804.

Jiang, J.H., Jin, C.M., Kim, Y.C., Kim, H.S., Park, W.C., Park, H., 2008. Anti-toxoplasmosis effects of oleuropein isolated from *Fraxinus rhychophylla*. Biol. Pharm. Bull. 31, 2273–2276.

Khoddami, A., Wilkes, M.A., Roberts, T.H., 2013. Techniques for analysis of plant phenolic compounds. Molecules 18, 2328–2375.

Korukluoglu, M., Sahan, Y., Yigit, A., Ozer, E.T., Gucer, S., 2010. Antibacterial activity and chemical constitutions of *Olea europaea* L. leaf extracts. J. Food Process. Preserv. 34, 383–396.

Kruk, I., Aboul-Enein, H.Y., Michalska, T., Lichszteld, K., Kladna, A., 2005. Scavenging of reactive oxygen species by the plant phenols genistein and oleuropein. Luminescence 20, 81–89.

Lafka, T.-I., Lazou, A.E., Sinanoglou, V.J., Lazos, E.S., 2013. Phenolic extracts from wild olive leaves and their potential as edible oils antioxidants. Foods 2, 18–31.

Lafka, T.-I., Sinanoglou, V., Lazos, E.S., 2007. On the extraction and antioxidant activity of phenolic compounds from winery wastes. Food Chem. 104, 1206–1214.

Le Floch, F., Tena, M.T., Ríos, A., Valcárcel, M., 1998. Supercritical fluid extraction of phenol compounds from olive leaves. Talanta 46, 1123–1130.

Lee-Huang, S., Huang, P.L., Zhang, D., Lee, J.W., Bao, J., Sun, Y., Chang, Y.T., Zhang, J., 2007a. Discovery of small-molecule HIV-1 fusion and integrase inhibitors oleuropein and hydroxytyrosol: Part I. Fusion [corrected] inhibition. Biochem. Biophys. Res. Commun. 354, 872–878.

Lee-Huang, S., Huang, P.L., Zhang, D., Lee, J.W., Bao, J., Sun, Y., Chang, Y.T., Zhang, J., 2007b. Discovery of small-molecule HIV-1 fusion and integrase inhibitors oleuropein and hydroxytyrosol: part II. Integrase inhibition. Biochem. Biophys. Res. Commun. 354, 879–884.

Leong, L.P., Shui, G., 2002. An investigation of antioxidant capacity of fruits in Singapore markets. Food Chem. 76, 69–75.

Luo, H., 2011. Extraction of antioxidante compounds from olive (*Olea europaea*) leaf. Master thesis, Massey University, Albany, New Zealand.

Luque de Castro, M.D., Capote, F.P., 2007. Analytical applications of ultrasound. Techniques and Instrumentation in Analytical Chemistry, vol. 26Elsevier, Amsterdam.

Luque-García, J.L., Luque de Castro, M.D., 2003a. Comparison of the static, dynamic and static-dynamic pressurised liquid extraction modes for the removal of nitrated polycyclic aromatic hydrocarbons from soil with on-line filtration-preconcentration. J. Chromatogr. A 1010, 129–140.

Luque-García, J.L., Luque de Castro, M.D., 2003b. Ultrasound: a powerful tool for leaching. Trends Anal. Chem. 22, 41–47.

Mandal, S., Mandal, V., Das, A., 2015. Essentials of Botanical Extraction, first ed. Academic Press, San Diego, CA.
Markin, D., Duek, L., Berdicevsky, I., 2003. In vitro antimicrobial activity of olive leaves. Mycoses 46, 132–136.
Mkaouar, S., Gelicus, A., Bahloul, N., Allaf, K.N., Kechaou, N., 2016. Kinetic study of polyphenols extraction from olive (*Olea europaea* L.) leaves using instant controlled pressure drop texturing. Sep. Purif. Technol. 161, 165–171.
Morales-Muñoz, S., Luque-Garcia, J.L., Luque de Castro, M.D., 2003. Approaches for accelerating sample preparation in environmental analysis. Crit. Rev. Environ. Sci. Technol. 33, 391–421.
Naczk, M., Shahidi, F., 2004. Extraction and analysis of phenolics in food. J. Chromatogr. A 1054, 95–111.
Nora, N.B., Hamid, K., Snouci, M., Boumedien, M., Abdellah, M., 2012. Antibacterial activity and phytochemical screening of *Olea europaea* leaves from Algeria. Open Conf. Proc. J. 3, 66–69.
Odiatou, E.M., Skaltsounis, A.L., Constantinou, A.I., 2013. Identification of the factors responsible for the in vitro pro-oxidant and cytotoxic activities of the olive polyphenols oleuropein and hydroxytyrosol. Cancer Lett. 330, 113–121.
Omar, S.H., 2010. Cardioprotective and neuroprotective roles of oleuropein in olive. Saudi Pharm. J. 18, 111–121.
Papagiannopoulos, M., Wollseifen, H.R., Mellenthin, A., Haber, B., Galensa, R., 2004. identification and quantification of polyphenols in carob fruits (*Ceratonia siliqua* L.) and derived products by HPLC-UV-ESI/MSn. J. Agric. Food Chem. 52, 3784–3791.
Pereda, S., Bottini, S.B., Brignole, E.A., 2008. Fundamentals of supercritical fluid technology. In: Martínez, J.L. (Ed.), Supercritical Fluid Extraction of Nutraceuticals and Bioactive Compounds. CRC Press, Boca Raton, pp. 1–24.
Pereira, A.P., Ferreira, I., Marcelino, F., Valentão, P., Andrade, P.B., Seabra, R., Estevinho, L., Bento, A., Pereira, J.A., 2007. Phenolic compounds and antimicrobial activity of olive (*Olea europaea* L Cv. Cobrançosa) leaves. Molecules 12, 1153–1162.
Plaza, M., Turner, C., 2015. Pressurized hot water extraction of bioactives. Trends Anal. Chem. 71, 39–54.
Quirantes-Piné, R., Lozano-Sánchez, J., Herrero, M., Ibáñez, E., Segura-Carretero, A., Fernández-Gutiérrez, A., 2013. HPLC-ESI-QTOF-MS as a powerful analytical tool for characterising phenolic compounds in olive-leaf extracts. Phytochem. Anal. 24, 213–223.
Raynie, D.E., 2000. Extraction. In: Wilson, I.D., Adlard, E.R., Cooke, M., Poolie, C.F. (Eds.), Encyclopedia of Separation Science. Academic Press, San Diego.
Şahin, S., Bilgin, M., 2012. Study on oleuropein extraction from olive tree (*Olea europaea*) leaves by means of SFE: comparison of water and ethanol as co-solvent. Sep. Sci. Technol. 47, 2391–2398.
Şahin, S., Samli, R., 2013. Optimization of olive leaf extract obtained by ultrasound-assisted extraction with response surface methodology. Ultrason. Sonochem. 20, 595–602.
Şahin, S., Ilbay, S., Kırbaşlar, S.I., 2015. Study on optimum extraction conditions for olive leaf extracts rich in polyphenol and flavonoid. Sep. Sci. Technol. 50, 1181–1189.
Salah, M.B., Abdelmelek, H., Abderraba, M., 2012. Study of phenolic composition and biological activities assessment of olive leaves from different varieties grown in Tunisia. Eur. J. Med. Chem. 2, 107–111.
Sánchez-Ávila, N., Priego-Capote, F., Ruiz-Jiménez, J., de Castro, L.M.D., 2009. Fast and selective determination of triterpenic compounds in olive leaves by liquid chromatography-tandem mass spectrometry with multiple reaction monitoring after microwave-assisted extraction. Talanta 78, 40–48.
Sanseverino, A.M., 2002. Micro-ondas em síntese orgânica. Quím. Nova 25, 660–667.
Sevim, D., Tuncay, O., 2012. Total phenolic contents and antioxidant activities of "ayvalik" and "memecik" olive leaves and olive fruits. J. Food 37, 219–226.
Silva, S., Gomes, L., Leitao, F., Coelho, A.V., Boas, L.V., 2006. Phenolic compounds and antioxidant activity of *Olea europaea* L fruits and leaves. Food Sci. Technol. Int. 12, 385–395.
Skerget, M., Kotnik, P., Hadolin, M., Hras, A.R., Simonic, M., Knez, Z., 2005. Phenols, proanthocyanidins, flavones and flavonols in some plant materials and their antioxidant activities. Food Chem. 89, 191–198.
Smith, R.M., 2006. Superheated water: the ultimate green solvent for separation science. Anal. Bioanal. Chem. 385, 419–421.
Soria, A.C., Villamiel, M., 2010. Effect of ultrasound on the technological properties and bioactivity of food: a review. Trends Food Sci. Technol. 21, 323–331.

Sovilj, M.N., Nikolovski, B.G., Spasojević, M.D., 2011. Critical review of supercritical fluid extraction of selected spice plant materials. Maced. J. Chem. Chem. Eng. 30, 197–220.

Stalikas, C.D., 2007. Extraction, separation, and detection methods for phenolic acids and flavonoids. J. Sep. Sci. 30, 3268–3295.

Sudjana, A.N., D'Orazio, C., Ryan, V., Rasool, N., Ng, J., Islam, N., Riley, T.V., Hammer, K.A., 2009. Antimicrobial activity of commercial *Olea europea* (olive) leaf extract. Int. J. Antimicrob. Agents 33, 461–463.

Suslick, K.S., Didenko, Y., Fang, M.M., Hyeon, T., Kolbeck, K.J., McNamara, III, W.B., Mdleleni, M.M., Wong, M., 1999. Acoustic cavitation and its chemical consequences. Philos. Trans. R. Soc. A 357, 335–353.

Syrpas, M., Van Hoed, V., Van Poucke, C., De Saeger, S. Kiritsakis, A., Verhé, R., 2010. Cultivar effect on the phenolics of olive leaves and their antioxidant activity (OTTO-008). 8th Euro Fed Lipid Congress, "Oils, Fats and Lipids: Health & Nutrition, Chemistry & Energy", 21-24 November 2010, Munich, Germany.

Talhaoui, N., Gómez-Caravaca, A.M., León, L., de la Rosa, R., Segura-Carretero, A., Fernández-Gutiérrez, A., 2014. Determination of phenolic compounds of "Sikitita" olive leaves by HPLC-DAD-TOF-MS. Comparison with its parents "Arbequina" and "Picual" olive leaves. LWT—Food Sci. Technol. 58, 28–34.

Tawaha, K., Alali, F.Q., Gharaibeh, M., Mohammad, M., El-Elimat, T., 2007. Antioxidant activity and total phenolic content of selected Jordanian plant species. Food Chem. 104, 1372–1378.

Toma, M., Vinatoru, M., Paniwnyk, L., Mason, T.J., 2001. Investigation of the effects of ultrasound on vegetal tissues during solvent extraction. Ultrason. Sonochem. 8, 137–142.

Tsimidou, M., Papadopoulos, G., Boskou, D., 1992. Phenolic compounds and stability of virgin olive oil: part I. Food Chem. 45, 141–144.

Tsukui, A., Rezende, C.M., 2014. Microwave assisted extraction and green chemistry. Rev. Virtual Quím. 6, 1713–1725.

Turkoglu, A., Duru, M.E., Mercan, N., Kivrik, I., Gezer, K., 2007. Antioxidant and antimicrobial activities of *Laetiporus sulphureus* (Bull) Murrill. Food Chem. 101, 267–273.

Veggi, P.C., Martinez, J., Meireles, M.A.A., 2013. Fundamentals of microwave extraction. In: Chemat, F., Cravotto, G. (Eds.), Microwave-Assisted Extraction for Bioactive Compounds: Theory and Practice Food Engineering, Series 4. Springer, New York, pp. 15–52.

Veličković, D.T., Milenović, D.M., Ristić, M.S., Veljković, V.B., 2006. Kinetics of ultrasonic extraction of extractive substances from garden (*Salvia officinalis* L.) and glutinous (Salvia glutinosa L.) sage. Ultrason. Sonochem. 13, 150–156.

Veličković, D.T., Milenović, D.M., Ristić, M.S., Veljković, V.B., 2008. Ultrasonic extraction of waste solid residues from the *Salvia* sp. essential oil hydrodistillation. Biochem. Eng. J. 42, 97–104.

Vinatoru, M., 2001. An overview of the ultrasonically assisted extraction of bioactive principles from herbs. Ultrason. Sonochem 8, 303–313.

Xynos, N., Papaefstathiou, G., Psychis, M., Argyropoulou, A., Aligiannis, N., Skaltsounis, A.-L., 2012. Development of a green extraction procedure with super/subcritical fluids to produce extracts enriched in oleuropein from olive leaves. J. Supercrit. Fluids 67, 89–93.

Xynos, N., Papaefstathiou, G., Gikas, E., Argyropoulou, A., Aligiannis, N., Skaltsounis, A.-L., 2014. Design optimization study of the extraction of olive leaves performed with pressurized liquid extraction using response surface methodology. Sep. Purif. Technol. 122, 323–330.

Yancheva, S., Mavromatis, P., Georgieva, L., 2016. Polyphenol profile and antioxidant activity of extracts from olive leaves. J. Cent. Eur. Agric. 17, 154–163.

Yang, Y.C., Wei, M.C., Huang, T.C., Lee, S.Z., Lin, S.S., 2013. Comparison of modified ultrasound-assisted and traditional extraction methods for the extraction of baicalin and baicalein from *Radix scutellariae*. Ind. Crops Prod. 45, 182–190.

Yateem, H., Afaneh, I., Al-Rimawi, F., 2014. Optimum conditions for oleuropein extraction from olive leaves. Int. J. Appl. Sci. Technol. 4, 153–157.

Zbidi, H., Salido, S., Altarejos, J., Perez-Bonilla, M., Bartegi, A., Rosado, J.A., Salido, G.M., 2009. Olive tree wood phenolic compounds with human platelet antiaggregant properties. Blood Cells Mol. Dis. 42, 279–285.

Zhao, G., Yin, Z., Dong, J., 2009. Antiviral efficacy against hepatitis B virus replication of oleuropein isolated from *Jasminum officinale* L. var. *grandiflorum*. J. Ethnopharmacol. 125, 265–268.

CHAPTER 12

Separation of Bioactive Whey Proteins and Peptides

Mohamed H. Abd El-Salam, Safinaz El-Shibiny

National Research Centre, Cairo, Egypt

1 Introduction

Whey is a major byproduct of the dairy industry, resulting mainly from cheese manufacture. It is estimated that the global production of whey reaches >200 million tons/year (Smithers, 2015). Whey consists of about 50% milk solids, including most of the lactose and water-soluble minerals and vitamins contents of milk, and about 20% milk proteins. However, whey has long been considered as a waste product representing a serious environmental problem. Direct disposal of untreated whey adds serious pollutants in water streams due to its high biological oxygen demand (BOD >35,000 ppm). Introduction of strict environmental regulations prevented the disposal of untreated whey, which represents an unavoidable cost to the dairy industry. This has encouraged the dairy industry and researchers to seek feasible ways to utilize whey. This trend has been boosted with (1) the advent of new technologies, in particular membrane separation; (2) the demand of the food industry for functional food ingredients to improve the functionality of conventional food products and to develop new products with tailored functionalities; (3) the growing scientific evidence about the physicochemical properties, nutritional, and therapeutic values of whey proteins and peptides; and (4) the more recent trend of using whey proteins as nanodelivery systems for nutraceuticals and drugs.

Sweet whey contains β-lactoglobulin (β-LG), α-lactalbumin (α-LA) as the major proteins, and much lower concentrations of immunoglobulins (Ig), blood serum albumin (BSA), and minor proteins, such as lactoferrin (LF), lactoperoxidase (LP), and growth factors (Smithers, 2008). Casein macropeptide (CMP) represents a significant part of whey proteins obtained from rennet-coagulated cheeses.

Since the 1970s the use of membrane separation processes in the production of whey protein concentrates has been considered a breakthrough in whey utilization. Several protein concentrates of different protein contents (35, 70, 80%) and functionalities (by modifying composition and technological treatments) have been developed. Later, whey proteins were

Ingredients Extraction by Physicochemical Methods in Food
http://dx.doi.org/10.1016/B978-0-12-811521-3.00012-0

fractionated to yield whey protein fractions rich in one of the main whey proteins, such as β-lactoglobulin rich fraction and α-lactalbumin rich fraction, which offer new whey protein products for more specific uses. For example, α-lactalbumin rich products has been used extensively in infant foods to simulate human milk, while β-lactoglobulin rich fraction blended with hydrocolloids has been used to develop a heat-set gel product with characteristics similar to solid fat that can be used as a fat mimetic in low-fat healthy foods (Smithers, 2008, 2015). Also, preparation of whey protein hydrolysates has received growing attention for special uses, particularly in sports nutrition and in humanized milk products. Whey protein hydolysates are characterized by high nutritional value, low antigenicity, and low bitterness, which make them ideal ingredients for the manufacture of humanized milk products. Whey protein hydrolysates rich in di- and tripeptides were reported to be superior to intact proteins or free amino acid mixtures in skeletal muscle anabolism (Manninen, 2009).

In addition to the major proteins, whey contains a large number of minor proteins with important biological and functional activities. However, exploitation of the industrial uses of these constituents is lagging behind those of the major whey proteins. This may be the result of their low concentration and the lack of cost-effective technology for the large-scale production of these constituents. During the last two decades significant progress has been achieved in this area. The same period has also witnessed progress in the fractionation of whey protein hydrolysates for the purpose of producing peptide fraction with specific functional activities.

The present chapter presents and discusses the different methods developed for the recovery of the minor whey proteins and peptides of different biological activities.

2 Separation of Lactoferrin and Lactoperoxidase

Separation of both lactoferrin (LF) and lactoperoxidase (LP) will be presented together as they share almost similar molecular weight and the high basic isoelectric. Simultaneous separation of LF and LP has been described in several studies.

Lactoferrin (LF) is an iron-containing protein present in high concentration in colostrum and much less concentrations in normal milk. Variations have been reported for LF content, homology, and glycosylation of LF isolated from the milk of different species (Lőnnerdal and Suzuki, 2013). Bovine LF is a single glycosylated polypeptide chain (689 amino acid residues) of a molecular weight 76–80 kDa depending on the degree of glycosylation. LF contains intramolecular disulfide bridges but no free sulfhydryl groups. LF polypeptide consists of two globular lobes of similar amino acid sequences and linked by a short α-helix peptide containing 10–15 amino acid residues (Baker and Baker, 2005). One iron-binding site and one glycan are present in each lobe of the LF molecule. However, slight differences have been found in the conformation and affinities for iron between the N and C terminals of the two lobes (Lőnnerdal and Suzuki, 2013). Each LF molecule can bind with high affinity

($K_d \sim 10^{-30}$) two ferric ions (Fe^{3+}) with concomitant incorporation of bicarbonate or carbonate ions. The iron containing LF molecule is termed holo-LF while the iron-depleted LF (<5% saturation) is termed apo-LF. The native bovine LF is 20%–30% saturated with respect to iron (Lőnnerdal and Suzuki, 2013). LF is a basic protein that has an isoelectric point of pH 8.7 (Lőnnerdal and Suzuki, 2013) and can interact with molecule of opposite charge.

Pasteurization has no marked effect on the structure or the biological activities of LF, marginal losses of LF occurred in spray-dried milk and apo-LF was more sensitive to heat treatment than holo-LF (Korhonen and Marnila, 2013). Therefore, LF retains its native form in whey.

The growing scientific evidences about the biological activities (Table 12.1) and safety of LF have expanded widely its potential application (Tomita et al., 2009). Toxicological studies suggested that cow milk LF can be regarded as a safe food ingredient (Tamano et al., 2008). The FDA has granted a "generally recognized as safe" (GRAS, GRN 67) status to cow LF for uses at defined levels in beef carcasses, subprimal, and finished cuts (Taylor et al., 2004). Nowadays, LF is used in food preservation, fortification of infant formula, health promoting

Table 12.1: Biological functions of LF (Gonzalez-Chavez et al., 2009).

Function	Target	Mode of Action
Antibacterial	Wide range of Gram-negative and Gram-positive bacteria	Bacteriostatic: depriving bacteria from Fe needed for their growth and activity Bactericidal: interaction with bacterial surface Prevention of the attachment of bacteria to host cell
Antiviral	Wide range of RNA and DNA viruses	Binding to and blocking glycosaminoglycan viral receptors, especially heparan sulfate
Antifungal	*Candida albicans* and *Candida krusei*, body tineas caused by *Trichophyton mentagrophytes*, *Aspergillus fumigatus*	Fe^{3+} sequestration, altering the permeability of the cell surface
Antiparasitic	Intestinal amebiasis caused by *Entamoeba histolytica*, toxoplasmosis caused by *Toxoplasma gondii*, hemoparasites *Babesia caballi*, and *Babesia equi*	Fe^{3+} sequestration, binding the membrane lipids causing membrane disruption and damage of the parasite
Immunomodulatory	Innate and acquired immune systems	Binding negatively charged molecules on the surface cells of the immune system that trigger signaling pathways
Antiinflammatory	Inflammation	Inhibition of proinflammatory cytokines, increases the number of natural killer cells, induces phagocytosis
Anticarcinogenic	Cancer	Inducing apoptosis and arresting tumor growth in vitro. It can also block the transition from G1 to S in the cycle of malignant cells
Enzymatic	Had slight amylase, DNAase, RNAase, and ATPase activities	Unknown

LF, Lactoferrin.

supplements, fish feeds, pharmaceuticals, health care, oral hygiene products, and cosmetics (Korhonen and Marnila, 2013). Also, LF has been used to fortify infant formula products to have a composition that is closer to breast milk (Smithers, 2015). Lactoferrin has been reported to enhance immune function and to reduce allergic reactions that can form the basis for its use in foods, personal care products, and immune support supplements. Clinical studies indicated that LF can be used as a possible treatment or prophylactic for a number of diseases, such as treatment against *Helicobacter pylori,* in which patients supplemented with bovine LF showed greater recovery from infection (Gonzalez-Chavez et al., 2009). LF has shown synergy with antiviral drugs and reduces the minimum inhibitory concentrations of antifungal agents against *C. albicans*. The most resistant strains of pathogens have been the most sensitive to these combinations (Gonzalez-Chavez et al., 2009). Also, LF has potent bone growth stimulation, which may be helpful when combined with other approaches in preventing and/or treating osteoporosis (Smithers, 2015).

Lactoperoxidase (LP, EC 1.11.1.7) is a peroxidase enzyme found in milk whey. LP represents about 0.5% of whey proteins. LP is a basic glycoprotein (isoelectric point 8.0–8.5) consisting of a single polypeptide chain (~80 kDa depending on the glycosylation ratio). LP is a major antimicrobial agent. In the presence of hydrogen peroxide, LP catalyzes the oxidation of thiocyanate to generate potent intermediate oxidants with a wide range of antibacterial activities. The efficacy of LP in inhibiting bacteria associated with gingivitis, reducing inflammation, promoting the healing of gums, and reducing halitosis has been shown from clinical trials (Boots and Floris, 2006). The availability of commercial quantities of LP has allowed for its use in several food and pharmaceutical applications. LP/thiocyanate system is recognized as safe as a natural food preservative, particularly in the dairy industry (FAO, 2006). Also, LP has been used in developing a wide range of health-care products (Boots and Floris, 2006).

The different methods developed and used for the recovery of LF and LP from whey have been based on their molecular structure and properties as follows:

1. *Molecular size.* Lactoferrin and lactoperoxidase have molecular weights of ~80 and ~66 kDa, respectively being higher than the major whey proteins namely: β-LG (36 kDa) and α-LA (14 kDa), but similar to or less than the minor whey proteins BSA (68 kDa) and IgG (150 kDa). Therefore, the capability of membrane processing methods to separate LF and LP is limited to the retention of LF and LP as more concentrated fraction.
2. *Charge.* Both LF and LP have basic isoelectric points (~pH 9), while all other whey proteins have lower isoelectric points (pH ~5.0). This has been the base of separation of LF and LP using ion exchange chromatography.
3. *Captured ion.* Lactoferrin is an iron containing protein. This has been used as a base for magnetic separation of LF from other whey proteins.

2.1 Using Column Chromatography

One of the most effective and widely used groups of techniques for recovery, separation, and purification of proteins are the column chromatographic methods. The interactions between proteins and the stationary phases are very complicated. Protein interaction with the solid adsorbent occurs with different strengths at more than one site on proteins' surfaces. On the other hand, the adsorbent carries a heterogeneous array of unevenly distributed binding sites. Thus, no single parameter can describe adequately the protein-matrix interaction except in highly specific affinity sites. Several materials can be used as stationary phases for the separation of proteins. The nature of these materials determines whether proteins are separated by difference in size, charge, or binding affinity.

2.1.1 Ion exchange chromatography

Ion exchange is the most popular chromatographic methods for the separation and purification of proteins. Ion exchangers are divided into two groups namely: anion exchangers and cation exchangers depending on the attached groups to the matrices. Both groups are subdivided into strong and weak exchanger depending on the pK_a of its attached charged group. Ion exchange separation is based on the ionic interaction of proteins with charged surface groups of the ion exchanger. Proteins are complex ampholytes carrying both negative and positive charges and the net charge of the protein can be controlled by changing the pH of surrounding media. This means that proteins can be separated on both anion and cation exchangers by selecting the suitable pH of the used buffer. To achieve good separation the pH of the used buffer should be at least one pH unit above or below the isoelectric point of the separated protein.

Separation of LF and LP from other whey proteins by ion exchange chromatography is based on the differences between their isoelectric points. All whey proteins have acidic isoelectric point except LF and LP, which have basic isoelectric point (~pH 9). Ion exchanger chromatography is the most frequently method used for extraction and purification of LF at industrial level using concentrated whey protein solution and, in some cases, diafiltered whey. In this way, yields between 50% and 96% LF can be obtained.

Lactoperoxidase and LF have been isolated from acid whey of bovine milk by chromatography on carboxy methyl-toyopearl cation exchange column where LP and LF were the only proteins retained on the column (Yoshida and Xiuyun, 1991). The two proteins were eluted from the column using phosphate buffer (0.05 M pH 7.7) and a linear NaCl gradient (0.0–0.55 M) where LP was recovered (91.1%) between 0.1 and 0.15 M NaCl, and LF between 0.4 and 0.55 M NaCl.

Lactoperoxidase and LF were recovered from rennet whey by passing through a strong cation exchange (SP-TP) column, followed by elution with phosphate buffer (0.05 M pH 6.5 and 0.16–0.19 M NaCl) to extract LP and with the same buffer but containing 0.42–0.55 M NaCl to extract LF (Ye et al., 2000). Cation exchanger SP-Sepharose Big Beads was used for direct recovery of LP and LF from raw milk (Fee and Chand, 2006). The performance of the column was not affected by the presence of fat in the raw milk. Approximately, 100-column volume can be loaded from raw milk before major leakage can be obtained from LF in the effluent.

Development of a model-based method to determine steric mass action (SMA) adsorption isotherm parameters of LP and LF was carried out using SP-Sepharose FF resin (Faraji et al., 2015). The equilibrium and dynamic binding capacity of the two proteins were determined, and adsorption of binary mixture of pure LP and LF was done to calibrate the steric mass action (SMA) model. The calibrated model was verified for various process operating conditions. The model parameters were evaluated using whey protein isolates (WPI) in real condition.

Reasonable agreement was found between the experimental results and the model predicted results.

Cation exchange composite cryogel embedded with macroporous cellulose beads has been prepared and used for the recovery of LP from cheese whey (Pan et al., 2015). Maximum recovery (92%) of highly purified LP (98.0%–99.8%) was obtained by capture of LP from whey using the prepared beads at pH 5.8, followed stepwise elution using 10 mM phosphate buffer containing 0.075 M NaCl, 0.15 M, and 1 M NaCl, respectively.

Lactoferrin and Ig G were effectively separated in high purity from crude whey using two sequential expanded beds (Du et al., 2014): the first was packed with a cationic exchanger (Fastline SP) and the second was packed with an anion exchanger (Streamline Direct CST-1). Lactoferrin was recovered (77.1%) at high purity (88.5%) from the first bed using phosphate buffer pH 7 containing 0.5 M NaCl, while immunoglobulin G was recovered (14.3%–63.7%) with high purity of 83.8%–92.4% from the second bed using 0.02 M sodium bicarbonate buffer containing 0.5 M NaCl.

2.1.2 Adsorption chromatography

Ng and Yoshitake (2010) described a simple method for separation of high-purity lactoferrin from whey in a single step. They used a commercial unit of mixed-mode chromatography–ceramic hydroxyapatite (CH). The column was first equilibrated with phosphate buffer 0.01 M, pH 6, whey was added and unadsorbed proteins were removed by the equilibration buffer. Both LP and LF were recovered by respective isocratic elution with phosphate buffer, 0.01 M, pH 7, and 0.4 M NaCl, and 0.6 M NaCl. Lactoferrin of high purity and free from LP was obtained.

2.1.3 Affinity chromatography

Affinity chromatography (AA) is based on the specific interactions of proteins with other molecules termed ligands. In AA, ligands are covalently bound to an inert chromatographic matrix. The fluid is passed through the AA column under conditions that permit specific and reversible binding of the target protein to the ligand while other constituents are washed away. The target protein is then recovered from the column by elution under conditions that favor the breakdown protein-ligand interaction.

Lactoferrin was recovered from colostrum and cheese whey by selective adsorption on heparin attached Sepharose (Al-Mashikhi and Naki, 1987) and then eluted with veronal-HCl buffer (5 mM pH 7.4) containing 0.5 M NaCl. The purity of the separated LF was confirmed by SDS-PAGE and immune elecrophoresis, but no quantitative estimate for the yield and purity were given.

Antibodies were separated from the yolk of hens immunized against LF and then immobilized on agarose beads. Affinity chromatography column was packed by the prepared antibody-agarose beads (Li-Chen et al., 1998). High-binding capacity of LF (80%) was achieved by the prepared column from phosphate buffer (0.01 M, pH 6). The retained LF can be recovered by elution with the same buffer but containing 0.5 NaCl.

The continuous phase chromatographic columns (cryogels) have been developed to overcome the limitation of the conventional column chromatography. The interconnected large-dimension pore structure of the column bed allows for purely convective flow through the pores and low resistance to mass transfer. Super macroporous ion affinity chromatography (MIAC) was prepared by loading polyacrylamide cryogels with immobilized Cu^{++} through binding to iminodiacetic acid (IDA). The thermodynamic parameters for adsorption affinity of LF on cryogel with immobilized Cu^{2+} ions were measured to maximize the recovery of LF from whey (Carvalho et al., 2013). The adsorption of the protein in the matrix was found to be spontaneous, entropy- and enthalpy-favored, and entropy-driven, and more tight binding occurred at higher temperature. The prepared MIAC has been used as an adsorbent for direct capture of LF from cheese whey (Carvalho et al., 2014). The high retention of LF by MIAC was attributed to the presence of two histidine residues in the primary structure of LF. The purity of the obtained LF was confirmed.

The triazine dye Yellow HE-4R has been immobilized on Sepharose 4B to act as pseudo bioaffinity ligand to LP (Baieli et al., 2014a). The prepared beads were soaked in defatted whey, stirred, washed with 0.5 M NaCl solution, and LF was extracted using 25% ethylene glycol solution containing 2 M NaCl. The yield of the recovered LF amounted to 71% of the original LF content in whey with 61% purity. The method was further developed (Baieli et al., 2014b) using the same dye but immobilized on chitosan microsphere (1.7 μm) for LF adsorption from defatted whey and subsequent recovery from the beads as described before. A yield of 77% with purity of >90 from whey LF was reported to achieve from one purification step, and that the beads can be used for three successive cycles without the need for regeneration.

Mercaptoethyl pyridine (MEP) HyperCel and phenyl propyl amine (PPA) HyperCel are two industrially chromatographic sorbents that have hydrophobic and ionic interactions capability.

They were tested for the recovery of IgG and LF from colostrum whey (Ravichandran et al., 2015). The highest recovery of LF (~91%) with 2.9-fold rise in purity was achieved using MEP HyperCel and 0.05 M phosphate.

Recently, different types of temperature-responsive agarose-based ion exchange resins have been developed (Maharjan et al., 2016). Under dynamic loading at 40°C, LF was highly retained (94%), which can be then recovered from the resin by lowering the temperature to 4°C without changing the buffer. Under these experimental conditions, β-LG was not retained by the resin. The method has been developed and tested using solutions of pure protein fractions. However, it can be considered a new and simple method for the recovery of LF from whey under mild conditions.

Despite the numerous methods developed for recovery of LF and LP from whey, conventional chromatographic processes have several disadvantages for fast and reliable recovery of LF and LP from whey. The large volume needed and the high protein concentrations of whey are responsible for many problems, such as fouling of chromatographic material, long cycle times, high pressure drops via chromatographic columns and complicated process control systems.

2.2 Using Membrane Filtration Coupled With Chromatography

Lactoferrin and IgG were recovered from cheese whey by binding to an affinity matrice (heparin-Sepharose, protein G-Sepharose, and protein G-bearing *Streptococcal* cells), and concentrated by microfiltration and diafiltration (Chen and Wang, 1991). Lactoferrin and IgG were recovered from the whey retentate by dilution with 5 mM veronal-HCl buffer (pH 7.4, containing 0.6 M NaCl) for recovery of LF, and 0.5 M glycine-HCl (pH 2.7) for recovery of IgG, and microfiltration. The prepared LF had 95% purity while the purity of the prepared IgG was 90% and 86% activity. Colostrum whey was first ultrafiltered using a membrane of 100 kDa MWCO and the obtained permeate was subjected to second ultrafiltration using membrane of 10 MWCO (Lu et al., 2007). The retentate from the second UF step contained the crude LF and was purified by loading on a cation exchange (SP-Sepharose fast flow), washing with phosphate buffer (pH 7, 0.05 M), followed by elution of retained LF with phosphate buffer containing 0.1 M NaCl. The eluted LF was desalted and concentrated by UF using 10 kDa membrane followed by freeze drying to give a product of high purity (>90%).

Lactoperoxidase was separated from whey using two different integrated processes, namely: partitioning between two aqueous phases and ultrasound-assisted ultrafiltration process for the purification and concentration of the extracted LP (Nandini and Rastogi, 2011). The two-phase system consisting of polyethylene glycol 6000 and potassium phosphate gave the maximum LP activity recovery of 150.70%, leading to 2.31-fold purity. Subsequent use of ultrafiltration increased the recovered activity and purification factor by 149.85% and 3.53-fold, respectively. The use of ultrasound during ultrafiltration increased the flux, but had no significant effect on activity recovery and purification factor of LP (Nandini and Rastogi, 2011).

2.3 Using Electrically Enhanced Microfiltration

Application of electrical field during microfiltration of whey protein solution was found to affect the transmission of proteins through the membrane (Brisson et al., 2007). When the cathode was placed on the retentate side and the electric field was set to 3333 V/m, the separation factors between LF and the two major whey proteins—β-LG and α-LA—were 3.0 and 9.1, respectively. Iron saturation of LF was found to affect the separation factor between LF and the major whey proteins. Also, under these conditions the permeation rate increased three times.

2.4 Using Simulated Moving Bed

The simulated moving bed (SMB) technology is a large-scale separation method using a continuous series of chromatographic columns simulating a countercurrent separation. The target analyte is partitioned between the solid and the liquid phase, of a series of columns or column movement in a carousel operated by valve switching. This method offers the advantages of higher productivity, smaller separation plants, higher product concentration, reduced buffer consumption, and more efficient use of raw material and higher target. Andersson and Mattiasson (2006) described a simplified SMB based on the use of 20 columns packed with cation exchanger (Streamline-SP) for the separation of LF and LP from whey protein concentrate. They reported an increase of 48% in productivity, a 4.3-times decrease in buffer consumption, and a 6.5-fold increase in target protein compared to conventional column chromatographic separation.

2.5 Using Membrane Chromatography

Membrane adsorber technology is a hybrid between chromatographic separation and membrane filtration process (Etzel and Arunkumar, 2015). Membrane adsorbers can be prepared in two ways. The first and most commercially applied is based on the attachment of functional groups to the inner surface of the microporous membranes. Thus ion exchange, affinity, and metal-chelating absorbers can be developed depending on the functional groups attached to the porous membrane. The second group of membrane adsorber is prepared by mixing fine ion exchanger beads in the polymer before casting the membrane. Commercial membrane adsorber systems are now available in different scales and functional groups. The nondiffusion limited adsorption can be considered as the main advantage of the membrane adsorber systems, which allow for high process fluxes.

A commercial cation exchange membrane (Sartobind membrane adsorber) was used for the recovery of LP and LF from cheese whey (Chiu and Etzel, 1997). The membrane was made of microporous (3–5 μm) cross-linked regenerated cellulose charged with $R—CH_2—SO_3$ moieties covalently bonded to its internal pore surfaces. Recovery of LP and LF was unaffected by the whey volume used up to eightfold or the number of processing cycles up to 12 cycles without the need for cleaning. The recoveries of LP and LF were 73.56% and 50.55%, respectively.

Lactoferrin and LP have been separated on both laboratory and large scale using a commercial strong cation exchange membrane adsorber from sweet cheese whey (Plate et al., 2006). The system was tested in a quasicontinuous mode in a twin module setup. The system was operated for eight cycles without the need for washing and 88% efficiency was obtained for the recovery of LF/cycle in the large-scale ($2 m^2$ membrane surface) experiment (Plate et al., 2006). The system was also used for the recovery of lactoferricin (potent antimicrobial peptide derived from LF) from pepsin treated whey.

Affinity hollow fiber membranes with high LF adsorption capacity were synthesized using Red HE-3B dye as legand. The prepared membrane had a maximum capacity of 111.0 mg LF/mL membrane (Wolman et al., 2007). In a batch system 91% of LF was retained by the membrane from whey and colostrum. A solution of 2 M NaCl in 25% ethylene glycol achieved 99% recovery of LF retained by the membrane with a purity of 94%. Compared with the Red HE-3B dye-agarose beads, the hollow fiber membrane showed better performance.

Saufi and Fee (2011a) prepared mixed matrix membrane (MMM) for recovery and fractionation of proteins from whey. The prepared MMM was based on the incorporation of anion and cation exchange resin in the membrane polymer solution before casting of the membrane. The ratio of the used anion (42.5%) and cation (7.5%) exchanger were determined relative to the relative contents of the acidic and basic proteins in whey. Whey was run through the membrane at pH 6, where proteins were retained and then recovered by elution with 1 M NaCl solution. Gradient elution suggested that fractionation of whey proteins can be achieved using MMM. A cationic mixed matrix membrane (MMM) of high affinity to LF was prepared by inclusion of grounded SP-Sepharose beads into the ethylene vinyl alcohol polymer used as the membrane base (Saufi and Fee, 2011b). The system was operated in cross-flow to enhance LF binding and minimizing membrane fouling. This resulted in high recovery (91%) of high purity LF from whey.

Voswinkel and Kulozik (2011, 2014) developed a method for the isolation and fractionation of β-LG, α-LA, BSA, IgG, LF, and LP on laboratory and pilot scales using Sartobind membranes. The process consisted of two steps. In the first step, whey was percolated through an anion exchanger where β-LG and BSA were retained, and the filtrate was adjusted to pH 4.6 and then eluted through a cation exchanger in the second step, in which LF, LP, and IgG were retained, leaving α-LA to pass in the filtrate. Proteins retained in the anion membrane were eluted using 0.01 phosphate buffer, pH 7, and 0.1 M NaCl for BSA, and 1 M NaCl for β-LG. IgG, LP, and LF were recovered from the cation exchanger using 0.01 acetate buffer containing 0.25, 0.35, and 1.0 M NaCl, respectively. Both laboratory and pilot scales gave comparable results with yields ranging from 80% to 97% except for LF, which gave a lower yield.

Factors affecting the separation of a binary mixture of LF and BSA using unmodified, and negatively and positively charged UF membrane were studied (Virginia Valin˘o et al., 2014). The negatively charged membrane was found to retain BSA completely at pH 9, while LF yielded the maximum permeation. Lactoferrin was rapidly and effectively separated from BSA using commercial membrane chromatography systems (Sartobind Q75 and S75). Complete separation of LF from BSA was achieved using Sartobind S75 and 5 mM phosphate buffer pH 6 as eluent while BSA was completely retained by the membrane (Teepakorn et al., 2015).

2.6 Using Molecularly Imprinted Polymer

Molecularly imprinted polymers (MIPs) are new synthetic materials able to recognize the target molecules in a similar way present in the living systems. Therefore, MIPs possess high selectivity and affinity for the target molecule (Jiménez-Guzmána et al., 2014). MIP was prepared using vinyl pyridine as functional monomer and pure LF as template to create a specific cavity in the polymer. The prepared MIP was selectively able to retain 34.5% of LF from a protein mixture rich in LF (81.9%) of total proteins. This study indicated that the retention of LF was selective in the cavity formed by the template.

2.7 Using Magnetic Micro Ion-Exchanger

This method is based on two steps: in the first step, the target species is captured by a functionalized magnetic support, and in the second step, the product-loaded sorbent is separated using high-gradient magnetic separation (HGMS) technology and the product was then eluted from the separated sorbent (Meyer et al., 2007). The recovery of LF from crude whey was carried out in three cycles to yield an average purification factor of 18.6, together with a concentration factor of 1.7 and product yield of about 47%. This process has the advantage of very rapid adsorption kinetics and used particles are less prone to fouling and can be easily cleaned.

Heparin-microsuper paramagnetic polyglycidyl methacrylate particles (PGMA-heparin) were prepared by polymerization (Chen et al., 2007). LF can be recovered from acid whey using PGMA-heparin particles with high effectiveness. The maximum binding capacity of LF was 164 mg/g.

A polyacrylic acid-fimbriated magnetic cation exchange supports was prepared. This support was characterized by ultrahigh binding capacity for basic proteins (Brown et al., 2013). This support coupled with a new rotor-stator was used to recover LF from crude sweet whey. The system can be operated for five cycles without any loss in efficiency with LF yield of 40%–49% with respective purification and concentration factors of 37- to 46-folds and 1.3- to 1.6-folds.

2.8 Using Electrodialysis

Whey enriched with LF was subjected to electrodialysis at pH 3, 4, and 5 using 500 kDa membrane (N'diaye et al., 2010). The highest migration yield (15%) of LF was obtained at pH 3, which can furnish an effective method for the recovery of LF from solutions. However, they concluded that difficulty in cleaning of the membrane should be solved before the method can be applied on industrial scale.

2.9 Using Hydrophobic Ionic Liquid

Lactoferrin was extracted with imidazolium-based ionic liquids as a novel separation method (Alvarez-Guerra and Irabien, 2012). Enhanced extraction efficiency (EE) of LF from its mixture with BSA was achieved at neutral pH, low ionic strength, and low concentration. However, the highest EE obtained was 20%. In addition, the used base may suffer from hydrolysis. Therefore, another system based on the use of 1-butyl-3-methylimidazolium trifluoro-methane sulfonate (BmimTfO)/phosphate was selected, which had the advantage of high LF retention (83%–99%) at the liquid–liquid interface and chemical stability (Alvarez-Guerra and Irabien, 2015). The highest LF recovery was obtained at pH 3.1–4.0 and low protein concentration.

2.10 Colloidal Gas Aphrons

This method is based on the partitioning of proteins between surfactant-stabilized microbubbles and liquid phase. The surfactant stabilized microbubbles are termed colloidal gas aphrons (CGAs), which are formed by intense stirring (8000 rpm/10 min) of anionic surfactant solution. Fuda et al. (2004) added the anionic surfactant AOT to whey and stirred to form the CGAs and to partition the whey proteins between CGAs and the remaining liquid. They evaluated the effect of pH, ionic strength, and surfactant/whey ratio on the recovery of LF and LP. They found that the LF and LP contents were 25-fold higher in the CGAs than in the liquid at pH 4 and 0.1 ionic strength.

3 Casein Macropeptide

The casein macropeptide (CMP) is a heterogeneous mixture of glycosylated and nonglycosylated peptides released by the action of chymosin on κ-casein during cheese manufacture and found in cheese whey (Abd El-Salam et al., 1996). All CMPs have the same amino acid sequence but differ in the degree of glycosylation (Manso and López-Fandiño, 2004). The carbohydrate-free CMP is a large peptide (69 amino acid residue) of unique amino acid composition being free of aromatic amino acid with special reference to Phe (Manso and López-Fandiño, 2004). Therefore, CMP is a suitable protein source for subjects suffering from phenylketonuria (Brody, 2000). Also, CMP is rich in branched chain amino acid and low in methionine suggesting that it can be beneficial for subjects suffering

Table 12.2: Physiological and biological activities of CMP (Abd El-Salam et al., 1996; Brody, 2000; Kvistgaard et al., 2014; Manso and López-Fandiño, 2004).

Activity	Description
Antimicrobial	• Carbohydrate mediated interaction with toxins, viruses, and pathogens • Inhibits the adhesion of cariogenic bacteria to the oral cavity • Growth inhibition against *S. mutans*, and *Escherichia coli* of the nonglycosylated monophosphorylated CMP
Immunomodulatory	• Inhibits mitogens from inducing the proliferation of spleen lymphocytes • Suppressing activity toward the production of IgG antibodies
Regulation of gut functions	• Growth-promoting activity on *Bifidobacterium* genus (healthy gut microflora) • Stimulates the release of CCK, the satiety hormone • Inhibits gastric secretions and slows down stomach contractions
Antithrombotic	• Several peptides derived from the sequence 106–116 of bovine κ-casein inhibit platelet aggregation and thrombosis.
ACE-inhibitory	Cow-, sheep-, and goat-derived CMPs have moderate activity

CCK, Cholecystokinin; CMPs, casein macropeptide.

from hepatic diseases. Wang et al. (2007) suggested that the sialic acid content of CMP may be beneficial in brain development and improving learning ability. CMP was reported to have several biological effects that can be summarized (Table 12.2)

CMP represents ~20% of whey proteins concentrates (WPC) prepared from sweet whey (Regester and Smithers, 1991). The presence of CMP in WPC affects its functional properties (Svanborg et al., 2016). Removal of CMP from WPC preparation can improve greatly the functional properties of CMP-free WPC products.

The different methods used in the separation and recovery of CMP from cheese whey have been based on the molecular structure and properties of CMP:

1. *Charge.* CMP is a mixture of acidic peptides without a single isoelectric point due to the variable glycosylation of its components. Glycosylated and nonglycosylated CMP were reported to have isoelectric points of 3.15 and 4.15, respectively (Kreuβ et al., 2009). This means that CMPs have negative charges at pH 7.0 while all other whey proteins carry positive charges under the same condition. This has been used as the base for the separation and recovery of CMP from whey using ion exchangers. Also, it has been used as a base for recovery of CMP from whey using cationic biopolymer, such as chitosan.
2. *Molecular size.* The theoretical molecular weight (MW) of the carbohydrate free bovine CMP peptide ranges from 6755 and 6787 Da according to the parent genetic variant of κ-casein. However, the apparent MW of CMP in whey is much larger than the theoretical one. The carbohydrate moiety of the CMP adds the CMP monomer up to 36.4% of its MW at the maximum degree of glycosylation (Kreuβ et al., 2009). It has been suggested that the MW of monomeric CMP ranges from 7 to 11 kDa depending on the degree of glycosylation. However, the reported MW of CMP is three to four times to the theoretical MW of the monomeric CMP, which has been explained by the formation of aggregates

(Farıias et al., 2010). Mikkelsen et al. (2005) demonstrated the molecular size of CMP to be pH dependent due to the aggregation of the monomeric CMP. The formed CMP aggregates resisted dissociation after their formation. On the other hand, changes in pH were reported (Kawasaki et al., 1993) to affect the molecular size of CMP, which has been used as the basis of separation of CMP by molecular exclusion chromatography and membrane filtration.

3. *Heat stability.* CMP remains in solution after high heat treatment of whey while all other whey proteins tend to form large aggregates that can be removed by filtration.

The different methods used in the separation and recovery of CMP can be classified according to the basis of separation as follows:

3.1 Methods Based on Removal of Whey Proteins by Heat Treatments

The differential heat stability of CMP and other whey proteins has been taken as the base of several methods developed for the preparation of CMP. Berrocal and Neeser (1991) patented a process for preparing CMP by heat treatment. A 10% solution of whey protein concentrate was heated (90°C/20 min) at pH 6.0, removing the curd, concentrating the remaining aqueous phase by ultrafiltration. Ethanol was then added to the retentate and pH was adjusted to 4.5 to remove the residual protein contaminant and the remaining solution was freeze dried. The product was reported to contain 84% CMP. Saito et al. (1991) heated 10% reconstituted whey powder (pH 6) at 98°C/60 min. Ethanol was added to the heated whey solution at the ratio of 1:1 and the formed precipitate was removed by centrifugation. CMP was recovered from the supernatant by adsorption on DEAE-Toyoperarl column and then eluted by ammonium bicarbonate solution and then dried. They reported a yield of 1.1 g CMP/100 g whey powder. Nielsen and Trombolt (1994) heated (90°C/15 min) reconstituted WPC solution (8% protein), cooled to 50°C, adjusted to pH 4.5, and the formed curd was removed by filtration. The residual proteins were removed from the filtrate by ultrafiltration using 100 kDa MW cut-off membrane and permeate containing CMP was concentrated by reverse osmosis (RO) membrane and spray dried. Martin-Diana and Fontecha (2002) utilized the high thermal stability of CMP to separate it from whey proteins. The described method involves complete denaturation and aggregation of whey proteins by the heat treatment of a rennet whey or WPC at 90°C for 1 h. The denatured proteins were then removed by centrifugation at 5200 g and 4°C for 15 min and the pH of the supernatant was adjusted to 7.0 and ultrafiltered using 10 kDa membrane. Rojas and Torres (2013) heated (90°C/60 min) sweet whey and the formed coagulum was removed by centrifugation. High percentage of CMP (34.08%) was recovered in the supernatant.

3.2 Methods Based on Membrane Ultrafiltration

Tanimoto et al. (1991) patented a process for the isolation of CMP from whey by adjusting the pH of whey to <4, followed by ultrafiltration using a membrane with a MWCO of 10–50 kDa followed by concentration of the permeate using a membrane of MWCO <50 kDa where the CMP retained in the concentrate. However, this method suffered from the low recovery of retained GMP. The permeability of CMP during ultrafiltration at different pH values through membrane filter was found to be pH dependent (Kawasaki et al., 1993). At pH 6.5, CMP was found to be retained by UF membrane filter of MWCO of 50 kDa or 20 kDa, but it passed through these membranes when the pH was adjusted to 3.5. Therefore, they described a method for the recovery of CMP, based on two steps of membrane filtration. In the first step, the whey protein concentrate solution (2%) was adjusted to pH 3.5 and then ultrafiltered using 50 kDa membrane and in the second step the pH of the obtained permeate was adjusted to pH 7.0 ultrafiltered using 20 kDa membrane and then diafiltered where CMP was retained by the membrane filter. The obtained CMP was reported to have a purity of ~80%.

Chatterton and Holst (2002) patented a process for the preparation of CMP from a solution (7.5%) of whey protein concentrate (WPC). The pH of the WPC solution was adjusted to 3.0 and ultrafiltration at <15°C using spiral-wound UF element having a membrane (20 kDa) coated with a suspension of calcium phosphate. They claimed that coating with calcium phosphate doubled the permeability of CMP. The pH of the permeate was adjusted to 6.7 and ultrafiltered and diafiltered using 5 kDa MWCO membrane. The retentate was then spray dried to give a product containing 70.9% protein, of which CMP represented 90%.

CMP was separated from other whey proteins by enzymatic treatment of whey with transglutaminase (TGase) followed by microfiltration (Tolkach and Kulozik, 2005). The open structure and the presence of glutamic and lysine residues in CMP composition allowed for the cross-linking of CMP when treated with TGase, while the globular structure of other whey proteins made them resistant to TGase cross-linking. Microfiltration of TGase-treated whey allowed for the retention of cross-linked CMP in the retentate, while other whey proteins were recovered in the permeate.

Whey protein isolates (WPI) solution was subjected to supercritical carbon dioxide treatment at 70°C to precipitate α-LA, which was then removed by centrifugation (Bonnaillie et al., 2014). The supernatant was then ultrafiltered using 30 kDa membrane. About 40% of the original CMP content in the WPI solution was recovered in the permeate with high purity (~94%).

3.3 Methods Based on Ion Exchange Chromatography

Whey protein isolates (WPI) and CMP were simultaneously recovered from cheese whey by first passing the whey through a cation exchange column followed by passing the effluent through an anion exchange column (Doultani et al., 2003). Nearly all the major whey proteins were retained on the cation exchange column, where they can be recovered in the form of WPI. The proeinaceous material recovered from the anion exchange column was almost free from other whey proteins and contained mainly CMP. About half the CMP in whey can be recovered by this process in almost pure form.

Etzel (2001) described a method for the preparation of high purity CMP based on two successive anion exchanges and metal affinity chromatography. In the first step, filtered whey was adjusted to pH ~5, passed through quaternary amino ethyl cellulose column operated at 40°C, where CMP was retained on the column while other whey constituents were found in the elute. CMP was then recovered by elution with 0.5 M NaCl. The elute was then adjusted to pH 7.15 and chromatographed on imino diacetic agarose beads immobilized with Cu^{++} ion. The elute was reported to contain substantially purified CMP.

Ayer et al. (2003) obtained highly purified CMP from 10% solution of whey protein concentrate (80% protein) using three successive steps of ion exchange chromatography. The WPC solution was adjusted to pH 4.9, stirred with anion exchange beads (QA GibcoCel), filtered and washed. Crude CMP was extracted from the washed beads at pH 2 and 100 mM NaCl, and then adjusted to 200 mM NaCl, stirred with cation exchange beads (SP GibcoCel), and then filtered. The filtrate was then chromatographed on anion exchange (QA GibcoCel) at pH 8.5. The obtained CMP was claimed to contain less than 0.5 mg/g phenylalanine.

The recovery and purity of the recovered CMP from mozzarella cheese whey using anion exchange chromatography were affected by operating conditions (Tek et al., 2005). Decreasing the conductivity and increasing the pH of whey increased substantially the recovery of CMP, while increasing the pH only from 4.0 to 6.0 decreased the purity of the recovered CMP. A minimum of 0.1 M NaCl was needed for complete recovery of the bound CMP from the column.

3.4 Methods Based on Combined Membrane Filtration and Ion Exchange Chromatography

Tanimoto et al. (1992), coagulated a casein solution with rennet, and the resultant whey was concentrated and desalted using RO and the concentrate (crude CMP) was freeze dried. The crude CMP was then purified using Q-Sepharose ion exchange chromatography.

A method has been described to prepare CMP low in Phe from commercially available CMP (LaClair et al., 2009). A solution of the commercial CMP was passed through a cation exchange resin column where the contaminated proteins were retained and the pure CMP was recovered and concentrated from the effluent by UF/DF using 3 kDa hollow-fiber membrane and then lyophilized.

3.5 Methods Based on the Use of Membrane Ion Exchanger

Kreuβ et al. (2008) compared the membrane ion exchange (MAC) with the conventional strong and weak ion exchanger columns in the separation of glycosylated CMP from the nonglycosylated CMP on a laboratory scale. Separation of the glycosylated and carbohydrate free CMPs was best achieved at pH 4.0–4.1 using 0.02 M sodium acetate buffer. They concluded that the MAC offered the advantage of being faster (four times) in comparison to the conventional ion exchanger chromatography. With the use of the MAC method it was possible to separate the glycosylated and the nonglycosylated CMPs in pure forms. A pilot scale separation of glycosylated CMP was carried out using MAC (Kreuβ and Kulozik, 2009). The use of commercial CMP preparation gave high binding (0.28 mg/cm^2) and purity (91%) of glycosylated CMP. However, the use of whey as a starting material gave lower binding and purity of the recovered glycosylated CMP, which they attributed to the presence of membrane foulants in whey. The use of ultrafiltered/diafiltered whey improved MAC performance during separation.

3.6 Methods Based on Binding to Basic Biopolymer

Chitosan was used to remove the glycosylated CMP under acidic conditions (Casal et al., 2005). At pH 5, glycosylated CMP was completely recovered from whey using 0.08 mg/mL of chitosan while 70% of the nonglycosylated CMP remained in solution. At higher pH (6.0, 6.6) more chitosan (0.19 and 0.34 mg/mL) was needed to ensure complete recovery of the glycosylated CMP from whey.

Modified chitosan beads of high adsorption affinity toward CMP were prepared (Li et al., 2010) by immobilization of β-cylodextrin on chitosan using 1,6-hexamethylene diisocyanate, CMP was selectively (90.27%) recovered from whey protein solution or from whey at pH 3. CMP was extracted from the modified chitosan beads using 1 M NaCl. The modified chitosan beads can be used repeatedly without loss in their capacity or selectivity.

4 Bioactive Peptides

Bioactive peptides are single or multi-functional peptides of specific amino acid sequence encrypted in the primary structure of natural proteins. The bioactivity of these peptides is inherent in their amino acid sequence and they can exhibit several physiological and biological effects, such as antihypertensive, antithrombotic, antimicrobial, antioxidative, immunomodulatory, and opioid molecules (Bazinet and Firdaous, 2009). They remain latent until they are released from proteins by enzymatic hydrolysis or microbial fermentation. Generally, the protein hydrolysate contains a mixture of biologically active and inactive peptides in addition to the undigested protein. Different methods can be followed to remove the undigested protein depending on the protein nature and its physicochemical properties while the mixture of the released peptides remains in solution (Fig. 12.1). Further fractionation of the peptide mixture is needed to obtain fractions of more homogeneous and concentrated bioactivities.

Despite the several attempts to produce bioactive peptides through direct chemical synthesis or using recombinant DNA technology, isolation of bioactive peptides from natural sources remains a cost effective and easy way to scale their production. However, bioprocessing

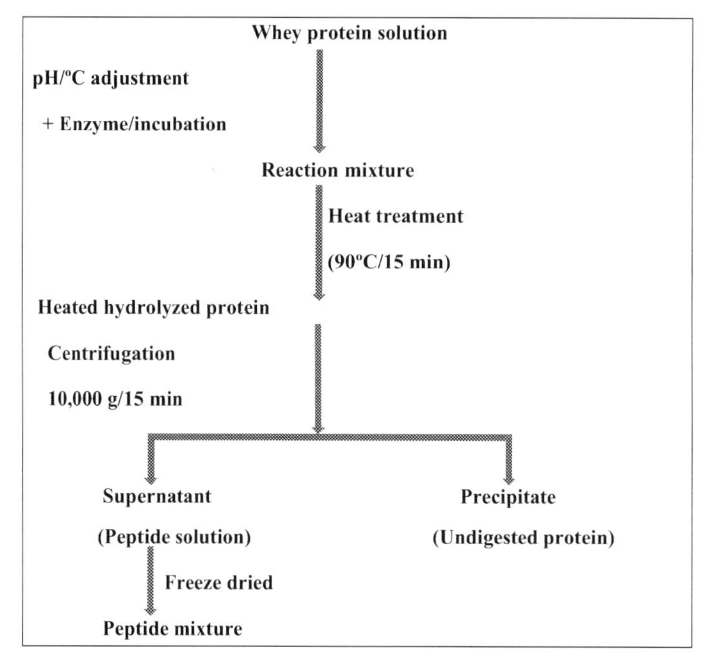

Figure 12.1: Preparation of Whey Protein Hydrolysate and Peptide Mixture.

of peptides remains a challenge due to the lack of commercial viable process for large-scale production of bioactive peptides. Each of the existing methods for the separation and purification of bioactive peptides has its advantages and limitations (Agyei et al., 2016).

Hydrolysis of whey proteins by digestive or microbial enzymes or microbial fermentation of whey has been reported to be important sources of bioactive peptides (Korhonen and Marnila, 2013). Preparation of bioactive peptides from the enzymatic protein hydrolysates has been preferred to the use of bacterial fermentation, as the enzymatic hydrolysis is more easily controlled. The type of the enzyme used and conditions of hydrolysis affects the hydrolysis and the bioactive peptides generated from whey proteins (Cheison et al., 2010; van der Ven et al., 2002). Hydrolysis of whey proteins under controlled and noncontrolled pH generated hydolysates of different activities (Le Maux et al., 2016). However, isolation of peptide fractions of a specific function from these protein hydrolysates represents a challenge.

4.1 Membrane Separation

Membrane separation processes (MSP) play an important role in the isolation and fractionation of bioactive peptides from protein hydrolysates. MSP include a wide range of processes that share the use of membranes as a basic element of these processes, but they differ in the driving forces used in separation. The MSP are classified in the following sections, according to the used driving forces.

4.1.1 Pressure-driven filtration

In the cross-filtration methods, the treated fluid is pumped along a fixed membrane in a flat, spiral wound, tubular, or hollow fiber modules. The occurring molecules are then fractionated according to their sizes and shapes by the used membranes, which can be asymmetric organic or inorganic porous or composite membrane with different pore sizes. Several advantages have been reported for these methods. The feed effluent is separated into two streams in cross-flow mode: (1) a retentate/concentrate containing the retained molecules, and (2) a permeate containing molecules that can pass the membrane pores. The cross-flow allows for the self-cleaning of the membrane surface from formed deposits and continuous nonstop operation for long time. Separation is carried out without phase changes and, therefore, low energy is needed for separation. Moreover, the applied mechanical separation offers mild conditions without any effect on the separated molecules. However, apparent gradual pressure drop occurs along the membrane which in turn reduces the transmembrane pressure (TMP) in the downstream part of the module and the average permeate flux. The pressure used in pressure driven membrane separation methods is inversely related to the membrane pore size.

The pressure-driven membrane separation methods are classified according to molecular weight of retained molecules into microfiltration (MF), ultrafiltration (UF), nanofiltration (NF), and reverse osmosis (RO). Only UF and NF are used in the separation, concentration, and fractionation of the bioactive peptides:

4.1.2 Ultrafiltration

In ultrafiltration, membranes have molecular weight cut-off (MWCO) of 10^3–10^6 Da. They are effectively used mainly to separate colloidal particles, such as proteins and large peptides. Ultrafiltration has been used extensively in the separation and fractionation of bioactive peptides from whey proteins in different ways (Muro et al., 2013), as follows:

- To separate the released peptide from high molecular weight molecules found in the protein hydrolysates. Membranes of high MWCO >20 kDa are used for this purpose.
- To fractionate of <7 kDa peptides using membranes of low MWCO <10 kDa. The peptides separated from WPH by UF through 5 and 2 kDa membranes along with the hydrophilic fraction isolated by solid-phase extraction were significantly more potent DPP-IV inhibitors than WPH (Nongonierma and FitzGerald, 2013).
- Combined hydrolysis and separation of released peptides using membrane bioreactor involving UF membranes in combination with nanofiltration (Fig. 12.2).

The first step of this process is the removal of high molecular proteinaceous material from the whey protein hydrolysates followed by fractionation of the low molecular weight peptides <1 kDa by using NF (Butylina et al., 2006). The size of peptides is not the only factor controlling their separation with UF/NF processes. Transmission of peptides through 5 kDa UF membrane has been reported to be governed mainly by charge mechanisms and reached

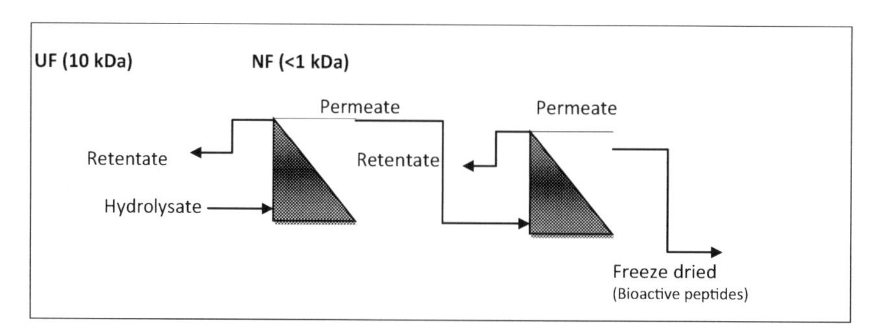

Figure 12.2: Ultrafiltration/Nanofiltration Separation of Bioactive Peptides From WP Hydrolysates.

its maximum value at the isoelectric point of the peptide (Fernändez et al., 2013). Lapointe et al. (2005) used an NF membrane with MWCO of 2500 Da to study factors affecting the separation of acid, neutral and basic peptides from tryptic digest of β-LG. Complete retention of acid peptides was observed at both pH 5 and 9, while separation between neutral and basic peptides was best achievable at pH 9. The fractionation between neutral and basic peptides was optimal when the transmembrane pressure was raised to intermediate or higher values. The composition of the peptide layer interacting with the used NF membrane and causing fouling was suggested to be pH dependent. The transmission of β-LG tryptic peptide hydrolysate through UF membranes were influenced by the membrane characteristics (Fernändez et al., 2014). When extremely hydrophilic membrane was used, the permeation rate was not affected by changes of pH from 4 to 10 even with presence of high peptide aggregates. Also, decreasing the MWCO of the membrane from 5 kDa to 1 kDa did not affect the selective transmission of the peptides through the membrane. Ultrafiltration/nanofiltration of tryptic whey protein concentrate hydrolysates was carried at pH 2, 6, and 8 using 1 and 5 kDa polyethersulfone membranes, respectively. The net charge, charge distribution, and size of peptide controlled their partition between the retentate and permeate (Arrutia et al., 2016).

4.1.3 Nanofiltration

In nanofiltration, membranes are used that have MWCO <1 kDa. The surfaces of NF membranes are slightly charged. They can be used for the separation of small peptides (e.g., di- and tripeptides) of molecular weight <1 kDa. Peptides are generally fractionated by NF according to size and charge (Schaep et al., 1998), on the interaction between the charged peptides and the membrane surface, peptide–peptide interaction, and the hydrodynamic parameter used (Muro et al., 2013). Table 12.3 summarizes the bioactive peptides separated from whey protein hydrolysates (WPH) by UF/NF membranes.

4.1.4 Dynamic filtration

Dynamic filtration or shear-enhanced filtration is a relatively recent membrane technology first introduced in 1990. It has the advantage of enhanced performance compared to the

Table 12.3: Separation of bioactive peptide enriched fractions from WPHs using UF and NF.

Type	Substrate	Membrane/Separation Conditions	Properties of Separated Fraction	References
Antihypertensive	Tryptic hydrolysate of α-lactalbumin and β-LG	Two-step UF using 30 kDa and 1 kDa membrane	The IC50 for 1–30 kDa and <1 kDa permeates ranged from 485 to 1134 μg/mL and 109–837 mg/mL	Pihlanto-Leppälä et al. (2000)
	Heated WPH using Corolase PP	Casscade of UF membranes 10, 5, 1 kDa followed by isoelectric focusing	1 kDa retentate exhibited highest ACE-inhibitory activity	O'Loughlin et al. (2014b)
	Whey concentrate (6% protein) hydrolyzed with combinations of thermolysin, proteinase K, Alcalase, Nutrase	Two-step UF, using 10 kDa membrane followed by 1 kDa membrane	1 kDa permeate exhibited 70% ACE-inhibitory	Ramos et al. (2012)
	Tryptic hydrolysate of whey proteins	Fractionation into <6 kDa, 6–10 kDa and >10 kDa using UF membrane 6 and 10 kDa. The <6 kDa was further fractionated by Gel chromatography on G-25 and G-10	The <6 kDa exhibited highest ACE-inhibitory activity, fraction with the lowest MW was the most potent and contain dipeptide LL	Pan et al. (2012)
Antimicrobial	Tryptic hydrolysate of WPI solution	UF 10 kDa MWCO/permeate fractionated by NF membrane 2.5 kDa MWCO (pH 7.0, 1.7 MPa, 50°C) retentate (NFR) and permeate (NFR) freeze dried	NFR was the most effective as inhibitor against *Listeria monocytogenes* and *E. coli* rich in anionic peptide >8 amino acid residues	Demers-Mathieu et al. (2013)
Dipeptidyl peptide inhibitor IV	Tryptic WPH solution (10%)	Successive UF using 5 kDa membrane followed by UF using 2 kDa membrane	Permeates from 5 and 2 kDa membranes had more potent DPP-IV inhibitor activity than the WP hydrolysate	Nongonierma and FitzGerald (2013)
	Tryptic WPH	NF membrane 200 Da	Permeate had more potent DPP-IV activity than WPH, contain short peptides, Val-Ala,Val-Leu, Try-Leu, Try-Ile	Le Maux et al. (2015)
Mixed bioactive peptides	β-LG tryptic hydrolysate	UF at pH 8.0 using two PES membranes having MWCO of 1 kDa and 5 kDa and stabilized 2 kDa cellulose membrane	Increased antioxidant, DPP-IV and ACE-inhibitory activities	Power et al. (2014)

(Continued)

Table 12.3: Separation of bioactive peptide enriched fractions from WPHs using UF and NF. (*cont.*)

Type	Substrate	Membrane/Separation Conditions	Properties of Separated Fraction	References
	WPH (heated, unheated) using Corolase PP (5, 10% degree of hydrolysis)	Cascade membrane fractionation using 0.14 μm, and 30, 10, 5, and 1 kDa MWCO membrane	Presence of multifunctional peptides; permeate from 1 kDa membrane exhibited the highest ACE-inhibitory activity from 10 DH hydrolysate of heated whey proteins	O'Loughlin et al. (2014a)
Antioxidant	β-Lg enriched WPH using Corolase PP and thermolysin	UF of thermolysin hydrolysate using 3 kDa hydrophilic membrane at 5°C	The highest antioxidant activity was obtained in permeate from hydrolysate after 8 h hydrolysis at 80°C	Contreras et al. (2011)
Immunomodulatory	Tryptic digest of β-LG	UF using 1, 5, and 2 kDa membranes	Fraction enriched in large and acid peptides induce Th1 responses, fractions containing short peptides activate monocytes increasing TNFa secretion	Rodríguez-Carrio et al. (2014)
Antithrombotic	Tryptic casein macropeptide hydrolysate	Membrane bioreactor followed by NF of permeate	NF retentate fraction containing small peptides (500–783 Da)	Bouhallab and Touzé (1995)
Antiulcerogenic	Whey proteins hydrolyzed using proteases from *Cynara cardunculus*	UF 3 kDa membrane	Total hydrolysate and UF permeate reduced ulcerative injuries by 37.4 and 68.5%, respectively	Tavares et al. (2011)

β-LG, β-Lactoglobulin; MWCO, molecular weight cut-off; NF, nanofiltration; PES, polyethersulfone; UF, ultrafiltration; WPH, whey protein hydrolysates; WPI, whey protein isolates cross-flow filtration techniques. In these methods, the shear rate at the membrane surface is created by (1) a rotating disk or rotor near a fixed membrane, (2) rotating membrane along a fixed shaft in a housing, (3) vibrations of a circular membranes stacked around a vertical axis, or (4) by longitudinal vibration of a hollow fiber (Ding et al., 2015). Compared to static cross-flow filtration, dynamic filtration has the advantage of increased permeation with the increase in transmembrane pressure (TMP) due to the decreased concentration polarization. Also, dynamic filtration is operated at lower feed velocity than that needed in cross-filtration as the shear rate is produced independently from the feed velocity. This means that less energy is needed in the feed pump. In addition, dynamic filtration can produce retentate of higher concentration than that obtained by cross-filtration. However, driven filtration is more expensive and more complex to operate as compared to the classic pressure driven membrane filtration.

4.1.5 Electric-driven filtration

Electric-driven filtration occurs in which electrically charged ionic species are migrated through perm-selective membranes, under the influence of an electric field (Bazinet, 2005; Bazinet and Firdaous, 2013). Three types of membranes are used namely: homopolar (ion exchange membrane permeable to only one species of ions), bipolar (membranes composed of an anion-exchange layer, a cation-exchange layer, and an intermediary hydrophilic layer at the center), and filtration membranes (Bazinet, 2015). Separation of the basic peptides from tryptic hydrolysate of β-LG was carried out using electrofiltration (EF), aiming to isolate the β-LG 142–148 (Lapointe et al., 2006), which is a potential antihypertensive peptide. The optimum conditions for the fractionation between peptides were assessed. The use of membrane with MWCO 2.5 kDa and pH 9 were found to affect the highest separation of basic peptides from neutral peptides. The relative concentration of the β-LG peptide 142–149 was increased from 3.5% in the hydrolysate to 38% in the permeate. The combined electrodialysis and ultrafiltration (EDUF) has been used to isolate bioactive compounds from several sources. The EDUF has the advantage of being able to fractionate both cationic and anionic peptides even when they have very close molecular weights. Stacking UF membrane in an electrodialysis cell was found to enhance the performance and characteristics of peptides separated from the tryptic digest of β-LG (Poulin et al., 2007). Increasing the UF membrane surface area increased the concentration of separated target antihypertensive peptide. Doyen et al. (2013) used electrodialysis cell stacked with UF membrane for simultaneous tryptic hydrolysis of β-LG and separation of the released anionic and cationic bioactive peptides. Comparing the separation of soy protein hydrolysates with nanofiltration and EDUF revealed that the two processes lead to different results: NF was more efficient in terms of mass flux than EDUF on the same basis of membrane area and process duration, while EDUF recovered peptides of a wide range of molecular weights and amount of polar amino acids (Langevin et al., 2012). This suggests that coupling the two processes in the same line would be more effective in separation of specific peptides.

4.2 Chromatography

Chromatographic methods are characterized by high selectivity in the separation of bioactive peptides. However, they suffer from the high cost and the low separation capacities. Therefore, they find limited use in downstream processing of effluents, such as whey. Pan and Guo (2010) used gel filtration on Sephadex G-75 and G-15 to separate ACE-inhibitory peptide rich fraction from whey protein hydrolysate produced by crude proteinases of *Lb. helveticus* LB10. The obtained fraction was further fractionated using reversed-phase, high-performance liquid chromatography to isolate a peptide with the RLSFNP sequence that had an ACE-inhibitory activity of 177.39 mM. Whey protein isolate (WPI) was hydrolyzed for 1 h using alcalase, protamex, and flavourzyme. Native WPI, hydrolyzed WPI and two commercial WPI hydrolysates were subjected to fractionation by size exclusion chromatography. For native and hydrolyzed WPI samples, the high molecular weight fraction showed a higher antioxidant activity with TBARS inhibition effect of 24%–27%. In contrast, for commercial WPI hydrolysates a higher antioxidant inhibitory effect was found in most of the lower molecular weight fractions (30%–55%).

Ion exchange membrane has been used for the separation of ACE-inhibitory peptides from the tryptic hydrolysate of β-LG (Leeb et al., 2014). The β-LG hydrolysate was first separated using anion exchange membrane, followed by separation of permeate concentrate through strong cation exchange membrane. The ACE-inhibitory activity of the fraction was found to be three- and sixfold higher than the original hydrolysate. The potent β-LG peptide (f 9–14) was recovered (yield 52%) in the separated fractions. A whey protein hydrolysate (WPH1) was prepared that had the characteristic of promoting a strong acute insulin secretion response in the presence of glucose from a clonal rat pancreatic beta cell line (BRIN-BD11) (Gaudel et al., 2013). This hydrolysate was fractionated by chromatographic techniques namely: solid-phase extraction and semipreparative reverse phase-high performance liquid chromatography (Nongonierma et al., 2013). The most potent insulinotropic fraction was enriched in free amino acids and contained relatively hydrophilic peptides indicating that they may be involved in the insulinotropic effect of WPH1.

5 Conclusions and Future Trends

Attention has been given during the last two decades to whey as a mine for several bioactive proteins and peptides. This trend has been boosted by the marked progress in the scientific evidence about their biological, physiological, and potential health benefits. In addition, several applications of these components have been developed in food and pharmaceutical industries. However, large-scale production and uses of these components represent a challenge for several reasons, but the main ones are their low concentrations of whey. Also, there is an absence of commercially and economically feasible protocols for their isolation, fractionation, and concentration without affecting their activities and bioavailability. Several methods have been developed for the separation of LF, LP, and CMP from whey, including membrane separation processes, and chromatographic methods. Each of these methods has its advantages and limitations.

Membrane separation processes are usually regarded as a good candidate for large-scale separation and fractionation of dairy streams. Membrane fouling is a major limitation to achieve this objective. Both the membrane characteristics and feed stream are involved in membrane fouling. Membrane fouling by whey has received much attention with the objective of its reduction to increase the efficacy of membrane filtration. A recent study (Steinhauer et al., 2015) showed that the presence of aggregated proteins can act as an initiator for membrane fouling during ultrafiltration of whey and their removal improved the performance of the used membrane. Along the same line, ultrasonic treatment of whey after heat exposure reduced greatly the pore-blocking formation and growth of foulant cake and improved the flux during ultrafiltration of whey (Koh et al., 2014).

The characteristic properties of the membrane surface play the main role in determining the overall performance of the membrane. Therefore, modification of the membrane surface has been found to overcome the disadvantages of polymeric membranes, such as high hydrophobicity, susceptibility to biofouling, low fluxes, and low mechanical strength (Ng et al., 2013). Several methods have been followed to modify the membrane surfaces in order to improve its performance. Introduction of a negative charge to the membrane surface (Arunkumar and Etzel, 2015) allowed the use of wide-pore UF membrane (100 MWCO) to double the flux during the UF of whey without any loss in the retention of whey proteins. Adjusting pH and ionic strength would allow these membranes to be used for fractionation of individual proteins. Interfacial polymerization is gaining rapid acceptance and use, particularly in the case of NF membranes in the formation of a thin layer on the surface of the membrane. This thin layer determines the membrane permeability, and solute retention as it increases the membrane hydrophilicity and the membrane is smoother compared to regular composite membranes (Mohammad et al., 2015). Also, modification of the membrane surface can be done by covalent attachment of selected macromolecule to the membrane surface (Kochkodan and Hilal, 2015). Several techniques can be followed to initiate surface grafting. These techniques include UV-irradiation, chemical reaction, and plasma treatment of the membrane surface. The characteristics of the membrane surface can be tailored for specific uses by the choice of the specific graft polymerization technique and grafted macromolecules.

A new generation of polymeric membranes based on the incorporation of metal nanoparticles has been developed to overcome the disadvantages of polymeric membranes, such as high hydrophobicity, susceptibility to biofouling, low fluxes, and low mechanical strength (Ng et al., 2013; Dong et al., 2015). Several inorganic nanoparticles have been used to modify the surface properties of the polymeric membranes, such as silver-, iron-, titanium-, zinc-, aluminum-, and magnesium-based nanoparticles. Incorporation of such inorganic nanoparticles markedly improved the characteristic performance of the different polymeric membranes in terms of permeability, selectivity, hydrophilicity, conductivity, mechanical strength, thermal stability, and the antiviral and antibacterial properties. The progress achieved in membrane manufactures offers unlimited choices for improved separation of bioactive proteins from whey using the new generation of membranes. However, there is a critical need for testing the performance of the modified membranes using whey and whey protein solution. This may lead to the development of a commercially and economically feasible method for large-scale separation of bioactive proteins from whey.

Also, newer bioseparation methods have been developed mainly on the laboratory scale or based on the use of pure protein solution. Application of these methods on whey on a large scale may be one of the important steps to study the feasibility of these trends.

The same situation is also found in the case of preparation, isolation, and fractionation of bioactive peptides from whey protein hydolysates. Selection of the appropriate enzyme used, degree of hydrolysis, and fractionation of the released bioactive peptides into fractions with specific activities still represent a challenge for commercial separation and utilization of bioactive peptides from whey protein hydrolysates. Well-defined approaches and methods are needed for large-scale exploitation of economic and commercial production of bioactive peptide fraction of unified and consistent properties and activities. Several new bioseparation approaches and methods have been proposed to achieve these objectives (Agyei et al., 2016). As an example, the new monolithic chromatographic methods offer better performance than the conventional chromatographic methods, due to their unique physical and chemical structures. Monolithic chromatographic columns have better hydrodynamics and can be scaled up. However, little has been done with respect to the application of these new methods in the separation of bioactive peptides particularly from whey protein hydrolysates.

It is hoped that the recently developed methods and those under development will offer better approaches for wider and more efficient techniques for the recovery of bioactive whey proteins and bioactive peptides.

References

Abd El-Salam, M.H., El-Shibiny, S., Buchheim, W., 1996. Characteristics and potential uses of the casein macropeptide. Int. Dairy J. 6, 327–341.

Agyei, D., Ongkudon, C.M., Wei, C.Y., Chanc, A.S., Danquah, M.K., 2016. Bioprocess challenges to the isolation and purification of bioactive peptides. Food Bioprod. Proc. 98, 244–256.

Al-Mashikhi, S.A., Naki, S., 1987. Isolation of bovine immunoglobulins and lactoferrin from whey proteins by gel filtration techniques. J. Dairy Sci. 70, 2486–2492.

Alvarez-Guerra, E., Irabien, A., 2012. Extraction of lactoferrin with hydrophobic ionic liquids. Sep. Purif. Technol. 98, 432–440.

Alvarez-Guerra, E., Irabien, A., 2015. Ionic liquid-based three phase partitioning (ILTPP) systems for whey protein recovery: ionic liquid selection. J. Chem. Technol. Biotechnol. 90, 939–946.

Andersson, J., Mattiasson, B., 2006. Simulated moving bed technology with a simplified approach for protein purification: separation of lactoperoxidase and lactoferrin from whey protein concentrate. J. Chromatogr. A 1107, 88–95.

Arrutia, F., Rubio, R., Riera, F.A., 2016. Production and membrane fractionation of bioactive peptides from a whey protein concentrate. J. Food Eng. 184, 1–9.

Arunkumar, A., Etzel, M.E., 2015. Negatively charged tangential flow ultrafiltration membranes for whey protein concentration. J. Membr. Sci. 475, 340–348.

Ayer, J. S., Coolbear, K. P., Elgar, D. F., Pritchard, M., 2003. Process for isolating glycomacropeptide from dairy products with a phenylalanine impurity of 0.5% w/w. US Patent 6,555,659.

Baieli, M.F., Urtasun, N., Miranda, M.V., Cascone, O., Wolman, F.J., 2014a. Bovine lactoferrin purification from whey using Yellow HE-4R as the chromatographic affinity ligand. J. Sep. Sci. 37, 484–487.

Baieli, M.F., Urtasun, N., Miranda, M.V., Cascone, O., Wolman, F.J., 2014b. Isolation of lactoferrin from whey by dye-affinity chromatography with Yellow HE-4R attached to chitosan mini-spheres. Int. Dairy J. 39, 53–59.

Baker, E.N., Baker, H.M., 2005. Molecular structure, binding properties and dynamics of lactoferrin. Cell Mol. Life Sci. 62, 2531–2539.

Bazinet, L., 2005. Electrodialytic phenomena and their applications in the dairy industry: a review. Crit. Rev. Food Sci. Nutr. 45, 307–326.

Bazinet, L., 2015. Electrodialysis applications on dairy ingredients separation. In: Hu, K., Dickson, J.M. (Eds.), Membrane Processing for Dairy Ingredient Separation. first ed. John Wiley & Sons, Hoboken, NJ.

Bazinet, L., Firdaous, L., 2009. Membrane processes and devices for separation of bioactive peptides. Rec. Pat. Biotechnol. 3, 61–72.

Bazinet, L., Firdaous, L., 2013. Separation of bioactive peptides by membrane processes: technologies and devices. Recent Pat. Biotechnol. 7, 9–27.

Berrocal, R., Neeser, J.R., 1991. Process for the production of κ-casein glycomacropeptide. Eur. Pat. Appl. EP 0453, 782.

Bonnaillie, L.M., Qi, P., Wickham, E., Tomasula, P.M., 2014. Enrichment and purification of casein glycomacropeptide from whey protein isolate using supercritical carbon dioxide processing and membrane ultrafiltration. Foods 3, 94–109.

Boots, J.-W., Floris, R., 2006. Lactoperoxidase: from catalytic mechanism to practical applications. Int. Dairy J. 16, 1272–1276.

Bouhallab, S., Touzé, C., 1995. Continuous hydrolysis of caseinomacropeptide in a membrane reactor: kinetic study and gram-scale production of antithrombotic peptides. Lait 75, 251–258.

Brisson, G., Britten, M., Pouliot, Y., 2007. Electrically enhanced cross-flow microfiltration for separation of lactoferrin from whey protein mixtures. J. Membr. Sci. 297, 206–216.

Brody, E.P., 2000. Biological activities of bovine glycomacropeptide. Br. J. Nutr. 84, S39–S46.

Brown, G.N., Müller, C., Theodosiou, E., Franzreb, M., Thomas, O.R.T., 2013. Multi-cycle recovery of lactoferrin and lactoperoxidase from crude whey using fimbriated high-capacity magnetic cation exchangers and a novel "rotor–stator" high-gradient magnetic separator. Biotechnol. Bioeng. 110, 1714–1725.

Butylina, S., Luque, S., Nyström, M., 2006. Fractionation of whey-derived peptides using a combination of ultrafiltration and nanofiltration. J. Membr. Sci. 280, 418–426.

Carvalho, B.M.A., Da Silva, L.H.M., Carvalho, L.M., Soaresd, A.M., Minim, L.A., Da Silva, S.L., 2013. Micro calorimetric study of the adsorption of lactoferrin in super macroporous continuous cryogel with immobilized Cu^{2+} ions. J. Chromatogr. A 1312, 1–9.

Carvalho, B.M.A., Carvalho, L.M., Silva, Jr., W.F., Minim, L.A., Soares, A.M., Carvalho, G.G.P., da Silva, S.L., 2014. Direct capture of lactoferrin from cheese whey on super macroporous column of polyacrylamide cryogel with copper ions. Food Chem. 154, 308–314.

Casal, E., Corzo, N., Moreno, F.J., Olano, A., 2005. Selective recovery of glycosylated casein macropeptide with chitosan. J. Agric. Food Chem. 53, 1201–1204.

Chatterton, D.E.W., Holst, H.H., 2002. Process for preparing a kappa-casein glycopeptide or a derivative thereof. US patent 6,462,181.

Cheison, S.C., Schmitt, M., Leeb, E., Letzel, T., Kulozik, U., 2010. Influence of temperature and degree of hydrolysis on the peptide composition of trypsin hydrolysates of β-lactoglobulin: analysis by LC–ESI-TOF/MS. Food Chem. 121, 457–467.

Chen, J.-P., Wang, C.-H., 1991. Microfiltration affinity purification of lactoferrin and immunoglobulin G from cheese whey. J. Food Sci. 53, 701–706.

Chen, L., Guo, C., Guan, Y., Liu, H., 2007. Isolation of lactoferrin from acid whey by magnetic affinity separation. Sep. Purif. Technol. 56, 168–174.

Chiu, C.K., Etzel, M.R., 1997. Fractionation of lactoperoxidase and lactoferrin from bovine whey using a cation exchange membrane. J. Food Sci. 62, 996–1000.

Contreras, M.D.M., Hernández-Ledesma, B., Amigo, L., Martín-Álvarez, P.J., Recio, I., 2011. Production of antioxidant hydrolyzates from a whey protein concentrate with thermolysin: optimization by response surface methodology. LWT Food Sci. Technol. 44, 9–15.

Demers-Mathieu, V., Gauthier, S.F., Britten, M., Fliss, I., Robitaille, G., Jean, J., 2013. Antibacterial activity of peptides extracted from tryptichydrolyzate of whey protein by nanofiltration. Int. Dairy J. 28, 94–101.

Ding, L., Jaffrin, M.Y., Luo, J., 2015. Dynamic filtration with rotating disks, and rotating or vibrating membranes. In: Tarleton, E.S. (Ed.), Progress in Filtration and Separation. Academic Press, Cambridge, MA, (Chapter 2).

Dong, L-x., Yang, H-w., Liu, S-t., Wang, X-m., Xie, Y.F., 2015. Fabrication and anti-biofouling properties of alumina and zeolite nanoparticle embedded ultrafiltration membranes. Desalination 365, 70–78.

Doultani, S., Turhan, K.N., Etzel, M.R., 2003. Whey protein isolate and glycomacropeptide recovery from whey using ion exchange chromatography. J. Food Sci. 68, 1389–1395.

Doyen, A., Husson, E., Bazinet, L., 2013. Use of an electrodialytic reactor for the simultaneous β-lactoglobulin enzymatic hydrolysis and fractionation of generated bioactive peptides. Food Chem. 136, 1193–1202.

Du, Q.-Y., Lin, D.-Q., Zhang, Q.-L., Shan-Jing Yao, S.-J., 2014. An integrated expanded bed adsorption process for lactoferrin and immunoglobulin G purification from crude sweet whey. J. Chromatogr. B 947–948, 201–207.

Etzel, M.A., 2001. Production of substantially pure casein macropeptide. US Patent 6,168,823.

Etzel, M.R., Arunkumar, A., 2015. Dairy protein fractionation and concentration using charged ultrafiltration membranes. In: Hu, K., Dickson, J.M. (Eds.), Membrane Processing for Dairy Ingredient Separation. first ed. John Wiley & Sons, Hoboken, NJ.

FAO, 2006. Benefits and potential risks of the lactoperoxidase system of raw milk preservation: report of an FAO/WHO technical meeting. FAO Headquarters, Rome, Italy, November 28–December 2, 2005. Available from: http://www.fao.org/ag/dairy.html

Faraji, N., Zhang, Y., Ray, A.K., 2015. Determination of adsorption isotherm parameters for minor whey proteins by gradient elution preparative liquid chromatography. J. Chromatogr. A 1412, 67–74.

Fariias, M.E., Martinez, M.J., Pilosof, A.M.R., 2010. Casein glycomacropeptide pH-dependent self-assembly and cold gelation. Int. Dairy J. 20, 78–88.

Fee, C.J., Chand, A., 2006. Capture of lactoferrin and lactoperoxidase from raw whole milk by cation exchange chromatography. Sep. Purif. Technol. 48, 143–149.

Fernändez, A., Suärez, A., Zhu, Y., FitzGerald, R.J., Riera, F.A., 2013. Membrane fractionation of a β-lactoglobulin tryptic digest: effect of the pH. J. Food Eng. 114, 83–89.

Fernändez, A., Suärez, A., Zhu, Y., FitzGerald, R.J., Riera, F.A., 2014. Membrane fractionation of a β-lactoglobulin tryptic digest: effect of the membrane characteristics. J. Chem. Technol. Biotechnol. 89, 508–515.

Fuda, E., Jauregi, P., Pyle, D.L., 2004. Recovery of lactoferrin and lactoperoxidase from sweet whey using colloidal gas aphrons (CGAs) generated from an anionic surfactant, AOT. Biotechnol. Prog. 20, 514–525.

Gaudel, C., Nongonierma, A.B., Maher, S., Flynn, S., Krause, M., Murray, B.A., et al., 2013. Effect of a whey protein hydrolysate promotes insulinotropic activity in a clonal pancreatic beta cell line and enhances glycemic function in ob/ob mice. J. Nutr. 143 (7), 1109–1114.

Gonzalez-Chavez, S.A., Arevalo-Gallegos, S., Rascon-Cruz, Q., 2009. Lactoferrin: structure, function and applications. Int. J. Antimicrob. Agents 33, 301.e1–301.e8.

Jiménez-Guzmána, J., Méndez-Palacios, I., López-Luna, A., Del Moral-Ramíreza, E., Bárzana, E., García-Garibaya, M., 2014. Development of a molecularly imprinted polymer for the recovery of lactoferrin. Food Bioprod. Process 92, 226–232.

Kawasaki, Y., Kawakami, H., Tanimoto, S., Dosako, S., Tomzawa, A., Kotake, M., m Nakajima, I., 1993. pH-dependent molecular weight changes of κ-casein glycomacropeptide and its preparation by ultrafiltration. Milchwissenschaft 48, 191–195.

Kochkodan, V., Hilal, N., 2015. A comprehensive review on surface modified polymer membranes for biofouling mitigation. Desalination 356, 187–207.

Koh, L.L.A., Nguyen, H.T.H., Chandrapala, J., Zisu, B., Ashokkumar, M., Kentish, S.E., 2014. The use of ultrasonic feed pretreatment to reduce membrane fouling in whey ultrafiltration. J. Membr. Sci. 453, 230–239.

Korhonen, H.J., Marnila, P., 2013. Milk bioactive proteins and peptides. In: Park, Y.W., Haenlein, G.F.W. (Eds.), Milk and Dairy Products in Human Nutrition: Production, Composition and Health. first ed. John Wiley & Sons, Hoboken, NJ.

Kreuβ, M., Kulozik, U., 2009. Separation of glycosylated caseinomacropeptide at pilot scale using membrane adsorption in direct-capture mode. J. Chromatogr. A 1216, 8771–8777.

Kreuβ, M., Krause, I., Kulozik, U., 2008. Separation of a glycosylated and nonglycosylated fraction of caseinomacropeptide using different anion-exchange stationary phases. J. Chromatogr. A 1208, 126–132.

Kreuβ, M., Strixner, T., Kulozik, U., 2009. The effect of glycosylation on the interfacial properties of bovine caseinomacropeptide. Food Hydro. 23, 1818–1826.

Kvistgaard, A.S., Schroder, J.B., Jensen, E., Setarehnejad, A., Kanekanian, A., 2014. Milk ingredients as functional foods. In: Kanekanian, A. (Ed.), Milk and Dairy Products as Functional Foods. first ed. John Wiley & Sons, Hoboken, NJ, pp. 198–217.

LaClair, C.E., Ney, D.M., MacLeod, E.L., Etzel, M.R., 2009. Purification and use of glycomacropeptide for nutritional management of phenylketonuria. J. Food Sci. 74, E199–E206.

Langevin, M.-E., Roblet, C., Moresoli, C., Ramassamy, C., Bazinet, L., 2012. Comparative application of pressure- and electrically driven membrane processes for isolation of bioactive peptides from soy protein hydrolysate. J. Membr. Sci. 403–404, 15–24.

Lapointe, J.-F., Gauthier, S.F., Pouliot, Y., Bouchard, C., 2005. Fouling of a nanofiltration membrane by a β-lactoglobulin tryptic hydrolysate: impact on the membrane sieving and electrostatic properties. J. Membr. Sci. 253, 89–102.

Lapointe, J.-F., Gauthier, S.F., Pouliot, Y., Bouchard, C., 2006. Selective separation of cationic peptides from a tryptic hydrolysate of β-lactoglobulin by electrofiltration. Biotechnol. Bioeng. 94, 223–233.

Le Maux, S., Nongonierma, A.B., Murray, B., Kelly, P.M., FitzGerald, R.J., 2015. Identification of short peptide sequences in the nanofiltration permeate of a bioactive whey protein hydrolysate. Food. Res. Int. 77, 534–539.

Le Maux, S., Nongonierma, A.B., Barre, C., FitzGerald, R.J., 2016. Enzymatic generation of whey protein hydrolysates under pH-controlled and non-pH-controlled conditions: impact on physicochemical and bioactive properties. Food Chem. 199, 246–251.

Leeb, E., Holder, A., Letzel, T., Cheison, S.C., Kulozik, U., Hinrichs, J., 2014. Fractionation of dairy-based functional peptides using ion-exchange membrane adsorption chromatography and cross-flow electro membrane filtration. Int. Dairy J. 38, 116–123.

Li, C., Song, X., Hein, S., Wang, K., 2010. The separation of GMP from milk whey using the modified chitosan beads. Adsorption 16, 85–91.

Li-Chen, E.C.Y., Ler, S.S., Kummer, A., Akita, E.M., 1998. Iolation of lactoferrin by immunoaffinity chromatography using yolk antibodies. J. Food Biochem. 22, 179–195.

Lőnnerdal, B., Suzuki, Y.A., 2013. Lactoferrin. In: McSweeney, P.L.H., Fox, P.F. (Eds.), Advanced Dairy Chemistry: Volume 1A: Proteins: Basic Aspects. fourth ed. Springer Science+Business Media, New York, NY.

Lu, R.R., Xu, S.Y., Wang, Z., Yang, R.J., 2007. Isolation of lactoferrin from bovine colostrum by ultrafiltration coupled with strong cation exchange chromatography on a production scale. J. Membr. Sci. 297, 152–161.

Maharjan, P., Campi, E.M., De Silva, K., Woonton, B.W., Jackson, W.R., Hearn, M.T.W., 2016. Studies on the application of temperature-responsive ion exchange polymers with whey proteins. J. Chromatogr. A 1438, 113–122.

Manninen, A.H., 2009. Protein hydrolysates in sports nutrition. Nutr. Metab. (Lond.) 6, 38.

Manso, M.A., López-Fandiño, R., 2004. κ-Casein macropeptides from cheese whey: physicochemical, biological, nutritional, and technological features for possible uses. Food. Rev. Int. 20, 329–355.

Martin-Diana, A.B., Fontecha, M.J.F.J., 2002. Isolation and characterization of caseinomacropeptide from bovine, ovine, and caprine cheese whey. Eur. Food Res. Technol. 214, 282–286.

Meyer, A., Berensmeier, S., Franzreb, M., 2007. Direct capture of lactoferrin from whey using magnetic micro-ion exchangers in combination with high-gradient magnetic separation. Reac. Func. Polym. 67, 1577–1588.

Mikkelsen, T.L., Frøkær, H., Topp, C., Bonomi, F., Iametti, S., Picariello, G., Ferranti, P., Barkholt, V., 2005. Caseinomacropeptide self-association is dependent on whether the peptide is free or restricted in κ-casein. J Dairy Sci. 88, 4228–4238.

Mohammad, A.W., Teow, Y.H., Ang, W.L., Chung, Y.T., Oatley-Radcliffe, D.L., et al., 2015. Nanofiltration membranes review: recent advances and future prospects. Desalination 356, 226–254.

Muro, C., Riera, F., Fernández, A., 2013. Advancements in the fractionation of milk biopeptides by means of membrane processes. InTech. (Chapter 10). Available from: http://creativecommons.org/licenses/by/3.0

N'diaye, N., Pouliot, Y., Sauciera, L., Beaulieua, L., Bazineta, L., 2010. Electroseparation of bovine lactoferrin from model and whey solutions. Sep. Purif. Technol. 74, 93–99.

Nandini, K.E., Rastogi, N.K., 2011. Integrated downstream processing of lactoperoxidase from milk whey involving aqueous two-phase extraction and ultrasound-assisted ultrafiltration. Appl. Biochem. Biotechnol. 163, 173–185.

Ng, P.K., Yoshitake, T., 2010. Purification of lactoferrin using hydroxyapatite. J. Chromatogr. B 878, 976–980.

Ng, L.Y., Mohammad, A.W., Leo, C.P., Hilal, N., 2013. Polymeric membranes incorporated with metal/metal oxide nanoparticles: a comprehensive review. Desalination 308, 15–33.

Nielsen, P.M., Trombolt, N., 1994. Method for production of a kappa-casein glycomacropeptide and use of a kappa-casein glycomacropeptide. WO, 94/15952.

Nongonierma, A.B., FitzGerald, R.J., 2013. Dipeptidyl peptidase IV inhibitory properties of a whey protein hydrolysate: influence of fractionation, stability to simulated gastrointestinal digestion and food-drug interaction. Int. Dairy J. 32, 33–39.

Nongonierma, A.B., Gaudel, C., Murray, B.A., Flynn, S., Kelly, P.M., Newsholme, P., FitzGerald, R.J., 2013. Insulinotropic properties of whey protein hydrolysates and impact of peptide fractionation on insulinotropic response. Int. Dairy J. 32, 163–168.

O'Loughlin, I.B., Murray, B.A., FitzGerald, R.J., Brodkorb, A., Kelly, P.M., 2014a. Pilot-scale production of hydrolysates with altered biofunctionalities based on thermally denatured whey protein isolate. Int. Dairy J. 34, 146–152.

O'Loughlin, I.B., Murray, B.A., Brodkorb, A., FitzGerald, R.J., Kelly, P.M., 2014b. Production of whey protein isolate hydrolysate fractions with enriched ACE-inhibitory activity. Int. Dairy J. 38, 101–103.

Pan, D., Guo, Y., 2010. Optimization of sour milk fermentation for the production of ACE-inhibitory peptides and purification of a novel peptide from whey protein hydrolysate. Int. Dairy J. 20, 472–479.

Pan, D., Cao, J., Guo, H., Zhao, B., 2012. Studies on purification and the molecular mechanism of a novel ACE inhibitory peptide from whey protein hydrolysate. Food Chem. 130, 121–126.

Pan, M., Shen, S., Chen, L., Dai, B., Xu, L., Yun, J., Yao, K., Lin, D.-Q., Yao, S.-J., 2015. Separation of lactoperoxidase from bovine whey milk by cation exchange composite cryogel embedded macroporous cellulose beads. Sep. Purif. Technol. 147, 132–138.

Pihlanto-Leppälä, A., Koskinen, P., Piilola, K., Tupasela, T., Korhonen, H., 2000. Angiotensin I-converting enzyme inhibitory properties of whey protein digests: concentration and characterization of active peptides. J. Dairy Res. 67, 53–64.

Plate, K., Beutel, S., Buchholz, H., Demmer, W., Fischer-Frühholz, S., Reif, O., Ulber, R., Scheper, T., 2006. Isolation of bovine lactoferrin, lactoperoxidase and enzymatically prepared lactoferricin from proteolytic digestion of bovine lactoferrin using adsorptive membrane chromatography. J. Chromatogr. A 1117, 81–86.

Poulin, J.F., Amiot, J., Bazinet, L., 2007. Improved peptide fractionation by electrodialysis with ultrafiltration membrane: Influence of ultrafiltration membrane stacking and electrical field strength. J. Membr. Sci. 299, 83–90.

Power, O., Fernández, A., Norris, R., Riera, F.A., FitzGerald, R.J., 2014. Selective enrichment of bioactive properties during ultrafiltration of a tryptic digest of β-lactoglobulin. J. Funct. Foods 9, 38–47.

Ramos, B.P., Rodriguez, A.P., Telle, N.E., Gonzales, J.P.F., Rodriguez, L.R., Castro, L.P., 2012. Optimized method for obtaining ACE-activity inhibitory peptides from whey, ACE inhibitory and food comprising them. US Patent 201,210,322,745 A1.

Ravichandran, R., Padmanabhan, V., Vijayalakhsmi, M.A., Jayaprakash, N.S., 2015. Studies on recovery of lactoferrin from bovine colostrum whey using mercapto ethyl pyridine and phenyl propyl amine HyperCel™ mixed mode sorbents. Biotechnol. Bioprocess Eng. 20, 148–156.

Regester, G.O., Smithers, G.W., 1991. Seasonal changes in the β-lactoglobulin, α-lactalbumin, glycomacropeptide, and casein contents of whey protein concentrates. J. Dairy Sci. 74, 796–802.

Rodríguez-Carrio, J., Fernández, A., Riera, F.A., Suárez, A., 2014. Immunomodulatory activities of whey β-lactoglobulin tryptic-digested fractions. Int. Dairy J. 34, 65–73.

Rojas, E., Torres, G., 2013. Isolation and recovery of glycomacropeptide from milk whey by means of thermal treatment. Food Sci. Technol. Campinas 33, 14–20.

Saito, T., Yamaji, A., Itoh, T., 1991. A new isolation method of caseinoglycopeptide from sweet cheese whey. J. Dairy Sci. 74, 2831–2837.

Saufi, S.M., Fee, C.J., 2011a. Simultaneous anion and cation exchange chromatography of whey proteins using a customizable mixed matrix membrane. J. Chromatogr. A 1218, 9003–9009.

Saufi, S.M., Fee, C.J., 2011b. Recovery of lactoferrin from whey using cross-flow cation exchange mixed matrix membrane chromatography. Sep. Purif. Technol. 77, 68–75.

Schaep, J., Van Der Bruggen, B., Vandecasteele, C., Wilms, D., 1998. Influence of ion size and charge in nanofiltration. Sep. Purif. Technol. 14, 155–162.

Smithers, G.W., 2008. Whey and whey proteins: from "gutter-to-gold". Int. Dairy J. 18, 695–704.

Smithers, G.W., 2015. Whey-ing up the options: yesterday, today and tomorrow. Int. Dairy J. 48, 2–14.

Steinhauer, T., Marx, M., Bogendörfer, K., Kulozik, U., 2015. Membrane fouling during ultra- and microfiltration of whey proteins at different environmental conditions: the role of aggregated whey proteins as fouling initiators. J. Membr. Sci. 489, 20–27.

Svanborg, S., Johansen, A.-G., Abrahamsen, R.K., Schüller, R.B., Skeie, S.B., 2016. Caseinomacropeptide influences the functional properties of a whey protein concentrate. Int. Dairy J. 60, 14–23.

Tamano, S., Sekine, K., Takase, M., Yamauchi, K., Iigo, M., Tsuda, H., 2008. Lack of chronic oral toxicity of chemopreventive bovine lactoferrin in F344/DuCrj rats. Asian Pac. J. Cancer Prev. 9, 313–316.

Tanimoto, M., Kawasaki, Y., Shinmoto, H., Dosako, S., Tomizawa, A., 1991. Process for producing kappacasein-glycomacropeptide. US Patent 5.075.424, 1991.

Tanimoto, M., Kawasaki, Y., Shinmoto, H., Dosako, S., Ahiko, K., 1992. Large-scale preparation of k-casein glycomacropeptide from rennet casein whey. Biosci. Biotechnol. Biochem. 56, 140–141.

Tavares, T.G., Monteiro, K.M., Possenti, A., Pintado, M.E., Carvalho, J.E., Malcata, F.X., 2011. Antiulcerogenic activity of peptide concentrates obtained from hydrolysis of whey proteins by proteases from *Cynara cardunculus*. Int. Dairy J. 21, 934–939.

Taylor, S., Brock, J., Kruger, C., Berner, T., Murphy, M., 2004. Safety determination for the use of bovine milk derived lactoferrin as a component of an antimicrobial beef carcass spray. Reg. Toxicol. Pharm. 39, 12–24.

Teepakorn, C., Fiaty, K., Charcosset, C., 2015. Optimization of lactoferrin and bovine serum albumin separation using ion-exchange membrane chromatography. Sep. Purif. Technol. 151, 292–302.

Tek, H.N., Turhan, K.N., Etzel, M.R., 2005. Effect of conductivity, pH, and elution buffer, salinity on glycomacropeptide recovery from whey using anion exchange chromatography. J. Food Sci. 70, S295–S300.

Tolkach, A., Kulozik, U., 2005. Fractionation of whey proteins and caseinomacropeptide by means of enzymatic cross linking and membrane separation techniques. J. Food Eng. 67, 13–20.

Tomita, M., Wakabayashi, H., Shin, K., Yamauchi, K., Yaeshima, T., Iwatsuki, K., 2009. Twenty-five years of research on bovine lactoferrin applications. Biochimie 91, 52–57.

van der Ven, C., Gruppen, H., de Bont, D.B.A., Voragen, A.G.J., 2002. Optimization of the angiotensin converting enzyme inhibition by whey protein hydrolysates using response surface methodology. Int. Dairy J. 12, 813–820.

Virginia Valin˜o, M., San Román, F., Iban˜ez, R., Ortiz, I., 2014. Improved separation of bovine serum albumin and lactoferrin mixtures using charged ultrafiltration membranes. Sep. Purif. Technol. 125, 163–169.

Voswinkel, L., Kulozik, U., 2011. Fractionation of whey proteins by means of membrane adsorption chromatography: 11th Int. Congr. Eng. Food (ICEF11). Proc. Food Sci. 1, 900–907.

Voswinkel, L., Kulozik, U., 2014. Fractionation of all major and minor whey proteins with radial flow membrane adsorption chromatography at lab and pilot scale. Int. Dairy J. 39, 209–214.

Wang, B., Yu, B., Karim, M., Hu, H.H., Sun, Y., McGreevy, P., Petocz, P., Held, S., Brandmiller, J., 2007. Dietary sialic acid supplementation improves learning and memory in piglets. Am. J. Clin. Nutr. 85, 561–569.

Wolman, F.J., Maglio, D.G., Grasselli, M., Cascone, O., 2007. One-step lactoferrin purification from bovine whey and colostrum by affinity membrane chromatography. J. Memb. Sci. 288, 132–138.

Ye, X., Yoshida, S., Ng, T.B., 2000. Isolation of lactoperoxidase, lactoferrin, α-lactalbumin, β-lactoglobulin B and β-lactoglobulin A from bovine rennet whey using ion exchange chromatography. Int. J. Biochem. Cell Biol. 32, 1143–1150.

Yoshida, S., Xiuyun, Y., 1991. Isolation of lactoperoxidase and lactoferrins from bovine milk acid whey by carboxymethyl cation exchange chromatography. J. Dairy Sci. 74, 1439–1444.

Further Reading

Pená-Ramos, E.A., Xiong, Y.L., Arteaga, G.E., 2004. Fractionation and characterization for antioxidant activity of hydrolyzed whey protein. J. Sci. Food Agric. 84, 1908–1918.

CHAPTER 13

Phytochemicals: An Insight to Modern Extraction Technologies and Their Applications

Priyanka Rao, Virendra Rathod
Institute of Chemical Technology, Mumbai, Maharashtra, India

1 Introduction

1.1 An Overview of Phytochemicals in Foods

Phytochemicals are plant chemicals identified from vegetables, fruits, beans, grains, nuts, and seeds. Phytochemical foods help in preventing various ailments and also in maintaining improved health (Saxena et al., 2013). Various phytochemicals are involved in an array of metabolic reactions, such as substrates, cofactors, inhibitors, intracellular receptors, scavengers, essential nutrients absorption enhancers, elective growth factors, and fermentation substrates for beneficial gastrointestinal bacteria, and selective inhibitors of injurious intestinal bacterial flora (Zhang et al., 2015). Natural biomolecules include a broad range of functional groups that deliver a wide choice of phtyochemicals for production of nutraceuticals, functional foods, and food additives. Phytochemicals in foods have diverse and complex chemical structures and are broadly classified into polyphenols, terpenoids, alkaloids, other nitrogen compounds, carbohydrates, and lipids (Kennedy and Wightman, 2011). Phenolics comprise the largest group of phytochemicals, which are secondary metabolites with wide distribution with a myriad of characteristics. They possess an aromatic ring bearing one or more hydroxyl groups and their structures may range from that of a simple phenolic molecule to that of a complex high-molecular weight polymer are often denoted as "polyphenols." Phenolic compounds as functional food render their effects via antioxidation and relief from oxidative stress and its consequences. The antioxidant action of phenolics in functional foods is due to the direct free radical scavenging activity, reducing activity, and an indirect effect arising from chelation of prooxidant metal ions (Nair, 2015). Some examples of phenolic compounds present in foods include quercetin in apples, red and yellow onions hesperidin in citrus fruits, catechins in white tea, green tea, grapes and cocoa, curcumins in turmeric. Terpenoids (isoprenoids) are another class of hydrocarbons of plant

Ingredients Extraction by Physicochemical Methods in Food
http://dx.doi.org/10.1016/B978-0-12-811521-3.00013-2

origin, which have unique antioxidant activity as they react with free radicals by partitioning themselves into fatty membranes by virtue their long carbon side chain. Carotenoids are a class of terpenoids that are fat-soluble natural pigments with antioxidant properties, with various other additional physiological functions, such as immunostimulation (Krinsky and Yeum, 2003). In humans, four carotenoids (α-, β-, and γ-carotene, and β-cryptoxanthin) have vitamin A activity, and these and other carotenoids can also act as antioxidants. Glucosinolates (GLS), a group of plant thioglucosides found among several vegetables, are a class of organic compounds, which are derived from glucose and an amino acid (Pazitna et al., 2014). Some economically important GLS containing plants are white mustard, brown mustard, radish, and broccoli (Devi and Thangam, 2010). GLS hydrolysis and metabolic products have proven chemoprotective properties against chemical carcinogens by blocking the initiation of tumors in various tissues. They exhibit their effect by inducing Phase I and II enzymes, inhibiting the enzyme activation, modifying the steroid hormone metabolism, and protecting against oxidative damage (Verkerk et al., 2009). Allicin is another important phytochemical, found abundantly in fresh garlic and in smaller amounts in onions, chives, and leeks. This chemical acts as an antioxidant similar to vitamins A, C, and E, and may help protect the body from free radicals. Allicin fights cancer by reacting with carcinogens and changing their structure so they can no longer initiate tumors or by speeding the death of cancer cells that have already been formed (Williams et al., 2013). Hundreds of phytochemical compounds, with several different biological functions, have been identified in plant-based foods. Therefore, consuming a variety of plant-based foods helps to ensure that individuals receive the optimum benefits from the fruits and vegetables consumed.

1.2 An Overview of Extraction Techniques Used in Food Industry

Recent technological advances and the development of new methods to improve the production and separation have revolutionized the screening of biomolecules and provided an opportunity to obtain natural extracts that could be potentially used. In addition, extraction techniques for the isolation of biomolecules have been developed to obtain highly purified products, rendering them useful in a wide range of applications. Extraction results in the isolation of that part of the plant material which shows pharmacological activity from the inert components, such as plant matrix, by using specific solvents. The extensively used extraction techniques of biomolecules include Soxhlet extraction, percolation, digestion, maceration, microwave-assisted extraction, ultrasound extraction, supercritical fluid extraction, continuous extraction, phytonic extraction, and distillation techniques (Handa, 2008). Use of appropriate solvents at higher temperature is the principle behind traditional solvent extraction. This accelerates the mass transfer and rate of reaction of the extraction process (Afoakwah et al., 2012). Conventional extraction methods require lengthy process time thus risking the stability of thermolabile biomolecules (Kadam et al., 2013). Hence, if novel techniques are explored scientifically, they can provide an efficient extraction technology for ensuring the quality of phyotchemicals worldwide. The

Table 13.1: An overview of recent extraction technologies used in food industry.

Extraction Systems	Types of Extraction	Mechanisms
Conventional extraction	Chemical extraction	Solubilization of plant cell walls by using solvents at high temperature
Three-phase partitioning	Chemical extraction	Salting out, isotonic precipitation, osmolytic and kosmotropic effect, protein hydration shifts, electrostatic forces
Hydrotropic extraction	Chemical extraction	Solubilization of plant matrix by hydrotropes
Accelerated solvent extraction	Chemical extraction	Elevated pressure and temperature to enhance extraction process
Aqueous two-phase extraction	Chemical extraction	Electrostatic interaction, salting out effect, hydrophobic forces, and steric hindrance interactions
Supercritical fluid extraction	Chemical extraction	Separating two components using supercritical CO_2 as the extracting solvent
Microwave-assisted extraction	Physical extraction	Electric and magnetic fields, ionic conduction, and dipole rotation
Ultrasound-assisted extraction	Physical extraction	Cavitation phenomenon
Enzyme-assisted extraction	Physical extraction	Binding of enzymes onto the active sites of plant matrix and its solubilization
Pulse electric field-assisted extraction	Physical extraction	Separation of components using discharge of high-voltage electric pulses in a few microseconds

aim of this chapter is to provide an insight to the sources, technologies, and methods that have been developed to improve the productivity and isolation of biomolecules in foods. Table 13.1 represents an overview of the recent extraction technologies in food industry along with an outline of their mechanisms.

2 Chemical Extraction Techniques

Chemical extraction relies greatly on the solvents used, amount of heat energy required, and agitation to improve the solubility and mass transfer. Bioactive extraction by chemical method is an inveterate method and easy to accomplish. In this section, various chemical extraction techniques in the food industry are explained alongside their applications.

2.1 Traditional Solvent Extraction

Solvent extraction has been used since ages in today's food industry. Water is the chosen solvent if the substance to be extracted or leached out is water-soluble. Organic and inorganic solvents are required otherwise and depending on the solubility characteristics of the extractable material, different solvents are selected. A large number of biomolecules are extracted from plants with organic solvents, comprising a vital part in the isolation of phytoconstituents by this technique. Solvents, such as hexane, acetone, chloroform, acetonitrile, and ethanol, are commonly used along with water for extraction of biomolecules.

Aforementioned solvents can be involved with the extraction of polar and nonpolar biomolecules, such as xanthanoids, lactones, phenols, aromatic hydrocarbons, essential oils, and fatty acids (Kaufmann and Christen, 2002). However, various solvents must be used with care as they are toxic for humans and dangerous for the environment, moreover, the extraction conditions are sometimes laborious. The solvent must be separated from the final extract before the product is to be used in food applications. Solvent extraction is advantageous compared to other methods due to low processing cost and ease of operation. Moreover, the possibility of thermal degradation of bioactives cannot be ignored due to the high temperatures of the solvents and the longer extraction time (Cabana et al., 2013). In general, the art of separation is improving, with new methods and procedures rapidly being developed. Solvent extraction has been improved by novel extraction methods, such as microwave or ultrasound extraction and supercritical fluid extraction, to obtain better yields in less time. Conventionally used solvent extraction methods in the food industry include maceration, percolation, and Soxhlet extraction techniques. The main disadvantage of maceration and percolation is that the process can be time-consuming, taking from a few hours up to several weeks (Margeretha et al., 2012). Maceration is a popularly used technique in the food industry for making red wine (Joscelyne, 2009). Soxhlet extraction is adequate for both initial and bulk extraction and it gives the advantage of stage-wise extraction, where complete extraction of the phytoconstituent of interest is possible. (Kumoroa et al., 2009). A recent extraction method of solvent extraction of plant materials by hydrofluorocarbon-134a is being used, which offers significant environmental advantages over conventional techniques. Fragrant components of essential oils, phytopharmacological extracts are mostly extracted using this technique. These extracts can be used directly without further physical or chemical treatment. The solvent used here is 1,1,2,2-tetrafluoroethane, also known as hydrofluorocarbon-134a (HFC-134a), it is not flammable and nontoxic (Kumar, 2014). Extraction using phytonics is advantageous because it uses a combination of solvents that can be employed along with HFC-134a. In this way various classes of phytoconstituents can be segregated from each other. The biological products obtained using phytonics have extremely low residual solvent. These solvents are neither acidic nor alkaline and, therefore, they have only minimal potential reaction effects on the botanical materials. The processing plant is totally sealed so that the solvents are continually recycled and completely recovered after each production cycle. The waste biomass from these plants is dry and ecofriendly to handle (Swapna et al., 2015). The phytonics process can be employed for extraction in biotechnology (e.g., for the production of antibiotics), in the herbal drug industry, in food, essential oil and flavor industries, and in the production of other pharmacologically active products. It is specifically utilized in the production of top-quality pharmaceutical-grade extracts, pharmacologically active intermediates, antibiotic extracts, and phytopharmaceuticals. The technique is being used in the extraction of high-quality essential oils, oleoresins, natural food colors, flavors, and aromatic oils from all manner of plant materials (Ghosh et al., 2011). Extraction of Tanshinone IIA from *Salvia miltiorrhizabunge* was done recently using phytosol solvent extraction by Dean et al. They studied the effect of extraction by varying the phytonic solvent mixture system using a mixture of butane and dimethyl ether along with 1,1,2,2-tetrafluoroethane. An increased yield of Tanshinone IIA, consuming less time, using the core solvent, 1,1,2,2-tetrafluoroethane was observed (Dean et al., 1998).

2.2 Extraction Using Three-Phase Partitioning

This technique employs an organic solvent like tertiary butanol and a salt, such as ammonium sulfate, to separate enzymes, and proteins from a crude mixture. Tertiary butanol is completely water miscible. On the addition of appropriate concentration of salt, the crude mixture separates out into two phases: a lower aqueous phase and an upper organic phase. Proteins in the aqueous solution get partitioned forming a third phase, between the lower aqueous and upper *t*-butanol phase (Dennison and Lovrien, 1997). This is the basis of three-phase partitioning (TPP). Fig. 13.1 represents the laboratory setup for TPP. In TPP, polar molecules get concentrated in the lower aqueous phase, whereas lipids, enzymes, and pigments are accumulated in upper organic phase. Separation of any compound by TPP is governed by an association of various factors, which include salting out, isotonic precipitation, cosolvent precipitation, osmolytic and kosmotropic effect, protein hydration shifts, and electrostatic forces (Shah et al., 2004). TPP can be conducted between microliter and liter levels. TPP does not require any pretreatment measures before being used with crude cell cultures. This technique has an edge over chromatography because it performs simutaneous purification and concentration of the biomolecules (Rachana and Jose, 2014).

Various parameters affect the extraction yield and time of biomolecules when three-phase extraction is used. Ammonium sulfate concentration has significant effect on the TPP system

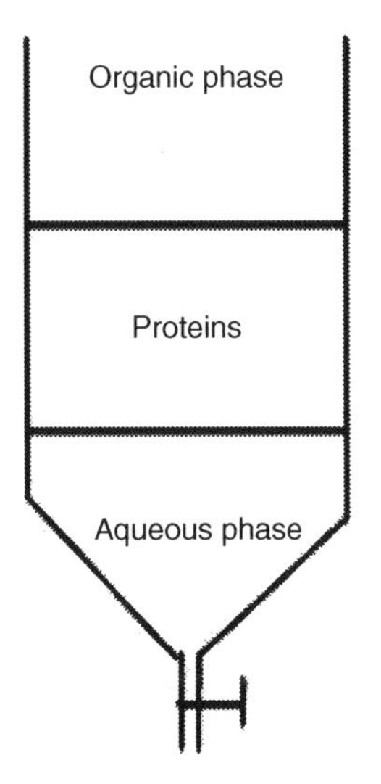

Figure 13.1: Three-Phase Partitioning Setup.

(Rao and Rathod, 2015b). It results in protein precipitation by salting out mechanism. Protein solubility is affected by the type of salt and its concentration. It has been reported that at higher salt concentration water molecules are attracted by salt ions resulting in an intense interaction between proteins, and the protein molecules further coagulate through hydrophobic interactions (Li et al., 2006a). Enzymes show different behavior in TPP systems on the basis of the pH of the system. The salting out of proteins depends on the net charge of the proteins, which is highly pH dependent. Proteins tend to precipitate most readily at their isoelectric point (pI). Electrostatic forces and binding of salts anion to cationic protein molecules, which promote macromolecular contraction and conformational shrinkage, are the main causes of the strong salt anion-pH dependency in salting out. Proteins tend to precipitate most readily at their pI (isoelectric point). Below the pI, proteins are positively charged and quantitatively precipitated out by TPP. On the other hand, negatively charged proteins are more soluble and not easily precipitated. Therefore, the selective pH should be carefully investigated to enhance the purification fold and recovery of target enzymes. Proteins are damaged by moderate or severe pH conditions, for example, exposure to acidic pHs or below, before the addition of sulfate or *t*-butanol may denature them (Gaur et al., 2007). Apart from pH, temperature for this operation is important because the use of low temperatures in solvent or salt precipitation dissipates the heat generated, ensuring minimal protein denaturation. Dogan and Tari also reported improved yields and purification of *exo*-polygalacturonase using TPP at 25°C rather than 37°C. Temperature of 25°C gave a 25.5% recovery of with a 6.7-fold purification, whereas only 8.8% recovery and 0.7-fold purification was obtained at 37°C (Dogan and Tari, 2008). However, Rajeeva and Lele showed that there was no significance in recovery and purification fold of laccase by TPP at 5°C compared to 35°C. Temperature of 5°C gave a 55% yield with a 4.2-fold purification, whereas a 53% recovery and 4.1-fold purification was obtained at 35°C (Rajeeva and Lele, 2011). TPP has been formerly used for the separation and purification of a number of biomolecules. TPP was used to recover alkaline proteases from farmed giant catfish viscera. Vidhate and Singhal (2013) reported the consequence of TPP for efficient extraction of kokum kernel fat (Vidhate and Singhal, 2013). TPP can be combined with other novel extraction techniques, such as microwave- and ultrasound-assisted extraction for improvement in yield and quality of extract. Rathod and Rao (2015) have recently separated and partially purified andrographolide from *Andrographis paniculata* using ultrasound assisted TPP.

2.3 Aqueous Two-Phase Extraction

Aqueous two-phase system was initially applied to the separation of plant organelles and viruses (Albertsson, 1958). Since then, attention has been directed toward widening its application scenario. During the last two decades, lot of work has been done to develop feasible separation techniques with aqueous two-phase systems for various biological materials, and proteins. Aqueous two-phase extraction (ATPE) involves the presence of either two immiscible mixing polymers, or one polymer with salt, which are water-soluble at certain concentrations (Goja et al., 2013). Optimization of several parameters, including polymer molecular weight and concentration, is essential. The presence of salts and the ionic strength of salts play a pivotal role in improving extraction yield. The partitioning of phytoconstituents can also be decided based on the pH and affinity of the biomolecule for the phase-forming polymer (Patil and Raghavarao, 2007). Some examples of biomolecule extraction by ATPE in the recent years are extraction from a variety of fruits, such as serine protease extracted from mango (Liu et al., 2012), bromelain from pineapple, and invertase from tomato (Goja et al., 2013). Electrostatic interactions, salting out effect, steric hindrance, van der Waals forces, and ionic interactions of molecules dominate the extraction by ATPE (Saravanan et al., 2008). Polymer (e.g., PEG) characteristics, such as its molecular size and weight and its quantity are elemental to decide the partitioning of the active components. PEG with low concentration is preferable for better separation (Yucekan and Onal, 2011). The PEG of low molecular weight has a hydrophilic moiety with short-polymer chains attached toward its end. This reduces the polymer hydrophobicity and an effective separation of biomolecules is observed (Rao and Nair, 2011). The use of PEG of higher molecular weight causes an overcrowding in the upper phase of the system. This leads to a movement of biomolecules in the lower phase thereby reducing the partition coefficient (Karkas and Önal, 2012). Researchers confirmed the dependency of polymer concentration on protein partition coefficient (K_P) and enzyme partition coefficient (K_E) by conducting a series of experiments. Polymer of intermediate concentration was suitable to achieve proper partitioning and purification. Using a polymer of high molecular weight and concentration could not separate the biomolecules adequately (Hemavathi and Raghavarao, 2011). Salt concentration is another parameter of great importance. The negatively charged proteins move to the upper phase of the system, whereas the ones having positive charge settle in the lower phase (Karkas and Önal, 2012). Salts that increase hydrophobic interactions among the active components are preferably in ATPS system. Salts, such as phosphates, sulphates, carbonates, and citrates have been widely experimented with. Negatively charged salts exhibit better partitioning ($SO4^{-2} > HPO4^{-2} >$ acetate) than the positively charged ones ($NH4^+ > K^+ > Na^+ > Mg^{2+} > Ca^{2+}$). Salts, such as phosphate and ammonium carbonate show better specificity, nontoxicity, and scalability. Hence, they are preferredover other salts (Goja et al., 2013). Isoelectric point of a protein (pI) greatly influences its separation within the system (Barbosa et al., 2011). A change in pH of the system alters the charge on the proteins to be separated, and this is followed by the partitioning of the desired proteins into the respective phases (Mohamadi et al., 2007). ATPE for most biomolecules are generally performed at neutral pH, as they are highly stable under such environment. ATPE can be implemented in the partitioning and purification of biomolecules in the food industry. However, the PEG/salt is unsuitable for scaling up the ATPE system owing to the environmental hazards from the salts and polymers, and the high cost incurred from recycling them. To vanquish these drawbacks, few alterations can be implemented to ATPE to purify high value phytochemicals in an economical and safe manner. Another modification in the ATPE system can be the development of a continuous process for the partitioning of proteins at a commercial level. This technique renders the purification of high value products in a time efficient manner, along with a better partition coefficient of proteins. Recently Zhang et al. (2016) utilized focused microwave extraction accompanied with aqueous two-phase extraction (FMAATPE) coupled with reversed micellar extraction for the recovery and purification of alkaloids from *Sophora flavescens* Ait. The alkaloids extracted by FMAATPE were further purified with reverse micelles of sodium dodecyl benzene sulfonate/isooctane/*n*-octanol. The relative purity of total alkaloids in the product was improved greatly from 58.30% to 88.75% (w/w), and purification recovery was higher than that of solvent extraction, ion exchange resin, and macroporous resin adsorption (Zhang et al., 2016). Negative-pressure cavitation (NPC) accompanied by ATPE was performed for enrichment of flavonoids and stilbenes from the pigeon pea leaves. The results demonstrated that NPC-ATPE could be an alternative method for simultaneous extraction and enrichment of natural compounds in a single step (Wang et al., 2015).

2.4 Extraction Using Hydrotrope Solutions

Hydrotropy refers to the ability of certain compounds termed hydrotropes to increase the solubility of sparingly soluble or water soluble organic solutes in aqueous solutions. It is a consequence of the tendency of amphiphilic hydrotrope molecules to aggregate among themselves and with other hydrophobic molecules. It is a molecular phenomenon whereby adding a second solute (hydrotrope) helps to increase the aqueous solubility of poorly soluble solutes (Nagarajan et al., 2016). The use of hydrotropes in extraction was investigated by few research groups because of its high selectivity in separation and easy recovery of natural compounds. The phenomena of hydrotropic extraction can be initiated by penetration of hydrotropes on the plant cell wall, followed by solubilization of the active compound. The hydrotrope solution should either partly dissolve the cell membrane or at least destabilize the cell wall structure during the process. Surfactant solutions at fairly high concentrations form lamellar crystalline structures in the solutions. Hydrotropes have a tendency to destabilize lamellar liquid crystalline phases of a conventional surfactant in aqueous solutions. The destabilizations of lamellar liquid crystalline structures of surfactants in aqueous solutions by hydrotropes have been extensively studied because of their similarity with the natural cell membrane structure (Mishra and Gaikar, 2009). The double phospholipid layers on the cell membrane resembles the lamellar crystalline structure of surfactant in aqueous solutions. Hence the hydrotrope, which shows the tendency to disrupt lamellar crystalline structure will also be capable of destabilizing the cell wall structure to extract bioactive compounds. The amphiphilic nature of the hydrotrope molecules also imparts the well-known characteristic of semipermeability to the cell membranes. Moreover, the hydrotrope helps the reduction of surface intensity on the cell wall and enhances the permeability and thus hydrotrope molecules penetrate into cell wall easily to extract bioactive compounds (Dandekar and Gaikar, 2003). The hydrotrope molecules penetrate the cell wall at higher concentration aggregate to form a stack-like structure. The nonpolar bioactive compounds would enter the hydrophobic layers of these assemblies, adding themselves between the layering molecules to form a stable structure. These hydrotropic solutions precipitates the bioactive compounds out of the solution on dilution with distilled water thus enabling the easy recovery of the extracted solute (Kumar et al., 2014). Temperature plays a vital role in influencing the phytochemical extraction efficiency. The rigidity of plant cell wall becomes much weaker at high temperatures, easing out the accessibility of hydrotropic solvents. The purity of the extract will also be affected due to higher temperatures. Apart from the target solute, dissolving ability of other intracellular compounds of the plant also increases rapidly, this then affects the final product purity. This situation was seen in the recovery of piperine from the sodium butyl monoglycolsulfate (NaBMGS) extraction solutions, where the extraction efficiency increased from 43% to 56% with a significant decrease in the purity when extraction was performed at a temperature range of 30–65°C. At the lower temperature (30°C), the liquid lamellar structure of the cell membranes destabilizes. These properties enable a more selective transport of piperine into the hydrotropic solutions (Raman and Gaikar, 2002). With sodium *n* butyl benzene sulfonate (NaNBBS), the extraction yield of embelin at 40°C was 95% with 65% purity. At 50°C, the maximum extraction was recorded, but with a further reduction in purity. At higher temperatures, the cell structure was disrupted, and the permeability of the hydrotrope solution increased. This facilitates increased diffusion of fatty acids and oleoresins into the solution and decreases the selectivity of the hydrotrope solution (Latha, 2006). Meanwhile, a contradictory behavior was exhibited by diosgenin, where solubility of diosgenin was found to decrease significantly above 60°C, suggesting the possibility of compound decomposition. A higher amount of sample loading in hydrotropic solvents makes the entire mixture too concentrated. This condition reduces the ability of the hydrotrope molecules to penetrate into the plant sample. The plant sample might absorb the hydrotrope solution becoming wet and swollen, without releasing the targeted solute, forming a semisolid mixture or slurry. Factors, such as the size of the hydrophobic portion and the number of $-CH_2$ groups in the structure, largely contribute to the extraction efficiency of the solute. In most hydrotropic extraction methods, higher extraction efficiency is obtained using aromatic sulfonates, such as NaNBBS or sodium cumene sulfonate (NaCS) (Mishra and Gaikar, 2004). In most cases, a large particle size results in a better extraction efficiency. A sample with a large particle size provides more surface area for hydrotrope molecule accessibility, increasing the penetration level. A significant decrease in the purity of extracted piperine from 98% to 89% was observed with decreasing particle size from 710 to 50 μm (Raman and Gaikar, 2002). Another study demonstrated the hydrotropic extraction of limonin from *Citrus aurantium* seed. Parameters, such as surfactant concentration, reaction temperature, and solid-to-solvent ratio influence the extraction system. Sodium salicylate (Na-Sal) and sodium cumene sulphonate (Na-CuS) were used for hydrotropic extraction. Highest limonin yield at 2M hydrotrope concentration, at 45°C and 10% solute to solvent ratio was observed. Hydrotropic extraction minimizes the use of organic solvents

(Dandekar et al., 2008). Embelin (2,5-dihydroxy-3-undecyl-*p*-benzoquinone) is an abundant phenolic compound found in Embeliaribes. Among the selected hydrotropes, recovery of embelin was insignificant with low purity from the sodium *P*-toluenesulfonate (NaPTS) and sodium xylenesulfonate (NaXS) extracts. NaNBBS and NaCS promoted the selective extraction of embelin, recovering 95% and 92% with high purity. Both NaBBS and NaCS have comparably high amounts of hydrophobic regions and low surface tension to enhance the level of penetration compared to NaPTS and NaXS (Latha, 2006).

2.5 Accelerated Solvent Extraction

Accelerated solvent extraction (ASE) or pressurized solvent extraction (PSE), or pressurized liquid extraction (PLE) is a technique that has been developed as an alternative to current extraction techniques, such as Soxhlet extraction, maceration, percolation or reflux, offering advantages with respect to extraction time, solvent consumption, extraction yields, and reproducibility. It uses organic solvents at elevated pressure and temperature to enhance the extraction yield (Shams et al., 2015). Increased temperature accelerates the extraction kinetics and elevated pressure keeps the solvent in the liquid state, thus enabling safe and rapid extractions. Furthermore, high pressure forces the solvent into the matrix pores and hence, facilitates extraction of analytes. High temperatures decrease the viscosity of the liquid solvent, allowing a better penetration of the matrix and weakened solute matrix interactions. Also, elevated temperatures enhance diffusivity of the solvent resulting in increased extraction speed (Plaza and Turner, 2015). Solvent selection depends on the polarity of the component being analyzed and compatibility with any postextraction processing steps and quantification equipment. Sometimes, the use of adsorbents in the sample cell along with sample may offer some highest degree of selectivity in ASE procedure. Typically, adsorbent is loaded into the sample cell first (outlet end) and the sample is loaded on the top of the adsorbent. This way flow of solvent during the extraction is such that undesired compounds may retain in the cell by adsorbent. However, these depend on types and physicochemical properties of solvents and target analytes, including undesired compounds in the samples (Mottaleb and Sarker, 2012). ASE operates by moving the extraction solvent through an extraction cell containing the sample. The sample cell is heated by unmediated connection with the oven. The extraction is carried out by contacting the sample with the hot solvent in both static and dynamic modes. When the extraction is complete, compressed nitrogen moves all of the solvent from the cell to the vial for analysis. The filtered extract is collected away from the sample matrix, ready for analysis. The elevated pressure serves to perpetuate the solvent in the liquid phase during separation process and ensures that the solvent remains in contact with the sample (Kassing et al., 2010). Some researchers reported that pressure had little effect on the extraction. However, others suggested that high pressure might increase extraction efficiency by forcing the solvent into the matrix pores demonstrated that the pressure effect was matrix-dependent. In their investigations, ASE carried out on tea leaves at temperature of 70–100°C resulted in lower caffeine yields when the pressure was increased above 100 bar. The elevated pressure might have caused compression of the soft tea matrix, reducing the efficiency of transport of the molecules out of the plant matrix and penetration of the solvent into its entire matrix (Rostango et al., 2004). ASE is generally performed at higher temperatures, to ameliorate the extraction kinetics by the interruption of plant matrix and analyte interactions, increased molecular motion of the solvent and greater analyte solubility in the extraction solvent due to elevated temperature. During the static phase of extraction, the diffusion of compounds from plant matrix into the solvent occurs without the outflow of solvent from the extraction vessel. The extract is finally accumulated by quickly providing the extraction cell with new solvent and nitrogen as an inert gas (Pitipanapong et al., 2007). An increase in the static extraction time generally increases the extraction yield until equilibrium is reached, increasing the static extraction time further does not result in further improvement on compound recovery (Xianwen et al., 1997). The flush volume, 40%–60% of the cell volume, is the amount of fresh solvent, that is used to sluice the sample after the completion of the process. While it was reported increased yields of active compounds from Angelica sinensis with reduced flush volumes (Li et al., 2006b), flush volume was reported to have no significant effect on extraction yields of Cortex Dictam (Jiang et al., 2006). Reduction of vessel void volume with an inert packing material ensures better solvent-matrix contact, reduces analyte oxidation due to the presence of air, and reduces solvent consumption. Diatomateous earth is commonly used especially if the material to be extracted exists as a fine powder. Neutral glass has also been used as a dispersant to reduce the volume of solvent used for extraction.

The obvious benefits of ASE cover rapid extraction for varied sample quantity, drastic reduction in the solvent quantities used, broad range of applications, and dealing with acidic and alkaline matrix. This technique proffers lesser cost per reaction, thus dramatically bringing down the solvent utilization (Khaled et al., 2015).

2.6 Supercritical Fluid Extraction

The biggest interest for the last decade has been the applications of supercritical carbon dioxide because it has a near ambient critical temperature (310°C), thus biological materials can be processed at temperatures around 350°C. The density of the supercritical CO_2 at around 200 bar pressure is close to that of hexane, and the solvation characteristics are also similar to hexane; thus, it acts as a nonpolar solvent. The major advantage is that a small reduction in temperature, or a slightly larger reduction in pressure, will result in almost the entire solute precipitating out as the supercritical conditions are changed or made subcritical. Supercritical fluids can produce a product with no solvent residues (De Melo et al., 2014). Examples of pilot and production scale products include decaffeinated coffee, cholesterol-free butter, low-fat meat, evening primrose oil, squalene from shark liver oil and so on. The solvation characteristics of supercritical CO_2 can be modified by the addition of an entrainer, such as ethanol, however, some entrainer remains as a solvent residue in the product, negating some of the advantages of the residue-free extraction (Sapkale et al., 2010).

A simplified process scale SFE system is shown in Fig. 13.2 and a typical batch extraction proceeds as follows. Raw material is charged in the extraction tank, which is equipped with temperature controllers and pressure valves at both ends to keep desired extraction conditions. The extraction tank is pressurized with the fluid by means of pumps, which are also needed for the circulation of the fluid in the system. From the tank the fluid and the solubilized components are transferred to the separator where the solvation power of the fluid is decreased by increasing the temperature, or more likely, decreasing the pressure of the system. The product is then collected via a valve located in the lower part of the separator (Capuzzo et al., 2013).

SC–CO_2 extraction has found application in the food industry, and today there are several products in the markets processed with SC–CO_2. Unusual vegetable oils, such as wheat germ oil, essential oils, green coffee oil, phospholipids, rice bran oil, fatty acids, and bioactive compounds have been extracted from fruits and vegetables using SC–CO_2. SC–CO_2 with water as cosolvent can be employed to selectively extract caffeine from coffee and green tea while avoiding the extraction of antioxidants (Ciftci, 2012). Supercritical extraction is used to add value to by-products of food industry. Extraction of by-products allows the removal of valuable compounds. These include extraction of polyphenols from rice wine lees and pomegranate seeds from juice production, and extraction of carotenoids from tomato and Sea buckthorn pomaces. Tsuda et al. (1995) demonstrated the supercritical CO_2 extraction of tamarind seed coat. At high reaction pressure and temperature, the extract showed improved antioxidant activity (Tsuda et al., 1995). Addition of cosolvents, such as ethanol, improved the antioxidant activity of the extracts (Herrero et al., 2010). This extraction method is extensively used for producing extracts of improved antioxidant effect. In another example, Yepez et al. (2002) demonstrated the ability of supercritical CO_2 extraction to produce extracts with improved antioxidant activity, which are odorless and flavorless extracts from coriander seeds

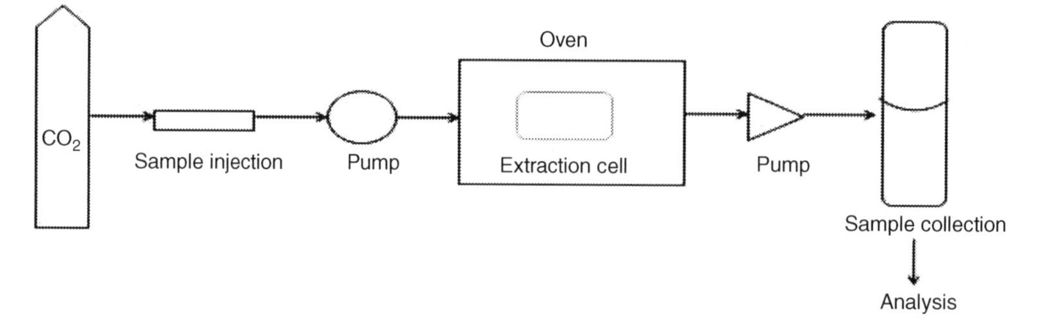

Figure 13.2: Supercritical CO_2 Extraction Setup.

(*Coriander sativum*) (Yepez et al. 2002). Due to the high antioxidant activity associated with vitamin E and its family of compounds, this has received an increasing interest in the field of extraction. Supercritical CO_2 extraction of vitamin E oil from a plant called Silybummarianum was demonstrated. It was observed that improved extraction yield of 19% was obtained at 60°C and 200 bar pressure (Hadolin et al., 2001). Natural pigment isolation is trending in the food industry because of the safety factor associated with natural ingredients and also because the associated antioxidant activities that can make them high-value products. Cadoni et al. (2000) demonstrated the extraction of lycopene and β-carotene from tomato from its skin and pulp. An improved product composition of 65% lycopene and 35% β-carotene was obtained at 275 bar pressure and a reaction temperature of 80°C. Both compounds exhibited separate solubility characteristics during the extraction process. Hence a development of two-step extraction process to segregate lycopene and β-carotene was performed. In the initial step, extraction was performed at 275 bar and 40°C, followed by the separation at 275 bar pressure and 80°C. They obtained a final product with 87% lycopene and 13% β-carotene composition. This was because β-carotene preferably got extracted under during the first step (Cadoni et al., 2000).

Subcritical water extraction (SWE) is a new and powerful technique at temperatures between 100°C and 374°C and pressure high enough to maintain the liquid state. Unique properties of water are its disproportionately high boiling point for its mass, high polarity, and a high dielectric constant. As the temperature rises, there is a marked and systematic reduction in permitivity with an enhanced diffusion rate and a decrease in the viscosity and surface tension. In consequence, more polar target materials with high solubility in water at ambient conditions are extracted most efficiently at lower temperatures, whereas moderately polar and nonpolar targets require a less polar medium induced by elevated temperature (HaghighiAsl and Khajenoori, 2013). Compared with supercritical fluid extraction, the required equipments for the subcritical water extraction (SWE) are relatively simpler and not much higher pressure is required. SWE has been employed to extract phytochemicals from a variety of plant or vegetable material. Rosemary (*Rosmarinus officinalis* L.) has been one of the most profoundly experimented subjects. Ibanez et al. performed the extraction of rosemary by SWE at a various temperatures. Extraction of antioxidant compounds was performed 25–200°C. It was concluded that the yield of antioxidants improved with temperature (Ibanez et al., 2003).

Even though SFE has an upper hand over conventional techniques, the equipment cost can be thought of as an obstacle to industrial scale commercialization of supercritical process. A safety risk due to high pressures is another concern. One of the ways to make the extraction processes economically feasible is the extraction of feed materials at higher pressures and temperatures and using cosolvents to extract most of the compounds, and then fractionate the extract to obtain different classes of high value fractions. SFE has been suggested as the method of choice for thermo labile compounds extraction. It has been used for developing an ever-expanding niche in the food industry whether it is used as a solvent for extraction or analyses (Capuzzo et al., 2013). Nowadays, SFE is seen as an advance method that can provide fast, reliable, clean, and cheap methods for routine analysis.

3 Physical Methods of Extraction

Physical methods of extraction involve alteration of the plant cell wall by rupturing them by physical means. In this section various physical extraction techniques in the food industry are explained alongside their applications.

3.1 Microwave-Assisted Extraction

The area of microwave-assisted extraction of bioactive compounds is at a benign stage. In the last two decades, new investigations have been prompted by an increasing demand of more efficient extraction techniques, amenable to automation. Lesser extraction times, reduced solvent consumption, energy and costs saved, were the main tasks pursued. Driven by these goals, advances in microwave extraction have resulted in a number of innovative techniques. Electric and magnetic fields oscillate perpendicular to each other and are responsible for producing microwaves. Ionic conduction and dipole rotation of solvent molecules cause its superboiling when exposed to microwave radiations. The solvent molecules attempt to position themselves along the changing electric field. But when the solvent molecules fail to realign them along the electric field, the solutions are heated up owing to the frictional resistance. Thus, it can be concluded that only dielectric materials with permanent dipoles can be utilized in microwave extraction (Tatke and Jaiswal, 2011). Microwave-assisted extraction requires less solvent consumption, has a shorter operational time, possess high recoveries and produces high extraction yield (Mandal et al., 2007). This technique outweighs SFE due to it being economical and easy to operate. Another merit of using microwave-assisted extraction is its reduced extraction time (Kaufmann and Christen, 2002).

The moisture present within plant cells heats up on exposure to microwave radiations. This causes evaporation of moisture and generates great amount of stress on the cell wall leading to its rupture. Phytoconstituents are released from the ruptured cells, hence increasing their extraction yield. The extraction yield of biomolecules is improved on soaking the plant material with solvents of high polarity for a particular time period. The cellulose within the plant cells is targeted and broken down in sometime (Tatke and Jaiswal, 2011). Elevated temperatures provide higher extraction yield as the biomolecule solubility improves, thus escalating the penetrability of solvents into plant cells. In one instance, the images of scanning electron micrographs (SEM) of normal plant material, heat-refluxed extract and MAE extract were collated. The structure of plant cell wall in the heat refluxed extract and normal extract remained intact. The cell walls of MAE treated extract was fully fragmented and broken (Jassie et al., 1997). Heat-reflux method doesn't rupture the cell walls because extraction here involves permeation and eventual solubilization processes to bring the phytoconstituents out of the substrate (Delhaes and Drillon, 1987). Parameters to be optimized in microwave-assisted extraction include the choice of solvent, microwave-irradiation time, microwave power, and solid-loading ratio. Correct choice of solvent is essential for any extraction reaction. Nonpolar solvents are transparent to microwave radiations and are unsuitable for microwave extraction. Polar solvents, such as water, have decent microwave absorbing capacity and get heated up faster and enhance the extraction process. Scientists have even experimented with a combination of different levels of microwave-absorbing solvents, to obtain high extraction yields (Chen et al., 2008). To promote green extraction technology, researchers designed solvent-free MAE systems, where no solvent was used and the water present within the cell walls aided the extraction process (Luque-Garcia and Luque de Castro, 2003). The solid loading is an essential factor in MAE. There must be enough solvent volume to submerge the plant material into it during the extraction reaction. A higher solvent volume gives greater extraction yields in case of traditional extraction. In MAE, a higher solvent volume might reduce the extraction yield because of the nonuniform subjection of the reaction mixture to microwaves. Heating time is another parameter, that is, studied MAE. Due to exposure of the reaction mixture to microwaves for a longer duration, there are chances of instability of heat sensitive components (Al-Harahsheh and Kingman, 2004). Microwave power and microwave reaction time are dependent on each other. Microwave reaction is usually operated at lesser or intermediate power for a longer time. The use of high microwave power is not suitable for thermally sensitive compounds. Researchers have utilized very high microwave power and observed that it has no major effects on the yield of flavanoids (Raner et al., 1993). At elevated temperatures there is a reduction in solvent viscosity with an increase in solubilizing and penetrability properties. Higher is the surface area of plant mixture more is the extraction yield. This is brought about by reducing the particle size of the plant matrix by milling, grinding, and homogenization. Kothari and Seshadri (2010) performed MAE on powdered seeds, to extract flavonoids from *Annona squamosa* and *Carica papaya* seeds. They observed an increase in the extraction yield of flavanoids because they had used finely ground seeds with low particle size (Kothari and Seshadri, 2010).

There are two forms of microwave system used these days—focused microwave ovens and multimode extraction vessel. A multimode system utilizes controlled pressure and temperature for extraction. Focused microwave-assisted system only focuses microwave irradiation on that area of extraction vessel carrying the reaction mixture. In closed microwave system, microwaves randomly reflect within the microwave vessel and every

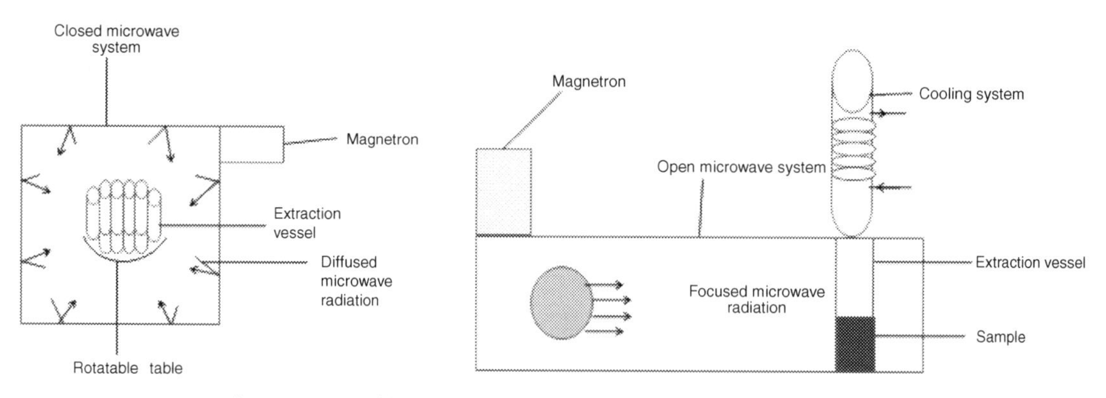

Figure 13.3: Diffused and Focused Microwave System Setup.

part of the vessel is equally irradiated (Tatke and Jaiswal, 2011). Fig. 13.3 shows the setup of an open as well as a closed microwave system. Several modified microwave-extraction systems have been developed. A few modifications incorporated are vacuum-induced microwave-assisted extraction of thermo labile compounds at gentle operating conditions (Chen et al., 2014), nitrogen-protected microwave-assisted extraction in which nitrogen is used to pressurize the extraction vessel to avoid the oxidation of oxygen sensitive molecules during extraction (Yu et al., 2009), ultrasonic microwave-assisted extraction, where a synergy of microwaves and ultrasonic waves augment the mass transfer process, which then acts on splintering the plant cells and segregating the active components into the extraction solvent, and dynamic microwave-assisted extraction, where the extraction process is operated in an uninterrupted manner and coupled to an online analytical step. Table 13.2 depicts the examples of synergistically used extraction techniques in the recent times. Combining microwave-assisted extraction with other extraction techniques, which may overcome the limitations of the single extraction technique and obtain the satisfactory extraction efficiency, has attracted more attention recently. A novel extraction method using microwave combined with ultrasound was established to isolate lycopene from tomato. The results showed that extraction rate of the new method was higher than that of ultrasonic-assisted extraction (97.4% vs. 89.4%) (Kumcuoglu et al., 2014). Most recently, concerns about environmental damage, resource depletion and food safety have fostered investigations into alternative ways to extract secondary metabolites from plants without the addition of solvents, especially organic ones like methanol (Zhang et al., 2011). Solvent-free microwave hydrodiffusion and gravity extraction of flavonols from onion was optimized. A protocol of pressurized solvent-free microwave-assisted extraction was developed for the isolation of antioxidants from the berries of Hippophaerhamnoides. Microwave techniques led to the highest antioxidant activity of the crude extract as compared to maceration, pressurized hot water extraction, and pressing extraction (Huma et al., 2009).

Table 13.2: Examples of synergistically used extraction techniques.

Extractions	Substrates for Extraction	Extraction Conditions	Yield	References
Ultrasound-assisted three-phase partitioning	Three-phase extraction with ultrasound on recovery of *Andrographis paniculata*	40% w/w ammonium sulfate, 32 min, pH 7, 1:1 slurry to *t*-butanol ratio, 30°C	35.28 mg/g of andrographolide	Rathod and Rao (2015)
Hydrotrope extraction coupled with ultrasound	Optimization of ultrasound-assisted extraction of defatted wheat germ proteins by reverse micelles	Power 363 W, ultrasonic time 24 min, and pulse mode 2.4 s on and 2 s off	Extraction efficiency of 45.6%	Zhu et al. (2009)
Microwave-assisted enzymatic extraction	Microwave-assisted aqueous enzymatic extraction of oil from pumpkin seeds	Enzyme concentration of 1.4%, w/w, at 44°C, 66 min, power 419 W	64.17% oil recovery	Jiao et al. (2014)
Ionic liquids-based enzyme-assisted extraction	Application of ionic liquids based enzyme-assisted extraction of chlorogenic acid from eucommiaulmoides leaves	1-alkyl-3-methylimidazolium ionic liquid, 0.5 M cellulose, 120 min, pH 3, 50°C	8.32 mg/g chlorogenic acid	Liu et al. (2016)
Ultrasound-assisted compound enzymatic extraction	Ultrasound-assisted compound enzymatic extraction of polysaccharides from black currant	Enzyme concentration 1.575%, pH 5.3, ultrasonic time 25.6 min	14.28% polysaccharides	Xu et al. (2015)
Ultrasound-assisted supercritical carbon dioxide	Ultrasound-assisted improved supercritical carbon dioxide to produce extracts enriched in oleanolic acid and ursolic acid from *Scutellaria barbata*	Particle size of 0.355 mm, 27.6 MPa, 55°C, CO_2 flow rate of 2.1 mL/min, 14.1% (v/v) aqueous ethanol solvent, 50 min extraction time	Oleanolic acid 14.142 μg/g; ursolic acid 59.275 μg/g	Yang et al. (2013)
Aqueous two-phase system coupled with ultrasound	Aqueous two-phase system coupled with ultrasound for the extraction of lignans from seeds of *Schisandra chinensis*	25% (w/w) $(NH_4)_2SO_4$ and 19% (w/w) ethanol, 20:1 of solvent: solid, 800 W, 61.1 min	13.1 mg/g schizandrin, 1.87 mg/g schisantherin A	Guo et al. (2013)

3.2 Ultrasound-Assisted Extraction

The field of biotechnology provides multitudinous opportunities for the use of ultrasound technology among the food researchers. Ultrasound-assisted extraction (UAE) is a process, which is time-efficient, improves the extraction yield along with better product quality. In the past few years numerous experiments on commercial application of ultrasound as a process-intensification technique in extraction of phytochemicals from plants, essential oils, and enzymes have been conducted. Ultrasound energy results in high shear forces with the reaction medium. When an ultrasound wave comes in contact with the solvent molecules, longitudinal waves form, further creating alternating compression and rarefaction waves within the medium (Petigny et al., 2013). In such areas, cavitation gas bubbles are found. During the expansion phase, the bubbles increase their surface area, which increases the penetration of gas. During the compression phase there comes a stage wherein the ultrasonic energy provided becomes insufficient to keep the gas bubbles in the medium intact. As a result, the bubbles burst, releasing great energy, increasing the temperature and pressure of the system. Due to severe collision of the bubbles with each other, shock waves are created within the system. When these cavitation bubbles strike on the plant material, microjets are formed and projected toward the solid surface. The high pressure and temperature involved in this process disrupt the cell wall of the plant matrix, and its contents are released into the medium. As a consequence, employing UAE has benefits in increased mass transfer, better solvent penetration, less dependence on solvent use, extraction at lower temperatures, faster extraction rates, and greater yields of product (Rao and Rathod, 2015a).

Ultrasound is renowned for its commercial application in the extraction of biomolecules from a variety of herbal extracts. UAE is known to enhance the extraction efficiency and yield of oils from seeds and flowers. Microfractures and rupture of cell walls in the seeds and plants provided more evidence for the mechanical effects of ultrasound, thus causing exudation of the biomolecules, unlike the traditional extraction techniques (Vilkhu et al., 2008). Ultrasound was used as a pretreatment technique prior to the extraction of oil from seeds of *Jatropha curcas* L. by aqueous enzymatic oil extraction, a process evaluated by Shah et al. (2005). Ultrasonic pretreatment of the almond seeds before aqueous oil extraction and aqueous enzymatic oil extraction provided significantly greater yield in a time efficient manner (Shah et al., 2005). Clove buds are used as a spice and in food flavoring. In a study by Wei et al., clove oil and α-humulene were extracted from cloves using supercritical carbon dioxide extraction with and without ultrasound assistance (USC–CO_2 and SC–CO_2, respectively) at different temperatures (32–50°C) and pressures (9.0–25.0 MPa). The results demonstrated that the USC–CO_2 extraction procedure may extract clove oil and α-humulene from clove buds with better yields and shorter extraction times than conventional extraction techniques while utilizing less severe operating parameters (Wei et al., 2016). It is very important to select optimum process parameters so that the overall process of ultrasonic extraction becomes industrially and economically viable. In ultrasound-assisted reactions, proper selection of operating conditions, such as ultrasound reactor configuration, ultrasound intensity, ultrasound frequency, and duty cycle in addition to reaction parameters, such as time, temperature, solute-to-solvent ratio, is very crucial. It is beneficial to select the right configuration setting of the ultrasound reactor. Sonochemical reactors are of two types: bath type and horn or probe type. Bath-type sonochemical reactors are the indirect contact type of reactors. In this type of ultrasound setup, the correct placement of the glass reactor and the level of water inside the sonochemical bath are very crucial (Khan and Rathod, 2015). The correct position of the glass reactor can be selected on the basis of earlier literature on the mapping studies on the same type of sonochemical reactions. Rao et al. have mapped an ultrasound bath for extraction of andrographolide from *A. paniculata* to identify the active and passive zones in the bath (Rao and Rathod, 2015a). Ultrasound technology is still in the developing stage and has abundant future scope to convert this technology to industrial scale.

3.3 Enzyme-Assisted Extraction

Enzymes extraction technology has created a buzz among researchers owing to the need for environmentally benign extraction technologies. Enzymes are target-specific and very efficient. Enzymes act on the active site of the cell wall, causing it to break and release the desired biomolecules with improved extraction yield. Enzymes offer the merits of enzyme reusability, without affecting the activity of biomolecules to a great extent (Puri et al., 2012). A variety of compounds, including carbohydrates, essential oils, natural colors, fragrances, and compounds of medicinal value have been extracted by enzyme-assisted extraction of their substrates (Sowbhagya and Chitra, 2010). The basic principle of enzyme-assisted extraction is that the enzymes hydrolyze the plant cell wall and disrupt it completely under optimum experimental conditions, to release the intracellular components. Enzymes act on the plant cell wall by binding onto its active site. This causes the enzyme to change its shape so as to fit into the active site of the substrate, causing maximum interaction between the enzyme and substrate. Change in the shape of enzyme leads to breakage of bonds of cell wall, thereby releasing the active constituents out of it (Meyer, 2010). The efficiency of extraction depends on system temperature, mode of action of the enzyme, extraction duration, enzyme loading, substrate availability, and pH of the system.

Enzyme extraction technology has several limitations. High enzyme cost and stability is one of the major hurdles to make this technology commercial. Enzymes present in the market these days are unable to completely hydrolyze the bonds in plants, thus limiting the reducing the release of active components from within it.

Enzyme technology is demonstrated for the extraction of an array of biomolecules, such as curcumin from turmeric, oil from grape seed, and mangiferin from mango leaves. Tchabo et al. (2015) used ultrasound radiation in synergy with enzymes for the extraction of phytochemicals from mulberry. Using this hybrid technology not only improved the product quality but also greatly reduced the time required for extraction (Tchabo et al., 2015). Additionally, enzymes are employed as a pretreatment technique of plant materials. Enzymes, such as papain, cellulases, lipases, and pectinases can be often employed to rupture the plant matrix, hence improving the yield of phytoconstituents from plants (Puri et al., 2012). Enzymes like α-amylase and amyloglucosidase were used in the pretreatment for ionic liquid extraction of curcumin from *Curcuma longa*. This enhanced the extraction yield of curcumin by a drastic 60% (Sahne et al., 2017). A recent case of the use of an enzyme system, is in the processing of pectin polysaccharide for enhancing extraction of antioxidants. Enzyme concentration of 0.1% w/w drastically boosted the extraction yield of antioxidants. In a second instance, the researchers performed enzyme-assisted extraction of lycopene from tomato tissues by employing cellulases and pectinases to significantly increase the lycopene yield (Choudhari and Ananthanarayan, 2007). Enzymes, such as cellulases, α-amylase and pectinase are generally employed for oil extraction. Oils extracted using hexane is of inferior quality as compared to oils extracted by applying enzymes (Ptichkina et al., 2008). In an instance of enzyme-assisted extraction of grapes skin during vinification process, higher pigment (anthocyanin) extraction was observed after extraction (Munoz et al., 2004). Oil extraction from defatted grape seed meal using enzymes rendered a 60% enhanced yield of phenolic compounds, unlike nonenzymatic process (Tobar et al., 2005). Enzyme-extraction technology is an emerging area in food research. A profound analysis of the cellular structure of the plant matrix with the employment of particular enzymes, which attack the active site of the substrate can be used as a strategy for improved hydrolysis.

3.4 Pulsed Electric Field Assisted Extraction

Pulsed electric field (PEF) assisted extraction is commonly understood as a fast, nonthermal, and highly effective method for extraction of intracellular compounds. This treatment involves discharge of high voltage electric pulses of a few microseconds into the food product, which is placed or passed between two electrodes (Angersbach et al., 2000). The extent of cell membrane disintegration is an important factor influencing the extraction procedure. Different physical, chemical, or biological treatments cause breakage of the cellular membrane. Therefore, among different nonthermal processing methods used in the food industry, pulsed-electric field (PEF) treatment was found to be a promising one and minimally invasive for breakage of cellular tissue (Yongguang et al., 2006).The application of electric fields for a short duration of a few to several hundred microseconds is capable of inducing cell membrane permeabilization through a phenomenon called electroporation (Asavasanti et al., 2011). Pore formation is a dynamic process and can be reversible or irreversible depending on the intensity of the pulse field treatment. The electric breakdown can be reversible if it is generated with PEF treatment of low intensity and when induced pores are smaller as compared to the membrane area. Enhancing the treatment intensity by increasing electric field strength (E) or treatment time (t) results in the formation of large pores, and reversible permeabilization will turn into irreversible disruption of the cell membrane (Barbosa-Cánovas et al., 2000). The irreversible permeabilization of the cell membrane in the plant tissue provides a wide range of process applications where disruption of the cell membrane is required, including extraction. Electric field strength is an important parameter that controls the efficiency of electroporation of the cellular tissue. Bazhal et al. (2003) presented classification of the PEF modes as low ($E \leq$ 100–200 V/cm), moderate (E = 300–1500 V/cm), and high ($E >$ 1500 V/cm). With low electric field strength, the treatment time should be longer for electroporation of the cellular membranes. It has been found experimentally that time needed for electroporation of cellular membrane of the different biological tissues is inversely proportional to the electric field strength by factor dependency (Bouzrara and Vorobiev, 2001). The material to be extracted is placed between a set of electrodes, which are connected to each other with a nonconductive material to avoid electrical conduction between them (Mohamed and Eissa, 2012). High voltage electrical pulses are generated from the electrodes. The pass the electrical pulse to the substance placed between the electrodes in the treatment chamber, which is responsible for the extraction (Zimmermann, 1986). Parameters that can affect extraction of biomolecules include the electric field strength, time of exposure, reaction temperature, and properties of the treatment substrate (Fincan et al., 2004). Application of PEF is also demonstrated to extract secondary plant metabolites, for example, sugar beets (Giri and Mangaraj, 2013). Extraction of antioxidants from various fruit juices, such as blueberry, apple, and cranberry juices have been dealt with PEF technology (Evrendilek et al., 2000; Qin et al., 1998). Extracts from grape by-products contain bioactive substances, such as anthocyanins, which could be used as natural antioxidants or colorants. The effect of heat treatment at 70°C combined with the effect of different emerging novel technologies, such as ultrasonics (35 KHz), high-hydrostatic pressure (600 MPa), and pulsed electric fields (3 kV cm^{-1}) (PEF) showed a great feasibility and selectivity for extraction purposes. After 1 h extraction, the total phenolic content of samples subjected to novel technologies was 50% higher than in the control samples (Corrales et al., 2008). Recently, the result of pulsed electric fields assisted extraction on antiinflammatory and cytotoxic activity of brown rice bioactive compounds, such as γ-oryzanol, polyphenols, and phenolic acids were studied. The results show that PEF-assisted extraction, enhancing the yield of bioactive compounds with respect to untreated extracts, significantly promotes their antioxidant activity, which is correlated with an increased HT29 cells cytotoxicity (Quagliariello et al., 2016).

4 Conclusions

By-products of plant origin represent an abundant source of bioactive compounds. However, to exploit these resources commercially relevant strategies for their extraction must be developed. There are many demerits to both chemical and physical extraction processes, which must be conquered. The common constraints are the pretreatment of raw materials prior to extraction, nonrecyclability of process thereby making the technology uneconomical, interference of hazardous by-products after extraction, observance in batch-to-batch variation in the system, and a variation in product quality due to the type of extraction method used. New technology for increasing the yield and limiting the use of solvents must be developed. As a result these extracts can be widely used without any negative effect on food properties. To conclude, the future holds great opportunities for modern extraction technologies in the area of food research on a commercial scale.

References

Afoakwah, A.N., Owusu, J., Adomako, C., Teye, E., 2012. Microwave assisted extraction (MAE) of antioxidant constituents in plant materials. Global J. Biosci. Biotechnol. 1, 132–140.

Albertsson, P.A., 1958. Partition of proteins in liquid polymer–polymer two-phase systems. Nature 182, 709–711.

Al-Harahsheh, M., Kingman, S.W., 2004. Microwave assisted leaching: a review. Hydrometallurgy 73, 189–193.

Angersbach, A., Heinz, V., Knorr, D., 2000. Effects of pulsed electric fields on cell membranes in real food systems. Innov. Food Sci. Emerg. Technol. 1, 135–149.

Asavasanti, S., Ristenpart, W., Stroeve, P., Barret, M., 2011. Permeabilization of plant tissue by monopolar pulsed electric fields: effect of frequency. J. Food Sci. 76, 98–111.

Barbosa, J.M.P., Souza, R.L., Fricks, A.T., Zanin, G.M., Soares, C.M.F., 2011. Purification of lipase produced by a new source of bacillus in submerged fermentation using an aqueous two-phase system. J. Chromatogr. B 879, 3853–3858.

Barbosa-Cánovas, G.V., Pierson, M.D., Zhang, Q.H., Schaffner, D.W., 2000. Pulsed electric fields. Food Sci. 65, 65–79.

Bazhal, M., Lebovka, N., Vorobiev, E., 2003. Optimisation of pulsed electric field strength for electroplasmolysis of vegetable tissue. Biosyst. Eng. 86, 339–345.

Bouzrara, H., Vorobiev, E., 2001. Non-thermal pressing and washing of fresh sugarbeet cossettes combined with a pulsed electrical field. Zuckerindustrie 126, 463–466.

Cabana, R., Silva, L., Valentao, P., Viturro, C.I., Andrade, P.B., 2013. Effect of different extraction methodologies on the recovery of bioactive metabolites from *Saturejaparvifolia* (Phil.) Epling (Lamiaceae). Ind. Crops Prod. 48, 49–56.

Cadoni, E., De Giorgi, M.R., Medda, E., Poma, G., 2000. Supercritical CO_2 extraction of lycopene and β-carotene from ripe tomatoes. Dyes. Pigm. 44, 27–32.

Capuzzo, A., Maffei, M.E., Occhipinti, A., 2013. Supercritical fluid extraction of plant flavors and fragrances. Molecules 18, 7194–7238.

Chen, L., Song, D., Tian, Y., Ding, L., Yu, A., Zhang, H., 2008. Application of on-line microwave sample-preparation techniques. Trends Anal. Chem. 27, 151–159.

Chen, F., Mo, K., Liu, Z., Yang, F., Hou, K., Li, S., Zu, Y., Yang, L., 2014. Ionic liquid-based vacuum microwave-assisted extraction followed by macroporous resin enrichment for the separation of the three glycosides salicin, hyperin and rutin from Populus bark. Molecules 19, 9689–9711.

Choudhari, S.M., Ananthanarayan, L., 2007. Enzyme aided extraction of lycopene from tomato tissues. Food Chem. 102, 77–81.

Ciftci, O.M., 2012. Supercritical fluid technology: application to food processing. J. Food Process. Technol. 3, 1–2.

Corrales, M., Toepfl, S., Butz, P., Knorr, D., Tauscher, B., 2008. Extraction of anthocyanins from grape by-products assisted by ultrasonics, high hydrostatic pressure or pulsed electric fields: a comparison. Innov. Food Sci. Emerg. Technol. 9, 85–91.

Dandekar, D.V., Gaikar, V.G., 2003. Hydrotropic extraction of curcuminoids from turmeric. Sep. Sci. Technol. 38, 1185–1215.

Dandekar, D.V., Jayaprakasha, G.K., Patil, B.S., 2008. Hydrotropic extraction of bioactive limonin from sour orange (*Citrusaurantium* L.) seeds. Food Chem. 109, 515–520.

De Melo, M.M.R., Silvestre, A.J.D., Silva, C.M., 2014. Supercritical fluid extraction of vegetable matrices. J. Supercrit. Fluids 92, 115–176.

Dean, J., Liu, B., Price, R., 1998. Extraction of tanshinone IIA from *Salvia miltiorrhizabunge* using supercritical fluid extraction and a new extraction technique, phytosol solvent extraction. J. Chromatogr. A 799, 343–348.

Delhaes, P., Drillon, M., 1987. Organic and Inorganic Low-Dimensional Crystalline Materials. Plenum, New York, NY.

Dennison, C., Lovrien, R., 1997. Three phase partitioning: concentration and purification of proteins. Protein Express Purif. 11, 149–161.

Devi, J.R., Thangam, E.B., 2010. Extraction and Separation of glucosinolates from brassica oleraceaevar rubra. Adv. Biol. Res. 4, 309–313.

Dogan, N., Tari, C., 2008. Characterization of three-phase partitioned exo-polygalacturonase from *Aspergillus sojae* with unique properties. Biochem. Eng. J. 39, 43–50.

Evrendilek, G.A., Jin, Z.T., Ruhlman, K.T., Qui, X., Zhang, Q.H., Ritcher, E.R., 2000. Microbial safety and shelf life of apple juice and cider processed by bench and pilot scale PEF systems. Innov. Food Sci. Emerg. Tech. 1 (1), 77–86.

Fincan, M., DeVito, F., Dejmek, P., 2004. Pulsed electric field treatment for solid–liquid extraction of red beetroot pigment. J. Food Eng. 64, 381–388.

Gaur, R., Sharma, A., Khare, S.K., Gupta, M.N., 2007. A novel process for extraction of edible oils: enzyme assisted three phase partitioning (EATPP). Biores. Technol. 98, 696–699.

Ghosh, U., Haq, M.B., Chakraborty, S., 2011. Application of systematic technologies for the extraction of novel phytoconstituents from pharmacologically important plants. Int. J. Chem. Anal. Sci. 2, 1153–1158.

Giri, S.K., Mangaraj, S., 2013. Application of pulsed electric field technique in food processing: a review. Asian J. Dairy. Food Res. 32, 1–12.

Goja, A.M., Yang, H., Cui, M., Li, C., 2013. Aqueous two-phase extraction advances for bioseparation. J Bioproces. Biotechniq. 4, 1–8.

Guo, Y.X., Han, J., Zhang, D.Y., Wang, L.H., Zhou, L.L., 2013. Aqueous two-phase system coupled with ultrasound for the extraction of lignans from seeds of *Schisandrachinensis* (turcz.) Baill. Ultrason. Sonochem. 20, 125–132.

Hadolin, M., Skerget, M., Knez, Z., Bauman, D., 2001. High pressure extraction of vitamin E-rich oil from Silybummarianum. Food Chem. 74, 355–364.

HaghighiAsl, A., Khajenoori, M., 2013. Subcritical water extraction. In: Nakajima, H. (Ed.), Mass Transfer: Advances in Sustainable Energy and Environment Oriented Numerical Modeling. Intech, Croatia, pp. 459–487.

Handa, S.S., 2008. An overview of extraction techniques for medicinal and aromatic plants. In: Handa, S.S., Khanuja, S.S., Longo, G., Rakesh, D. (Eds.), Extraction Technologies for Medicinal and Aromatic Plants. International Centre for Science and Technology (ICS-UNIDO), Trieste, Italy, pp. 21–25.

Hemavathi, A.B., Raghavarao, K. S.M.S., 2011. Differential partitioning of β-galactosidase and β-glucosidase using aqueous two phase extraction. Process Biochem. 46, 649–655.

Herrero, M., Mendiola, J.A., Cifuentes, A., Ibáñez, E., 2010. Supercritical fluid extraction: recent advances and applications. J. Chromatogr. A 1217, 2495–2511.

Huma, Z., Vian, M., Maingonnat, J.F., Chemat, F., 2009. Clean recovery of antioxidant flavonoids from onions: optimising solvent free microwave extraction method. J. Chromatogr. A 1216, 7700–7707.

Ibanez, E., Kuvatov, A., Senorans, F.J., Cavero, S., Reglero, G., Hawthorne, S.B., 2003. Subcritical water extraction of antioxidant compounds from rosemary plants. J. Agric. Food Chem. 51, 375–382.

Jassie, L., Revesz, R., Kierstead, T., Hasty, E., Matz, S., 1997. Microwave-assisted solvent extraction. In: Kingston, H.M., Haswell, S.J. (Eds.), Microwave-Enhanced Chemistry: Fundamentals, Sample Preparation, Applications. American Chemical Society, Washington, DC, pp. 569–610.

Jiang, Y., Li, S.P., Chang, H.T., Wang, Y.T., Tu, P.F., 2006. Pressurized liquid extraction followed by high-performance liquid chromatography for determination of seven active compounds in Cortex Dictamni. J. Chromatr. A 1108, 268–272.

Jiao, J., Li, Z., Gai, Q., Li, X., Wei, F., Fu, Y., Ma, W., 2014. Microwave-assisted aqueous enzymatic extraction of oil from pumpkin seeds and evaluation of its physicochemical properties, fatty acid compositions and antioxidant activities. Food Chem. 147, 17–24.

Joscelyne, V., 2009. Consequences of extended maceration for red wine colour and phenolics. PhD Thesis. University of Adelaide, Adelaide, Australia.

Kadam, S.U., Tiwari, B.K., Donnell, P.C., 2013. Application of novel extraction technologies for bioactives from marine algae. J. Agric. Food Chem. 61, 4667–4675.

Karkas, T., Önal, S., 2012. Characteristics of invertase partitioned in poly (ethylene glycol)/magnesium sulfate aqueous two-phase system. Biochem. Eng. J. 60, 142–150.

Kassing, M., Jenelten, U., Schenk, J., Strube, J., 2010. A new approach for process development of plant-based extraction processes. Chem. Eng. Technol. 33, 377–387.

Kaufmann, B., Christen, P., 2002. Recent extraction techniques for natural products: microwave-assisted extraction and pressurised solvent extraction. Phytochem. Anal. 13, 105–113.

Kennedy, D.O., Wightman, E.L., 2011. Herbal extracts and phytochemicals: plant secondary metabolites and the enhancement of human brain function. Adv. Nutr. 2, 32–50.

Khaled, A.S., Nahla, S.A., Ibrahim, A.S., Mohamed-Elamir, F.H., Mostafa, M.E., Faiza, M.H., 2015. Green technology: Economically and environmentally innovative methods for extraction of medicinal and aromatic plants (MAP) in Egypt. J. Chem. Pharm. Res. 7, 1050–1074.

Khan, N., Rathod, V.K., 2015. Enzyme catalyzed synthesis of cosmetic esters and its intensification: a review. Process Biochem. 50., 1793–1806.

Kothari, V., Seshadri, S., 2010. Antioxidant activity of seed extracts of *Annona squamosa* and *Carica papaya*. Nutr. Food Sci. 40, 403–408.

Krinsky, N.I., Yeum, K.J., 2003. Carotenoid-radical interactions. Biochem. Biophys. Res. Comm. 305, 754–760.

Kumar, S., 2014. Design and development of extraction process in the isolation of phytopharmaceuticals from plant sources. Int. J. Med. Health Prof. Res. 1, 28–38.

Kumar, S.V., Raja, C., Jayakumar, C., 2014. A review on solubility enhancement using hydrotropic phenomena. Int. J. Phar. Phar. Sci. 6, 1–7.

Kumcuoglu, S., Yilmaz, T., Tavman, S., 2014. Ultrasound assisted extraction of lycopene from tomato processing wastes. J. Food Sci. Technol. 51, 4102–4107.

Kumoroa, A.C., Hasana, M., Singh, H., 2009. Effects of solvent properties on the Soxhlet extraction of diterpenoid lactones from *Andrographis paniculata* leaves. Sci. Asia 35, 306–309.

Latha, C., 2006. Selective extraction of embelin from embeliaribes by hydrotropes. Sep. Sci. Technol. 16, 3721–3729.

Li, B.B., Smith, B., Hossain, M., 2006a. Extraction of phenolics from citrus peels: II. Enzyme-assisted extraction method. Sep. Purif. Tech. 48, 189–196.

Li, P., Li, S.P., Lao, S.C., Fu, C.M., Kan, K.K.W., Wang, Y.T., 2006b. Optimization of pressurized liquid extraction for Z-ligustilide, Z-butylidenephthalide and ferulic acid in *Angelica sinensis*. J. Pharm. Biomed. Anal. 40, 1073–1079.

Liu, Y., Wu, Z., Zhang, Y., Yuan, H., 2012. Partitioning of biomolecules in aqueous two-phase systems of polyethylene glycol and nonionic surfactant. Biochem. Eng. J. 69, 93–99.

Liu, T., Sui, X., Li, L., Zhang, J., Liang, X., Li, W., Zhang, H., Fu, S., 2016. Application of ionic liquids based enzyme-assisted extraction of chlorogenic acid from Eucommiaulmoides leaves. Anal. Chim. Acta 90, 91–99.

Luque-Garcia, J.L., Luque de Castro, M.D., 2003. Where is microwave based analytical treatment for solid sample pre-treatment going? Trends Anal. Chem. 22, 90–99.

Mandal, V., Mohan, Y., Hemalatha, S., 2007. Microwave assisted extraction: an innovative and promising extraction tool for medicinal plant research. Pharmacogn. Rev. 1, 7–18.

Margeretha, I., Suniarti, D.F., Herda, E., Masud, Z., 2012. Optimization and comparative study of different extraction methods of biologically active components of Indonesian propolis *Trigona* spp. J. Nat. Prod. 5, 233–242.

Meyer, A.S., 2010. Enzyme technology for precision functional food ingredient processes. Ann. NY Acad. Sci. 1190, 126–132.

Mishra, S.P., Gaikar, V.G., 2004. Recovery of diosgenin from dioscorea rhizomes using aqueous hydrotropic solutions of sodium cumene sulfonate. Ind. Eng. Chem. Res. 43, 5339–5346.

Mishra, S.P., Gaikar, V.G., 2009. Hydrotropic extraction process for recovery of forskolin from coleus forskohlii roots. Ind. Eng. Chem. Res. 48, 8083–8090.

Mohamadi, H., Omidinia, E., Dinarvand, R., 2007. Evaluation of recombinant phenylalanine dehydrogenase behavior in aqueous two-phase partitioning. Process Biochem. 42, 1296–1301.

Mohamed, M.E., Eissa, A.H., 2012. Pulsed electric fields for food processing technology. In: Eissa, A.H. (Ed.), Structure and Function of Food Engineering. Intech, Dammam, Saudi Arabia, pp. 275–306.

Mottaleb, M.A., Sarker, S., 2012. Accelerated solvent extraction for natural products isolation. In: Walker, ., John, M. (Eds.), Methods in Molecular Biology. Humana, New Jersey, pp. 75–87.

Munoz, O., Sepulveda, M., Shwart, M., 2004. Effects of enzymatic treatment on anthocyanic pigments from grapes skin from Chilean wine. Food Chem. 87, 487–490.

Nagarajan, J., Heng, W., Galanakis, C.M., Ramanan, R.N., Eshwaraiah, R.M., Sun, J., Ismail, A., Ti, T., Krishnamurthy, N.P., 2016. Extraction of phytochemicals using hydrotropic solvents. Sep. Sci. Tech. 51 (7), 1151–1165.

Nair, D.G., 2015. Use of phytochemicals as functional food: an overview. PARIPEX Ind. J. Res. 4, 1–3.

Patil, G., Raghavarao, K.S.M.S., 2007. Aqueous two phase extraction for purification of C-phycocyanin. Biochem. Eng. J. 34, 156–164.

Pazitna, A., Dzurova, J., Spanik, I., 2014. Enantiomer distribution of major chiral volatile organic compounds in selected types of herbal honeys. Chirality 26, 670–674.

Petigny, L., Périno-Issartier, S., Wajsman, J., Chema, F., 2013. Batch and continuous ultrasound assisted extraction of boldo leaves (*Peumusboldus* Mol.). Int. J. Mol. Sci. 14, 5750–5764.

Pitipanapong, J., Chitprasert, S., Goto, M., Jiratchariyakul, W., Mitsuru, S.M., Shotipruk, A., 2007. New approach for extraction of charantin from *Momordica charantia* with pressurized liquid extraction. Sep. Purif. Technol. 52, 416–422.

Plaza, M., Turner, C., 2015. Pressurized hot water extraction of bioactives. Trends Anal. Chem. 71, 39–54.

Ptichkina, N.M., Markina, O.A., Rumyantseva, G.N., 2008. Pectin extraction from pumpkin with the aid of microbial enzymes. Food Hydrocolloids 22, 192–195.

Puri, M., Sharma, D., Barrow, C.J., 2012. Enzyme-assisted extraction of bioactives from plants. Trends Biotech. 30, 37–44.

Qin, B.L., Barbosa-Cánovas, G., Swanson, B., Pedrow, P.D., Olsen, R.G., 1998. Inactivating microorganisms using a pulsed electric field continuous treatments system. IEEE Trans. Ind. Appl. 34, 43–50.

Quagliariello, V., Iaffaioli, R.V., Falcone, M., Ferrari, G., Pataro, G., Donsì, F., 2016. Effect of pulsed electric fields: assisted extraction on anti-inflammatory and cytotoxic activity of brown rice bioactive compounds. Food Res. Int. 87, 115–124.

Rachana, C.R., Jose, L.V., 2014. Three phase partitioning: a novel protein purification method. Int. J. ChemTech Res. 6, 3467–3472.

Rajeeva, S., Lele, S.S., 2011. Three-phase partitioning for concentration and purification of laccase produced by submerged cultures of Ganoderma sp. Biochem. Eng. J. 54, 103–110.

Raman, G., Gaikar, V.G., 2002. Extraction of piperine from *Piper nigrum* (black pepper) by hydrotropic solubilization. Ind. Eng. Chem. Res. 41, 2966–2976.

Raner, K.D., Strauss, C.R., Vyskoc, F., Mokbel, L., 1993. A comparison of reaction kinetics observed under microwave irradiation and conventional heating. J. Org. Chem. 58, 950–995.

Rao, J.R., Nair, B.U., 2011. Novel approach towards recovery of glycosaminoglycans from tannery wastewater. Biores. Technol. 102, 872–878.

Rao, P.R., Rathod, V.K., 2015a. Mapping study of an ultrasonic bath for the extraction of andrographolide from *Andrographis paniculata* using ultrasound. Ind. Crops Prod. 66, 312–318.

Rao, P.R., Rathod, V.K., 2015b. Rapid extraction of andrographolide from *Andrographis paniculata* Nees by three phase partitioning and determination of its antioxidant activity. Biocatal. Agr. Biotechnol. 4, 586–593.

Rathod, V.K., Rao, P.R., 2015. Effect of three phase extraction with ultrasound on recovery and antioxidant activity of *Andrographis paniculata*. J. Biol. Active Prod. Nat. 5, 264–275.

Rostango, M.A., Palma, M., Barroso, C.G., 2004. Pressurized liquid extraction of isoflavones from soybeans. AnalyticachimicaActa 522, 169–177.

Sahne, F., Mohammadi, M., Najafpour, G.D., Moghadamnia, A., 2017. Enzyme-assisted ionic liquid extraction of bioactive compound from turmeric (*Curcuma longa* L.): isolation, purification and analysis of curcumin. Ind. Crops Prod. 95, 686–694.

Sapkale, G.N., Patil, S.M., Surwase, U.S., Bhatbhage, P.K., 2010. Supercritical fluid extraction: a review. Int. J. Chem. Sci. 8, 729–743.

Saravanan, S., Rao, J.R., Nair, B.U., Ramasami, T., 2008. Aqueous two-phase poly(ethylene glycol)–poly (acrylic acid) system for protein partitioning: influence of molecular weight, ph and temperature. Process Biochem. 43, 905–911.

Saxena, M., Saxena, J., Nema, R., Singh, D., Gupta, A., 2013. Phytochemistry of medicinal plants. J. Pharmacogn. Phytochem. 1, 168–182.

Shah, S., Sharma, A., Gupta, M.N., 2004. Extraction of oil from *Jatropha curcas* L. seed kernels by enzyme assisted three phase partitioning. Ind. Crops Prod. 20, 275–279.
Shah, S., Sharma, A., Gupta, M.N., 2005. Extraction of oil from Jatropha curcas L. seed kernels by combination of ultrasonication and aqueous enzymatic oil extraction. Biores. Technol. 96, 121–123.
Shams, K.A., Abdel-Azim, N.S., Saleh, I.A., Hegazy, M.F., El-Missiry, M.M., Hammouda, M.F., 2015. Green technology: economically and environmentally innovative methods for extraction of medicinal & aromatic plants (MAP) in Egypt. J. Chem. Pharm. Res. 7, 1050–1074.
Sowbhagya, H.B., Chitra, V.N., 2010. Enzyme-assisted extraction of flavorings and colorants from plant materials. Crit. Rev. Food Sci. Nutr. 50, 146–161.
Swapna, G., Jyothirmai, T., Lavanya, V., Swapnakumari, S., Sri Lakshmi Prasanna, A., 2015. Extraction and characterization of bioactive compounds from plant extracts: a review. Eur. J. Pharm. Sci. Res. 2, 1–6.
Tatke, P., Jaiswal, Y., 2011. An overview of microwave assisted extraction and its applications in herbal drug research. Res. J. Med. Plants 5, 21–31.
Tchabo, W., Ma, Y., Engmann, F.N., Zhang, H., 2015. Ultrasound-assisted enzymatic extraction (UAEE) of phytochemical compounds from mulberry (*Morus nigra*) must and optimization study using response surface methodology. Ind. Crops Prod. 63, 214–225.
Tobar, P., Soto, M., Chamy, R., Zuniga, M.E., 2005. Winery solid residue revalorization into oil and antioxidant with nutraceutical properties by an enzyme assisted process. Water Sci. Technol. 51, 47–52.
Tsuda, T., Mizuno, K., Ohshima, K., Kawakishi, S., Osawa, T., 1995. Supercritical carbon dioxide extraction of antioxidative components from tamarind (*Tamarindus indica* L.) seed coat. J. Agr. Food Chem. 43, 2803–2806.
Verkerk, R., Schreiner, M., Krumbein, A., Ciska, E., Holst, B., Rowland, I., Schrijver, R., Hansen, M., Gerhuser, C., Mithen, R., Dekker, M., 2009. Glucosinolates in Brassica vegetables: the influence of the food supply chain on intake, bioavailability and human health. Mol. Nutr. Food Res., 1.
Vidhate, G., Singhal, R.S., 2013. Extraction of cocoa butter alternative from kokum (*Garcinia indica*) kernel by three phase partitioning SI: extraction and encapsulation. J. Food Eng. 117, 464–466.
Vilkhu, K., Mawson, R., Simons, L., Bates, D., 2008. Applications and opportunities for ultrasound assisted extraction in the food industry: a review. Innov. Food Sci. Emerg. Technol. 9, 161–169.
Wang, X., Wei, W., Zhao, C., Li, C., Luo, M., Wang, W., Zu, Y., Efferth, T., Fu, Y., 2015. Negative-pressure cavitation coupled with aqueous two-phase extraction and enrichment of flavonoids and stilbenes from the pigeon pea leaves and the evaluation of antioxidant activities. Sep. Purf. Technol. 156, 116–123.
Wei, M.C., Xiao, J., Yang, Y., 2016. Extraction of α-humulene-enriched oil from clove using ultrasound assisted supercritical carbon dioxide extraction and studies of its fictitious solubility. Food Chem. 210, 172–181.
Williams, D.J., Edwards, D., Hamernig, I., Jian, L., James, A.P., Johnson, S.K., Tapsell, L.C., 2013. Vegetables containing phytochemicals with potential anti-obesity properties: a review. Food Res. Int. 52, 323–333.
Xianwen, L., Janssen, H., Cramers, C.A., 1997. Parameters affecting the accelerated solvent extraction of polymeric samples. Anal. Chem. 69, 1598–1603.
Xu, Y., Zhang, L., Yang, Y., Song, X., Yu, Z., 2015. Optimization of ultrasound-assisted compound enzymatic extraction and characterization of polysaccharides from blackcurrant. Carbohydr. Polym. 117, 895–902.
Yang, Y., Wei, M., Hong, S., Huang, T., Lee, S., 2013. Development/optimization of a green procedure with ultrasound-assisted improved supercritical carbon dioxide to produce extracts enriched in oleanolic acid and ursolic acid from *Scutellariabarbata* D. Don. Ind. Crops Prod. 49, 542–553.
Yepez, B., Espinosa, M., Lopez, S., Bolanos, G., 2002. Producing antioxidant fractions from herbaceous matrices by supercritical fluid extraction. Fluid Phase Equilib. 197, 879–884.
Yongguang, Y., Yuzhu, H., Yong, H., 2006. Pulsed electric field extraction of polysaccharide from *Rana temporariachensinensis* David. Int. J. Pharm. 312, 33–36.
Yu, Y., Chen, B., Chen, Y., Xie, M., Duan, H., Li, Y., Duan, Y., 2009. Nitrogen-protected microwave-assisted extraction of ascorbic acid from fruit and vegetables. J. Sep. Sci. 32, 4227.
Yucekan, I., Onal, S., 2011. Partitioning of invertase from tomato in poly (ethylene glycol)/sodium sulfate aqueous two-phase systems. Process Biochem. 46, 226–232.

Zhang, H., Yang, X., Wang, Y., 2011. Microwave assisted extraction of secondary metabolites from plants: current status and future directions. Trends Food Sci. Technol. 22, 672–688.

Zhang, Y., Gan, R.Y., Li, S., Zhou, Y., Li, A.N., Xu, D.P., Li, H.B., 2015. Antioxidant phytochemicals for the prevention and treatment of chronic diseases. Molecules 20, 21138–21156.

Zhang, W., Liu, X., Fan, H., Zhu, D., Wu, X., Huang, X., Tang, J., 2016. Separation and purification of alkaloids from *Sophora flavescens* Ait. by focused microwave-assisted aqueous two-phase extraction coupled with reversed micellar extraction. Ind. Crops Prod. 86, 231–238.

Zhu, K., Sun, X., Zhou, H., 2009. Optimization of ultrasound-assisted extraction of defatted wheat germ proteins by reverse micelles,J. Cereal Sci 50, 266–271.

Zimmermann, U., 1986. Electric breakdown, electropermeabilization and electrofusion. Rev. Phys. Biochem. Pharmacol. 105, 196–256.

CHAPTER 14

Extraction Technologies and Solvents of Phytocompounds From Plant Materials: Physicochemical Characterization and Identification of Ingredients and Bioactive Compounds From Plant Extract Using Various Instrumentations

Ida I. Muhamad, Nor D. Hassan, Siti N.H. Mamat, Norazlina M. Nawi, Wahida A. Rashid, Nuraimi A. Tan

University of Technology Malaysia, Johor Bahru, Johor, Malaysia

Diverse ingredients and bioactive compounds are present in different plant materials, such as leaves, stem, roots, flower, fruits, and seeds. The development of advance and modern instrumentation techniques make the identification and characterization of ingredients and bioactive compounds analysis easier. Phytocompounds are beneficial ingredients (phenolic, vitamin, amino acids, minerals) in the various plant materials, which show the potential therapeutic for various biological activities, such as treatment and prevention of cancer, cardiovascular, and other chronic diseases. This chapter aims to review the several physicochemical extraction techniques, including conventional and advanced techniques, such as solvent extraction, microwave-assisted extraction (MAE), ultrasonic-assisted extraction (UAE), aqueous extraction, enzymatic extraction, and supercritical fluid extraction (SFE). Different solvent system could give different chemical compounds. Polar solvents (water, ethanol, and methanol) are commonly solvent for phenolic compounds extraction, while nonpolar solvent (hexane, chloroform, petroleum ether) are used for oil and fats extraction. Additionally, the impacts of the extraction operations parameters and different polarity of solvent are highlighted. The comparisons of each technique and solvents are also summarized to facilitate better understanding of phytocompounds from plant material extracts. Physical analysis of plant extract, includes pH, color, moisture, flavor, and appearance from which was determined by using pH meter, colorimeter, and moisture analyzer, respectively. While, chemical analysis of plant extract includes fatty

Ingredients Extraction by Physicochemical Methods in Food
http://dx.doi.org/10.1016/B978-0-12-811521-3.00014-4

acids, heavy metals, amino acids, phytochemical compounds, such as volatile compounds (VOCs) and non-VOCs, and others. The chromatographic methods, such as liquid chromatography-mass spectrometry (LC-MS), high-performance liquid chromatography (HPLC), and ultraperformance liquid chromatography (UPLC) used for both VOC and non-VOCs, while gas chromatography (GC), gas chromatography (GC-MS), and gel permeation chromatography (GPC) used for VOCs, spectrometric techniques for specific known compounds. The nutritional value (carbohydrate, fat, protein, fiber, moisture) was assessed using proximate analysis. These physicochemical characterizations of ingredients and bioactive compounds using various instrumentations could provide informative and scientific reference for diverse potential uses of plant extracts especially for nutraceuticals and functional food applications.

1 Extraction of Colorant From Plant Parts

1.1 Method of Evaluation of Natural Colorant

In general, overall appearance and pigment content analysis are two approaches that need to be evaluated in a suitable formulation to study the potential of natural colorant (Francis, 1989). Natural colors can be described as organic colorant that are derived from agricultural, biological, or edible sources using food-preparation procedures, for example, curcumin (from turmeric), bixin (from annatto seeds), and anthocyanins (from red fruits). The refined form of natural colors are used as food additives, which are the green pigment chlorophyll, the caretonoids, which give yellow to red colors, and the flavonoids, with their principal subclass the anthocyanins, which give flowers and fruits their red to blue colors (Hendry and Houghton, 1992).

Changes in texture, particle size, microbial growth, and precipitation would have to be done visually to see overall appearance. Natural colorants are beneficial in improving visual color that can be measured either by spectrophotometry or tristimulus colorimetry. Pigment content analysis is necessary to understand the cause of the color changes. Tinctorial power of colorant extract can be measured by absorption data at one or two wavelengths. Dilution is necessary if the color of extracts is highly concentrated but this should be avoided because the solutions may not follow Beers law due to copigmentation effects. Generally, qualitative of anthocyanins involves three main steps in systematically approaches; extraction, purification, and identification (Jackman et al., 1987a).

1.2 Extraction Method of Anthocyanins

Extraction is one of important steps in getting the crude extract from plant (Chan et al., 2011). The purpose of the extraction and the nature of the constituent anthocyanins themselves determine the method of anthocyanins extractions from any of their various sources (Jackman and Smith, 1992).

It is also necessary to consider any factors that influence anthocyanins structure and stability. If the extracted pigments are to be analyzed subsequently, either qualitatively or quantitatively, it is desirable to choose a method that retains the pigments as close to their natural state (i.e., in vivo) as possible. Alternatively, if the extracted pigments are to be used as colorant/food ingredient that maximum pigment yield, tinctorial strength, and stability are of greater concern. It is also important that the extraction and cleanup procedures are not too complex, time-consuming, or costly (Jackman and Smith, 1992).

Studies on natural colorant extraction techniques were attempted by using both conventional water bath steaming, Soxhlet, maceration method, as well as nonconventional methods, such as ultrasonic (Cilek et al., 2012) and microwave (Zhang et al., 2014). In raising the efficiency and valuable extracts of bioactive compounds from plants, several techniques of extraction had been utilized, such as ultrasonic waves, supercritical fluids, or microwave (Parthasarathi et al., 2013). Different methods of extraction to get extracts, which are rich in anthocyanin, depend on different solvents used, such as methanol, ethanol, acetone, water, and mixture (Ahmed et al., 2011). Anthocyanins can be extracted from plant sources and carried out in suitable form depending on the application used.

1.3 Solvent Etraction

Extraction of anthocyanins with ethanol containing 200–2000 ppm. SO_2 has been reported to increase pigment yield and to result in concentrates possessing up to twice the tinctorial strength and flavor intensity as anthocyanin concentrates obtained from extraction with ethanol alone. Extraction of anthocyanins with aqueous SO_2 or bisulfate solutions also led to concentrates of higher purity and pigment stability than those obtained from hot water extraction. However, aqueous SO_2 extraction of anthocyanins would appear to be less selective than extraction with methanol-HCl followed by purification using ion-exchange chromatography (Hendry and Houghton, 1992).

1.3.1 Solvent selection for extraction of anthocyanins

Francis (1989) found that in most fruits and vegetables the anthocyanin pigments are located in cells near the surface of exterior. These water-soluble plant pigments are usually dissolved in the cell sap rather than in lipoidal bodies. Increasing hydroxylation or methoxylation of the aglycone makes it more violet. Anthocyanins are not stable in neutral or alkaline solution. Thus, extraction procedures have generally involved the use of acidic solvents, which denature the membranes of cell tissue and simultaneously dissolve the pigments, followed by ether precipitation (Markakis, 1982; Simpson, 1985). The traditional and most common means of anthocyanin extraction involves maceration or soaking of plant material in a low boiling point alcohol containing a small amount of mineral acid (e.g., $\leq 1\%$ HCl). Methanol is often used as the extracts are more easily concentrated prior to the application to paper (Harborne, 1958); however, due its toxicity, acidified ethanol may be preferred with food sources, although it is a less-effective extractant and more difficult to concentrate due to its higher boiling point (Markakis, 1982).

Acidification with hydrochloric acid serves to maintain a low pH, thereby providing a favorable medium for the formation flavylium chloride salts (Markakis, 1982; Simpson, 1985) from simple anthocyanins (Hendry and Houghton, 1992). The acid tends to stabilize anthocyanins but the use of mineral acids, such as HCl, may, however, alter the native form of complex pigments by breaking associations with metals, copigments, and so on (Hendry and Houghton, 1992; Moore et al., 1982). The solvent systems generally used for the extraction purposes are by no means specific for anthocyanins (Markakis, 1974). The loss of labile acyl and sugar residues may also occur during subsequent concentration. Thus, to obtain anthocyanins closer to their natural state, neutral solvents have been suggested for initial pigment extraction, including 60% methanol, *n*-butanol, ethylene glycone, propylene glycol, cold acetone, acetone/methanol/water mixtures, and simply, boiling water. In addition, weaker organic acids, mainly formic acid but also acetic, citric, and tartaric acids, have been used in extraction solvents (Hendry and Houghton, 1992).

In part of the study, 10% of formic acid was added to the methanol before extraction, creating a low pH environment for anthocyanin stabilization (Su et al., 2016). Sample preparation of anthocyanins extraction is the crucial step for producing product quality in food processing. The common method for extraction of anthocyanins included combining the material with water–organic solvent mixture (acetone-ethanol-methanol) at the optimum temperature, time, and solid–solvent ratio for anthocyanins extraction were at 80°C, 60 min, and 1:32, respectively (Truong et al., 2012).

1.4 Microwave-Assisted Extraction

From the variety of extraction techniques, MAE had drawn significant research particularly in food processing due to special heating mechanisms, moderate capital cost (Chan et al., 2011), simple and efficient (Parthasarathi et al., 2013). MAE is a method to selectively extract target compound from various raw material. MAE is a friendly technique since it shortens the time of extraction and has an inherent capability to quickly heat the sample-solvent mixture and thus speed up the extraction process. As compared to conventional method, MAE has many advantages, which offer much lowered extraction time (Dhobi et al., 2009; Parthasarathi et al., 2013), lesser solvent extraction, higher extraction rate, enhanced efficiency, as well as cost effective (Dhobi et al., 2009).

In previous studies, there have been reported that the extraction yield will increase by increasing the time of extraction, but there was not much increment of yield with further increase in time (Routray and Orsat, 2012). Generally, it has been examined that increasing microwave power leads to increasing the yield of extracted compounds. However, if the microwave power is too high, it can decrease the yield of extraction by damaging

Figure 14.1: Purple Sweet Potatoes Contain Anthocyanins.

or breakdown the compound of the product (Routray and Orsat, 2012). The key factors influencing microwave extraction, include effect of solvent system, time of application of microwave, microwave power level, temperature, and contact surface area (Routray and Orsat, 2012) (Fig. 14.1).

1.4.1 Mechanism of microwave extraction

MAE is a new process that applies microwave energy to heat the solvent and discharge the contents of the plant materials into the liquid phase (Parthasarathi et al., 2013). Basic mechanisms of heat and mass transfer in MAE are illustrated in Fig. 14.2. Although the mass transfer for MAE and convention occur in same direction, it differs because volumetric heating effect occurs in MAE process, which means that the heat is dissipated volumetrically inside the irradiated medium (Veggi et al., 2013). In addition, microwave heating speeds up

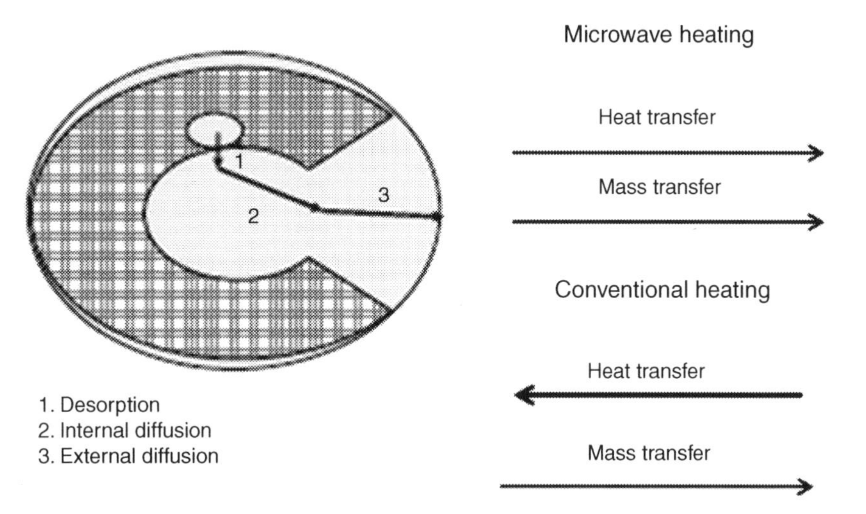

Figure 14.2: Mechanism of Microwave-Assisted Extraction of Natural Products. *Adapted from Perino-Issartier, S., Maingonnat, J.-F., Chemat, F., 2011. Microwave food processing. In: Proctor, A. (Ed.), Alternatives to Conventional Food Processing. Royal Society of Chemistry, United Kingdom, pp. 415–458.*

the rasing temperature, which depends on the microwave power and the dielectric loss factor of the irradiated material (Al Bittar et al., 2013).

According to Veggi et al. (2013), there are several steps that must occur during interaction of the solid–solvent extraction during MAE process:

1. penetration of the solvent into the solid matrix;
2. solubilization and/or breakdown of the component;
3. transport of the solute out of the solid matrix;
4. migration of the extracted solute from the external surface of the solid into the bulk solution;
5. movement of the extract with respect to the solid;
6. separation and discharge of the extract and solid.

Therefore, the solvent penetrates into the solid matrix by diffusion (effective), and the solute is dissolved until reaching a concentration limited by the characteristics of the solid (Veggi et al., 2013).

The extraction process takes place in three different steps: an equilibrium phase, transition phase, and diffusion phase (Fig. 14.3). During equilibrium phase, solubilization and partition govern the process in which the substrate is removed from the outer surface of the particle. Intermediary transition phases occur prior to diffusion. The resistance to mass transfer begins to appear in the solid–liquid interface; in this period the mass transfer by convection and diffusion prevails. In the last phase, the solute must overcome the interactions that bind it to the matrix and diffuse into the extracting solvent (Veggi et al., 2013).

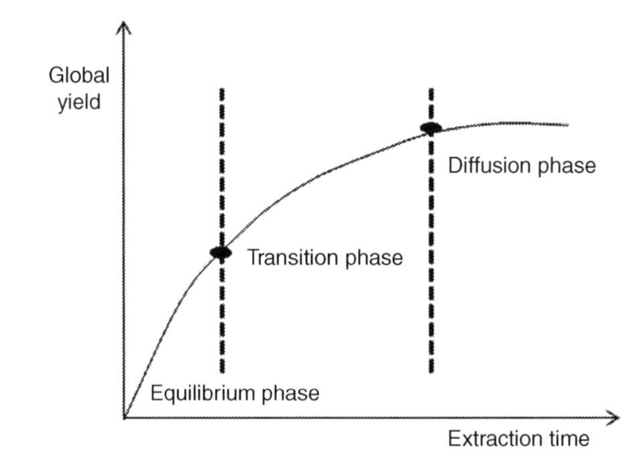

Figure 14.3: Schematic Representation of Yield Versus Time in Extraction Process.
Adapted from Veggi et al. (2013).

1.4.2 MAE instrumentation

Microwave ovens can be monomode or multimode cavity. The monomode cavity can generate a frequency, which excites only one mode of resonance. The sample can be placed at the maximum of the electrical field as the distribution of the field is known. The multimode cavity is large and the incident wave is able to affect several modes of resonance. This superimposition of modes allows the homogenization of the field. Systems, such as rotating plates are included for homogenization (Routray and Orsat, 2012). Presently, two types of MAE systems have been developed: open system and closed system. Open system is classified as a focused-mode system (monomode), which means that the system allows focused microwave radiation on a restricted zone in cavity. Meanwhile, closed system is classified as a multimode system, which allows random dissipation of microwave radiation in cavity by mode stirrer (Routray and Orsat, 2012).

1.4.3 Important parameters in microwave-assisted extraction

The efficiency of the MAE process is directly related to the selection of operating conditions and parameters, which affect the extraction yields and mechanism. A great attention should be given in understanding the effects and interactions of the factors, including solvent composition, solvent-to-feed ratio, extraction temperature and time, microwave power, and sample characteristics on the MAE process.

1.4.4 Effect of solvent compositions

The selection of an appropriate solvent is the most essential factor that has an effect on the MAE process. The solvent selection highly relies on the solubility of the compounds of interest, solvent penetration, and its interaction with the sample matrix and its dielectric constant (Veggi et al., 2013). Normally, the solvent has a high capacity to absorb microwave energy when the solvent has higher dielectric constant and dielectric loss, which causes faster rates of heating of the solvent as regards the plant material (Routray and Orsat, 2012). Due to these factors, water results as the best solvent of MAE (Spigno and De Faveri, 2009).

1.4.5 Solvent-to-feed ratio

The solvent-to-feed ratio also plays the vital role in extraction of MAE process. According to Afoakwah et al. (2012) the solvent volume must be sufficient to ensure that the sample is fully immersed in the solvent through the entire irradiation process. In conventional extraction method, a higher ratio of solvent to feed gives better extraction yield because of large volume of solvent raising the extraction recovery. In contrast, MAE process did not give better yield due to nonuniform distribution and exposure to electromagnetic wave (Veggi et al., 2013).

1.4.6 Effect of extraction time and cycle

The extraction time of MAE process is very short compared to conventional techniques. It requires heating up only for a few minutes to a half-hour. The increase of extraction time usually tends to increase the extraction yield but it faces the thermal degradation and oxidation during heating. Overheating occurs due to the high dielectric properties of the solvent, such as ethanol and methanol (Veggi et al., 2013), and thus further dilution with water enhances the heat capacity of the solvent combination. Irradiation time is also influenced by the dielectric properties of the solvent. Consequently, prolonging the microwave exposure may lead to degradation of the target compound with overheating of the solute/solvent system (Routray and Orsat, 2012).

If longer extraction time is required, it risks the thermolabile constituents. But it can be reduced via the extraction cycle to retain getting the high yield and thus preventing prolong heating using the same volume of solvent. This can be done by repeating the same extraction step until the process of extraction is completed. The number of cycles will be depending on the type of solvent and solid (Chan et al., 2011)

1.4.7 Effect of microwave power and extraction temperature

In general, higher extraction yield will be increased when exposed to higher microwave power until it becomes insignificant or declines. However, high microwave power might result in poor extraction yield due to thermal degradation (Routray and Orsat, 2012). Also, rapid rupture of the cell wall takes place at a higher temperature when using higher power, and resulting impurities leaches out into solvent together with the desired solutes (Chan et al., 2011). Therefore, power intensity must be chosen exactly to maximize yields, selectivity, stability of the target solutes (Chan et al., 2011). On the other hand, it also requires minimization of the extraction time until reaching the temperature needed, avoiding the "bumping" phenomenon in temperature during extraction (Veggi et al., 2013) and overpressure and excessive temperature (Flórez et al., 2014).

Temperature and microwave power are correlated with each other as elevated microwave power can bring up the extraction temperature, which enhances extraction yields, consequently increasing diffusivity of the solvent into the matrix and desorption and partition of the solute into the solvent (Flórez et al., 2014). Temperature is closely related to the power intensity and the choice of temperature greatly relies on the stability and extraction yield of the desired bioactive compound (Chan et al., 2011).

1.4.8 Advantages and drawbacks of MAE

MAE is increasingly employed for the extraction of the natural product as a promising alternative sample preparation technique for a numerous of applications (Kubra et al., 2013)

The use of MAE technology has gained several advantages:

1. Shorter extraction time. Microwave extraction takes much less time compared to conventional extraction. For instance, Soxhlet extraction takes time at least 1 h and occasionally 12–24 h for most extraction to utilize a lot of solvent. High temperature exposure and extended

extraction time affect the quality of the extract. In MAE process, it takes about 30 min or less; in consequence, considerably decreasing extraction time is required. The process is short and faster giving a higher yield of bioactive compound (Routray and Orsat, 2012).

2. The heating system can be turned on or off instantly.
3. High extraction speed. Accelerating the speed of heating because of heating directly to the solvent via microwave (Gadkari and Balaraman, 2014).
4. Less solvent. MAE enables a significant reduction in the consumption of solvent during extraction process (Gadkari and Balaraman, 2014). Consuming a low volume of solvent assists in decreasing the time required for solvent evaporation (Routray and Orsat, 2012).
5. Less environmental pollution. Increasing efficiency and clean transfer to energy lead to green environmental process (Veggi et al., 2013).
6. Small sample size. Small sample size as low as 0.1 g helps in maintaining the compound concentration in extracts besides overcoming the loss sensitivity caused by dilution (Routray and Orsat, 2012).
7. Moderate capital cost. The equipment setup for microwave extraction is much cheaper than nonconventional extraction methods, such as SFE (Chan et al., 2011).
8. Improved extraction yield and superior product quality. Material can be heated rapidly (Veggi et al., 2013).
9. High efficiency. MAE has the uniqueness of microwave heating as its interaction with extraction system and has led to enhancing the mass transfer (Chan et al., 2011).
10. Energy saving. The process occurs at low temperature, as well as the material absorbs the energy directly (Veggi et al., 2013).
11. Minor risk. MAE has low risk and minor safety issues because most of the extraction process is generated under atmospheric condition (Chan et al., 2011).

Nevertheless, some drawbacks and limitations are associated with MAE processing.

1. Nonpolar solvent should be avoided in MAE as they are poor absorbents for microwave heating. However, the addition of modifier into nonpolar solvent can help to solve this problem (Chan et al., 2011).
2. Low selectivity. MAE is highly dependent on the solvent nature and extraction temperature (Chan et al., 2011).
3. Additional filtration or centrifuge is required to remove the solid residue after the MAE process (Veggi et al., 2013).
4. Degradation of heat-sensitive bioactive compound occurs if exposed to high temperatures (Veggi et al., 2013).

1.5 Purification of Anthocyanins

Chromatographic technique, especially paper chromatography, has been carried out primarily in the purification of anthocyanins. Quantification of anthocyanins can be done by HPLC.

Jackman et al. (1987b) suggested a large number of developing solvents that have been used for purification purposes, which the selections depend on the nature of the crude anthocyanin sample. HPLC methodology has limited application because the use of authentic standards is required in conventional HPLC chromatograms to identify anthocyanins. The separation and purification of anthocyanin in large quantities leads to development of several column chromatograhic techniques, such as thin-layer chromatography (TLC), acetate precipitation, and ion-exchange resin column chromatoghraphy to concentrate and purify the anthocyanins. Paper chromatography and TLC were the most popular among chromatograpic techniques used in anthocyanin identification (Jackman et al., 1987a).

1.6 Identification

Identification of anthocyanin involves some application techniques, such as well-developed chromatographic and spectrophotometric techniques. The absorption spectra of anthocyanins and their aglycones in the visible and ultraviolet region have proven useful in the identification of these pigments (Jackman et al., 1987a).

Absorption data at one or two wavelengths has been used as a measure of tinctorial power of colorant extracts. Total anthocyanins concentration is determined by measuring the absorbance of the sample, which is usually diluted with acidified alcohol at the appropriate wavelength. Most applications involve a 1 cm cell path but usually dilution is necessary because extract usually is highly colored. It is important that Beer's law be obeyed in the concentration range under study: self-association and copigmentation effects cause deviations in Beer's law and can lead to inaccurate estimates of total anthocyanin concentration. Therefore, sample dilution should be avoided because solutions may not follow Beer's law due to copigmentation effects.

The substractive method always involves measurement of the sample at the visible maximum, which the substraction provides the absorbance reading due to anthocyanin and it can be converted to anthocyanin concentration with reference to a calibration curve by standard pigment. The quantitative analysis of individual anthocyanins is usually carried out by chromatography for separation and purification from mixture. Spectral measurements usage is limited because most anthocyanins give similar absorption readings. The ability to obtain pure pigments from complex mixture is difficult to separate by conventional paper chromatography especially for acylated pigments. The development of automated preparative HPLC solved this problem. Preparative columns easily provide sufficient pigment for pigment identification by conventional paper chromatography and spectrophotometry. The use of mass spectrometry is essential for more complicated and highly acylated pigments (Francis, 1989).

LC coupled electrospray ionization tandem mass spectrometry (LC-MS/MS) was used to carry out anthocyanin identification and quantification (Hassan et al., 2014; Su et al., 2016).

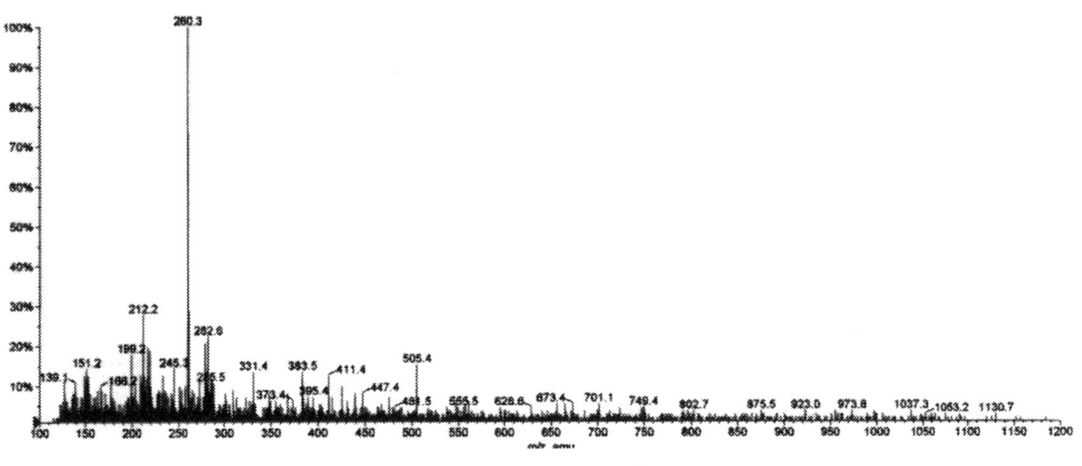

Figure 14.4: LC/MS/MS Spectrum of *Clitorea ternatea* Extract.

Hassan et al. (2014) identified anthocyanins in blue butterfly pea flowers by LC/MS/MS. The LC/MS/MS spectrum is shown in Fig. 14.4. The flavonoid glycoside detected were: quercetin 3-O-(2″-O-α-rhamnosyl-6″-O-malonyl)-β-glucoside (*m*/*z* 699.465, MW 698.465), kaempferol 3-2^{G}-rhamnosylrutinoside(*m*/*z* 741.305, MW 740.305), kaempferol 3-glucoside(*m*/*z* 449.292, MW 448.292), and myricetin 3-glucoside(*m*/*z* 481.425, MW 448.292). Anthocyanin was detected as delphinidin 3-glucoside at *m*/*z* 465.411 (MW 464.411) (Fig. 14.5).

2 *Extraction of* Piper betle *Oil*

Piper betle Linn from the family of *Piperaceae* is a perennial dioceious, semiwoody climber. Betle leaves are widely used in India, Sri Lanka, Malaysia, Thailand, and other Southeast Asia countries. Commonly known as betle (in English), sireh (in Bahasa Indonesia),

Figure 14.5: The Flowers of Butterfly Pea (*C. Ternatea*).

Figure 14.6: *Piper betle* Leaves.

paan (in India), and phlu (in Thai), betle have been used traditionally from generation to generation as a medicinal substance. The leaves (Fig. 14.6) have been widely used in traditional medicine as antiseptic, antifungal, antibacterial agent, carminative, and stimulant. Because of the strong pungent aromatic flavor, betle leaves are used by the Asian people as masticatory. In India, traditional healers from different remote communities claim that their medicine obtained from these betle leaves is cheaper and more effective than modern medicine (Arani et al., 2011). The use of plant extracts and phytochemicals, both with known antimicrobial properties, can be of great significance in therapeutic treatments (Subashkumar et al., 2013).

2.1 Chemical Constituent

This leaf is effective toward gram-positive bacteria. Betle oil, whose composition varies among varieties, is a medicinal substance of volatile oil produced by *P. betel*. The important content in the volatile oil of betel leaf is kavikol, methyl kavikol (estragol), eugenol, and kavibetol (betelfenol, isomer of eugenol). The kavikol provides a distinctive smell of betle and is able to kill bacteria 5 times more than ordinary ethanol. In addition, the oil of betle leaves and roots also has tumor-inhibiting properties, which give it the ability of an antiseptic. Therefore, betle can be of great benefit in treating diseases caused by fungi and bacteria. Previous studies on the betle leave, roots, and whole extract (mixture of VOC and non-VOC) of the green variety showed a very strong antimicrobial activity (Jenie, 2001).

It has been described that the *P. betle* leaf has piperol-A, piperol-B, methyl piper betlol, and they also have been isolated. The betle leaves have starch, sugars, diastases, and an essential oil composing of terpinen-4-ol, safrole, allyl pyrocatechol monoacetate, eugenol, eugenyl acetate, hydroxyl chavicol, eugenol, piper betol, and the betle oil contains cadinene carvacrol, allyl catechol, chavicol, *p*-cymene, caryophyllene, chavibetol, cineole, estragol, and so on, as the key components (Sugumaran et al., 2011). Phytochemical analysis on leaves revealed the presence of alkaloids, tannins, carbohydrate, amino acids, and steroidal components. The chief component of the leaves is a volatile oil in the leaves from different countries, called betle oil and contains two phenols, betle phenol (chavibetol and chavicol). Codinene has also been found (Prabodh and William, 2012).

2.2 Function

In modern era, various studies have been done to prove the ability or advantages of betle leaves as medicinal substances. Many researchers have studied the properties of the steam, leaves, and bark of *P. betle* to investigate their properties. Arani et al. (2011) had run a study to determine the antimicrobial properties of *P. betle* leaf against clinical isolate of bacteria. Different kinds of bacteria from Gram-positive and Gram-negative had been used in their work to study the evaluation of the constituent of phytochemical of the ethanol extract to the betel leaf. They also investigated the efficacy of the same as an antimicrobial agent on the different pathogenic bacteria species used. Their study showed that crude ethanol extract of *P. betle* showed strong antimicrobial activity against all the tested pathogenic bacterial strains used.

An antimicrobial agent is a substance that kills or inhibits the growth of microorganisms. It may be categorized on the basis of their antibacterial activity as either bacteriostatic or bactericidal. Some medicinal plant extracts have antifungal activity. These are agents capable of destroying or inhibiting the growth of fungi. These agents may either be fungicidal or fungistatic. The study was conducted to prove that the active compounds on *P. betle* is the chief constituent that causes its antimicrobial activity.

2.3 Soxhlet Extraction

There are many techniques to recover antioxidants from plants, such as Soxhlet extraction, maceration, SFE, subcritical water extraction, and ultrasound-assisted extraction (Quy et al., 2014). The Soxhlet extraction system was developed in 1879 by Franz von Soxhlet for the determination of fat in milk. Today, the Soxhlet extractor is not limited to fate extraction but becomes the most used tool for solid–liquid extraction in many fields, such as food, herbal, pharmaceutical, and environmental industry. This kind of extraction was applied in the extraction of *P. betle* to obtain the crude extract for further analysis of chemical constituent.

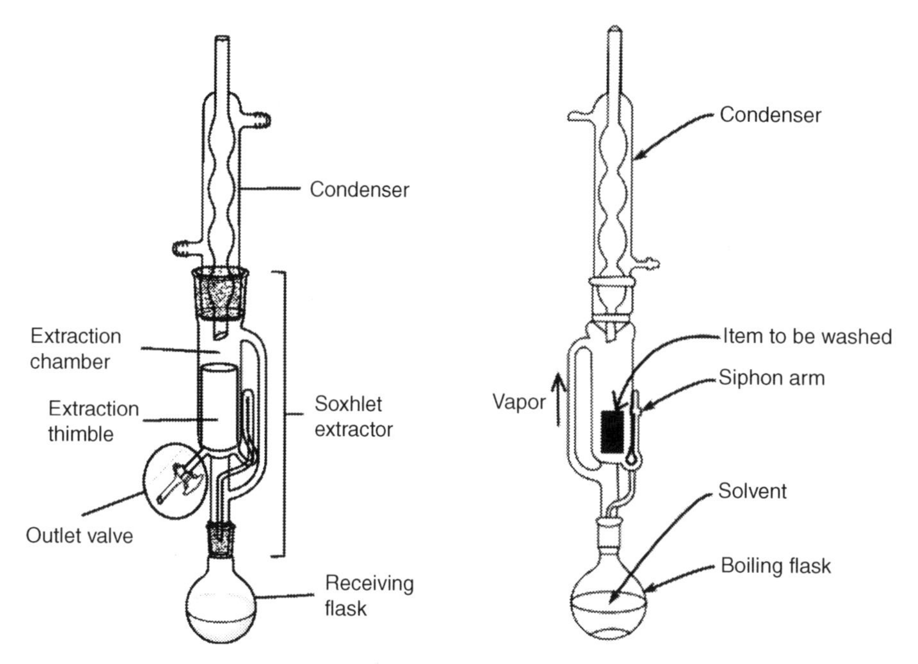

Figure 14.7: Soxhlet Apparatus.

The Soxhlet apparatus is built in three parts, shown in Fig. 14.7: condenser, Soxhlet extractor, and solvent flask. The extracting solvent, which is located in the flask, is set to heat to its boiling point. Solvent in the vapor phase will condense back to the liquid state by moving upward through the tube on the far right side of the apparatus, and then reaches the condenser. The solvent then passes a poros cellulose filter thimble, which held the sample, collecting the upper reservoir. The solvent and extracted analytes are siphoned back to the lower reservoir once the solvent in the upper reservoir reaches the return tube's upper bend. The analyte concentrations in the lower reservoir will increase over time. Extract of *P. betle* leaves is shown in Fig. 14.8.

Figure 14.8: Extract of *P. betle* leaves.

2.4 Choices of Solvent

Extraction yield and antioxidant activity not only depend on the extraction method but also on the solvent used for extraction. A good solvent must have low toxicity and have a preservative action on extract. Other than that, the solvent should evaporate at low heat, have an ability to quickly absorb to the extract, and cannot make the abstract dissociate. The specific nature of the targeted bioactive compound influenced the selection of solvent system. Bioactive compound can be extracted from natural products by different solvent systems. The extraction of hydrophilic compounds uses polar solvents, such as methanol, ethanol, or ethyl-acetate. For extraction of more lipophilic compounds, dichloromethane, or a mixture of dichloromethane/methanol in ratio of 1:1 are used. In some instances, extraction with hexane is used to remove chlorophyll (Cosa et al., 2006).

Water is a universal solvent, commonly used to extract plant compounds. Previous studies reported that uses of aqueous acetone for extraction of tannins and other phenolic compound are better in aqueous methanol. The presence of various antioxidant compounds with different chemical characteristics and polarities may or may not be soluble in a particular solvent (Turkmen et al., 2006). Polar solvents are frequently used for recovering polyphenols from plant matrices. The most suitable solvents are aqueous mixtures containing ethanol, methanol, acetone, and ethyl acetate. Ethanol has been known as a good solvent for polyphenol extraction and is safe for human consumption. Methanol has been generally found to be more efficient in the extraction of lower molecular weight polyphenols, whereas aqueous acetone is good for extraction of higher molecular weight flavanols. The maximum total phenolic content was obtained from barley flour by extraction using a mixture of ethanol and acetone (Bonoli et al., 2004).

Foo et al. (2015) in their study found that 10 g of dried *P. betle* leaves can produce of 1.57 g of extract in aqueous extraction and 1.23 g of extract in ethanol (70%) extraction. The yield of aqueous extraction better than ethanol (70%) extraction because of polarity water is greater than ethanol and the polar compounds are easier to be extracted compared with nonpolar compounds. Water and ethanol contain hydroxyl group and can form hydrogen bonding with the bioactive compounds. Aqueous extraction is more effective than ethanol extraction in antimicrobial compound because water has higher polarity and shorter chain than ethanol (Pin et al., 2010).

2.5 Phytochemical Test

The phytohemical tests of extracts from *P. betle* leaves were performed and the results were presented in Table 14.1. Separation and detection of bioactive compounds from the dry leaves extract can be achieved by GC-MS. GC-MS can separate more than 50 individual components in one sample and is consider one of the best techniques for identifying the

Table 14.1: GCMS data of importance bioactive compound with their function in aqueous extraction.

	RT (min)	Peak Area (%)	Name	Compound Nature	Functions
1.	3.02	2.45	Isopropyl isothiocyanate	Esther	Antimicrobial, flavoring agent, antifungal
2.	3.78	2.24	3-Heptanone, 6-methyl	Ketones	Flavoring agent, fragrance agent
3.	5.73	0.48	4H-pyran-4-1,2,3-dihydro-3,5-dihydroxy-6-methyl	Flavonoid compound	Antimicrobial, antiinflammatory, antioxidant agent
4.	6.50	0.19	1,2,3-Propanetriol, monoacetate	Fatty acid and their ester	Antimicrobial and antifungal properties at low pH
5.	7.25	0.18	Triacetin	Ester	Antifungal, Flavoring agent, humectants, plasticizer, and as solvent
6.	7.64	0.84	Pyrene, 4,5,9,10-tetrahydro	Pyrene and its derivatives	Dye and dye precursors
7.	7.70	0.34	Pyrimidin-4(3H)-1,2-amino-6-(3-fluorophenyl)	Heterocyclic aromatic organic compounds	DNA repair for cancer in research
8.	7.86	0.75	1-Pentene-3-ol	Alcoholic compound	Antimicrobial, antifungal
9.	8.15	91.51	Benzoic acid, 2,5-dimethyl	Aromatic carboxylic acid/ benzoic acid compound	Antimicrobial, food preservation
10.	8.51	0.11	Benzoic acid, 3,5-dimethyl	Aromatic carboxylic acid/ benzoic acid compound	Antimicrobial, food preservation
11.	8.53	0.19	Benzoic acid, 2,3-dimethyl	Aromatic carboxylic acid/ benzoic acid compound	Antimicrobial, food preservation
12.	8.79	0.21	*trans*-3-(2-Nitrovinyl)pyridine	Pyridine derivatives/alkaloid	—
13.	10.01	0.10	2-Methoxy-4-vinylphenol (synonyms: *p*-vinylguaiacol)	Phenolic compound	Antimicrobial, antioxidant, antiinflammatory, analgesic
14.	13.55	0.20	2,5-Dimethoxy-4-ethoxybenzonitrile	Benzoic acid amides	—
15.	13.63	0.20	Indole-2-1,2,3-dihydro-*N*-hydroxy-4-methoxy-3,3-dimethyl	Alkaloid/aromatic compound	Antimicrobial
16.	13.96	0.03	2-Ethylacridine	Alkaloid	Antimicrobial

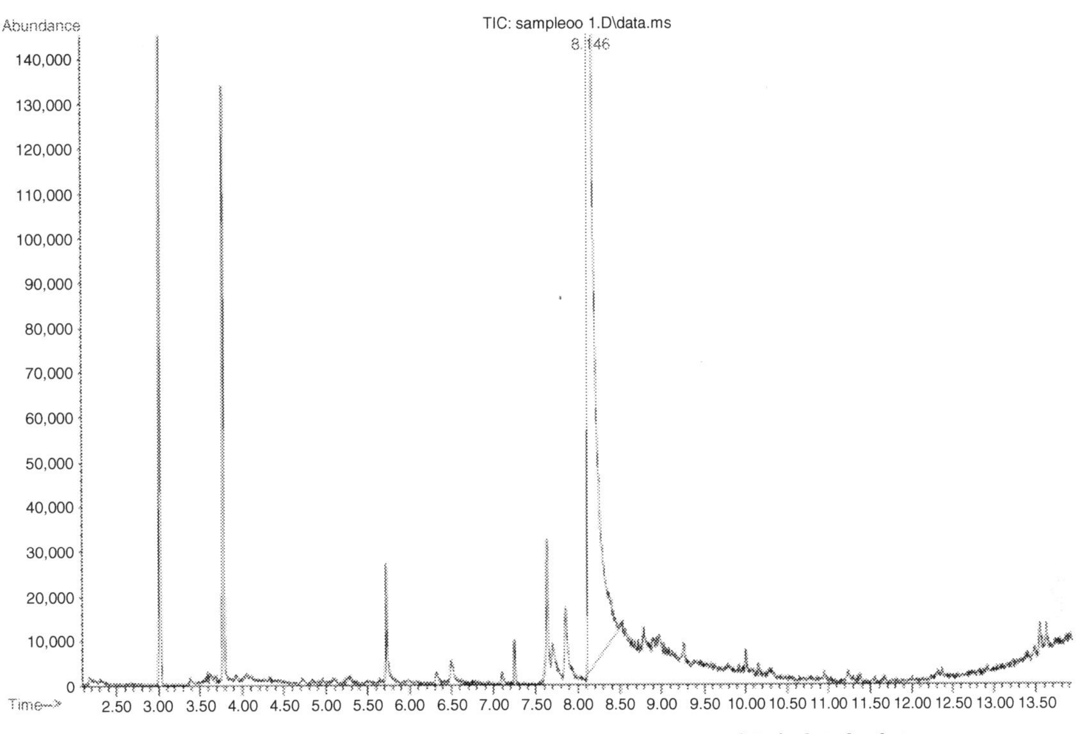

Figure 14.9: GCMS Chromatogram of Aqueous Extract of Dried *P. betle* Leaves.

bioactive compound of long chain hydrocarbon, alcohols, acids esters, alkaloids, steroids, amino acids, and nitro compounds (Muthulakshmi et al., 2012).

The bioactive compound present in the aqueous extract and ethanol (70%) extract of dried leaves of *P. betle* were identified by GC-MS (Figs. 14.9 and 14.10). GC-MS data of important bioactive compounds with their function in aqueous extraction and ethanol (70%) extraction were shown in Table 14.2. Mostly, antimicrobial compounds in both extractions were extracted out in the GC-MS retention time range 1-–10 min.

GC-MS analysis showed the presence of Benzoic acid, 2,5-dimethyl as a main compound in both extractions. The aqueous extraction showed 91.51% and ethanol (70%) extraction showed 76.22% of benzoic acid, 2,5-dimethyl. The presence of various bioactive compounds from both plants justified the use of various treatments by local traditional practitioners. Phytochemical screening of extracts in GC-MS showed the presence of various bioactive compounds, for example, alkaloids, fatty acids, phenolic compounds, alcoholic compounds, flavonoids compounds, terpenes compounds, coumaran compounds, and organic acids (Foo et al., 2015). The dried *P. betle* leaves extract is a complex with many bioactive components (Arani et al., 2011), which shows significant biological activities. The phenolic compounds, terpenes, alcohols compounds, aldehydes compounds, ketone, acid compounds, and flavonoid compounds have been known with antimicrobial properties for a long time

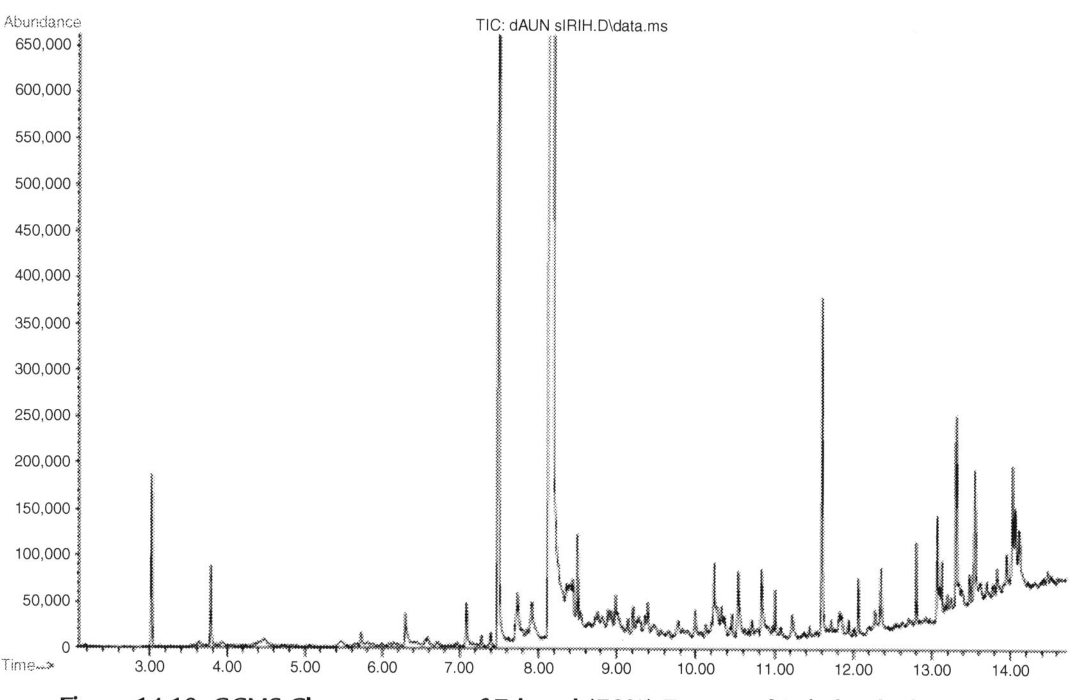

Figure 14.10: GCMS Chromatogram of Ethanol (70%) Extract of Dried *P. betle* Leaves.

(Tiwari et al., 2009). The aqueous extraction of *P. betle* plant mainly showed antimicrobial properties. Other than antimicrobial, *P. betle* extract in aqueous extraction also showed antifungal, antioxidant, and antiinflammatory properties. The ethanol (70%) extraction of dried *P. betle* leaves showed antimicrobial activity, antiinflammatory activity, antifungal, antitumor properties or cancerpreventive, antidiabetic, hepotoprotective, and so on. Ethanol and water mixtures are commonly used for the extraction of phenolic compound from plant materials. This is because water had its limited ability to extract oil-based components, such as eugenol (phenolic compound). This is the reason why some bioactive compounds that are only soluble in organic solvent could not be detected in aqueous extract in the GC-MS result. However, based on the GC-MS result, both methods of extraction (water and ethanol [70%]) were reported to show significant inhibition of bacterial activities.

3 Extraction of Vegetable Oil/Seed Oil

3.1 Function

Nowadays, healthier and more nutritious food products are the main criteria chosen by consumers. At the moment, many food products are fortified with oils rich in essential fatty acids: omega-3 and omega-6, as well as other bioactive compounds, such as antioxidants. The demand for these oils is increasing every year with the increase of human population

Table 14.2: GCMS data of importance bioactive compound with their function in ethanol (70%) extraction.

	RT (min)	Peak Area (%)	Name	Compound Nature	Function
1.	3.03	0.46	Isopropyl isothiocyanate	Ester	Antimicrobial, flavoring agent, antifungal
2.	5.726	0.07	4H-pyran-4-1,2,3-dihydro-3,5,dihydroxy-6-methyl	Flavonoid compound	Antimicrobial, antiinflammatory, antioxidant agent
3.	6.30	0.19	Benzofuran, 2,3-dihydro	Coumaran	Antimicrobial, antiinflammatory
4.	6.71	0.03	Hydroquinone	Phenolic compound	Antimicrobial, antioxidant, antimalarial, flavor
5.	6.97	0.03	Indole	Alkaloid	Antimicrobial, anticariogenic, antiseptic, cancer-preventive, perfumery
6.	7.09	0.18	2-Methoxy-4-vinyphenol	Phenolic compound	Antioxidant, antimicrobial, antiinflammatory
7.	7.28	0.05	Triacetin	Ester	Antifungal
8.	7.50	4.88	Phenol, 2-methoxy-4-(1-propenyl)-(isoeugenol)	Phenolic compound	Antimicrobial
9.	7.60	0.03	Quinolin-5(6H)-1,7,8-dihydro-2-hydroxy-4,7,7-trimethyl	Quinolines compound/alkaloid	Anticancer, antimicrobial, anticonvulsant, antiinflammatory, and cardiovascular activity
10.	7.85	0.10	*trans*-Cinnamic acid	Phenolic compound	Antitumor activity, antioxidant
11.	8.18	76.22	Benzoic acid, 2,5-dimethyl	Aromatic carboxylic acid/benzoic acid compound	Antimicrobial, food preservation
12.	8.41	0.40	Benzoic acid, 2,3-dimethyl	Aromatic carboxylic acid/benzoic acid compound	Antimicrobial, food preservation
13.	8.44	0.38	Benzoic acid, 3,5-dimethyl	Aromatic carboxylic acid/benzoic acid compound	Antimicrobial, food preservation
14.	8.50	0.41	Eugenol	Phenolic compound	Antimicrobial, antimutagenic, antiseptic, antiviral, insecticide, ulcerogenic, fungicide
15.	9.28	0.19	Napthalene, 1,2,3,5,6,7,8,8a-octahydro-1,8a-dimethyl-7-(1-methylethenyl)-[1R-(1.alpha, 7.beta, 8a.alpha)]	Alkenes/terpenes	Antimicrobial
16.	9.65	0.03	Phytol	Diterpene	Antimicrobial, anticancer, antiinflammatory, antidiuretic, and antidiabetic
17.	9.67	0.05	4-Chloromanol	Alcohol	Antioxidant
18.	10.84	0.33	*n*-Hexadecanoic acid	Fatty acid	Antiinflammatory, antioxidant
19.	11.83	0.11	9,12-Octadecanoic acid (Z,Z)	Fatty acid	Hepotoprotective
20.	11.85	0.12	6-Octadecenoic acid	Steric acid	Cancer preventive
21.	12.55	0.04	Oleic acid	Fatty acid	Antioxidant, cancer preventive
22.	13.12	0.18	Undecanal, 2-methyl	Aldehyde compound	Antimicrobial

associated with their health beneficial effects, such as immune response disorders, hypertension, arthritis, and reduced risk of coronary diseases (Rubio-Rodríguez et al., 2010).

A range of fatty acids is a lucrative compound in natural seed oils that are always preferred to be a part of the ingredient that has been used in the production of cosmetic and personal care products, include lubricants, soaps, and hair shampoo. Linoleic acid is the most favored fatty acid used in cosmetic and personal care products as it cannot be synthesised by the body and always used as emollients or moisturiser for skin, nails, and hair. The deficiency of linoleic acid will cause drying and scaly skin, crack nails, hair loss, and increase transepidermal water loss (Vermaak et al., 2011).

3.2 Sources

Edible oil from different natural sources is sought after, owing to its interesting compound, such as high content of unsaturated fatty acid, tocopherols, and antioxidant. Rural community-utilized plant-based oils for centuries are a good source of food, medicine, cosmetics, and fuel (Vermaak et al., 2011). Therefore, the interest to explore the potential of various plants as the sources of nutritious oil has increased. To date, there are a number of reports that study various types of vegetable oil, including walnut oil, evening primrose oil, canola oil, sunflower oil, olive oil, soybean oil, perah seed oil (Yong and Salimon, 2006), pomegranate seed oil (Liu et al., 2009), flaxseed oil (Carneiro et al., 2013), almond oil, sesame oil, avocado oil, apricot kernel oil, rapeseed oil, linseed oil, sunflower seed oil, and palm oil (Vermaak et al., 2011).

3.3 Extraction of Oil

Traditional methods of solid–liquid extraction techniques have been applied for many decades to get the oil, including Soxhlet extraction, maceration, percolation, turbo-extraction (high-speed mixing), and sonication. However, several disadvantages of these techniques have been identified, such as time-consuming and required nonfood-grade solvents (Kaufmann and Christen, 2002). Fast and efficient modern extraction techniques, including pressurised solvent extraction (PSE), SFE, and MAE for extracting analyses from solid matrixes, becomes more attractive in research nowadays.

In general, extraction of vegetable oils can be grouped into two different approaches, which are mechanical oil extraction (physical process) and solvent extraction (chemical process). There are two types of mechanical oil extraction: cold press, where no heat is applied and hot press, where external heat is applied (Katkar et al., 2015).

3.3.1 Expeller

Expeller pressing is the method of extracting oil with mechanical pressing equipment. No further purification is required because there are no solvent residues in the oil. The oil yield is clean pure oil, which is high in natural colors and flavors.

A helical screw is used in oil expellers for continuous mechanical pressing to move and compress the material (seeds) to get the oil and produce the cake. The cake is compressed and yields the output of oil by the linings in the equipment as the transition takes place through the barrel or the casing. This results in the volume of the forced oil seeds mass being reduced, at the same time the deoiled cake is discharged through the annular ending of the process unit. The oil expeller can be operated by diesel, electric motor, or even human-powered flywheel (Katkar et al., 2015).

3.3.2 Microwave extraction

Microwave extraction can be applied to a broader range of applications because it uses microwave dielectric heating as the heating mode. With frequency 2.46 GHz of microwave energy, is well known to have a significant effect on the rate of various processes in the chemical and food industry. Microwaves use electromagnetic waves, which penetrate into certain materials to provide volumetric heating through ionic conduction and dipole rotation (Chan et al., 2014). One of the biggest benefits is realized in the reduced amount of time the extraction takes for multiple samples in a single-extraction operation. Additionally, simplified manipulation and higher purity of the final product made the application of microwave extraction gains more attention. Numerous classes of compounds, such as fats and oil, essential oils, aromas, pesticides, phenols, dioxin, and other organic compounds have been successfully extracted from a variety of matrices, including soils, sediments, animals' tissues, food, or plant material (Virot et al., 2008). All the reported applications have revealed that MAE is a viable alternative to conventional techniques for such matrixes.

3.3.3 Ultrasound extraction (sonication)

Ultrasonic or ultrasound extraction is produced by an elastic mechanical wave with 20–2000 kHz to increase cell wall permeability and to produce cavitation. The ultrasound equipment provides simple operation with high efficiency. It significantly reduces extraction time and temperature, requires less solvent quantity, and improves oil yield compared to conventional extraction (Hu et al., 2010). However, the disadvantages of this process are the high cost of operation and could generate free radicals from plant constituents due to the effect of the ultrasound energy (Handa et al., 2008).

3.3.4 Supercritical fluid extraction

SFE is an attractive alternative separation technology to the traditional separation method. According to Doker et al. (2010), SFE eliminates the disadvantages of conventional solvent extraction (CSE), which leads to thermal degradation of heat labile compound and leaving toxic solvent (hexane) in the extracted product. It reduces the usage of organic solvent but at the same time increases the extraction product. Certain parameters need to be controlled, such as temperature, pressure, sample volume, analyte collection, modifier (cosolvent), flow, and pressure control and restrictors (Handa et al., 2008).

Generally, the component recovery rate increases with increasing the pressure. CO_2 is always preferred as solvent as it is cheap, environmentally acceptable, safe, operates at low temperature. The beneficial transport properties of fluids near their critical points provide deeper penetration into the solid plant matrix and makes the extraction process more efficient and faster compared to conventional organic solvents (Setapar et al., 2011).

Chatterjee and Bhattacharjee (2012) have optimized different methods to extract eugenol from clove buds, such as liquid, subcritical, and supercritical carbon dioxide (SC-CO_2) extractions, steam distillation, and solvent extraction. They concluded that SC-CO_2 clove extracts have the best interactions of therapeutic and phytochemical properties. Moreover, SFE are widely utilized in the separation of essential oil especially from herb matter (seed, leaves, etc). Its nutritional derivative use in food, nutraceutical, pharmaceutical, and cosmetics applications, due to the concern of consumer health risk and environment contamination associated with the use of chemical solvent in conventional method.

3.3.5 Phytonic extraction

This type of extraction involves new solvent bases on hydroflurocarbon-134 and new technology to optimize the plant material. This process provides several advantages, such as environmental friendly technology and health extraction, produce a high quality of natural fragrant oil, flavor, and biological extracts and do not require any further physical and chemical treatment (Handa et al., 2008).

3.4 Physiochemical Properties Analysis of Elateriospermum tapos *Seed*

E. tapos seed (ETS), which is also known as perah seed (Fig. 14.11) is one of new local underutilized seeds that have recent studies of rich omega 3 contents. Studies made by

Figure 14.11: Perah Seed.

Yong and Salimon (2006), 17.14% of alpha linoleic acid content, which is recognized as a precursor of omega 3 from plants and was found in ETS oil.

The physicochemical of ETS oil has been studied by Yong and Salimon (2006) by normal lipid extraction. However, there are no studies on physicochemical of ETO extracted by MAE and soxhlet extraction. Soxhlet extraction is a conventional method that can be considered as reference to other extraction techniques, which are almost used in extraction of plant seed oil (Ofori-Boateng et al., 2012). According to Pradhan et al. (2010), Soxhlet extraction can result in up to 99% of extraction recovery. Meanwhile, microwave-assisted extraction (MAE) is green extraction technology due to the use of aqueous water based as a solvent for this study. Moreover, it is known as a superb fast and efficient extraction method as it uses electromagnetic fields as a heating mode where electromagnetic waves penetrate into certain materials to provide volumetric heating through ionic conduction and dipole rotation (Chan et al., 2014).

Therefore, the physiochemical properties of ETO by Soxhlet extraction and MAE extraction had to determine the review of the oil quality properties. Acid value (AV), free fatty acid (FFA), peroxide value (PV), iodine value (IV), saponification value (SV) test of ETS oil extracted by Soxhlet, and MAE were conducted by PORIM method.

Among the methods extraction, the microwave method gives the lowest yield. This might due to the fact that aqueous cannot extract more nonpolar compounds as compared to Soxhlet that use hexane as a solvent for extraction; however, the MAE was efficient and the fast extraction method can extract an amount of yield within only 5 min extraction time. From the results tabulated in Table 14.3, acid value, free fatty acid value, and peroxide value of perah oil extracted by microwave is low compared to Soxhlet extraction. This shows that microwaves can reduce the spoilage of the oil, which can enhance the oxidative stability of the oil. As stated by Gai et al. (2013), low acid value shows the oil less rancidity. The amount of iodide value (141.47 ± 0.53) in perah seed oil using microwave extraction was higher compared to iodine value by Soxhlet extraction (129.21 ± 0.25) and lipid extraction (106.77 ± 0.37),

Table 14.3: Comparison of physiochemical analysis of extracted ETS oil by Soxhlet and MAE.

	Soxhlet	MAE	Standard Lipid Extraction (Yong and Salimon, 2006)
Yield (%)	51.81 ± 0.47	20.1	38.59 ± 1.77
Acid value	15.9941 ± 0.0655	13.99	8.21 ± 0.06
Free fatty acid	8.0372 ± 0.0329	7.03	4.12 ± 0.03
Peroxide value	4.3998 ± 0.1109	2.79	0.46 ± 0.16
Iodine value	129.2137 ± 0.2473	141.47 ± 0.53	106.77 ± 0.37
Saponification value	185.4058 ± 0.7583	182.69	150.90 ± 0.32

MAE, Microwave-assisted extraction.

Figure 14.12: Extract of Perah Seed Oil Extracted by Difference Types of Solvent.
(A) Hexane, (B) isopropanol, and (C) methanol.

reported by Yong and Salimon (2006). Iodine value measures the content of unsaturated or the average number of double bond in a fatty acid (Rui et al., 2009). Higher iodine value results in high unsaturated bond in oil. Hence, the iodine value in perah oil extracted by microwave shows the high content of unsatrated fatty acid as compared to Soxhlet and method extraction by Yong and Salimon (2006).

Azlan Hadi Tan et al. (2013) investigated chemical constituents along with chemical properties in perah seed oil extracted by different solvents through Soxhlet extraction using fourier transform infrared (FTIR), GC-MS, and GC. Fig. 14.12 shows the extract of perah seed oil using different solvents.

The FTIR spectra for perah seed oil using hexane and isopropanol as extraction solvents were similar with typical absorption bands in the 3600–3200, 3300–3000, 2960–2850, 1760–1670, 1650–1580, and 1340–1020 cm^{-1} regions, which associated with alcohols (3600–3200), alkanes groups (2960–2850), aldehydes, ketones, carboxylic acids, esters (1760–1670), amide bend (1650–1580), and amide stretch (1340–1020) as shown in Fig. 14.13. Using methanol as extraction solvent showed different structures of the oil possibly caused by the polarity of the solvent and polymerization due to high extraction temperature of the oil. Broad peaks at absorbance from 2700 to 3700 cm^{-1}, probably due to the forming of polymers during the extraction process, reinforces why extraction using methanol gives a small yield.

In the GS-MC analysis as shown in Fig. 14.14, most compounds in perah seed oil were detected by using hexane as extraction solvent (37), followed by methanol (23), and isopropanol (13) possibly due to its polarity (Azlan Hadi Tan et al., 2013).

4 *Extraction of Sabah Snake Grass (*Clinacanthus nutans*)*

4.1 Clinacanthus nutans' *Background*

C. nutans (Burm. f.) Lindau commonly known as Sabah Snake Grass has been traditionally used as a medicinal plant since ancient times. This plant (shown in Fig. 14.15), which can be

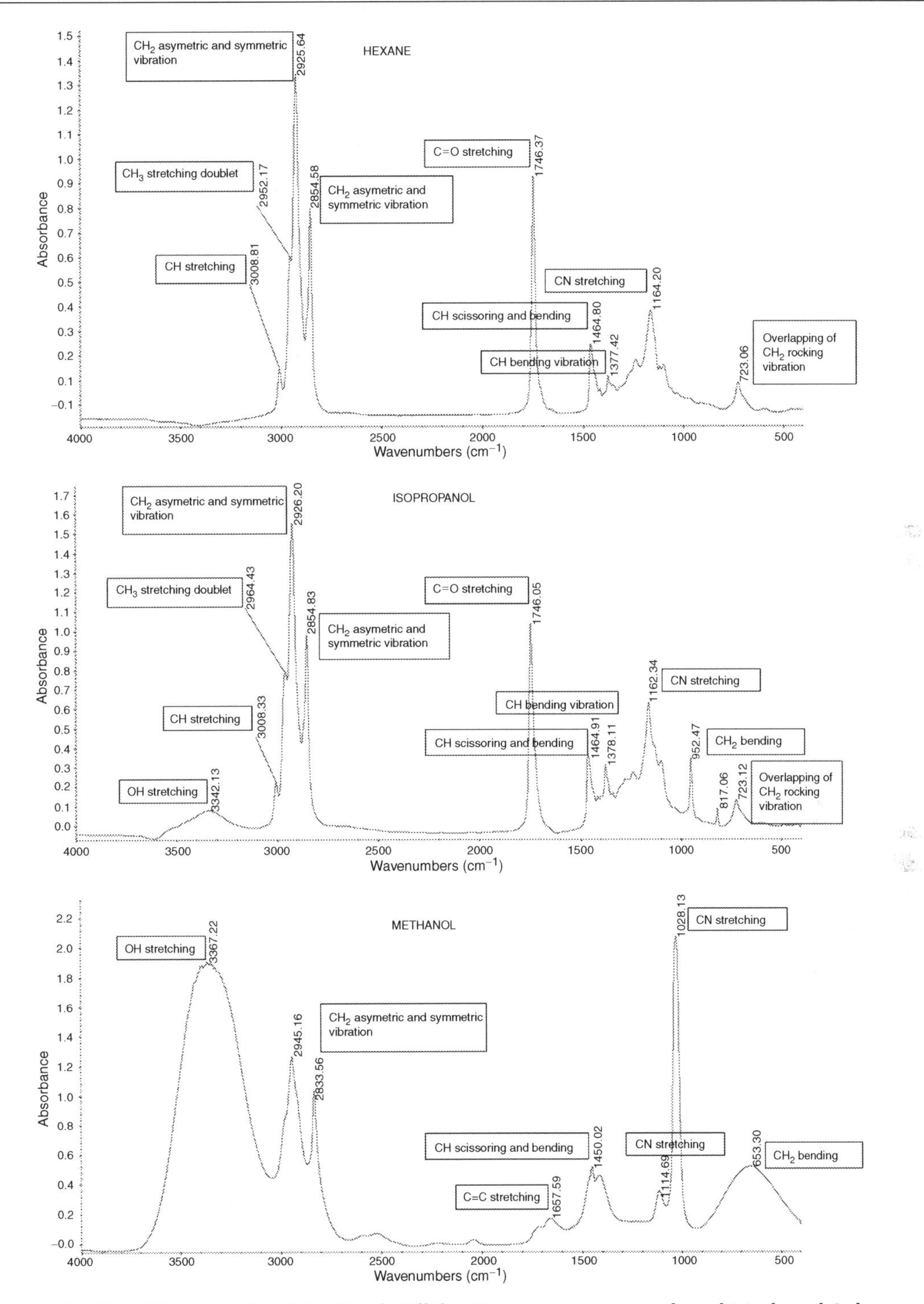

Figure 14.13: FTIR Analysis of the Perah Oil for Hexane, Isopropanol, and Methanol Solvent.

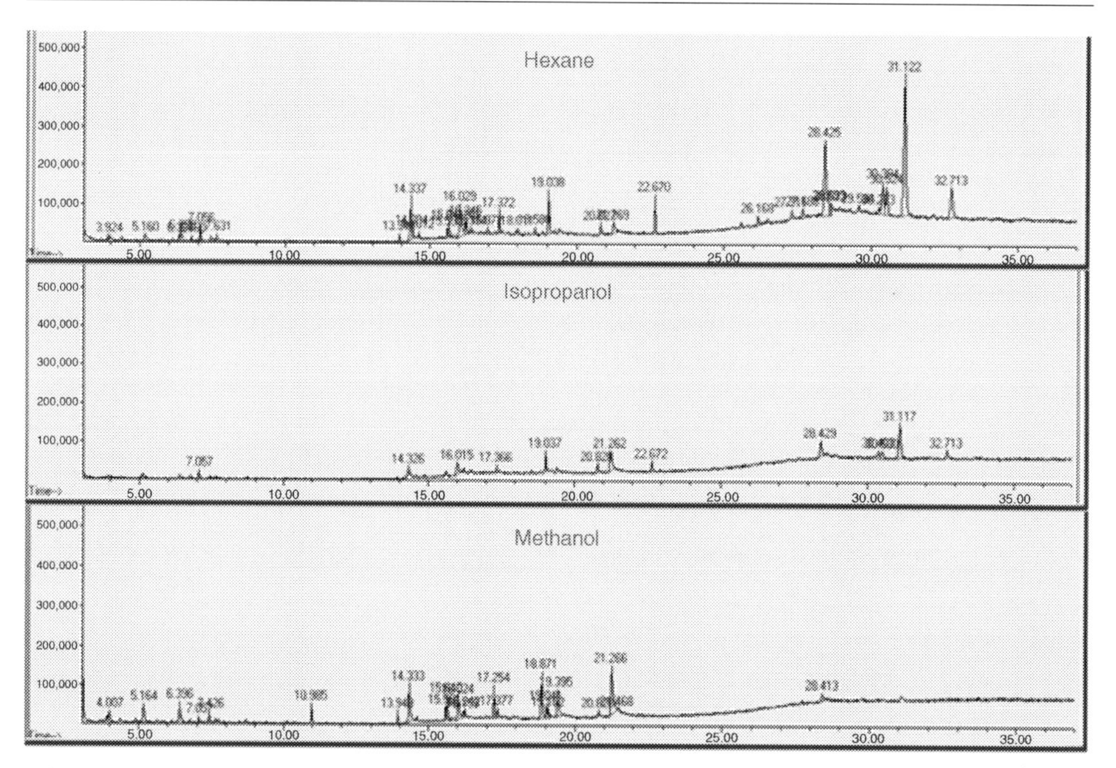

Figure 14.14: GC-MS Analysis of the Perah Oil for Hexane, Isopropanol, and Methanol Solvent.

found in tropical Asia countries, such as Malaysia, Indonesia, and Thailand belongs to the family of Acanthaceae (Wanikiat et al., 2008). The common names of this plant all around the world are Phaya Yo/Phaya Plong Thong (Thailand), Belalai Gajah (Malay), Dandang Gendis (Indonesia), and E Zui Hua (Chinese). The synonyms, including *C. angustus*, *C. burmanni*, *C. spirei*, and *C. siamensis* (Fig. 14.16).

Figure 14.15: *Clinacanthus nutans* (Burm.f.) Leaves (left) and Extraction Powder (right).

Figure 14.16: Percolator.

4.2 Morphology

Traditionally, the leaves of *C. nutans* are the most applied in health treatments. According to Sakdarat et al. (2009), fresh leaves of *C. nutans* have the ability to treat many infections, including skin rashes, herpes simplex virus, varicella-zoster virus lesions, and is an antidote for snakebites. Furthermore, it is reported that *C. nutans* is a shrub or perennial herbs that can grow up to 1 m in height with young branches and belongs to Acanthaceae family (Wanikiat et al., 2008). This family is mostly tropical herbs and epiphytes. Morphologically, they can be easily recognized where the leaves are simple and narrowly elliptic-oblong with 2.5–13 cm long × 0.5–1.5 cm wide (Panyakom, 2006). The leaves have acute shapes with a short sharp apex angled less than 90 degree. When young, both surfaces of leaves are pubescent then it tends to become glabrous or glabrescent. The leaf bases are obtuse rounded or truncated, often oblique. The petiole, which is the transition between the stem and the leaf blade is 3–15 mm long. As recorded by Panyakom (2006), the flowers are basely yellow or greenish yellow. They are in dense cymes at the top of branches and branchlets, always covered with 5-alpha cymules. The calyx of the flower is about 1 cm long with granular-pubescent; corolla dull red with a green base, about 3–4.2 cm. The stamen is exerted from the throat of corolla. The ovary is compressed into two cells and each cell has two ovules. The styles are filiform with shortly bidentate. The capsule is oblong basally wrapped into 4-seeded short stalks.

4.3 Bioactive Compounds of **Clinacanthus nutans**

Bioactive compounds are essential to human health. The bioactive compounds in *C. nutans* contain potential cytotoxic, antioxidant, and antimicrobial agents. In previous studies, a series of flavonoids, steroids, triterpenoids, cerebrosides, glycoglycero-lipids, glycerides, and sulfur containing glycosides were isolated from this plant. *C. nutans* extract are rich with six known C-glycosyl flavones, such as, vitexin, isovitexin, shaftoside, isomollupentin 7-O-β–glucopyranoside, orientin, and isoorientin (Gaitan et al., 1995; Yoosook et al., 1999). Vitexin and isovitexin have been found to have ability as antioxidant and strong antithyroid effect, which prevent thyroid glands malfunction (Gaitan et al., 1995).

4.3.1 Vitexin and isovitexin

Vitexin and isovitexin are C-glycosyl flavones that have high benefits to human health. Studies reported antitumor efficacy of vitexin in preclinical models of ectopic growth of several cancer cells, including breast, prostate, liver, and cervical cancer cells. The C-linked glycosidase flavones vitexin and isovitexin have been shown to inhibit alpha-glucosidase, an enzyme that is responsible for the breakdown of carbohydrate to sugar. According to Gaitan et al. (1995), vitexin and isovitexin have been found to have ability as antioxidant and strong antithyroid effect.

4.3.2 Total polysaccharide

Polysaccharides are complex carbohydrate polymers consisting of more than two monosaccharides linked together covalently by glycosidic linkages in a condensation reaction. Being comparatively large macromolecules, polysaccharides are most often insoluble in water. Polysaccharide is a natural macromolecule located in the primary cell walls of plants. It was built from hundreds to thousands of monosaccharide combination through dehydration synthesis. Starch, cellulose, and glycogen are some examples of polysaccharides. In the food industry, the addition of polysaccharides acts as dietary fiber and stabilizers. Polysaccharides are also formed as products of bacteria, for example, in yogurt production). Polysaccharides from plants have pharmacological effects and are beneficial for health. The pharmacological activities of polysaccharides are antitumor activity, antivirus activity, antibacterial activity, immune activating activity, and hypoglycemic. Furthermore, polysaccharides have been reported to exhibit several biological activities, such as antiinflammation, antioxidation, anticomplement, antifatigue, anticoagulation, and enhancement of probiotic bacteria growth (Zheng et al., 2010).

For separation, detection, and identification of polysaccharides, several methods have been developed, such as HPLC-MS, GC-MS, NMR, CE-DAD (Zheng et al., 2010). The measurement using this instrumentation was applied if the sample has major impurities of different natures. The amount of polysaccharides can be measured accurately and convenient methods using colorimetric analysis, which the transformation of selective chemicals will produce color. The absorbance of the color is directly proportional with the presence of polysaccharides.

4.3.3 Total protein

Protein is essential to human health because it is source of energy and contains necessary amino acids, such as tryptophan, lysine, methionine, leucine, valine, and isoleucine, which the body cannot synthesize. Protein is the polymer of amino acids that are bonded together by peptide linkages. Protein and amino acids are important for continuous repair and maintenance of body tissue. In addition, protein protects the body against diseases when the body detects invading antigens.

Moreover, acids and bases within the body fluids could be balanced and maintained with the presence of protein by accepting and releasing hydrogen ions. Besides protein being important for body growth, repair, and maintenance, it also helps increase glucose content when needed (National Coordinating Committee on Food and Nutrition, 2005).

Two main sources of protein in the human diet are animal protein and plant protein. Animal protein is referred to as complete protein as it consists of all the necessary amino acids that are needed by the body. Meanwhile, plant protein is an incomplete protein because it misses one or more essential amino acids. However, the mixture of two incomplete proteins can be changed into a complete protein. For example, mixture of legumes and grains or legumes and nuts can produce a complete protein. Although the amount of protein from plant sources is smaller than animal protein, it can be a sufficient substitute when it is eaten with other foods (National Coordinating Committee on Food and Nutrition, 2005). Total protein assays are applied to evaluate the products in industrial, biotechnology, and agricultural. Issues concerned with performing total protein assays are the sensitivity and technique of the method used, clear units' definition, interfering compounds, removal of interfering substances before assaying samples, and correlation of information from various methods. Several methods were developed for total protein determination, such as Biuret assay, Hartree-Lowry assay, bicinchinonic acid (BCA) assay, acid digestion-ninhydrin method, ultraviolet absorption coomassie dye-binding assay (Bradford assay), and dry weight determination.

4.3.4 Total glycosaponins

Glycosaponins are secondary metabolites in plants with the form of steroids, glycosides of triterpenes, and sometimes alkaloids. Therefore, glycosaponins can be categorized as steroidal, triterpenoidal, or alkaloidal depending on the nature of aglycone. The aglycone of glycosaponins is known as sapogenin, whereas glycone of glycosaponins is referred to as oligosaccharides (Kareru et al., 2007). In nature, glycosaponins are widely distributed and their presences have been reported in at least 500 genera of plants. They are freely soluble in water because they are polar in nature, but insoluble in nonpolar solvents. In shaken water, they produce a soapy lather, thus they are added to shampoos, toothpastes, liquid detergents, and beverages. Glycosaponins have been used in several industries, such as in food and cosmetics industry as additives, in agriculture and photographic industry as wetting agents and in the pharmaceutical industry as adjuvants. They are surface-active glycosides with wetting, detergent, foaming, and emulsifying properties owing to the presence of water-soluble sugar chain and lipid soluble aglycone in their structure (Negi et al., 2011).

Glycosaponins in plants have been used for medical purposes. They have various physical, chemical, and biological properties due to their complex structure. Several biological activities of glycosaponins are immunostimulant, antioxidant, antihepatotoxic, anticarcinogenic, antibacterial, antiulcerogenic, antidiarrheal, antioxytocic, anticoagulant, antiinflammatory, hypocholesterolemic, hypoglycemic, hepatoprotective, neuroprotective, inhibition of dental caries, useful in diabetic retinopathy, and platelet aggregation (Negi et al., 2011). They are also useful as antiseptic, in treatment of liver, kidney, and cardiovascular disorders, and in enhancing mood. According to Kareru et al. (2007), glycosaponins are antimutagenic, antiviral, and have cytotoxicity effects.

The amount of glycosaponins in plants is influenced by several factors, such as the age, cultivating, physiological state, and geographical location of the plant. Different plant locations might create different composition and quantity of glycosaponins. Many methods have been developed for qualitative and quantitative determination of glyosaponins in plants. For example, haemolysis, piscicidal activity, gravimetry, spectrophotometry, TLC, GC, and HPLC (Negi et al., 2011).

4.4 *Extraction of* Clinacanthus nutans

Selection of suitable *C. nutans* extraction method is important in order to get high qualitative and quantitative bioactive compounds from *C. nutans*. Extraction is the first step in medicinal plant study and plays an important role in producing a good final result. Although modern spectrometric and chromatographic techniques have been developed to ease bioactive compound analysis, they are still dependent on the extraction methods, parameters effects and also exact nature of plant parts (Azmir et al., 2013). In pharmaceutical, extraction process consists of separation of essential parts or necessary compounds of animal or plant tissues from the inert or inactive components through preferred solvents. Products manufactured from the extraction process came in various forms, such as in impure liquids, semisolids, or powders and they are used in oral or external practice.

There are many types of extraction process for plant ingredients. General techniques of extraction for medicinal plants are infusion, maceration, digestion, percolation, hot continuous extraction (Soxhlet), decoction, counter current extraction (CCE), aqueous-alcoholic extraction by fermentation, ultrasound extraction (sonication), MAE, phytonic extraction with hydroflouro-carbon solvents) and SFE.

Maceration involves the powdered crude drug are placed in a stoppered container for at least 3 days at room temperature with frequent agitation until the soluble matter has dissolved. The mixture then is strained together with liquid from pressed damp solid material and clarified by filtration or decantation after standing (Handa et al., 2008).

Percolation is a technique often used in the extraction of active ingredients in the preparation of tinctures and fluid extract. After the mass is packed in a closed container of which the top of the percolator is closed, certain amounts of solvent moistens the solid ingredients for about 4 h. Additional solvent is added until it reaches above the mass to form a layer and the mixture is allowed to macerate in the closed percolator for 24 h. The liquid contained therein is allowed to drip slowly after it opens the outlet of the percolator. Additional solvent is added until the percolate measures about three-quarters of the required volume of the finished product. The expressed liquid is mixed together to the percolate and the last step is clarifying the mixed liquid by filtration (Handa et al., 2008).

For aromatic plants, the extractions methods involve are hydrodistillation either water distillation, steam distillation, or steam and water distillation and hydrolytic maceration followed by distillation method (Handa et al., 2008).

To understand extraction selectivity from different natural sources, extraction techniques used should be varied in different conditions but with the same objectives:

1. To get targeted bioactive compounds extracted from complex plant sample,
2. To get high selectivity of analytical methods,
3. To get high concentration of targeted compounds to increase sensitivity of bioassay,
4. To change the extraction compounds in other forms, which are suitable for detection and separation,
5. To provide strong extraction methods that are suitable for any independent variables of extracted sample.

4.4.1 Parameters that affect extraction process

Process parameters, such as solvent characteristics, solvent-to-solid ratio, temperature of extraction, duration of extraction, particle sizes of raw materials, and agitation rates are important parameters involved in *C. nutans* extraction process. Most of these parameters can have individual or combined effects on the yield extract and composition of the extract.

4.4.1.1 Solvents

Important things to consider in choosing the solvent are the solubility of the targeted bioactive compounds, diffusivity of the compounds in the solvent, and the sample's characteristics. In order to get high purity and selectivity of extract, the selected compound should have higher solubility in the solvent compared to the other compounds. Besides that, the other aspects, such as safety, economy, and sustainability should be considered in the choice of solvent. Nonharmful and less toxic solvents are to be preferred. Water, acetates, alcohols, and glycol are common solvents used for plant extraction. Water is a suitable solvent to extract polar, moderately polar, and nonpolar organic compounds by increasing the temperature during extraction (Mustafa and Turner, 2011; Azmir et al., 2013).

4.4.1.2 Solvent-to-solid ratio

Solvent-to-solid ratio is the ratio of solvent per feed used in the extraction process. To dissolve solute and transfer it to the exterior of the solid matrix, an adequate amount of solvent is important. Higher ratios lead to enhance extraction kinetics due to greater concentration between solute trapped inside the raw material and those located at the surface. However, using solvents in excess is not recommended because it will require higher energy and heat capacity (Vuong et al., 2013).

4.4.1.3 Temperature

Efficiency and selectivity of extraction process were also affected by temperature during extraction. High temperatures used enhance the extraction efficiency due to disruption of analyte-sample matrix interaction caused by hydrogen bonding, van der Waals forces, and dipole attraction. Moreover, using elevated temperature in extraction process could decrease the surface tensions of the solvent, solutes, and matrix, and thus helps the solvent wetting of the sample. High extraction temperatures also decrease solvent viscosity and improve diffusion rate. Using extraction temperatures close to the boiling point of the solvent could increase bioproduct yield. However, too high temperatures may cause solvents to be lost to evaporation, damage sensitive equipment, and may extract undesirable constituents (Mustafa and Turner, 2011).

4.4.1.4 Duration of extraction

Duration of extraction is crucial in minimizing energy and cost of the extraction process. Moreover, duration of processing influence the degradation of targeted compounds. Decreasing extraction duration as much as possible could reduce the degradation effect. Moreover, increased duration of extraction will increase the loss of solvent by evaporation. Duration of extraction no longer than 3 h is suggested (Mustafa and Turner, 2011).

4.4.1.5 Particle size

Particle size of raw materials influences the rate of extraction and its efficiency. Smaller raw materials particle sizes enhance the efficiency of extraction due to shorter internal diffusion path lengths over the extracted solutes and must travel to reach the fluid phase. Moreover, higher surface areas available in smaller particles could increase mass transfer rate, subsequently reducing the time needed for initial soaking of plant matrix. Higher rate of extraction could also be achieved in higher fluid–solid contact areas of smaller particle size. However, too small of a particle size would reduce the effectiveness of surface area available because of agglomeration of fine particles, and thus, makes it difficult for the solvent to permeate (Vuong et al., 2013).

4.4.1.6 Agitation rate

In extraction, increasing agitation rate increases the dissolution, however, agitation above its maximum has little effect or no benefit. In poorly mixed systems, increasing the rate of agitation may have significant effect and improved homogeneity in the bulk solution. Though, in well-mixed systems the effect is less significant due to the bulk solution being more homogenous and more difficult to decrease diffusion by agitation alone. Thus, the agitation is sometimes unjustifiable (Marsden and House, 2006).

4.5 Spray Drying in *Clinacanthus nutans* Production

C. nutans extract in aqueous solution is highly exposed to harmful bacteria and fungi. For industrial purposes, *C. nutans* extract have to go through a drying process to have longer shelf life, higher stability, and easier handling, transportation, standardization, and storage. However, drying process typically affects the chemical and physical properties of extracts. Since quality control becomes a big issue in the drying process, an appropriate drying method should be considered to maintain the good quality of dried product (Zuhaili et al., 2012). In herbal processing, spray drying is a commonly used method to obtain dried extract (Zuhaili et al., 2012). However, high temperatures used in a drying system become the limitation to its application when processing thermal sensitive materials. Thus, the spray-drying method is chosen in *C. nutans* production with optimum conditions to produce dried *C. nutans* in good quality and low cost product.

4.5.1 Spray drying

Spray drying is a very commonly used technical process to dry organic or aqueous solutions, suspension, dispersion, and emulsion in industrial chemistry and food industry. Since 1940s, spray drying has been used successfully in pharmaceutical industry to produce drug substances and various excipients, such as analgesics, antibiotics, vitamins, and antacids. Since late 1950s, spray-drying encapsulation has been used in the food industry for protection from degradation and oxidation of oil flavor and production of powders from liquids. Spray drying is an appropriate process for drying heat-sensitive biological materials, such as pharmaceutical proteins and enzymes, with slight activity losses (Phisut, 2012; Zuhaili et al., 2012).

The dried spray-drying products can be in powders, agglomerates, or in granular form that are influenced by the chemical and physical characteristics of the feed solution, design of the dryer, and also the desired characteristics of the final product (Patel et al., 2009). Formation and drying of particles are involved in spray drying the feed and is converted from fluid state to droplets. Then, it becomes dried particles that are usually collected in a cyclone or drum. Meanwhile, the liquid was converted into a hot vapor stream and vaporized. The nozzle that was used to spray the liquid input stream has made the droplets as small as possible for greater heat transfer and water vaporization rate. Spray drying presents low operational costs, short contact time and weight, as well as volume reduction (Patel et al., 2009; Phisut, 2012).

5 Conclusions

The use of plant extracts for foods, cosmetics, and pharmaceuticals nowadays is becoming increasingly important in the developed countries due to more consumer-friendly labels, health benefits, medicinal and nutraceutical value. This study reviewed extraction of several plants that are known for their health benefits, including butterfly pea flowers (natural colorant), *P. betle* leaves (antimicrobe), perah seed (oil), and snake grass (anticancer), and that would help to get better understanding of extraction technologies. Recent development of extraction technologies has been introduced to get bioactive compounds extracted from plant parts, such as tubers, seed, leaves, seed, and flowers by different extraction method, including the use of enzymes, MAE, SFE, and UAE. The use of recent extraction technologies has been reported as a faster, more highly efficient, and solvent-saving technique compared to CSE methods.

References

Afoakwah, A.N., Owusu, J., Adomako, C., Teye, E., 2012. Microwave assisted extraction (MAE) of antioxidant constituents in plant materials. Global J. Biosci. Biotechnol. 1 (2), 132–140.

Ahmed, M., Akter, M.S., Eun, J.-B., 2011. Optimization conditions for anthocyanin and phenolic content extraction form purple sweet potato using response surface methodology. Int. J. Food Sci. Nutr. 62 (1), 91–96.

Al Bittar, S., Périno-Issartier, S., Dangles, O., Chemat, F., 2013. An innovative grape juice enriched in polyphenols by microwave-assisted extraction. Food Chem. 141 (3), 3268–3272.

Arani, D., Shreya, G., Mukesh, S., 2011. Antimicrobial property of piper betel leaf against clinical isolates of bacteria. Int. J. of Pharma Sci. Res. 2 (3), 104–109.

Azlan Hadi Tan, N., Siddique, B.M., Muhamad, I.I., Mad Salleh, L., Hassan, N.H., 2013. Perah oil: a new opportunity in health and skincare wellness. Int. J. Biotechnol. Well. Ind. 2, 22–28.

Azmir, J., Zaidul, I.S.M., Rahman, M.M., Sharif, K.M., Mohamed, A., Sahena, F., Jahurul, M.H.A., Ghafoor, K., Norulaini, N.A.N., Omar, A.K.M., 2013. Techniques for extraction of bioactive compounds from plant materials: a review. J. Food Eng. 117, 426–436.

Bonoli, M., Verardo, V., Marconi, E., 2004. Antioxidant phenols in barley (*Hordeum vulgare* L.) flour: comparative spectrophotometric study among extraction methods of free and bound phenolic compounds. J Agric Food Chem 52, 5195–5200.

Carneiro, H.C.F., Tonon, R.V., Grosso, C.R.F., Hubinger, M.D., 2013. Encapsulation efficiency and oxidative stability of flaxseed oil microencapsulated by spray drying using different combinations of wall materials. J. Food Eng. 115, 443–451.

Chan, C.H., Yusoff, R., Ngoh, G.C., Kung, F.W.L., 2011. Microwave-assisted extractions of active ingredients from plants. J. Chromatogr. A 1218 (37), 6213–6225.

Chan, C.-H., Yusoff, R., Ngoh, G.-C., 2014. Optimization of microwave-assisted extraction based on absorbed microwave power and energy. Chem. Eng. Sci. 111, 41–47.

Chatterjee, D., Bhattacharjee, P., 2012. Supercritical carbon dioxide extraction of eugenol from clove buds: Process optimization and packed bed characterization. Food Bioprocess Technol. 6 (10), 2587–2599.

Cilek, B., Luca, A., Hasirci, V., Sahin, S., Sumnu, G., 2012. Microencapsulation of phenolic compounds extracted from sour cherry pomace: effect of formulation, ultrasonication time and core to coating ratio. Eur. Food Res. Technol. 235 (4), 587–596.

Cosa, P., Vlietinck, A.J., Berghe, D.V., Maes, L., 2006. Anti-infective potential of natural products: how to develop a stronger in vitro "proof-of-concept". J. Ethnopharmacol. 106, 290–302.

Dhobi, M., Mandal, V., Hemalatha, S., 2009. Optimization of microwave assisted extraction of bioactive flavonolignan-silybinin. J. Chem. Metrol. 3 (1), 13–23.

Doker, O., Salgin, U., Yildiz, N., Aydogmus, M., Alimli, A.C., 2010. Extraction of sesame seed oil using supercritical CO_2 and mathematical modelling. J. Food Eng. 97, 360–366.
Flórez, N., Conde, E., Domínguez, H., 2014. Microwave assisted water extraction of plant compounds. J. Chem. Technol. Biotechnol. 90, 1–18.
Foo, L.W., Salleh, E., Mamat, S.N.H., 2015. Extraction and qualitative analysis of piper betel leaves for antimicrobial activities. Int. J. Eng. Technol. Sci. Res. 2, 1–8.
Francis, F.J., 1989. Food colourants: anthocyanins. Critical Rev. Food Sci. Nutr. 28, 273–314.
Gadkari, P.V., Balaraman, M., 2014. Catechins: sources, extraction and encapsulation: a review. Food Bioprod. Proc. 93, 122–138.
Gai, Q.Y., Jiao, J., Mu, P.S., Wang, W., Luo, M., Li, C.Y., Fu, Y.J., 2013. Microwave-assisted aqueous enzymatic extraction of oil from Isatis indigotica seeds and its evaluation of physicochemical properties, fatty acid compositions and antioxidant activities. Ind. Crops Prod. 45, 303–311.
Gaitan, E., Cooksey, R.C., Legan, J., Lindsay, R.H., 1995. Antithyroid effects in vivo and in vitro of vitex: a C-glucosylflavone in millet. J. Clin Endocrinol Met 80, 1144–1151.
Handa, S.S., Khanuja, S.P.S., Longo, G., Rakesh, D.D., 2008. Extraction Technologies for Medicinal and Aromatic Plants. ICS Unido, Trieste.
Harborne, J.B., 1958. The chromatoghraphic identification of anthocyanin pigments. J. Chromatogr. 1, 473–488.
Hassan, N.D., Muhamad, I.I., Sarmidi, M.R., 2014. The effect of copigmentation on the stability of butterfly pea extract. Key Eng. Mater. 594–595, 245–249.
Hendry, G.A.F., Houghton, J.D., 1992. Natural Food Colourants. Blackie and Son Ltd, New York.
Hu, A., Feng, Q., Zheng, J., Hu, X., Wu, C., Liu, C., 2010. Kinetic model and technology of ultrasound extraction of safflower seed oil. J. Food. Proc. Eng. 35, 278–294.
Jackman, R.L., Smith, J.L., 1992. Anthocyanins and Betalains. In: Hendry, G.A.F., Houghton, J.D. (Eds.), Natural Food Colourants. Blackie and Son Ltd, New York, pp. 183–207.
Jackman, R.L., Yada, R.Y., Tung, M.A., 1987a. A Review: separation and chemical properties of anthocyanin used for qualitative and quantitative analysis. J. Food Biochem. 11, 279–308.
Jackman, R.L., Yada, R.Y., Tung, M.A., Speers, R.A., 1987b. Anthocyanin as food colourants: a review. J. Food Biochem. 11, 201–247.
Jenie, B.S.L., 2001. Antimicrobial activity of *Piper betle* Linn extract towards foodborne pathogens and food spoilage microorganisms. FT Annual Meeting, New Orleans, LA.
Kareru, P.G., Kenji, G.M., Gachanja, A.N., Keriko, J.M., Mungai, G., 2007. Traditional medicines among the Embu and Mbeere peoples of Kenya. Afri J. Trad Compl Alter Med. 4, 75–86.
Katkar, A.L., Sakhale, C.N., Undirwade, S.K., 2015. Design and mechanization of oil expeller. Int. J. New Technol. Sci. Eng. 2 (4), 7–13.
Kaufmann, B., Christen, P., 2002. Recent extraction techniques for natural products: microwave-assisted extraction and pressurised solvent extraction. Phytochem. Anal. 13 (2), 105–113.
Kubra, I.R., Kumar, D., Rao, L.J.M., 2013. Effect of microwave-assisted extraction on the release of polyphenols from ginger (*Zingiber officinale*). Int. J. Food Sci. Technol. 48, 1828–1833.
Liu, G., Xiang Xu, X., QinfengHao, Q., YanxiangGao, Y., 2009. Supercritical CO_2 extraction optimization of pomegranate (*Punicagranatum* L.) seed oil using response surface methodology. LWT—Food Sci. Technol. 42, 1491–1495.
Markakis, P., 1974. Anthocyanins and their stability in foods. CRC Crit. Rev. Food Technol. 4, 437–456.
Markakis, P., 1982. Stability of Anthocyanins in Food. In: Markakis, P. (Ed.), Anthocyanins as Food Colours. Academic Press, New York, pp. 163–178.
Marsden, J.O., House, C.I., 2006. The Chemistry of Gold Extraction, 2nd ed. SME Littleton, Colorado, USA.
Moore, A.B., Francis, F.J., Clydesdale, F.M., 1982. Changes in chromathographic profile of anthocyanins of red onion during extraction. J. Food Prot. 45, 738–743.
Mustafa, A., Turner, C., 2011. Pressurized liquid extraction as a green approach in food and herbal plants extraction: a review. Anal. Chimica Acta 703 (1), 8–18.
Muthulakshmi, A., Margret, J.R., Mohan, V.R., 2012. GC-MS analysis of bioactive components of feoniaelephantumcorrea (*Rutaceae*). J. App. Pharm. Sci. 2 (2), 69–74.

National Coordinating Committee on Food and Nutrition, 2005. Recommended Nutrient Intakes of Malaysia. Ministry of Health Malaysia, Putrajaya.

Negi, J.S., Pramod, S., Bipin, R., 2011. Chemical constituents and biological importance of swertia: a review. Curr. Res. Chem. 3, 1–15.

Ofori-Boateng, C., Keat Teong, L., Jit Kang, L., 2012. Comparative exergy analyses of *Jatropha curcas* oil extraction methods: solvent and mechanical extraction processes. Energy Convers. Manage. 55, 164–171.

Panyakom, K., 2006. Structural elucitation of bioactive compounds of *Clinacanthus nutans* (Burm. f.) Lindau leaves. Master thesis, Suranaree University of Technology, Nakhon Ratchasima, Thailand.

Parthasarathi, S., Ezhilarasi, P.N., Jena, B.S., Anandharamakrishnan, C., 2013. A comparative study on conventional and microwave-assisted extraction for microencapsulation of garcinia fruit extract. Food Bioprod. Proc. 91 (2), 103–110.

Patel, R.P., Patel, P.M., Suthar, M.A., 2009. Spray drying technology: an overview. Indian J. Sci. Technol. 10, 44–47.

Phisut, N., 2012. Spray drying technique of fruit juice powder. Int. Food Res. J. 19 (4), 1297–1306.

Pin, K.Y., LuqmanChuah, A., AbdullRashih, A., Mazura, M.P., Fadzureena, J., Vimala, S., Rasadah, M.A., 2010. Antioxidant and anti-inflammatory activities of extracts of betel leaves (*Piper betle*) from solvents with different polarities. J. Trop. Forest Sci. 22, 448–455.

Prabodh, S., William, S.N., 2012. Chemical composition and biological activities of Nepalese *Piper betle* L. IJPHA 1 (2), 23–26.

Pradhan, R.C., Meda, V., Rout, P.K., Naik, S., Dalai, A.K., 2010. Supercritical CO_2 extraction of fatty oil from flaxseed and comparison with screw press expression and solvent extraction processes. J. Food Eng. 98 (4), 393–397.

Quy, Diem Do., Artik, E.A., Phuong Lan, T.-N., Lien Huong, H., Felycia, E.S., Suryadi, I., Yi-Hsu, J., 2014. Effect of extraction solvent on total phenol content, total flavonoid content, and antioxidant activity of *Limnophila aromatica*. J. Food Drug Anal. 22 (3), 296–302.

Routray, W., Orsat, V., 2012. Microwave-assisted extraction of flavonoids: a review. Food Bioproc. Technol. 5 (2), 409–424.

Rubio-Rodríguez, N., Beltrán, S., Jaime, I., de Diego, S.M., Sanz, M.T., Rovira-Carballido, J., 2010. Production of u-3 polyunsaturated fatty acid concentrates: a review. Inn. Food Sci. Emerg. Technol. 2010 (11), 1–12.

Rui, H., Zhang, L., Li, Z., Pan, Y., 2009. Extraction and characteristics of seed kernel oil from white pitaya. J. Food Eng. 93 (4), 482–486.

Sakdarat, S., Shuyprom, A., Pientong, C., Ekalaksananan, T., Thongchai, S., 2009. Bioactive constituents from the leaves of *Clinacanthus nutans* Lindau. Bioorg. Med. Chem. 17 (5), 1857–1860.

Setapar, S.H., Lee, N.Y., Mohd, A., 2011. Comparison of physico-chemical properties and fatty acid compostion of *Elateriospermum Tapos* (Buah perah), palm oil and soybean oil. World Academy of Science. Eng. Technol. 5 (9), 81–165.

Simpson, K.L., 1985. Chemical Changes in Natural Food Pigments. In: Richardson, T., Finley, J.W. (Eds.), Chemical Changes in Food During Processing. AVI Publishing Company, Westport, CT, pp. 409–441.

Spigno, G., De Faveri, D.M., 2009. Microwave-assisted extraction of tea phenols: a phenomenological study. J. Food Eng. 93 (2), 210–217.

Su, X., Jianteng Xu, J., Rhodes, D., Shen, Y., Song, W., Katz, B., Tomich, J., Wang, W., 2016. Identification and quantification of anthocyanins in transgenic purple tomato. Food Chem. 202, 184–188.

Subashkumar, R., Sureshkumar, M., Babu, S., Thayumanavan, T., 2013. Antibacterial effect of crude aqueous extract of *Piper betel* L. against pathogenic bacteria. Int. J. Res. Pharm. Biomed. Sci. 4 (1), 42–46.

Sugumaran, M., Poornima, M., Venkatraman, S., Lakshmi, M., 2011. Srinivasansethuvani. Chemical composition and antimicrobial activity of sirugamani variety of *Piper betel* Linn leaf oil. J. Pharm. Res. 4 (10), 3424–3426.

Tiwari, B., Valdramidis, V.V., O' Donnell, C.P., Muthukumarappan, K., Cullen, P.J., 2009. Application of natural antimicrobials for food preservation. J. Agr. Food Chem. 57 (14), 5987–6000.

Truong, V.D., Hu, Z., Thompson, R.L., Yencho, G.C., Pecota, K.V., 2012. Pressurized liquid extraction and quantification of anthocyanins in purple-fleshed sweet potato genotypes. J. Food Comp. Anal. 26 (1–2), 96–103.

Turkmen, N., Sari, F., Velioglu, Y.S., 2006. Effects of extraction solvents on concentration and antioxidant activity of black and black mate tea polyphenols determined by ferrous tartrate and Folin–Ciocalteu methods. Food Chem. 99, 835–841.

Veggi, P.C., Martinez, J., Meireles, M.A.A., 2013. Fundamentals of Microwave Extraction. In: Chemat, F., Cravotto, G. (Eds.), Microwave-assisted Extraction for Bioactive Compounds. Springer Science, London, pp. 15–52.

Vermaak, I., Kamatou, G.P.P., Komane-Mofokeng, B., Viljoen, A.M., Beckett, K., 2011. African seed oils of commercial importance: cosmetic applications. S. Afr. J. Bot. 77, 920–933.

Virot, M., Tomao, V., Ginies, C., Visinoni, F., Chemat F, 2008. Microwave-integrated extraction of total fats and oils. J. Chromatogr. A 1196–1197, 57–64.

Vuong, Q.V., Hirun, S., Roach, P.D., Bowyer, M.C., Phillips, P.A., Scarlett, C.J., 2013. Effect of extraction conditions on total phenolic compounds and antioxidant activities of Carica papaya leaf aqueous extracts. J. Herbal Med. 3, 104–111.

Wanikiat, P., Panthong, A., Sujayanon, P., Yoosook, C., Rossi, A.G., Reutrakul, V., 2008. The anti-inflammatory effects and the inhibition of neutrophil responsiveness by *Barleria lupulina* and *Clinacanthus nutans* extracts. J. Ethnopharmacol. 116 (2), 234–244.

Yong, O.Y., Salimon, J., 2006. Characteristics of *Elateriospermum tapos* seed oil as a new source of oilseed. Ind. Crops Prod. 24 (2), 146–151.

Yoosook, C., Panpisutchai, Y., Chaichana, S., Santisuk, T., Reutrakul, V., 1999. Evaluation of anti-HSV-2 activities of *Barleria lupulina* and *Clinacanthus nutans*. J. Ethnopharmacol. 67 (2), 179–187.

Zhang, H., Wang, R., Wang, J., Dong, Y., 2014. Microwave-assisted synthesis and characterization of acetylated corn starch. Starch 66, 515–523.

Zheng, S.Q., Jiang, F., Gao, H.Y., Zheng, J.G., 2010. Preliminary observations on the antifatigue effects of longan (Dimocarpus longanLour.) seed polysaccharides. Phytother. Res. 24, 622–624.

Zuhaili, I., Ida, I.M., Mohd, R.S., 2012. Degradation kinetics and colour stability of spray dried encapsulated anthocyanins from *Hibiscus sabdariffa* L. J. Food Process Eng. 35 (4), 522–542.

Further Reading

Bangash, F.A., Hashmi, A.N., Mahboob, A., Zahid, M., Hamid, B., Muhammad, S.A., Shah, Z.U., Afzaal, H., 2012. In-vitro antibacterial activity of *Piper betel* leaf extracts. J. Appl. Pharm. 03 (04), 639–646.

Chan, K.W., Khong, N.M.H., Iqbal, S., Ismail, M., 2012. Simulated gastrointestinal pH condition improves antioxidant properties of wheat and rice flours. Int. J. Mol. Sci. 13 (6), 7496–7507.

Charuwichitratana, S., Wongrattanapasson, N., Timpatanapong, P., Bunjob, M., 1996. Herpes zoster: treatment with *Clinacanthus nutans* cream. Int. J. Dermatol. 35 (9), 665–666.

Chemat, F., Cravotto, G. (Eds.), 2013. Microwave-Assisted Extraction for Bioactive Compounds. first ed. Springer, New York.

Dai, R., Mumper, J., 2015. Plant phenolics: extraction, analysis and their antioxidant and anticancer properties. Molecules 15, 7313–7352.

Datta, A., Ghoshdastidar, S., Singh, M., 2011. Antimicrobial property of *Piper betle* leaf against clinical isolates of bacteria. Int. J. Pharm. Sci. Res. 2 (3), 104–109.

Do, Q.D., Angkawijaya, A.E., Tran-Nguyen, P.L., Huynh, L.H., Soetaredjo, F.E., Suryadi Ismadji, S., Ju, Y.H. Effect of extraction solvent on total phenol content, total flavonoid content, and antioxidant activity of *Limnophila aromatic*. J. Food Drug. Anal. 22 (3), 296–302.

Hu, F., Yi, B.W., Zhang, W., 2012. Carotenoids and breast cancer risk: a meta-analysis and meta-regression. Breast Cancer Res. Treat. 131 (1), 239–253.

Klaunig, J.E., Kamendulis, L.M., 2004. The role of oxidative stress in carcinogenesis. Ann. Rev. Pharmacol. Toxicol. 44, 239–267.

Kongkaew, C., Chaiyakunapruk, N., 2011. Efficacy of *Clinacanthus nutans* extracts in patients with herpes infection: systematic review and meta-analysis of randomised clinical trials. Compl. Thera. Med. 19 (1), 47–53.

Kumari, O.S., Rao, N.B., 2015. Phyto chemical analysis of *Piper betel* leaf extract. World J. Pharm. Pharm. Sci. 4 (1), 699–703.

Vachirayonstien, T., Promkhatkaew, D., Bunjob, M., Chueyprom, A., Chavalittumrong, P., Sawanpanyalert, P., 2010. Molecular evaluation of extracellular activity of medicinal herb *Clinacanthus nutans* against herpes simplex virus type-2. Nat. Prod. Res. 24 (3), 236–245.

Valko, M., Rhodes, C.J., Moncol, J., Izakovic, M., Mazur, M., 2006. Free radicals, metals and antioxidants in oxidative stress-induced cancer. Chem. Biol. Inter..

CHAPTER 15

An Energy-Based Approach to Scale Up Microwave-Assisted Extraction of Plant Bioactives

Chung-Hung Chan*, Rozita Yusoff, Gek Cheng Ngoh****

**Malaysian Palm Oil Board, Kajang, Selangor, Malaysia; **University of Malaya, Kuala Lumpur, Malaysia*

1 Introduction

The extraction of bioactive compounds from plants is being conducted intensively in various fields, such as agricultural analysis, food processing, and drug delivery. The extracted bioactive compounds possess valuable therapeutic effects that are good for human health. For instance, quercetin, catechin, and kaempherol give antioxidant and hypoglycemic effects, which are good for diabetes treatment (Sultana and Anwar, 2008). These compounds have potential to replace synthetic drugs, such as thiazolidinedione, which often burdens patients with undesirable secondary effects, namely edema, abnormal water retention, and coronary heart disease (Benbow et al., 2001). In order to obtain bioactive compounds, extraction process is crucial and the conventional techniques, such as soaking, maceration, and Soxhlet extraction are commonly employed. The overall steps involved in the extraction of bioactive compounds from plants are illustrated in Fig. 15.1. In the conventional practice, dried and grinded plant materials are immersed in a solvent for a specific time to allow the bioactive compounds within to diffuse out. Such a process usually requires a lengthy heating to attain the equilibrium condition and this could risk the degradation of bioactive compounds. The shortcomings of the conventional extraction can be alleviated with the incorporation of assisted means, such as microwave heating.

Microwave heating is associated with ionic conduction and dipole rotation mechanisms (Sparr Eskilsson and Björklund, 2000). Basically, microwave, an electromagnetic wave with frequency of 0.3 to 300 GHz, is able to penetrate nonpolar substance and to interact with polar substances to generate heat. The heating efficiency is depended very much on the dielectric properties of the substance. One of the properties is known as dissipation factor of

Ingredients Extraction by Physicochemical Methods in Food
http://dx.doi.org/10.1016/B978-0-12-811521-3.00015-6

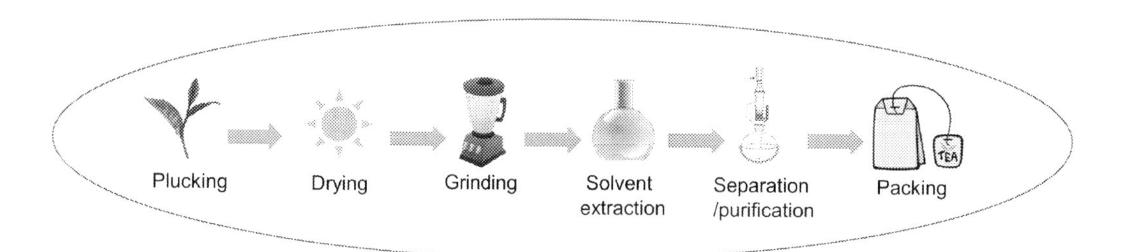

Figure 15.1: Steps Involved in the Extraction of Bioactive Compounds From Plants.

material or dielectric loss tangent, tan δ. This property indicates the ability of a substance to absorb microwave energy in relative to its ability to dissipate heat (Ayappa et al., 1991):

$$\tan\,\delta = \frac{\varepsilon''}{\varepsilon'} \tag{15.1}$$

where ε'' is the dielectric loss and ε' is the dielectric constant. The properties ε'' and ε' indicate the ability of the material to convert microwave energy into heat and its ability to absorb microwave energy, respectively. The power dissipation of the substance can be calculated using the following equation (Ayappa et al., 1991).

$$P = Kf\varepsilon' E^2 \tan\delta \tag{15.2}$$

where P is the dissipated power per unit volume, K is a constant, f is the frequency of microwave, and E is the strength of the electric field.

Microwave-assisted extraction (MAE) technique has high efficiency in extracting plant bioactives. It requires shorter time, lower solvent consumption, moderate capital cost, and milder extraction conditions due to the superior effect of microwave heating (Barrera Vázquez et al., 2014; Chan et al., 2011; Mandal et al., 2007; Tanongkankit et al., 2013; Wang and Weller, 2006; Zhang et al., 2011). MAE has been used to extract bioactives from various plants, such as red beet (Cardoso-Ugarte et al., 2014), *Schisandra chinensis* Bail fruit (Cheng et al., 2015b), lettuce (Périno et al., 2016), and others (Dahmoune et al., 2014; Dong et al., 2014; Setyaningsih et al., 2012). Besides, it is also employed to recover valuable compounds from plant-based waste residues, such as fruit skins or peels (Franco-Vega et al., 2016; Liazid et al., 2011; Nayak et al., 2015; Pandit et al., 2015; Thirugnanasambandham et al., 2014), Eucalyptus wood veneer trimmings (Fernández-Agulló et al., 2015), and so on (Milutinović et al., 2014, 2015; Pandit et al., 2015).

There are various issues pertaining to optimization and scaling up of MAE, which have restricted the commercialization of this technique. In general, MAE is difficult to scale up. This is due to the inconsistency of microwave absorption capability of extraction mixture at various extraction scales and the variation of heating efficiency of microwave systems. In other words, using different extraction scales and microwave extractors would give different extraction performances even though the same heating conditions, such as microwave power and heating time, are applied. To date, most of the optimum operating conditions and kinetic models reported in the literature are applicable only for specific scale and microwave system, making their applications limited for comparison study to assess impacts of various extraction techniques (Dahmoune et al., 2014; Dong et al., 2014; Li et al., 2014) and for analytical chemistry to characterize bioactive compounds (Aeenehvand et al., 2016; Chen et al., 2015; Cheng et al., 2015a; Gao et al., 2012; Shen et al., 2014; Sternbauer et al., 2013). These operational issues can be resolved by calibrating the actual power and energy requirement of the extraction system.

The actual power and energy requirement of the MAE can be calibrated based on energy-based parameters, that is, absorbed power density (APD) and absorbed energy density (AED). These parameters were employed to replace nominal microwave power and extraction time in optimization study (Chan et al., 2014b). They were also adopted in kinetic models to predict and simulate extraction profiles at various conditions, including larger scales and different microwave setups (Chan et al., 2013, 2015a,b). This energy-based approach provides a practical solution to commercialize MAE technique in extracting bioactive compounds from plants. With that, this chapter discusses the optimization and modeling methods based on APD and AED parameters to scale-up and commercialize MAE. Topics covered include the extraction mechanism and operating parameters, the theory related to the energy-based parameters and their applications in optimization and modeling. The concept and strategy to commercialize the MAE technique are also included.

2 Extraction Kinetics and Mechanism

In general, the extraction of plant bioactives is comprised of two steps, a fast-washing step and a slow extraction step as illustrated in Fig. 15.2 (Franco et al., 2007; So and Macdonald, 1986). In the first step, the extraction of bioactive compounds is initiated by the penetration of solvent into plant matrices and subsequently the washing of the bioactive compounds from the exposed cytoplasm layer of broken cells (Crossley and Aguilera, 2001). This occurs at the beginning of the extraction with an extremely fast rate (Franco et al., 2007; Rakotondramasy-Rabesiaka et al., 2009). In the second step, the bioactive compounds from the intact cells diffuse slowly and will then dissolve in the solvent. This diffusion step is usually facilitated by a thermal treatment. The extraction yield associated to this step is greatly dependent on the amount of intact plant cells after sample preparation process, such as mechanical grinding of plant sample (Crossley and Aguilera, 2001). The kinetic mechanism of this diffusion step depends on the extraction technique. For instance, conventional extraction techniques involve diffusion of bioactive

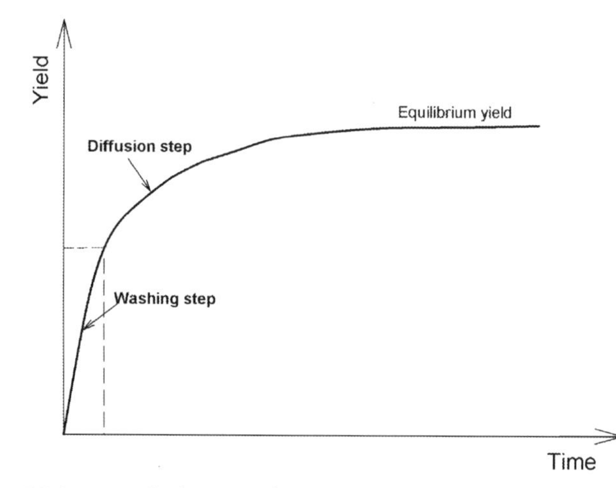

Figure 15.2: A Typical Extraction Curve of Plant Bioactive Extraction.

compounds due to the concentration gradient whereas in assisted extraction techniques, such as MAE, the extraction mechanism involves the disruption of plant sample followed by the elution of bioactive compounds to the surrounding solvent (Dahmoune et al., 2014; Zhou and Liu, 2006). Eventually, the extraction reaches equilibrium yield after sufficient extraction time.

Extraction mechanisms of MAE of bioactive compounds from plants are detailed in Fig. 15.3. These mechanisms are based on the microwaved plant cell rupture model (Chan et al., 2016). The dominant mechanism in MAE is the disruption of plant sample via cell rupture mechanism. The conventional mass transfer mechanism, namely concentration-driven diffusion mechanism, is not significant in the MAE. At the beginning of the extraction, penetration of solvent into plant matrices takes place. Then, during the microwave heating, the intracellular moisture of the plant sample and the solvent absorb microwave energy. As a result, the internal pressure within the plant matrices is gradually built up due to vaporization of the intracellular moisture in the cell vacuoles. The pressure expands the vacuoles and stretches the cell membrane to the cell wall. Subsequently, the cell wall stretches together during the expansion and ruptures when the internal pressure is raised beyond the strength of the cell wall. The disruption of plant sample has been evidenced in the MAE of *Pistacia lentiscus* leaves (Dahmoune et al., 2014), *Schisandra chinensis* fruits (Ma et al., 2012), *Sophora flavescens* roots (Zhang et al., 2015), and so on (Chan et al., 2014b; Kong et al., 2010). Once the cell is ruptured, the bioactive compounds elute and dissolve into the surrounding solvent. The rate-limiting mechanism of MAE is the disruption of plant sample through microwave heating. Whereas, the rate of solvent penetration and elution of bioactive compounds are relatively fast. It should be noted that all the mechanisms mentioned earlier are responsible for the contribution and characterization of the overall diffusion step of the MAE.

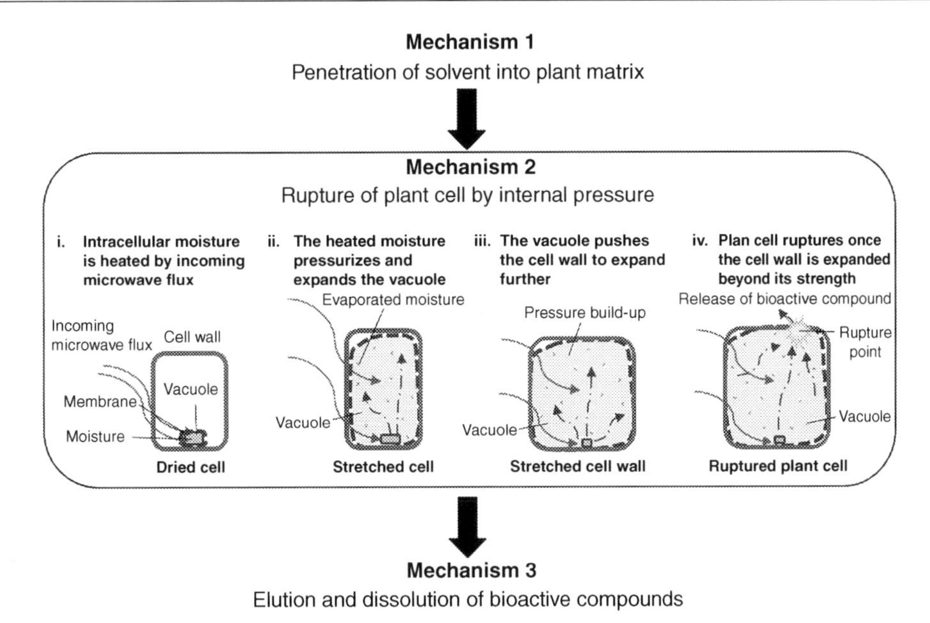

Figure 15.3: Schematic of Extraction Mechanism of Microwave-Assisted Extraction *(MAE)* of Plant Bioactives. *Cell rupture mechanism is adapted with permission from Chan, C.-H., Yeoh, H.K., Yusoff, R., Ngoh, G.C., 2016. A first-principles model for plant cell rupture in microwave-assisted extraction of bioactive compounds. J. Food Eng. 188, 98–107.*

3 Influencing Parameters

Many groups of operating parameters affect every single mechanism of MAE, and in turn characterize the overall extraction yield and kinetics (Chan et al., 2014b). The relationship between the MAE mechanisms and their influencing operating parameter is shown in Fig. 15.4. Theoretically, general extraction parameters, such as solvent type, solvent-to-sample ratio, sample particle size, and temperature affect the penetration of solvent into plant matrix and the elution of bioactive compounds from disrupted structure as stated in mechanism 1 and 3 in Fig. 15.4, respectively. These parameters only affect the extraction yield of the MAE and they do not have direct impact on the rate of limiting mechanism (i.e., disruption of plant cell structure). On the contrary, operating parameters, such as type of microwave system, heating mode, microwave power, and heating time affect the microwave heating efficiency and also have direct impacts on the disruption of plant structure or the rate of limiting mechanism. These parameters influence both the extraction yield and the extraction kinetics. The individual effects of the stated operating parameters are further discussed in the following subsections.

3.1 Solvent

Factors to be considered in the selection of suitable extraction solvent are molecular size, polarity, and the solubility of the compounds of interest. Selection of solvent for MAE

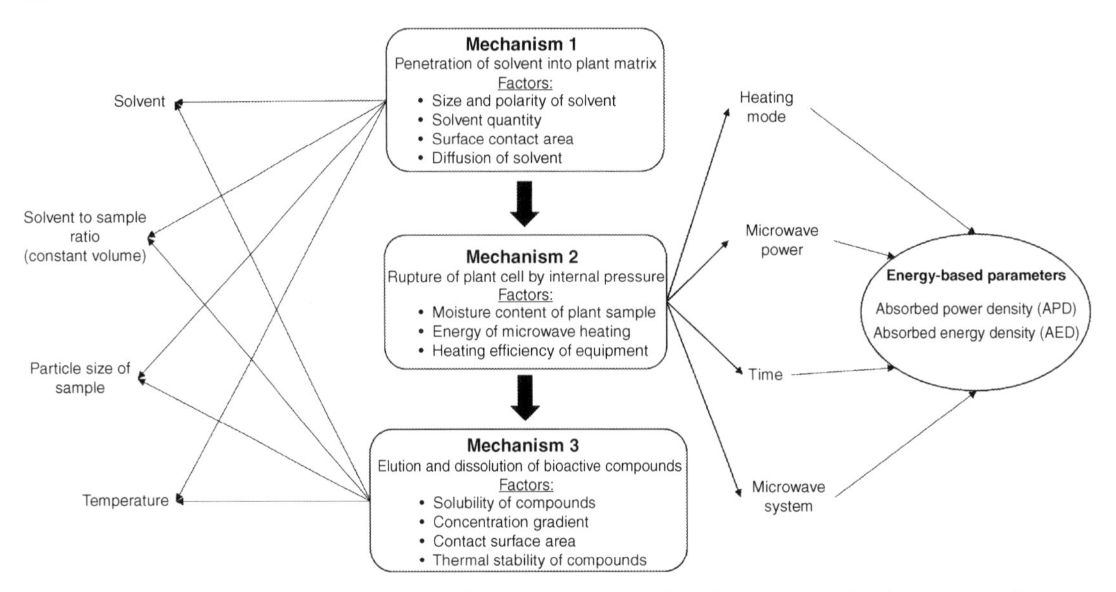

Figure 15.4: Influencing Parameters in MAE of Plant Bioactives. *Adapted with permission from Chan, C.-H., Yusoff, R., Ngoh, G.-C., 2014b. Optimization of microwave-assisted extraction based on absorbed microwave power and energy. Chem. Eng. Sci. 111, 41–47.*

cannot be deduced based on the performance of the specific solvent in conventional extraction techniques as the requirements are different in both techniques. Solvents with smaller molecular size and higher polarity, such as methanol, give better performance in terms of extraction yield, for example; in the MAE of phenolic compounds from grape skins and seeds (Casazza et al., 2010). However, ethanol is widely adopted in extraction in preference over methanol due to safety concern (Zhou and Liu, 2006). For better heating efficiency, a modifier, such as water, can be added into the extraction solvent because it has good dielectric properties (Alfaro et al., 2003). The addition of solvent modifier is useful in cases where a less polar solvent, such as hexane or acetone, has to be employed for the extraction. Usually, aqueous organic solvent is widely employed in MAE, such as in the bioactive extraction from *Fumaria officinalis* L. (Rakotondramasy-Rabesiaka et al., 2010), chestnut tree wood (Gironi and Piemonte, 2011), and grape seeds (Bucic-Kojic et al., 2007).

3.2 Solvent-to-Sample Ratio

Solvent-to-sample ratio can be manipulated either at constant solvent volume or sample mass, which affects the extraction kinetics differently. For instance, increasing solvent-to-sample ratio without changing the solvent volume decreases the mass transfer barrier. Hence, it facilitates the penetration of solvent into the plant matrix and elution of bioactive compounds to improve extraction yield. Once this ratio is increased above its optimum point, the improvement on yield will not be significant and this will cause solvent wastage instead (Karabegović et al., 2013; Qu et al., 2010; Spigno and De Faveri, 2009). On the other hand, increasing solvent-to-sample ratio at constant sample mass affects both the mass transfer barrier and the absorption of microwave energy (Kingston and Jassie, 1988). Increasing this ratio beyond its optimum values causes poor extraction yield due to insufficient supplied energy to complete the extraction (Mandal and Mandal, 2010; Spigno and De Faveri, 2009). Therefore, for the ease of optimization, solvent-to-sample ratio at constant solvent volume should be considered. This ratio does not exhibit interaction with microwave power and heating time on the cell disruption mechanism. Looking into the reported optimum solvent-to-sample ratio in the literature, most of the MAE employed 5–50 mL/g (Chan et al., 2011).

3.3 Particle Size of Sample

Extraction of bioactive compounds from an intact plant part is difficult and inefficient without the aid of mechanical treatment, such as cutting and grinding of plant sample. Basically, particle size of sample characterizes the amount of disrupted and intact cells present in the sample, which determines the extractable yield during the washing step and the diffusion step as previously mentioned in Section 2. A slight change in particle size of plant sample exerts predominant effects on the extraction yield. For instance, smaller size of pomegranate marc (Qu et al., 2010) and *Hibiscus sabdariffa* calyces (Cissé et al., 2012) improve the extraction yield in the washing step. In this case, the diffusivity of the bioactive compounds increases with smaller particles due to larger contact surface area with the solvent and shorter average diffusion path of the compounds from the plant sample to the bulk solvent (Cissé et al., 2012; Herodež et al., 2003; Hojnik et al., 2008). Nevertheless, excessive particle size reduction leads to several adverse effects, such as high extraction yields of undesired compounds and difficulty in separating the plant residue with the extract (Cissé et al., 2012; Qu et al., 2010). Despite the fact that the effect of particle size is obvious, the aforementioned effects are not significant for plant sample with plate geometry as the relevant dimension for the diffusion of bioactive compounds is the thickness of leaves (Wongkittipong et al., 2004). The effects are noticeable only when the particle size is reduced below its leaf thickness.

3.4 Temperature

Control of extraction temperature is vital to avoid degradation or decomposition of thermal sensitive compounds. It should be monitored at all cost as it affects the rate of degradation of bioactive compounds (Xiao et al., 2012). According to the first-principle model (Chan et al., 2016), extraction temperature does not significantly affect the extraction kinetics of the diffusion step in MAE. Nevertheless, high extraction temperature promotes better solvent penetration into plant matrix and diffusivity of bioactive compounds during elution and dissolution into the surrounding solvent (Dahmoune et al., 2014; Tan et al., 2011). Besides, the increase in solvent temperature enhances the solvation power to allow more bioactive compound to be dissolved in the solvent and hence it improves the extraction yield (Rakotondramasy-Rabesiaka et al., 2007, 2010; Tsubaki et al., 2010). When thermal labile compounds is involved, mild extraction temperature in the range of 50–70°C should be considered. Temperature control can be achieved by several ways. One can employ low boiling point solvents, such as methanol and ethanol, or adopt different heating modes, such as intermittent and temperature-controlled microwave heating. When extracting thermally stable compounds, MAE can be performed using simple constant-power heating mode to allow temperature to reach the boiling point of the solvent.

3.5 Microwave Power

Microwave power controls the heating rate of the extraction mixture, and it acts as the driving force to disrupt the plant sample in the MAE. Conducting the MAE at higher microwave power can result in several consequences. To begin with, the MAE can reach equilibrium extraction yield in shorter time (Dong et al., 2014; Mandal and Mandal, 2010). Besides, higher equilibrium extraction yield can be achieved provided that the extraction temperature is below the degradation temperature of the bioactive compounds (Biesaga, 2011; Chemat et al., 2005; Kwon et al., 2003; Mandal and Mandal, 2010; Xiao et al., 2008). At high operating power, the temperature of extraction solvent elevates in a much faster rate, evaporating large amounts of solvent during boiling. This might result in poor extraction yield (Chan et al., 2013; Mustapa et al., 2015). Nevertheless, the solvent evaporation can be minimized by conducting extraction in closed systems or by incorporating an external condenser to cool down the evaporated solvent back to the system. When MAE is performed at extremely high microwave power, for example, 800 W, it subjects the plant sample to possible carbonization or burning (Cardoso-Ugarte et al., 2014). Conversely, when conducting MAE at low microwave power, longer extraction time is required to complete the extraction, and also using low power, might not be sufficient to disrupt the cell plant sample (Ma et al., 2011; Song et al., 2011). Having said that, there is no standard microwave power level for the MAE because different plant samples have different characteristics and so the power requirement. The selection of microwave power level is also depended very much on the working volume of extraction solvent and the desired extraction time.

3.6 Extraction Time

Extraction time is usually associated with microwave power level as these two parameters demonstrate an inverse relationship. This means that when microwave power is increased, the required optimum extraction time is decreased and vice versa. At each power level, there is an optimum extraction time that gives an equilibrium extraction yield (Chan et al., 2013). Once the extraction time exceeds the optimum condition, overexposure to microwave heating even at low power might degrade the chemical structure of the bioactive compounds, if the extraction temperature exceeded the degradation temperature of the compounds (Cardoso-Ugarte et al., 2014; Hao et al., 2002; Švarc-Gajić et al., 2013; Wang et al., 2009). Otherwise, extraction yield is stable across a period of time if the solvent loss due to evaporation is not significant. Normally, the extraction time of MAE varies from a few minutes up to half an hour with the exception of solvent-free MAE whereby longer extraction time of 1 h is needed to complete the essential oil extraction (Chan et al., 2011).

3.7 Operational Mode of Heating

The selection of heating mode in MAE system is dependent on many factors, such as the stability of bioactive compounds, the characteristic of plant sample, the types of extraction solvent, and the type of microwave system. The MAE heating modes include constant-power heating, intermittent-power heating, and constant-temperature heating. In constant-power heating, persistent microwave heating at a specific power level is engaged. This heating mode is widely adopted to extract thermally stable bioactive compounds (Terigar et al., 2010). On the other hand, pulsed microwave heating in intermittent-power MAE is preferable when extracting thermal-labile compounds (Hiranvarachat and Devahastin, 2014). It is able to control the extraction temperature to prevent thermal degradation and to minimize evaporation of solvent. Last, temperature control feature in constant temperature MAE is able to extract highly degradable active compounds (Tsukui et al., 2014). The differences in heating modes might complicate the optimization study and thus careful consideration should be given especially in the selection of optimizing parameters. In view of that, constant-power heating involves microwave power and heating time (Song et al., 2011; Švarc-Gajić et al., 2013); intermittent-power heating involves microwave power, intermittency ratio, and heating time (Ahmad and Langrish, 2012; Rodríguez-Rojo et al., 2012) whereas constant-temperature heating involves microwave power during ramping, extraction temperature, and heating time (Li et al., 2011; Setyaningsih et al., 2015).

3.8 Microwave System

Most of the microwave systems in MAE are in batch operation, despite that continuous operational mode has been mushrooming in recent years (Leone et al., 2014; Terigar et al., 2011; Wu et al., 2016). Among them, domestic microwave system is widely employed due to its low cost and simplicity. To cater for specific requirements in MAE, some commercial microwave systems from companies, such as Milestone, CEM, and Sineo offer various high throughput extraction vessel, pressurized system, temperature control and monitoring, vessel cooling system, overhead condensing unit, and so on (Chan et al., 2011). Domestic microwave systems can also be modified to include these features (Chan et al., 2011). Different microwave systems have different heating efficiencies. Hence, the extraction result of MAE is difficult to be reproduced despite similar optimum operating conditions, such as when microwave power and heating time are applied. Under this circumstance, effects of operating parameters on MAE need to be ascertained when different microwave systems are employed.

4 Energy-Based Parameters

Based on the parametric effects detailed previously in an earlier section, one may notice that the parameters driving the disruption mechanism of the MAE exhibit complex interaction effects. For example, the reported optimum MAE conditions in the literature are applicable only for a specific type of microwave system, because different microwave systems have different heating efficiencies regardless of using similar operating conditions. As such, different microwave systems would give different extraction performances. Besides that, the optimum microwave power and extraction time can only cater to a specific extraction scale. This is because changing the working volume of the extraction solvent affects the absorption of microwave energy in the system. Therefore, in most cases, the optimum MAE conditions can only serve as a reference for new extraction if a similar type of extractor is used, or reproduction of the similar extraction at different microwave systems is attempted.

Energy-based parameters can be incorporated into the study of MAE to overcome the mentioned operating issues. They are able to characterize the heating performance of the microwave system intrinsically. The idea of energy-based parameters started since the term "energy density" was first introduced to indicate microwave irradiation power for a given unit of extraction volume (Alfaro et al., 2003). This parameter was claimed to be more applicable than microwave power level in the optimization of MAE (Li et al., 2012). As the applied power is not related to the actual power absorbed to disrupt the plant sample, Alfaro's energy density cannot be used to scale-up the process (Chan et al., 2015b). Further intuitive ways to explore the energy-based parameters is through the determination of absorbed microwave power and energy per unit solvent volume, also known as APD and AED, respectively. These energy-based parameters can be used not only to optimize and scale up the MAE (Chan et al., 2014b), but also to model and predict the extraction behavior (Chan et al., 2013, 2015a). The determinations of APD and AED parameters are elaborated in the next subsections.

4.1 Determination of Energy-Based Parameters

The heating efficiency of a microwave system at any extraction condition can be characterized based on the absorbed power of solvent or the APD parameter. APD is defined as the absorbed heating power per unit volume of solvent (W/mL) and it can be determined experimentally. Considering extraction solvent with volume V (mL) is heated under a microwave system at any heating power for a duration of t_H (s), its APD value can be quantified from the heat absorbed during the duration, Q(J) as follows:

$$\mathrm{APD} = \frac{Q}{V \cdot t_{\mathrm{H}}} \tag{15.3}$$

In Eq. (15.3), the heat Q is calculated from the temperature profile of the solvent using calorimetric method (Incropera, 2007). The calculation depends on the heating cases as follows:

Case A: heating of solvent without boiling

$$Q = m_{\mathrm{L}} C_{\mathrm{p}} \Delta T \tag{15.4}$$

Case B: boiling of solvent with evaporation

$$Q = m_{\mathrm{L}} C_{\mathrm{p}} \Delta T + m_{\mathrm{V}} \Delta H_{\mathrm{vap}} \tag{15.5}$$

where m_L is the initial mass of the solvent, C_p is the heat capacity of the solvent, ΔT is the temperature difference after heating, m_v is the mass of the evaporated solvent, H_{vap} is the heat of vaporization of the solvent. Note that Eqs. (15.4) and (15.5) neglect the heat loss to surrounding. In practice, a representative APD value of the extraction system at certain conditions can be obtained by repeating the calculation of Q at different heating time t_H.

AED parameter, which is defined as the absorbed microwave energy in the solvent (J/mL), can be used to extend the capability of APD parameter. AED parameter is mathematically related to APD parameter through the following relationship:

$$\mathrm{AED} = \mathrm{APD} \times t \tag{15.6}$$

where t is the extraction time. The example of APD calculation is given as follows.

4.1.1 Example of APD calculation

Considering an MAE, which is carried out at 100 W microwave power and 100 mL of 85% (v/v) aqueous ethanol. The increment of solvent temperature is about 51°C during the first 5 min heating time. The solvent is boiled and 17 mL of solvent is evaporated after 27 min of heating. Taking the initial temperature of the solvent as 29°C and the boiling point of the solvent as 70°C. The APD value of the MAE system at this extraction conditions can be determined using the following solution.

4.1.1.1 Solution

The extraction solvent is assumed to be ideal solution and the vapor–liquid equilibrium is formed during vaporization of solvent. The density of ethanol (ρ_{EtOH}) and water (ρ_{water}) are 0.789 g/mL and 1 g/mL, respectively (Miller et al., 1976). The heat capacity of ethanol ($C_{p.EtOH}$) and water ($C_{p.water}$) are found to be 2.63 $Jg^{-1}\ K^{-1}$ and 4.2 $Jg^{-1}\ K^{-1}$, respectively (Miller et al., 1976). The latent heat of vaporization for the extraction solvent is about 40.7 kJ/mol (Tamir, 1982). The volume of ethanol (V_{EtOH}) and water (V_{water}) are 85 mL and 15 mL, respectively. The temperature rise during 5 min of heating (ΔT_1) is 22K and during 27 min (ΔT_2) is 41K. When 17 mL (V_{vap}) of the extraction solvent vaporizes, ethanol and water have a mass fraction of 0.91 (y_{EtOH}) and 0.09 (y_{water}) in the vapor, respectively (Smith et al., 1996). Based on ideal solution assumption, the density of the liquid solvent mixture at 0.91 mass fraction of ethanol (ρ_{vap}) is 0.808 g/mL.

Case A: heating of solvent without boiling

$$\begin{aligned} Q_1 &= m_L C_p \Delta T_1 \\ &= V_{EtOH}\rho_{EtOH}C_{p.EtOH}\Delta T_1 + V_{water}\rho_{water}C_{p.water}\Delta T_1 \\ &= (85\,mL)\left(0.789\frac{g}{mL}\right)\left(2.63\frac{J}{gK}\right)(22K) + (15\,mL)\left(1\frac{g}{mL}\right)\left(4.2\frac{J}{gK}\right)(22K) \\ &= 5266\,J \end{aligned}$$

$$\begin{aligned} APD_1 &= \frac{Q_1}{V \cdot t_{H_1}} \\ &= \frac{(5266.38\,J)}{(100\,mL)\ 5\,min \times 60\frac{s}{min}} \\ &= 0.175\ \frac{W}{mL} \end{aligned}$$

Case B: boiling of solvent with evaporation

$$\begin{aligned} m_L C_p \Delta T_2 &= V_{EtOH}\rho_{EtOH}C_{p.EtOH}\Delta T_2 + V_{water}\rho_{water}C_{p.water}\Delta T_2 \\ &= (85\,mL)\left(0.789\frac{g}{mL}\right)\left(2.63\frac{J}{gK}\right)(41K) + (15\,mL)\left(1\frac{g}{mL}\right)\left(4.2\frac{J}{gK}\right)(41K) \\ &= 9815\,J \end{aligned}$$

$$\begin{aligned}
m_v H_{vap} &= \left[\left(\frac{\rho_{vap} V_{vap} y_{EtOH}}{M_{EtOH}} \right) + \left(\frac{\rho_{vap} V_{vap} y_{water}}{M_{water}} \right) \right] H_{vap} \\
&= \left[\rho_{vap} V_{vap} \left(\frac{y_{EtOH}}{M_{EtOH}} + \frac{y_{water}}{M_{water}} \right) \right] H_{vap} \\
&= \left[0.808 \frac{\text{g}}{\text{mL}} \cdot 17\,\text{mL} \left(\frac{0.91}{46\,\text{g/mol}} + \frac{0.09}{18\,\text{g/mol}} \right) \right] 40.7 \times 10^3 \frac{\text{J}}{\text{mol}} \\
&= 13855\,\text{J}
\end{aligned}$$

$$\begin{aligned}
Q_2 &= m_L C_p \Delta T_2 + m_v H_{vap} \\
&= 9814.62\,\text{J} + 13851.42\,\text{J} \\
&= 23669\,\text{J}
\end{aligned}$$

$$\begin{aligned}
\text{APD}_2 &= \frac{Q_2}{V \cdot t_{H_2}} \\
&= \frac{(23669.47\,\text{J})}{(100\,\text{mL})\ 27\,\text{min} \times 60 \frac{\text{s}}{\text{min}}} \\
&= 0.146 \frac{\text{W}}{\text{mL}}
\end{aligned}$$

$$\text{APD}_{av} = \frac{\text{APD}_1 + \text{APD}_2}{2} = 0.161\,\text{W/mL}$$

∴ The average APD value, APD_{av} is 0.16 W/mL when 100 W microwave power and 100 mL of 85% EtOH is used.

4.2 Effect of Absorbed Power Density

Absorbed power density (APD) parameter is used to indicate the heating efficiency of extraction system. It can be treated as a lumped parameter, representing the effects of solvent nature, microwave system, solvent loading, and nominal microwave power on the microwave energy absorption. In practice, the influences of solvent loading and nominal microwave power on APD parameter are vital especially to scale-up MAE. From their response surface curve in Fig. 15.5, it shows that APD increases exponentially with the increase in microwave power or the decrease in solvent loading (Chan, 2013).

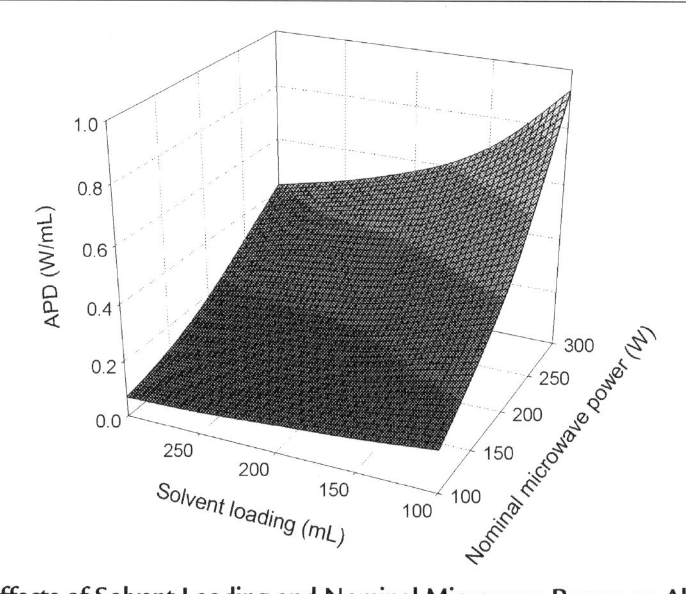

Figure 15.5: Effects of Solvent Loading and Nominal Microwave Power on Absorbed Power Density *(APD)* Parameter.

The APD values were calculated based on 85 (v/v) EtOH as extraction solvent and domestic microwave oven as extraction system. *Adapted from Chan, C.-H., 2013. Optimization and modeling of microwave-assisted extraction of active compounds from cocoa leaves. Ph.D., University of Malaya.*

APD parameter has strong influence on the extraction kinetics of MAE. Higher APD condition implies a higher rate of heating and hence the MAE requires shorter extraction time to achieve equilibrium yield. For better illustration, Fig. 15.6 presents the extraction kinetics of MAE under the influences of microwave power and solvent loading. It can be seen that the effect of APD on extraction kinetic is similar with that of microwave power, and there is an optimum extraction time at each APD condition. When the MAE is conducted at larger scale, higher power or APD condition should be engaged to sustain the extraction, otherwise longer heating time is required. Note that the extraction yield at higher APD condition (e.g., >0.5 W/mL) in Fig. 15.6 is slightly lower. This is due to microwave overheating, causing the extraction temperature to increase beyond its normal boiling point and subsequently triggered the adverse effect of the temperature parameter (Chan et al., 2013). As a summary, APD characterizes both the extraction yield and the optimum extraction time regardless of the heating condition and extraction scale.

4.3 Effect of Absorbed Energy Density

Absorbed energy density (AED) parameter is able to track the progress of MAE approaching equilibrium, using the absorbed energy in the solvent as the indicator. Theoretically, microwave energy is utilized to disrupt the plant sample and a certain amount of energy

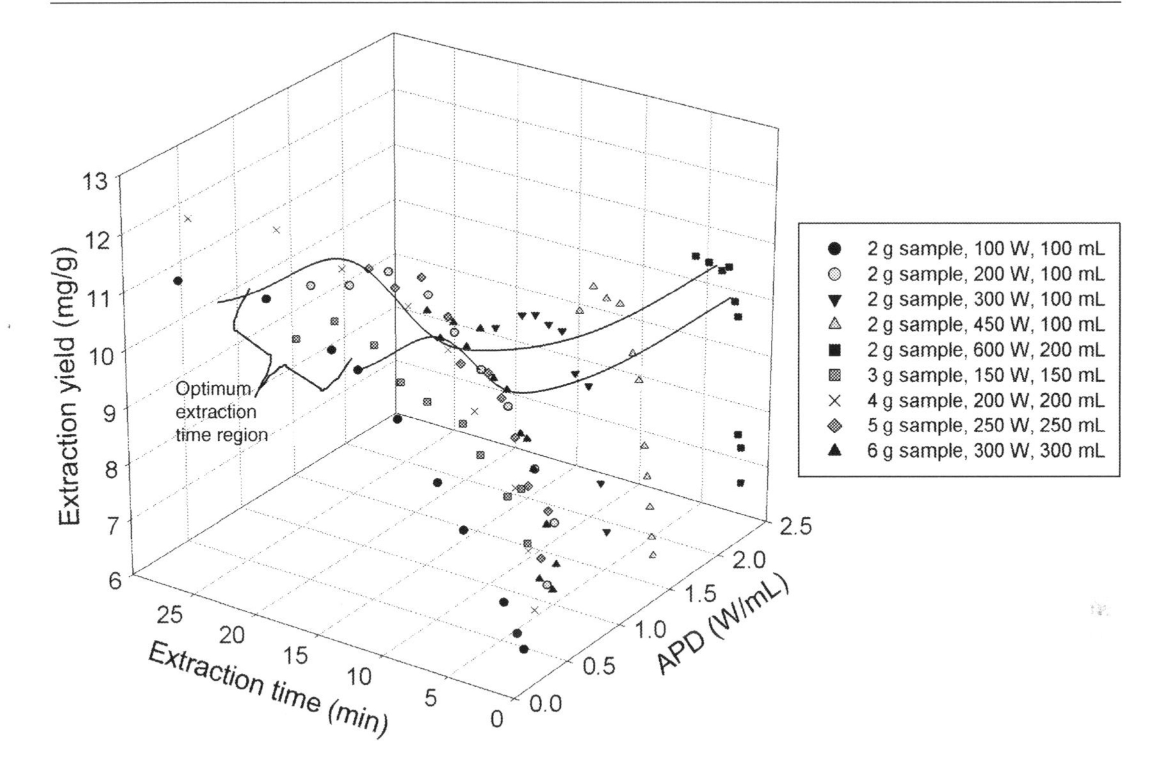

Figure 15.6: The Effect of APD on MAE Performance.
MAE conditions: domestic microwave oven with constant-power heating, 85% (v/v) EtOH solvent, sample particle size of 0.25–0.60 mm, 50 mL/g solvent-to-sample ratio. Optimum extraction time region is defined as the extraction time that can achieve 80%–95% of extraction yield during the diffusion step period. *Adapted with permission from MAE of bioactive compounds from cocoa (*Theobroma cacao *L.) leaves by Chan, C.-H., Yusoff, R., Ngoh, G.-C., 2015b. Assessment of scale-up parameters of microwave-assisted extraction via the extraction of flavonoids from cocoa leaves. Chem. Eng. Technol. 38, 489–496.*

delivered to the extraction system is associated with a certain degree of extraction completion (Chan et al., 2013). To highlight the impact of AED, Fig. 15.7 shows the extraction profile of MAE at various heating conditions in the basis of AED. It can be seen that the scattered extraction profiles of the MAE exhibit a similar trend, despite conducting at different conditions. Although AED is calculated based on the APD values, there is no interaction between APD and AED on MAE kinetics. All the extraction profiles reached equilibrium yield at the same AED values, though they were conducted at different APD values. After all, the progress of MAE to reach equilibrium yield is strongly dependent on AED, in which three distinct extraction regions can be observed. In the first region, the extraction proceeds with constant rate, indicating that a steady diffusion of bioactive compounds takes place. After some time, the diffusion rate decreases when the extraction approaching equilibrium yield. Beyond the equilibrium yield, a further increase in microwave energy would not improve the

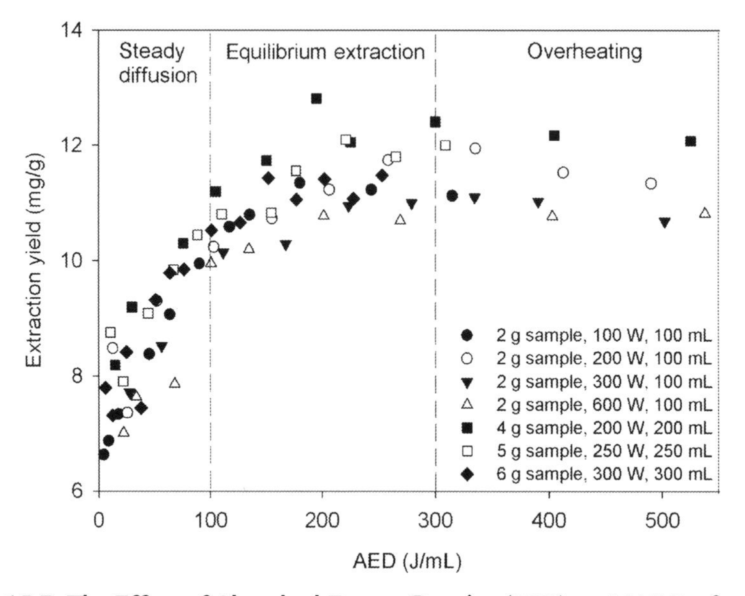

Figure 15.7: The Effect of Absorbed Energy Density *(AED)* **on MAE Performance.**
MAE conditions: domestic microwave oven with constant-power heating, 85% (v/v) EtOH solvent, sample particle size of 0.25–0.60 mm, 50 mL/g solvent-to-sample ratio. Different AED value is achieved by changing the extraction time. *Adapted with permission from MAE of bioactive compounds from cocoa (*Theobroma cacao *L.) leaves by Chan, C.-H., Yusoff, R., Ngoh, G.-C., 2013. Modeling and prediction of extraction profile for microwave-assisted extraction based on absorbed microwave energy. Food Chem. 140, 147–153.*

extraction yield substantially. According to Fig. 15.7, 100–300 J/mL AED is required to drive the MAE toward equilibrium stage.

5 Energy-Based Optimization Method

This method utilizes APD and AED parameters as the intensive optimum conditions to optimize, scale-up the MAE and also to reproduce the MAE outputs. Basically, it converts the scale-dependent parameters, namely microwave power and extraction time to scale-independent parameters, such as APD and AED and then optimize these parameters accordingly. The optimization strategy is described in the following section.

5.1 Optimization Strategy

Response surface methodology (RSM) is a well-used optimization method. It is normally employed to optimize microwave power and extraction time, together with other interactive parameters, such as solvent-to-sample ratio at constant sample mass (Chen et al., 2010; Yang and Zhai, 2010). The drawbacks are that this method requires screening of suitable ranges of the operating parameters and the optimization result is restricted to certain extraction scales and microwave systems. In this section, a systematic way to optimize operating parameters of the MAE based on its kinetic mechanism is presented in Fig. 15.4. Since the mechanism 1 and mechanism 3, which respectively involve solvent penetration and bioactive compounds elution, are not rate limiting, their associated operating parameters, such as solvent type, solvent-to-sample ratio at constant volume, sample particle size and temperature should be optimized prior. These parameters are not interacting with each other. Hence, they can be optimized using single factor experiment. However, in common practice, extraction solvent is predetermined based on the solubility of the bioactive compounds and dielectric properties of the mixture. Besides, the selection of sample particle size is dependent on the efficiency of the mechanical pretreatment, such as grinding. Therefore, only solvent-to-sample ratio at constant solvent is required for optimization. For extraction temperature, it is optional and only important when extracting thermal-sensitive compound.

The following procedure aims to optimize the rate-limiting mechanism, which is the disruption of plant structure. The influencing parameters involved are microwave power, extraction time, heating mode, and microwave system, and their combined effects are addressed based on APD and AED parameters. As independent variables, APD and AED can be optimized individually using a single-factor experiment. The optimum APD and AED are then used to determine their corresponding optimum microwave power and extraction time, at any extraction scale and microwave system. This is possible by tuning the nominal power and heating time of an MAE system at a particular extraction scale (working volume of solvent) such that its APD and AED parameters attain the optimum values. The optimum APD and AED conditions together with other scale-independent parameters, such as type of solvent, particle size of sample, extraction temperature, and solvent-to-sample ratio at constant volume are the intensive optimum conditions. They can be used for scale-up and reproducibility study of extraction at different microwave system.

5.2 Case Study of Optimization

This section demonstrates the energy-based optimization method based on the MAE using a case study (Chan et al., 2014b). Considering extraction of antidiabetic compounds, such as isoquercitrin, (–)-epicatechin and rutin from cocoa (*Theobroma cacao* L.) leaves is performed using domestic microwave oven with adjustable nominal power output. Prior to the optimization study, the type of solvent and particle size of sample are specified to be 85% (v/v) EtOH and 0.25–0.60 mm, respectively. As the bioactive compounds of interest are relatively stable under 100°C (Liazid et al., 2007; Rohn et al., 2007), the extraction temperature is not controlled and hence it is capped on the solvent boiling point (approximately 70°C). In this optimization study, only solvent-to-sample ratio, microwave power, and extraction time are the focus.

5.2.1 Optimization of solvent-to-sample ratio

Single factor experiments can be used to optimize the solvent-to-sample ratio at constant solvent volume. At an arbitrary heating condition, the effect of the solvent-to-sample ratio on the extraction yield is shown in Fig. 15.8. Generally, the extraction yield increases with the solvent-to-sample ratio up to an optimum point. In this case, the solvent-to-sample ratio of 50 mL/g is the optimum value and the increment of extraction yield beyond this point is not significant. The solvent-to-sample ratio of 50 mL/g is selected as the optimum value after considering both performance and solvent consumption of MAE.

5.2.2 Optimization of AED

The optimum AED is investigated by changing the extraction time, while keeping other parameters, such as the optimum solvent-to-sample and microwave power, unchanged in the single-factor experiment. The AED values can be calculated by multiplying their corresponding extraction time with the APD value of the condition using Eq. (15.6). When comparing the extraction profile of MAE at time basis with the one in AED basis, both profiles give similar trends as observed in Fig. 15.9. But, their implications are different. AED parameter denotes the required amount of microwave energy to attain certain extent of equilibrium yield regardless of extraction scale and microwave power (or APD), while extraction time is merely the heating time. In this case study, AED of 300 J/mL is the optimum energy requirement for the MAE.

5.2.3 Optimization of APD

Optimization of APD can be carried out using single-factor experiment by changing the microwave power at the optimum values of both solvent-to-sample ratio and AED. As similar AED value has to be applied to isolate the effect of APD, the extraction time for each microwave power level are calculated through Eq. (15.6). This is to ensure that all the extractions in the evaluation reach the same extent of completion. It is not sensible if the effect of heating power (APD) is evaluated at similar extraction time because the optimum extraction time required for higher microwave power is shorter. From the illustration shown in Fig. 15.10, the optimum extraction time decreases exponentially with the increase of APD. Higher equilibrium yield can be achieved at low APD condition (i.e., <0.35 W/mL). The extraction yield at higher APD condition is slightly lower due to thermal degradation of bioactive compounds. As constant-power heating is used in this case, microwave overheating normally occurs, causing the solvent temperature to increase beyond its normal boiling temperature. Concerning to the optimum APD condition, 0.30 W/mL is selected.

5.2.4 Intensive optimum conditions

The ultimate aim of this optimization is to determine the intensive optimum conditions of the MAE. In this particular case study, the intensive optimum conditions obtained are 85% (v/v)

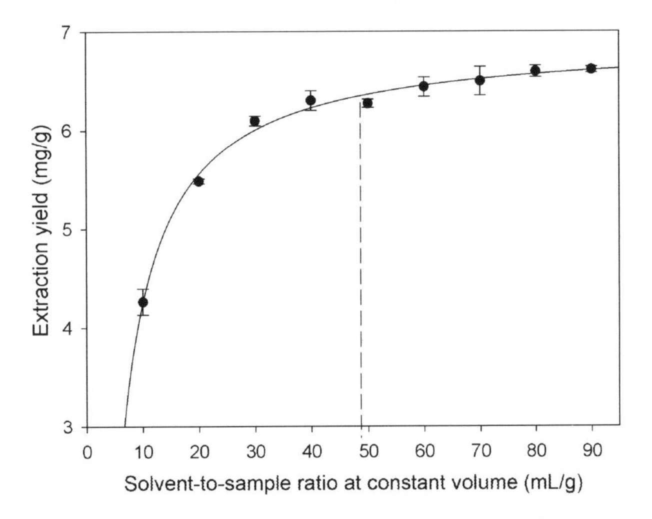

Figure 15.8: Single Factor Optimization on Solvent-to-Sample Ratio at Constant Mass for the MAE of Bioactive Compounds From Cocoa (*T. cacao* L.) Leaves.

MAE conditions: domestic microwave oven with constant-power heating, 50 mL of 85% (v/v) EtOH solvent, sample particle size of 0.25–0.60 mm, 100 W, 10 min. *Reproduced with permission from Chan, C.-H., Yusoff, R., Ngoh, G.-C., 2014b. Optimization of microwave-assisted extraction based on absorbed microwave power and energy. Chem. Eng. Sci. 111, 41–47.*

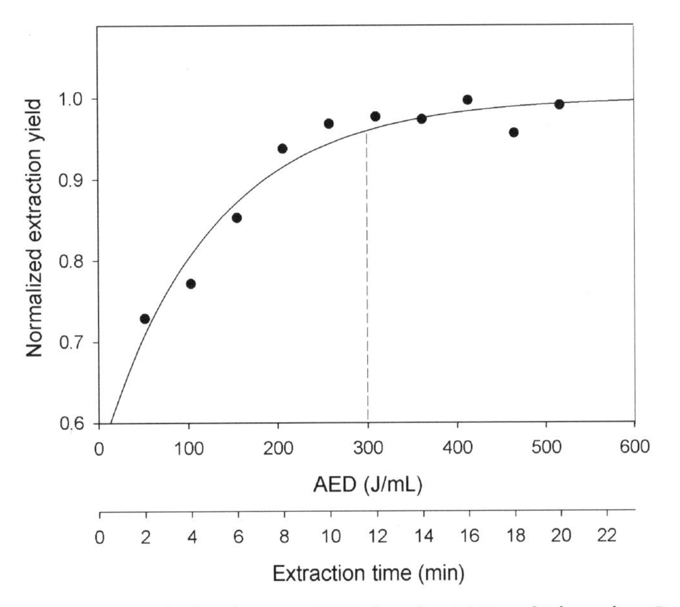

Figure 15.9: Single Factor Optimization on AED for the MAE of Bioactive Compounds From Cocoa (*T. cacao* L.) Leaves.

MAE conditions: domestic microwave oven with constant-power heating, 50 mL of 85% (v/v) EtOH solvent, sample particle size of 0.25–0.60 mm, 50 mL/g, 100 W (APD of 0.43 W/mL). *Reproduced with permission from Chan, C.-H., Yusoff, R., Ngoh, G.-C., 2014b. Optimization of microwave-assisted extraction based on absorbed microwave power and energy. Chem. Eng. Sci. 111, 41–47.*

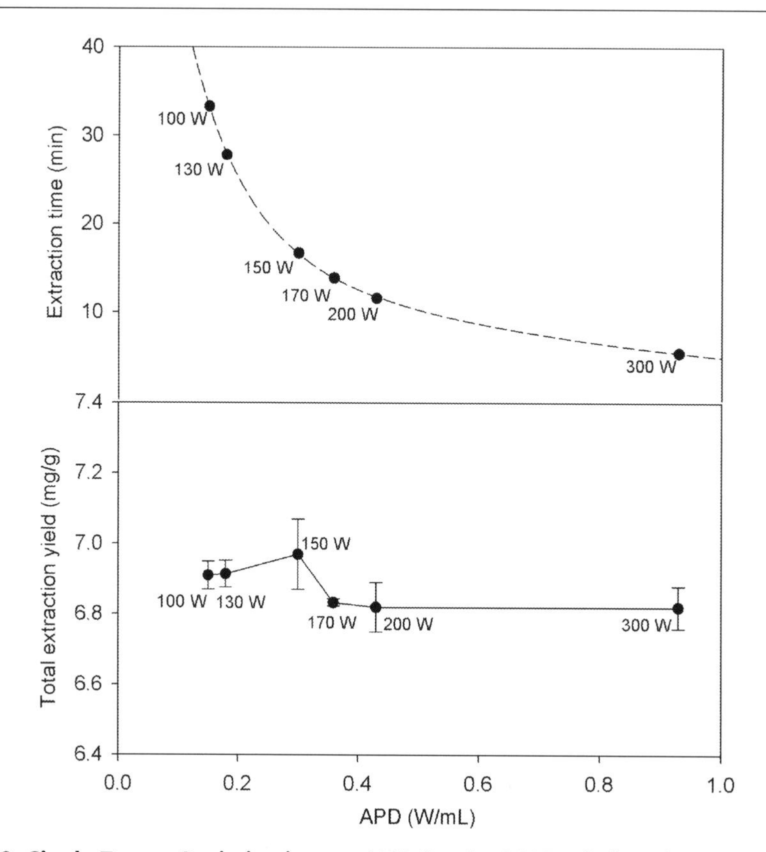

Figure 15.10: Single Factor Optimization on APD for the MAE of Bioactive Compounds From Cocoa (*T. cacao* L.) Leaves (Chan et al., 2014a).

MAE conditions: domestic microwave oven with constant-power heating, 100 mL of 85% (v/v) EtOH solvent, sample particle size of 0.25–0.60 mm, 50 mL/g. Extraction time at each APD condition is calculated based on AED of 300 J/mL. *Reproduced with permission from Chan, C.-H., Yusoff, R., Ngoh, G.-C., 2014a. Modeling and kinetics study of conventional and assisted batch solvent extraction. Chem. Eng. Res. Des. 92, 1169–1186; Chan, C.-H., Yusoff, R., Ngoh, G.-C., 2014b. Optimization of microwave-assisted extraction based on absorbed microwave power and energy. Chem. Eng. Sci. 111, 41–47.*

EtOH as solvent, 50 mL/g solvent-to-sample ratio, 0.25–0.60 mm sample particle size, 0.30 W/mL APD, and 300 J/mL AED. These intensive conditions enable MAE to be scaled up and reproduced at any extraction scale and microwave system, by determining their corresponding optimum operating conditions, such as nominal microwave power and heating time. A detailed demonstration is presented in Table 15.1. To scale up MAE, the optimum microwave power of the MAE at various solvent loading (150–300 mL) is adjusted such that the APD of the system is at 0.30 W/mL. If the adjustment is difficult to be achieved due to the power setting of the microwave system, the nearest optimum APD value in the range of 0.30 ± 0.04 W/mL obtained from the best tuning can be considered. Upon determination of microwave power, Eq. (15.6) is used to calculate the optimum extraction time based on AED of 300 J/mL and the obtained APD

Table 15.1: Determination of optimum microwave power and extraction time at various solvent loading based on intensive optimum conditions (50 mL/g, 300 J/mL, 0.3 W/mL).

Solvent loading (mL)	100	150	200	250	300
Optimum microwave power (W)[a]	150	200	220	220	260
APD (W/mL)	0.30	0.32	0.34	0.27	0.31
Optimum extraction time (min)[b]	16.7	15.6	14.7	18.5	16.1
Total extraction yield (mg/g)	6.97 ± 0.11	6.82 ± 0.06	7.15 ± 0.19	7.01 ± 0.09	6.97 ± 0.08
IQ yield (mg/g)	1.05 ± 0.06	1.01 ± 0.01	1.04 ± 0.01	1.03 ± 0.01	1.02 ± 0.04
EC yield (mg/g)	1.38 ± 0.03	1.30 ± 0.01	1.47 ± 0.09	1.40 ± 0.04	1.40 ± 0.03
RT yield (mg/g)	4.49 ± 0.06	4.51 ± 0.06	4.64 ± 0.09	4.58 ± 0.05	4.55 ± 0.06
Percentage difference of total extraction yield (%)[c]	/	2.2	2.6	0.6	0

[a]Determined based on APD of 0.3 W/mL.
[b]Determined based on AED of 300 J/mL.
[c]Determined based on the total extraction yields at solvent loading of 100 mL.
Source: Reproduced with permission from Chan, C.-H., Yusoff, R., Ngoh, G.-C., 2014b. Optimization of microwave-assisted extraction based on absorbed microwave power and energy. Chem. Eng. Sci. 111, 41–47.

obtained. From the tabulated results in Table 15.1, the MAE conducted at the intensive optimum conditions give consistent extraction yield. As a whole, the energy-based optimization method is reliable to determine the optimum MAE parameters at various extraction scales. Besides, it is also applicable for outputs reproduction at different microwave system.

6 Energy-Based Modeling Method

Modeling of MAE is essential for the prediction and simulation of extraction behavior. Most of the extraction models that are commonly used in MAE are empirical. These models do not have predictive capability to simulate the extraction behavior beyond the fitted conditions. To expand the predictive capability of MAE model, energy-based parameters, that is, APD and AED can be incorporated into the empirical extraction model to predict the extraction profile of MAE at various heating conditions.

6.1 Development of Energy-Based Predictive Models

Most of the extraction models focus on the diffusion step as this step determines the equilibrium extraction yield.

6.1.1 Film theory model

Film theory (Stanisavljević et al., 2007; Velickovic et al., 2008) is the simplest empirical model to simulate typical extraction curve (yield vs. extraction time) as follows:

$$Y = \frac{y}{y_s} = 1 - (1 - b)\exp(-k \cdot t) \tag{15.7}$$

where Y is the normalized yield, y is the yield at time t (mg/g), y_s is the equilibrium yield (mg/g), b indicates the kinetics of the washing step (dimensionless), and k denotes the kinetics of the diffusion step (min^{-1}).

6.1.2 AED-film theory model

In order to get a simple energy-based model (Chan et al., 2013), the original film theory model in Eq. (15.7) is adapted in AED basis as follows:

$$Y = \frac{y}{y_s} = 1 - (1 - b)\exp(-k' \cdot \text{AED}_t) \tag{15.8}$$

where k' is the energy-based kinetic constant for the diffusion step (mL/J) and AED_t indicates the amount of AED in the extraction solvent during the extraction time t. Note that the coefficient b in both Eqs. (15.7) and (15.8) are the same since they characterize the normalized yield during the washing step. This step is normally affected by sample pretreatment, such as grinding. Eq. (15.8) can be used to simulate AED-based extraction profile, in which its trend is similar to those in time basis as illustrated previously in Fig. 15.9. The only difference is that Eq. (15.8) simulates the profile based on the absorbed microwave energy accumulated in the extraction system. Besides, the simulated normalized profile, that is, yield versus AED in Eq. (15.8) is generally applicable for any heating modes and heating conditions of the MAE (Chan et al., 2013, 2015a). This is because the operating parameters, such as solvent loading, applied microwave power that influence the absorption of microwave energy, are already imposed on the AED_t value. In view of the model coefficients, the coefficients b and k' are affected by the type of plant sample, solvent-to-sample ratio, and particle size of sample, and these parameters are not related to microwave heating efficiency.

6.1.3 Predictive AED-film theory model

To form a predictive model, Eq. (15.8) is rewritten in term of APD as follow:

$$Y = \frac{y}{y_s} = 1 - (1 - b)\exp(-k' \cdot \text{APD} \cdot t) \tag{15.9}$$

Comparing the model coefficient of Eq. (15.9) with Eq. (15.7), it can be observed that the diffusion coefficient of the original film theory, k is equal to $k' \times \text{APD}$. This signifies that the model in Eq. (15.9) can be used to predict the time-based normalized extraction profile at any microwave power, solvent loading, and microwave system, once the APD of the extraction condition and the energy-based kinetic constant k' are calibrated.

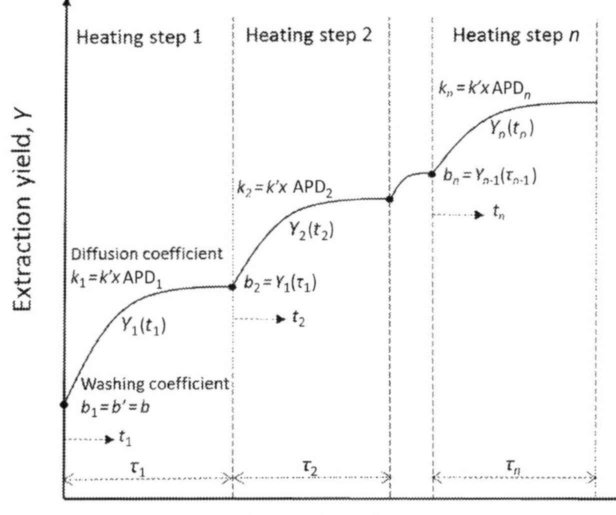

Figure 15.11: Schematic Diagram of the Development of Generalized Predictive AED-Film Theory Model. *Reproduced with permission from Chan, C.-H., Lim, J.-J., Yusoff, R., Ngoh, G.-C., 2015a. A generalized energy-based kinetic model for microwave-assisted extraction of bioactive compounds from plants. Sep. Purif. Technol. 143, 152–160.*

6.1.4 Generalized predictive AED-film theory model

The applicability and the performance of this model can be extended to other heating modes namely, two-steps-power, intermittent-power, and constant-temperature MAE. This is achievable by applying Eq. (15.9) to each heating steps involved in the heating mode as illustrated in Fig. 15.11. As such, a generalized energy-based MAE model (Chan et al., 2015a) can be developed to capture the normalized yield of MAE at the *i*th heating step, Y_i as presented in Eq. (15.10).

$$Y_i(t) = \frac{y}{y_s} = 1 - (1 - b_i)\exp\left[-k_i \cdot \left(t - \sum_{j=0}^{i-1} \tau_j\right)\right] \tag{15.10}$$

where τ is the period associated in a heating step, b_i and k_i are the washing and diffusion coefficient at *i*th heating step, respectively. These coefficients in each heating steps can be calculated using Eqs. (15.11) and (15.12), respectively.

$$b_i = Y_{i-1}(\tau_{i-1}) = 1 - (1 - b_{i-1})\exp(-k' \cdot \mathrm{APD}_{i-1} \cdot \tau_{i-1}), \quad \text{where} \quad b_1 = b \tag{15.11}$$

$$k_i = k' \times \mathrm{APD}_i \tag{15.12}$$

The coefficient b_i is the final normalized yield obtained at previous $(i - 1)$th heating step. On the other hand, the coefficient k_i is computed based on the APD associated to the *i*th heating step.

6.2 Case Study of Modeling

A case study is presented to demonstrate the energy-based modeling method. In this case study, MAE of antidiabetic compounds from cocoa (*T. cacao* L.) leaves (Chan et al., 2015a) was conducted at constant-power heating, two-steps-power, intermittent-power, and constant-temperature heating modes. The earlier MAE was performed using 85% (v/v) EtOH as solvent, sample particle size of 0.25–0.60 mm and solvent-to-sample ratio of 50 mL/g at constant solvent volume of 100 mL.

6.2.1 Calibration of model coefficients

Calibration of the initial washing coefficient b and the energy-based diffusion coefficient k' in Eq. (15.8) are required for the construction of MAE predictive model. This can be done by curve-fitting the equation with an experimental extraction profile. Fig. 15.12 demonstrates the determination of the coefficients of film theory-AED model by curve fitting the extraction profile at constant-power heating. Generally, the calibrated coefficient k' (0.01279 mL/J) is consistent, regardless of the heating modes, microwave power, microwave system, and extraction scale. On the other hand, the coefficient b in Fig. 15.12 denotes the extraction yield (56%) obtained at the beginning of the extraction before microwave heating.

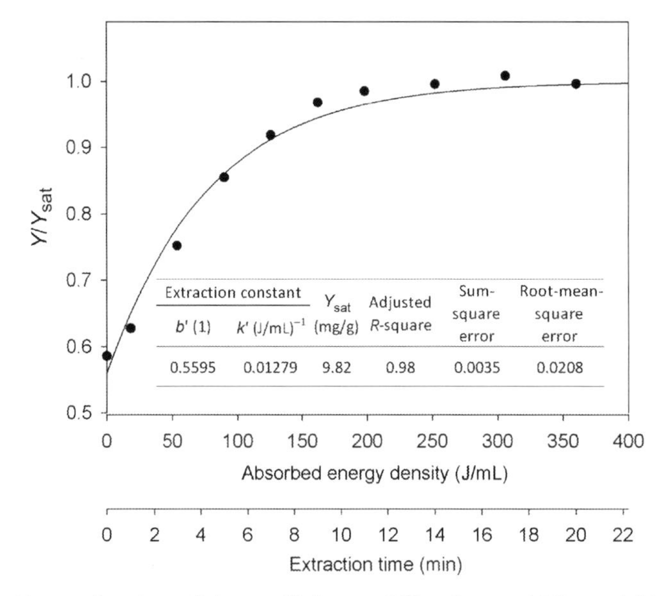

Extraction constant		Y_{sat}	Adjusted	Sum-square error	Root-mean-square error
b' (1)	k' (J/mL)$^{-1}$	(mg/g)	R-square		
0.5595	0.01279	9.82	0.98	0.0035	0.0208

Figure 15.12: Determination of the coefficients of film theory-AED model based on curve fitting of extraction profile for the MAE of bioactive compounds from cocoa (*T. cacao* L.) leaves (Chan et al., 2015a) at conditions: constant-power heating of 150 W and APD of 0.3 W/mL; · experimental extraction yield; — fitted extraction curve. *Reproduced with permission from Chan, C.-H., Lim, J.-J., Yusoff, R., Ngoh, G.-C., 2015a. A generalized energy-based kinetic model for microwave-assisted extraction of bioactive compounds from plants. Sep. Purif. Technol. 143, 152–160.*

6.2.2 Prediction of extraction profiles

By substituting the calibrated coefficient (Fig. 15.12) to the model in Eqs. (15.10)–(15.12), the extraction profiles of MAE at various heating modes can be predicted. The predictive parameter of this model is the APD values associated in each heating step. The considered heating steps in constant-power heating, two-steps-power, intermittent-power, and constant-temperature heating modes are shown in Fig. 15.13. Among all, the heating steps at constant-temperature MAE is not obvious. Therefore, the APD values at constant-temperature MAE are computed based on the average microwave powers to ramp the extraction temperature reaching the set point, and to maintain the set point temperature across the extraction duration. After the APD at each heating step of the heating modes are determined, their respective washing (b_i) and diffusion coefficients (k_i) can be computed as tabulated in Table 15.2.

After substituting the model coefficients (Table 15.2) in the generalized model in Eq. (15.10), the extraction profiles of the MAE at various heating modes can be predicted.

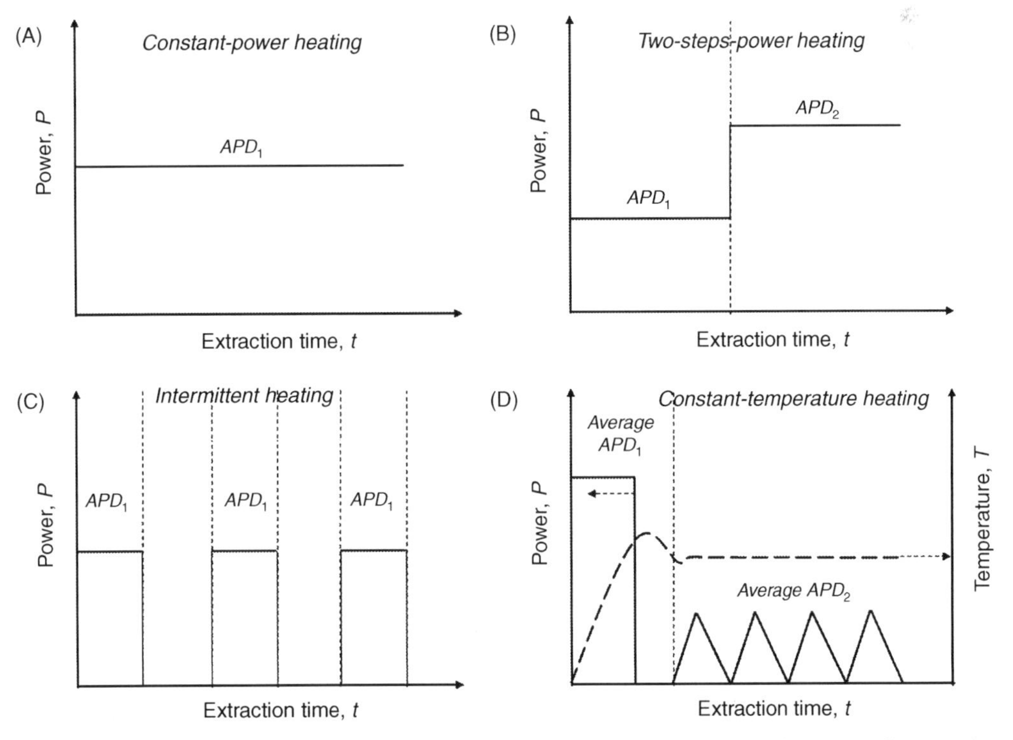

Figure 15.13: APD Parameters Associated to the Heating Steps of Various Heating Modes of MAE.

(A) Constant-power heating, (B) two-steps power heating, (C) intermittent-power heating, (D) constant-temperature heating. *Reproduced with permission from Chan, C.-H., Lim, J.-J., Yusoff, R., Ngoh, G.-C., 2015a. A generalized energy-based kinetic model for microwave-assisted extraction of bioactive compounds from plants. Sep. Purif. Technol. 143, 152–160.*

Table 15.2: Predictive capability of generalized predictive AED-film theory model.

Operational Heating Mode	Heating Steps	Predicted Extraction Constant		Average Relative Error (%)
		b (1)	k (min)$^{-1}$	
Two-steps power heating (Samsung MW718 microwave system)	(1) 100 W (0.15 W/mL), 13.46 min	b_1 = 0.5595	k_1 = 0.1151	3.18
	(2) 300 W (0.93 W/mL), 4.14 min	b_2 = 0.9064	k_2 = 0.7137	
	(1) 300 W (0.93 W/mL), 3.20 min	b_1 = 0.5595	k_1 = 0.7137	2.93
	(2) 100 W (0.15 W/mL), 16.40 min	b_2 = 0.9551	k_2 = 0.1151	
Intermittent power heating (Samsung MW718 microwave system)	(1) 150 W (0.30 W/mL), 4 min	b_1 = 0.5595	k_1 = 0.2302	2.52
	(2) 0 W, 4 min	b_2 = 0.8246	k_2 = 0	
	(3) 150 W (0.30 W/mL), 4 min	b_3 = 0.8246	k_3 = 0.2302	
	(4) 0 W, 4 min	b_4 = 0.9302	k_4 = 0	
	(5) 150 W (0.30 W/mL), 4 min	b_5 = 0.9302	k_5 = 0.2302	
	(6) 0 W, 4 min	b_6 = 0.9722	k_6 = 0	
	(7) 150 W(0.30 W/mL), 4 min	b_7 = 0.9722	k_7 = 0.2302	
	(8) 0 W, 4 min	b_8 = 0.9889	k_8 = 0	
	(1) 300 W (0.93 W/mL), 1 min	b_1 = 0.5595	k_1 = 0.7137	2.10
	(2) 0 W, 3 min	b_2 = 0.7842	k_2 = 0	
	(3) 300 W (0.93 W/mL), 1 min	b_3 = 0.7842	k_3 = 0.7137	
	(4) 0 W, 3 min	b_4 = 0.8943	k_4 = 0	
	(5) 300 W (0.93 W/mL), 1 min	b_5 = 0.8943	k_5 = 0.7137	
	(6) 0 W, 3 min	b_6 = 0.9482	k_6 = 0	
	(7) 300 W (0.93 W/mL), 1 min	b_7 = 0.9482	k_7 = 0.7137	
	(8) 0 W, 3 min	b_8 = 0.9746	k_8 = 0	
Constant temperature heating (Milestone RotoSYNTH)	Extraction temperature of 50°C	b_1 = 0.5595	k_1 = 0.4604	4.09
	(1) Average 101 W (0.60 W/mL), 2 min	b_2 = 0.8246	k_2 = 0.0844	
	(2) Average 38 W (0.08 W/mL), 28 min			
	Extraction temperature of 70°C	b_1 = 0.5595	k_1 = 0.9669	3.35
	(1) Average 209 W (1.26 W/mL), 2 min	b_2 = 0.9363	k_2 = 0.1305	
	(2) Average 71 W (0.20 W/mL), 13 min			

Source: Reproduced with permission from Chan, C.-H., Lim, J.-J., Yusoff, R., Ngoh, G.-C., 2015a. A generalized energy-based kinetic model for microwave-assisted extraction of bioactive compounds from plants. Sep. Purif. Technol. 143, 152–160.

Fig. 15.14 shows that the predicted extraction profiles match with the experimental profiles with average relative error less than 4%. For two-step heating mode, both the extraction and temperature profiles exhibit two distinct regions with different growth rates. The growth rate is higher at higher APD condition. Two-step heating mode in the MAE is able to produce higher extraction yield than other heating modes. Regardless of the order of power configuration (e.g., low-high or high-low power levels), additional 5% of extraction yield can be achieved (Chan et al., 2015a). This is probably due to the effectiveness of this mode to disrupt the plant cell structure. As for the intermittent power MAE, it produces an extraction profile with stepwise increment trends. This is because the extraction of

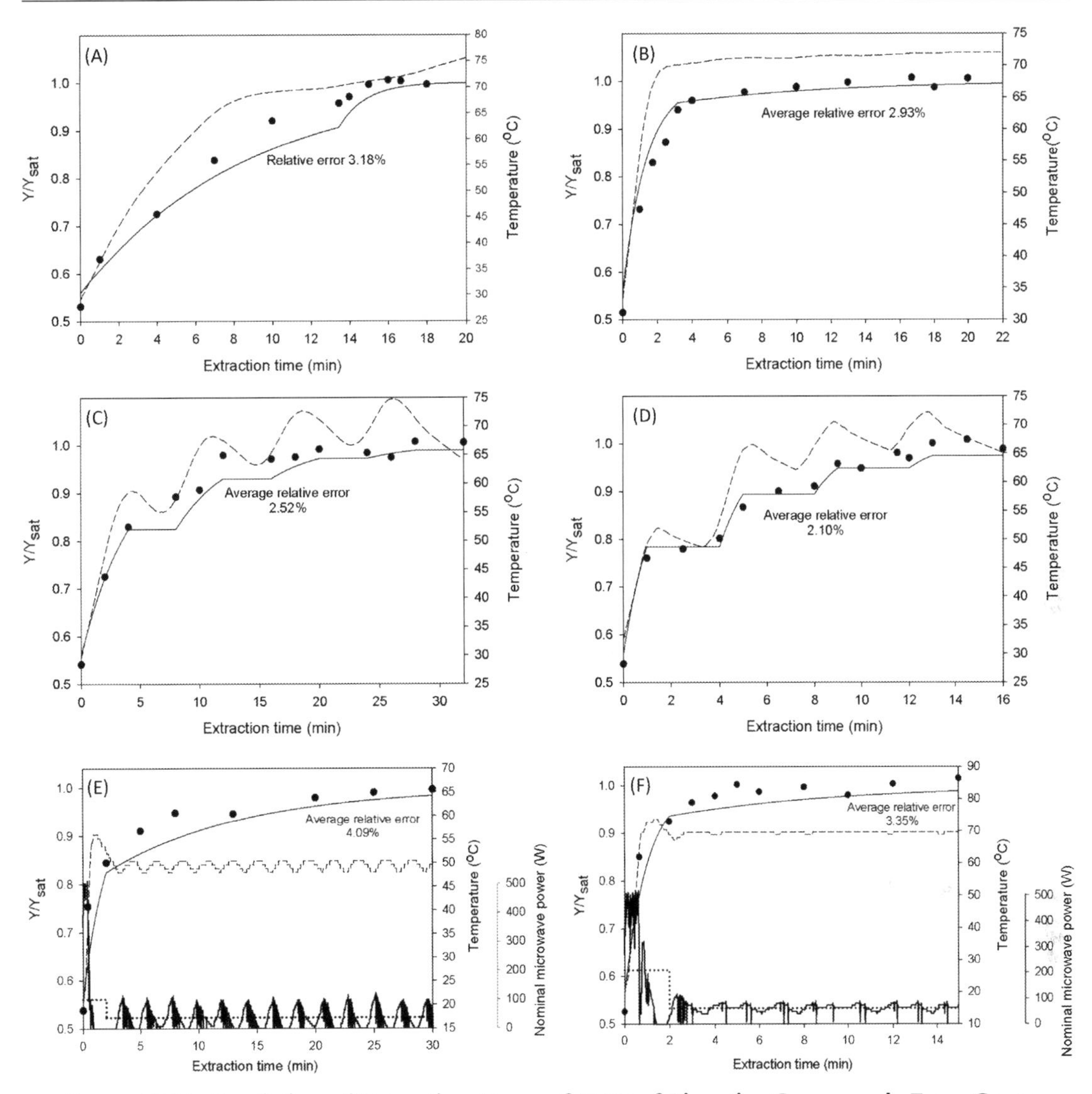

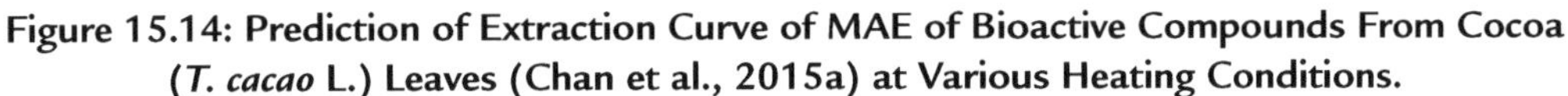

Figure 15.14: Prediction of Extraction Curve of MAE of Bioactive Compounds From Cocoa (*T. cacao* L.) Leaves (Chan et al., 2015a) at Various Heating Conditions.

(A) Two-steps power heating, 100 W for 13.46 min followed by 300 W for 4.14 min; (B) two-steps power heating 300 W for 3.20 min followed by 100 W for 16.40 min; (C) intermittent power heating, 150 W, $\alpha = 0.50$ (on: 4 min, off: 4 min) for 32 min; (D) intermittent power heating, 300 W, $\alpha = 0.25$ (on: 1 min, off: 3 min) for 16 min; (E) constant temperature heating, 500 W (ramping for 25 s), 50°C, 30 min; (F) constant temperature heating, 500 W (ramping for 40 s), 70°C, 15 min. · experimental extraction yield; — predicted extraction curve; - - solvent temperature; ▬ nominal power; ··· average nominal power. *Reproduced with permission from Chan, C.-H., Lim, J.-J., Yusoff, R., Ngoh, G.-C., 2015a. A generalized energy-based kinetic model for microwave-assisted extraction of bioactive compounds from plants. Sep. Purif. Technol. 143, 152–160.*

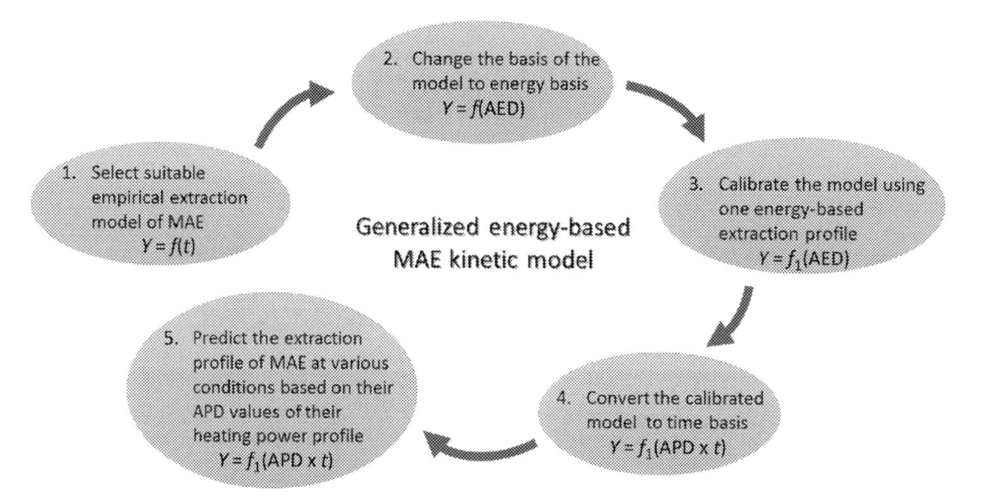

Figure 15.15: General Procedure to Develop Energy-Based Predictive Model for MAE.

bioactive compounds from plants occurs during the presence of microwave heating and is not due to the effect of temperature. Moreover, constant-temperature MAE gives extraction profile with two distinct growth rates, as the heating mode consists of ramping and control steps.

In summary, this generalized model predicts MAE kinetics based on suitable diffusion coefficients according to the heating power in the extraction. Calibration of the model coefficients, such as the initial washing coefficient b and the energy-based diffusion coefficient k' are prerequisite when attempting new extraction or when the solvent type, solvent-to-sample ratio or sample particle size is changed. As far as energy-based modeling is concerned, the modeling scheme presented in Fig. 15.15 is applicable to other empirical models. However, the most suitable one is the film theory model due to its consistent normalized response. Other models use the real extraction yield as response are not recommended because the modeling outcome would be inconsistent if different batches of sample are involved.

6.3 Comparison With Other Models

This energy-based model is the first MAE model to have such predictive capability. Other empirical models, such as film theory, chemical rate law and species balance equation cannot be used for any prediction as they are constructed entirely from fitting through experimental data and only capable to simulate extraction behavior within the fitted conditions (Amarni and Kadi, 2010; Chan et al., 2014a; Chumnanpaisont et al., 2014; Xiao et al., 2012). Besides, statistical-based empirical models, namely RSM and artificial neutral network (ANN) can be used to simulate the extraction behavior and to optimize the operating parameters of MAE

(Sinha et al., 2012, 2013). The predictive performance of these models are usually restricted to the range of the operating parameters used in the model calibration.

7 Concept and Strategy of Commercialization

Successful commercialization of MAE can potentially save operational cost because MAE requires shorter extraction time and lower solvent usage as compared to other techniques. To facilitate that, energy-based parameters, that is, APD and AED, are the keys as they can be used to characterize the extraction kinetics and to indicate the degree of extraction completion, respectively. They can be used as references for optimum extraction time prediction, scale-up, optimization, reproduction, or even as performance indicator to deduce the extraction efficiency of an MAE at any extraction scale and microwave setup. In the following section, the operating framework for MAE is detailed.

7.1 Operating Framework for MAE

Operating framework based on energy-based parameters, namely APD and AED, enables operational scalability, reproducibility, adaptability, flexibility, and extensibility of the MAE. The operating framework for MAE in this section refers to the applications of intensive optimum conditions and predictive models obtained from the optimization and modeling methods.

The intrinsic criteria for optimum MAE is completely described by the intensive optimum extraction conditions, namely, solvent type, sample particle size, solvent-to-sample ratio, APD, AED, and extraction temperature (optional). These intensive parameters allow process scale-up, outputs reproduction, and flexibility in operation. As scale-up and reproduction of MAE process has been demonstrated previously, this section elaborates on the operational flexibility aspect of the intensive optimum conditions. Basically, only simple tuning or modification of the intensive optimum conditions is required to achieve the specific objective of the optimization. This procedure saves the effort to reinvestigate the interactions between the parameters. To demonstrate this, the extraction case in Section 5.2 is considered. Despite that the optimum solvent-to-sample ratio of 50 mL/g has been determined for the extraction, one can use higher ratio (at constant volume), such as 80 mL/g while keeping other parameters, such as APD and AED, unchanged to maximize the extraction yield. Apart from the solvent consumption, one may alter APD and AED to give a different extraction performance. In general, nine performance regimes of MAE can be identified under different combinations of APD and AED as illustrated in Fig. 15.16. The optimum region in the case study is bounded within APD of 0.25–0.35 W/mL and AED of 250–350 J/mL. When conducting the MAE below optimum AED conditions, the extraction is incomplete due to the insufficient heating time. On the other hand, there is a risk of thermal degradation when

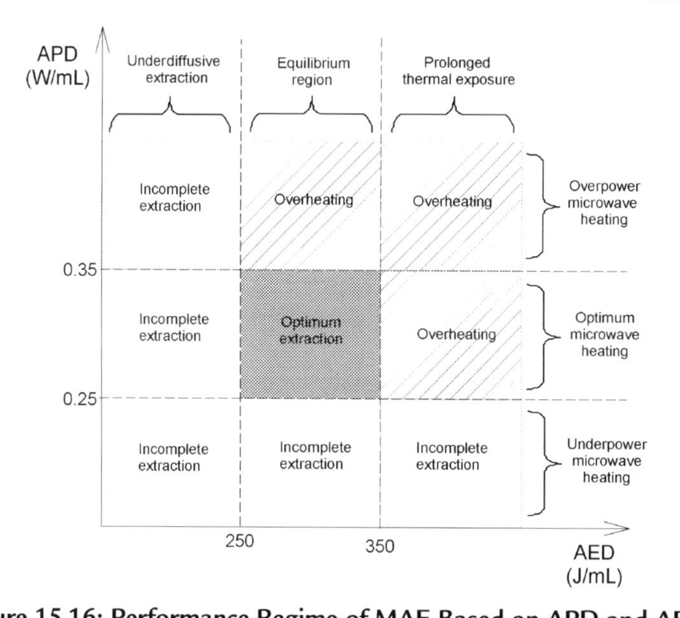

Figure 15.16: Performance Regime of MAE Based on APD and AED.
The bounded optimum region for the MAE of bioactive compounds from cocoa (*T. cacao* L.) leaves. *Reproduced with permission from Chan, C.-H., Yusoff, R., Ngoh, G.-C., 2014b. Optimization of microwave-assisted extraction based on absorbed microwave power and energy. Chem. Eng. Sci. 111, 41–47.*

prolonging the extraction above the optimum AED. With regard to the APD effect, MAE conducted below optimum APD gives poor extraction yield due to insufficient heating power to disrupt the plant sample effectively. Higher APD, that is, beyond the optimum value, may subject the extraction to a high temperature, resulting in thermal degradation due to microwave overheating. Overall, operating framework based on APD and AED is compiled with the green extraction principles (Chemat et al., 2012) as it is able to adapt for different types of plant extraction and produce desired extraction efficiency without engaging excessive power or time.

Apart from the application of intensive optimum conditions, the AED-based predictive model also plays an important role to reproduce the extraction outputs and to adapt the extraction at different microwave systems with different heating modes. The model can be used to generate an energy-performance curve, which correlates the normalized extraction yield with the AED of the extraction system. The example of the energy-performance curve for MAE of bioactive compounds from cocoa (*T. cacao* L.) leaves (Chan et al., 2015a) is shown in Fig. 15.17. The performance curve as in the figure is associated with three regions, which are steady diffusion (AED <100 J/mL), equilibrium region (100 J/mL $<$ AED $<$ 300 J/mL), and overheating (AED >300 J/mL). It denotes the degree of equilibrium of MAE based on the amount of energy accumulated in the system. This performance curve facilitates the prediction of extraction profiles and the determination of optimum extraction time of MAE at any heating

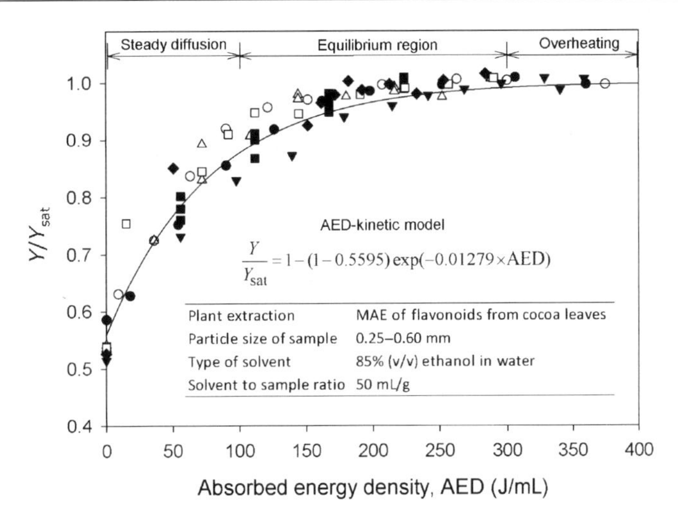

Figure 15.17: Energy-performance curve of MAE flavonoid compounds from cocoa leaves; • constant-power heating at 150 W (20 min); ^ two-steps power heating at 100 W (13.46 min) followed by 300 W (4.14 min); ▼ two-steps power heating at 300 W (3.20 min) followed by 100 W (6.40 min); Δ intermittent power heating at 150 W, $\alpha = 0.50$ (32 min); ■ intermittent power heating at 300 W, $\alpha = 0.25$ (16 min); □ constant temperature heating at 50°C (30 min); ♦ constant temperature heating at 70°C (15 min). *Reproduced with permission from Chan, C.-H., Lim, J.-J., Yusoff, R., Ngoh, G.-C., 2015a. A generalized energy-based kinetic model for microwave-assisted extraction of bioactive compounds from plants. Sep. Purif. Technol. 143, 152–160.*

condition, microwave system, and extraction scale. Note that the performance curve is not valid when the sample particle size, solvent, and solvent-to-sample ratio changed.

7.2 Current Scenario and Challenges in Commercialization of MAE

Although the applications of the energy-based parameters are useful for MAE, they are only applicable for batch operation and more work needs to be done to extend their applicability to continuous microwave systems. From an industrial perspective, continuous operation is preferable due to the ease of control and scale-up. Nevertheless, there is no operational framework to commercialize continuous MAE systems, despite various prototypes of the system that have been developed (Leone et al., 2014; Setyaningsih et al., 2012; Terigar et al., 2011).

APD and AED can be used as potential equipment design parameters for industrial microwave extractors. They can be employed to characterize the extraction vessel configuration and microwave power output because they are able to address the microwave absorption capability, the required power, and the energy for optimum extraction. After all, they serve as the important criteria for design specifications in ensuring high efficiency of the MAE process. More studies are needed to design an operational framework for continuous

MAE system by investigating the relationship of operating parameters in continuous flow, such as flow rate and flow modes (e.g., packed bed flow, slurry flow) with the energy-based parameters.

8 Conclusions

The energy-based approach using APD and AED parameters serves as tools and practical guidelines to perform MAE of bioactive compounds from plants. It covers optimization of extraction yields, modeling, and prediction of extraction behavior, scale-up, and also reproduction of extraction result. Besides, it offers operational flexibility to perform the MAE effectively according to user-defined specifications and enables operational adaptability for different types of plant samples. Despite the earlier features and advantages, this energy-based approach has only been employed for laboratory-scale MAE processes particularly in batch operations. To commercialize MAE techniques, more works have to be done to validate and refine this approach for pilot and industrial-scale MAEs, especially for continuous microwave system.

References

Aeenehvand, S., Toudehrousta, Z., Kamankesh, M., Mashayekh, M., Tavakoli, H.R., Mohammadi, A., 2016. Evaluation and application of microwave-assisted extraction and dispersive liquid–liquid microextraction followed by high-performance liquid chromatography for the determination of polar heterocyclic aromatic amines in hamburger patties. Food Chem. 190, 429–435.

Ahmad, J., Langrish, T.a.G., 2012. Optimisation of total phenolic acids extraction from mandarin peels using microwave energy: the importance of the Maillard reaction. J. Food Eng. 109, 162–174.

Alfaro, M.J., Belanger, J.M.R., Padilla, F.C., Pare, J.R.J., 2003. Influence of solvent, matrix dielectric properties, and applied power on the liquid-phase microwave-assisted processes (MAP (TM)) extraction of ginger (*Zingiber officinale*). Food Res. Int. 36, 499–504.

Amarni, F., Kadi, H., 2010. Kinetics study of microwave-assisted solvent extraction of oil from olive cake using hexane: comparison with the conventional extraction. Innov. Food Sci. Emerg. Technol. 11, 322–327.

Ayappa, K.G., Davis, H.T., Davis, E.A., Gordon, J., 1991. Analysis of microwave heating of materials with temperature-dependent properties. AIChE J. 37, 313–322.

Barrera Vázquez, M.F., Comini, L.R., Martini, R.E., Núñez Montoya, S.C., Bottini, S., Cabrera, J.L., 2014. Comparisons between conventional, ultrasound-assisted and microwave-assisted methods for extraction of anthraquinones from *Heterophyllaea pustulata* Hook f. (Rubiaceae). Ultrason. Sonochem. 21, 478–484.

Benbow, A., Stewart, M., Yeoman, G., 2001. Thiazolidinediones for type 2 diabetes. All glitazones may exacerbate heart failure. BMJ 322, 236.

Biesaga, M., 2011. Influence of extraction methods on stability of flavonoids. J. Chromatogr. A 1218, 2505–2512.

Bucic-Kojic, A., Planinic, M., Tomas, S., Bilic, M., Velic, D., 2007. Study of solid-liquid extraction kinetics of total polyphenols from grape seeds. J. Food Eng. 81, 236–242.

Cardoso-Ugarte, G.A., Sosa-Morales, M.E., Ballard, T., Liceaga, A., San Martín-González, M.F., 2014. Microwave-assisted extraction of betalains from red beet (*Beta vulgaris*). LWT Food Sci. Technol. 59, 276–282.

Casazza, A.A., Aliakbarian, B., Mantegna, S., Cravotto, G., Perego, P., 2010. Extraction of phenolics from *Vitis vinifera* wastes using non-conventional techniques. J. Food Eng. 100, 50–55.

Chan, C.-H., 2013. Optimization and modeling of microwave-assisted extraction of active compounds from cocoa leaves. Ph.D., University of Malaya.

Chan, C.-H., Lim, J.-J., Yusoff, R., Ngoh, G.-C., 2015a. A generalized energy-based kinetic model for microwave-assisted extraction of bioactive compounds from plants. Sep. Purif. Technol. 143, 152–160.

Chan, C.-H., Yusoff, R., Ngoh, G.-C., Kung, F.W.-L., 2011. Microwave-assisted extractions of active ingredients from plants. J. Chromatogr. A 1218, 6213–6225.

Chan, C.-H., Yusoff, R., Ngoh, G.-C., 2013. Modeling and prediction of extraction profile for microwave-assisted extraction based on absorbed microwave energy. Food Chem. 140, 147–153.

Chan, C.-H., Yusoff, R., Ngoh, G.-C., 2014a. Modeling and kinetics study of conventional and assisted batch solvent extraction. Chem. Eng. Res. Des. 92, 1169–1186.

Chan, C.-H., Yusoff, R., Ngoh, G.-C., 2014b. Optimization of microwave-assisted extraction based on absorbed microwave power and energy. Chem. Eng. Sci. 111, 41–47.

Chan, C.-H., Yusoff, R., Ngoh, G.-C., 2015b. Assessment of scale-up parameters of microwave-assisted extraction via the extraction of flavonoids from cocoa leaves. Chem. Eng. Technol. 38, 489–496.

Chan, C.-H., Yeoh, H.K., Yusoff, R., Ngoh, G.C., 2016. A first-principles model for plant cell rupture in microwave-assisted extraction of bioactive compounds. J. Food Eng. 188, 98–107.

Chemat, S., Ait-Amar, H., Lagha, A., Esveld, D.C., 2005. Microwave-assisted extraction kinetics of terpenes from caraway seeds. Chem. Eng. Process. 44, 1320–1326.

Chemat, F., Vian, M.A., Cravotto, G., 2012. Green extraction of natural products: concept and principles. Int. J. Mol. Sci. 13, 8615–8627.

Chen, Y.Y., Gu, X.H., Huang, S.Q., Li, J.W., Wang, X., Tang, J., 2010. Optimization of ultrasonic/microwave assisted extraction (UMAE) of polysaccharides from *Inonotus obliquus* and evaluation of its anti-tumor activities. Int. J. Biol. Macromol. 46, 429–435.

Chen, F., Du, X., Zu, Y., Yang, L., 2015. A new approach for preparation of essential oil, followed by chlorogenic acid and hyperoside with microwave-assisted simultaneous distillation and dual extraction (MSDDE) from *Vaccinium uliginosum* leaves. Ind. Crops Prod. 77, 809–826.

Cheng, G.-J.-S., Li, G.-K., Xiao, X.-H., 2015a. Microwave-assisted extraction coupled with counter-current chromatography and preparative liquid chromatography for the preparation of six furocoumarins from Angelica Pubescentis Radix. Sep. Purif. Technol. 141, 143–149.

Cheng, Z., Song, H., Yang, Y., Liu, Y., Liu, Z., Hu, H., Zhang, Y., 2015b. Optimization of microwave-assisted enzymatic extraction of polysaccharides from the fruit of *Schisandra chinensis* Baill. Int. J. Biol. Macromol. 76, 161–168.

Chumnanpaisont, N., Niamnuy, C., Devahastin, S., 2014. Mathematical model for continuous and intermittent microwave-assisted extraction of bioactive compound from plant material: extraction of β-carotene from carrot peels. Chem. Eng. Sci. 116, 442–451.

Cissé, M., Bohuon, P., Sambe, F., Kane, C., Sakho, M., Dornier, M., 2012. Aqueous extraction of anthocyanins from *Hibiscus sabdariffa*: Experimental kinetics and modeling. J. Food Eng. 109, 16–21.

Crossley, J.I., Aguilera, J.M., 2001. Modeling the effect of microstructure on food extraction. J. Food Process. Eng. 24, 161–177.

Dahmoune, F., Spigno, G., Moussi, K., Remini, H., Cherbal, A., Madani, K., 2014. *Pistacia lentiscus* leaves as a source of phenolic compounds: microwave-assisted extraction optimized and compared with ultrasound-assisted and conventional solvent extraction. Ind. Crops Prod. 61, 31–40.

Dong, Z., Gu, F., Xu, F., Wang, Q., 2014. Comparison of four kinds of extraction techniques and kinetics of microwave-assisted extraction of vanillin from *Vanilla planifolia* Andrews. Food Chem. 149, 54–61.

Fernández-Agulló, A., Freire, M.S., González-Álvarez, J., 2015. Effect of the extraction technique on the recovery of bioactive compounds from eucalyptus (*Eucalyptus globulus*) wood industrial wastes. Ind. Crops Prod. 64, 105–113.

Franco, D., Sineiro, J., Pinelo, M., Núñez, M.J., 2007. Ethanolic extraction of *Rosa rubiginosa* soluble substances: oil solubility equilibria and kinetic studies. J. Food Eng. 79, 150–157.

Franco-Vega, A., Ramírez-Corona, N., Palou, E., López-Malo, A., 2016. Estimation of mass transfer coefficients of the extraction process of essential oil from orange peel using microwave assisted extraction. J. Food Eng. 170, 136–143.

Gao, S., You, J., Wang, Y., Zhang, R., Zhang, H., 2012. On-line continuous sampling dynamic microwave-assisted extraction coupled with high performance liquid chromatographic separation for the determination of lignans in Wuweizi and naphthoquinones in Zicao. J. Chromatogr. B 887–888, 35–42.

Gironi, F., Piemonte, V., 2011. Temperature and solvent effects on polyphenol extraction process from chestnut tree wood. Chem. Eng. Res. Des. 89, 857–862.

Hao, J.-Y., Han, W., Huang, S.-D., Xue, B.-Y., Deng, X., 2002. Microwave-assisted extraction of artemisinin from *Artemisia annua* L. Sep. Purif. Technol. 28, 191–196.

Herodež, Š.S., Hadolin, M., Škerget, M., Knez, Ž., 2003. Solvent extraction study of antioxidants from Balm (*Melissa officinalis* L.) leaves. Food Chem. 80, 275–282.

Hiranvarachat, B., Devahastin, S., 2014. Enhancement of microwave-assisted extraction via intermittent radiation: extraction of carotenoids from carrot peels. J. Food Eng. 126, 17–26.

Hojnik, M., Skerget, M., Knez, Z., 2008. Extraction of lutein from Marigold flower petals: experimental kinetics and modelling. LWT Food Sci. Technol. 41, 2008–2016.

Incropera, F.P., 2007. Fundamentals of Heat and Mass Transfer, sixth ed. John Wiley, Hoboken NJ.

Karabegović, I.T., Stojičević, S.S., Veličković, D.T., Nikolić, N.Č., Lazić, M.L., 2013. Optimization of microwave-assisted extraction and characterization of phenolic compounds in cherry laurel (*Prunus laurocerasus*) leaves. Sep. Purif. Technol. 120, 429–436.

Kingston, H.M., Jassie, L.B., 1988. Introduction to Microwave Sample Preparation: Theory and Practice. American Chemical Society, Washington, DC.

Kong, Y., Zu, Y.-G., Fu, Y.-J., Liu, W., Chang, F.-R., Li, J., Chen, Y.-H., Zhang, S., Gu, C.-B., 2010. Optimization of microwave-assisted extraction of cajaninstilbene acid and pinostrobin from pigeonpea leaves followed by RP-HPLC-DAD determination. J. Food Compos. Anal. 23, 382–388.

Kwon, J.H., Belanger, J.M.R., Pare, J.R.J., Yaylayan, V.A., 2003. Application of the microwave-assisted process (MAP) to the fast extraction of ginseng saponins. Food Res. Int. 36, 491–498.

Leone, A., Tamborrino, A., Romaniello, R., Zagaria, R., Sabella, E., 2014. Specification and implementation of a continuous microwave-assisted system for paste malaxation in an olive oil extraction plant. Biosyst. Eng. 125, 24–35.

Li, Y., Han, L., Ma, R., Xu, X., Zhao, C., Wang, Z., Chen, F., Hu, X., 2012. Effect of energy density and citric acid concentration on anthocyanins yield and solution temperature of grape peel in microwave-assisted extraction process. J. Food Eng. 109, 274–280.

Li, M., Ngadi, M.O., Ma, Y., 2014. Optimisation of pulsed ultrasonic and microwave-assisted extraction for curcuminoids by response surface methodology and kinetic study. Food Chem. 165, 29–34.

Li, Y., Skouroumounis, G.K., Elsey, G.M., Taylor, D.K., 2011. Microwave-assistance provides very rapid and efficient extraction of grape seed polyphenols. Food Chem. 129, 570–576.

Liazid, A., Guerrero, R.F., Cantos, E., Palma, M., Barroso, C.G., 2011. Microwave assisted extraction of anthocyanins from grape skins. Food Chem. 124, 1238–1243.

Liazid, A., Palma, M., Brigui, J., Barroso, C.G., 2007. Investigation on phenolic compounds stability during microwave-assisted extraction. J. Chromatogr. A 1140, 29–34.

Ma, C.-H., Liu, T.-T., Yang, L., Zu, Y.-G., Chen, X., Zhang, L., Zhang, Y., Zhao, C., 2011. Ionic liquid-based microwave-assisted extraction of essential oil and biphenyl cyclooctene lignans from *Schisandra chinensis* Baill fruits. J. Chromatogr. A 1218, 8573–8580.

Ma, C.-H., Yang, L., Zu, Y.-G., Liu, T.-T., 2012. Optimization of conditions of solvent-free microwave extraction and study on antioxidant capacity of essential oil from *Schisandra chinensis* (Turcz.) Baill. Food Chem. 134, 2532–2539.

Mandal, V., Mandal, S.C., 2010. Design and performance evaluation of a microwave based low carbon yielding extraction technique for naturally occurring bioactive triterpenoid: oleanolic acid. Biochem. Eng. J. 50, 63–70.

Mandal, V., Mohan, Y., Hemalatha, S., 2007. Microwave assisted extraction: an innovative and promising extraction tool for medicinal plant research. Pharmacogn. Rev. 1, 7–18.

Miller, J.W., Shah, P.N., Yaws, C.L., 1976. Correlation constants for chemical compounds: heat capacity. Chem. Eng. 83, 129.

Milutinović, M., Radovanović, N., Ćorović, M., Šiler-Marinković, S., Rajilić-Stojanović, M., Dimitrijević-Branković, S., 2015. Optimisation of microwave-assisted extraction parameters for antioxidants from waste Achillea millefolium dust. Ind. Crops Prod. 77, 333–341.

Milutinović, M., Radovanović, N., Rajilić-Stojanović, M., Šiler-Marinković, S., Dimitrijević, S., Dimitrijević-Branković, S., 2014. Microwave-assisted extraction for the recovery of antioxidants from waste *Equisetum arvense*. Ind. Crops Prod. 61, 388–397.

Mustapa, A.N., Martin, A., Gallego, J.R., Mato, R.B., Cocero, M.J., 2015. Microwave-assisted extraction of polyphenols from *Clinacanthus nutans* Lindau medicinal plant: energy perspective and kinetics modeling. Chem. Eng. Process. 97, 66–74.

Nayak, B., Dahmoune, F., Moussi, K., Remini, H., Dairi, S., Aoun, O., Khodir, M., 2015. Comparison of microwave, ultrasound and accelerated-assisted solvent extraction for recovery of polyphenols from *Citrus sinensis* peels. Food Chem. 187, 507–516.

Pandit, S.G., Vijayanand, P., Kulkarni, S.G., 2015. Pectic principles of mango peel from mango processing waste as influenced by microwave energy. LWT Food Sci. Technol. 64, 1010–1014.

Périno, S., Pierson, J.T., Ruiz, K., Cravotto, G., Chemat, F., 2016. Laboratory to pilot scale: microwave extraction for polyphenols lettuce. Food Chem. 204, 108–114.

Qu, W., Pan, Z., Ma, H., 2010. Extraction modeling and activities of antioxidants from pomegranate marc. J. Food Eng. 99, 16–23.

Rakotondramasy-Rabesiaka, L., Havet, J.-L., Porte, C., Fauduet, H., 2007. Solid-liquid extraction of protopine from *Fumaria officinalis* L.: analysis determination, kinetic reaction and model building. Sep. Purif. Technol. 54, 253–261.

Rakotondramasy-Rabesiaka, L., Havet, J.L., Porte, C., Fauduet, H., 2009. Solid-liquid extraction of protopine from *Fumaria officinalis* L.: kinetic modelling of influential parameters. Ind. Crops Prod. 29, 516–523.

Rakotondramasy-Rabesiaka, L., Havet, J.-L., Porte, C., Fauduet, H., 2010. Estimation of effective diffusion and transfer rate during the protopine extraction process from *Fumaria officinalis* L. Sep. Purif. Technol. 76, 126–131.

Rodríguez-Rojo, S., Visentin, A., Maestri, D., Cocero, M.J., 2012. Assisted extraction of rosemary antioxidants with green solvents. J. Food Eng. 109, 98–103.

Rohn, S., Buchner, N., Driemel, G., Rauser, M., 2007. Thermal degradation of onion quercetin glucosides under roasting conditions. J. Agric. Food Chem. 55, 1568.

Setyaningsih, W., Palma, M., Barroso, C.G., 2012. A new microwave-assisted extraction method for melatonin determination in rice grains. J. Cereal Sci. 56, 340–346.

Setyaningsih, W., Saputro, I.E., Palma, M., Barroso, C.G., 2015. Optimisation and validation of the microwave-assisted extraction of phenolic compounds from rice grains. Food Chem. 169, 141–149.

Shen, Y., Lin, S., Han, C., Zhu, Z.O., Hou, X., Long, Z., Xu, K., 2014. Rapid identification and quantification of five major mogrosides in *Siraitia grosvenorii* (Luo-Han-Guo) by high performance liquid chromatography–triple quadrupole linear ion trap tandem mass spectrometry combined with microwave-assisted extraction. Microchem. J. 116, 142–150.

Sinha, K., Saha, P.D., Datta, S., 2012. Response surface optimization and artificial neural network modeling of microwave assisted natural dye extraction from pomegranate rind. Ind. Crops Prod. 37, 408–414.

Sinha, K., Chowdhury, S., Saha, P.D., Datta, S., 2013. Modeling of microwave-assisted extraction of natural dye from seeds of *Bixa orellana* (Annatto) using response surface methodology (RSM) and artificial neural network (ANN). Ind. Crops Prod. 41, 165–171.

Smith, J.M., Van Ness, H.C., Abbott, M.M., 1996. Introduction to Chemical Engineering Thermodynamics. McGraw-Hill, New York, NY.

So, G.C., Macdonald, D.G., 1986. Kinetics of oil extraction from Canola (rapeseed). Can. J. Chem. Eng. 64, 80–86.

Song, J., Li, D., Liu, C., Zhang, Y., 2011. Optimized microwave-assisted extraction of total phenolics (TP) from *Ipomoea batatas* leaves and its antioxidant activity. Innov. Food Sci. Emerg. Technol. 12, 282–287.

Sparr Eskilsson, C., Björklund, E., 2000. Analytical-scale microwave-assisted extraction. J. Chromatogr. A 902, 227–250.

Spigno, G., De Faveri, D.M., 2009. Microwave-assisted extraction of tea phenols: a phenomenological study. J. Food Eng. 93, 210–217.
Stanisavljević, I.T., Lazić, M.L., Veljković, V.B., 2007. Ultrasonic extraction of oil from tobacco (*Nicotiana tabacum* L.) seeds. Ultrason. Sonochem. 14, 646–652.
Sternbauer, L., Hintersteiner, I., Buchberger, W., Standler, A., Marosits, E., 2013. Evaluation of a microwave assisted extraction prior to high performance liquid chromatography for the determination of additives in polyolefins. Polym. Test. 32, 901–906.
Sultana, B., Anwar, F., 2008. Flavonols (kaempeferol, quercetin, myricetin) contents of selected fruits, vegetables and medicinal plants. Food Chem. 108, 879–884.
Švarc-Gajić, J., Stojanović, Z., Segura Carretero, A., Arráez Román, D., Borrás, I., Vasiljević, I., 2013. Development of a microwave-assisted extraction for the analysis of phenolic compounds from *Rosmarinus officinalis*. J. Food Eng. 119, 525–532.
Tamir, A., 1982. Prediction of latent heat of vaporization of multicomponent mixtures. Fluid Phase Equilibr. 8, 131–147.
Tan, S.N., Yong, J.W.H., Teo, C.C., Ge, L., Chan, Y.W., Hew, C.S., 2011. Determination of metabolites in *Uncaria sinensis* by HPLC and GC–MS after green solvent microwave-assisted extraction. Talanta 83, 891–898.
Tanongkankit, Y., Sablani, S.S., Chiewchan, N., Devahastin, S., 2013. Microwave-assisted extraction of sulforaphane from white cabbages: effects of extraction condition, solvent and sample pretreatment. J. Food Eng. 117, 151–157.
Terigar, B.G., Balasubramanian, S., Boldor, D., Xu, Z., Lima, M., Sabliov, C.M., 2010. Continuous microwave-assisted isoflavone extraction system: design and performance evaluation. Bioresour. Technol. 101, 2466–2471.
Terigar, B.G., Balasubramanian, S., Sabliov, C.M., Lima, M., Boldor, D., 2011. Soybean and rice bran oil extraction in a continuous microwave system: from laboratory- to pilot-scale. J. Food Eng. 104, 208–217.
Thirugnanasambandham, K., Sivakumar, V., Prakash Maran, J., 2014. Process optimization and analysis of microwave assisted extraction of pectin from dragon fruit peel. Carbohydr. Polym. 112, 622–626.
Tsubaki, S., Sakamoto, M., Azuma, J., 2010. Microwave-assisted extraction of phenolic compounds from tea residues under autohydrolytic conditions. Food Chem. 123, 1255–1258.
Tsukui, A., Santos Júnior, H.M., Oigman, S.S., De Souza, R.O.M.A., Bizzo, H.R., Rezende, C.M., 2014. Microwave-assisted extraction of green coffee oil and quantification of diterpenes by HPLC. Food Chem. 164, 266–271.
Velickovic, D.T., Milenovic, D.M., Ristic, M.S., Veljkovic, V.B., 2008. Ultrasonic extraction of waste solid residues from the *Salvia* sp. essential oil hydrodistillation. Biochem. Eng. J. 42, 97–104.
Wang, L., Weller, C.L., 2006. Recent advances in extraction of nutraceuticals from plants. Trends Food Sci. Technol. 17, 300–312.
Wang, J.L., Zhang, J., Wang, X.F., Zhao, B.T., Wu, Y.Q., Yao, J., 2009. A comparison study on microwave-assisted extraction of *Artemisia sphaerocephala* polysaccharides with conventional method: molecule structure and antioxidant activities evaluation. Int. J. Biol. Macromol. 45, 483–492.
Wongkittipong, R., Prat, L., Damronglerd, S., Gourdon, C., 2004. Solid-liquid extraction of andrographolide from plants: experimental study, kinetic reaction and model. Sep. Purif. Technol. 40, 147–154.
Wu, L., Hu, M., Li, Z., Song, Y., Yu, C., Zhang, H., Yu, A., Ma, Q., Wang, Z., 2016. Dynamic microwave-assisted extraction combined with continuous-flow microextraction for determination of pesticides in vegetables. Food Chem. 192, 596–602.
Xiao, W., Han, L., Shi, B., 2008. Microwave-assisted extraction of flavonoids from Radix Astragali. Sep. Purif. Technol. 62, 614–618.
Xiao, X., Song, W., Wang, J., Li, G., 2012. Microwave-assisted extraction performed in low temperature and in vacuo for the extraction of labile compounds in food samples. Anal. Chim. Acta 712, 85–93.
Yang, Z., Zhai, W., 2010. Optimization of microwave-assisted extraction of anthocyanins from purple corn (*Zea mays* L.) cob and identification with HPLC-MS. Innov. Food Sci. Emerg. Technol. 11, 470–476.

Zhang, H.-F., Yang, X.-H., Wang, Y., 2011. Microwave assisted extraction of secondary metabolites from plants: current status and future directions. Trends Food Sci. Technol. 22, 672–688.

Zhang, W., Zhu, D., Fan, H., Liu, X., Wan, Q., Wu, X., Liu, P., Tang, J.Z., 2015. Simultaneous extraction and purification of alkaloids from *Sophora flavescens* Ait. by microwave-assisted aqueous two-phase extraction with ethanol/ammonia sulfate system. Sep. Purif. Technol. 141, 113–123.

Zhou, H.-Y., Liu, C.-Z., 2006. Microwave-assisted extraction of solanesol from tobacco leaves. J. Chromatogr. A 1129, 135–139.

Index

C

D

E

N

O

P

T

U

V

W

X

Z

Printed in the United States
By Bookmasters